VOLUME FOUR HUNDRED AND FIFTY-EIGHT

METHODS IN ENZYMOLOGY

Complex Enzymes in Microbial Natural Product Biosynthesis, Part A: Overview Articles and Peptides

METHODS IN ENZYMOLOGY

Editors-in-Chief

JOHN N. ABELSON AND MELVIN I. SIMON

*Division of Biology
California Institute of Technology
Pasadena, California, USA*

Founding Editors

SIDNEY P. COLOWICK AND NATHAN O. KAPLAN

VOLUME FOUR HUNDRED AND FIFTY-EIGHT

METHODS IN ENZYMOLOGY

Complex Enzymes in Microbial Natural Product Biosynthesis, Part A: Overview Articles and Peptides

EDITED BY

DAVID A. HOPWOOD
Department of Molecular Microbiology
John Innes Centre
Norwich, UK

AMSTERDAM • BOSTON • HEIDELBERG • LONDON
NEW YORK • OXFORD • PARIS • SAN DIEGO
SAN FRANCISCO • SINGAPORE • SYDNEY • TOKYO
Academic Press is an imprint of Elsevier

ELSEVIER

Academic Press is an imprint of Elsevier
525 B Street, Suite 1900, San Diego, CA 92101-4495, USA
30 Corporate Drive, Suite 400, Burlington, MA 01803, USA
32 Jamestown Road, London NW1 7BY, UK

First edition 2009

Copyright © 2009, Elsevier Inc. All Rights Reserved.

No part of this publication may be reproduced, stored in a retrieval system or transmitted in any form or by any means electronic, mechanical, photocopying, recording or otherwise without the prior written permission of the publisher

Permissions may be sought directly from Elsevier's Science & Technology Rights Department in Oxford, UK: phone (+44) (0) 1865 843830; fax (+44) (0) 1865 853333; email: permissions@elsevier.com. Alternatively you can submit your request online by visiting the Elsevier web site at http://elsevier.com/locate/permissions, and selecting *Obtaining permission to use Elsevier material*

Notice
No responsibility is assumed by the publisher for any injury and/or damage to persons or property as a matter of products liability, negligence or otherwise, or from any use or operation of any methods, products, instructions or ideas contained in the material herein. Because of rapid advances in the medical sciences, in particular, independent verification of diagnoses and drug dosages should be made

> For information on all Academic Press publications visit our website at elsevierdirect.com

ISBN: 978-0-12-374588-0
ISSN: 0076-6879

Printed and bound in United States of America
09 10 11 12   10 9 8 7 6 5 4 3 2 1

Working together to grow
libraries in developing countries

www.elsevier.com | www.bookaid.org | www.sabre.org

ELSEVIER   BOOK AID International   Sabre Foundation

Contents

Contributors xiii
Preface xix
Volumes in Series xxiii

Section I. Overview Articles 1

1. Approaches to Discovering Novel Antibacterial and Antifungal Agents 3

Stefano Donadio, Paolo Monciardini, and Margherita Sosio

1. Introduction 4
2. Strains and Samples 7
3. Targets and Assays for Antibacterial and Antifungal Programs 13
4. Screening 16
5. Hit Follow-up 19
6. Databases, Operations, and Costs 23
7. Perspectives 25
Acknowledgment 25
References 25

2. From Microbial Products to Novel Drugs that Target a Multitude of Disease Indications 29

Flavia Marinelli

1. Microbial Diversity and Biotechnological Products 30
2. Secondary Metabolites 30
3. Conclusions 53
Acknowledgments 54
References 54

3. Discovering Natural Products from Myxobacteria with Emphasis on Rare Producer Strains in Combination with Improved Analytical Methods — 59
Ronald O. Garcia, Daniel Krug, and Rolf Müller

1. Introduction — 60
2. The Search for Novel Myxobacteria and Their Metabolites—Basic Considerations — 63
3. Methods for Isolation, Purification, and Preservation of Novel Myxobacteria — 71
4. Fermentation and Screening for Known and Novel Metabolites — 78
Acknowledgment — 88
References — 88

4. Analyzing the Regulation of Antibiotic Production in Streptomycetes — 93
Mervyn Bibb and Andrew Hesketh

1. Introduction — 94
2. The Regulation of Antibiotic Production in Streptomycetes — 94
3. Identifying Regulatory Genes for Antibiotic Biosynthesis — 97
4. Characterizing Regulatory Genes for Antibiotic Biosynthesis — 104
Acknowledgments — 113
References — 113

5. Applying the Genetics of Secondary Metabolism in Model Actinomycetes to the Discovery of New Antibiotics — 117
Gilles P. van Wezel, Nancy L. McKenzie, and Justin R. Nodwell

1. Introduction — 118
2. Actinomycetes as Antibiotic Factories — 119
3. Effects of Culture Conditions and Metabolism — 120
4. Molecular Genetic Factors that Regulate Antibiotic Production — 125
5. Applications for New Antibiotic Screening Technologies — 129
6. Future Prospects — 133
Acknowledgments — 134
References — 134

6. Regulation of Antibiotic Production by Bacterial Hormones — 143
Nai-Hua Hsiao, Marco Gottelt, and Eriko Takano

1. Introduction — 144
2. Rapid Small-Scale γ-Butyrolactone Purification — 145
3. Antibiotic Bioassay — 146

4.	Kanamycin Bioassay	148
5.	Identification of γ-Butyrolactone Receptors	150
6.	Identification of the γ-Butyrolactone Receptor Targets	151
7.	Gel Retardation Assay to Detect Target Sequences of the γ-Butyrolactone Receptors	152
8.	Conclusions	156
	Acknowledgments	156
	References	156

7. Cloning and Analysis of Natural Product Pathways   159
Bertolt Gust

1.	Introduction	160
2.	Cloning and Identification of Biosynthetic Gene Clusters	161
3.	Analysis of Natural Product Pathways by PCR-Targeted Gene Replacement	163
4.	*In Vitro* Transposon Mutagenesis	170
5.	Heterologous Expression of Biosynthetic Gene Clusters	173
6.	Reassembling Entire Gene Clusters by "Stitching" Overlapping Cosmid Clones	174
7.	Conclusions	177
	Acknowledgments	177
	References	177

8. Methods for *In Silico* Prediction of Microbial Polyketide and Nonribosomal Peptide Biosynthetic Pathways from DNA Sequence Data   181
Brian O. Bachmann and Jacques Ravel

1.	Introduction	182
2.	Converting Type I PKSs to Structural Elements	191
3.	Converting NRPS Domain Strings to Structural Elements	205
4.	Concluding Remarks	212
	Acknowledgments	214
	References	214

9. Synthetic Probes for Polyketide and Nonribosomal Peptide Biosynthetic Enzymes   219
Jordan L. Meier and Michael D. Burkart

1.	Introduction	220
2.	Synthetic Probes of PKS and NRPS Mechanism	221
3.	Synthetic Probes of PKS and NRPS Structure	233

4. Synthetic Probes for Proteomic Identification of PKS
    and NRPS Enzymes — 241
 5. Conclusions — 250
 References — 250

10. Using Phosphopantetheinyl Transferases for Enzyme Posttranslational Activation, Site Specific Protein Labeling and Identification of Natural Product Biosynthetic Gene Clusters from Bacterial Genomes — 255

Murat Sunbul, Keya Zhang, and Jun Yin

 1. Introduction — 256
 2. Experimental Procedures — 263
 3. Conclusion — 271
 References — 271

11. Sugar Biosynthesis and Modification — 277

Felipe Lombó, Carlos Olano, José A. Salas, and Carmen Méndez

 1. Introduction — 278
 2. Deoxysugar Biosynthesis — 279
 3. Deoxysugar Transfer — 283
 4. Modification of the Glycosylation Pattern through Gene Inactivation — 284
 5. Modification of the Glycosylation Pattern through Heterologous Gene Expression — 288
 6. Modification of the Glycosylation Pattern through Combinatorial Biosynthesis — 290
 7. Gene Cassette Plasmids for Deoxysugar Biosynthesis — 292
 8. Generation of Glycosylated Compounds — 299
 9. Tailoring Modifications of the Attached Deoxysugars — 301
 10. Detection of Glycosylated Compounds — 303
 Acknowledgments — 303
 References — 303

12. The Power of Glycosyltransferases to Generate Bioactive Natural Compounds — 309

Johannes Härle and Andreas Bechthold

 1. Introduction — 310
 2. Application of GTs in Producing Unnatural Bioactive Molecules — 322
 References — 328

Section II. Peptides — 335

13. Nonribosomal Peptide Synthetases: Mechanistic and Structural Aspects of Essential Domains — 337
M. A. Marahiel and L.-O. Essen

1. Introduction — 338
2. Mechanistic and Structural Aspects of Essential NRPS Domains — 339
3. Structural Insights into an Entire Termination Module — 347
References — 349

14. Biosynthesis of Nonribosomal Peptide Precursors — 353
Barrie Wilkinson and Jason Micklefield

1. Introduction — 354
2. Precursors from Amino Acid Metabolism — 355
3. Fatty Acid Precursor Biosynthesis — 360
4. Polyketide Precursors — 364
5. Glycosyl Building Blocks — 371
6. Conclusion — 372
References — 373

15. Plasmid-Borne Gene Cluster Assemblage and Heterologous Biosynthesis of Nonribosomal Peptides in *Escherichia coli* — 379
Kenji Watanabe, Alex P. Praseuth, Mike B. Praseuth, and Kinya Hotta

1. Introduction — 380
2. Biosynthetic Pathway of Nonribosomal Peptides — 382
3. Echinomycin Biosynthetic Pathway — 383
4. Construction of A Multigene Assembly on Expression Vectors — 387
5. Heterologous Gene Expression and NRP Biosynthesis in *E. coli* — 390
6. Self-Resistance Mechanism — 391
7. Stability of Transformants Carrying Multiple Very Large Plasmids — 392
8. Engineering of Heterologous NRP Biosynthetic Pathways in *E. coli* — 393
9. Conclusion — 395
References — 396

16. Enzymology of β-Lactam Compounds with Cephem Structure Produced by Actinomycete — 401
Paloma Liras and Arnold L. Demain

1. Introduction — 402
2. Biosynthesis of Cephamycins: Enzymes and Genes — 405
3. Early Steps Specific for Cephamycin Biosynthesis — 405

4. Common Steps in Cephamycin-Producing Actinomycetes and Penicillin- or Cephalosporin-Producing Filamentous Fungi	409
5. Specific Steps for Tailoring the Cephem Nucleus in Actinomycetes	420
6. Regulation of Cephamycin C Production	422
Acknowledgments	423
References	423

17. Siderophore Biosynthesis: A Substrate Specificity Assay for Nonribosomal Peptide Synthetase-Independent Siderophore Synthetases Involving Trapping of Acyl-Adenylate Intermediates with Hydroxylamine — 431

Nadia Kadi and Gregory L. Challis

1. Introduction	432
2. NRPS-Dependent Pathways for Siderophore Biosynthesis	434
3. NRPS-Independent Pathway for Siderophore Biosynthesis	442
4. Hybrid NRPS/NIS Pathway for Petrobactin Biosynthesis	448
5. Hydroxamate-Formation Assay for NIS Synthetases	450
References	455

18. Molecular Genetic Approaches to Analyze Glycopeptide Biosynthesis — 459

Wolfgang Wohlleben, Evi Stegmann, and Roderich D. Süssmuth

1. Structural Classification of Glycopeptide Antibiotics	460
2. Methods for Analyzing Glycopeptide Biosynthesis	462
3. Investigation of Glycopeptide Biosynthetic Steps	466
4. Regulation, Self-Resistance, and Excretion	477
5. Linking Primary and Secondary Metabolism	478
6. Approaches for the Generation of New Glycopeptides	479
Acknowledgments	480
References	480

19. *In Vitro* Studies of Phenol Coupling Enzymes Involved in Vancomycin Biosynthesis — 487

Dong Bo Li, Katharina Woithe, Nina Geib, Khaled Abou-Hadeed, Katja Zerbe, and John A. Robinson

1. Introduction	488
2. Peptide Synthesis	491
3. Peptide Thioesters	499
4. *In Vitro* Assays with OxyB	502
5. Production and Purification of Enzymes	503
References	507

20. Biosynthesis and Genetic Engineering of Lipopeptides in *Streptomyces roseosporus* — 511
Richard H. Baltz

1. Introduction — 512
2. Biosynthesis and Genetic Engineering of Daptomycin in S. rosesoporus — 514
3. Sources of Genes for Combinatorial Biosynthesis — 523
4. Genetic Engineering of Novel Lipopeptides — 524
5. Concluding Remarks — 527
Acknowledgments — 528
References — 528

21. *In Vitro* Studies of Lantibiotic Biosynthesis — 533
Bo Li, Lisa E. Cooper, and Wilfred A. van der Donk

1. Introduction — 534
2. Mining Microbial Genomes for Novel Lantibiotics — 537
3. Expression and Purification of Lantibiotic Precursor Peptides (LanAs) — 538
4. Expression, Purification, and Assay of LanM Enzymes — 542
5. Expression, Purification, and Assays of LanC Cyclases — 547
6. The Protease Domain of Class II Lantibiotic Transporters — 552
7. Additional Posttranslational Modifications in Lantibiotics — 553
References — 554

22. Whole-Cell Generation of Lantibiotic Variants — 559
Jesús Cortés, Antony N. Appleyard, and Michael J. Dawson

1. Introduction — 559
2. Variant Generation — 561
3. Conclusions — 571
References — 571

23. Cyanobactin Ribosomally Synthesized Peptides—A Case of Deep Metagenome Mining — 575
Eric W. Schmidt and Mohamed S. Donia

1. Introduction — 576
2. Some Remaining Questions — 583
3. Obtaining *Prochloron* Cells and DNA — 583
4. Chemical Analysis — 586
5. Cyanobactin Gene Cloning and Identification — 587
6. Heterologous Expression in *E. coli* — 590

7.	Deep Metagenome Mining	591
8.	Enzymatic Analysis of Cyanobactin Biosynthesis	593
9.	Applying Deep Metagenome Mining: Pathway Engineering	593
	Acknowledgments	595
	References	595

Author Index   *597*
Subject Index   *627*

Contributors

Khaled Abou-Hadeed
Institute of Organic Chemistry, University of Zürich, Zürich, Switzerland

Antony N. Appleyard
Novacta Biosystems Ltd, BioPark Hertfordshire, Welwyn Garden City, United Kingdom

Brian O. Bachmann
Department of Chemistry, Vanderbilt Institute for Chemical Biology, Vanderbilt University, Nashville, Tennessee, USA

Richard H. Baltz
Cubist Pharmaceuticals Inc, Lexington, Massachusetts, USA

Andreas Bechthold
Institut für Pharmazeutische Wissenschaften, Lehrstuhl für Pharmazeutische Biologie und Biotechnologie, Albert-Ludwigs-Universität Freiburg, Freiburg, Germany

Mervyn Bibb
Department of Molecular Microbiology, John Innes Centre, Norwich Research Park, Colney Lane, Norwich, United Kingdom

Michael D. Burkart
Department of Chemistry and Biochemistry, University of California, San Diego, La Jolla, California, USA

Gregory L. Challis
Department of Chemistry, University of Warwick, Coventry, United Kingdom

Lisa E. Cooper
Departments of Chemistry and Biochemistry and the Howard Hughes Medical Institute, University of Illinois, Urbana, Illinois, USA

Jesús Cortés
Novacta Biosystems Ltd, BioPark Hertfordshire, Welwyn Garden City, United Kingdom

Michael J. Dawson
Novacta Biosystems Ltd, BioPark Hertfordshire, Welwyn Garden City, United Kingdom

Arnold L. Demain
Research Institute for Scientists Emeriti (R.I.S.E.), Drew University, Madison, New Jersey, USA

Stefano Donadio
KtedoGen and NAICONS, Via Fantoli, Milano, Italy

Mohamed S. Donia
Department of Medicinal Chemistry, University of Utah, Salt Lake City, Utah, USA

L.-O. Essen
Biochemistry-Department of Chemistry, Philipps-University Marburg, Marburg, Germany

Ronald O. Garcia
Department of Pharmaceutical Biotechnology, Saarland University, Saarbrücken, Germany

Nina Geib
Institute of Organic Chemistry, University of Zürich, Zürich, Switzerland

Marco Gottelt
Department of Microbial Physiology, GBB, University of Groningen, Haren, The Netherlands

Bertolt Gust
Pharmazeutische Biologie, Pharmazeutisches Institut, Eberhard-Karls-Universität Tübingen, Tübingen, Germany

Johannes Härle
Institut für Pharmazeutische Wissenschaften, Lehrstuhl für Pharmazeutische Biologie und Biotechnologie, Albert-Ludwigs-Universität Freiburg, Freiburg, Germany

Andrew Hesketh
Department of Molecular Microbiology, John Innes Centre, Norwich Research Park, Colney Lane, Norwich, United Kingdom

Kinya Hotta
Department of Biological Sciences, National University of Singapore, Singapore

Nai-Hua Hsiao
Department of Microbial Physiology, GBB, University of Groningen, Haren, The Netherlands

Nadia Kadi
Department of Chemistry, University of Warwick, Coventry, United Kingdom

Daniel Krug
Department of Pharmaceutical Biotechnology, Saarland University, Saarbrücken, Germany

Bo Li
Departments of Chemistry and Biochemistry and the Howard Hughes Medical Institute, University of Illinois, Urbana, Illinois, USA

Dong Bo Li
Institute of Organic Chemistry, University of Zürich, Zürich, Switzerland

Paloma Liras
Biotechnological Institute INBIOTEC, Scientific Park of León, and Área de Microbiología, Facultad de Ciencias Biológicas y Ambientales, Universidad de León, León, Spain

Felipe Lombó
Departamento de Biología Funcional and Instituto Universitario de Oncología del Principado de Asturias (I.U.O.P.A), Universidad de Oviedo, Oviedo, Spain

M. A. Marahiel
Biochemistry-Department of Chemistry, Philipps-University Marburg, Marburg, Germany

Flavia Marinelli
Department of Biotechnology and Molecular Sciences, University of Insubria, Varese, Italy

Nancy L. McKenzie
Michael G. DeGroote Centre for Infectious Disease Research, Department of Biochemistry and Biomedical Sciences, McMaster University, Hamilton, Ontario, Canada

Carmen Méndez
Departamento de Biología Funcional and Instituto Universitario de Oncología del Principado de Asturias (I.U.O.P.A), Universidad de Oviedo, Oviedo, Spain

Jordan L. Meier
Department of Chemistry and Biochemistry, University of California, San Diego, La Jolla, California, USA

Jason Micklefield
School of Chemistry and Manchester Interdisciplinary Biocentre, The University of Manchester, Manchester, United Kingdom

Paolo Monciardini
KtedoGen, Via Fantoli, Milano, Italy

Rolf Müller
Department of Pharmaceutical Biotechnology, Saarland University, Saarbrücken, Germany

Justin R. Nodwell
Michael G. DeGroote Centre for Infectious Disease Research, Department of Biochemistry and Biomedical Sciences, McMaster University, Hamilton, Ontario, Canada

Carlos Olano
Departamento de Biología Funcional and Instituto Universitario de Oncología del Principado de Asturias (I.U.O.P.A), Universidad de Oviedo, Oviedo, Spain

Alex P. Praseuth
Department of Pharmacology and Pharmaceutical Sciences, University of Southern California, Los Angeles, California, USA

Mike B. Praseuth
Department of Pharmacology and Pharmaceutical Sciences, University of Southern California, Los Angeles, California, USA

Jacques Ravel
Institute for Genomic Sciences, Department of Microbiology and Immunology, University of Maryland School of Medicine, Baltimore, Maryland, USA

John A. Robinson
Institute of Organic Chemistry, University of Zürich, Zürich, Switzerland

José A. Salas
Departamento de Biología Funcional and Instituto Universitario de Oncología del Principado de Asturias (I.U.O.P.A), Universidad de Oviedo, Oviedo, Spain

Eric W. Schmidt
Department of Medicinal Chemistry, University of Utah, Salt Lake City, Utah, USA

Margherita Sosio
KtedoGen, Via Fantoli, Milano, Italy

Evi Stegmann
Institut für Mikrobiologie, Mikrobiologie/Biotechnologie, Universität Tübingen, Tübingen, Germany

Roderich D. Süssmuth
Institut für Chemie, Technische Universität Berlin, Berlin, Germany

Murat Sunbul
Department of Chemistry, The University of Chicago, Chicago, Illinois, USA

Eriko Takano
Department of Microbial Physiology, GBB, University of Groningen, Haren, The Netherlands

Wilfred A. van der Donk
Departments of Chemistry and Biochemistry and the Howard Hughes Medical Institute, University of Illinois, Urbana, Illinois, USA

Gilles P. van Wezel
Molecular Genetics, Leiden Institute of Chemistry, Gorlaeus Laboratories, The Netherlands

Kenji Watanabe
Research Core for Interdisciplinary Sciences, Okayama University, Okayama, Japan

Barrie Wilkinson
Biotica, Chesterford Research Park, Little Chesterford, Essex, United Kingdom

Wolfgang Wohlleben
Institut für Mikrobiologie, Mikrobiologie/Biotechnologie, Universität Tübingen, Tübingen, Germany

Katharina Woithe
Institute of Organic Chemistry, University of Zürich, Zürich, Switzerland

Jun Yin
Department of Chemistry, The University of Chicago, Chicago, Illinois, USA

Katja Zerbe
Institute of Organic Chemistry, University of Zürich, Zürich, Switzerland

Keya Zhang
Department of Chemistry, The University of Chicago, Chicago, Illinois, USA

Preface

The complex structures of microbial natural products have fascinated chemists for decades. As the tools of chemistry and biochemistry were sharpened, huge advances in understanding natural product biosynthesis were made, but there were still barriers to a satisfactory understanding. Many such impediments were due to the instability of intermediates in the biosynthetic pathways, which hampered chemical analysis. At the same time, a frequent inability to obtain active cell-free preparations severely limited the success of biochemical approaches. A striking example of these limitations is provided by the polyketides, the largest and most important family of secondary metabolites. Chemistry and biochemistry had deduced the relationships between polyketide and fatty acid biosynthesis and had revealed the basic biochemical reactions involved, but there was little understanding of the "programming" of the enzymes, that is control of the variables that make the polyketides such a varied class of chemicals: choice of starter and extender units for carbon chain building, and control of chain length, degree of reduction of keto groups, and chirality of carbon and hydroxyl branches. Isolation of the actinomycete gene clusters that encode the polyketide synthases, their sequencing, and their manipulation into unnatural combinations in the early 1990s changed the landscape almost overnight. There followed a period in which genetics provided a primary stimulus to much of the research in natural product biosynthesis. Now, chemistry, genetics, enzymology, and structural studies are working synergistically to reveal the details of biochemical control.

This two-volume set of *Methods in Enzymology* reflects these developments in the study of natural product biosynthesis. As expressed by Mel Simon in his invitation to edit the set, it is especially timely in view of the increasing need for novel bioactive natural products, especially antibiotics and anticancer drugs, and the new possibilities for addressing this need by carrying out "chemistry through genetics" and by studying the gamut of potential natural products revealed by the sequencing of microbial genomes.

We begin Volume A with the isolation and screening of various kinds of microorganisms, to provide the raw material for subsequent fundamental studies or for the development of natural products as drugs. Then come three chapters dealing with the regulation of secondary metabolite production in actinomycetes – the group of filamentous soil bacteria that are preeminent secondary metabolite producers – and how an understanding of such regulation can furnish compounds that would otherwise be hard to obtain.

Next are chapters covering the cloning and analysis of biosynthetic pathway genes and computer-based methods for predicting the products encoded by gene sets for two key classes of secondary metabolites, the polyketides and nonribosomal peptides, from DNA sequence data, as well as articles describing innovative approaches to probing their biosynthesis. Two final chapters in the first section deal with the biosynthesis of sugars and their attachment to secondary metabolite aglycones, thereby conferring biological activity.

The section on peptide natural products begins with an overview of nonribosomal peptide biosynthesis, followed by a detailed description of methods for studying the biosynthesis of the amino acids and other precursors that function as building blocks in their assembly, as well as a chapter on the heterologous expression of nonribosomal peptide synthetase genes. Next come chapters on a specific class of compounds in this super-family, the cephem beta-lactams, on a special type of iron-chelating siderophore, and on the important glycopeptide and lipopeptide families of antibiotics. Moving to ribosomally synthesised peptide natural products, two chapters cover the lantibiotics, a topic of increasing current focus in the search for antibiotics effective against resistant pathogens. We end Volume B with another example of ribosomally synthesised peptides, this time coupled with techniques for metagenomics mining.

Volume B is dominated by the polyketides, reflecting their pre-eminence as natural products. Kira Weissman introduces polyketide synthesis and the different types of polyketide synthases, and puts the 16 chapters in this section elegantly into context, making redundant any further remarks here, except to note the absence of a chapter on the type III polyketide synthases, an omission stemming from the last-minute withdrawal of the author chosen for this topic. The section on aminocoumarins contains a single chapter that provides a particularly fine example of the application of molecular genetics to another class of compounds, with considerable potential for the generation of "unnatural natural products" by genetic engineering techniques first developed for the polyketides. The volume ends with a section on carbohydrate-type natural products, with two chapters on aminoglycosides and one on the biosynthesis of the TDP-deoxysugars that play such a crucial role in conferring biological activity on a whole range of secondary metabolites, harking back to the chapters on sugar biosynthesis in Volume A.

Inevitably, the choice of topics to include in these volumes is somewhat arbitrary. The peptides and polyketides chose themselves because of their importance amongst natural products, especially as antibiotics, and because of the huge amount of recent research devoted to them. Historically, the aminoglycosides were centre-stage in the early days of antibiotic discovery – streptomycin was the first important actinomycete antibiotic to be described, and only the second, after penicillin, from any source to be a medical marvel – and they probably still make up the third largest chemical family of antibiotics, earning them a place in Volume B. Several other classes of microbial natural products were contenders for inclusion: aminocoumarins, terpenoids, and tetrapyrroles amongst others. However, space constraints precluded inclusion of all of them, and in the end only the aminocoumarins made the cut. Hopefully, other classes will take their place in a further volume in due course, along with a fuller coverage of natural product production by a wider range of microorganisms outside of the actinomycetes.

I am most grateful for the enthusiastic response that greeted my invitations to contribute to this project. Inevitably, leaders in the field have many calls on their time, but it was most gratifying that nearly all my invitees either accepted or offered suggestions for alternative authors. I am especially grateful to Greg Challis, Chaitan Khosla, Tom Simpson, and Chris Walsh for their insightful ideas. To those who accepted – as well as to the many co-authors who were recruited to the writing – thank you for the time and effort that went into the preparation of the chapters and to the friendly way in which you all responded to my – usually minor – editorial suggestions, making my task a very pleasant one.

DAVID A. HOPWOOD

Methods in Enzymology

Volume I. Preparation and Assay of Enzymes
Edited by Sidney P. Colowick and Nathan O. Kaplan

Volume II. Preparation and Assay of Enzymes
Edited by Sidney P. Colowick and Nathan O. Kaplan

Volume III. Preparation and Assay of Substrates
Edited by Sidney P. Colowick and Nathan O. Kaplan

Volume IV. Special Techniques for the Enzymologist
Edited by Sidney P. Colowick and Nathan O. Kaplan

Volume V. Preparation and Assay of Enzymes
Edited by Sidney P. Colowick and Nathan O. Kaplan

Volume VI. Preparation and Assay of Enzymes *(Continued)*
Preparation and Assay of Substrates
Special Techniques
Edited by Sidney P. Colowick and Nathan O. Kaplan

Volume VII. Cumulative Subject Index
Edited by Sidney P. Colowick and Nathan O. Kaplan

Volume VIII. Complex Carbohydrates
Edited by Elizabeth F. Neufeld and Victor Ginsburg

Volume IX. Carbohydrate Metabolism
Edited by Willis A. Wood

Volume X. Oxidation and Phosphorylation
Edited by Ronald W. Estabrook and Maynard E. Pullman

Volume XI. Enzyme Structure
Edited by C. H. W. Hirs

Volume XII. Nucleic Acids (Parts A and B)
Edited by Lawrence Grossman and Kivie Moldave

Volume XIII. Citric Acid Cycle
Edited by J. M. Lowenstein

Volume XIV. Lipids
Edited by J. M. Lowenstein

Volume XV. Steroids and Terpenoids
Edited by Raymond B. Clayton

VOLUME XVI. Fast Reactions
Edited by KENNETH KUSTIN

VOLUME XVII. Metabolism of Amino Acids and Amines (Parts A and B)
Edited by HERBERT TABOR AND CELIA WHITE TABOR

VOLUME XVIII. Vitamins and Coenzymes (Parts A, B, and C)
Edited by DONALD B. MCCORMICK AND LEMUEL D. WRIGHT

VOLUME XIX. Proteolytic Enzymes
Edited by GERTRUDE E. PERLMANN AND LASZLO LORAND

VOLUME XX. Nucleic Acids and Protein Synthesis (Part C)
Edited by KIVIE MOLDAVE AND LAWRENCE GROSSMAN

VOLUME XXI. Nucleic Acids (Part D)
Edited by LAWRENCE GROSSMAN AND KIVIE MOLDAVE

VOLUME XXII. Enzyme Purification and Related Techniques
Edited by WILLIAM B. JAKOBY

VOLUME XXIII. Photosynthesis (Part A)
Edited by ANTHONY SAN PIETRO

VOLUME XXIV. Photosynthesis and Nitrogen Fixation (Part B)
Edited by ANTHONY SAN PIETRO

VOLUME XXV. Enzyme Structure (Part B)
Edited by C. H. W. HIRS AND SERGE N. TIMASHEFF

VOLUME XXVI. Enzyme Structure (Part C)
Edited by C. H. W. HIRS AND SERGE N. TIMASHEFF

VOLUME XXVII. Enzyme Structure (Part D)
Edited by C. H. W. HIRS AND SERGE N. TIMASHEFF

VOLUME XXVIII. Complex Carbohydrates (Part B)
Edited by VICTOR GINSBURG

VOLUME XXIX. Nucleic Acids and Protein Synthesis (Part E)
Edited by LAWRENCE GROSSMAN AND KIVIE MOLDAVE

VOLUME XXX. Nucleic Acids and Protein Synthesis (Part F)
Edited by KIVIE MOLDAVE AND LAWRENCE GROSSMAN

VOLUME XXXI. Biomembranes (Part A)
Edited by SIDNEY FLEISCHER AND LESTER PACKER

VOLUME XXXII. Biomembranes (Part B)
Edited by SIDNEY FLEISCHER AND LESTER PACKER

VOLUME XXXIII. Cumulative Subject Index Volumes I-XXX
Edited by MARTHA G. DENNIS AND EDWARD A. DENNIS

VOLUME XXXIV. Affinity Techniques (Enzyme Purification: Part B)
Edited by WILLIAM B. JAKOBY AND MEIR WILCHEK

Volume XXXV. Lipids (Part B)
Edited by John M. Lowenstein

Volume XXXVI. Hormone Action (Part A: Steroid Hormones)
Edited by Bert W. O'Malley and Joel G. Hardman

Volume XXXVII. Hormone Action (Part B: Peptide Hormones)
Edited by Bert W. O'Malley and Joel G. Hardman

Volume XXXVIII. Hormone Action (Part C: Cyclic Nucleotides)
Edited by Joel G. Hardman and Bert W. O'Malley

Volume XXXIX. Hormone Action (Part D: Isolated Cells, Tissues, and Organ Systems)
Edited by Joel G. Hardman and Bert W. O'Malley

Volume XL. Hormone Action (Part E: Nuclear Structure and Function)
Edited by Bert W. O'Malley and Joel G. Hardman

Volume XLI. Carbohydrate Metabolism (Part B)
Edited by W. A. Wood

Volume XLII. Carbohydrate Metabolism (Part C)
Edited by W. A. Wood

Volume XLIII. Antibiotics
Edited by John H. Hash

Volume XLIV. Immobilized Enzymes
Edited by Klaus Mosbach

Volume XLV. Proteolytic Enzymes (Part B)
Edited by Laszlo Lorand

Volume XLVI. Affinity Labeling
Edited by William B. Jakoby and Meir Wilchek

Volume XLVII. Enzyme Structure (Part E)
Edited by C. H. W. Hirs and Serge N. Timasheff

Volume XLVIII. Enzyme Structure (Part F)
Edited by C. H. W. Hirs and Serge N. Timasheff

Volume XLIX. Enzyme Structure (Part G)
Edited by C. H. W. Hirs and Serge N. Timasheff

Volume L. Complex Carbohydrates (Part C)
Edited by Victor Ginsburg

Volume LI. Purine and Pyrimidine Nucleotide Metabolism
Edited by Patricia A. Hoffee and Mary Ellen Jones

Volume LII. Biomembranes (Part C: Biological Oxidations)
Edited by Sidney Fleischer and Lester Packer

VOLUME LIII. Biomembranes (Part D: Biological Oxidations)
Edited by SIDNEY FLEISCHER AND LESTER PACKER

VOLUME LIV. Biomembranes (Part E: Biological Oxidations)
Edited by SIDNEY FLEISCHER AND LESTER PACKER

VOLUME LV. Biomembranes (Part F: Bioenergetics)
Edited by SIDNEY FLEISCHER AND LESTER PACKER

VOLUME LVI. Biomembranes (Part G: Bioenergetics)
Edited by SIDNEY FLEISCHER AND LESTER PACKER

VOLUME LVII. Bioluminescence and Chemiluminescence
Edited by MARLENE A. DELUCA

VOLUME LVIII. Cell Culture
Edited by WILLIAM B. JAKOBY AND IRA PASTAN

VOLUME LIX. Nucleic Acids and Protein Synthesis (Part G)
Edited by KIVIE MOLDAVE AND LAWRENCE GROSSMAN

VOLUME LX. Nucleic Acids and Protein Synthesis (Part H)
Edited by KIVIE MOLDAVE AND LAWRENCE GROSSMAN

VOLUME 61. Enzyme Structure (Part H)
Edited by C. H. W. HIRS AND SERGE N. TIMASHEFF

VOLUME 62. Vitamins and Coenzymes (Part D)
Edited by DONALD B. MCCORMICK AND LEMUEL D. WRIGHT

VOLUME 63. Enzyme Kinetics and Mechanism (Part A: Initial Rate and Inhibitor Methods)
Edited by DANIEL L. PURICH

VOLUME 64. Enzyme Kinetics and Mechanism
(Part B: Isotopic Probes and Complex Enzyme Systems)
Edited by DANIEL L. PURICH

VOLUME 65. Nucleic Acids (Part I)
Edited by LAWRENCE GROSSMAN AND KIVIE MOLDAVE

VOLUME 66. Vitamins and Coenzymes (Part E)
Edited by DONALD B. MCCORMICK AND LEMUEL D. WRIGHT

VOLUME 67. Vitamins and Coenzymes (Part F)
Edited by DONALD B. MCCORMICK AND LEMUEL D. WRIGHT

VOLUME 68. Recombinant DNA
Edited by RAY WU

VOLUME 69. Photosynthesis and Nitrogen Fixation (Part C)
Edited by ANTHONY SAN PIETRO

VOLUME 70. Immunochemical Techniques (Part A)
Edited by HELEN VAN VUNAKIS AND JOHN J. LANGONE

VOLUME 71. Lipids (Part C)
Edited by JOHN M. LOWENSTEIN

VOLUME 72. Lipids (Part D)
Edited by JOHN M. LOWENSTEIN

VOLUME 73. Immunochemical Techniques (Part B)
Edited by JOHN J. LANGONE AND HELEN VAN VUNAKIS

VOLUME 74. Immunochemical Techniques (Part C)
Edited by JOHN J. LANGONE AND HELEN VAN VUNAKIS

VOLUME 75. Cumulative Subject Index Volumes XXXI, XXXII, XXXIV–LX
Edited by EDWARD A. DENNIS AND MARTHA G. DENNIS

VOLUME 76. Hemoglobins
Edited by ERALDO ANTONINI, LUIGI ROSSI-BERNARDI, AND EMILIA CHIANCONE

VOLUME 77. Detoxication and Drug Metabolism
Edited by WILLIAM B. JAKOBY

VOLUME 78. Interferons (Part A)
Edited by SIDNEY PESTKA

VOLUME 79. Interferons (Part B)
Edited by SIDNEY PESTKA

VOLUME 80. Proteolytic Enzymes (Part C)
Edited by LASZLO LORAND

VOLUME 81. Biomembranes (Part H: Visual Pigments and Purple Membranes, I)
Edited by LESTER PACKER

VOLUME 82. Structural and Contractile Proteins (Part A: Extracellular Matrix)
Edited by LEON W. CUNNINGHAM AND DIXIE W. FREDERIKSEN

VOLUME 83. Complex Carbohydrates (Part D)
Edited by VICTOR GINSBURG

VOLUME 84. Immunochemical Techniques (Part D: Selected Immunoassays)
Edited by JOHN J. LANGONE AND HELEN VAN VUNAKIS

VOLUME 85. Structural and Contractile Proteins (Part B: The Contractile Apparatus and the Cytoskeleton)
Edited by DIXIE W. FREDERIKSEN AND LEON W. CUNNINGHAM

VOLUME 86. Prostaglandins and Arachidonate Metabolites
Edited by WILLIAM E. M. LANDS AND WILLIAM L. SMITH

VOLUME 87. Enzyme Kinetics and Mechanism (Part C: Intermediates, Stereo-chemistry, and Rate Studies)
Edited by DANIEL L. PURICH

VOLUME 88. Biomembranes (Part I: Visual Pigments and Purple Membranes, II)
Edited by LESTER PACKER

VOLUME 89. Carbohydrate Metabolism (Part D)
Edited by WILLIS A. WOOD

VOLUME 90. Carbohydrate Metabolism (Part E)
Edited by WILLIS A. WOOD

VOLUME 91. Enzyme Structure (Part I)
Edited by C. H. W. HIRS AND SERGE N. TIMASHEFF

VOLUME 92. Immunochemical Techniques (Part E: Monoclonal Antibodies and General Immunoassay Methods)
Edited by JOHN J. LANGONE AND HELEN VAN VUNAKIS

VOLUME 93. Immunochemical Techniques (Part F: Conventional Antibodies, Fc Receptors, and Cytotoxicity)
Edited by JOHN J. LANGONE AND HELEN VAN VUNAKIS

VOLUME 94. Polyamines
Edited by HERBERT TABOR AND CELIA WHITE TABOR

VOLUME 95. Cumulative Subject Index Volumes 61–74, 76–80
Edited by EDWARD A. DENNIS AND MARTHA G. DENNIS

VOLUME 96. Biomembranes [Part J: Membrane Biogenesis: Assembly and Targeting (General Methods; Eukaryotes)]
Edited by SIDNEY FLEISCHER AND BECCA FLEISCHER

VOLUME 97. Biomembranes [Part K: Membrane Biogenesis: Assembly and Targeting (Prokaryotes, Mitochondria, and Chloroplasts)]
Edited by SIDNEY FLEISCHER AND BECCA FLEISCHER

VOLUME 98. Biomembranes (Part L: Membrane Biogenesis: Processing and Recycling)
Edited by SIDNEY FLEISCHER AND BECCA FLEISCHER

VOLUME 99. Hormone Action (Part F: Protein Kinases)
Edited by JACKIE D. CORBIN AND JOEL G. HARDMAN

VOLUME 100. Recombinant DNA (Part B)
Edited by RAY WU, LAWRENCE GROSSMAN, AND KIVIE MOLDAVE

VOLUME 101. Recombinant DNA (Part C)
Edited by RAY WU, LAWRENCE GROSSMAN, AND KIVIE MOLDAVE

VOLUME 102. Hormone Action (Part G: Calmodulin and Calcium-Binding Proteins)
Edited by ANTHONY R. MEANS AND BERT W. O'MALLEY

VOLUME 103. Hormone Action (Part H: Neuroendocrine Peptides)
Edited by P. MICHAEL CONN

VOLUME 104. Enzyme Purification and Related Techniques (Part C)
Edited by WILLIAM B. JAKOBY

VOLUME 105. Oxygen Radicals in Biological Systems
Edited by LESTER PACKER

VOLUME 106. Posttranslational Modifications (Part A)
Edited by FINN WOLD AND KIVIE MOLDAVE

VOLUME 107. Posttranslational Modifications (Part B)
Edited by FINN WOLD AND KIVIE MOLDAVE

VOLUME 108. Immunochemical Techniques (Part G: Separation and Characterization of Lymphoid Cells)
Edited by GIOVANNI DI SABATO, JOHN J. LANGONE, AND HELEN VAN VUNAKIS

VOLUME 109. Hormone Action (Part I: Peptide Hormones)
Edited by LUTZ BIRNBAUMER AND BERT W. O'MALLEY

VOLUME 110. Steroids and Isoprenoids (Part A)
Edited by JOHN H. LAW AND HANS C. RILLING

VOLUME 111. Steroids and Isoprenoids (Part B)
Edited by JOHN H. LAW AND HANS C. RILLING

VOLUME 112. Drug and Enzyme Targeting (Part A)
Edited by KENNETH J. WIDDER AND RALPH GREEN

VOLUME 113. Glutamate, Glutamine, Glutathione, and Related Compounds
Edited by ALTON MEISTER

VOLUME 114. Diffraction Methods for Biological Macromolecules (Part A)
Edited by HAROLD W. WYCKOFF, C. H. W. HIRS, AND SERGE N. TIMASHEFF

VOLUME 115. Diffraction Methods for Biological Macromolecules (Part B)
Edited by HAROLD W. WYCKOFF, C. H. W. HIRS, AND SERGE N. TIMASHEFF

VOLUME 116. Immunochemical Techniques (Part H: Effectors and Mediators of Lymphoid Cell Functions)
Edited by GIOVANNI DI SABATO, JOHN J. LANGONE, AND HELEN VAN VUNAKIS

VOLUME 117. Enzyme Structure (Part J)
Edited by C. H. W. HIRS AND SERGE N. TIMASHEFF

VOLUME 118. Plant Molecular Biology
Edited by ARTHUR WEISSBACH AND HERBERT WEISSBACH

VOLUME 119. Interferons (Part C)
Edited by SIDNEY PESTKA

VOLUME 120. Cumulative Subject Index Volumes 81–94, 96–101

VOLUME 121. Immunochemical Techniques (Part I: Hybridoma Technology and Monoclonal Antibodies)
Edited by JOHN J. LANGONE AND HELEN VAN VUNAKIS

VOLUME 122. Vitamins and Coenzymes (Part G)
Edited by FRANK CHYTIL AND DONALD B. MCCORMICK

Volume 123. Vitamins and Coenzymes (Part H)
Edited by Frank Chytil and Donald B. McCormick

Volume 124. Hormone Action (Part J: Neuroendocrine Peptides)
Edited by P. Michael Conn

Volume 125. Biomembranes (Part M: Transport in Bacteria, Mitochondria, and Chloroplasts: General Approaches and Transport Systems)
Edited by Sidney Fleischer and Becca Fleischer

Volume 126. Biomembranes (Part N: Transport in Bacteria, Mitochondria, and Chloroplasts: Protonmotive Force)
Edited by Sidney Fleischer and Becca Fleischer

Volume 127. Biomembranes (Part O: Protons and Water: Structure and Translocation)
Edited by Lester Packer

Volume 128. Plasma Lipoproteins (Part A: Preparation, Structure, and Molecular Biology)
Edited by Jere P. Segrest and John J. Albers

Volume 129. Plasma Lipoproteins (Part B: Characterization, Cell Biology, and Metabolism)
Edited by John J. Albers and Jere P. Segrest

Volume 130. Enzyme Structure (Part K)
Edited by C. H. W. Hirs and Serge N. Timasheff

Volume 131. Enzyme Structure (Part L)
Edited by C. H. W. Hirs and Serge N. Timasheff

Volume 132. Immunochemical Techniques (Part J: Phagocytosis and Cell-Mediated Cytotoxicity)
Edited by Giovanni Di Sabato and Johannes Everse

Volume 133. Bioluminescence and Chemiluminescence (Part B)
Edited by Marlene DeLuca and William D. McElroy

Volume 134. Structural and Contractile Proteins (Part C: The Contractile Apparatus and the Cytoskeleton)
Edited by Richard B. Vallee

Volume 135. Immobilized Enzymes and Cells (Part B)
Edited by Klaus Mosbach

Volume 136. Immobilized Enzymes and Cells (Part C)
Edited by Klaus Mosbach

Volume 137. Immobilized Enzymes and Cells (Part D)
Edited by Klaus Mosbach

Volume 138. Complex Carbohydrates (Part E)
Edited by Victor Ginsburg

VOLUME 139. Cellular Regulators (Part A: Calcium- and Calmodulin-Binding Proteins)
Edited by ANTHONY R. MEANS AND P. MICHAEL CONN

VOLUME 140. Cumulative Subject Index Volumes 102–119, 121–134

VOLUME 141. Cellular Regulators (Part B: Calcium and Lipids)
Edited by P. MICHAEL CONN AND ANTHONY R. MEANS

VOLUME 142. Metabolism of Aromatic Amino Acids and Amines
Edited by SEYMOUR KAUFMAN

VOLUME 143. Sulfur and Sulfur Amino Acids
Edited by WILLIAM B. JAKOBY AND OWEN GRIFFITH

VOLUME 144. Structural and Contractile Proteins (Part D: Extracellular Matrix)
Edited by LEON W. CUNNINGHAM

VOLUME 145. Structural and Contractile Proteins (Part E: Extracellular Matrix)
Edited by LEON W. CUNNINGHAM

VOLUME 146. Peptide Growth Factors (Part A)
Edited by DAVID BARNES AND DAVID A. SIRBASKU

VOLUME 147. Peptide Growth Factors (Part B)
Edited by DAVID BARNES AND DAVID A. SIRBASKU

VOLUME 148. Plant Cell Membranes
Edited by LESTER PACKER AND ROLAND DOUCE

VOLUME 149. Drug and Enzyme Targeting (Part B)
Edited by RALPH GREEN AND KENNETH J. WIDDER

VOLUME 150. Immunochemical Techniques (Part K: *In Vitro* Models of B and T Cell Functions and Lymphoid Cell Receptors)
Edited by GIOVANNI DI SABATO

VOLUME 151. Molecular Genetics of Mammalian Cells
Edited by MICHAEL M. GOTTESMAN

VOLUME 152. Guide to Molecular Cloning Techniques
Edited by SHELBY L. BERGER AND ALAN R. KIMMEL

VOLUME 153. Recombinant DNA (Part D)
Edited by RAY WU AND LAWRENCE GROSSMAN

VOLUME 154. Recombinant DNA (Part E)
Edited by RAY WU AND LAWRENCE GROSSMAN

VOLUME 155. Recombinant DNA (Part F)
Edited by RAY WU

VOLUME 156. Biomembranes (Part P: ATP-Driven Pumps and Related Transport: The Na, K-Pump)
Edited by SIDNEY FLEISCHER AND BECCA FLEISCHER

VOLUME 157. Biomembranes (Part Q: ATP-Driven Pumps and Related Transport: Calcium, Proton, and Potassium Pumps)
Edited by SIDNEY FLEISCHER AND BECCA FLEISCHER

VOLUME 158. Metalloproteins (Part A)
Edited by JAMES F. RIORDAN AND BERT L. VALLEE

VOLUME 159. Initiation and Termination of Cyclic Nucleotide Action
Edited by JACKIE D. CORBIN AND ROGER A. JOHNSON

VOLUME 160. Biomass (Part A: Cellulose and Hemicellulose)
Edited by WILLIS A. WOOD AND SCOTT T. KELLOGG

VOLUME 161. Biomass (Part B: Lignin, Pectin, and Chitin)
Edited by WILLIS A. WOOD AND SCOTT T. KELLOGG

VOLUME 162. Immunochemical Techniques (Part L: Chemotaxis and Inflammation)
Edited by GIOVANNI DI SABATO

VOLUME 163. Immunochemical Techniques (Part M: Chemotaxis and Inflammation)
Edited by GIOVANNI DI SABATO

VOLUME 164. Ribosomes
Edited by HARRY F. NOLLER, JR., AND KIVIE MOLDAVE

VOLUME 165. Microbial Toxins: Tools for Enzymology
Edited by SIDNEY HARSHMAN

VOLUME 166. Branched-Chain Amino Acids
Edited by ROBERT HARRIS AND JOHN R. SOKATCH

VOLUME 167. Cyanobacteria
Edited by LESTER PACKER AND ALEXANDER N. GLAZER

VOLUME 168. Hormone Action (Part K: Neuroendocrine Peptides)
Edited by P. MICHAEL CONN

VOLUME 169. Platelets: Receptors, Adhesion, Secretion (Part A)
Edited by JACEK HAWIGER

VOLUME 170. Nucleosomes
Edited by PAUL M. WASSARMAN AND ROGER D. KORNBERG

VOLUME 171. Biomembranes (Part R: Transport Theory: Cells and Model Membranes)
Edited by SIDNEY FLEISCHER AND BECCA FLEISCHER

VOLUME 172. Biomembranes (Part S: Transport: Membrane Isolation and Characterization)
Edited by SIDNEY FLEISCHER AND BECCA FLEISCHER

VOLUME 173. Biomembranes [Part T: Cellular and Subcellular Transport: Eukaryotic (Nonepithelial) Cells]
Edited by SIDNEY FLEISCHER AND BECCA FLEISCHER

VOLUME 174. Biomembranes [Part U: Cellular and Subcellular Transport: Eukaryotic (Nonepithelial) Cells]
Edited by SIDNEY FLEISCHER AND BECCA FLEISCHER

VOLUME 175. Cumulative Subject Index Volumes 135–139, 141–167

VOLUME 176. Nuclear Magnetic Resonance (Part A: Spectral Techniques and Dynamics)
Edited by NORMAN J. OPPENHEIMER AND THOMAS L. JAMES

VOLUME 177. Nuclear Magnetic Resonance (Part B: Structure and Mechanism)
Edited by NORMAN J. OPPENHEIMER AND THOMAS L. JAMES

VOLUME 178. Antibodies, Antigens, and Molecular Mimicry
Edited by JOHN J. LANGONE

VOLUME 179. Complex Carbohydrates (Part F)
Edited by VICTOR GINSBURG

VOLUME 180. RNA Processing (Part A: General Methods)
Edited by JAMES E. DAHLBERG AND JOHN N. ABELSON

VOLUME 181. RNA Processing (Part B: Specific Methods)
Edited by JAMES E. DAHLBERG AND JOHN N. ABELSON

VOLUME 182. Guide to Protein Purification
Edited by MURRAY P. DEUTSCHER

VOLUME 183. Molecular Evolution: Computer Analysis of Protein and Nucleic Acid Sequences
Edited by RUSSELL F. DOOLITTLE

VOLUME 184. Avidin-Biotin Technology
Edited by MEIR WILCHEK AND EDWARD A. BAYER

VOLUME 185. Gene Expression Technology
Edited by DAVID V. GOEDDEL

VOLUME 186. Oxygen Radicals in Biological Systems (Part B: Oxygen Radicals and Antioxidants)
Edited by LESTER PACKER AND ALEXANDER N. GLAZER

VOLUME 187. Arachidonate Related Lipid Mediators
Edited by ROBERT C. MURPHY AND FRANK A. FITZPATRICK

VOLUME 188. Hydrocarbons and Methylotrophy
Edited by MARY E. LIDSTROM

VOLUME 189. Retinoids (Part A: Molecular and Metabolic Aspects)
Edited by LESTER PACKER

VOLUME 190. Retinoids (Part B: Cell Differentiation and Clinical Applications)
Edited by LESTER PACKER

VOLUME 191. Biomembranes (Part V: Cellular and Subcellular Transport: Epithelial Cells)
Edited by SIDNEY FLEISCHER AND BECCA FLEISCHER

VOLUME 192. Biomembranes (Part W: Cellular and Subcellular Transport: Epithelial Cells)
Edited by SIDNEY FLEISCHER AND BECCA FLEISCHER

VOLUME 193. Mass Spectrometry
Edited by JAMES A. MCCLOSKEY

VOLUME 194. Guide to Yeast Genetics and Molecular Biology
Edited by CHRISTINE GUTHRIE AND GERALD R. FINK

VOLUME 195. Adenylyl Cyclase, G Proteins, and Guanylyl Cyclase
Edited by ROGER A. JOHNSON AND JACKIE D. CORBIN

VOLUME 196. Molecular Motors and the Cytoskeleton
Edited by RICHARD B. VALLEE

VOLUME 197. Phospholipases
Edited by EDWARD A. DENNIS

VOLUME 198. Peptide Growth Factors (Part C)
Edited by DAVID BARNES, J. P. MATHER, AND GORDON H. SATO

VOLUME 199. Cumulative Subject Index Volumes 168–174, 176–194

VOLUME 200. Protein Phosphorylation (Part A: Protein Kinases: Assays, Purification, Antibodies, Functional Analysis, Cloning, and Expression)
Edited by TONY HUNTER AND BARTHOLOMEW M. SEFTON

VOLUME 201. Protein Phosphorylation (Part B: Analysis of Protein Phosphorylation, Protein Kinase Inhibitors, and Protein Phosphatases)
Edited by TONY HUNTER AND BARTHOLOMEW M. SEFTON

VOLUME 202. Molecular Design and Modeling: Concepts and Applications (Part A: Proteins, Peptides, and Enzymes)
Edited by JOHN J. LANGONE

VOLUME 203. Molecular Design and Modeling: Concepts and Applications (Part B: Antibodies and Antigens, Nucleic Acids, Polysaccharides, and Drugs)
Edited by JOHN J. LANGONE

VOLUME 204. Bacterial Genetic Systems
Edited by JEFFREY H. MILLER

VOLUME 205. Metallobiochemistry (Part B: Metallothionein and Related Molecules)
Edited by JAMES F. RIORDAN AND BERT L. VALLEE

VOLUME 206. Cytochrome P450
Edited by MICHAEL R. WATERMAN AND ERIC F. JOHNSON

VOLUME 207. Ion Channels
Edited by BERNARDO RUDY AND LINDA E. IVERSON

VOLUME 208. Protein–DNA Interactions
Edited by ROBERT T. SAUER

VOLUME 209. Phospholipid Biosynthesis
Edited by EDWARD A. DENNIS AND DENNIS E. VANCE

VOLUME 210. Numerical Computer Methods
Edited by LUDWIG BRAND AND MICHAEL L. JOHNSON

VOLUME 211. DNA Structures (Part A: Synthesis and Physical Analysis of DNA)
Edited by DAVID M. J. LILLEY AND JAMES E. DAHLBERG

VOLUME 212. DNA Structures (Part B: Chemical and Electrophoretic Analysis of DNA)
Edited by DAVID M. J. LILLEY AND JAMES E. DAHLBERG

VOLUME 213. Carotenoids (Part A: Chemistry, Separation, Quantitation, and Antioxidation)
Edited by LESTER PACKER

VOLUME 214. Carotenoids (Part B: Metabolism, Genetics, and Biosynthesis)
Edited by LESTER PACKER

VOLUME 215. Platelets: Receptors, Adhesion, Secretion (Part B)
Edited by JACEK J. HAWIGER

VOLUME 216. Recombinant DNA (Part G)
Edited by RAY WU

VOLUME 217. Recombinant DNA (Part H)
Edited by RAY WU

VOLUME 218. Recombinant DNA (Part I)
Edited by RAY WU

VOLUME 219. Reconstitution of Intracellular Transport
Edited by JAMES E. ROTHMAN

VOLUME 220. Membrane Fusion Techniques (Part A)
Edited by NEJAT DÜZGÜNEŞ

VOLUME 221. Membrane Fusion Techniques (Part B)
Edited by NEJAT DÜZGÜNEŞ

VOLUME 222. Proteolytic Enzymes in Coagulation, Fibrinolysis, and Complement Activation (Part A: Mammalian Blood Coagulation Factors and Inhibitors)
Edited by LASZLO LORAND AND KENNETH G. MANN

VOLUME 223. Proteolytic Enzymes in Coagulation, Fibrinolysis, and Complement Activation (Part B: Complement Activation, Fibrinolysis, and Nonmammalian Blood Coagulation Factors)
Edited by LASZLO LORAND AND KENNETH G. MANN

VOLUME 224. Molecular Evolution: Producing the Biochemical Data
Edited by ELIZABETH ANNE ZIMMER, THOMAS J. WHITE, REBECCA L. CANN, AND ALLAN C. WILSON

VOLUME 225. Guide to Techniques in Mouse Development
Edited by PAUL M. WASSARMAN AND MELVIN L. DEPAMPHILIS

VOLUME 226. Metallobiochemistry (Part C: Spectroscopic and Physical Methods for Probing Metal Ion Environments in Metalloenzymes and Metalloproteins)
Edited by JAMES F. RIORDAN AND BERT L. VALLEE

VOLUME 227. Metallobiochemistry (Part D: Physical and Spectroscopic Methods for Probing Metal Ion Environments in Metalloproteins)
Edited by JAMES F. RIORDAN AND BERT L. VALLEE

VOLUME 228. Aqueous Two-Phase Systems
Edited by HARRY WALTER AND GÖTE JOHANSSON

VOLUME 229. Cumulative Subject Index Volumes 195–198, 200–227

VOLUME 230. Guide to Techniques in Glycobiology
Edited by WILLIAM J. LENNARZ AND GERALD W. HART

VOLUME 231. Hemoglobins (Part B: Biochemical and Analytical Methods)
Edited by JOHANNES EVERSE, KIM D. VANDEGRIFF, AND ROBERT M. WINSLOW

VOLUME 232. Hemoglobins (Part C: Biophysical Methods)
Edited by JOHANNES EVERSE, KIM D. VANDEGRIFF, AND ROBERT M. WINSLOW

VOLUME 233. Oxygen Radicals in Biological Systems (Part C)
Edited by LESTER PACKER

VOLUME 234. Oxygen Radicals in Biological Systems (Part D)
Edited by LESTER PACKER

VOLUME 235. Bacterial Pathogenesis (Part A: Identification and Regulation of Virulence Factors)
Edited by VIRGINIA L. CLARK AND PATRIK M. BAVOIL

VOLUME 236. Bacterial Pathogenesis (Part B: Integration of Pathogenic Bacteria with Host Cells)
Edited by VIRGINIA L. CLARK AND PATRIK M. BAVOIL

VOLUME 237. Heterotrimeric G Proteins
Edited by RAVI IYENGAR

VOLUME 238. Heterotrimeric G-Protein Effectors
Edited by RAVI IYENGAR

VOLUME 239. Nuclear Magnetic Resonance (Part C)
Edited by THOMAS L. JAMES AND NORMAN J. OPPENHEIMER

VOLUME 240. Numerical Computer Methods (Part B)
Edited by MICHAEL L. JOHNSON AND LUDWIG BRAND

VOLUME 241. Retroviral Proteases
Edited by LAWRENCE C. KUO AND JULES A. SHAFER

VOLUME 242. Neoglycoconjugates (Part A)
Edited by Y. C. LEE AND REIKO T. LEE

VOLUME 243. Inorganic Microbial Sulfur Metabolism
Edited by HARRY D. PECK, JR., AND JEAN LEGALL

VOLUME 244. Proteolytic Enzymes: Serine and Cysteine Peptidases
Edited by ALAN J. BARRETT

VOLUME 245. Extracellular Matrix Components
Edited by E. RUOSLAHTI AND E. ENGVALL

VOLUME 246. Biochemical Spectroscopy
Edited by KENNETH SAUER

VOLUME 247. Neoglycoconjugates (Part B: Biomedical Applications)
Edited by Y. C. LEE AND REIKO T. LEE

VOLUME 248. Proteolytic Enzymes: Aspartic and Metallo Peptidases
Edited by ALAN J. BARRETT

VOLUME 249. Enzyme Kinetics and Mechanism (Part D: Developments in Enzyme Dynamics)
Edited by DANIEL L. PURICH

VOLUME 250. Lipid Modifications of Proteins
Edited by PATRICK J. CASEY AND JANICE E. BUSS

VOLUME 251. Biothiols (Part A: Monothiols and Dithiols, Protein Thiols, and Thiyl Radicals)
Edited by LESTER PACKER

VOLUME 252. Biothiols (Part B: Glutathione and Thioredoxin; Thiols in Signal Transduction and Gene Regulation)
Edited by LESTER PACKER

VOLUME 253. Adhesion of Microbial Pathogens
Edited by RON J. DOYLE AND ITZHAK OFEK

VOLUME 254. Oncogene Techniques
Edited by PETER K. VOGT AND INDER M. VERMA

VOLUME 255. Small GTPases and Their Regulators (Part A: Ras Family)
Edited by W. E. BALCH, CHANNING J. DER, AND ALAN HALL

VOLUME 256. Small GTPases and Their Regulators (Part B: Rho Family)
Edited by W. E. BALCH, CHANNING J. DER, AND ALAN HALL

VOLUME 257. Small GTPases and Their Regulators (Part C: Proteins Involved in Transport)
Edited by W. E. BALCH, CHANNING J. DER, AND ALAN HALL

VOLUME 258. Redox-Active Amino Acids in Biology
Edited by JUDITH P. KLINMAN

VOLUME 259. Energetics of Biological Macromolecules
Edited by MICHAEL L. JOHNSON AND GARY K. ACKERS

VOLUME 260. Mitochondrial Biogenesis and Genetics (Part A)
Edited by GIUSEPPE M. ATTARDI AND ANNE CHOMYN

VOLUME 261. Nuclear Magnetic Resonance and Nucleic Acids
Edited by THOMAS L. JAMES

VOLUME 262. DNA Replication
Edited by JUDITH L. CAMPBELL

VOLUME 263. Plasma Lipoproteins (Part C: Quantitation)
Edited by WILLIAM A. BRADLEY, SANDRA H. GIANTURCO, AND JERE P. SEGREST

VOLUME 264. Mitochondrial Biogenesis and Genetics (Part B)
Edited by GIUSEPPE M. ATTARDI AND ANNE CHOMYN

VOLUME 265. Cumulative Subject Index Volumes 228, 230–262

VOLUME 266. Computer Methods for Macromolecular Sequence Analysis
Edited by RUSSELL F. DOOLITTLE

VOLUME 267. Combinatorial Chemistry
Edited by JOHN N. ABELSON

VOLUME 268. Nitric Oxide (Part A: Sources and Detection of NO; NO Synthase)
Edited by LESTER PACKER

VOLUME 269. Nitric Oxide (Part B: Physiological and Pathological Processes)
Edited by LESTER PACKER

VOLUME 270. High Resolution Separation and Analysis of Biological Macromolecules (Part A: Fundamentals)
Edited by BARRY L. KARGER AND WILLIAM S. HANCOCK

VOLUME 271. High Resolution Separation and Analysis of Biological Macromolecules (Part B: Applications)
Edited by BARRY L. KARGER AND WILLIAM S. HANCOCK

VOLUME 272. Cytochrome P450 (Part B)
Edited by ERIC F. JOHNSON AND MICHAEL R. WATERMAN

VOLUME 273. RNA Polymerase and Associated Factors (Part A)
Edited by SANKAR ADHYA

VOLUME 274. RNA Polymerase and Associated Factors (Part B)
Edited by SANKAR ADHYA

VOLUME 275. Viral Polymerases and Related Proteins
Edited by LAWRENCE C. KUO, DAVID B. OLSEN, AND STEVEN S. CARROLL

VOLUME 276. Macromolecular Crystallography (Part A)
Edited by CHARLES W. CARTER, JR., AND ROBERT M. SWEET

VOLUME 277. Macromolecular Crystallography (Part B)
Edited by CHARLES W. CARTER, JR., AND ROBERT M. SWEET

VOLUME 278. Fluorescence Spectroscopy
Edited by LUDWIG BRAND AND MICHAEL L. JOHNSON

VOLUME 279. Vitamins and Coenzymes (Part I)
Edited by DONALD B. MCCORMICK, JOHN W. SUTTIE, AND CONRAD WAGNER

VOLUME 280. Vitamins and Coenzymes (Part J)
Edited by DONALD B. MCCORMICK, JOHN W. SUTTIE, AND CONRAD WAGNER

VOLUME 281. Vitamins and Coenzymes (Part K)
Edited by DONALD B. MCCORMICK, JOHN W. SUTTIE, AND CONRAD WAGNER

VOLUME 282. Vitamins and Coenzymes (Part L)
Edited by DONALD B. MCCORMICK, JOHN W. SUTTIE, AND CONRAD WAGNER

VOLUME 283. Cell Cycle Control
Edited by WILLIAM G. DUNPHY

VOLUME 284. Lipases (Part A: Biotechnology)
Edited by BYRON RUBIN AND EDWARD A. DENNIS

VOLUME 285. Cumulative Subject Index Volumes 263, 264, 266–284, 286–289

VOLUME 286. Lipases (Part B: Enzyme Characterization and Utilization)
Edited by BYRON RUBIN AND EDWARD A. DENNIS

VOLUME 287. Chemokines
Edited by RICHARD HORUK

VOLUME 288. Chemokine Receptors
Edited by RICHARD HORUK

VOLUME 289. Solid Phase Peptide Synthesis
Edited by GREGG B. FIELDS

VOLUME 290. Molecular Chaperones
Edited by GEORGE H. LORIMER AND THOMAS BALDWIN

VOLUME 291. Caged Compounds
Edited by GERARD MARRIOTT

VOLUME 292. ABC Transporters: Biochemical, Cellular, and Molecular Aspects
Edited by SURESH V. AMBUDKAR AND MICHAEL M. GOTTESMAN

VOLUME 293. Ion Channels (Part B)
Edited by P. MICHAEL CONN

VOLUME 294. Ion Channels (Part C)
Edited by P. MICHAEL CONN

VOLUME 295. Energetics of Biological Macromolecules (Part B)
Edited by GARY K. ACKERS AND MICHAEL L. JOHNSON

VOLUME 296. Neurotransmitter Transporters
Edited by SUSAN G. AMARA

VOLUME 297. Photosynthesis: Molecular Biology of Energy Capture
Edited by LEE MCINTOSH

VOLUME 298. Molecular Motors and the Cytoskeleton (Part B)
Edited by RICHARD B. VALLEE

VOLUME 299. Oxidants and Antioxidants (Part A)
Edited by LESTER PACKER

VOLUME 300. Oxidants and Antioxidants (Part B)
Edited by LESTER PACKER

VOLUME 301. Nitric Oxide: Biological and Antioxidant Activities (Part C)
Edited by LESTER PACKER

VOLUME 302. Green Fluorescent Protein
Edited by P. MICHAEL CONN

VOLUME 303. cDNA Preparation and Display
Edited by SHERMAN M. WEISSMAN

VOLUME 304. Chromatin
Edited by PAUL M. WASSARMAN AND ALAN P. WOLFFE

VOLUME 305. Bioluminescence and Chemiluminescence (Part C)
Edited by THOMAS O. BALDWIN AND MIRIAM M. ZIEGLER

VOLUME 306. Expression of Recombinant Genes in Eukaryotic Systems
Edited by JOSEPH C. GLORIOSO AND MARTIN C. SCHMIDT

VOLUME 307. Confocal Microscopy
Edited by P. MICHAEL CONN

VOLUME 308. Enzyme Kinetics and Mechanism (Part E: Energetics of Enzyme Catalysis)
Edited by DANIEL L. PURICH AND VERN L. SCHRAMM

VOLUME 309. Amyloid, Prions, and Other Protein Aggregates
Edited by RONALD WETZEL

VOLUME 310. Biofilms
Edited by RON J. DOYLE

VOLUME 311. Sphingolipid Metabolism and Cell Signaling (Part A)
Edited by ALFRED H. MERRILL, JR., AND YUSUF A. HANNUN

VOLUME 312. Sphingolipid Metabolism and Cell Signaling (Part B)
Edited by ALFRED H. MERRILL, JR., AND YUSUF A. HANNUN

VOLUME 313. Antisense Technology (Part A: General Methods, Methods of Delivery, and RNA Studies)
Edited by M. IAN PHILLIPS

VOLUME 314. Antisense Technology (Part B: Applications)
Edited by M. IAN PHILLIPS

VOLUME 315. Vertebrate Phototransduction and the Visual Cycle (Part A)
Edited by KRZYSZTOF PALCZEWSKI

VOLUME 316. Vertebrate Phototransduction and the Visual Cycle (Part B)
Edited by KRZYSZTOF PALCZEWSKI

VOLUME 317. RNA–Ligand Interactions (Part A: Structural Biology Methods)
Edited by DANIEL W. CELANDER AND JOHN N. ABELSON

VOLUME 318. RNA–Ligand Interactions (Part B: Molecular Biology Methods)
Edited by DANIEL W. CELANDER AND JOHN N. ABELSON

VOLUME 319. Singlet Oxygen, UV-A, and Ozone
Edited by LESTER PACKER AND HELMUT SIES

VOLUME 320. Cumulative Subject Index Volumes 290–319

VOLUME 321. Numerical Computer Methods (Part C)
Edited by MICHAEL L. JOHNSON AND LUDWIG BRAND

VOLUME 322. Apoptosis
Edited by JOHN C. REED

VOLUME 323. Energetics of Biological Macromolecules (Part C)
Edited by MICHAEL L. JOHNSON AND GARY K. ACKERS

VOLUME 324. Branched-Chain Amino Acids (Part B)
Edited by ROBERT A. HARRIS AND JOHN R. SOKATCH

VOLUME 325. Regulators and Effectors of Small GTPases (Part D: Rho Family)
Edited by W. E. BALCH, CHANNING J. DER, AND ALAN HALL

VOLUME 326. Applications of Chimeric Genes and Hybrid Proteins (Part A: Gene Expression and Protein Purification)
Edited by JEREMY THORNER, SCOTT D. EMR, AND JOHN N. ABELSON

VOLUME 327. Applications of Chimeric Genes and Hybrid Proteins (Part B: Cell Biology and Physiology)
Edited by JEREMY THORNER, SCOTT D. EMR, AND JOHN N. ABELSON

VOLUME 328. Applications of Chimeric Genes and Hybrid Proteins (Part C: Protein–Protein Interactions and Genomics)
Edited by JEREMY THORNER, SCOTT D. EMR, AND JOHN N. ABELSON

VOLUME 329. Regulators and Effectors of Small GTPases (Part E: GTPases Involved in Vesicular Traffic)
Edited by W. E. BALCH, CHANNING J. DER, AND ALAN HALL

VOLUME 330. Hyperthermophilic Enzymes (Part A)
Edited by MICHAEL W. W. ADAMS AND ROBERT M. KELLY

VOLUME 331. Hyperthermophilic Enzymes (Part B)
Edited by MICHAEL W. W. ADAMS AND ROBERT M. KELLY

VOLUME 332. Regulators and Effectors of Small GTPases (Part F: Ras Family I)
Edited by W. E. BALCH, CHANNING J. DER, AND ALAN HALL

VOLUME 333. Regulators and Effectors of Small GTPases (Part G: Ras Family II)
Edited by W. E. BALCH, CHANNING J. DER, AND ALAN HALL

VOLUME 334. Hyperthermophilic Enzymes (Part C)
Edited by MICHAEL W. W. ADAMS AND ROBERT M. KELLY

VOLUME 335. Flavonoids and Other Polyphenols
Edited by LESTER PACKER

VOLUME 336. Microbial Growth in Biofilms (Part A: Developmental and Molecular Biological Aspects)
Edited by RON J. DOYLE

VOLUME 337. Microbial Growth in Biofilms (Part B: Special Environments and Physicochemical Aspects)
Edited by RON J. DOYLE

VOLUME 338. Nuclear Magnetic Resonance of Biological Macromolecules (Part A)
Edited by THOMAS L. JAMES, VOLKER DÖTSCH, AND ULI SCHMITZ

VOLUME 339. Nuclear Magnetic Resonance of Biological Macromolecules (Part B)
Edited by THOMAS L. JAMES, VOLKER DÖTSCH, AND ULI SCHMITZ

VOLUME 340. Drug–Nucleic Acid Interactions
Edited by JONATHAN B. CHAIRES AND MICHAEL J. WARING

VOLUME 341. Ribonucleases (Part A)
Edited by ALLEN W. NICHOLSON

VOLUME 342. Ribonucleases (Part B)
Edited by ALLEN W. NICHOLSON

VOLUME 343. G Protein Pathways (Part A: Receptors)
Edited by RAVI IYENGAR AND JOHN D. HILDEBRANDT

VOLUME 344. G Protein Pathways (Part B: G Proteins and Their Regulators)
Edited by RAVI IYENGAR AND JOHN D. HILDEBRANDT

VOLUME 345. G Protein Pathways (Part C: Effector Mechanisms)
Edited by RAVI IYENGAR AND JOHN D. HILDEBRANDT

VOLUME 346. Gene Therapy Methods
Edited by M. IAN PHILLIPS

VOLUME 347. Protein Sensors and Reactive Oxygen Species (Part A: Selenoproteins and Thioredoxin)
Edited by HELMUT SIES AND LESTER PACKER

VOLUME 348. Protein Sensors and Reactive Oxygen Species (Part B: Thiol Enzymes and Proteins)
Edited by HELMUT SIES AND LESTER PACKER

VOLUME 349. Superoxide Dismutase
Edited by LESTER PACKER

VOLUME 350. Guide to Yeast Genetics and Molecular and Cell Biology (Part B)
Edited by CHRISTINE GUTHRIE AND GERALD R. FINK

VOLUME 351. Guide to Yeast Genetics and Molecular and Cell Biology (Part C)
Edited by CHRISTINE GUTHRIE AND GERALD R. FINK

VOLUME 352. Redox Cell Biology and Genetics (Part A)
Edited by CHANDAN K. SEN AND LESTER PACKER

VOLUME 353. Redox Cell Biology and Genetics (Part B)
Edited by CHANDAN K. SEN AND LESTER PACKER

VOLUME 354. Enzyme Kinetics and Mechanisms (Part F: Detection and Characterization of Enzyme Reaction Intermediates)
Edited by DANIEL L. PURICH

VOLUME 355. Cumulative Subject Index Volumes 321–354

VOLUME 356. Laser Capture Microscopy and Microdissection
Edited by P. MICHAEL CONN

VOLUME 357. Cytochrome P450, Part C
Edited by ERIC F. JOHNSON AND MICHAEL R. WATERMAN

VOLUME 358. Bacterial Pathogenesis (Part C: Identification, Regulation, and Function of Virulence Factors)
Edited by VIRGINIA L. CLARK AND PATRIK M. BAVOIL

VOLUME 359. Nitric Oxide (Part D)
Edited by ENRIQUE CADENAS AND LESTER PACKER

VOLUME 360. Biophotonics (Part A)
Edited by GERARD MARRIOTT AND IAN PARKER

VOLUME 361. Biophotonics (Part B)
Edited by GERARD MARRIOTT AND IAN PARKER

VOLUME 362. Recognition of Carbohydrates in Biological Systems (Part A)
Edited by YUAN C. LEE AND REIKO T. LEE

VOLUME 363. Recognition of Carbohydrates in Biological Systems (Part B)
Edited by YUAN C. LEE AND REIKO T. LEE

VOLUME 364. Nuclear Receptors
Edited by DAVID W. RUSSELL AND DAVID J. MANGELSDORF

VOLUME 365. Differentiation of Embryonic Stem Cells
Edited by PAUL M. WASSAUMAN AND GORDON M. KELLER

VOLUME 366. Protein Phosphatases
Edited by SUSANNE KLUMPP AND JOSEF KRIEGLSTEIN

VOLUME 367. Liposomes (Part A)
Edited by NEJAT DÜZGÜNEŞ

VOLUME 368. Macromolecular Crystallography (Part C)
Edited by CHARLES W. CARTER, JR., AND ROBERT M. SWEET

VOLUME 369. Combinational Chemistry (Part B)
Edited by GUILLERMO A. MORALES AND BARRY A. BUNIN

VOLUME 370. RNA Polymerases and Associated Factors (Part C)
Edited by SANKAR L. ADHYA AND SUSAN GARGES

VOLUME 371. RNA Polymerases and Associated Factors (Part D)
Edited by SANKAR L. ADHYA AND SUSAN GARGES

VOLUME 372. Liposomes (Part B)
Edited by NEJAT DÜZGÜNEŞ

VOLUME 373. Liposomes (Part C)
Edited by NEJAT DÜZGÜNEŞ

VOLUME 374. Macromolecular Crystallography (Part D)
Edited by CHARLES W. CARTER, JR., AND ROBERT W. SWEET

VOLUME 375. Chromatin and Chromatin Remodeling Enzymes (Part A)
Edited by C. DAVID ALLIS AND CARL WU

VOLUME 376. Chromatin and Chromatin Remodeling Enzymes (Part B)
Edited by C. DAVID ALLIS AND CARL WU

VOLUME 377. Chromatin and Chromatin Remodeling Enzymes (Part C)
Edited by C. DAVID ALLIS AND CARL WU

VOLUME 378. Quinones and Quinone Enzymes (Part A)
Edited by HELMUT SIES AND LESTER PACKER

VOLUME 379. Energetics of Biological Macromolecules (Part D)
Edited by JO M. HOLT, MICHAEL L. JOHNSON, AND GARY K. ACKERS

VOLUME 380. Energetics of Biological Macromolecules (Part E)
Edited by JO M. HOLT, MICHAEL L. JOHNSON, AND GARY K. ACKERS

VOLUME 381. Oxygen Sensing
Edited by CHANDAN K. SEN AND GREGG L. SEMENZA

VOLUME 382. Quinones and Quinone Enzymes (Part B)
Edited by HELMUT SIES AND LESTER PACKER

VOLUME 383. Numerical Computer Methods (Part D)
Edited by LUDWIG BRAND AND MICHAEL L. JOHNSON

VOLUME 384. Numerical Computer Methods (Part E)
Edited by LUDWIG BRAND AND MICHAEL L. JOHNSON

VOLUME 385. Imaging in Biological Research (Part A)
Edited by P. MICHAEL CONN

VOLUME 386. Imaging in Biological Research (Part B)
Edited by P. MICHAEL CONN

VOLUME 387. Liposomes (Part D)
Edited by NEJAT DÜZGÜNEŞ

VOLUME 388. Protein Engineering
Edited by DAN E. ROBERTSON AND JOSEPH P. NOEL

VOLUME 389. Regulators of G-Protein Signaling (Part A)
Edited by DAVID P. SIDEROVSKI

VOLUME 390. Regulators of G-Protein Signaling (Part B)
Edited by DAVID P. SIDEROVSKI

VOLUME 391. Liposomes (Part E)
Edited by NEJAT DÜZGÜNEŞ

VOLUME 392. RNA Interference
Edited by ENGELKE ROSSI

VOLUME 393. Circadian Rhythms
Edited by MICHAEL W. YOUNG

VOLUME 394. Nuclear Magnetic Resonance of Biological Macromolecules (Part C)
Edited by THOMAS L. JAMES

VOLUME 395. Producing the Biochemical Data (Part B)
Edited by ELIZABETH A. ZIMMER AND ERIC H. ROALSON

VOLUME 396. Nitric Oxide (Part E)
Edited by LESTER PACKER AND ENRIQUE CADENAS

VOLUME 397. Environmental Microbiology
Edited by JARED R. LEADBETTER

VOLUME 398. Ubiquitin and Protein Degradation (Part A)
Edited by RAYMOND J. DESHAIES

VOLUME 399. Ubiquitin and Protein Degradation (Part B)
Edited by RAYMOND J. DESHAIES

VOLUME 400. Phase II Conjugation Enzymes and Transport Systems
Edited by HELMUT SIES AND LESTER PACKER

VOLUME 401. Glutathione Transferases and Gamma Glutamyl Transpeptidases
Edited by HELMUT SIES AND LESTER PACKER

VOLUME 402. Biological Mass Spectrometry
Edited by A. L. BURLINGAME

VOLUME 403. GTPases Regulating Membrane Targeting and Fusion
Edited by WILLIAM E. BALCH, CHANNING J. DER, AND ALAN HALL

VOLUME 404. GTPases Regulating Membrane Dynamics
Edited by WILLIAM E. BALCH, CHANNING J. DER, AND ALAN HALL

VOLUME 405. Mass Spectrometry: Modified Proteins and Glycoconjugates
Edited by A. L. BURLINGAME

VOLUME 406. Regulators and Effectors of Small GTPases: Rho Family
Edited by WILLIAM E. BALCH, CHANNING J. DER, AND ALAN HALL

VOLUME 407. Regulators and Effectors of Small GTPases: Ras Family
Edited by WILLIAM E. BALCH, CHANNING J. DER, AND ALAN HALL

VOLUME 408. DNA Repair (Part A)
Edited by JUDITH L. CAMPBELL AND PAUL MODRICH

VOLUME 409. DNA Repair (Part B)
Edited by JUDITH L. CAMPBELL AND PAUL MODRICH

VOLUME 410. DNA Microarrays (Part A: Array Platforms and Web-Bench Protocols)
Edited by ALAN KIMMEL AND BRIAN OLIVER

VOLUME 411. DNA Microarrays (Part B: Databases and Statistics)
Edited by ALAN KIMMEL AND BRIAN OLIVER

VOLUME 412. Amyloid, Prions, and Other Protein Aggregates (Part B)
Edited by INDU KHETERPAL AND RONALD WETZEL

VOLUME 413. Amyloid, Prions, and Other Protein Aggregates (Part C)
Edited by INDU KHETERPAL AND RONALD WETZEL

VOLUME 414. Measuring Biological Responses with Automated Microscopy
Edited by JAMES INGLESE

VOLUME 415. Glycobiology
Edited by MINORU FUKUDA

VOLUME 416. Glycomics
Edited by MINORU FUKUDA

VOLUME 417. Functional Glycomics
Edited by MINORU FUKUDA

VOLUME 418. Embryonic Stem Cells
Edited by IRINA KLIMANSKAYA AND ROBERT LANZA

VOLUME 419. Adult Stem Cells
Edited by IRINA KLIMANSKAYA AND ROBERT LANZA

VOLUME 420. Stem Cell Tools and Other Experimental Protocols
Edited by IRINA KLIMANSKAYA AND ROBERT LANZA

VOLUME 421. Advanced Bacterial Genetics: Use of Transposons and Phage for Genomic Engineering
Edited by KELLY T. HUGHES

VOLUME 422. Two-Component Signaling Systems, Part A
Edited by MELVIN I. SIMON, BRIAN R. CRANE, AND ALEXANDRINE CRANE

VOLUME 423. Two-Component Signaling Systems, Part B
Edited by MELVIN I. SIMON, BRIAN R. CRANE, AND ALEXANDRINE CRANE

VOLUME 424. RNA Editing
Edited by JONATHA M. GOTT

VOLUME 425. RNA Modification
Edited by JONATHA M. GOTT

VOLUME 426. Integrins
Edited by DAVID CHERESH

VOLUME 427. MicroRNA Methods
Edited by JOHN J. ROSSI

VOLUME 428. Osmosensing and Osmosignaling
Edited by HELMUT SIES AND DIETER HAUSSINGER

VOLUME 429. Translation Initiation: Extract Systems and Molecular Genetics
Edited by JON LORSCH

VOLUME 430. Translation Initiation: Reconstituted Systems and Biophysical Methods
Edited by JON LORSCH

VOLUME 431. Translation Initiation: Cell Biology, High-Throughput and Chemical-Based Approaches
Edited by JON LORSCH

VOLUME 432. Lipidomics and Bioactive Lipids: Mass-Spectrometry–Based Lipid Analysis
Edited by H. ALEX BROWN

VOLUME 433. Lipidomics and Bioactive Lipids: Specialized Analytical Methods and Lipids in Disease
Edited by H. ALEX BROWN

VOLUME 434. Lipidomics and Bioactive Lipids: Lipids and Cell Signaling
Edited by H. ALEX BROWN

VOLUME 435. Oxygen Biology and Hypoxia
Edited by HELMUT SIES AND BERNHARD BRÜNE

VOLUME 436. Globins and Other Nitric Oxide-Reactive Protiens (Part A)
Edited by ROBERT K. POOLE

VOLUME 437. Globins and Other Nitric Oxide-Reactive Protiens (Part B)
Edited by ROBERT K. POOLE

VOLUME 438. Small GTPases in Disease (Part A)
Edited by WILLIAM E. BALCH, CHANNING J. DER, AND ALAN HALL

VOLUME 439. Small GTPases in Disease (Part B)
Edited by WILLIAM E. BALCH, CHANNING J. DER, AND ALAN HALL

VOLUME 440. Nitric Oxide, Part F Oxidative and Nitrosative Stress in Redox Regulation of Cell Signaling
Edited by ENRIQUE CADENAS AND LESTER PACKER

VOLUME 441. Nitric Oxide, Part G Oxidative and Nitrosative Stress in Redox Regulation of Cell Signaling
Edited by ENRIQUE CADENAS AND LESTER PACKER

VOLUME 442. Programmed Cell Death, General Principles for Studying Cell Death (Part A)
Edited by ROYA KHOSRAVI-FAR, ZAHRA ZAKERI, RICHARD A. LOCKSHIN, AND MAURO PIACENTINI

VOLUME 443. Angiogenesis: *In Vitro* Systems
Edited by DAVID A. CHERESH

VOLUME 444. Angiogenesis: *In Vivo* Systems (Part A)
Edited by DAVID A. CHERESH

VOLUME 445. Angiogenesis: *In Vivo* Systems (Part B)
Edited by DAVID A. CHERESH

VOLUME 446. Programmed Cell Death, The Biology and Therapeutic Implications of Cell Death (Part B)
Edited by ROYA KHOSRAVI-FAR, ZAHRA ZAKERI, RICHARD A. LOCKSHIN, AND MAURO PIACENTINI

VOLUME 447. RNA Turnover in Bacteria, Archaea and Organelles
Edited by LYNNE E. MAQUAT AND CECILIA M. ARRAIANO

VOLUME 448. RNA Turnover in Eukaryotes: Nucleases, Pathways and Analysis of mRNA Decay
Edited by LYNNE E. MAQUAT AND MEGERDITCH KILEDJIAN

VOLUME 449. RNA Turnover in Eukaryotes: Analysis of Specialized and Quality Control RNA Decay Pathways
Edited by LYNNE E. MAQUAT AND MEGERDITCH KILEDJIAN

VOLUME 450. Fluorescence Spectroscopy
Edited by LUDWIG BRAND AND MICHAEL L. JOHNSON

VOLUME 451. Autophagy: Lower Eukaryotes and Non-Mammalian Systems (Part A)
Edited by DANIEL J. KLIONSKY

VOLUME 452. Autophagy in Mammalian Systems (Part B)
Edited by DANIEL J. KLIONSKY

VOLUME 453. Autophagy in Disease and Clinical Applications (Part C)
Edited by DANIEL J. KLIONSKY

VOLUME 454. Computer Methods (Part A)
Edited by MICHAEL L. JOHNSON AND LUDWIG BRAND

VOLUME 455. Biothermodynamics (Part A)
Edited by MICHAEL L. JOHNSON, JO M. HOLT, AND GARY K. ACKERS

VOLUME 456. Mitochondrial Function, Part A: Mitochondrial Electron Transport Complexes and Reactive Oxygen Species
Edited by WILLIAM S. ALLISON AND IMMO E. SCHEFFLER

VOLUME 457. Mitochondrial Function, Part B: Mitochondrial Protein Kinases, Protein Phosphatases and Mitochondrial Diseases
Edited by WILLIAM S. ALLISON AND ANNE N. MURPHY

VOLUME 458. Complex Enzymes in Microbial Natural Product Biosynthesis, Part A: Overview Articles and Peptides
Edited by DAVID A. HOPWOOD

SECTION ONE

OVERVIEW ARTICLES

CHAPTER ONE

Approaches to Discovering Novel Antibacterial and Antifungal Agents

Stefano Donadio,[*,†] Paolo Monciardini,[*] and Margherita Sosio[*]

Contents

1. Introduction	4
1.1. Objectives of a screening program	4
1.2. Screening strategy and novelty of the program	5
2. Strains and Samples	7
2.1. Strain isolation, purification, and storage	8
2.2. Example: Isolation of acidophilic actinomycetes from soil	9
2.3. Example: Purification and storage of actinomycetes	10
2.4. Sample preparation	11
3. Targets and Assays for Antibacterial and Antifungal Programs	13
3.1. Target identification and validation	13
3.2. Assay development	14
3.3. Secondary assays	15
4. Screening	16
4.1. Example: Screening for bacterial cell-wall inhibitors	16
5. Hit Follow-up	19
5.1. Activity confirmation	20
5.2. Novelty evaluation	20
5.3. Profiling	22
6. Databases, Operations, and Costs	23
7. Perspectives	25
Acknowledgment	25
References	25

Abstract

The need for novel antibiotics to fight multidrug-resistant pathogens calls for a return to natural product screening, but novel approaches must be implemented to increase the chances of discovering novel compounds. This chapter illustrates strategic considerations and the required ingredients for screening programs:

[*] KtedoGen, Via Fantoli, Milano, Italy
[†] NAICONS, Via Fantoli, Milano, Italy

microbial diversity, samples for screening, targets and assays, assay development and implementation, hit identification and follow-up. When appropriate, we highlight the impact that chemical diversity consisting of mixtures of different compounds, amid a large background of known antibiotics, has on the screening process. Examples of detailed procedures are described for strain isolation and preservation, sample preparation, primary and secondary assays, and extract fractionation. While these limited examples are not sufficient to organize a complete screening program, they provide a basis for understanding the details of microbial product screening in the anti-infective field.

1. Introduction

There is a need for novel and more effective antibiotics to combat multidrug resistant pathogens. Due to aging, immunosuppression and invasive surgical procedures, an ever-larger population is at risk of contracting severe infections, while bacterial pathogens are becoming increasingly resistant to currently available antibiotics. As a consequence, infections caused by drug-resistant bacteria are associated with increased morbidity, mortality, and health care costs.

It is now widely accepted that, despite the plethora of novel targets provided by the sequenced genomes of bacterial pathogens, combinatorial chemistry and high throughput screening (HTS) have failed to provide novel drug candidates in the anti-infective field, resulting in a virtually empty pipeline of compounds under development (Donadio et al., 2005a; Payne et al., 2007). This has prompted many players in the field to advocate a return to screening natural products, which constitute most of the clinically used antibiotics and the bulk of compounds under development (Butler, 2008).

During the golden era of antibiotic discovery in the 1950s and 1960s, the then-untapped diversity of soil actinomycetes provided most of the antibiotics known today. Over several decades, tens of millions of soil microorganisms were screened for anti-infective activity (Baltz, 2005). Thus, most low-hanging fruits have probably been picked and the discovery of a new antibiotic is nowadays a rare event. Furthermore, newly discovered compounds must possess advantages over the many antibiotics in clinical use, which implies the early recognition of potentially valuable compounds.

1.1. Objectives of a screening program

The objective of a screening program for anti-infectives is the discovery of a new, patentable chemical entity possessing desired properties, such as antimicrobial spectrum, molecular weight range, solubility, and preferred route of administration. Obviously, the higher the bar is set in terms of desired properties, the lower the probability of identifying the desired compound(s). Thus, a compromise is reached by defining the minimal properties a compound should have to be pursued and characterized. To reduce costs, it is also important that these properties can be established using a small amount of compound. For some classes of antibiotics (e.g., β-lactams, macrolides, glycopeptides, aminoglycosides, and polyenes) many variants are known, either as the product of different strains or because they have undergone extensive chemical analoguing. There is thus a small probability of discovering improved variants through screening in these classes. Consequently, screening programs usually aim at discovering either a novel class or an improved variant of a poorly explored class.

Operationally, screening microbial products requires: (1) a library of microbial strains that produce a diverse and biologically relevant set of compounds; (2) appropriate tests to detect compounds of potential value; (3) samples derived from the strains; (4) instrumentation and data capture tools adequate to the size of the screening program; (5) identification of the molecules responsible for the activity and recognition of known compounds; (6) additional tests to profile the molecules; and (7) ability to supply increasing amounts of active compounds. These ingredients, schematized in Fig. 1.1, are strongly intertwined as, for example, the tests must take into account the known compounds produced by the screened strains. Screening also uses its own jargon and relevant terminology is summarized in Table 1.1.

1.2. Screening strategy and novelty of the program

Once the objective is set, the screening strategy must consider chemical diversity (how many different strains are available? how frequently do they produce bioactive compounds? do these belong to different chemical classes?), the test system (selectivity and sensitivity of the assay, hit rate), and the known natural products detected by the assay. This knowledge helps delineate the screening program in terms of size, screening algorithm, and hit dereplication (Table 1.1).

A key factor for success is to introduce elements of novelty with respect to previous screening campaigns. Although many details are not known, a reasonable assumption is that easy-to-isolate strains (e.g., streptomycetes and other easily retrieved actinomycetes) were screened for antibacterial and antifungal agents by simple growth inhibition tests, prioritizing the many positives by potency, selective effect on a macromolecular synthesis and/or lack of toxicity. Consequently, most discovered compounds are produced by relatively abundant species within the sampled genera, act on canonical targets (e.g., cell-wall synthesis, transcription, and translation), and are rather potent and not obviously toxic (or coproduced with toxic compounds).

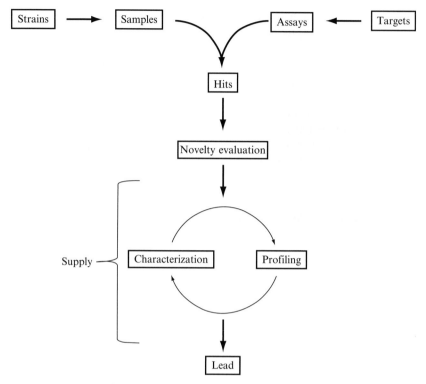

Figure 1.1 Elements of a screening process.

Table 1.1 Common terms used in microbial product screening

Term	Meaning
Primary assay	A test used to screen all samples in the library
Secondary assay	One or more tests used only on the positives
Screening algorithm	The sequence of primary and secondary assays with the corresponding threshold values for a sample being positive or negative
Positive	A sample giving a signal above the established threshold in a primary assay
Hit	A sample passing the selectivity criteria of the screening algorithm
Dereplication	The procedure of selecting one representative only among identical strains or among strains producing an identical compound
Extract	A sample used in screening obtained by processing a fermentation broth or a cell culture

The novelty of a program can thus result from a relatively unexplored source of microbial diversity, from a sensitive assay that detects a subset of growth inhibition events, and from a screening algorithm (Table 1.1) that effectively filters out most known compounds. These factors greatly contribute to the probability that a screening campaign will identify novel compounds at a reasonable cost.

The following sections describe the elements of a screening process, providing also selected experimental details. They are mostly based on the authors' experience at previous companies (Biosearch Italia and Vicuron Pharmaceuticals), with modifications implemented at their current affiliations.

2. STRAINS AND SAMPLES

Chemical diversity is obtained by the simplified steps illustrated in Fig. 1.2. Several methods allow culturing of microorganisms from environmental samples, which are then purified as single colonies and grown in

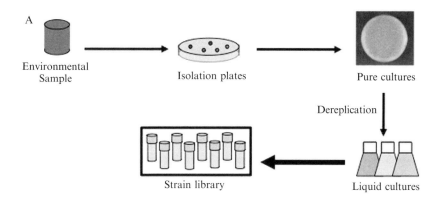

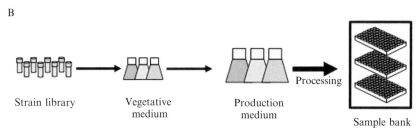

Figure 1.2 Strains and samples. (A) activities needed to generate a strain library. (B) activities for generation of a sample bank from the strain library.

liquid media for inclusion in the strain library (Fig. 1.2A). A sample bank can be obtained by culturing the strains in appropriate media, followed by sample processing (Fig. 1.2B). Although samples can be immediately screened, it is preferable to store them as separate aliquots for comparing results obtained with different tests.

The key element in generating a sample bank is diversification, which results from the diversity of the strains and their growth under appropriate conditions. This ideally requires the *ad hoc* development of methods for strain cultivation. However, since newly isolated strains are, by definition, uncharacterized, new fermentation or processing methods are usually developed through a pilot study on a limited number of strains, and then systematically applied to the entire library.

The results from the screening establish the true value of the samples, in terms of diversity of isolated strains, extent of replicas and appropriate growth conditions. However, the centralization of HTS programs has resulted in the need for large sample banks, which are then rapidly screened on different targets. This has substantially increased the time interval between strain isolation and the corresponding screening results, so that a substantial investment will have been made in strain library and sample bank before their actual value is established.

2.1. Strain isolation, purification, and storage

Antibiotics are produced by a variety of microorganisms, with fungi and actinomycetes representing the most intensively screened groups. While strain diversity is critical for identifying diverse molecules (Baltz, 2005; Bull *et al.*, 2000), the ability to produce bioactive metabolites is not uniformly distributed in the microbial world (Bull *et al.*, 2000; Donadio *et al.*, 2007), microbial products in randomly screened strains are observed at frequencies that can be several orders of magnitude apart (Baltz, 2005), and the ease of strain isolation and cultivation varies greatly for different groups. It is therefore important to gather a library of strains that produce anti-infective compounds at an appreciable frequency, produce different classes of compounds, and can be reasonably cultivated with existing methodologies and equipment. Strain diversity and novelty can derive from underexplored sources (e.g., Jensen *et al.*, 2005; Lam, 2007), from novel isolation procedures (e.g., Busti *et al.*, 2006; Davis *et al.*, 2005), or simply by the painstaking recognition of the few uncommon strains present on the isolation plates. In any case, the generation of a diversified strain library must rely on different sources, on a variety of isolation methods and, in the end, on the recognition of truly novel strains (first step in Fig. 1.2A).

Only selected strains present on the isolation plates are retrieved in pure culture (second step in Fig. 1.2A), and a skilled microbiologist can rapidly recognize filamentous strains by morphological features such as shape and arrangement of hyphae, spores, or sporangia. In our experience, taxon assignment by 16S rRNA gene sequencing can effectively train the microbiologist's eye in recognizing relevant morphological features for previously uncultured actinomycete taxa (Donadio et al., 2005b).

Along with purification, strains are dereplicated, a process that aims at storing a single representative of sibling strains. Dereplication occurs before the liquid culture phase in Fig. 1.2A, and can be accomplished by evaluation of morphological characters, an inexpensive and effective tool limited, however, to strains isolated and analyzed within a relatively short time period. Quantitative tools for dereplication include fingerprinting of DNA (Busch and Nitschko, 1999; Bull et al., 2000) or of other cellular components (Bull et al., 2000; Larsen et al., 2005). Dereplication aims at discarding those strains that have a *low probability* of producing additional novel metabolites. However, this must be done judiciously considering, for example, the sample from where the strain was isolated or the coverage of the corresponding taxon. Furthermore, dereplication must be cost-effective in comparison to downstream screening operations, since true dereplication occurs only after hit identification (see Fig. 1.1).

2.2. Example: Isolation of acidophilic actinomycetes from soil

Published protocols are available (e.g., Hunter-Cevera and Belt, 1999; Otoguro et al., 2001) which enhance the percentage of desired taxa on isolation plates, but no method is absolute. Furthermore, most taxa are likely to consist of both abundant and rare species, hence the need for effective dereplication. An example is reported of an effective method for isolating novel *Actinomycetales* taxa from soil (Busti et al., 2006).

2.2.1. Procedure

1. Sieve collected soil sample through a 1–2 mm mesh and air-dry for 1 week at room temperature.
2. Prepare soil extract by autoclaving 100 g fresh soil in 500 ml distilled water, followed by filtration through sterile gauze.
3. Prepare agar plates by mixing 500 ml soil extract, 10 g gellan gum, 333 mg $CaCl_2$ in 1 l distilled H_2O, adjusting to pH 5.5 with HCl. After autoclaving, add 1 ml vitamin solution,[1] and 5 ml of 10 mg/ml cycloheximide in 50% ethanol, and dispense 25–30 ml per 90-mm petri dish. Dry plates for about 30 min in a sterile hood.

---

[1] A filter-sterilized solution contains per liter 25 mg thiamin·HCl, 250 mg calcium panthotenate, 250 mg nicotinic acid, 500 mg biotin, 1.25 g riboflavin, 6 mg vitamin B_{12}, 25 mg p-aminobenzoic acid, 500 mg folic acid, 500 mg pyridoxal·HCl.

4. Suspend 0.5 g of air-dried soil in 5 ml sterile water, vortex briefly, and prepare serial 1:10 dilutions in water.
5. Use a sterile rod to spread 0.1 ml aliquots from the serial dilutions on the plates, dry them briefly and incubate at 28–30 °C.

Different actinomycetes will appear at different times, with *Streptomyces* strains requiring ca. 7 days, while members of other taxa may require up to several weeks. Useful isolation plates for strain purification contain a few 100 colonies, are devoid of fungal contaminants and represent a significant number of desired strains. The appropriate dilution of the soil suspension strongly depends on the soil type and the isolation method, and is empirically determined.

2.3. Example: Purification and storage of actinomycetes

Desired strains are transferred to new plates, usually containing a rich medium, with possible restreaking to purify away contaminants. In contrast to the many different isolation methods, strain purification and storage rely on few, established procedures that work with most actinomycetes.

2.3.1. Procedure

1. Using a 40× Long Working Distance objective, identify interesting strains on the isolation plates. Mark chosen colonies on the back of the plate.
2. Pick chosen colonies with a sterile needle under a dissection microscope.
3. Deposit colony into a 100-μl H_2O drop placed on a sterile surface, such as the inside of the lid of a petri dish, and fragment mycelium with a metal rod (2–3 mm diameter).
4. Transfer the liquid to an appropriate medium[2] in a 60-mm petri dish and incubate at 28–30 °C until good growth is observed (1–3 weeks).
5. Check purity of the strain, repeating steps 2–4 if necessary.[3]
6. Dereplicate apparently identical strains, and scrape sufficient mycelium from each selected strain, depositing it into a 300-μl H_2O drop and fragmenting as in step 3.

---

[2] Suitable media can be oatmeal agar [prepared by boiling 60 g oatmeal in 1 l H_2O for 20 min, filtering through cheese cloth, then adding 18 g agar, 1 ml of mineral solution (1 g/l each of $FeSO_4 \cdot 7H_2O$, $MnCl_2 \cdot 7H_2O$, and $ZnSO_4 \cdot 7H_2O$), adjusting to the desired pH and autoclaving]; the isolation medium itself; or 0.5 × BTT (5 g glucose, 0.5 g yeast extract, 0.5 g beef extract, 1 g casitone, 20 g agar, made up to 1 l with distilled H_2O, adjusted to the desired pH and autoclaved).

[3] In case of contaminants, it may be advisable to repeat streaking on the isolation medium.

7. Transfer the liquid into 15 ml of suitable medium[4] in a 50-ml baffled flask and incubate at 28–35 °C on an orbital shaker (200 rpm) for 4–10 days.
8. Control culture purity, add glycerol to 10% and transfer 1-ml aliquots to 2-ml screw-cap cryovials. An unequivocal strain code is best assigned at this stage. Vials are then stored at −80 °C.

A strain may be streaked on different media to enhance the probability of obtaining good growth and different liquid media can be used. The examples reported should work with most actinomycetes.

2.4. Sample preparation

Strains are grown under suitable conditions, which usually include a vegetative medium for biomass accumulation, followed by a production medium, where the switch to secondary metabolism is expected to occur. As growth conditions are empirically established on a limited number of strains, rarely are they optimal, and probably no single medium allows production of all metabolites encoded by a strains' genome. The same strain may be grown under multiple conditions, in an attempt to stimulate production of different metabolites.

After growth, cultures may be either tested directly or processed. Fermentation broths are complex mixtures consisting of media residues and byproducts, cellular components (secreted enzymes and residues from lysed cells), in addition to secondary metabolites. In each sample, the relative concentrations of these components may vary greatly. While fermentation broths can be used as samples in simple tests (e.g., growth inhibition), they are not compatible with many assays and not amenable to long-term storage. Culture processing to produce extracts (Table 1.1) obviates these problems, but adds substantial cost to sample preparation. Processing may also reduce sample complexity, by performing a prefractionation of the extracts, to help with assay design and to speed up hit dereplication (see below).

Extract generation may involve: separating the fermentation broth from the biomass by centrifugation or filtration; solid-phase extraction of the fermentation broth, followed by washing and solvent elution; solvent extraction of the biomass; solvent extraction of the entire culture. The resulting extracts may be prefractionated by partition in different organic phases or by collecting a small number of fractions from an HPLC column.

---

[4] Suitable media can be: AF (20 g dextrose, 8 g soybean meal, 2 g yeast extract, 4 g $CaCO_3$, 1 g NaCl, made up to 1 l with distilled H_2O, adjusted to the desired pH, and autoclaved); TSB (17 g tryptone, 3 g soytone, 5 g NaCl, 2.5 g K_2HPO_4, 2.5 g glucose, made up to 1 l with distilled H_2O, adjusted to the desired pH, and autoclaved).

The higher the extent of processing, the lower the complexity (i.e., the number of different compounds present) of the resulting samples.

Samples are stored in appropriate microtiter plates (Fig. 1.2B) and, in order to reduce variability, independent aliquots of the same sample are distributed into equivalent wells of different microtiter plates, so that an extract bank actually consists of twin plates, each containing 80 different samples and 16 empty wells for the positive and negative controls.[5]

The steps necessary for preparation of a sample bank are schematized in Fig. 1.2B. One procedure is described below for the preparation of processed fermentation broths and mycelial extracts from actinomycetes.

2.4.1. Procedure

1. Inoculate a 1.5-ml aliquot of a frozen culture into 15 ml of AF medium[4] (or other medium used for storage) in a 50-ml baffled flask and incubate at 28–30 °C on an orbital shaker (200 rpm) for 3–5 days.
2. Transfer 5 ml into 95 ml production medium[6] in a 500-ml baffled flask and incubate as above for 5–10 days.
3. Transfer each culture into two 50-ml tubes and separate cells from supernatant by centrifugation for 10 min at 3000g.
4. Add 2 volumes of ethanol to biomass and incubate for 90 min at room temperature with shaking.
5. Remove residue by centrifugation and pool liquids from the two tubes.
6. Distribute 0.2 ml aliquots of each sample from step 5 into equivalent positions of twin 96-well microtiter plates, dry under vacuum, and cold storage it.
7. Mix 50 ml of the supernatant from step 3 with 5 g of the polystyrene resin HP20 (Mitsubishi Chemical Co.), incubate for 2 h at room temperature, wash the resin with 20 ml H_2O, and elute with 60 ml methanol.
8. Distribute 0.2 ml aliquots as in step 6 above.

To speed up operations, strains with similar growth characteristics should be processed in parallel. Furthermore, extracts prepared with different methods should be stored in different plates. This compartmentalization allows screening only of selected portions of the sample bank. Before screening, samples are resuspended in 0.1 ml 10% DMSO and 5 or 10 μl are dispensed in the equivalent position of the assay plate.

---

[5] We refer here to 96-well plates only. The reader is referred to specialized reports on the challenges of working with very small volumes in high density plates (Mishra et al., 2008; Wölcke and Ullmann, 2001).
[6] Examples of productive media are: M8 (20 g starch, 10 g glucose, 4 g hydrolyzed casein, 2 g yeast extract, 2 g meat extract, 3 g $CaCO_3$, made up to 1 l with distilled water, pH adjusted to 7.0 before autoclaving); G1/0 (10 g glucose, 10 g maltose, 8 g soybean meal, 4 g $CaCO_3$, 2 g yeast extract, made up to 1 l with distilled water, pH adjusted to 7.2 before autoclaving); and INA5 (30 g glycerol, 15 g soybean meal, 2 g NaCl, 5 g $CaCO_3$, made up to 1 l with distilled water, pH adjusted to 7.2 before autoclaving).

3. Targets and Assays for Antibacterial and Antifungal Programs

The samples are screened with specific tests, which in turn are based on modulating the activity of a desired target. The increased use of HTS as a drug discovery tool has resulted in a vast literature describing target identification and validation, assay formats, assay development and automation, and the reader is referred to general reviews on these topics (Chan and Hueso-Rodríguez, 2002; Janzen, 2002). The main focus of this section is on the features relevant to anti-infective programs that use mixtures of microbial products as chemical diversity, stressing the elements of novelty that target and assays can provide. The main steps necessary to progress from target to screening are schematized in Fig. 1.3.

3.1. Target identification and validation

In the antibacterial and antifungal fields, a validated target is a cellular component essential for viability or disease, conserved in the pathogens of interest, where it plays an equivalent role, and absent or significantly divergent in humans. The genomic revolution and the systematic analysis of essential genes in model organisms have led to catalogs of essential bacterial and fungal genes, their conservation in other microbes and their homology to human sequences (e.g., Hu *et al.*, 2007; Sakharkar *et al.*, 2008).

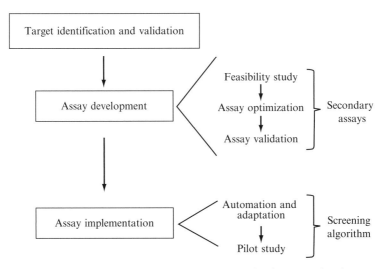

Figure 1.3 From targets to screening. Main steps involved in assay development and implementation.

The number of validated bacterial targets is significant (>100 genes), and examples exist of new antibacterial targets and ingenious assays introduced into HTS programs (Brandi et al., 2007; Chalker and Lunsford, 2002; DeVito et al., 2002; Singh et al., 2007). In contrast, the antifungal field is characterized by a dearth of potential targets (about one order of magnitude lower than for bacteria) and choices are consequently limited (Roemer et al., 2003; Seringhaus et al., 2006).

New anti-infective agents must act on drug-resistant pathogens. Since drug-resistance mechanisms are often class-specific and the drugs in clinical use affect few bacterial targets, there is ample choice of validated targets for antibacterial programs. Furthermore, most targets present multiple sites for potential inhibition, further increasing the likelihood that a new antibiotic class acting on an established target will not exhibit cross-resistance. In this case, however, the major challenge is to devise primary assays insensitive to the frequent known compounds or secondary assays to effectively recognize them.

3.2. Assay development

Once the target has been identified, the assay concept (how inhibitors will be detected) is developed, and the assay format established (Fig. 1.3). Two fundamental choices are available: cell-free assays, which directly measure the biological activity on one or more targets; and cell-based assays, in which the biological activity of the target is determined by monitoring a cellular response. Cell-free assays have the advantage of generally being more sensitive and specific than cell-based assays. Furthermore, because of their limited use, they are more likely to discover new classes of inhibitors. A serious drawback is that they also detect compounds unable to enter microbial cells, and these inhibitors may be prohibitively hard to transform into antibiotics. On the other hand, the challenge with cell-based assays is to make them specific, that is, responsive to inhibition of the desired target(s) only, and more sensitive than simple growth inhibition assays. It is beyond the scope of this chapter to describe all assay formats reported in the literature, and the many sophisticated detection technologies. Among the cell-based assays, general formats amenable to multiple targets include reporter (Fischer et al., 2004; Urban et al., 2007) and antisense (DeBacker et al., 2001; Ji et al., 2001) assays.

It is important to have a rough idea of the expected positivity rate of the assay in the screening, which depends on the assay type and its sensitivity, on the number of targets affected, and on the characteristics of the sample bank. For example, a sensitive cell-free assay in which multiple targets can be inhibited (e.g., coupled transcription/translation; Dermyer et al., 2007) may have a positivity rate 2–3 orders of magnitude higher than an assay based on the inhibition of a single target within a Gram-negative cell (e.g., Fossum et al., 2008).

Generally, assay development involves a feasibility study, assay optimization, and assay validation (Fig. 1.3). The feasibility study aims at experimental demonstration that the assay is actually specific for the desired target and reasonably sensitive. This requires at least one compound acting on the target (a positive control), which should generate the expected readout in a reasonable time and at a concentration compatible with the available chemical diversity (see above). [In the absence of a small-molecule positive control, genetic constructs can be used to validate the assay.] In addition, the assay should not be influenced by compounds commonly found in the samples (negative controls). During this phase, one must also consider the assay reagents (enzymes, substrates, engineered strains), their cost and stability, and the amenability of the assay to miniaturization and automation. The ideal test should be performed in a single well and require only addition of reagents, that is, a "mix and measure" assay, with a detection system not requiring separation of the reaction product(s) from substrate(s).

During the optimization phase, the assay parameters are varied to establish the best signal-to-noise ratio, using positive and negative controls. Samples representative of the extract bank are also used, with and without spiking with the positive control(s), to establish the presence of any interference. Further negative controls are tested at this stage. Depending on the results obtained, secondary assays are devised.

Assay validation consists of executing the test on a relevant number of samples (from a few hundred to a few thousand) representative of the entire collection, in terms of producing strains and extraction methods. Its objective is to establish whether the assay is robust enough to withstand the inevitable variability associated with the assay procedure and with the samples, and whether the positivity rate is within the expected range. In our experience, it is advisable to perform assay validation manually, before spending resources in adapting and automating the assay. During assay validation, the threshold for positivity is usually set.

Finally, the assay is automated, adapted to the available HTS instrumentation and routines, and implemented. This is followed by a pilot study (Fig. 1.3), as described below.

3.3. Secondary assays

Rarely is a single assay sufficient to identify a hit. The positives identified with the primary assay must undergo further evaluation to confirm their activity and to discard unwanted signals. The succession of primary and secondary assays, with the corresponding thresholds for accepting or discarding a sample, constitutes the screening algorithm (Table 1.1). In microbial product screening, identification of the active molecule in a positive sample is a lengthy and laborious process, and effective screening algorithms must be implemented to identify only the most promising hits. Consequently, secondary assays are usually performed to recognize known antibiotic classes or undesirable biological properties, and rarely are they used to independently confirm the activity observed with the primary assay. Secondary assays may involve dose-ranging experiments *vis-à-vis* the primary assay, to establish the selectivity window.

With simply processed samples obtained from good metabolite producers (e.g., actinomycetes, fungi, myxobacteria), a single sample may well contain two or more classes of bioactive metabolites. This requires a judicious choice of primary and secondary assays of comparable sensitivity and specificity and compatible with sample complexity. For example, selective inhibitors of translation can be identified by comparing the effect observed on bacterial enzymes with that observed on eukaryotic enzymes [see Brandi *et al.* (2007) for an example].

4. Screening

With the design of a screening algorithm, the program can enter its true test with the execution of a pilot study, during which the assay is run on a significant number of samples (a few thousand), performing all the steps of a true HTS program, as specified in Fig. 1.4: execution of primary assays, identification of positives, cherry-picking of an independent sample of the positives, repetition of the primary assay and execution of the secondary assay(s), data analysis, and finally, hit declaration. The purpose of the pilot study is to "road-test the engine", making sure no unexpected results are encountered. This is particularly relevant for microbial product screening, as different strains, growth media and extraction procedures result in heterogeneous samples and lower signal-to-noise ratios. Furthermore, comparison with other screening programs may help identify possible false positives and suggest strategies to recognize them. Often, hits are processed to confirm that the active molecules correspond to expectations. Additional secondary assays may be implemented after examining the results of the pilot study.

4.1. Example: Screening for bacterial cell-wall inhibitors

As an example of a screening program, we report a set of primary and secondary assays used for identifying bacterial cell-wall inhibitors using a reporter assay. In a *Bacillus subtilis* strain expressing the *Enterococcus faecium vanRS* genes and a *vanH–lacZ* fusion, β-galactosidase activity appears in response to cell-wall inhibitors and lysozyme (Ulijasz *et al.*, 1996). During assay validation, it was observed that some membrane-interfering compounds elicited a signal (De Pascale *et al.*, 2007; Falk *et al.*, 2007). Thus, a secondary assay based on the same reporter strain was devised to recognize these unwanted compounds on the basis of their fast response.

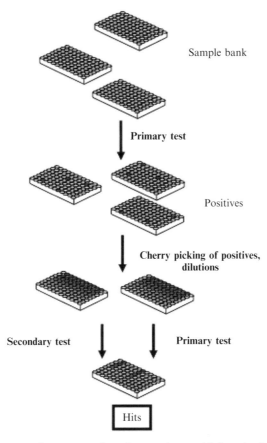

Figure 1.4 The screening process. Samples meeting positivity criteria in the primary tests (red) are transferred to twin plates, diluted as appropriate and tested with primary and secondary assays for hit identification. (See Color Insert.)

4.1.1. Procedure

1. Inoculate 10 µl of an 8×10^8 spore/ml suspension of *B. subtilis* BAU102 (Ulijasz *et al.*, 1996) in 30 ml TSB[7] containing 10 µg/ml erythromycin and 34 µg/ml chloramphenicol and incubate for 16 h at 37 °C with moderate shaking.
2. Dilute cells 1:100 in 30 ml of fresh medium and incubate 2–4 h at 37 °C to an OD_{600} of 0.2.
3. Centrifuge culture and resuspend cells in 30 ml fresh TSB. Use within 1–2 h.

---

[7] TSB is described in footnote 4.

4. Prepare screening plates, by dispensing 10 μl of each sample in the wells of a 96-well microtiter plate. Add positive and negative controls.[8]
5. Dispense 90 μl of the cell suspension in each well and incubate 1 h at 37 °C.
6. Add 10 μl per well of 400-μg/ml 4-methylumbelliferyl-β-D-galactopyranoside in DMSO and further incubate for 2 h at room temperature.
7. Measure fluorescence at excitation and emission wavelengths of 360 and 455 nm, respectively, and express results as

$$S_i = 100(F_i - F_c)/(F_p - F_c),$$

where S_i is the normalized signal of sample i, with F_i representing the fluorescence values of sample i, and F_c and F_p the average of negative and positive controls, respectively.

8. Identify positives as samples possessing an $S_i > 60$. Prepare two identical plates, each containing 10 μl aliquots of 10 different positives in the upper row, and serial 1:2 dilutions of each sample in the remaining rows.
9. Use one plate to repeat steps 5–7, determining for each sample and its dilutions the corresponding S_i.
10. To the other plate, add cells as in step 5, and incubate for 15 min at room temperature, then repeat step 6 as above.
11. For each sample and its dilutions from step 10, measure fluorescence as in step 7, expressing results as

$$N_i = 100(R_i - R_c)/(R_p - R_c)$$

where N_i is the normalized signal of sample i, with R_i representing the fluorescence values of sample i, and R_c and R_p the average of the negative and positive controls, respectively.[9]

12. Determine for each sample D_S and D_N (the highest dilutions giving S and N signals above 60 and 50, respectively).
13. Identify as hits those samples confirming an S value >60 and showing a D_S/D_N ratio ≥ 8.[10]

---

[8] Positive and negative controls consist of 10 μl and 25 μg/ml ramoplanin in 10% DMSO and 10 μl 10% DMSO, respectively.
[9] Positive and negative controls are represented by 5 μg/ml benzyldimethylhexadecyl-ammonium chloride and 10% DMSO, respectively.
[10] It should be noted that the probability that a sample retains a signal above threshold decreases with its dilution, so in many cases the measured D_S/D_N ratio may only be ≥ 2. These samples may be included among the hits, bearing in mind that they may not confirm their specificity upon further testing.

If the amount of sample is not limiting, further tests may be introduced before hit identification. In the example above, positives may be tested for: (1) growth inhibition of a desired Gram-positive strain; (2) the same test as in (1), but in the presence of 2 mg/ml D-alanyl-D-alanine; (3), the same test as in (1), but with the sample preincubated with a cocktail of β-lactamases[11] for 1 h at 37 °C. These tests establish if growth inhibition is due to a D-alanyl-D-alanine-binding antibiotic (e.g., a glycopeptide) or a β-lactamase-sensitive compound, respectively. Only samples active under (1), (2) and (3) become hits.

5. Hit Follow-up

The ultimate goal of microbial product screening is to establish the chemical identity of the hits and their biological properties. Since this may be a lengthy and expensive process, further filters are introduced to dedicate resources only to the most promising hits. These filters involve dereplication, novelty evaluation and profiling.

Dereplication requires establishing whether different hits actually represent the same compound. For example, different samples derived from the same strain (through the use of different production media or extraction procedures) may contain the same compound, or the same strain may have been independently added to the library several times. Depending on the information available in the database, dereplication may be performed *in silico* or along with novelty evaluation (see below). After dereplication, the sample showing the highest potency or specificity ratio is selected.

Depending on sample availability, its complexity (weight per culture volume after extraction) and potency (i.e., highest dilution giving a measurable signal), novelty evaluation may be directly performed on the sample bank. This is the preferred route in the presence of a high hit rate or of many microbial product classes detected by the screening algorithm. On the other hand, when dealing with low-potency or high-complexity samples, it may be preferable to confirm the activity by reprocessing the producing strain before performing novelty evaluation. Indeed, hit characterization and profiling depend on the supply of adequate amounts of samples (Fig. 1.1). Different preparations are likely to yield different concentrations of the active compound and of interfering substances. Thus, each sample preparation must be validated with respect to the biological selection criteria, and consistency of the results must be thoroughly checked. For this reason, all profiling tests should be performed, whenever possible, on the same sample batch.

---

[11] 0.001 U/ml type 1 and 0.002 U/ml type 2 penicillases from *Bacillus cereus* (Sigma), and 0.0025 U/ml type 3 and 0.5 U/ml type 4 penicillases from *Enterobacter cloacae* (Sigma).

5.1. Activity confirmation

Eventually, the activity observed during screening must be reconfirmed by growing the producer strain and preparing a new sample under conditions identical to those that yielded the original sample. At this stage, the strain is also characterized to establish likely genus or species. A typical procedure is described below.

5.1.1. Procedure

1. Thaw a frozen vial of the original strain, streak on appropriate medium and ensure purity and identity.
2. Inoculate strain in vegetative medium as in the original procedure.
3. From the vegetative medium, inoculate into three identical flasks of production medium as in the original procedure.
4. Harvest the three flasks at 24-h intervals, starting from 1 day before the original harvest time.
5. Process culture as in the original procedure but preparing a 2× concentrated sample, dividing it into a smaller aliquot for testing and a larger one for novelty evaluation.
6. Repeat primary and secondary assays on the smaller aliquot.
7. Process through novelty evaluation the larger aliquot of the best sample.[12]

The use of different growth times helps establishing a rough production curve, while the preparation of a 2× concentrated sample increases the chance of reproducing the original activity and of measuring specificity when the ratio in the original sample was just ≥ 2. If the original sample was not prefractionated, it may be advisable to perform a prefractionation, determining the best fraction to use for novelty evaluation.

5.2. Novelty evaluation

A critical step in microbial product screening is determining the likelihood that the observed bioactivity is due to a novel compound. This is a continuous process, as it is based on negative evidence (i.e., the compound under investigation does not appear to resemble any known molecule), and must be repeated until convincing proof of the compound novelty is established. Novelty evaluation involves separating the active molecule(s) from other sample components and acquiring data on physico-chemical (molecular mass, UV spectrum, chemical class, presence of functional groups, molecular formula, etc.) and biological (antimicrobial spectrum, producer strain

---

[12] The best sample is the one with the highest potency and/or the highest specificity ratio, or that responds to other biological criteria.

identity) properties. These data are then compared with those available in the literature until a tentative match is established (Fig. 1.5). Relevant information may not be obtained at each step and, depending on prioritization, not all hits may undergo a complete set of analyses (dashed arrows in Fig. 1.5).

Rigorous proof that a hit corresponds to a new molecule requires comparison with a standard or establishing its chemical structure. However, this will frequently be too expensive and usually congruence of all chemical and biological data is sufficient to establish correspondence between a hit and a known compound. A general protocol for hit HPLC fractionation is described below.

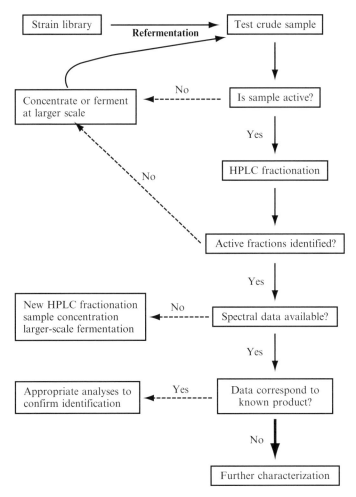

Figure 1.5 Novelty evaluation. Striped arrows indicate iterative tasks that may or may not be performed on all hits.

5.2.1. Procedure

1. Inject 0.1 ml of a 10-mg/ml sample onto a SymmetryShield RP18 column (Waters).
2. Apply a gradient in which phase B (20 mM HCOONH$_4$ pH 4.5, CH$_3$CN 1:19) increases in 28 min from 10 to 90%, with a corresponding decrease of phase A (20 mM HCOONH$_4$ pH 4.5, CH$_3$CN 19:1), at a flow rate of 1 ml/min.
3. Collect 1-ml fractions, acquiring UV–vis spectra with a diode array spectrophotometer and mass data by diverting 5% of the HPLC effluent into an electrospray ionization mass detector.
4. Evaporate solvent from the collected fractions under vacuum and dissolve residues in 100 μl 10% DMSO.
5. Test each fraction with the primary assay and identify active peak(s).
6. Retrieve UV–vis and mass data on the active peak(s) and search literature for comparable data.
7. Match other criteria of the active peaks (producer strain, target, profile) with those reported in the literature.

A first HPLC run may not lead to the identification of active fractions because of sample instability, dilution or interference from the eluent. Similarly, active peak(s) may be identified, but multiple or no spectral data may be present (Fig. 1.5). In these situations, prefractionation of the sample and/or the use of alternative eluents may be necessary. This may require a more concentrated sample, and possibly a new preparation.

5.3. Profiling

Hits are characterized for their activity on different microbial strains. Depending on the extent of purification, samples may contain more than one antibiotic and the molecule of interest may be present at low purity. Consequently, care must be taken in evaluating the observed spectrum and MIC data for low-purity samples.

At the early stages, a microbial panel may include isogenic antibiotic-resistant and -sensitive strains to help in novelty evaluation. As the interest in the hit increases because of its potential novelty, purer samples become available, and microbial panels include relevant pathogens with clinically relevant mechanisms of resistance. At this stage, the cytotoxicity of the sample is also evaluated and relevant animal models of infection are used to test *in vivo* efficacy. With the caveat of performing experiments with compounds of unknown structure and (relatively) unknown concentration, these profiling activities are independent of the origin of the compound (fermentation, semi-synthesis, or totally synthetic), are tightly linked to the objective of the discovery program, and will not be described here.

6. Databases, Operations, and Costs

The screening of large sample banks on different targets requires investment in automation, data acquisition and data management, as described in specialized articles (Koehn, 2008; Ling, 2008). Extremely relevant to microbial product screening are databases linking the strains to the compounds identified in the samples. In contrast to databases for chemical libraries, databases related to microbial products start with no initial information on the compounds present in the samples and must accommodate the possibility of more than one chemical class being produced by a single strain. Furthermore, since compound identification during novelty evaluation is often tentative, this information needs to be amended over time. Finally, the identification of the same compound in different samples does not necessarily imply that the samples are identical, since they are likely to differ in additional molecules. These peculiarities often require the adaptation of commercial software to the specific needs.

A database is necessary for handling the strain library (Fig. 1.6). Its complexity can vary, but it should contain, for each strain, at least the

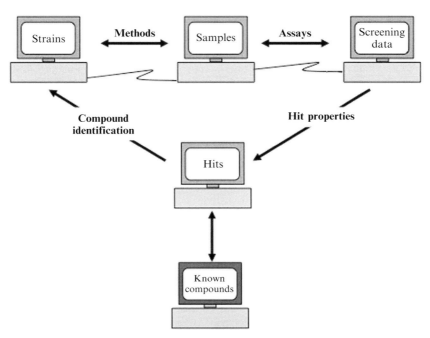

Figure 1.6 Databases. In-house generated data are inserted in four interlinked databases. The Hits database must effectively interface with a database describing known molecules.

following information: a unique code identifier; physical localization; origin; medium and conditions used for isolation, purification, and cryovial preparation; and taxonomic classification.[13] Further information may include method of isolation, morphological description, genus/species identification, phylotype assignment, and identified compound(s). Some of these data may not be uniformly available for the entire collection. A database must also be implemented for the sample bank (Fig. 1.6), with at least the following information: unique code, physical location, corresponding strain, vegetative medium and conditions, production medium and conditions, and extraction method. This information is essential for repreparing an equivalent sample after hit identification. It is also important to implement stock management for tracking the number of intact twin plates and those in use (Chan and Hueso-Rodríguez, 2002). An additional database deals with the screening data and associates to each sample the results of standardized primary and secondary assays, including hit identification (Fig. 1.6). Finally, information on processing of the hits (active peaks, spectral data, profiling, chemical identification), including different hit preparations, must also be electronically available (Fig. 1.6). These databases must be interlinked, to enable efficient retrieval of all necessary information.

Another relevant aspect concerns access to databases on known microbial products. When most pharmaceutical companies were involved in natural product screening, each had built its customized database of known natural products, gathering literature information and reporting relevant data on the chemical and biological properties of the described compounds. Each database addressed the needs of the natural product chemists involved, and was often integrated with external compendia (e.g., www.nih.go.jp/~jun/NADB/search.html). These databases are particularly valuable for old compounds discovered when analytical tools were less sophisticated.

Finally, we would like to conclude this section with considerations on operational aspects, including costs. As mentioned, the activities necessary for screening are highly connected, and decisions must be made on the best strategies and approaches to achieve the overall objective. In order to execute many screening campaigns on a large number of samples, the throughput of sample preparation (and hence of strain management) is extremely important. However, throughput strongly depends on standardization of procedures, which contrasts with the need for diversification. Therefore, it is extremely important to strike a proper balance between the quality of the samples (in terms of richness and diversity of the molecules

---

[13] Classification should be functional to sample preparation, allowing the use of the growth conditions established for each taxonomic group.

they contain) and their number. Quality controls should be introduced to address whether newly isolated strains are adding diversity to an existing strain library and whether the chosen cultivation and processing methods result in valuable samples. Hits should be prioritized on the basis of observed properties and likelihood to be novel. All these operational aspects have a profound effect on the cost of microbial product screening and hence on its commercial viability.

7. Perspectives

We have tried to provide an overview of the identification of novel anti-infective agents from microbial sources, focusing on key elements of the process and combining strategic considerations with selected detailed procedures. There are significant challenges in this endeavor, but also opportunities offered by newly discovered strains, by a better understanding of the large genomic potential of the known strains to produce metabolites and by the possibility of designing effective assays to identify novel compounds. The recent discovery of novel chemical classes effective on microbial targets (Brandi *et al.*, 2006; Riedlinger *et al.*, 2004; Scott *et al.*, 2008; Wang *et al.*, 2006) bodes well for the future.

Screening requires the integration of different skills and experiences. Valuable insights can be obtained in this respect from profound retrospectives on past successes (Baltz, 2005; Lancini, 2006; Weinstein, 2004). In our opinion, a key element for success is a team of highly motivated scientists, each with a sufficient knowledge of chemistry and microbiology to understand all the required steps and activities, who share the overall objective of the program.

ACKNOWLEDGMENT

This work was partially supported by a grant from the Italian Ministry of Research (FIRB RBIP06NSSX).

REFERENCES

Baltz, R. H. (2005). Antibiotic discovery from actinomycetes: Will a renaissance follow the decline and fall? *SIM News* **55,** 186–196.

Brandi, L., Fabbretti, A., La Teana, A., Abbondi, M., Losi, D., Donadio, S., and Gualerzi, C. O. (2006). Specific, efficient, and selective inhibition of prokaryotic translation initiation by a novel peptide antibiotic. *Proc. Natl. Acad. Sci. USA* **103,** 39–44.

Brandi, L., Fabbretti, A., Milon, P., Carotti, M., Pon, C. L., and Gualerzi, C. O. (2007). Methods for identifying compounds that specifically target translation. *Methods Enzymol.* **431,** 229–267.

Bull, A. T., Ward, A. C., and Goodfellow, M. (2000). Search and discovery strategies for biotechnology: The paradigm shift. *Microbiol. Mol. Biol. Rev.* **64,** 573–606.

Busch, U., and Nitschko, H. (1999). Methods for the differentiation of microorganisms. *J. Chromatogr. B Biomed. Sci. Appl.* **722,** 263–278.

Busti, E., Cavaletti, L., Monciardini, P., Schumann, P., Rohde, M., Sosio, M., and Donadio, S. (2006). *Catenulispora acidiphila* gen. nov., sp. nov., a novel, mycelium-forming actinomycete, and proposal of *Catenulisporaceae* fam. nov. *Int. J. Syst. Evol. Microbiol.* **56,** 1741–1746.

Butler, M. S. (2008). Natural products to drugs: Natural product-derived compounds in clinical trials. *Nat. Prod. Rep.* **25,** 475–516.

Chalker, A. F., and Lunsford, R. D. (2002). Rational identification of new antibacterial drug targets that are essential for viability using a genomics-based approach. *Pharmacol. Ther.* **95,** 1–20.

Chan, J. A., and Hueso-Rodríguez, J. A. (2002). Compound library management. *Methods Mol. Biol.* **190,** 117–127.

Davis, K. E. R., Joseph, S. J., and Janssen, P. H. (2005). Effects of growth medium, inoculum size, and incubation time on culturability and isolation of soil bacteria. *Appl. Environ. Microbiol.* **71,** 826–834.

De Pascale, G., Grigoriadou, C., Losi, D., Ciciliato, I., Sosio, M., and Donadio, S. (2007). Validation for high-throughput screening of a VanRS-based reporter gene assay for bacterial cell wall inhibitors. *J. Appl. Microbiol.* **103,** 133–140.

DeBacker, M. D., Nelissen, B., Logghe, M., Viaene, J., Loonen, I., Vandoninck, S., de Hoogt, R., Dewaele, S., Simons, F. A., Verhasselt, P., Vanhoof, G., Contreras, R., *et al.* (2001). An antisense-based functional genomics approach for identification of genes critical for growth of *Candida albicans*. *Nat. Biotechnol.* **19,** 235–241.

Dermyer, M., Wise, S. C., Braden, T., and Holler, T. P. (2007). Simultaneous screening of multiple bacterial tRNA synthetases using an *Escherichia coli* S30-based transcription and translation assay. *Assay Drug Dev. Technol.* **5,** 515–521.

DeVito, J. A., Mills, J. A., Liu, V. G., Agarwal, A., Sizemore, C. F., Yao, Z., Stoughton, D. M., Cappiello, M. G., Barbosa, M. D., Foster, L. A., *et al.* (2002). An array of target-specific screening strains for antibacterials discovery. *Nat. Biotechnol.* **20,** 478–483.

Donadio, S., Brandi, L., Serina, S., Sosio, M., and Stinchi, S. (2005a). Discovering novel antibacterial agents by high throughput screening. *Front. Drug Design Discov.* **1,** 3–16.

Donadio, S., Busti, E., Monciardini, P., Bamonte, R., Mazza, P., Sosio, M., and Cavaletti, L. (2005b). Sources of polyketides and nonribosomal peptides. In "Biocombinatorial Approaches for Drug Finding" (W. Wohlleben, T. Spelling, and B. Müller-Thiemann, eds.) pp. 19–41. Ernst Schering Research Foundation, Springer, Berlin.

Donadio, S., Monciardini, P., and Sosio, M. (2007). Polyketide synthases and nonribosomal peptide synthetases: The emerging view from bacterial genomics. *Nat. Prod. Rep.* **24,** 1073–1109.

Falk, S. P., Ulijasz, A. T., and Weisblum, B. (2007). Differential assay for high-throughput screening of antibacterial compounds. *J. Biomol. Screen.* **12,** 1102–1108.

Fischer, H. P., Brunner, N. A., Wieland, B., Paquette, J., Macko, L., Ziegelbauer, K., and Freiberg, C. (2004). Identification of antibiotic stress-inducible promoters: A systematic approach to novel pathway-specific reporter assays for antibacterial drug discovery. *Genome Res.* **14,** 90–98.

Fossum, S., De Pascale, G., Weigel, C., Messer, W., Donadio, S., and Skarstad, K. (2008). A robust screen for novel antibiotics: Specific knockout of the initiator of bacterial DNA replication. *FEMS Microbiol. Lett.* **281,** 210–214.

Hu, W., Sillaots, S., Lemieux, S., Davison, J., Kauffman, S., Breton, A., Linteau, A., Xin, C., Bowman, J., Becker, J., *et al.* (2007). Essential gene identification and drug target prioritization in *Aspergillus fumigatus*. PLoS Pathog. **3,** e24.

Hunter-Cevera, J. C., and Belt, A. (1999). Isolation of cultures. *In* "Manual of Industrial Microbiology and Biotechnology" (A. L. Demain and J. E. Davies, eds.), 2nd ed. pp. 3–20. ASM Press, Washington, DC.

Janzen, W. P. (2002). HighThroughput Screening: Methods and Protocols. Vol. 190. Humana Press.

Jensen, P. R., Gontang, E., Mafnas, C., Mincer, T. J., and Fenical, W. (2005). Culturable marine actinomycete diversity from tropical Pacific Ocean sediments. *Environ. Microbiol.* **7,** 1039–1048.

Ji, Y., Zhang, B., Van Horn, S. F., Warren, P., Woodnutt, G., Burnham, M. K., and Rosenberg, M. (2001). Identification of critical staphylococcal genes using conditional phenotypes generated by antisense RNA. *Science* **293,** 2266–2269.

Koehn, F. E. (2008). High impact technologies for natural products screening. *Prog. Drug Res.* **65,** 177–210.

Lam, K. S. (2007). New aspects of natural products in drug discovery. *Trends Microbiol.* **15,** 279–289.

Lancini, G. (2006). Forty years of antibiotic discovery at Lepetit: A personal journey. *SIM News* **56,** 192–212.

Larsen, T. O., Smedsgaard, J., Nielsen, K. F., Hansen, M. E., and Frisvad, J. C. (2005). Phenotypic taxonomy and metabolite profiling in microbial drug discovery. *Nat. Prod. Rep.* **22,** 672–695.

Ling, X. B. (2008). High throughput screening informatics. *Comb. Chem. High Throughput Screen.* **11,** 249–257.

Mishra, K. P., Ganju, L., Sairam, M., Banerjee, P. K., and Sawhney, R. C. (2008). A review of high throughput technology for the screening of natural products. *Biomed. Pharmacother.* **62,** 94–98.

Otoguro, M., Hayakawa, M., Yamazaki, T., and Iimura, Y. (2001). An integrated method for the enrichment and selective isolation of *Actinokineospora* spp. in soil and plant litter. *J. Appl. Microbiol.* **91,** 118–130.

Payne, D. J., Gwynn, M. N., Holmes, D. J., and Pompliano, D. L. (2007). Drugs for bad bugs: Confronting the challenges of antibacterial discovery. *Nat. Rev. Drug Discov.* **6,** 29–40.

Riedlinger, J., Reicke, A., Zähner, H., Krismer, B., Bull, A. T., Maldonado, L. A., Ward, A. C., Goodfellow, M., Bister, B., Bischoff, D., Süssmuth, R. D., and Fiedler, H. P. (2004). Abyssomicins, inhibitors of the *para*-aminobenzoic acid pathway produced by the marine *Verrucosispora* strain AB-18-032. *J. Antibiot.* **57,** 271–279.

Roemer, T., Jiang, B., Davison, J., Ketela, T., Veillette, K., Breton, A., Tandia, F., Linteau, A., Sillaots, S., Marta, C., Martel, N., Veronneau, S., *et al.* (2003). Large-scale essential gene identification in *Candida albicans* and applications to antifungal drug discovery. *Mol. Microbiol.* **50,** 167–181.

Sakharkar, K. R., Sakharkar, M. K., and Chow, V. T. (2008). Biocomputational strategies for microbial drug target identification. *Method Mol. Med.* **142,** 1–9.

Scott, J. J., Oh, D. C., Yuceer, M. C., Klepzig, K. D., Clardy, J., and Currie, C. R. (2008). Bacterial protection of beetle–fungus mutualism. *Science* **322,** 63.

Seringhaus, M., Paccanaro, A., Borneman, A., Snyder, M., and Gerstein, M. (2006). Predicting essential genes in fungal genomes. *Genome Res.* **16,** 1126–1135.

Singh, S. B., Phillips, J. W., and Wang, J. (2007). Highly sensitive target-based whole-cell antibacterial discovery strategy by antisense RNA silencing. *Curr. Opin. Drug Discov. Devel.* **10,** 160–166.

Ulijasz, A. T., Grenader, A., and Weisblum, B. (1996). A vancomycin-inducible *lacZ* reporter system in *Bacillus subtilis*: Induction by antibiotics that inhibit cell wall synthesis and by lysozyme. *J. Bacteriol.* **178,** 6305–6309.

Urban, A., Eckermann, S., Fast, B., Metzger, S., Gehling, M., Ziegelbauer, K., Rübsamen-Waigmann, H., and Freiberg, C. (2007). Novel whole-cell antibiotic biosensors for compound discovery. *Appl. Environ. Microbiol.* **73,** 6436–6443.

Wang, J., Soisson, S. M., Young, K., Shoop, W., Kodali, S., Galgoci, A., Painter, R., Parthasarathy, G., Tang, Y. S., Cummings, R., Ha, S., Dorso, K., *et al.* (2006). Platensimycin is a selective FabF inhibitor with potent antibiotic properties. *Nature* **441,** 358–361.

Weinstein, M. P. (2004). *Micromonospora* antibiotic discovery at Schering-Plough (1961–1973): A personal reminiscence. *SIM News* **54,** 56–66.

Wölcke, J., and Ullmann, D. (2001). Miniaturized HTS technologies uHTS. *Drug Discov. Today* **6,** 637–646.

CHAPTER TWO

From Microbial Products to Novel Drugs that Target a Multitude of Disease Indications

Flavia Marinelli

Contents

1. Microbial Diversity and Biotechnological Products	30
2. Secondary Metabolites	30
2.1. Source of new microbial drugs	34
2.2. Overview of bioactivities and assays	35
2.3. Successful drugs	40
2.4. "Old" and "novel" antitumor antibiotics	41
2.5. DNA-targeting antitumor antibiotics	41
2.6. Tubulin-targeting anticancer drugs	47
2.7. Discarded antifungals as agents for organ transplantation	49
2.8. The best-selling drugs from natural products: Fungal statins	51
3. Conclusions	53
Acknowledgments	54
References	54

Abstract

More than 20,000 bioactive, so-called microbial secondary metabolites are known. In nature, they can play many different roles as antibiotics, toxins, ionophores, bioregulators, and in intra- and interspecific signaling. Their most versatile producers are differentiating filamentous fungi and actinomycetes, followed by other bacteria such as *Bacillus, Pseudomonas*, Myxobacteria, and Cyanobacteria. From a biotechnological point of view, bioactive metabolites have been mainly studied as potential anti-infectives (antibacterials, antifungals, antivirals, and antiparasitics). Many of them, originally discovered for their antibiotic activity, were developed further to become leading anticancer drugs, immunosuppressive agents for organ transplantation, and successful pharmaceuticals targeting metabolic and cardiovascular diseases. Old and novel antitumor antibiotics can be divided into two groups spanning diverse chemical

Department of Biotechnology and Molecular Sciences, University of Insubria, Varese, Italy classes: those causing some damage to the DNA such as Mitomycins (quinones), Bleomycins (glycopeptides), Actinomycins (peptides), Anthracyclines (aromatic polyketides), Pentostatin (nucleoside), Enediynes (polyketides), and Rebeccamycin derivatives (indolocarbazole glycosides)—all produced by *Streptomyces* strains or related genera; and the group of tubulin-targeting molecules such as plant/fungal Taxanes (terpenes), myxobacterial Epothilones (macrocyclic polyketides), and the revisited Cryptophycins (peptolides) produced by cyanobacteria. Immunosuppressive agents used in clinical practice include Cyclosporin A (cyclopeptide) and an ester of Mycophenolic acid (small aromatic polyketide), both produced by fungi, and two streptomycete macrocyclic polyketides, Sirolimus (rapamycin) and Tacrolimus (FK506). Statins include a group of fungal polyketides (Compactin or Mevastatin, Lovastatin) and their derivatives, which specifically inhibit cholesterologenesis in liver, and represent the best-selling drugs. They have been used to prevent cardiovascular diseases.

1. Microbial Diversity and Biotechnological Products

Biotechnology is currently considered a key enabling technology that has an impact on major global problems such as disease, environmental pollution, and malnutrition and that can provide innovative and economically competitive solutions to long-standing and emerging problems in industrial processes. To date, the major impact of biotechnology has been in pharmaceuticals although it is now progressively penetrating other industrial sectors, too. Microbes in all their diversity represent a major source for biotechnological exploitation for various categories of relevant products, that is, biomass, products from anaerobic metabolism (ethanol, butanol, lactic acid) or from incomplete oxidation (citric acid, acetic acid), primary and secondary metabolites, enzymes, biocatalysts, and, more recently, recombinant proteins (Demain and Adrio, 2008). In this chapter, a review of those valuable bioactive small molecules (the so-called secondary metabolites) of microbial origin targeting human diseases is accompanied by a description of classic and novel approaches being pursued to discover and develop new drugs.

2. Secondary Metabolites

Bu'Lock (1961) first explicitly introduced the term "secondary metabolite" in microbiology, taking it from previous studies of German plant physiologists on cellular components not essential for cell life and not found in every growing cell. Microbial secondary metabolites are generally defined as metabolic products found as differentiation compounds in restricted taxonomic groups. They are not essential for vegetative growth of the producing organisms, at least under laboratory conditions, and they are biosynthesized from one or more general (primary) metabolites by a wider variety of pathways than those involved in general (primary) metabolism. Microbial secondary metabolites are compounds of low molecular weight (<3000) and are frequently accumulated after vegetative or exponential growth has ceased, as families (complexes) of structurally related components (congeners). They are mainly produced by a relatively restricted group of bacteria and fungi, but their intergeneric, interspecific, and intraspecific variation is extremely high. Many of them are endowed with biological activities, such as antibiotics, toxins, ionophores, bioregulators, and intra- and interspecific signaling (Demain and Adrio, 2008; Fajardo and Martinez, 2008). Among these molecules, antibiotics have been studied most extensively. It is generally assumed that the majority of them are important as weapons in intermicrobial competition. Most of the known antibiotics are produced by differentiating microbes such as filamentous fungi and actinomycetes, which have in common a nonmotile, saprophytic life-style in a complex habitat such as terrestrial soil, where competition with other inhabitants is high. Examples of protocols to produce bioactive secondary metabolites from filamentous actinomycetes and fungi in laboratory conditions are described at the end of this section.

Antibiotics are typically produced when the growth of vegetative mycelium slows down as nutrients are exhausted and secondary structures evolve at the expense of nutrients released by the breakdown of vegetative cells. Thus, it has been proposed that they defend the food source from competitors (Challis and Hopwood, 2003). In some cases, secondary metabolites have been suggested to be possible waste products or shunt metabolites, but this is hard to reconcile with the complexity of their biosynthetic pathways and of the underlying genetic information (Vining, 1990). For some other secondary metabolites, a function in cell-to-cell or species-to-species communication has been demonstrated: they may act at the low concentrations present in the environment as chemical signals to modulate metabolic processes in bacteria by stimulating or depressing gene expression at the transcription level and thereby influencing population structure and dynamics (Davies *et al.*, 2006; Demain and Adrio, 2008).

Example: Cultivation of filamentous actinomycetes to produce bioactive secondary metabolites (for further details see the author's references (Brandi et al., 2006; Castiglione et al., 2007, 2008)).

- From the Master Cell Bank (MCB) (lyophilised vials, usually supplied by Culture Collections), prepare the Working Cell Bank (WCB) (frozen

vegetative stock at $-80\ °C$ in 15% glycerol at a biomass concentration of ~ 0.08–1.2 g/ml dry weight) of the producer strain.
- For vegetative seed (or WCB) culture preparations, inoculate 500-ml baffled Erlenmeyer flasks containing 100 ml seed medium[1] with one glycerol stock vial (for the first time with the MCB lyophilised powder) and incubate for 72–96 h at 28–30 °C on a rotary shaker (200–250 rpm) till abundant biomass is produced.
- Check biomass growth and colony morphology by plating aliquots of seed culture on plates containing suitable agar media.[2] Check morphology and purity by phase-contrast microscope at ×400.
- Use seed culture for inoculating (at 10% v/v) 100-ml aliquots of production medium[3] in 500-ml baffled Erlenmeyer flasks and incubate at 28–30 °C on a rotary shaker (200–250 rpm) for 120–144 h.
- Alternatively, after the seed stage in shaken flasks, further vegetative and production stages may be carried out in mechanically stirred fermenters two-thirds full of the appropriate media, aerated by an air-flow of 0.5 vvm (volume of air per volume of medium per minute), stirred at 150–700 rpm according to the vessel size, and kept at 28–30 °C by circulating cold water.
- Twice a day, measure biomass growth as packed mycelium volume (PMV%) by centrifuging 10 ml of the sampled culture in graduated vials, measure pH of the supernatant and prepare extracts (for extract preparation methods see examples described in the following section) to assay the bioactivity or check secondary metabolite production by HPLC or LC–MS, if an analytical method and/or a pure standard sample are available.

Example: Cultivation of filamentous fungi to produce bioactive secondary metabolites (Sponga et al., 1999; Stefanelli et al., 1996).

Procedures to cultivate secondary metabolite-producing fungi are approximately the same than those for filamentous actinomycetes, except for the lower incubation temperature (22–25 °C), lower agitation speed in

---

[1] Two examples of suitable vegetative media are: AF/MS (in g/l: dextrose 20, yeast extract 2, soybean meal 8, NaCl 1, and $CaCO_3$ 4, distilled water, pH adjusted to 7.3 with NaOH and autoclaved) (Castiglione et al., 2007, 2008); V6 (in g/l:glucose 20, meat extract 5, peptone 5, yeast extract 5, casein hydrolysate 3, NaCl 1.5, distilled water, pH adjusted to 7.4 with NaOH), (Brandi et al., 2006).

[2] Two examples of suitable agar media are: Salt Medium Agar (in g/l: glucose 10 g, Bacto-peptone 4 g, Bacto-yeast extract 4 g, $MgSO_4 \cdot 7H_2O$ 0.5 g, KH_2PO_4 2 g, K_2HPO_4 4 g, distilled water, solidified with agar 15 g, autoclaved) (Beltrametti et al., 2006); Oat Meal Agar (used also as isolation medium, see Chapter 1 for preparation and composition).

[3] Two examples of suitable production media are: M8 (in g/l: starch 20, glucose 10, yeast extract 2, hydrolyzed casein 4, meat extract 2, and $CaCO_3$ 3, distilled water, pH adjusted to 7.0 and autoclaved) (Castiglione et al., 2007, 2008); INA5 (in g/l: glycerol 30, soybean meal 15, $CaCO_3$ 5 and NaCl 2, distilled water, pH adjusted to 7.3 and autoclaved) (Brandi et al., 2006).

shaken flasks (150 rpm) and in reactors (150–500 rpm), and composition of solid,[4] liquid vegetative,[5] and production media.[6]

Example: Recovery of a secondary metabolite from actinomycetal and fungal cultures (Brandi et al., 2006; Castiglione et al., 2007, 2008).

A secondary metabolite can be produced intracellularly or extracellularly.

In the first case:

- Extract the mycelium by adding an equal volume of methanol and recover the metabolite after vacuum evaporation of the solvent. Acid or base solution can be added to facilitate the extraction, taking into account the chemical properties of the molecule (i.e., its stability at different pHs and at different temperatures).
- Concentrate under vacuum and redissolve in an appropriate solvent.

In the second case:

- Filter the harvest broth at the end of fermentation, by tangential filtration or centrifugation. Acid or base solution can be added according to the chemical properties of the molecule.
- Stir the filtrate with 2.5% (v/v) HP 20 polystyrene resin (Mitsubishi Chemical Co.). Recover the resin, wash it batch-wise with methanol–water 1:1 (v/v), and elute it with methanol–n-butanol–water 9:1:1 (v/v/v).
- Pool the fractions containing the secondary metabolites (tested by biological assay or HPLC/LC–MS methods) and concentrate them under vacuum to a residue of raw material, then dissolve it in n-butanol. Extract this solution with water, or precipitate the metabolite by adding a solvent such as acetone, then dissolve it in an appropriate solvent.

In both cases, the crude secondary metabolite solution so far achieved can be further purified by HPCL preparative chromatography set up according to the chemical properties of the secondary metabolite. Typically, isocratic or gradient elution are performed with a Phase A consisting of 25–40 mM ammonium formate buffer pH 4.5, acetonitrile 95:5 (v/v), and pure acetonitrile as Phase B. The fractions from repeated chromatographic runs are pooled, concentrated under vacuum, and lyophilized sequentially 2–3 times to yield the purified secondary metabolite.

---

[4] Two examples of suitable vegetative media are: AF/MS as for actinomycetes and Tomato Medium (in g/l: tomato paste 40, corn steep powder 2.5, glucose 10, oat flour 10 and 10 ml/l of trace solution described below, distilled water, pH adjusted to 6.8 and autoclaved).
[5] Potato Carrot Agar (in g/l: mashed potato 20, mashed carrot 20, agar 20, tap water, autoclaved (Stefanelli *et al.*, 1996) or Potato Dextrose Agar (Difco).
[6] Two examples of suitable production media are: KFO (in g/l: glucose 10, potato starch 10, barley 20 g/l, corn steep liquor 5, trace element 10 ml/l, distilled water, pH adjusted to 6.8 and autoclaved) and Potato Dextrose Broth (Difco). Trace element solution in g/l: $FeSO_4$ 1, $MnSO_4$ 1, $CuCl_2$ 0.025, $CaCl_2$ 0.1, H_3BO_3 0.056, $(NH_4)_6Mo_7O_{24}$ 0.019, $ZnSO_4$ 0.2, dissolved in 1 l HCl 0.6N.

2.1. Source of new microbial drugs

A wide range of potential disease-targeting activities of secondary metabolites has already been described. Since the "Golden Age" of antibiotic discovery in the 1940s and 1950s, thousands of bioactive secondary metabolites have been identified. In 1995, it was estimated that almost 12,000 bioactive secondary metabolites were discovered in those decades, and about 160 of them had by that time reached clinical use as natural products *per se* or as chemical derivatives of natural scaffolds (Demain and Adrio, 2008). Indeed, 55% were produced by the genus *Streptomyces*, 11% from other actinomycetes, 12% from nonfilamentous bacteria, and 22% from filamentous fungi. According to Berdy (2005), ca. 20,000 and 22,000 bioactive microbial secondary metabolites were described in the scientific and patent literature by the end of 2000 and 2002, respectively. About 38% of these molecules are produced by filamentous fungi, whereas the largest group (45%) derives from actinomycetes (7600 metabolites from *Streptomyces* and 2500 from the so-called rare filamentous actinomycetes). The remaining 17% is produced by other bacteria such as *Bacillus, Pseudomonas,* Myxobacteria, and Cyanobacteria. Examples of protocols to cultivate and prepare extracts for the screening of myxobacteria and cyanobacteria are described here (see Gaspari *et al.*, 2005 and Biondi *et al.*, 2008 for details and additional procedures for strain isolation and characterization).

Example: Cultivation and extract preparation in myxobacteria.

- Plate aliquots of WCB (stock solutions frozen at $-80\ °C$ in nutrient glycerol 10%) on CY or VY/2 agar (Gerth *et al.*, 1996) at 28 °C.
- Inoculate with selected colonies, 20 ml vegetative medium CYE Broth,[7] in 50-ml Erlenmeyer flasks and incubate them for 24–72 h at 28 °C.
- Inoculate 80 ml of production Starch Soybean Glucose (SSG) medium[8] in 500-Erlenmeyer flasks and incubate them at 28 °C and 200 rpm for 48–120 h.
- Centrifuge 50 ml of the harvested broth at 2000g for 15 min and load 12 ml of supernatant onto a 3-ml Solid Phase Extraction column (63 × 9 mm) filled with 1.2 ml of HP-20 resin (Mitsubishi Chemical Co., Tokyo, Japan), previously washed with 6 ml of methanol and then conditioned with 6 ml of distilled water.
- Wash the columns with 5.8 ml of water and elute them with 5.8 ml of a mixture of methanol–water 3:1 (v/v).
- Distribute the eluates in a 96-well plate (200 μl/well). This procedure can be performed on a robotic workstation. Concentrate the samples to

---

[7] CYE Broth (in g/l: casitone 10, yeast extract 1, $CaCl_2 \cdot 2H_2O$ 1, distilled water, autoclaved).
[8] SSG (in g/l: potato starch 8, soybean meal 3, glucose 2, yeast extract 1.5, $MgSO_4 \cdot 7H_2O$ 1, $CaCl_2 \cdot 2H_2O$ 1, EDTA-Na-Fe (III) 0.008, distilled water, pH adjusted to 7.2, autoclaved).

dryness under vacuum and store the plates at $-10\ °C$. These samples could be used, once resuspended in DMSO and water, for different screening assays.

Example: Cultivation and extract preparation in cyanobacteria.

- Inoculate with selected colonies grown on slants, 500–1100 ml glass tubes bubbled with air/CO_2 (98/2, v/v) or in 1000-ml Erlenmeyer flasks (400 ml culture volume) kept in an orbital shaker flushed with air/CO_2 (95/5, v/v) containing mineral media such as BG11 (Rippka, 1988) or ASNIII (Waterbury and Stanier, 1981) at 20 °C and under continuous light provided by daylight fluorescent tubes.
- Check for bacterial contamination of the cultures by observation in an optical microscope.
- After 3–7 days of growth, harvest biomass by filtration on a nylon net and carefully wash it with saline solution under vacuum.
- Extract 1 g biomass (dry weight) with 50 ml of ethyl acetate overnight.
- Filter the suspension through paper and evaporate the solvent under vacuum.
- Dissolve the dry residues in 50 μl of DMSO–H_2O 1:9 (v/v) and use them in different screenings.

2.2. Overview of bioactivities and assays

Similar results on bioactivity distribution among microbial producers were obtained by authors who queried the natural product ABL database (Lazzarini *et al.*, 2001) to collect literature and patent information: among more than 31,600 microbial products discovered from 1900 onwards, ca. 20,200 possess some biological activity (Marinelli, 2003). Here, 35% were produced from filamentous fungi, 48% from actinomycetes, and 17% from other bacteria. As shown in Fig. 2.1, nearly 50% of these active metabolites were detected through antibacterial and antifungal screenings (see Chapter 1 describing antibacterial and antifungal screenings). At the end of 1999, ca. 8000 potentially anti-infective drugs, that is, molecules endowed with antifungal or antibacterial activity, but devoid of any reported cytotoxicity, had been described (Lazzarini *et al.*, 2001), and this number increased to 9500 in 2003, with an average rate of 300–400 novel antibiotics having been discovered per year (Marinelli, 2003). About 1000 small microbial molecules identified through antiviral screenings (Fig. 2.1, Table 2.1) should be added to the anti-infective pool. Most of them are nucleoside derivatives or protease inhibitors.

A considerable proportion of those bioactive molecules identified as toxic antibiotics early on and that were discarded as therapeutically valuable anti-infectives, were (starting in the 1960s and 1970s to the present) retested as anticancer drugs because of their inhibitory effect toward rapid proliferating eukaryotic cells (see Table 2.1 for anticancer screening assays and

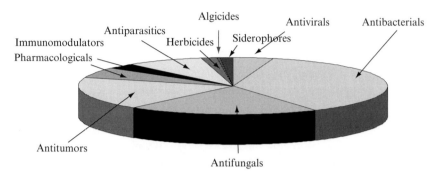

Figure 2.1 Grouping of 20,200 bioactive microbial secondary metabolites according to their biological activities in screening assays. Data from ABL queries (Lazzarini et al., 2001; Marinelli, 2003). Pharmacologicals are those substances that act on the physiology of higher animals; they usually do not exhibit general cytotoxicity, even if their pharmacological action may make them extremely toxic. They include inhibitors of receptor–ligand binding, enzyme inhibitors, analogs of animal hormones. Immunomodulators, which are here considered separately, belong to the pharmacological group of activities, too.

references). Figure 2.1 reports ca. 4400 microbial small molecules possessing cytotoxic activity in *in vitro* assays. This group includes the first widely used antitumor drugs Doxorubicin, Daunorubicin, Mitomycin, and Bleomycin, all of which were produced from actinomycetes, and the promising, more recently developed Epothilones produced from myxobacteria.

Another group of metabolites initially often isolated in antibiotic screening and then "rediscovered" between the 1970s and 1990s for other useful pharmacological activity in animals are those showing an immunomodulating property (ca. 670 in Fig. 2.1, Table 2.1 for screening approaches); these include those currently used as agents for organ transplantation such as Mycophenolic acid, Cyclosporin, Sirolimus (Rapamycin), and Tacrolimus (FK506). The first two are produced from filamentous fungi, whereas the rest come from members of the genus *Streptomyces*. In the same period, the introduction of cell-free assays in pharmaceutical screenings based on the inhibition of receptor–ligand binding or of key enzymes playing a role in recognized human disease (Table 2.1) led to the discovery of selective, nontoxic pharmacological activities (ca. 1400 in Fig. 2.1). In some successful cases such as that of the statins (Compactin, Lovastatin, etc.), fungal metabolites acting as cholesterol-lowering agents due to their inhibition of eukaryotic 3-hydroxy-3-methylglutaryl-coenzyme A reductase, the activities detected *in vitro* were transformed into successful drugs showing physiological activity *in vivo*. Another example is Acarbose, a pseudo-tetrasaccharide produced by an *Actinoplanes* sp., which inhibits intestinal α-glucosidases and is thus used as an antihyperglycemic drug. Finally, ca. 310 molecules are grouped in Fig. 2.1 as siderophores. Among them

Table 2.1 Screening assays for discovering drugs to be used in therapies against cancer, and in anti-hypertensive, anti-inflammatory, and antiviral indications. Examples of screening for insecticide and antiparasitic compounds are also reported.

Class of activity	Assay system	Examples	References
Anticancer	Cell-based: cytotoxicity versus rapidly proliferating cells/tumoral cell lines	Cytotoxicity/growth inhibition versus mouse L929 fibroblasts, human T-24 bladder carcinoma cells, leukaemia L1210 cells Microbial prescreening	Gerth et al. (1996), Newman and Shapiro (2008), Reichenbach and Höfle (2008), Wani et al. (1971)
Immunosuppressant	In vivo in animals Cell-based: inhibition of immune response Cell-free: inhibition of receptor–ligand binding	Haemagglutinin test in mice Suppression of mixed lymphocyte reaction in mouse cell lines Inhibition of interleukin-2 production in mouse cell lines Immunoassay based on the cyclosporin binding to cyclophilin	Borel et al. (1976), Mann (2001), Quesniaux et al. (1987), Reynolds and Demain (1997)
Anticholesterolemic	Cell-free: enzyme inhibition	Inhibition of 3-hydroxy-3-methylglutaryl-coenzyme A reductase	Alberts et al. (1980), Endo (1980), Endo et al. (1976)
Anti-inflammatory	Cell-free: inhibition of receptor–ligand binding	Immobilized-ligand IL-1 receptor binding assay Particle concentration fluorescence receptor binding assay	Stefanelli et al. (1997)

(*continued*)

Table 2.1 (continued)

Class of activity	Assay system	Examples	References
Anti-hyperglycemic	Cell-free: enzyme inhibition; genetic screening	Inhibition of α-glucosidases PCR based method to detect sedo-heptulose 7-phosphate cyclase	Hyun et al. (2005)
Antiviral	Cell-free: enzyme inhibition of viral enzymes	Inhibiton of HIV-1 integrase	Singh et al. (2003)
Antiparasitic	In vivo in animals	Activity in mice against the nematode *Nematosporides dubius*	Vining (1990)
Insecticide	In vivo in insects	Microbial solid cultures supplied as food to larvae and to adults of *Musca domestica*	Fabre et al. (1988)

some molecules, such as Desferrioxamine (Desferal) produced by *Streptomyces pilosus*, have found medical application in iron diseases (hemochromatosis) and kidney aluminium overload in dialysis patients as a result of their high level metal-binding activity (see Chapter 17 in this volume).

Since the 1970s secondary metabolites have also been increasingly used in veterinary medicine and agriculture. The group of insecticides and antiparasitic activities amounted to almost 2000 molecules in 2003 (Fig. 2.1), whereas the group of herbicides and algicides (agricultural antibiotics such as Bialaphos or antifouling agents) included ca. 500 molecules. A major veterinary problem has been worm infections in farm animals. Screening for metabolites inhibiting nematodes, cestodes, and protozoa (see Table 2.1 for screening procedures) found new products such as Avermectins, which are potent and specific inhibitors of invertebrates and lack antibiotic activity. One of the major economic diseases of poultry is coccidiosis, which is caused by species of the parasitic protozoan *Eimeria*. For years only synthetic products were screened for coccidiostatic activity but through their use resistance rapidly developed. Narrow-spectrum and human toxic polyether antibiotics (Monensin, Lasalocid, and Salinomycin) proved to be extremely active against coccids, but great effort was needed to improve production by genetic and fermentation tools in order to reduce the cost and enhance their appeal for such nonmedical applications. Bialaphos, avermectins, and polyether antibiotics are all produced by actinomycetes (Berdy, 2005; Demain and Adrio, 2008).

Table 2.1 lists examples of screening assays used to detect the classes of compounds described above, excluding those used in discovering antibacterial and antifungal agents (see Chapter 1). An example of an easy to handle cytotoxic assay protocol that can be used to rapidly check antitumor potential in a group of microbial extracts is reported at the end of this section.

Considerations and ingredients for microbial product screenings, unravelled by Donadio *et al.* (Chapter 1), are valid irrespective of the type of activity being looked for. The success of a screening program depends on the novelty of the microbial diversity, on the quality of the chemical library (the collection of extracts prepared by actinomycetes, fungi, myxobacteria, and cyanobacteria in screening format), as well as on the development of a proper set of assays to effectively sort out the novel molecules active against the preselected target. As shown by the examples in Table 2.1, in the non anti-infective area, screening can be performed *in vivo* by directly injecting extracts into animals or supplying them as a food to invertebrates, or in cell-based or cell-free formats, as in the case of more traditional antibacterial and antifungal screenings. Rapid dereplication of known molecules, additional tests to profile the biological activity, structure elucidation, identification of molecular mechanism of action, and, overall, the capacity to supply increasing amounts of pure active principles, represent the crucial steps to support drug development.

Example: Cytotoxic assays of microbial extracts.

- Keep human cervix epitheloid carcinoma cells (HeLa cells) in continuous culture as a mono-layer in 75 cm^2 tissue culture grade flasks (Costar) in a complete medium (RPMI 1640 supplemented with 10% fetal calf serum, penicillin 100 units/ml, and streptomycin 100 μg/ml) at 37 °C and 5% CO_2. Every 3 days, the cell mono-layer is removed from the flask by the addition of trypsin, washed by centrifugation at 1500 rpm for 10 min, and transferred to another flask containing fresh complete medium (subcloning).
- Seed cells, after 1 day of subcloning, for testing in microtiter plates at a density of 10^5 cells/well.
- Add 90 μl of complete medium plus 10 μl of the microbial extract or control DMSO–H_2O solution.
- Incubate plates at 37 °C and 5% CO_2 for 24 h.
- Wash the cells with 200 μl/well of a phosphate buffer solution at pH 7.3 and pulse them for 4 h with ^{3}H-thymidine 0.1 μCi of in 100 μl/well of complete medium without serum.
- Trypsinize cells and harvested them on a glass fiber filter with a semi-automated cell harvester.
- Measure incorporated radioactivity using a Betaplate scintillation counter. Those extracts able to inhibit 40% cell thymidine uptake relative to the control are flagged as cytotoxic. A broth microdilution method may then be used to confirm positive broths and to assay their potency (Biondi et al., 2008).

2.3. Successful drugs

Berdy reported that 150 compounds from microbial sources (less than 1% of the known bioactive metabolites) are being used in human and veterinary medicine and agriculture nowadays. One hundred of them are used in human therapy. Recent investigations demonstrate that ca. 50% of the small molecules approved as drugs in the years 2000–2006 were still produced from natural products (Newman and Cragg, 2007). The major area in which natural products have found success in the pharmaceutical industry continues to be infectious diseases, followed by emerging therapies against cancer, anti-hypertensives, and anti-inflammatory indications. The most approved drugs from natural sources are antibacterials and antifungals, but this numerical trend does not correlate with the value as measured by sales since the best-selling drugs of all are the hypochlolesterolemic statins (Newman and Cragg, 2007).

In the remainder of this chapter, examples of discovery and development strategies leading to the drugs targeting human diseases in the emerging fields of cancer, cardiovascular, and inflammatory therapies will be investigated, with the conviction that many other interesting and useful drugs are hidden and waiting to be fished from the pool of secondary metabolites already known or to a major extent from the yet-to-be discovered chemical diversity produced by an estimated 99% of bacteria and 95% of fungi that could not yet be isolated and cultivated (Demain and Adrio, 2008).

2.4. "Old" and "novel" antitumor antibiotics

The recent review of Newman and Cragg (2007) covers the anticancer drugs developed from 1940 to July 2006. Of the 175 available agents in the West and Japan, ca. 80% are small molecules, the rest being peptides or proteins (>45 residues) such as antibodies, growth factors, cytokines, and vaccines produced by biotechnological means in a surrogate host. More than 60% of the approved small molecule drugs are either natural products or natural product derivatives. Currently, more than 30 compounds of microbial origin are in various stages of clinical development as anticancer agents (Lam, 2007).

2.5. DNA-targeting antitumor antibiotics

Mitomycins, Bleomycins, Actinomycins, Anthracyclines, Pentostatin, Enediynes, and Rebeccamycin derivatives (Fig. 2.2) are structurally and biosynthetically diverse microbial metabolites all produced by actinomycetes, mainly *Streptomyces* strains. They possess a marked antibacterial activity and have the common feature of being toxic DNA-damaging agents. Toxicity limits their clinical use as antineoplastic drugs and has stimulated research in preparing novel derivatives by medicinal chemistry and combinatorial biosynthesis. Novel derivatives are also needed to overcome tumor cell resistance mechanisms such as drug inactivation, increased drug efflux, and enhanced DNA repair systems. From the biotechnological point of view, self-toxicity in the producing actinomycetes strongly limits their production, but for the majority of them the only feasible supply process is fermentation, total synthesis being too complicated or too expensive. Thus, understanding their mechanisms of biosynthesis, resistance, efflux, and regulation continues to be critical for the success of these molecules *per se* or for their use as starting material for more innovative drugs.

The oldest FDA-approved anticancer drug (1956) seems to be Mitomycin C (**I** in Fig. 2.2), a potent antibacterial that belongs to the family of antitumor quinones (see the review by Galm *et al.* (2005) and references cited therein). Mitomycin A, B, and C were first isolated form *Streptomyces caespitosus*. The clinically used **I** showed superior activity against solid tumors and reduced toxicity as compared to natural counterparts. It is used to treat gastric, colorectal, and lung cancer in particular (source National

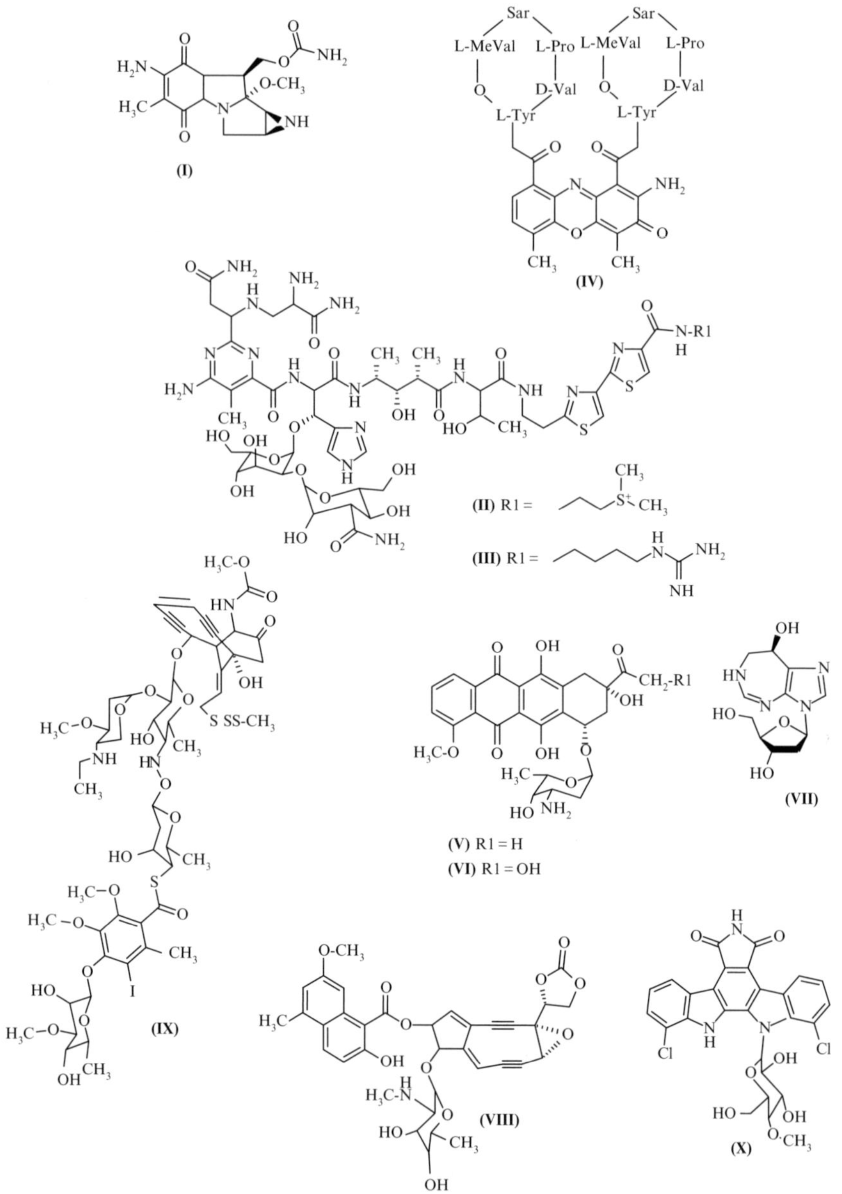

Figure 2.2 Chemical structures of DNA-damaging antitumor antibiotics: Mitomycin C (I), Bleomycin A2 (II) and B2 (III), Actinomycin (IV), Daunorubicin (V) and Doxorubicin (VI), Pentostatin (VII), Neocarzinostatin (VIII), Calicheamicin (IX) and Rebeccamycin (X).

Cancer Institute (NCI) USA web site: http://www.cancer.gov/). This molecule is comprised of aziridine, quinone, and carbamate moieties arranged in a compact pyrroloindole structure. Precursor-feeding studies showed that it is derived from 3-amino-5-hydroxybenzoic acid (AHBA synthesized through a variant of the shikimate pathway), D-glucosamine, L-methionine, and carbamoyl phosphate (Mao *et al.*, 1999). It forms covalent linkages with the DNA and functions as an alkylating agent. It is a natural agent for the approach termed bioreductive therapy since **I** is a prodrug activated through an enzymatic reduction that preferentially proceeds in the absence of oxygen. This quality makes it a valuable drug able to attack the hypoxic regions of solid tumors.

The discovery of the alkaloid FR900482 from *Streptomyces sandaensis* in 1987 revealed a structurally related mitomycinoid with a mode of action analogous to that of **I**. This representative of a new class of antitumor agents displays markedly lower hematotoxicity, and semisynthetic derivatives have shown promising activity in human trials. Currently, the usefulness of **I** and its derivatives is limited by the acquired or intrinsic resistance of tumor-cell populations. In recent years, a specific mechanism of resistance in cancer cells has been proposed that involves detoxification by reoxidizing the reduced cytotoxic intermediate of **I** and that is based on the self-protection mechanism described for the mitomycin industrial producer *Streptomyces lavendulae*. Cloning the microbial 54-kDa flavoprotein called MRCA (mitomycin C resistance associated), which acts as a hydroquinone oxidase and protects DNA from cross-linking by oxidizing the toxin mitomycin hydroquinone in the producer organism, confers resistance to Chinese hamster ovary (CHO) cells (Belcourt *et al.*, 1999). Two other resistance genes located within the mitomycin biosynthetic gene cluster (*mrd* and *mct*) code for a protein able to bind and sequester **I** and for a specific transporter gene, respectively, together forming an efficient drug-binding export system (Galm *et al.*, 2005). Targeted manipulation of a putative pathway regulator in the mitomycin biosynthetic gene cluster led to a substantial increase in drug production (Mao *et al.*, 1999).

Bleomycins are a family of glycopeptide-derived antibiotics isolated from several *Streptomyces* species (Galm *et al.* (2005) and references cited therein). They exert their biological effects through sequence-selective, metal-dependent oxidative cleavage of DNA and RNA in the presence of oxygen. Structurally and biosynthetically related metabolites are the phleomycins and the tallysomycins. Their complex glycosylated, linear, hybrid peptide–polyketide backbone is assembled by a megasynthetase that consists of both nonribosomal peptide synthetase (NRPS) and polyketide synthase (PKS) modules (Du *et al.*, 2000). The commercial products were introduced into clinics in 1966 as a mixture of Bleomycin A2 and B2 (**II** and **III** in Fig. 2.2) and they are still used in combination with a number of other agents for the treatment of several types of tumors, notably squamous cell carcinomas and malignant lymphomas (NCI, http://www.cancer.gov/). Unlike most anticancer drugs, **II** and **III** do not cause myelosuppression, promoting their wide application in combination chemotherapy.

Early development of drug resistance and cumulative pulmonary toxicity represent the major limitations. Structural complexity has limited most modification attempts at either the C-terminal amine or the N-terminal β-aminoalaninamide moiety by either directed biosynthesis or semisynthesis (Galm et al., 2005). Total chemical synthesis is expensive, thus limiting the practicality of novel derivatives with better clinical efficacy and reduced toxicity in pharmaceutical applications. The *in vivo* manipulation of the producing *Streptomyces verticillus* ATCC15003 has been hampered for nearly two decades because it is refractory to all means of introducing plasmid DNA into its cells. Only very recently, several years after cloning and identification of the bleomycin biosynthetic cluster, was it possible to manipulate this pharmaceutically relevant producer by RED-mediated PCR-targeting mutagenesis and intergeneric *Escherichia coli–Streptomyces* conjugation (see Gust et al. (2004) and Chapter 20, in this volume), paving the way for combinatorial biosynthesis of improved derivatives (Galm et al., 2008). As in the case of mitomycin, a comparison of resistance mechanisms within the producing organism with those predominant among tumor cells could offer some clues about the evolution of improved derivatives. Two resistance genes, *blmA* and *blmB*, have been characterized in the producer *S. verticillus*. They code for a bleomycin-binding protein, BlmA, whose mode of binding is interestingly similar to the Mrd from the mitomycin producer, and for an *N*-acetyltransferase, BlmB, which can inactivate bleomycin with remarkable substrate specificity for the α-amine of the β-aminoalanine moiety (Du et al., 2000; Sugiyama and Kumagai, 2002). Although bleomycins have never been used as antibacterials, many clinically isolated Methicillin Resistant *Staphylococcus aureus* (MRSA) and *Klebsiella pneumoniae* strains were found to be resistant to this drug at high levels. Mobile genes (on plasmids or transposons) coding for bleomycin-binding proteins, which act via drug sequestering, have been identified in these pathogens; they are likely recruited from the producer organisms. No naturally occurring bleomycin-binding proteins have been identified in eukaryotic cells so far, but this protein should also be considered as a possible emerging mechanism of resistance in tumor cells in the future (Galm et al., 2005; Sugiyama and Kumagai, 2002).

Among the other "old" antitumor agents produced by different species of actinomycetes, Actinomycin D or Dactinomycin (IV in Fig. 2.2) belongs to the class of antitumor peptides. Actinomycins produced by *Streptomyces parvulus* and other streptomycetes have historical importance since they were the first antibiotics isolated from a *Streptomyces* strain. Their structure contains a phenoxazinone chromophore and two cyclic peptides. The assembly of the acyl peptide lactone precursor of actinomycin occurs via an NRPS mechanism (Schauwecker *et al.*, 2000), whereas the chromophore derives from the condensation of two tryptophan residues. Too cytotoxic to be used as anti-infective drug, IV has been being used parenterally alone or in combination with other antineoplastic drugs since 1964 to treat Wilms' tumor, rhabdomyosarcoma, Ewing's sarcoma, and trophoblastic neoplasms. It acts as an intercalating agent in the DNA by stabilizing cleavable complexes of topoisomerases I and II with DNA (NCI, http://www.cancer.gov/).

Several members of the Anthracycline family have been used as antitumor agents since the 1960s. Daunorubicin (V in Fig. 2.2) and Doxorubicin (VI in Fig. 2.2), also known as Adriamicin, are produced from *Streptomyces peucetius* and several other streptomycetes. They were isolated as antibacterials but then developed into useful drugs to treat leukemia, lymphoma, breast, ovarian, and lung cancer. Chemically, anthracyclines are hydroxyanthraquinones carrying one or more sugar substituents. They are synthesized by an iterative type II PKS, using propionate as the starter unit for the polyketide chain. Doxorubicin-overproducing strains of *S. peucetius* ATCC 29050 can be obtained by manipulating the genes in the region of the doxorubicin gene cluster (Lomovskaya *et al.*, 1999). By a similar approach of introducing heterologous genes into blocked mutants, novel derivatives have been produced by fermentation (Madduri *et al.*, 1998). As bleomycins, V and VI intercalate DNA and inhibit the activity of topoisomerase II. Unfortunately, these drugs show dose-limiting toxicity, causing cardiac damage, for example (NCI, http://www.cancer.gov/). Thus, a variety of delivery systems have been tried. Their liposomally formulated variants have been introduced for the treatment of AIDS-related Kaposi's sarcoma. Different chemically or biotechnologically modified generations of doxorubicin (Idarubicin and Epirubicin) have prompted the development of the last one introduced into the market (2002): Amrubicin hydrochloride, a synthetic 9-amino-anthracycline, which shows higher levels of antitumor activity than conventional anthracycline drugs without exhibiting any indication of cumulative cardiac toxicity (Newman and Cragg, 2007).

Among the more than 200 nucleoside antibiotics produced by streptomycetes, one of the few that has found clinical application is Pentostatin (VII in Fig. 2.2), also known as covidarabine, which is produced from *Streptomyces antibioticus*. It is a purine nucleotide analog which binds to and inhibits adenine deaminase (ADA), an enzyme essential for purine metabolism; ADA activity is greatest in cells of the lymphoid system, with T cells having higher activity than B cells and T-cell malignancies higher ADA activity than B-cell malignancies. Here, VII inhibition of ADA appears to result in elevated intracellular levels of dATP, which may block DNA synthesis by inhibiting ribonucleotide reductase. It is used to treat particular forms of leukemias (NCI, http://www.cancer.gov/).

Another group of potent anticancer agents, the polyketide Enediynes, are produced from *Streptomyces, Micromonospora,* and *Actinomadura* species, or related genera and cause single-stranded or double-stranded DNA (and in some cases RNA) damage (see the review by Galm *et al.* (2005) and Volume II of this series). They are structurally characterized by an unsaturated core with two acetylenic groups conjugated to a double bond or incipient double bond and have been categorized in two subfamilies: 9-membered ring chromophore cores (Neocarzinostatin VIII as model structure in Fig. 2.2) or 10-membered rings (Calicheamicin IX as a model structure in Fig. 2.2). Enediynes are synthesized by a novel iterative type I PKS. The gene clusters of five of them have recently been cloned and sequenced, providing a foundation for better understanding their biosynthesis. Recently, 11 cryptic gene clusters encoding enediyne biosynthesis from 70 actinomycetes that were previously not known as enediyne producers have been identified, suggesting that many other enediynes can be discovered by employing a genomics-guided approach. Armed with this genomic information, antibiotic production in the strains was verified by optimizing medium and fermentation conditions (Zazopoulos *et al.*, 2003). All members have very potent anticancer activity (5–8000 times more potent than adriamycin) and antibiotic activity, but their clinical application is currently limited by a delayed toxicity. To overcome the systemic side effects, enediynes have been conjugated with tumor-directed monoclonal antibodies to provide a localized exposure to the cytotoxic agent. Mylotarg (NCI, http://www.cancer.gov/), the first antibody-targeted, cytotoxic small molecule agent approved by FDA, is IX prepared as a conjugate with a humanized monoclonal antibody specific for CD33 antigen in acute myeloid leukemia cells. The concept established for Mylotarg opened a new therapeutic approach, immunotherapy, which represents one of the most promising areas of cancer research.

Among those DNA-targeting microbial products not yet approved for clinical use, Becatecarin is one of the most advanced candidates; it is an NCI drug presently in the completing phase II of clinical trials for metastatic carcinomas, neuroblastomas, and leukemias (NCI, http://www.cancer.gov/). Becatecarin is a synthetic diethylaminoethyl analog of Rebeccamycin (X in Fig. 2.2), an indolocarbazole glycoside antibiotic discovered in 1987 in the fermentation broth of the actinomycete *Saccharotrix aerocolonigenes,* recently reclassified as *Lechevalieria aerocolonigenes*. The indocarbazole core is formed by decarboxylative fusion of two tryptophan-derived units, whereas the sugar moiety is derived from glucose. Strong DNA intercalation represents its primary mechanism of action, resulting in the potent catalytic inhibition of both topoisomerases I and II. The biosynthetic pathway for X has been dissected and reconstituted in the heterologous host strain *Streptomyces albus,* which provides an environment capable of supplying precursors without the need for further genetic manipulation. By co-expressing different combinations of genes isolated from the X-producing microorganism, more than 30 novel indocarbazole derivatives have been obtained in yields comparable to those from the original producer (Sanchez et al., 2005). Although the total synthesis route is available for becatecarin, combinatorial biosynthesis may offer new tools for discovering and developing improved derivatives.

2.6. Tubulin-targeting anticancer drugs

Tubulin is an antitumor target that continues to attract significant attention for drug discovery and development. The first destabilizing tubulin agents, the plant vinca alkaloids (Vinblastine, Vincristine), were approved as chemotherapeutics in the early 1960s. Their depolymerization action on microtubules causes cell cycle arrest at the G2/M transition and apoptosis follows. More recently developed and successful non-actinomycetic drugs are the Taxanes and Epothilones (Fig. 2.3), which act by blocking depolymerization of microtubules and promoting tubulin polymerization, which is opposite to the mechanism of vinca alkaloids. However, the end result of their effect is equivalent, that is, cell cycle arrest and apoptosis.

Diterpene Taxol (Paclitaxel) (XI in Fig. 2.3), originally isolated from plants (Wani et al., 1971), was later found also to be a microbial metabolite

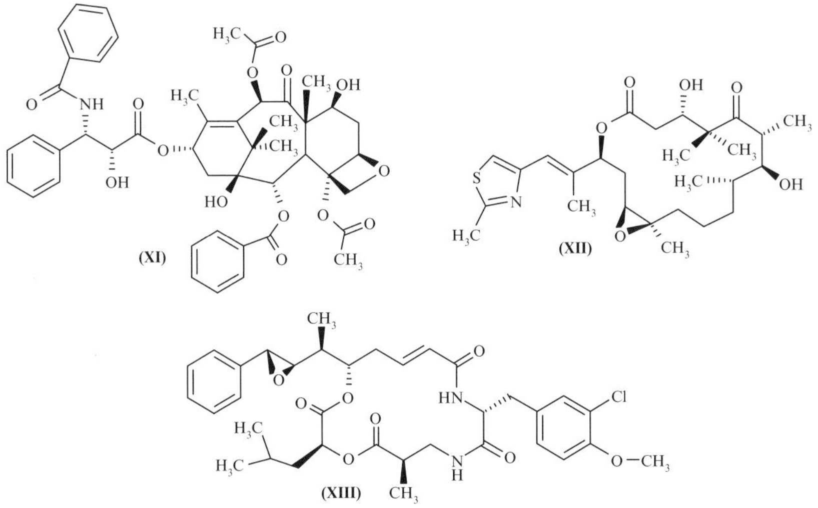

Figure 2.3 Chemical structures of tubulin-targeting agents: Taxol (XI), Epothilone B (XII), and Cryptophycin 1 (XIII).

produced by an endophyte fungus of Pacific yew, called *Taxomyces andreanae* (Stierle et al., 1993). It was approved in 1993 for the treatment of breast and ovarian cancer. Here, XI represents an interesting example of alternative strategies aimed to overcome the problem of natural drug supply, as reviewed in Cragg (1998). The original source of XI, the bark of the yew tree *Taxus brevifolia,* was finite. It took some years to develop a semisynthetic analog, which was then launched in 1995 as Docetaxel (Taxotere). Docetaxel is derived from 10-deacetylbaccatin III that is contained in a renewable source, the leaves of *Taxus baccata.* In the meantime, endophytic fungi able to produce very low amounts of XI have been isolated. Even if none of them gave rise to a feasible fermentative production of XI, these findings renewed interest in isolating endophytic and symbiotic microbes as bioactive metabolite producers (Strobel, 2006). Currently, total synthesis has been achieved for both XI and taxotere, and drug supply is no longer a problem. In 2005, a novel form of the agent ensuring better water solubility, Paclitaxel nanoparticles, was introduced into the clinics (Newman and Cragg, 2007).

The 16-membered macrolides, Epothilones (Epothilone B XII in Fig. 2.3), isolated from the myxobacterium *Sorangium cellulosum,* have the advantage that they are cytotoxic for cells overexpressing the P-glycoprotein drug efflux pump, which are resistant to taxanes. Moreover, they are more water soluble than taxane. An orally bioavailable semisynthetic analog (lactam) of epothilone B, Ixabepilone, was approved by the FDA in October 2007 for the treatment of aggressive metastatic or locally advanced breast cancer no longer responding to currently available chemotherapy regimens (NCI, http://www.cancer.gov/). Epothilone A and B were discovered in the resin-extracted fermentation broths of *S. cellulosum* for their very narrow and selective antifungal activity against the zygomycete *Mucor hiemalis,* coupled with a strong cytotoxic activity against eukaryotic cells (Gerth et al., 1996). A decade of intense research activities followed. The biosynthetic gene cluster was identified at nearly the same time by scientists at Novartis (Molnar et al., 2000) and Kosan (Tang et al., 2000). Epothilones, like many other peculiar myxobacteria metabolites, are biosynthesized as a hybrid system of PKSs and NRPSs (see Chapter 3 in this volume). Since *S. cellulosum* cannot be genetically manipulated, produces only 20 mg/l, and shows doubling times of 16–24 h in liquid cultures, various successful attempts were pursued to integrate in a stepwise fashion the whole epothilone cluster into more industrially convenient heterologous hosts such as *Streptomyces coelicolor* (Tang et al., 2000) or *Myxococcus xanthus* (Julien and Shah, 2002). The original strain or the recombinant ones were then improved in productivity by using classical mutagenesis and optimizing the medium (Gong et al., 2007). Recombinant strains were used for combinatorial biosynthesis and preparation of novel derivatives (Tang et al., 2005).

More than a dozen novel microtubule-destabilizing agents are presently under clinical assessment and, interestingly, most of them are produced by cyanobacteria (blue-green algae) or by soft-body marine invertebrates (sponges, molluscs, tunicates, etc.) (Berdy, 2005; Lam, 2007; Mann, 2001). Recently revisited drugs are Cryptophycins, a group of more than 25 cyanobacterial peptolides whose prototype (XIII in Fig. 2.3) is produced by the symbiotic *Nostoc* sp. ATCC 53789. These were discovered to be potent tubulin-destabilizing agents that are also active on P-glycoprotein overexpressing cells (Smith *et al.*, 1994). Clinical trials of the cryptophycins have been limited by a supply problem since no large-scale biotechnological production method exists for the producer cyanobacteria and chemical synthesis is complex and expensive. This fact, together with some toxic side effects, has blocked the development of the synthetic analog crytophycin 52 at phase 2 of clinical trials and hampered the progression of other derivatives with improved performance. Intensive research work elucidating XIII biosynthesis, cloning the biosynthetic gene cluster consisting of two modular type-I PKSs and two NRPS genes, and producing 30 novel analogs by precursor-directed biosynthesis, including cryptophycin 52 (Magarvey *et al.*, 2006), may reattract pharmaceutical interest to these drugs. The identification of the tailoring enzyme P450 oxygenase responsible for the formation of epoxide in the desired β-stereochemistry, which is difficult to achieve by chemical steps, has paved the way for chemoenzymatic synthesis of XIII analogs (Ding *et al.*, 2008).

2.7. Discarded antifungals as agents for organ transplantation

The most recently approved immunosuppressant agent Mycophenolate mofetil is the ester of the very first secondary metabolite of microbial origin to be discovered. Gosio in 1896 isolated and crystallized a mycotoxin from *Penicillium glaucum* (*Penicillium glaucoma*) that was later identified as Mycophenolic acid (XIV in Fig. 2.4). It was then used locally as a broad spectrum topical antibiotic, for example, to treat psoriasis and related infections. This low-molecular-weight polyketide was then recovered from many *Penicillium* strains often associated with cheese and other conserved foods. The ester acts as a prodrug, which is hydrolyzed to XIV in the body. It inhibits the synthesis of GDP, GTP, and dGTP. Most cells have a salvage pathway for these nucleotides but lymphocytes do not; therefore, administration of XIV inhibits DNA synthesis in lymphocytes (Mann, 2001). It was approved in combination with cyclosporin A and corticosteroids for kidney transplantation in 1995 and for heart transplants in 1998 (Demain and Adrio, 2008).

Cyclosporin A (XV in Fig. 2.4), a cyclic undecapeptide produced by the fungus *Tolypocladium nivenum* (or *inflatum*), was the first immunosuppressive drug of microbial origin to be launched (1978); it has had a dramatic impact

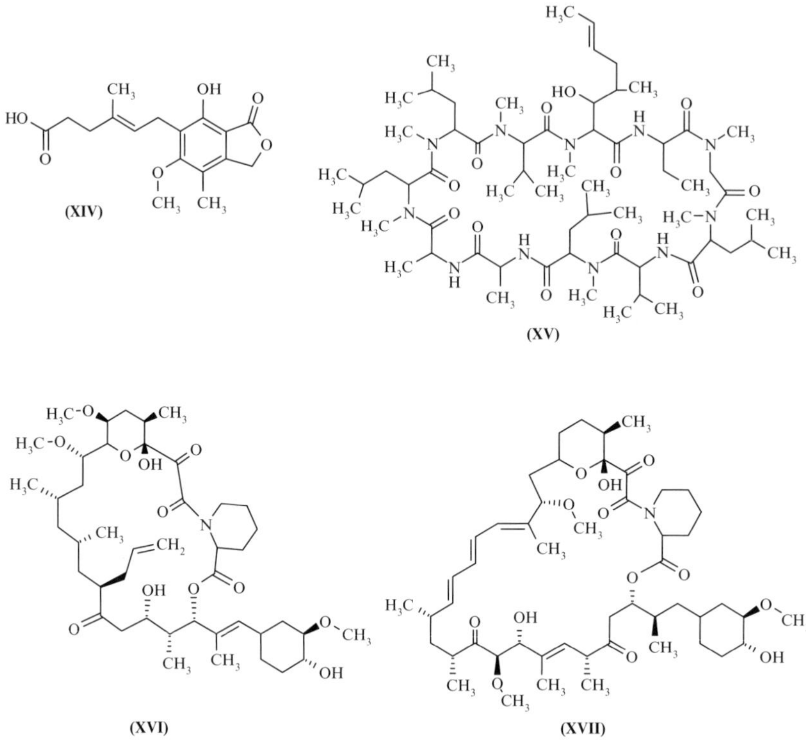

Figure 2.4 Chemical structures of immunosuppressive agents: Mycophenolic acid (XIV), Cyclosporin A (XV), Tacrolimus (XVI), and Sirolimus (XVII).

in clinical medicine and is widely used to prevent and treat graft rejection and graft-versus-host disease following both solid organ and bone marrow transplants. Originally isolated at Sandoz Laboratories in Basel for its antifungal activity, it was fortunate that this group tested all of their new compounds for antiviral, cytostatic, and immunosuppressive activity. In the latter tests, XV proved to inhibit specifically clonal proliferation of T-lymphocytes and lacked general cytotoxicity (Borel et al., 1976). Biosynthetically, it can be produced by a NRPS (Lawen and Zocher, 1990). Many natural or precursor-directed analogs have been produced but XV is still the most active (see the review by Mann (2001) and references cited therein).

Although cyclosporin was the only immunosuppressant product on the market for many years, two other products produced by actinomycetes have been developed more recently that are 100-fold more active and less toxic. Tacrolimus (formerly known as FK506) (XVI in Fig. 2.4) and Sirolimus (known as Rapamycin) (XVII in Fig. 2.4) are polyketide macrolactones produced from *Streptomyces tsukubaensis* and *Streptomyces hygroscopicus*, respectively (Kino et al., 1987; Vezina et al., 1975). They are both antifungal agents, but they also inhibit T-cell activation and proliferation by interacting with intracellular proteins and blocking signal transduction, involving IL-2 and other cytokines as effector molecules (Mann, 2001). It is worth mentioning that most of the complex and not completely elucidated cascade of events that follow antigen binding to T-cells and leading to an immune response have been unveiled in connection with investigations of the differential mode of action of cyclosporin, tacrolimus, and sirolimus. Tacrolimus was discovered in 1984 at Fusjisawa Pharmaceutical Company in a systematic screening program for agents with activity similar to that of cyclosporin. Its structure was finally elucidated in 1987 and it subsequently entered clinical use in 1991. It is a 23-member macrolide lactone including an *N*-heterocyclic ring. Sirolimus was discovered as a potent anticandida agent at Wyeth-Ayerst Pharmaceuticals in 1975: subsequent studies revealed impressive antitumor and immunosuppressive activity and it was approved as an immunosuppressive drug in 1999. The major portion of the molecule is a 31-member macrolide containing three conjugated double bonds and a heterocyclic group. The part of the molecule containing the heterocyclic group and the substituted cyclohexane ring is identical for tacrolimus and sirolimus. In both molecules, the polyketide chain is assembled by modular type I PKSs; the substituted cyclohexane ring derives from shikimic acid and the heterocyclic aromatic ring from L-lysine via pipecolic acid (see the review by Reynolds and Demain (1997) and references cited therein). Notwithstanding the substantial structural similarity, their mechanism of action is different. Tacrolimus, similarly to the structurally unrelated cyclosporin, acts preferentially by disrupting the activation pathway of IL-2 gene transcription, whereas sirolimus interferes in a later step of T-cell activation, disturbing the cascade process triggered by IL-2 binding to T-cell receptors (Mann, 2001). Sirolimus and its derivatives bind to a cytosolic protein, FK506 binding protein (FKBP), which subsequently inhibits mTOR (mammalian target of rapamycin), resulting in decreased expression of mRNAs necessary for cell-cycle progression and arresting cells in the G1 phase. This effect occurs not only in both T cells and B cells, but also has been demonstrated in many tumor cell lines. For this reason new anticancer agents are being developed by semisynthetic modification of the sirolimus polyketide scaffold. An ester derivative, Temsirolimus, has recently been approved for renal cell carcinoma treatment and it is undergoing clinical trials for the treatment of many other types of tumors (NCI, http://www.cancer.gov/).

2.8. The best-selling drugs from natural products: Fungal statins

Since the discovery of penicillin from *Penicillium chrysogenum*, scientists have continued to search for new antibiotics produced by filamentous fungi belonging to the phylum *Ascomycota*. More recently, the potential of fungi as producers of mammalian enzyme inhibitors has been recognized. Fungal statins (XVIII in Fig. 2.5), which act as competitive inhibitors of 3-hydroxy-3-methylglutaryl coenzyme A (HMG-CoA)-reductase, the regulatory and rate-limiting enzyme of cholesterol biosynthesis in liver, have been very successful in this regard. The structural similarity and high affinity of the acid form of natural statins and HMG, the natural substrate of the

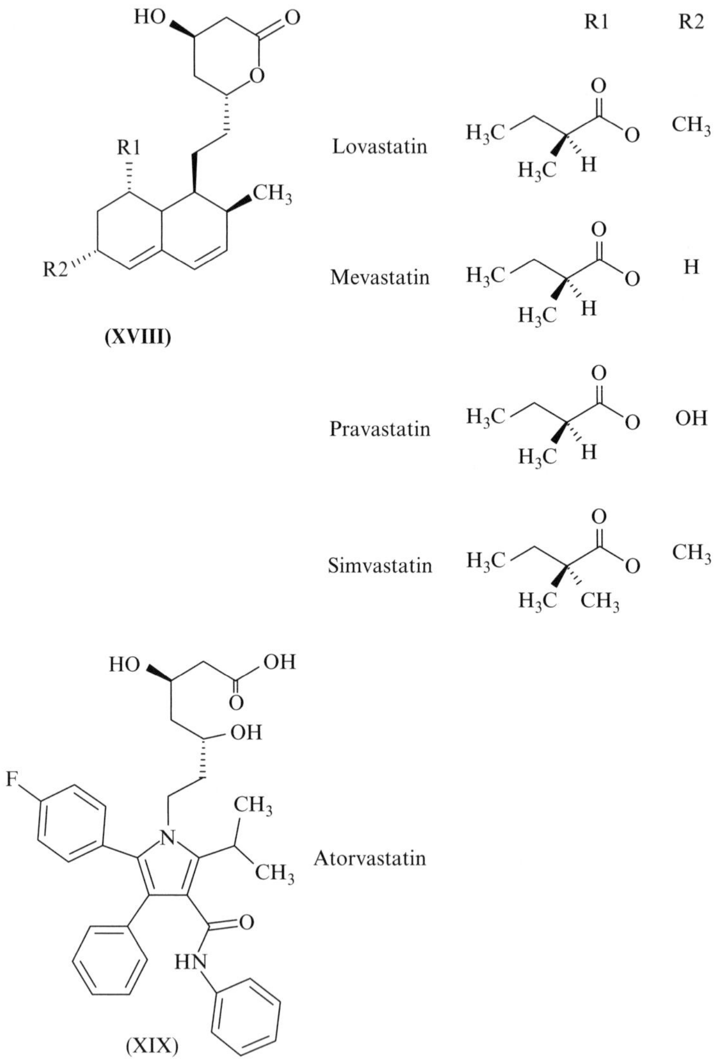

Figure 2.5 Chemical structures of statins: basic structure of natural and semisynthetic statins (XVIII) with side chains at C8 (R1) and C6 (R2), and synthetic Atorvastatin (XIX).

enzyme, afford specific and effective inhibition of HMG-CoA-reductase, blocking the synthesis of mevalonic acid.

The market for statins is very large, in the 15 billion dollar range (Demain and Adrio, 2008). Statins represent the first-line drug therapy for lowering cholesterol levels to prevent coronary artery disease. The best-selling such drug is the synthetic Atorvastatin (XIX in Fig. 2.5) (Lipitor), which sold over 11 billion dollars in 2004 and is at or above this level even today (Newman and Cragg, 2007). All members of the natural group possess a common polyketide portion, a hydroxy-hesahydro naphthalene ring system, to which different side chains are linked. Compactin (Fig. 2.5), also called Mevastatin, was first discovered in 1976 as an antifungal product of *Penicillium brevicompactum* (Brown *et al.*, 1976) and, at about the same time, as a specific inhibitor of cholesterolgenesis in *Penicillium citrum* (Endo *et al.*, 1976). Later, the more active methylated form of compactin known as Lovastatin (Fig. 2.5) (also named Monacolin K or Mevinolin and traded as Mevacor) was discovered independently at Sankyo (Endo, 1980) and at Merck (Alberts *et al.*, 1980) in fermentation broths of *Monascus ruber* and *Aspergillus terreus,* respectively. Lovastatin is produced by fermentation and was FDA-approved in 1987. Gene cloning confirmed the biosynthetic pathways hypothesized in early investigations (see the review by Manzoni and Rollini (2002) and references cited therein). The lovastatin biosynthetic cluster contains two PKSs—one involved in the assembly of the nonaketide skeleton of the hexahydro-naphthalene and another diketide synthase for the formation of the methylbutyric side chain—plus accessory proteins modulating their activities (Kennedy *et al.*, 1999). Pravastatin, launched 2 years after lovastatin, is a more *in vivo* active and tissue-selective derivative of mevastatin; it is obtained by microbial hydroxylation (biotransformation) from *Streptomyces carbophilus*. This new statin was first shown to be a minor but potent urinary metabolite during mevastatin experimentation in dogs. A screening for microbial hydrolases followed and a two-step fermentation process was developed, first to produce mevastatin and then to biotransform the drug. Simvastatin is a semisynthetic derivative of lovastatin and was the best-selling drug from 1998 to 2001 before the advent of atorvastatin, which represents a novel generation of synthetic statins quite different in structure from the natural ones but which strongly interact with the catalytic site of the human HMG-CoA reducatase (Istvan and Deisenhofer, 2001).

3. Conclusions

In the past, countless compounds that first were not believed to be truly active or proved toxic were rediscovered in later investigations by applying more specific and sensitive screening methods. Isolation of novel organisms, but also reinvestigation of known microbial products and producers, has proven to be fruitful, especially in the light of the expanding opportunities offered by combinatorial biosynthesis, heterologous expression, chemoenzymatic synthesis, and of the acquired knowledge about the genomes. This seems particularly true for the non anti-infective microbial drugs such as anticancer, immunosuppressive agents, and metabolic disease-targeting molecules. In these therapeutic areas, the real problem appears quite often not to be whether we are able to discover further new useful microbial leads, but rather how we can optimize and quickly and effectively apply the chances derived from these new discoveries to develop and produce clinically valuable drugs.

ACKNOWLEDGMENTS

The author thanks Giorgio Toppo for Natural Product Database (ABL) queries and Giancarlo Lancini for useful discussions and for his valuable help in identifying and drawing the chemical structures.

REFERENCES

Alberts, A. W., Chen, J., Kuron, G., Hunt, V., Huff, J., Hoffman, C., Rothrock, J., Lopez, M., Joshua, H., Harris, E., Patchett, A., Monaghan, R., et al. (1980). Mevinolin: A highly potent competitive inhibitor of hydroxymethylglutaryl-coenzyme A reductase and a cholesterol-lowering agent. *Proc. Natl. Acad. Sci. USA* **77,** 3957–3961.

Belcourt, M. F., Penketh, P. G., Hodnick, W. F., Johnson, D. A., Sherman, D. H., Rockwell, S., and Sartorelli, A. C. (1999). Mitomycin resistance in mammalian cells expressing the bacterial mitomycin C resistance protein MCRA. *Proc. Natl. Acad. Sci. USA* **96,** 10489–10494.

Beltrametti, F., Rossi, R., Selva, E., and Marinelli, F. (2006). Antibiotic production improvement in the rare actinomycete *Planobsipora rosea* by selection of resistance mutations to the aminoglycosides streptomycin and gentamycin and to rifamycin. *J. Ind. Microbiol. Biotechnol.* **34,** 283–288.

Berdy, J. (2005). Bioactive microbial metabolites. *J. Antibiot. (Tokyo)* **58,** 1–26.

Biondi, N., Tredici, M. R., Taton, A., Wilmotte, A., Hodgson, D. A., Losi, D., and Marinelli, F. (2008). Cyanobacteria from benthic mats of Antarctic lakes as a source of new bioactivities. *J. Appl. Microbiol.* **105,** 105–115.

Borel, J. F., Feurer, C., Gubler, H. U., and Stahelin, H. (1976). Biological effects of cyclosporin A: A new antilymphocytic agent. *Agents Actions* **6,** 468–475.

Brandi, L., Lazzarini, A., Cavaletti, L., Abbondi, M., Corti, E., Ciciliato, I., Gastaldo, L., Marazzi, A., Feroggio, M., Fabbretti, A., Maio, A., Colombo, L., et al. (2006). Novel tetrapeptide inhibitors of bacterial protein synthesis produced by a *Streptomyces* sp. *Biochemistry* **45,** 3692–3702.

Brown, A. G., Smale, T. C., King, T. J., Hasenkamp, R., and Thompson, R. H. (1976). Crystal and molecular structure of compactin, a new antifungal metabolite from *Penicillium brevicompactum*. *J. Chem. Soc. [Perkin 1]* 1165–1170.

Bu'Lock, J. D. (1961). Intermediary metabolism and antibiotic synthesis. *Adv. Appl. Microbiol.* **3,** 293–342.

Castiglione, F., Cavaletti, L., Losi, D., Lazzarini, A., Carrano, L., Feroggio, M., Ciciliato, I., Corti, E., Candiani, G., Marinelli, F., and Selva, E. (2007). A novel lantibiotic acting on bacterial cell wall synthesis produced by the uncommon actinomycete *Planomonospora* sp. *Biochemistry* **46,** 5884–5895.

Castiglione, F., Lazzarini, A., Carrano, L., Corti, E., Ciciliato, I., Gastaldo, L., Candiani, P., Losi, D., Marinelli, F., Selva, E., and Parenti, F. (2008). Determining the structure and mode of action of microbisporicin, a potent lantibiotic active against multiresistant pathogens. *Chem. Biol.* **15,** 22–31.

Challis, G. L., and Hopwood, D. A. (2003). Synergy and contingency as driving forces for the evolution of multiple secondary metabolite production by *Streptomyces* species. *Proc. Natl. Acad. Sci. USA* **100,** 14555–14561.

Cragg, G. M. (1998). Paclitaxel (Taxol): A success story with valuable lessons for natural product drug discovery and development. *Med. Res. Rev.* **18,** 315–331.

Davies, J., Spiegelman, G. B., and Yim, G. (2006). The world of subinhibitory antibiotic concentrations. *Curr. Opin. Microbiol.* **9,** 445–453.

Demain, A. L., and Adrio, J. L. (2008). Contributions of microorganisms to industrial biology. *Mol. Biotechnol.* **38,** 41–55.

Ding, Y., Seufert, W. H., Beck, Z. Q., and Sherman, D. H. (2008). Analysis of the cryptophycin P450 epoxidase reveals substrate tolerance and cooperativity. *J. Am. Chem. Soc.* **130,** 5492–5498.

Du, L., Sanchez, C., Chen, M., Edwards, D. J., and Shen, B. (2000). The biosynthetic gene cluster for the antitumor drug bleomycin from *Streptomyces verticillus* ATCC15003 supporting functional interactions between nonribosomal peptide synthetases and a polyketide synthase. *Chem. Biol.* **7,** 623–642.

Endo, A. (1980). Monacolin K, a new hypocholesterolemic agent that specifically inhibits 3-hydroxy-3-methylglutaryl coenzyme A reductase. *J. Antibiot. (Tokyo)* **33,** 334–336.

Endo, A., Kuroda, M., and Tanzawa, K. (1976). Competitive inhibition of 3-hydroxy-3-methylglutaryl coenzyme A reductase by ML-236A and ML-236B fungal metabolites, having hypocholesterolemic activity. *FEBS Lett.* **72,** 323–326.

Fabre, B., Armau, E., Etienne, G., Legendre, F., and Tiraby, G. (1988). A simple screening method for insecticidal substances from actinomycetes. *J. Antibiot. (Tokyo)* **41,** 212–219.

Fajardo, A., and Martinez, J. L. (2008). Antibiotics as signals that trigger specific bacterial responses. *Curr. Opin. Microbiol.* **11,** 161–167.

Galm, U., Hager, M. H., Van Lanen, S. G., Ju, J., Thorson, J. S., and Shen, B. (2005). Antitumor antibiotics: Bleomycin, enediynes, and mitomycin. *Chem. Rev.* **105,** 739–758.

Galm, U., Wang, L., Wendt-Pienkowski, E., Yang, R., Liu, W., Tao, M., Coughlin, J. M., and Shen, B. (2008). *In vivo* manipulation of the bleomycin biosynthetic gene cluster in *Streptomyces verticillus* ATCC15003 revealing new insights into its biosynthetic pathway. *J. Biol. Chem.* **283,** 28236–28245.

Gaspari, F., Paitan, Y., Mainini, M., Losi, D., Ron, E. Z., and Marinelli, F. (2005). Myxobacteria isolated in Israel as potential source of new anti-infectives. *J. Appl. Microbiol.* **98,** 429–439.

Gerth, K., Bedorf, N., Hofle, G., Irschik, H., and Reichenbach, H. (1996). Epothilons A and B: Antifungal and cytotoxic compounds from *Sorangium cellulosum* (Myxobacteria). Production, physico-chemical and biological properties. *J. Antibiot. (Tokyo)* **49,** 560–563.

Gong, G. L., Sun, X., Liu, X. L., Hu, W., Cao, W. R., Liu, H., Liu, W. F., and Li, Y. Z. (2007). Mutation and a high-throughput screening method for improving the production of epothilones of *Sorangium*. *J. Ind. Microbiol. Biotechnol.* **34,** 615–623.

Gust, B., Chandra, G., Jakimowicz, D., Yuqing, T., Bruton, C. J., and Chater, K. F. (2004). Lambda red-mediated genetic manipulation of antibiotic-producing *Streptomyces*. *Adv. Appl. Microbiol.* **54,** 107–128.

Hyun, C. G., Kim, S. Y., Hur, J. H., Seo, M. J., Suh, J. W., and Kim, S. O. (2005). Molecular detection of alpha-glucosidase inhibitor-producing actinomycetes. *J. Microbiol.* **43,** 313–318.

Istvan, E. S., and Deisenhofer, J. (2001). Structural mechanism for statin inhibition of HMG-CoA reductase. *Science* **292,** 1160–1164.

Julien, B., and Shah, S. (2002). Heterologous expression of epothilone biosynthetic genes in *Myxococcus xanthus*. *Antimicrob. Agents Chemother.* **46,** 2772–2778.

Kennedy, J., Auclair, K., Kendrew, S. G., Park, C., Vederas, J. C., and Hutchinson, C. R. (1999). Modulation of polyketide synthase activity by accessory proteins during lovastatin biosynthesis. *Science* **284,** 1368–1372.

Kino, T., Hatanaka, H., Hashimoto, M., Nishiyama, M., Goto, T., Okuhara, M., Kohsaka, M., Aoki, H., and Imanaka, H. (1987). FK-506, a novel immunosuppressant isolated from a *Streptomyces*. I. Fermentation, isolation, and physico-chemical and biological characteristics. *J. Antibiot. (Tokyo)* **40,** 1249–1255.

Lam, K. S. (2007). New aspects of natural products in drug discovery. *Trends Microbiol.* **15,** 279–289.

Lawen, A., and Zocher, R. (1990). Cyclosporin synthetase. The most complex peptide synthesizing multienzyme polypeptide so far described. *J. Biol. Chem.* **265,** 11355–11360.

Lazzarini, A., Cavaletti, L., Toppo, G., and Marinelli, F. (2001). Rare genera of actinomycetes as potential producers of new antibiotics. *Antonie Van Leeuwenhoek* **79,** 399–405.

Lomovskaya, N., Otten, S. L., Doi-Katayama, Y., Fonstein, L., Liu, X. C., Takatsu, T., Inventi-Solari, A., Filippini, S., Torti, F., Colombo, A. L., *et al.* (1999). Doxorubicin overproduction in *Streptomyces peucetius*: Cloning and characterization of the *dnrU* ketoreductase and *dnrV* genes and the *doxA* cytochrome P-450 hydroxylase gene. *J. Bacteriol.* **181,** 305–318.

Madduri, K., Kennedy, J., Rivola, G., Inventi-Solari, A., Filippini, S., Zanuso, G., Colombo, A. L., Gewain, K. M., Occi, J. L., MacNeil, D. J., *et al.* (1998). Production of the antitumor drug epirubicin (4′-epidoxorubicin) and its precursor by a genetically engineered strain of *Streptomyces peucetius*. *Nat. Biotechnol.* **16,** 69–74.

Magarvey, N. A., Beck, Z. Q., Golakoti, T., Ding, Y., Huber, U., Hemscheidt, T. K., Abelson, D., Moore, R. E., and Sherman, D. H. (2006). Biosynthetic characterization and chemoenzymatic assembly of the cryptophycins. Potent anticancer agents from cyanobionts. *ACS Chem. Biol.* **1,** 766–779.

Mann, J. (2001). Natural products as immunosuppressive agents. *Nat. Prod. Rep.* **18,** 417–430.

Manzoni, M., and Rollini, M. (2002). Biosynthesis and biotechnological production of statins by filamentous fungi and application of these cholesterol-lowering drugs. *Appl. Microbiol. Biotechnol.* **58,** 555–564.

Mao, Y., Varoglu, M., and Sherman, D. H. (1999). Molecular characterization and analysis of the biosynthetic gene cluster for the antitumor antibiotic mitomycin C from *Streptomyces lavendulae* NRRL 2564. *Chem. Biol.* **6,** 251–263.

Marinelli, F. (2003). Industrial relevance of microbial diversity. *In* "Abstract Book 1st FEMS Congress of European Microbiologists," p. 8. Ljubljana, Slovenia.

Molnar, I., Schupp, T., Ono, M., Zirkle, R., Milnamow, M., Nowak-Thompson, B., Engel, N., Toupet, C., Stratmann, A., Cyr, D. D., Gorlach, J., Mayo, J. M., *et al.* (2000). The biosynthetic gene cluster for the microtubule-stabilizing agents epothilones A and B from *Sorangium cellulosum* So ce90. *Chem. Biol.* **7,** 97–109.

Newman, D. J., and Cragg, G. M. (2007). Natural products as sources of new drugs over the last 25 years. *J. Nat. Prod.* **70,** 461–477.

Newman, D. J., and Shapiro, S. (2008). Microbial prescreens for anticancer activity. *SIM News* **58,** 132–150.

Quesniaux, V. F., Schreier, M. H., Wenger, R. M., Hiestand, P. C., Harding, M. W., and Van Regenmortel, M. H. (1987). Cyclophilin binds to the region of cyclosporine involved in its immunosuppressive activity. *Eur. J. Immunol.* **17,** 1359–1365.

Reichenbach, H., and Hofle, G. (2008). Discovery and development of the epothilones: A novel class of antineoplastic drugs. *Drugs R D* **9,** 1–10.

Reynolds, K. A., and Demain, A. L. (1997). Rapamycin, FK506, and ascomycin-related compounds. *In* "Biotechnology of Antibiotics" (R. K. A. and D. A. L. , eds.), 2nd ed., pp. 497–520. Marcel Dekker, NewYork.

Rippka, R. (1988). Isolation and purification of cyanobacteria. *Methods Enzymol.* **167,** 3–27.

Sanchez, C., Zhu, L., Brana, A. F., Salas, A. P., Rohr, J., Mendez, C., and Salas, J. A. (2005). Combinatorial biosynthesis of antitumor indolocarbazole compounds. *Proc. Natl. Acad. Sci. USA* **102,** 461–466.

Schauwecker, F., Pfennig, F., Grammel, N., and Keller, U. (2000). Construction and *in vitro* analysis of a new bi-modular polypeptide synthetase for synthesis of *N*-methylated acyl peptides. *Chem. Biol.* **7,** 287–297.

Singh, S. B., Jayasuriya, H., Dewey, R., Polishook, J. D., Dombrowski, A. W., Zink, D. L., Guan, Z., Collado, J., Platas, G., Pelaez, F., Felock, P. J., and Hazuda, D. J. (2003). Isolation, structure, and HIV-1-integrase inhibitory activity of structurally diverse fungal metabolites. *J. Ind. Microbiol. Biotechnol.* **30,** 721–731.

Smith, C. D., Zhang, X., Mooberry, S. L., Patterson, G. M., and Moore, R. E. (1994). Cryptophycin: A new antimicrotubule agent active against drug-resistant cells. *Cancer Res.* **54,** 3779–3784.

Sponga, F., Cavaletti, L., Lazzarini, A., Losi, D., Borghi, A., Ciciliato, I., and Marinelli, F. (1999). Biodiversity and potential of marine-derived microrganisms. *J. Biotechnol.* **70,** 65–69.

Stefanelli, S., Corti, E., Montanini, N., Denaro, M., and Sarubbi, E. (1997). Inhibitors of type-I interleukin-1 receptor from microbial metabolites. *J. Antibiot. (Tokyo)* **50,** 484–489.

Stefanelli, S., Sponga, F., Ferrari, P., Sottani, C., Corti, E., Brunati, C., and Islam, K. (1996). Inhibitors of myo-inositol monophosphatase, ATCC 20928 factors A and C. Isolation, physico-chemical characterization and biological properties. *J. Antibiot. (Tokyo)* **49,** 611–616.

Stierle, A., Strobel, G., and Stierle, D. (1993). Taxol and taxane production by *Taxomyces andreanae*, an endophytic fungus of Pacific yew. *Science* **260,** 214–216.

Strobel, G. (2006). Harnessing endophytes for industrial microbiology. *Curr. Opin. Microbiol.* **9,** 240–244.

Sugiyama, M., and Kumagai, T. (2002). Molecular and structural biology of bleomycin and its resistance determinants. *J. Biosci. Bioeng.* **93,** 105–116.

Tang, L., Chung, L., Carney, J. R., Starks, C. M., Licari, P., and Katz, L. (2005). Generation of new epothilones by genetic engineering of a polyketide synthase in *Myxococcus xanthus*. *J. Antibiot. (Tokyo)* **58,** 178–184.

Tang, L., Shah, S., Chung, L., Carney, J., Katz, L., Khosla, C., and Julien, B. (2000). Cloning and heterologous expression of the epothilone gene cluster. *Science* **287,** 640–642.

Vezina, C., Kudelski, A., and Sehgal, S. N. (1975). Rapamycin (AY-22,989), a new antifungal antibiotic. I. Taxonomy of the producing streptomycete and isolation of the active principle. *J. Antibiot. (Tokyo)* **28,** 721–726.

Vining, L. C. (1990). Functions of secondary metabolites. *Annu. Rev. Microbiol.* **44,** 395–427.

Wani, M. C., Taylor, H. L., Wall, M. E., Coggon, P., and McPhail, A. T. (1971). Plant antitumor agents. VI. The isolation and structure of taxol, a novel antileukemic and antitumor agent from *Taxus brevifolia*. *J. Am. Chem. Soc.* **93,** 2325–2327.

Waterbury, J. B., and Stanier, R. Y. (1981). Isolation and growth of cyanobacteria from marine and hypersaline environments. *In* "The Prokaryotes: A Handbook on Habitats, Isolation, and Identification of Bacteria," (H. G. Schlegel, ed.), pp. 221–223. Springer-Verlag, Berlin, Germany.

Zazopoulos, E., Huang, K., Staffa, A., Liu, W., Bachmann, B. O., Nonaka, K., Ahlert, J., Thorson, J. S., Shen, B., and Farnet, C. M. (2003). A genomics-guided approach for discovering and expressing cryptic metabolic pathways. *Nat. Biotechnol.* **21,** 187–190.

CHAPTER THREE

Discovering Natural Products from Myxobacteria with Emphasis on Rare Producer Strains in Combination with Improved Analytical Methods

Ronald O. Garcia, Daniel Krug, *and* Rolf Müller

Contents

1. Introduction	60
1.1. Myxobacteria as proficient producers of bioactive compounds	62
2. The Search for Novel Myxobacteria and Their Metabolites—Basic Considerations	63
2.1. Prospects for the discovery of novel myxobacteria	64
2.2. Choice of material to favor the discovery of novel myxobacteria	65
2.3. A short survey of previous isolation efforts	66
2.4. Influence of geographical and environmental factors on the isolation of novel myxobacteria	68
2.5. Genetic characterization of novel strains	68
2.6. The issue of "unculturability"	69
3. Methods for Isolation, Purification, and Preservation of Novel Myxobacteria	71
3.1. Baiting method for the isolation of myxobacteria	72
3.2. Important factors for successful isolation of novel strains	72
3.3. Swarming and fruiting body recognition	75
3.4. A simple and effective method for the purification of myxobacteria	76
3.5. Culture maintenance and preservation	77
4. Fermentation and Screening for Known and Novel Metabolites	78
4.1. Cultivation conditions and sample workup procedure	79
4.2. Bioassays using crude extracts and prepurified fractions	79
4.3. Analysis of myxobacterial metabolite profiles using high-resolution mass spectrometry	81

Department of Pharmaceutical Biotechnology, Saarland University, Saarbrücken, Germany

4.4. A diversity-oriented approach to mining myxobacterial
   secondary metabolomes   84
4.5. Summary and outlook   87
Acknowledgment   88
References   88

Abstract

Myxobacteria produce a range of structurally novel natural products which exhibit unusual or unique modes of action, attracting significant interest from both the academic and drug discovery communities. Efforts to discover new strains with the potential to biosynthesize novel molecules have revealed that myxobacterial diversity and natural products are far from exhausted. We describe here a general, nonselective approach to unearth further myxobacterial strains, in order to mine them for compounds with potential as medicines. Sample collection from locations world-wide has shown that environments which exhibit significant biological complexity yield the highest probability of isolating novel myxobacterial strains. Here, we illustrate the details of simple and efficient strain purification techniques, which lead systematically to the identification of new and promising myxobacteria. Compound identification is then facilitated by molecular biological approaches, coupled with sophisticated high resolution mass spectrometry, statistical analysis, and bioassays.

1. INTRODUCTION

Myxobacteria are fascinating prokaryotes, which exhibit a complex developmental life cycle mimicking that of the eukaryotic Dictyosteliomycetes—an apparent example of convergent evolution. Growth is characterized by the aggregation of solitary, gliding vegetative rods into a pseudoplasmodium-like slimy cell mass, which finally transforms into a multicellular fruiting body bearing propagative spores (Fig. 3.1). Myxobacteria have also emerged as a promising alternative to more conventional sources of microbial secondary metabolites, such as the actinomycetes and fungi. Of particular note, compounds isolated from myxobacteria often exhibit unique modes of action against a broad range of both prokaryotic and eukaryotic cells. Compounds isolated from myxobacterial strains which colonize the dung of herbivores usually exhibit antagonistic activity against competing bacteria (generally Gram-positive spore-formers), whereas natural products found in strains isolated from decaying wood and plant materials inhibit yeasts and molds (Reichenbach, 1999b). Approximately 54% of known myxobacterial natural products are active against fungi (Gerth et al., 2003). In addition, the compounds are likely to have a function in the complex and micropredatory lifestyle of the myxobacteria (Meiser et al., 2006), making them of great academic biological interest.

Figure 3.1 Myxobacterial fruiting bodies. (A) *Stigmatella erecta* on the surface of minimal casitone agar; (B) *Cystobacter* sp. on VY/2 agar; (C) *Chondromyces apiculatus*, top view on rabbit dung bait; (D) *Melittangium boletus*, early stage of development, on a piece of plant material adhering to rabbit dung; (E) *Sorangium cellulosum* on filtered yeast agar (cVY/2); (F) *Myxococcus* sp. on casitone agar.

In order to discover new secondary metabolites, we aim to identify novel myxobacterial species, genera, and families, with the expectation that such strains are likely to produce as yet uncharacterized metabolites. Strain isolation is carried out by standard methods (Dawid, 2000; Shimkets *et al.*, 2006), although we highlight here several features which are key to the success of these experiments. We also discuss methods to recognize novel strains, as well as to obtain them as pure isolates. A new mass spectrometric approach is also reviewed which, in combination with bioinformatics methods, can be employed to detect new compounds from myxobacteria at unprecedented selectivity and sensitivity.

1.1. Myxobacteria as proficient producers of bioactive compounds

Due in large part to the work of the Reichenbach and Höfle research groups at the HZI Braunschweig, Germany (Helmholtz Centre for Infection Research; formerly the GBF, German Research Centre for Biotechnology), more than 7500 different myxobacterial strains have been isolated to date, most of which are currently housed in the "Deutsche Sammlung von Mikroorganismen und Zellkulturen" (DSMZ), in Braunschweig (Reichenbach, 2005). The majority of the isolates deposited in the open collection belong to the suborders *Cystobacterineae* and *Sorangiineae*, with members of the genus *Sorangium* most common in the latter. Although the number of actinomycetes discovered to date far exceeds that of the myxobacteria, known myxobacterial strains have yielded at least 100 natural product core structures, and some 500 derivatives. Secondary metabolism is unequally distributed among the myxobacteria, as the majority of compounds have been isolated from selected species, including strains of *Sorangium cellulosum*, *Myxococcus*, and *Chondromyces* species. However, these findings may reflect productivity under laboratory culture conditions, and not the true genetic potential of the strains.

A number of the most valuable compounds discovered from myxobacteria act on the cytoskeleton of eukaryotic cells. This mechanism of action, shown for example by epothilone, tubulysin (Fig. 3.2A), and chondramide (Mulzer, 2008; Sasse *et al.*, 1998) is observed only rarely among compounds of bacterial origin (Gronewold *et al.*, 1999). Epothilone, the most promising of the metabolites, has already been developed as an antitumor agent due to its paclitaxel-type activity (Müller, 2009; Mulzer, 2009; Reichenbach and Höfle, 1999). Epothilone is active against diverse tumor cell lines, including breast carcinoma, HeLa cervical carcinoma, Burkitt's lymphoma, colon carcinoma, ovarian carcinoma, and neuroblastoma (Reichenbach and Höfle, 1999). Last year, the United States Food and Drug Administration (FDA) approved the first epothilone derivative, ixabepilone, as a drug for the treatment of breast cancer. In contrast to epothilone which acts to stabilize microtubules, tubulysin inhibits the polymerization of tubulin (Khalil *et al.*, 2006; Steinmetz *et al.*, 2004) (Fig. 3.2B), and additionally shows antiangiogenic activity (Kaur *et al.*, 2006).

The myxobacterial compound argyrin (Fig. 3.2A), isolated from *Archangium gephyra*, acts as a potent immunosuppressant, targeting T-cell-independent antibodies (Sasse *et al.*, 2002). More recently, it was discovered to

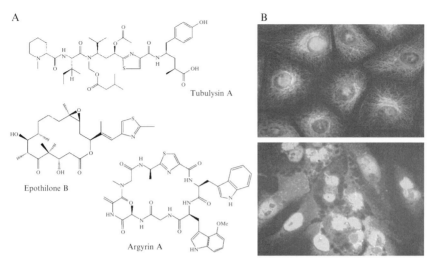

Figure 3.2 (A) Chemical structures of some biologically active myxobacterial natural products. (B) Influence of tubulysin A on the microtubule cytoskeleton of Ptk2 potoroo cells. The cells were fixed and immunostained for tubulin. The upper picture shows microtubules of control cells (green, microtubule network; blue, cell nuclei), while the lower picture was taken after incubation with tubulysin A for 24 h Following treatment, only diffuse tubulin fluorescence remains, and the early stages of degradation of the cell nucleus are apparent. Photograph taken with permission from Sandmann *et al.* (2004). (See Color Insert.)

exhibit antitumor activity (Nickeleit *et al.*, 2008). Argyrin A acts by stabilizing the protein p27^{kip1}, ultimately leading to inhibition of the 20S proteasome and so to apoptosis. Loss of p27^{kip1} was sufficient to confer resistance to argyrin A.

This section has provided only an incomplete survey of the broad range of activities exhibited by myxobacterial compounds. The reader is referred to several more comprehensive reviews for further information (Bode and Müller, 2007; Reichenbach and Höfle, 1999). See also Chapter 2 in this volume.

2. THE SEARCH FOR NOVEL MYXOBACTERIA AND THEIR METABOLITES—BASIC CONSIDERATIONS

The search for myxobacteria over the last three decades has led to the isolation of thousands of strains, producing hundreds of new compounds. This observation prompts the following questions. Is it reasonable still to expect to find novel myxobacteria? What are the prospects for finding new compounds in novel myxobacterial strains? What are the best ways of identifying novel compounds from new isolates? We address these questions in turn in the following sections.

2.1. Prospects for the discovery of novel myxobacteria

A brief survey of the literature reveals that novel myxobacterial strains continue to be isolated from locations world-wide, suggesting that strain diversity has yet to be exhausted. For example, Dawid and coworkers (Dawid, 2000) found at least 10 new species during a world-wide search, including psychrophilic ("cold-loving") strains belonging to the suborder of *Nannocystineae* and the family *Polyangiaceae*, as well as to unassigned genera of myxobacteria. Such strains are likely to be underrepresented in the current collections, due to their unusual physiological growth requirements, which make them resistant to laboratory cultivation.

One obvious prerequisite for discovering new strains is the ability to grow the bacteria under laboratory conditions. In our experience, efforts to isolate and cultivate from within the suborder *Cystobacterineae* are likely to be the most successful. Indeed, the *Cystobacterineae* comprise 55% of the HZI collection (Gerth *et al.*, 2003). These strains tend to grow faster than isolates from the other two suborders of the myxobacteria, the *Nannocystineae* and the *Sorangiineae*, facilitating attempts to purify the strains away from contaminating species. However, several species, including members of the genera *Cystobacter*, *Archangium*, and *Stigmatella*, produce a viscous slime, which often traps other bacterial species within the colony. In general, the steps necessary to obtain pure (axenic) cultures vary from species to species and depend on the degree of contamination. For *Myxococcus* (also a member of the *Cystobacterineae*), for example, single-step purification is often enough to obtain a pure culture.

Finding new species within the *Cystobacterineae*—a discovery obviously correlated with a higher probability of identifying novel metabolites—would appear to be possible, in principle. However, a new genus has not been added to the *Cystobacterineae* over the last five years. For example, although the genus *Hyalangium* was recently inaugurated (Reichenbach, 2005), the type species NOCB2 and NOCB4 (New Organism of the *Cystobacterineae*-type) were identified a decade ago on the basis of their 16S rDNA sequence (Spröer *et al.*, 1999). Similarly, *Angiococcus* (e.g. *A. disciformis* An d1) has already been known for more than a decade, but lately reclassified as *Pyxidicoccus* (Reichenbach, 2005) on the main basis of 16S rDNA (Spröer *et al.*, 1999). The situation appears more promising, however, for the suborder *Nannocystineae*, as additions in the last 6 years include the marine isolates *Plesiocystis*, *Enhygromyxa* (Iizuka *et al.*, 2003a,b), *Haliangium* (Fudou *et al.*, 2002) and *Kofleria flava*, a reclassification of *Polyangium vitellinum* Pl vt1 (Reichenbach, 2005).

We believe that the suborder *Sorangiineae* also holds great promise for the discovery of novel myxobacteria. As these bacteria are notoriously difficult to culture and purify, many of the strains are likely to have been missed in initial screening programs. Indeed, certain described members of this group were never re-isolated from either crude cultures or natural field samples (Reichenbach, 1999a, 2005). The genus *Polyangium* is an illustrative example: among its seven described species, only three have type cultures deposited in the collections (Reichenbach, 2005). The *Polyangium* sp. used in the first reported phylogenetic tree for myxobacteria (Shimkets and Woese, 1992) is no longer present in the open collections for validation studies. Five other species in the genus were also described once by early investigators, but the absence of type specimens or cultures renders verification of their identities extremely difficult. If the strains still exist, they are probably deposited in herbaria that cannot be accessed straightforwardly (Reichenbach, 2005).

In contrast to this bleak picture, we have been fortunate to isolate and cultivate a sizeable number of both rare and entirely novel isolates in the last 8 years, all of which belong to the suborder *Sorangiineae* (R. G. and R. M., unpublished). We elaborate below on the fundamentals of this success, as well as detailing the striking features of the novel isolates. We also describe methods to preserve novel strains. In the last section, we explain the screening process used to identify known and novel compounds using a method which correlates them to their bioactivity.

2.2. Choice of material to favor the discovery of novel myxobacteria

Myxobacteria colonize a wide variety of environments. Standard samples employed for laboratory-based strain isolation include soil, the bark of trees, decaying plant materials, and the dung of herbivores. Myxobacteria can also be found growing directly on top of natural substrata in the field; in such cases, the cells exhibit a more characteristic and elaborate fruiting body structure than that which is typically observed under laboratory growth conditions. To our surprise, we discovered different *Chondromyces* species on leaves and on the hymenial surfaces of a basidiomycetous bracket fungus, *Ganoderma* (R. G., I. J. Dogma and R. M., unpublished). The latter material has not, to our knowledge, been previously reported as a substrate for the isolation of myxobacteria.

Forest soil matter found in tropical and subtropical regions often yield novel myxobacterial strains. Decaying plant materials serve as the best substrate for many cellulose-degraders, and animal droppings are productive sources for bacteriolytic and/or proteolytic species. The type strain of the new species *Phaselicystis flava* was isolated from a mountain forest soil sample in the Philippines (Garcia *et al.*, 2009). Other novel strains (which represent the type species for new genera) similarly originated from forest soil samples rich in humus and decaying plant debris (R. G. and R. M., unpublished).

We have also rediscovered a rare species of myxobacteria, which had previously been described only on the basis of its appearance on decaying wood substrate (Reichenbach, 2005). Culturing was never achieved, and therefore only an incomplete strain description had been possible. We have now obtained this isolate in pure culture (Fig. 3.3D), and deposited it in the collection at our institute.

A further novel species of myxobacteria was obtained from a remote forest in Africa, following growth on the mineral salts agar (ST21) used for cellulose-degraders (Shimkets et al., 2006). The strain appears to be unique, as it forms fruiting bodies which consist of long, robust stalks with large, cup-like sporangioles. Unusually, the sporangioles formed a mid-apical crown of short and fine spikes. We have never before encountered this combination of phenotypic features in our screening efforts.

Taken together, our data suggest that tropical and subtropical forest samples represent the best source for discovering new myxobacterial strains. The high complexity of the forest environment, coupled with the need to outcompete other species (or strains within the same species) for common resources, has apparently favored the development of a rich secondary metabolic armory in forest strains.

2.3. A short survey of previous isolation efforts

In this section, we discuss previous efforts to isolate myxobacteria, many of which were targeted to certain genera (e.g., *Sorangium* and *Myxococcus*). The motivation here was the proven capability of these genera as secondary metabolite producers. However, this selective approach is likely to have missed many important myxobacterial strains.

Strains belonging to the genus *Sorangium* are reported to comprise 5.5% of myxobacterial species worldwide (Dawid, 2000). Specifically searching for *Sorangium* species yielded a hit rate of ~23% (Gerth et al., 2003). Given their inherently slow growth (Gerth et al., 2003, 2007), it can take as long as 1–2 years to eliminate all contaminants from a *Sorangium* culture (Reichenbach, 1999a). Cocultivating microbes commonly include bacilli, amoebae, molds, actinomycetes, other gliding bacteria, and even nematodes. To overcome the contamination problem, cocktails of antibiotics (e.g., levamisole and cycloheximide) are commonly used in the purification of *Sorangium* (Gerth et al., 2007) and *Myxococcus* (Karwowski et al., 1996). *Sorangium* can usually tolerate high doses of kanamycin (Reichenbach, 1999a) to as much as 1000 μg/ml (Reichenbach, 2006). However, antibiotics used in the isolation and purification process may counter-select against novel species. For example, in an antibiotic sensitivity test employing three of our novel strains, growth inhibition by kanamycin (50 μg/ml) varied from partial to total. This observation illustrates that it would be worth investigating the general effects of such compounds on myxobacteria.

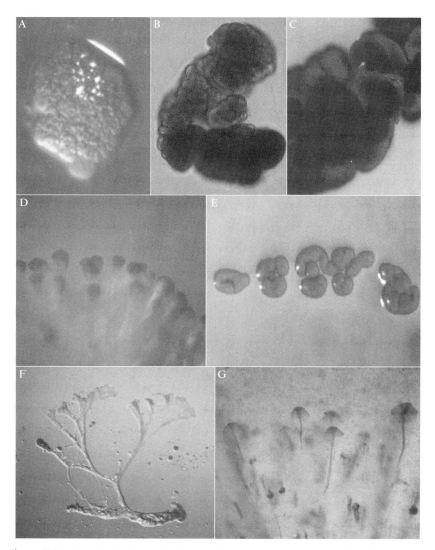

Figure 3.3 Fruiting body morphology and swarming patterns of novel and rare myxobacterial isolates. (A) Cluster of golden sporangioles. (B) Sporangioles tightly packed in sori. (C) Mass of minute and dense sporangioles. (D) Migrating bands of swarm produced under the agar by a rare strain. (E) The bean-shaped sporangioles of *Phaselicystis flava* SBKo001^T. (F) Fan-shaped and slimy migrating colony on the surface of the agar. (G) Independently migrating colony showing comet-like structures, which are produced deep within the agar. The images in panels A and E were obtained under a dissecting microscope, while those in B and C were generated using a fluorescent laser scanning microscope. D, F, and G constitute dissecting micrographs of swarm colonies. (See Color Insert.)

In the meantime, we argue that the isolation of novel myxobacteria should be favored by the absence of antibiotics in the growth medium. To date, all of our novel strains were isolated and brought into pure culture in the absence of antibiotics. Myxobacteria which are good candidates for production of bioactive compounds should be able to survive in the midst of a wide range of contaminants, and therefore it should not be necessary to supplement with antibiotics during the isolation process.

2.4. Influence of geographical and environmental factors on the isolation of novel myxobacteria

Sampling for myxobacteria has been performed worldwide, covering diverse environments and geographical locations. Probing of extreme environments, including the Antarctic, resulted in the isolation of some unusual psychrophilic myxobacteria (Dawid, 2000), while soil samples collected in countries with semi-arid climates have yielded moderately thermophilic strains (Gerth and Müller, 2005). Furthermore, exploration of the Japanese coastal marine environment resulted in the identification of several halophilic myxobacterial species (Fudou et al., 2002; Iizuka et al., 2003a,b). Molecular analyses of these halophilic isolates strongly indicate that the marine environment is likely to be a reservoir for other as yet uncharacterized marine myxobacteria, as has recently been found for actinomycetes too.

Soil samples collected at high elevation yielded novel strains with physicochemical characteristics different from those of related groups within the suborder. These strains form swarm colonies of comparable diameters at both 18 °C and 30 °C; hence, the strains grow optimally in a broader range of temperatures than normal psychrophiles and mesophiles (organisms with optimum growth at 30 °C). Whether or not this feature is associated with adaptation to the high altitude environment remains to be demonstrated. We believe that many more rare and wholly novel strains remain to be discovered, by specifically targeting as yet unexplored and isolated environments in tropical and subtropical zones. The warm climate in these regions may support myxobacterial diversity, and thus it seems very likely that novel strains will be found. Dawid and coworkers also highlighted these regions as boasting the largest range of myxobacterial species (Dawid, 2000).

2.5. Genetic characterization of novel strains

Once a putative novel strain is isolated, it is important to establish that the bacterium is truly novel. Morphological characterization is regarded as the foundation for classification and strain identification and is usually (but not always) supported by genetic data (e.g., 16S rDNA).

Based on the complete 16S rDNA sequences, all of the novel strains mentioned in this chapter show 94–96% homology to cellulose-degrading myxobacteria. Neighbor-joining trees however, constructed from similar sequences identified by BLASTn, reveal that these strains form a separate cluster within the δ-Proteobacteria, located in unexpectedly close proximity to sequences associated with as yet uncultureable bacterial clones of the δ-branch, and not with the cellulose-degrading myxobacteria (Fig. 3.4). This finding supports our hypothesis that these strains represent unexplored groups of myxobacteria.

One of the new species (representing a new genus) under investigation is *P. flava,* which is in fact the founding member of a new family, *Phaselicystidaceae* (Garcia *et al.*, 2009). 16S rDNA reveals 94% identity to cellulose degraders such as *Sorangium* and *Byssovorax*. Based on the constructed phylogenetic tree, both type and reference strains give rise to an independent branch of the *Polyangiaceae*, consistent with the existence of a novel family (Fig. 3.5) (Garcia *et al.*, 2009). Several other novel strains also exhibit homology to a *Polyangium* sp. (M94280). However, large sequence differences in 16S rDNA and the fact that they occupy an independent branch in the phylogenetic tree, coupled with unique physicochemical characteristics and morphology, suggest that these strains should be assigned to new genera (*R. G. and R. M., unpublished*).

As these examples illustrate, comparison to published 16S rDNA proved instrumental in establishing that a particular newly isolated strain was indeed novel, and also helped to assign it to a new species, genus, or even family. In addition, the sequence comparisons showed that some species are closely related to uncultureable bacterial clones which form specific branches within the myxobacterial tree. Thus, we have been able to culture myxobacteria which were previously thought to be "uncultureable". We believe that many of these clones represent novel groups of myxobacteria, which may give rise to yet unidentified metabolites. Accordingly, these species will shortly be assigned to new taxa (*R. G. and R. M., unpublished*).

2.6. The issue of "unculturability"

The presumption of "unculturability" may simply correlate to the unknown physiology of the respective organism. Novel strains are generally difficult to handle because their nutritional and metabolic growth requirements are initially unknown, and must be determined for each strain. Unlike faster growing bacteria, myxobacteria cannot be isolated from high-nutrient media. Instead, nutrient-lean media are preferable as they enable the germination of myxospores, and later support swarming of the vegetative cells. Thus, minimal media reduce the risk that myxobacteria will be outcompeted by faster growing microbes. Myxobacteria are ubiquitous organisms in the environment, and thus it should in principle be possible to isolate

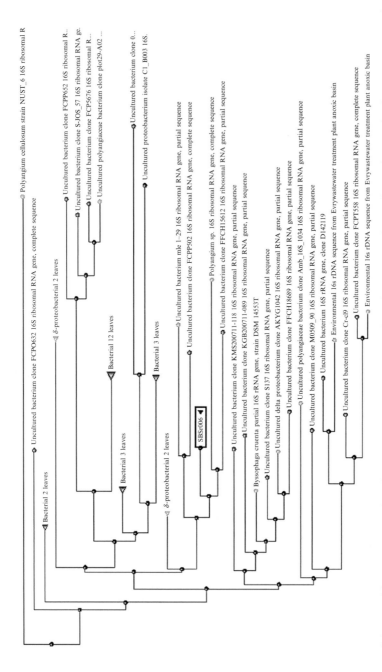

Figure 3.4 Neighbor-joining tree based on 16S rDNA sequences, revealing the position of a novel myxobacterial isolate (SBSr006) within a phylogenetic branch that contains sequences from "uncultured" bacterial clones.

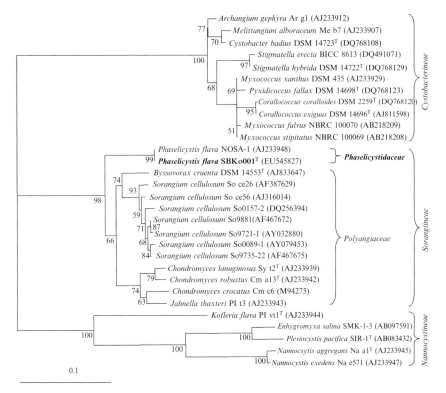

Figure 3.5 Neighbor-joining tree based on 16S rRNA gene sequences, showing the phylogenetic position of *P. flava* strain SBKo001^T.

them from any representative soil sample, assuming appropriate conditions can be identified. In many cases, it may only be a matter of recognizing them.

3. METHODS FOR ISOLATION, PURIFICATION, AND PRESERVATION OF NOVEL MYXOBACTERIA

Isolation methods for novel myxobacteria are designed primarily to meet nutritional demands, and adapted to fit their predatory lifestyle and ability to degrade biomacromolecules. The methods described here for cultivation also take into account the ecological niches of the myxobacteria. As most myxobacteria are predatory, live microorganisms are commonly used as bait to induce their growth (from swarm or fruiting bodies) in agar culture. In contrast, ordinary bacteria need not be baited and can simply be isolated and grown by streaking on nutrient-rich agar medium. A lean medium is known to promote swarming and is typically used in the isolation of myxobacteria; this approach has the advantage of producing thin and film-like swarms on the surface of the agar. Other members of the group (suborder *Sorangiineae*), including cellulose-degrading genera, can be isolated and purified by baiting with filter paper.

3.1. Baiting method for the isolation of myxobacteria

Samples obtained from the field are added onto a lean water agar medium (0.1% $CaCl_2 \cdot 2H_2O$, 5 mM HEPES, 1.5% agar) and a mineral salts agar medium (Shimkets et al., 2006), both adjusted to pH 7.0 before autoclaving using KOH. In our experience, these media induce swarming, spore germination, and fruiting body formation, all of which facilitate the isolation of myxobacteria. Live bacteria (*Escherichia coli*) and filter paper aid in the appearance and development of a myxobacterial colony because the cells lyse and degrade these bait items, which then also serve as nutrient sources for the colony's growth. Isolation plates are incubated at room temperature for at least a month, especially if the filter paper is placed on agar, as many *Sorangiineae*-type fruiting bodies do not immediately appear in the first weeks of incubation. In fact, we only noticed very small and pale fruiting bodies from one of our novel isolates upon viewing the plate using a stereoscopic microscope, and with illumination from below.

Some strains produce only a few swarms, which remain scattered about the plate. Attempts to isolate such strains are hampered by contamination with amoebae and other gliding bacteria. Therefore, agar plates should be inspected daily for swarms and fruiting bodies, beginning at the third day of the isolation procedure. Advancing clean swarms should immediately be cut out from the plate and transferred onto fresh agar medium. Subsequent transfers can be used to isolate a pure culture.

3.2. Important factors for successful isolation of novel strains

In the following section, we discuss the growth conditions that we have found to be optimal for the isolation of new strains of myxobacteria. We also try to enumerate difficulties encountered in the isolation, and make several suggestions as to how to solve these problems.

3.2.1. Medium

Water agar with bacterial (or yeast) bait and mineral salt agar with a filter paper overlay are standard media for the isolation of myxobacteria (Reichenbach and Dworkin, 1992; Shimkets et al., 2006). In our experience, use of these two media leads to the discovery of novel bacteriolytic myxobacteria. It may be possible by changing the bait to other macromolecules (e.g., chitin) to isolate additional novel groups.

Some strains can only be isolated using certain media. For example, fruiting bodies of *Chondromyces* cannot be found in water agar, but often grow on mineral salts medium. It appears that inorganic compounds are required for some groups of myxobacteria, including several of the novel strains described.

3.2.2. pH
Isolation of novel strains is performed with minimal medium, buffered to pH 7.0–7.2. We found that HEPES [4-(2-hydroxyethyl)-1-piperazineethanesulfonic acid] is a good substitute for inorganic buffers. In a test for pH tolerance, pH 5–8 was preferred by at least two of our novel strains. Colony swarms spread faster and more fruiting bodies are formed in slightly acidic media (pH 5–6). However, growth was severely depressed between pH 8–10. This surprising result shows that certain groups of myxobacteria prefer acidic conditions, which may mimic the pH of soil samples rich in decaying plant materials. Preparing the medium to imitate the natural environment may be a generally useful strategy for the isolation of new myxobacteria.

3.2.3. Temperature
All of our novel strains were isolated following incubation at room temperature (23 °C). Two of the strains produced swam colonies with essentially the same diameter, following incubation at both 18 °C and 30 °C. However, such temperature tolerance is not likely to be universal, and isolation attempts may fail if the temperature is elevated beyond 30 °C. A single isolate incubated at room temperature formed masses of cellular aggregates, presumed to be an early stage of fruiting body development. Subsequent overnight incubation at 30 °C led, however, to a collapse of the cell mound and a color change (golden yellow to pale), which was likely indicative of cell lysis. Thus, incubation temperature, in addition to its effect on the moisture level of the agar medium, is a crucial factor in the growth stages of myxobacteria.

3.2.4. Light
Some novel isolates show improved growth in the light. In the darkness, swarms and cell aggregates turn pale and do not spread to any significant extent over the agar surface. Light has also been shown to influence fruiting body formation in *Chondromyces* (Shimkets *et al.*, 2006). Based on this observation, it is probably best to incubate isolation plates under white light. Carotenoid biosynthesis in *Myxococcus xanthus* is induced by light (Elias-Arnanz *et al.*, 2007), so some novel myxobacterial strains may easily be recognized by the light-promoted production of colored pigments.

3.2.5. Air

The discovery of *Anaeromyxobacter* (Sanford *et al.*, 2002) necessitates revision of the concept of strict aerobiosis for myxobacteria as a group. It is not surprising that there are anaerobic strains in the δ-branch of the Proteobacteria, as they are related phylogenetically to the anaerobes (Shimkets and Woese, 1992). In test tube cultures of our three novel strains, swarm colonies penetrated deeply into the agar, and formed fruiting bodies (Fig. 3.6). This finding leads us to the conclusion that it may be worthwhile to perform strain isolation under reduced air set-up, in order to identify possible facultative anaerobes or microaerophiles among the myxobacteria.

3.2.6. Extended incubation

If isolation plates are not heavily contaminated with filamentous molds, incubation can be performed for a month or longer. During this period, the use of antifungal agents is recommended to suppress fungal growth (Shimkets *et al.*, 2006). Nevertheless, many molds tolerate or are resistant to cycloheximide (actidione) even at high concentrations.

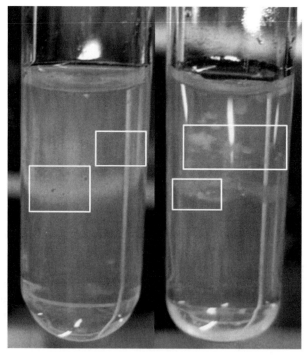

Figure 3.6 Unusual deep swarming and fruiting body formation inside the agar by two novel myxobacterial strains (highlighted by white boxes), indicating their uncommon microaerophilic behavior. (See Color Insert.)

Myxococcus, *Corallococcus*, *Archangium*, and *Cystobacter* of the *Myxococcaceae* generally appear during the first 2 weeks of incubation. These myxobacteria are easily recognized due to their fast swarming and large numbers of fruiting bodies. Prolonged incubation of the isolation plate may reveal a succession of other slow growers such as members of the *Sorangiineae*. When using mineral salt agar, many strains of *Chondromyces* and *Stigmatella* appear after more than 3 weeks of incubation, often preceded by other genera such as *Nannocystis* and *Melittangium*. *Sorangium* fruiting bodies are strikingly pigmented. In large numbers they impart orange, black, or yellow coloration to the decomposing filter paper. *Polyangium* may form scattered or dense aggregates of fruiting bodies under the medium, which also appear after more than 3 weeks of incubation. Several of the novel myxobacterial isolates described here were found following nearly a month of incubation, in the form of mature fruiting bodies. This observation shows that myxobacteria appear at different incubation times, which we presume to correlate with growth rate differences of the different genera and suborders, at least under laboratory conditions.

3.3. Swarming and fruiting body recognition

Isolation of myxobacteria requires familiarity with the taxonomy of the group. To date, all of the novel strains we isolated were readily recognizable by the unique morphology of their swarm colonies and fruiting bodies. Even in the absence of fruiting body formation, novel strains can be isolated by selecting migrating swarms. These are detected as waves of cells on the agar surface, and occasionally as band-shaped veins (Fig. 3.3D) which efficiently degrade the agar medium. Several novel myxobacteria are also striking for the fan-like (Fig. 3.3F) and comet-like (Fig. 3.3G) architecture of the swarms. Samples of either of these are suitable starting material for purification. Cells swarming under the agar have been purified after careful inversion of the agar, excision of the target migrating cells, and transfer to fresh medium. Depending on the age of the culture, this purification method requires two or three repetitions before purity is achieved. On the plus side, the method minimizes opportunity for contamination with surface bacteria and both trophozoites and amoebic cysts.

Familiarity with myxobacterial fruiting body morphology is fundamental to identifying new isolates. In one case, we observed small, pale fruiting bodies which closely resembled the abundant amoebic cysts present on isolation plates. In pure culture, the migrating cells of another strain aggregated to form unusual, white mounds, which could easily have been mistaken for compact colonies of ordinary bacteria. The mounds later developed the characteristic deep orange to reddish color of a mature fruiting body, which is microscopically composed of dense small sporangioles (Fig. 3.3C). One isolate is remarkable for its hump-like cell aggregates resembling the fruiting bodies of *Myxococcus*. Later in development, it differentiates into golden clusters of sporangioles (Fig. 3.3A). Remarkably, one strain forms yellowish-orange fruiting bodies which are often produced as clusters of chains of sori under the agar (Fig. 3.3B). The chain arrangement of fruiting is similar to that of *P. flava* SBKo001^T, although it differs by the formation of bean shaped sporangioles as opposed to *P. flava* (Fig. 3.3E). Many of the novel strains also produced fruiting bodies in and under the agar, which were clearly visible after inversion of the agar plate and examination with bright field microscope.

One of the species found only rarely was easily identified by the overall morphology of its huge and elegant fruiting body. In fact, members of this genus (*Chondromyces*) have attained the highest degree of morphological sophistication in the prokaryotic kingdom. However, because of the unusual fruiting body structure, which includes a stalk and apical clusters of sporangioles, these myxobacteria are often mistaken for fungi (e.g., the zygomycete *Cunninghamella*).

3.4. A simple and effective method for the purification of myxobacteria

Success in purification of novel myxobacteria relies mainly on two important growth stages: swarm colony and fruiting body. Either stage can be purified by washing with sterile distilled water. Mature fruiting bodies can be removed carefully from the isolation medium using sterile fine syringes or glass needles. Many adhering contaminants can then be washed away using sterile water. Repeated washing significantly decreases the load of adhering microbial contaminants, and thus shortens the overall purification protocol. Washed sporangioles are inoculated into the same medium used in the isolation procedure. Alternatively, washed sporangioles can be transferred to a drop of sterile distilled water on a glass slide, and then the tough protective wall can be cracked open using fine sterile needles. This step releases the myxospores which can then be seeded onto the agar using a glass or wire loop. The emerging swarm, if still contaminated, can be purified further by cutting out a small piece from the distal end of the farthest swarm growth and transferring it to the same agar medium. Depending on the degree of contamination and growth rate of the isolate, two or more transfers onto the same type of medium may be required before a pure culture is obtained.

Sporangioles formed by noncellulose degrading strains can be isolated using needles, and then washed several times with sterile distilled water. The strains can then be transferred to the same type of medium or to water agar containing spots of live *E. coli*. Given the unknown nutritional requirements of novel myxobacteria, it is always advisable to subculture any strains in the same medium used to obtain the original isolate.

Some myxobacteria may require very specific isolation conditions. For example, a rarely discovered myxobacterium was initially isolated on mineral salts agar (ST21). Fruiting bodies formed directly on a piece of decaying wood and on a filter paper bait, but growth to fructification stage did not extend to the surrounding agar surface. We therefore hypothesized that a cellulosic substrate was required to induce fruiting body development. This idea was also supported by our finding in pure culture that fruiting bodies are often formed in the presence of baited wood material. To isolate this strain, the topmost sporangiole was plucked off and its myxospores eventually germinated on fresh agar containing an overlaid piece of filter paper. The resulting swarm cells were used as the starting material for subsequent purification, which was achieved by cutting out a portion of a swarm colony at its most distal advancing edge using a sterile needle, and transferring the cells to the same medium. In general, repeated transfer of the swarm edge often leads to pure culture.

Purification of novel myxobacteria can also be performed from actively spreading colonies (swarm) especially if they produce mound-, ridge- or roll-shaped structures on the surface of the agar. A series of washing steps with sterile distilled water helps to decrease the microbial contaminants adhering on the surface of the swarm. Often these migrating cells are compacted so that washing with water does not destroy their intact structure. Alternatively, the smooth top surface of the swarm, which contains fewer contaminants, can be scraped or dabbed carefully with a fine wire needle. Vegetative cells adhering to the swarm can be suspended in a drop of sterile water and the cell suspension streaked on water agar. Streaking should progressively eliminate the remaining contaminants and also separate the myxobacterial cells. A good final transfer medium is a buffered yeast agar [VY/2 agar: 5 g baker's yeast, 1 g $CaCl_2 \cdot 2H_2O$, 5 mM HEPES, 10 g Bacto Agar, 0.5 μg/ml vitamin B_{12} (filter-sterilized and added after autoclaving of medium), 1 l distilled water, pH adjusted to 7.0 with KOH before autoclaving]. This medium supports germination of myxospores, promotes rapid spreading of the swarm colony, and formation of fruiting bodies.

Taken together, these data illustrate that careful inspection of the culture set-up can enable the discovery and isolation of novel myxobacteria even in the absence of recognizable fruiting bodies. In contrast, large and elegant fruiting bodies are usually easy to identify, especially in the case of the rarely found myxobacteria.

3.5. Culture maintenance and preservation

Once a strain is purified, a stock culture can be maintained in lean medium. A particular useful medium for preservation at room temperature is a buffered yeast agar. Cell viability is maintained for at least 2 months if fruiting bodies are present. In contrast, storage at 4 °C usually leads to rapid loss of viability (Reichenbach and Dworkin, 1992). Alternatively, the novel myxobacterial isolates can be maintained in water agar containing autoclaved *E. coli* cells which serve as a nutrient source for growth and fruiting body formation (Shimkets *et al.*, 2006). On high-nutrient medium, myxobacterial strains can be stored as long as 3–4 weeks. Prolonged storage, however, results in a rapid decrease in viability, which probably results from build up of metabolic end-products.

Cultures are normally preserved at −80 °C, and in fact myxobacteria have been shown to survive deep freezing for 8–10 years (Shimkets *et al.*, 2006). Transfer of a small amount (ca. 50 µl) of the frozen culture to yeast agar or a suitable liquid medium is enough to revive the cells. Visible growth is often observed after 3–5 days of incubation at 30 °C.

An alternate and cheaper method of preservation is to desiccate the fruiting bodies. Mature fruiting bodies are cut or scraped from the agar using needles or a loop, and placed on sterile filter paper (10 × 5 mm) in sterile vials. The sterile vials are then capped loosely, and stored in a desiccator jar for 2 weeks to allow the specimens to dry. Vials with dried materials are sealed by tightening and wrapping the cap with parafilm. Desiccation-resistant myxospores of specimens treated in this way and stored at room temperature remain viable for 5–15 years (Shimkets *et al.*, 2006), and sometimes as long as 20 years (Reichenbach, 1999a). The first batches of our known and novel strains preserved in this way are still viable after 8 years, and were also found to produce fruiting bodies after subsequent inoculation to fresh medium.

4. FERMENTATION AND SCREENING FOR KNOWN AND NOVEL METABOLITES

In order to investigate the potential of novel myxobacteria for the production of new natural products, crude extracts derived from small-scale cultivation in liquid medium are initially subjected to various bioassays. Extracts of interest are then fractionated to narrow down the number of candidate compounds, until a specific biological activity can be unambiguously correlated to a single substance ("micro-purification"). At every stage of the discovery process, LC-coupled high-resolution mass spectrometry is used to reveal known myxobacterial metabolites and to propose molecular formulae for as yet unidentified compounds. In Section 4.4, we review the implementation of statistical tools to mine mass spectrometric data for the presence of putative novel compounds, regardless of their biological function. We also illustrate how this approach can be employed to estimate species-wide secondary metabolite richness.

4.1. Cultivation conditions and sample workup procedure

Once pure cultures of novel myxobacterial strains are established, the next challenge is to optimize their growth in liquid medium. For many members of the suborder *Cystobacterineae*, it has not proven difficult to obtain high cell density. In fact, high density growth is typically achieved after a few successive transfers in high nutrient medium [e.g., CTT (Shimkets *et al.*, 2006) or M-medium (Müller and Gerth, 2006)]. In contrast, members of the suborder *Sorangiineae* are more problematic to grow in liquid medium. Initial growth occurs in clumps, and so the strains need to be shaken longer than other myxobacteria. As illustrated by our experience with the novel strains, however, such problems can usually be overcome by extending the incubation time to yield a reasonable cell density as well as homogeneous growth. Increasing the inoculum sometimes also helps to improve the growth characteristics of novel isolates.

Cultivation in liquid begins with the addition of several pieces (2–3) of agar containing swarming cells into a 100 ml Erlenmeyer flask containing 20 ml of liquid medium (e.g., CTT or MD1 medium (Shimkets *et al.*, 2006)). This small volume culture (preculture) is normally shaken for 3–4 days in the case of fast-growing strains, but incubation is extended up to several weeks for slow-growing myxobacteria. The addition of sugar (e.g., 0.35% glucose) can sometimes improve the growth, but it may also delay it. The choice of peptone source is also an important consideration. Marcor soy peptone (Müller and Gerth, 2006) supported the growth of many of our novel isolates, an observation which may be attributed to the high vitamin and carbohydrate content of the soy plant from which soy peptone is derived.

To prepare samples for screening of both known and novel compounds, myxobacteria are routinely grown in 50 ml liquid medium with 2% (w/v) XAD adsorber resin (Amberlite XAD-16, Sigma) (Gerth *et al.*, 2003). Cultures are shaken at 170 rpm and 30 °C for 5 days (*Cystobacterineae*), or up to 10 days (*Sorangiineae* and *Nannocystineae*). At the end of fermentation, both biomass and resin are collected together by centrifugation (10,000 rpm, 4 °C, 15 min). The combined pellet of cells and resin is stirred for 15 min in a solvent mixture consisting of equal volumes of methanol and acetone in order to break the cells and simultaneously elute compounds from the XAD adsorber resin. This crude extract is filtered, evaporated to dryness and then dissolved in 1 ml methanol.

4.2. Bioassays using crude extracts and prepurified fractions

Initial screening for bioactive compounds is performed using samples of the crude extract, which are subjected to bioassays using a set of indicator bacteria (both Gram-positive and Gram-negative), fungi, and cancer cell lines. A sample volume of 100 μl is usually enough for a wide range of tests to determine inhibitory effects with sets of standardized microorganisms in a 96-well plate format. The samples are loaded and dried in the wells prior to the addition of agar medium and inoculation of the test microorganisms. Extracts which exhibit inhibitory effects are investigated further by standard microbiological agar diffusion techniques, using an extended selection of potentially sensitive organisms. In parallel, cytotoxicity tests (MTT) are conducted using cancer cell lines (Elnakady *et al.*, 2004; Sandmann *et al.*, 2004). We prefer at the outset to use lymphoma (U937) and cervical carcinoma (KB3.1) cells to test our crude extracts, as these are, in our hands, the most reproducibly sensitive cell lines.

Bioassays of crude myxobacterial extracts are useful to quickly evaluate the overall ability of a novel strain to produce bioactive compounds. The specific substances responsible for bioactivity are then elucidated by retesting the fractionated samples against the sensitive organism, or in the respective assays. Sample fractionation is conveniently achieved by small-scale reverse-phase HPLC separation and automated fraction collection, while the content of every fraction is monitored simultaneously via high-resolution mass spectrometry (Fig. 3.7). This procedure facilitates the rapid and reliable identification of known metabolites and their correlation to an observed bioactivity, as well as the definition of novel candidate compounds for larger-scale production, purification, and structure elucidation. The mass spectrometric analysis of myxobacterial metabolite profiles is discussed in more detail in the following section.

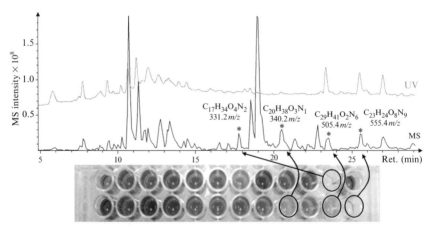

Figure 3.7 Correlation of biological activity (MTT assay in 96-well plate format) to fractions obtained by LC-fractionation of a 10-μl aliquot from a crude myxobacterial extract (arrows), and proposed molecular formulae for putative novel compounds (marked with asterisks).

4.3. Analysis of myxobacterial metabolite profiles using high-resolution mass spectrometry

A considerable number of myxobacterial secondary metabolites have been identified in the past by activity-guided screening methods. Bioassays represent a convenient screening method because they directly deliver candidate compounds for a specific biological activity. However, they are inherently biased by the choice of assays and targets. Additional myxobacterial compounds were discovered during workup of large-scale fermentations by HPLC-coupled diode array detector-based analysis, due to their "eye-catching" UV–vis absorption characteristics. Currently, LC-coupled ESI mass spectrometry is considered the most valuable analytical technique for the characterization of bacterial secondary metabolite profiles, as it can in principle visualize a highly diverse range of chemical compounds and facilitates the detection of even minor constituents in a sample.

Comprehensive profiling of extracts derived from novel myxobacterial isolates is fundamental to successful dereplication in order to identify the bioactive component. This step is always required, as production of certain myxobacterial natural products has been shown to be a strain-specific rather than species-specific property (Reichenbach and Höfle, 1999), and therefore it cannot be assumed that a particular strain will produce a given metabolite. For example, not all strains of *M. xanthus* which were screened in a recent study (Krug *et al.*, 2008a) produced myxovirescin and cittilin (39 out of 98 and 69 out of 98, respectively), while conversely, myxochelin has been identified in extracts from a variety of myxobacterial groups, including *Sorangium*, *Stigmatella*, and *Myxococcus* species (Schneiker *et al.*, 2007; Silakowski *et al.*, 2000).

Technical advances in recent years have afforded MS devices based on Time-of-Flight (ToF) or Fourier-Transform (FT) technology which are now in widespread use and can routinely perform analysis at a mass accuracy around 1 ppm. These instruments also combine high-resolution measurements with improved sensitivity and dynamic range. The acquired LC–MS data can be employed for the reliable identification of distinct chemical compounds in highly complex mixtures, whether they are crude extracts from myxobacterial cultivations or prepurified fractions. Notably, the minimum input for a high-resolution target screening database is the molecular formula and the retention time determined under standardized chromatographic conditions (Krug *et al.*, 2008b; Ojanperä *et al.*, 2006). As a first step, accurate m/z values for pseudomolecular ions $[M + H]^+$ are calculated from the molecular formulae and extracted ion chromatograms (EICs) are created in the dataset under investigation, with a low m/z tolerance (e.g., 3 mDa). A chromatographic peak is subsequently detected using a peakfinder algorithm, and mass spectra found in the respective peak are averaged and compared to the theoretical spectrum, derived from the molecular formula. The deviation of mass positions is evaluated and a match factor (termed "sigma value", σ) is calculated based on the deviations of measured signal intensities from the theoretical isotope pattern.

This targeted screening procedure is illustrated in Fig. 3.8A, using an example from (Krug et al., 2008a). Mass deviation ($\Delta m/z = 1.4$ ppm) and sigma value ($\sigma = 0.01$) are small and thereby indicate with high confidence the presence of cittilin A in a sample derived from *M. xanthus*. Furthermore, integration of the generated EIC traces enables the immediate semi-quantitative comparison of production yield in a number of strains (Fig. 3.8B). We routinely use this method for the characterization of metabolite profiles of novel myxobacterial strains and find that it reliably reveals the presence of known natural products which could otherwise easily be overlooked. Critical parameters for targeted screening of high-resolution LC–MS data are summarized in the following list.

High-resolution MS target screening checklist

1. *Chromatography*: Stable retention times and fast separations without compromising separation efficiency are achieved using an ultra-performance liquid chromatography (UPLC) system and sub-2 μm particle columns (e.g., Waters Acquity RP-C18 BEH column, 50 × 2.1 mm, 1.7 μm particle size). A mobile phase gradient using water and acetonitrile (5–95%), each containing 0.1% formic acid, is commonly employed. For optimal analysis, samples should be diluted prior to injection in a solvent mixture that matches the initial mobile phase composition.
2. *Electrospray ionization*: Use of a nanoESI source (with appropriate flow-splitting, e.g., Advion NanoMate) can substantially reduce adduct formation as well as unwanted in-source fragmentation and is also advantageous for increasing sensitivity and minimizing ion suppression effects.
3. *Sampling frequency*: The speed of the attached MS device should allow coverage of chromatographic peaks with 5–10 data points (e.g., sampling rate of 2 Hz or more on a typical ESI-ToF device). On combined ion-trap-FT machines (e.g., LTQ Orbitrap), tandem-MS measurements should not be performed simultaneously, in order to devote all sampling time to peak coverage.
4. *Mass precision and resolution*: By means of internal or external calibration, a mass precision around 1 ppm is within reach for state-of-the-art instrumentation, and should be the goal when data is recorded for target screening. A resolution higher than 10,000 (FWHM, full width at half maximum) is desirable, but resolutions exceeding 30,000 should only be used if not compromising sampling frequency.
5. Isotope pattern accuracy should be regarded as equally important as mass precision, as the authenticity of the isotopic distribution directly impacts

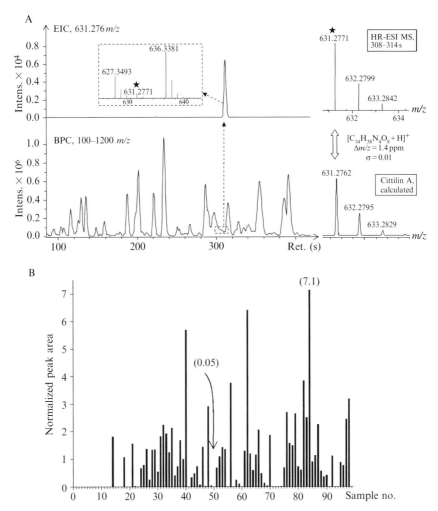

Figure 3.8 (A) The targeted screening procedure enables the identification of natural products in samples from very weak producers, using data from high-resolution (HR) ESI-TOF measurements. Evaluation of both mass position and isotope pattern is an important key to reliable compound identification. The analysis was carried out with the TargetAnalysis software (Bruker Daltonik, Bremen, Germany). (B) Semiquantitative comparison of production yield for cittilin in 98 *M. xanthus* strains. Numbers in parentheses are yields relative to the average level from all 98 strains (which was set as 1). From Krug *et al.* (2008a) with permission.

the outcome of the comparison between measured and theoretical spectra in the targeted screening procedure. ToF instruments have been shown to perform especially well in this respect. For routine screening, a threshold σ value of 0.03 is generally applied during data evaluation (Krug *et al.*, 2008b; Ojanperä *et al.*, 2006).

6. The dynamic range should be determined in the course of analytical performance verification, carried out with both a standard solution and an authentic sample. Sample stability and chromatographic reproducibility must be monitored when running longer sequences of samples in order to enable quantitative data treatment.

4.4. A diversity-oriented approach to mining myxobacterial secondary metabolomes

Target screening based on high-resolution MS measurements is a powerful tool for the characterization of myxobacterial secondary metabolite production profiles. However, it only readily identifies known compounds already present in the database, while failing to highlight potentially novel metabolites. The discovery of new compounds is therefore typically accomplished by manual comparison of chromatograms obtained from new strains to the established metabolite profiles of reference strains. This procedure is highly time-consuming even for an experienced analytical researcher, and is likely to represent a substantial bottleneck in a screening project aimed at the identification of novel metabolites. Thus, the implementation of statistical tools for the extraction of significant differences between a number of LC–MS datasets is highly desirable. We have recently demonstrated that Principal Component Analysis (PCA) is well-suited to this purpose (Krug et al., 2008a,b). PCA is an unsupervised pattern recognition technique which creates an overview of a multivariate dataset and thereby reveals groups of observations, trends and outliers (Lavine, 2000). In the context of LC–MS, PCA is applied as a statistical filter which highlights compounds responsible for significant variance in a set of samples (Idborg-Bjorkman et al., 2003). The input variables for PCA are retention time—m/z pairs with assigned intensity values, and the output of a PCA model consists of two-dimensional representations from which a grouping pattern of samples and the observations which are responsible for this pattern can be derived (see Fig. 3.9 and legend text).

However, data from LC–MS measurements present a significant challenge to statistical treatment, as every sample contains a large number of analytes with extremely variable properties, along with a considerable degree of chemical noise. Additional complicating factors include "biological" variation among replicate samples introduced during the cultivation process and variability in retention time during LC analysis. Therefore, LC–MS data need to be thoroughly preprocessed prior to statistical evaluation. Such data preparation can be achieved by implementation of a compound-finding algorithm. The basic idea behind this strategy is to differentiate all compounds with a chromatographic elution profile from randomly dispersed background noise (Sturm et al., 2008). All m/z signals

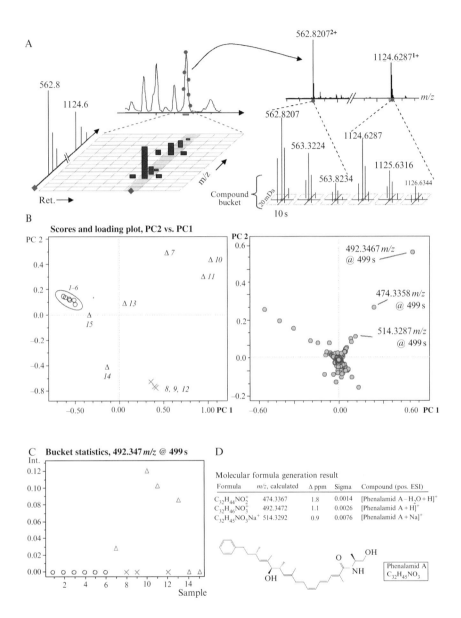

Figure 3.9 (A) Preparation of high-resolution LC–MS data prior to evaluation by PCA, via a compound-based bucketing approach. (B) Analysis of variation in a set of myxobacterial extracts derived from DK1622 (samples 1–6) and nine novel *Myxococcus* isolates (samples 7–15) by a PCA model, calculated with the ProfileAnalysis software package (Bruker Daltonik). The scores and loadings plot reveals accurate mass values and retention times of compounds (right diagram) which are responsible for the

that originate from one compound (including isotope peaks and multiply charged ions) are then merged into a noise-free mass spectrum, preserving the retention time as well as chromatographic peak area and intensity (Fig. 3.9A). Tolerances for RT and m/z values are subsequently applied and appropriate signals are added to build a "compound bucket" which is then considered for the PCA model calculation (Krug et al., 2008b).

In a recent study, we have employed this compound-based PCA approach to reveal distinct compounds with a nonubiquitous distribution in a set of samples derived from cultivation of nine novel *Myxococcus* spp. isolates (Krug et al., 2008b). High-resolution ESI-ToF MS data acquisition was performed according to the criteria mentioned in the above checklist. The presence of other secondary metabolites in addition to those identified by targeted screening was anticipated in this sample set, and PCA was successfully used to describe variation among production patterns (Fig. 3.9B and C). A list of candidate compounds which represent strain-specific differences was compiled, molecular formulae were proposed and one compound was subsequently identified as a known myxobacterial natural product which was not included in the screening database (Fig. 3.9D). Thus, the generation of molecular formulae for putative novel compounds with high confidence due to evaluation of both exact mass position and isotopic pattern was demonstrated as an important key for dereplication and prioritization of candidates for further characterization.

In another study, we compared the metabolic profiles of 98 *M. xanthus* strains isolated from locations worldwide (Krug et al., 2008a). We aimed to determine whether the metabolic inventory of the species *M. xanthus* is largely present in the standard, genome-sequenced strain DK1622, or whether there is significant potential for novel compound discoveries in other strains of the same species. Our screen revealed a strikingly high level of intraspecific diversity in the secondary metabolome of *M. xanthus*: the identification of 37 nonubiquitous candidate compounds greatly exceeded the eight secondary metabolite classes previously known to derive from this species. These results suggest that strains of *M. xanthus* should be regarded as a promising source of as yet unidentified natural products.

Furthermore, these findings emphasize the enormous genetic potential of individual *M. xanthus* strains for the production of secondary metabolites (Goldman et al., 2006). For example, *M. xanthus* DK1622 is known to produce five classes of natural products, yet the strain contains 18 biosynthetic gene clusters devoted to secondary metabolite biosynthesis.

---

grouping pattern of samples (left diagram). (C) The corresponding bucket statistics graph provides information on the presence or absence of a selected compound in all samples, as well as quantitative differences. (D) Identification of ions derived from phenalamid A by generation of high-confidence molecular formulae. From Krug et al. (2008b) with permission.

Encouragingly, a recent proteomics study (Schley et al., 2006), demonstrated that most of the pathways (e.g., of the polyketide synthase and/or nonribosomal peptide synthetase-type) are active, even if the metabolites have not yet been detected.

M. xanthus DK1622 was found to produce 10 candidates from the list of 37 putative novel compounds reported in the MS-based intraspecific screen, and subsequent studies to identify the corresponding biosynthetic gene clusters by targeted gene inactivation experiments are beginning to correlate genes to compound masses (*N. Cortina, D. K. and R. M., unpublished*). Thus, our diversity-oriented screening approach for the *M. xanthus* secondary metabolome is helping to uncover the products assembled by biosynthetic pathways which could not previously be correlated to a specific secondary metabolite. Furthermore, the procedure described here is unbiased with respect to the specific biological function of any putative metabolite and therefore has the potential to reveal compounds which are not easily identified by standard activity-based screens.

Currently, the application of statistical methods for the mining of LC–MS data in screening projects aimed at the discovery of novel natural products from bacteria appears to be underutilized. In contrast, tools for sophisticated statistical analyses have been employed thoroughly in a multitude of routine applications in the genomics and proteomics fields. The examples discussed here demonstrate that the development of methods for proper conditioning of high-resolution LC–MS data constitutes a crucial prerequisite for the applicability of techniques such as PCA, which in turn have the potential to significantly enable efficient data mining from a screening project. In addition to comparative metabolite profiling of related myxobacterial strains or species, the use of these methods extends to a wide range of metabolomics-based experiments enabling the discovery and identification of compounds which give rise to distinct differences in a set of highly complex samples. A library of bacterial clones, generated by targeted knockout or random mutagenesis, could be analyzed for the absence (or unexpected appearance) of as yet unidentified metabolites relative to the wild type strain, without defining *a priori* the specific signals to be monitored. It should also be possible to assess quantitative changes in secondary metabolomes in order to study regulatory aspects of natural product formation, or to comprehensively assay changes in a metabolite production pattern in response to different environmental conditions, in an unbiased fashion.

4.5. Summary and outlook

Over the last three decades, myxobacteria have transitioned from highly exotic organisms which attracted attention due to their unusual morphological characteristics and complex lifestyle, to valuable sources of natural products with significant potential in medical applications. The particular utility of myxobacterial compounds stems from their uncommon or even unique modes of action. In addition, the enzymes responsible for biosynthesis of these compounds often carry out unusual or unprecedented chemical reactions. Thus, these catalysts make useful additions to the growing "tool box" of enzymes available for genetic engineering of natural product biosynthetic pathways.

Further advances in this research area will depend on the continued discovery and characterization of novel myxobacterial families, genera and species as sources of new metabolites. Extensive sampling of geographically and ecologically promising habitats is an obvious strategy to identifying such strains. However, improving isolation and purification procedures to accommodate the specific and often unusual requirements of novel and rare myxobacteria is considered to be equally important. These microbiological efforts must be underpinned by advances in analytical technology and method development, which make it possible to identify novel metabolites, even if they are produced at very low levels. Encouragingly, modern liquid chromatography-coupled mass spectrometry can efficiently capture microbial metabolic diversity with ever-improving sensitivity and accuracy. Therefore, its implementation in combination with improved tools for statistical data evaluation looks set to enable the mining of the natural product inventory of many myxobacterial species with previously unmatched success. We anticipate that myxobacterial genome mining will continue to reveal metabolites with immense promise for human therapy.

ACKNOWLEDGMENT

The authors express their sincere thanks to Kira J. Weissman for her help in improving this manuscript. Research in the laboratory of RM was funded by the "Deutsche Forschungsgemeinschaft" (DFG) and the "Bundesministerium für Bildung und Forschung" (BMBF).

REFERENCES

Bode, H. B., and Müller, R. (2007). Secondary metabolism in myxobacteria. *In* "Myxobacteria: Multicellularity and differentiation" (D. Whitworth, ed.), pp. 259–282. ASM Press.

Dawid, W. (2000). Biology and global distribution of myxobacteria in soils. *FEMS Microbiol. Rev.* **24,** 403–427.

Elias-Arnanz, M., Fontes, M., and Padmanabhan, S. (2007). Carotenogenesis in *Myxococcus xanthus*: A Complex Regulatory Network. *In* "Myxobacteria: Multicellularity and Differentiation" (D. Whitworth, ed.), pp. 211–225. ASM Press.

Elnakady, Y. A., Sasse, F., Lünsdorf, H., and Reichenbach, H. (2004). Disorazol A1, a highly effective antimitotic agent acting on tubulin polymerization and inducing apoptosis in mammalian cells. *Biochem. Pharmacol.* **67,** 927–935.

Fudou, R., Jojima, Y., Iizuka, T., and Yamanaka, S. (2002). *Haliangium ochraceum* gen. nov., sp nov and *Haliangium tepidum* sp nov.: Novel moderately halophilic myxobacteria isolated from coastal saline environments. *J. Gen. Appl. Microbiol.* **48,** 109–115.

Garcia, R. O., Reichenbach, H., Ring, M. W., and Müller, R. (2009). *Phaselicystidaceae*, fam. nov., *Phaselicystis flava*, gen. nov., sp. nov., an arachidonic acid-containing soil myxobacterium. *Int. J. Syst. Evol. Microbiol.* in press.

Gerth, K., and Müller, R. (2005). Moderately thermophilic myxobacteria: Novel potential for production of natural products. *Environ. Microbiol.* **7,** 874–880.

Gerth, K., Perlova, O., and Müller, R. (2007). Sorangium cellulosum. In "Myxobacteria: Multicellularity and Differentiation" (D. Whitworth, ed.), pp. 329–348. ASM Press.

Gerth, K., Pradella, S., Perlova, O., Beyer, S., and Müller, R. (2003). Myxobacteria: Proficient producers of novel natural products with various biological activities—past and future biotechnological aspects with the focus on the genus *Sorangium*. *J. Biotechnol.* **106,** 233–253.

Goldman, B. S., Nierman, W. C., Kaiser, D., Slater, S. C., Durkin, A. S., Eisen, J., Ronning, C. M., Barbazuk, W. B., Blanchard, M., Field, C., Halling, C., Hinkle, G., et al. (2006). Evolution of sensory complexity recorded in a myxobacterial genome. *Proc. Natl. Acad. Sci. USA* **103,** 15200–15205.

Gronewold, T. M., Sasse, F., Lunsdorf, H., and Reichenbach, H. (1999). Effects of rhizopodin and latrunculin B on the morphology and on the actin cytoskeleton of mammalian cells. *Cell Tissue Res.* **295,** 121–129.

Idborg-Bjorkman, H., Edlund, P. O., Kvalheim, O. M., Schuppe-Koistinen, I., and Jacobsson, S. P. (2003). Screening of biomarkers in rat urine using LC/electrospray ionization-MS and two-way data analysis. *Anal. Chem.* **75,** 4784–4792.

Iizuka, T., Jojima, Y., Fudou, R., Hiraishi, A., Ahn, J. W., and Yamanaka, S. (2003a). *Plesiocystis pacifica* gen. nov., sp. nov., a marine myxobacterium that contains dihydrogenated menaquinone, isolated from the pacific coasts of Japan. *Int. J. Syst. Evol. Microbiol.* **53,** 189–195.

Iizuka, T., Jojima, Y., Fudou, R., Tokura, M., Hiraishi, A., and Yamanaka, S. (2003b). *Enhygromyxa salina* gen. nov., sp. nov., a slightly halophilic myxobacterium isolated from the coastal areas of Japan. *Syst. Appl. Microbiol.* **26,** 189–196.

Karwowski, J. P., Sunga, G. N., Kadam, S., and McAlpine, J. B. (1996). A method for the selective isolation of *Myxococcus* directly from soil. *J. Ind. Microbiol.* **16,** 230–236.

Kaur, G., Hollingshead, M., Holbeck, S., Schauer-Vukasinovic, V., Camalier, R. F., Domling, A., and Agarwal, S. (2006). Biological evaluation of tubulysin A: A potential anticancer and antiangiogenic natural product. *Biochem. J.* **396,** 235–242.

Khalil, M. W., Sasse, F., Lunsdorf, H., Elnakady, Y. A., and Reichenbach, H. (2006). Mechanism of action of tubulysin, an antimitotic peptide from myxobacteria. *ChemBioChem* **7,** 678–683.

Krug, D., Zurek, G., Revermann, O., Vos, M., Velicer, G. J., and Müller, R. (2008a). Discovering the hidden secondary metabolome of *Myxococcus xanthus*: A study of intra-specific diversity. *Appl. Environ. Microbiol.* **74,** 3058–3068.

Krug, D., Zurek, G., Schneider, B., Garcia, R., and Müller, R. (2008b). Efficient mining of myxobacterial metabolite profiles enabled by liquid-chromatography—electrospray ionization—time-of-flight mass spectrometry and compound-based principal component analysis. *Anal. Chim. Acta* **624,** 97–106.

Lavine, B. K. (2000). Clustering and classification of analytical data. In "Encyclopedia of Analytical Chemistry: Applications, Theory, and Instrumentation" (R. A. Meyers, ed.), pp. 1–20. Wiley.

Meiser, P., Bode, H. B., and Müller, R. (2006). DKxanthenes: Novel secondary metabolites from the myxobacterium *Myxococcus xanthus* essential for sporulation. *Proc. Natl. Acad. Sci. USA* **103,** 19128–19133.

Müller, R. (2009). Biosynthesis and Heterologous Production of Epothilones. *In* "The Epothilones—An Outstanding Family of Anti-Tumor Agents" (J. Mulzer, ed.). Springer in press.

Müller, R., and Gerth, K. (2006). Development of simple media which allow investigations into the global regulation of chivosazol biosynthesis with *Sorangium cellulosum* So ce56. *J. Biotechnol.* **121,** 192–200.

Mulzer, J. (2009). The Epothilones—An Outstanding Family of Anti-Tumour Agents: From Soil to the Clinic" Springer, Wien/New York.

Nickeleit, I., Zender, S., Sasse, F., Geffers, R., Brandes, G., Sörensen, I., Steinmetz, H., Kubicka, S., Carlomagno, T., Menche, D., Gütgemann, I., Buer, J., *et al.* (2008). Argyrin A reveals a critical role for the tumor suppressor protein p27^{kip1} in mediating antitumor activities in response to proteasome inhibition. *Cancer Res.* **14,** 23–35.

Ojanperä, S., Pelander, A., Pelzing, M., Krebs, I., Vuori, E., and Ojanperä, I. (2006). Isotopic pattern and accurate mass determination in urine drug screening by liquid chromatography/time-of-flight mass spectrometry. *Rapid Commun. Mass Spectrom.* **20,** 1161–1167.

Reichenbach, H. (1999a). Myxobacteria. *In* "Encyclopedia of Bioprocess Technology: Fermentation, Biocatalysis, and Bioseparation" (M. C. Flickinger and S. W. Drew, eds.), pp. 1823–1832. Wiley.

Reichenbach, H. (1999b). The ecology of the myxobacteria. *Environ. Microbiol.* **1,** 15–21.

Reichenbach, H. (2005). Order VIII. Myxococcales. Tchan, Pochon and Pre'vot 1948, 398AL. *In* "Bergey's Manual of Systematic Bacteriology" (D. J. Brenner, N. R. Krieg, and J. T. Staley, eds.), pp. 1059–1144. Springer.

Reichenbach, H., and Dworkin, M. (1992). The Myxobacteria. *In* "The Procaryotes" (A. Balows, H. G. Trüper, M. Dworkin, W. Harder, and K. H. Schleifer, eds.), pp. 3416–3487. Springer.

Reichenbach, H., and Höfle, G. (1999). Myxobacteria as producers of secondary metabolites. *In* "Drug Discovery from Nature" (S. Grabley and R. Thiericke, eds.), pp. 149–179. Springer.

Reichenbach, H., Lang, E., Schumann, P., and Sproer, C. (2006). *Byssovorax cruenta* gen. nov., sp nov., nom. rev., a cellulose-degrading myxobacterium: Rediscovery of '*Myxococcus cruentus*' Thaxter 1897. *Int. J. Syst. Evol. Microbiol.* **56,** 2357–2363.

Sandmann, A., Sasse, F., and Müller, R. (2004). Identification and analysis of the core biosynthetic machinery of tubulysin, a potent cytotoxin with potential anticancer activity. *Chem. Biol.* **11,** 1071–1079.

Sanford, R., Cole, J., and Tiedje, J. (2002). Characterization and description of *Anaeromyxobacter dehalogenans* gen. nov., sp. nov., an aryl-halorespiring facultative anaerobic Myxobacterium. *Appl. Environ. Microbiol.* **68,** 893–900.

Sasse, F., Kunze, B., Gronewold, T. M., and Reichenbach, H. (1998). The chondramides: Cytostatic agents from myxobacteria acting on the actin cytoskeleton. *J. Nat. Cancer. Inst.* **90,** 1559–1563.

Sasse, F., Steinmetz, H., Schupp, T., Petersen, F., Memmert, K., Hofmann, H., Heusser, C., Brinkmann, V., von Matt, P., Höfle, G., and Reichenbach, H. (2002). Argyrins, immunosuppressive cyclic peptides from myxobacteria. I. Production, isolation, physico-chemical and biological properties. *J. Antibiot.* **55,** 543–551.

Schley, C., Altmeyer, M. O., Swart, R., Müller, R., and Huber, C. G. (2006). Proteome analysis of *Myxococcus xanthus* by off-line two-dimensional chromatographic separation using monolithic poly-(styrene-divinylbenzene) columns combined with ion-trap tandem mass spectrometry. *J. Proteome Res.* **5,** 2760–2768.

Schneiker, S., Perlova, O., Kaiser, O., Gerth, K., Alici, A., Altmeyer, M. O., Bartels, D., Bekel, T., Beyer, S., Bode, E., Bode, H. B., Bolten, C. J., *et al.* (2007). Complete genome sequence of the myxobacterium *Sorangium cellulosum*. *Nat. Biotechnol.* **25,** 1281–1289.

Shimkets, L., Dworkin, M., and Reichenbach, H. (2006). The Myxobacteria. *In* "The Prokaryotes" (M. Dworkin, ed.), pp. 31–115. Springer.

Shimkets, L., and Woese, C. R. (1992). A phylogenetic analysis of the myxobacteria: Basis for their classification. *Proc. Natl. Acad. Sci. USA* **89,** 9459–9463.

Silakowski, B., Kunze, B., Nordsiek, G., Blöcker, H., Höfle, G., and Müller, R. (2000). The myxochelin iron transport regulon of the myxobacterium *Stigmatella aurantiaca* Sg a15. *Eur. J. Biochem.* **267,** 6476–6485.

Spröer, C., Reichenbach, H., and Stackebrandt, E. (1999). The correlation between morphological and phylogenetic classification of myxobacteria. *Int. J. Syst. Bacteriol.* **49,** 1255–1262.

Steinmetz, H., Glaser, N., Herdtweck, E., Sasse, F., Reichenbach, H., and Höfle, G. (2004). Isolation, crystal and solution structure determination, and biosynthesis of tubulysins-powerful inhibitors of tubulin polymerization from myxobacteria. *Angew. Chem. Int. Ed.* **43,** 4888–4892.

Sturm, M., Bertsch, A., Gropl, C., Hildebrandt, A., Hussong, R., Lange, E., Pfeifer, N., Schulz-Trieglaff, O., Zerck, A., Reinert, K., and Kohlbacher, O. (2008). OpenMS—An open-source software framework for mass spectrometry. *BMC Bioinformatics* **9,** 163–173.

CHAPTER FOUR

Analyzing the Regulation of Antibiotic Production in Streptomycetes

Mervyn Bibb *and* Andrew Hesketh

Contents

1. Introduction	94
2. The Regulation of Antibiotic Production in Streptomycetes	94
3. Identifying Regulatory Genes for Antibiotic Biosynthesis	97
3.1. Random generation of antibiotic nonproducing or overproducing mutants by UV, NTG, transposon, and insertion mutagenesis	97
3.2. Identifying regulatory genes from nonproducing mutants	99
3.3. Identifying antibiotic regulatory genes by overexpression	101
3.4. Genome scanning for regulatory genes for antibiotic biosynthesis	102
3.5. Confirming the nature of antibiotic regulatory genes	102
4. Characterizing Regulatory Genes for Antibiotic Biosynthesis	104
4.1. Pathway-specific regulatory genes	104
4.2. Pleiotropic regulatory genes	108
4.3. Concluding remarks	113
Acknowledgments	113
References	113

Abstract

This chapter outlines the approaches and techniques that can be used to analyze the regulation of antibiotic production in streptomycetes. It describes how to isolate antibiotic nonproducing and overproducing mutants by UV, nitrosoguanidine (NTG), transposon, and insertion mutagenesis, and then how to use those mutants to identify regulatory genes. Other approaches to identify both pathway-specific and pleiotropic regulatory genes include overexpression and genome scanning. A variety of methods used to characterize pathway-specific regulatory genes for antibiotic biosynthesis are then covered,

Department of Molecular Microbiology, John Innes Centre, Norwich Research Park, Colney Lane, Norwich, United Kingdom including transcriptional analysis and techniques that can be used to distinguish between direct and indirect regulation. Finally, genome-wide approaches that can be taken to characterize pleiotropic regulatory genes, including microarray and ChIP-on-Chip technologies, are described.

1. INTRODUCTION

This chapter will focus on experimental procedures for analyzing the transcriptional regulation of antibiotic production in streptomycetes, although in general the same procedures can be used to study the control of expression of any gene in any microorganism. The actinomycetes, the family to which the streptomycetes belong, are responsible for the production of over two-thirds of known antibiotics (defined as compounds produced by one microorganism that inhibit the growth of another), and the biosynthesis of many of these compounds is discussed elsewhere in this series. While of immense fundamental interest, insights into the regulation of antibiotic production provide new opportunities for knowledge-based approaches for strain improvement to complement the classical and undoubtedly successful strategy of mutation and screening for improved productivity. Many of the approaches described here are also theoretically applicable to other actinomycetes, but since many of the tools used are available only for streptomycetes, we have confined our considerations to this genus. This contribution is not intended to review our current knowledge of the regulation of antibiotic production in these organisms (see Chapter 5 in this volume for a more complete coverage), but instead to provide a primer for those interested in developing an understanding of the regulation of synthesis of a particular compound. To assist in this endeavor, we have quoted references that demonstrate the use of many of the techniques described. Frequently, these stem from our own studies, and we apologize for not providing a comprehensive list of publications that utilize these technologies.

2. THE REGULATION OF ANTIBIOTIC PRODUCTION IN STREPTOMYCETES

Antibiotics are the products of complex biosynthetic pathways that utilize primary metabolites as building blocks. Apparently without exception, all of the genes that are required for the production of a particular compound are clustered together in the chromosome, or sometimes on a plasmid, of the producing organism, markedly facilitating the isolation and analysis of entire antibiotic biosynthetic gene clusters. In addition to containing the genes encoding the biosynthetic enzymes, such clusters (which can vary in size from around 15 kb to over 100 kb) frequently contain pathway-specific regulatory genes that control the onset of biosynthesis, as well as genes involved in antibiotic export and self-resistance.

Antibiotic production in streptomycetes usually starts at the onset of stationary phase in liquid grown cultures, and coincides with the beginning of morphological differentiation in agar-grown cultures (see Bibb, 2005, for a review). Where they exist, pathway-specific regulatory genes play a key role in triggering stationary-phase expression, and indeed there is evidence to suggest that for at least some gene clusters the level of the pathway-specific regulatory protein is the only limiting factor in determining the onset of antibiotic biosynthesis. Constitutive expression of such regulatory genes may lead to precocious antibiotic synthesis during rapid growth, and their overexpression can result in marked increases in the level of antibiotic production (Gramajo et al., 1993; Takano et al., 1992).

Many streptomycetes produce several compounds with antimicrobial activity, and often the onset of production of some or all of these antibiotics is coordinately controlled by pleiotropic regulatory genes. Consequently, mutations in these genes abolish or impair the synthesis of several of the antibiotics made by a particular strain. It is striking that while pathway-specific regulatory genes, such as the *Streptomyces* Antibiotic Regulatory Proteins (SARPs) (Bibb, 2005), are absolutely required for antibiotic production, most of the pleiotropic regulatory genes characterized thus far are only required under particular growth conditions. This suggests that they serve to sense different environmental parameters, such as various nutritional limitations, and as a consequence mutants deficient in a particular pleiotropic regulatory gene often have wild-type phenotypes when grown on some growth media and mutant phenotypes on others. An example of such a pleiotropic regulatory gene is *relA* of *Streptomyces coelicolor* A3(2), which encodes synthesis of the intracellular signaling molecule (p)ppGpp. This highly phosphorylated guanine nucleotide plays a key role in regulating stationary phase gene expression in *Escherichia coli* (reviewed in Magnusson et al., 2005), and so it is perhaps not too surprising that it has also been shown to play a crucial role in triggering antibiotic biosynthesis in *S. coelicolor* (Chakraburtty and Bibb, 1997; Sun et al., 2001). However, it appears to be required only under conditions of nitrogen limitation; growth of a *relA* mutant under conditions of phosphate limitation results in a wild-type phenotype. Interestingly, induction of (p)ppGpp synthesis in the wild-type strain under conditions of nutritional sufficiency appears to directly activate the transcription of pivotal pathway-specific regulatory genes, suggesting that there are no intermediary steps in the regulatory cascade (Hesketh et al., 2001, 2007a). Binding of (p)ppGpp to RNA polymerase may allow it to adopt a configuration that favors transcription initiation at the promoters of these genes, or to select a particular σ factor that preferentially recognizes their promoter sequences. DasR, a DNA binding protein of *S. coelicolor*, is another example of a pleiotropic regulatory protein that links antibiotic production to the nutritional status of the cell (Rigali et al., 2008). DasR inhibits the production of two different antibiotics by binding to operator sequences upstream of the pathway-specific regulatory genes of each of the biosynthetic gene clusters, thus repressing their transcription. Repression is relieved in the presence of the metabolite glucosamine-6-phosphate, the abundance of which indicates levels of extracellular *N*-acetylglucosamine, an important environmental source of carbon and nitrogen.

Among the many genes that have been identified as having a pleiotropic influence on antibiotic biosynthesis, there are a subset, exemplified by the *bld* (for bald) genes of *S. coelicolor*, which in addition to being pleiotropically deficient in antibiotic production are also unable to erect aerial hyphae (hence the term bald) and to undergo normal morphological development. While the effect on antibiotic biosynthesis for some of these mutants may be indirect and reflect gross changes in physiology, it is apparent that some (e.g., *bldA* of *S. coelicolor*) play a direct regulatory role in controlling both morphological development and antibiotic biosynthesis. *bldA* encodes the only tRNA in *S. coelicolor* that can translate the rare leucine codon UUA; the presence of such UUA codons in many pathway-specific regulatory genes (Chater and Chandra, 2008) provides a means for implementing translational control over antibiotic biosynthesis, the physiological basis of which is not currently understood.

Many antibiotics are also regulated by small diffusible extracellular signaling molecules, the γ-butyrolactones (see Takano, 2006, for a review and Chapter 6 in this volume) and 2-alkyl-4-hydroxymethylfuran-3-carboxylic acids (Corre et al., 2008). While most of these compounds appear to regulate the production of specific antibiotics, at least one, the γ-butyrolactone A-factor made by *Streptomyces griseus*, regulates the production of several secondary metabolites and the onset of morphological differentiation (Ohnishi et al., 2005). Although these signaling molecules are reminiscent of the *N*-acyl homoserine lactones involved in quorum sensing in many Gram-negative bacteria, it is not clear that they always play a similar role in streptomycetes, where their synthesis may instead be triggered at least in part by environmental or nutritional signals.

In the following sections, we have attempted to describe the methodologies that can be used to identify and analyze genes that are involved in regulating antibiotic biosynthesis in streptomycetes. Where detailed protocols have been reported previously, we have directed the reader to specific references.

3. IDENTIFYING REGULATORY GENES FOR ANTIBIOTIC BIOSYNTHESIS

3.1. Random generation of antibiotic nonproducing or overproducing mutants by UV, NTG, transposon, and insertion mutagenesis

The classical approach to identifying genes, and consequently regulatory genes, involved in antibiotic biosynthesis is to carry out UV and/or NTG mutagenesis (see Kieser *et al.*, 2000, for detailed protocols). If UV mutagenesis is carried out, care must be taken to avoid photoreactivation which compromises the efficacy of the mutagenic treatment in at least some streptomycetes. Mutagenesis is then followed by screening for loss of or overproduction of the compound of interest. If the compound is pigmented (e.g., actinorhodin of *S. coelicolor*), then visual screening for changes in production can be readily accomplished. If not, then an activity screen is generally conducted using a sensitive microorganism to detect antibiotic nonproducing or overproducing mutants. This can be accomplished in a number of ways, just two of which are described below:

1. The survivors of mutagenesis, which have been grown on a medium that promotes antibiotic production, are replicated to agar to yield a set of master plates; the original survivors of mutagenesis are then overlaid generally in soft nutrient agar with a sensitive assay microorganism and screened for loss of inhibition of the background lawn of growth. Any candidate nonproducers can then be retrieved from the set of master plates, and loss of antibiotic production confirmed by bioassay and physical techniques, such as High Performance Liquid Chromatography (HPLC) and/or mass spectrometry. Alternatively, although more laborious, small cylinders of agar containing the survivors of mutagenesis can be extracted using a suitable cork borer and arrayed on a 20 cm x 20 cm assay plate and embedded (but not submerged) in a thin layer of soft nutrient agar containing the assay organism. Potential nonproducing mutants can then be isolated directly from the top of the agar cylinders and analyzed further (e.g., see Wright and Hopwood, 1976).

2. The survivors of mutagenesis are picked to 96-well agar plates, preferably with the use of robotics. Once grown, the arrayed survivors can be inoculated (using robotics or pin-tools) into 96-well or 32-deep-well plates containing a suitable liquid growth medium, and samples of the culture supernatants used in standard assays for bioactivity.

Antibiotic overproducing mutants can be identified using the same procedures by screening for enhanced levels of inhibition of the assay organism.

Since many streptomycetes make more than one antibiotic, such simple activity screens may not suffice. While it may be possible to choose a growth medium that favors the production of one antibiotic over others, or to use an indicator organism that is sensitive only to the compound of interest among those made by the streptomycete, it may nevertheless be necessary to resort to physical means (e.g., high throughput HPLC or Matrix Assisted Laser Desorption Ionisation-Time of Flight (MALDI-ToF) mass spectrometry analysis to detect the loss or overproduction of a particular compound.

Another approach that can be used to obtain antibiotic nonproducing and overproducing mutants is mutational cloning (Chater and Bruton, 1983; Kieser et al., 2000). In its initial form, this involved the use of derivatives of the temperate phage ΦC31 from which the attachment site *attP* had been deleted and into which had been inserted an antibiotic resistance gene selectable in the streptomycete of interest. Relatively small fragments (about 1–2 kb) of the latter's genomic DNA are inserted into the phage vector, and the phage library introduced into the antibiotic producing streptomycete by protoplast transfection (if this technology is not available for the streptomycete of interest, the initial phage library can be made in, e.g., *Streptomyces lividans,* and introduced into the antibiotic producer by phage infection; about two-thirds of streptomycetes are sensitive to ΦC31). Stable antibiotic resistant lysogens can only be obtained by homologous recombination between the insert cloned in the phage vector and homologous sequences present in the host's genome. If the cloned insert is internal to a transcription unit, then this will generally result in loss of function of the disrupted gene. Screening of lysogens for loss of antibiotic production thus serves to identify insertional mutants in the biosynthetic gene cluster of interest. In principle, a similar procedure could now be adopted using a plasmid vector. For example, a small-insert library of genomic DNA from the strain of interest could be generated in *E. coli* in a conjugative plasmid (such as pSET151; Bierman et al., 1992) that lacks any site-specific integration system and that is selectable in the *Streptomyces* strain of interest. Stable ex-conjugants can only be obtained by homologous recombination between the cloned inserts and homologous sequences present in the host. Screening for loss of antibiotic production should again reveal the required insertional mutants.

More recently, it has been possible to contemplate the use of transposon mutagenesis to identify genes involved in antibiotic production. Although several transposon systems have been described for use in streptomycetes, the resulting phenotypes have often been found not to be linked to the transposon insertion, suggesting that introduction of the transposon and/or its delivery vector into the host may itself promote random mutagenesis. Nevertheless, the pKay1 system developed by Fowler (personal communication, for further information contact Bibb), which is based on Tn4560, has been used to identify genes for antibiotic production in *S. coelicolor,* although again, not all of the non-producing mutants are transposon-linked (Fowler, Hesketh, and Bibb, unpublished results).

3.2. Identifying regulatory genes from nonproducing mutants

While antibiotic overproducing mutations are likely to act by derepressing expression of the cluster or increasing precursor or cofactor availability, the majority of mutations that result in loss of antibiotic production will lie in genes encoding biosynthetic enzymes rather than regulatory proteins. Here, we consider ways in which these nonproducing mutants can be used to identify the corresponding regulatory genes.

3.2.1. Inability to undergo cosynthesis

Since antibiotics are generally the products of stepwise linear biosynthetic pathways, it is often possible to carry out cosynthesis tests between mutants blocked at different points in the pathway. This assumes that any intermediates that accumulate in the blocked mutants are freely diffusible and capable of being taken up by other mutants grown in close proximity. In principle then, if we take two mutants A and B, where mutant B is blocked later in the pathway than mutant A, mutant B will accumulate an intermediate that may be able to enter mutant A and provide a missing intermediate for continued synthesis of the antibiotic in the latter strain. Mutant B would thus be a "secretor" and mutant A a "convertor." Mutants that are unable to act as "secretors" or "convertors" are candidates for regulatory mutants, since they are likely not to express any of the biosynthetic enzymes. (However, note that mutants unable to produce an extracellular signaling molecule are also likely to be classified as convertors in such tests, and potentially eliminated as regulatory candidates; see O'Rourke *et al.* (2009), who described such a system involved in methylenomycin biosynthesis.)

3.2.2. Transposon mutants

The advantage of transposon mutagenesis is that it may be possible to rapidly identify the gene into which the transposon has inserted, which by definition is involved in antibiotic production. This can be accomplished either by cloning the transposon out into a tractable host such as *E. coli*, by selecting for an antibiotic resistance gene carried by the transposon and then sequencing flanking regions, or (more efficiently) by Ligation-Mediated-Polymerase Chain Reaction (LM-PCR; Guilfoyle *et al.*, 1997) of DNA isolated from the mutant followed by nucleotide sequencing. In either case, if a genome sequence for the organism of interest is available, the precise point of transposon insertion can be determined and the putative antibiotic biosynthetic gene identified. The transposon insertion may have disrupted a pleiotropic or a pathway-specific regulatory gene, but in a nonproducing mutant it is more likely to have occurred in one of the more numerous genes encoding biosynthetic enzymes. In the latter case, the sequences adjacent to the point of transposon insertion can be used as hybridization probes with cosmid or BAC libraries of wild-type streptomycete DNA to identify clones that singly or collectively contain the entire biosynthetic gene cluster by "genome walking," and hence any pathway-specific regulatory genes contained therein. Alternatively, LM-PCR combined with nucleotide sequencing could be used to "walk" towards the required regulatory gene(s).

3.2.3. Insertional mutants

It is possible to rapidly identify antibiotic biosynthetic genes from insertional mutants in a similar fashion. If phage-based mutational cloning has been used to create an antibiotic nonproducing mutant, phages released from the antibiotic nonproducing lysogen can be amplified in *S. lividans*, their DNA isolated (Kieser et al., 2000) and the mutagenic DNA fragment that lies internal to a transcription unit identified. If a plasmid-based approach has been used, growth in liquid culture in the absence of selection should lead to the excision of some of the integrated plasmids. Transformation of *E. coli* with a crude cell lysate from such a culture may be sufficient to recover the mutagenic plasmid. Alternatively, restriction endonuclease digestion of genomic DNA obtained from the mutant with an enzyme that does not cut within the plasmid vector, followed by ligation and transformation of *E. coli*, should lead to isolation of the required DNA sequence. For both phage- and plasmid-based mutational cloning, the nucleotide sequence of part or all of the mutagenic insert can be determined and used to isolate the remainder of the biosynthetic gene cluster by "genome walking."

3.2.4. Genetic complementation

Once antibiotic nonproducing mutants have been isolated, be they regulatory mutants or not, then the corresponding wild-type genes can be identified by genetic complementation. This requires methodology to introduce wild-type DNA efficiently back into the mutant host, either by protoplast transformation or by conjugation from *E. coli* (see Kieser et al., 2000, for both procedures). If the latter approach is adopted, it may be necessary to use a methylation-deficient *E. coli* donor strain to avoid the methyl-specific restriction systems possessed by some streptomycetes (e.g., *S. coelicolor* and *Streptomyces avermitilis*). Complementing clones can be obtained by cloning relatively short fragments (around 5 kb) of wild-type DNA into the mutant strain on autonomously replicating plasmid vectors (that may be transferred by conjugation from *E. coli*) and screening for restoration of antibiotic production. Plasmid DNA can then be isolated from the complementing clones and analyzed by nucleotide sequencing.

Should the complementing clone not contain a pathway-specific regulatory gene, but one of the genes encoding a biosynthetic enzyme, the former can be obtained using the procedure of "genome walking" outlined above for transposon mutagenesis. However, given that the mutation (if not in a pleiotropic regulatory gene) will almost certainly lie within a large biosynthetic gene cluster, maximal information will be gained by cloning (and subsequently sequencing) large segments of wild-type DNA that might conceivably contain the entire biosynthetic gene cluster, including the desired regulatory gene(s). This could be achieved by using a low copy number *Streptomyces* plasmid vector based on SCP2 that may be conjugable from *E. coli* (Kieser *et al.*, 2000). Although, to our knowledge, not demonstrated thus far in streptomycetes, an alternative approach would be to make a library of wild-type DNA in a cosmid vector containing an antibiotic resistance gene that can be selected in streptomycetes and that has been modified to allow conjugation from *E. coli* into the nonproducing mutant. Selection for the resistance gene carried on the cosmid vector will result in integration of the cosmid clones into the streptomycete genome by homologous recombination. Screening for restoration of antibiotic biosynthesis will identify cosmid clones containing the required wild-type gene. Since the streptomycete containing the complementing clones now possesses two duplicated segments of genomic DNA of ~40 kb separated only by the cosmid vector, relaxation of antibiotic selection should lead to a second crossover that liberates the integrated cosmid vector together with part or all of either the wild-type or mutant biosynthetic gene cluster. Transformation of *E. coli* with a cell lysate of the streptomycete culture after such nonselective growth and selection for the antibiotic resistance gene carried on the cosmid vector should permit recovery of the required cosmid clone, which can then be characterized by nucleotide sequencing. Ultimately this approach, perhaps combined with additional "genome walking," should lead to the identification of putative pathway-specific regulatory genes.

3.3. Identifying antibiotic regulatory genes by overexpression

As indicated previously, the enhanced or constitutive expression of pathway-specific regulatory genes can increase markedly the level of antibiotic biosynthesis, and the same is true for at least some pleiotropic regulatory genes. Consequently, in principle members of both classes of regulatory gene can be isolated by cloning wild-type DNA in a high copy number plasmid vector and screening for elevated levels of antibiotic production. Alternatively, such wild-type DNA could be cloned in an expression vector with a strong constitutive promoter with the aim of achieving high levels of expression of the regulatory gene throughout growth without the requirement of any additional potentially developmentally controlled regulatory elements. While high copy number vectors would appear to be the obvious choice, integrative expression vectors may achieve the same end; indeed there is evidence that just a single additional copy of a pathway-specific regulatory gene is sufficient to cause a marked increase in antibiotic production (Hopwood et al., 1985). DNA sequencing of the cloned inserts should then reveal the required regulatory gene(s). (Note that if the cloned regulatory gene is present in a stably integrated vector that cannot readily be recovered from the streptomycete genome, then it will be necessary to first clone it out into E. coli using a selectable marker present in the integrated element, as outlined earlier in this chapter.)

3.4. Genome scanning for regulatory genes for antibiotic biosynthesis

Given the high-throughput nature and decreasing costs of nucleotide sequencing delivered by the next-generation sequencing technologies, such as Solexa, 454, and Solid, it is now feasible to consider identifying antibiotic biosynthetic gene clusters, and their corresponding pathway-specific and indeed pleiotropic regulatory genes, by a procedure known as genome scanning. In this approach, genomic DNA of the streptomycete of interest is subject to high-throughput sequencing to yield around fivefold coverage of the genome of interest. While this is not sufficient to yield a single genomic contig, it provides a database of contigs that can then be interrogated *in silico* to identify genes that are potentially components of the gene cluster of interest. These sequences then provide a means to generate high-fidelity PCR probes from genomic DNA that can be used to probe cosmid or BAC libraries of the antibiotic producing streptomycete of interest. Sequencing of the corresponding cosmid or BAC clones, perhaps with additional rounds of library screening, should ultimately reveal putative pathway-specific regulatory genes, the nature of which can then be confirmed by targeted mutagenesis. Similarly, the contig database can also be interrogated *in silico* to identify potential pleiotropic regulatory genes, which can be further analyzed by mutagenesis and overexpression.

3.5. Confirming the nature of antibiotic regulatory genes

Regardless of how a putative antibiotic regulatory gene has been identified, but particularly in the case of transposon-induced mutations and genes identified by overexpression, further mutational analysis is generally required to show conclusively that the gene is a true regulator of antibiotic biosynthesis. In the case of transposon mutagenesis, given the frequent lack of linkage of the mutant phenotype to the transposon insertion, it is crucial to either complement the mutant with the wild-type gene or to inactivate the identified gene by targeted mutagenesis to confirm its role in antibiotic production. Similarly, for genes identified by apparent genetic complementation or overexpression, it is important to show for the former that restoration of antibiotic production does not reflect suppression (rather then genetic complementation) of the original mutation, and for the latter that it reflects a true regulatory role for the cloned gene(s) rather than a perturbation of physiology that is able to restore antibiotic biosynthesis.

For UV and NTG mutagenesis, and potentially for some transposon-induced mutations, it may be important to determine the effect of a null mutation in the gene of interest. The original mutation may not have completely abolished gene function, giving a partially deficient antibiotic phenotype and suggesting a nonessential role in antibiotic biosynthesis, whereas the corresponding null-mutation may reveal complete dependency. Indeed, in the case of one antibiotic regulatory gene, point and null mutations gave completely opposite phenotypes (see McKenzie and Nodwell, 2007, and references therein).

There are several ways in which targeted mutations can be made in actinomycetes:

3.5.1. PCR targeting
This technology relies on the use of the lambda *red* recombination system to rapidly create defined mutations in cloned DNA present in *E. coli* that can then be introduced into the wild-type strain of interest and integrated into the host genome, at the same time replacing the wild-type copy of the gene with a null mutant allele (Gust *et al.*, 2004 and Chapter 7 in this volume). This is most efficiently carried out on a sequenced cosmid clone containing the gene of interest. The mutated cosmid is then further modified to allow transfer by conjugation into the pertinent streptomycete, where double-crossover recombination mediated through long flanking sequences is used to replace the wild-type gene with the mutant allele.

3.5.2. Insertion and deletion mutagenesis by homologous recombination
If appropriate clones and the lambda *red* recombination system are not available, then targeted mutations (either insertions or deletions) can be made by creating a mutant allele in *E. coli*, transferring it by conjugation or transformation into the wild-type streptomycete, and selecting or screening for the desired mutant, which can then be characterized for antibiotic production. This can be achieved by simply inserting an antibiotic resistance gene into the streptomycete gene of interest at a position likely to result in a null allele, by replacing part or all of the gene with an antibiotic resistance cassette, or by creating a precise and preferably in-frame deletion (to avoid possible polar effects on downstream genes). All of these modifications can be achieved by routine cloning and/or by using PCR.

4. Characterizing Regulatory Genes for Antibiotic Biosynthesis

4.1. Pathway-specific regulatory genes

If the gene that has been identified is a pathway-specific regulatory gene, it will be necessary to determine which of the genes in the cluster it regulates (it may not regulate all of them), and ultimately whether it does so directly or indirectly. This can be accomplished in a number of ways.

4.1.1. Northern analysis

In principle, Northern analysis (Kieser *et al.*, 2000) remains the simplest procedure to monitor the effects on transcription of a gene set of interest of a mutation in a putative pathway-specific regulatory gene. By using radioactively labeled probes corresponding to the entire biosynthetic gene cluster (made, e.g., by nick-translation (Sambrook and Russell, 2001) of a cosmid clone containing the entire biosynthetic gene cluster), it is possible in one experiment to assess the role of a particular pathway-specific regulatory gene. RNA is isolated from the wild-type and mutant strains, size fractionated by denaturing gel electrophoresis, transferred to a suitable membrane (often positively charged nylon), and hybridized with labeled DNA corresponding to the entire biosynthetic gene cluster. Autoradiography is then used to determine the number and size of transcripts present in the wild-type strain, and to identify any that are missing or changed in abundance in the mutant. Given that antibiotic production very often occurs in a growth phase-dependent manner, ideally Northern analysis should be carried out using RNA samples obtained at different stages of growth (or at stages of growth shown previously to be associated with production).

4.1.2. Low-resolution S1 nuclease protection analysis

Unfortunately, given the relatively short half life of bacterial mRNA compared to eukaryotic transcripts, autoradiographs from Northern analyzes often reveal smears of degraded mRNA that can be difficult to interpret. As a consequence, low resolution S1 nuclease protection analysis is often used instead to assess the effects of mutation of a putative pathway-specific regulatory gene on transcription of an antibiotic biosynthetic gene cluster (Kieser *et al.*, 2000). In this procedure, DNA segments internal to or overlapping the predicted termini of individual transcripts derived from the cluster are hybridized to total RNA isolated from the wild-type and mutant strains, and the resulting DNA–RNA hybrids treated with S1 nuclease to remove unhybridized single strands of DNA and RNA. The protected double-stranded DNA fragments are resolved by agarose gel electrophoresis, blotted onto a suitable membrane (traditionally nitrocellulose, more recently nylon), and hybridized with a labeled probe. The resulting pattern of bands serves to identify any transcripts that are altered in the mutant and to localize the approximate start- and end-points of individual transcripts.

4.1.3. High-resolution S1 nuclease protection analysis

S1 nuclease protection analysis can be used not only to determine whether or not a particular gene in the cluster is transcribed in the mutant strain, but when used in high resolution mode it can identify, at the nucleotide level, the precise site of transcription initiation. This may provide information that can be used subsequently to identify consensus binding sites for a pathway-specific transcriptional activator. In this procedure, a DNA fragment is labeled on the 5′-end of the strand that is complementary to the transcript of interest and which extends beyond the likely transcriptional start site. The labeled probe (now generally made by PCR using an oligonucleotide primer that has been labeled on its 5′-end with ^{32}P using polynucleotide kinase) is hybridized to total RNA and the resulting DNA–RNA hybrids are treated with S1 nuclease and resolved on a sequencing gel adjacent to a sequencing ladder derived from either the same labeled PCR product or by using the oligonucleotide used to generate the PCR product as a sequencing primer. Consequently, it is generally possible to locate, within 1–3 nucleotides, the precise transcriptional start site of a particular gene. The addition of sequences at the 3′-end of the labeled DNA strand of the probe that are not homologous to genomic DNA (accomplished when designing the unlabeled primer used in the PCR generation of the probe) enables the detection of read-through transcription (as opposed to reannealing of the DNA probe; Kieser et al., 2000), and thus provides additional information about overall transcript organization.

4.1.4. Primer extension

Primer extension is an alternative to high resolution S1 nuclease mapping (Kieser et al., 2000). It uses a 5′-end-labeled synthetic oligonucleotide and reverse transcriptase to make a cDNA copy of the 5′-end of a transcript of interest, the length of which localizes the transcriptional start site to within a couple of nucleotides when resolved on a denaturing acrylamide gel adjacent to a sequence ladder derived from the same synthetic oligonucleotide. This procedure is often used in streptomycetes instead of high resolution S1 nuclease mapping and is less labor-intensive, but it can be less reliable (premature termination, which appears to occur more frequently with GC-rich DNA, can lead to incorrectly predicted transcriptional start sites). While primer extension is used to localize transcriptional start sites, it is not generally used to detect the presence or absence of a particular mRNA.

4.1.5. In vitro transcription and dinucleotide priming

Although not used on a routine basis, should there be a need, it is possible to confirm the results of S1 nuclease mapping or primer extension by carrying out *in vitro* run-off transcription analysis (Kieser *et al.*, 2000). Ideally this should be done using RNA polymerase isolated from the streptomycete of interest. However, should an alternative σ factor be implicated in the transcription of an antibiotic biosynthetic gene, and if that σ factor can be purified, then it is possible to carry out *in vitro* reconstitution experiments using commercially available *E. coli* core RNA polymerase (Fujii *et al.*, 1996). In this procedure, labeled transcripts are produced *in vitro* from a purified DNA template that terminates before the predicted end of the transcript (hence the term "run-off transcript"). The sizes of the run-off transcripts are determined by denaturing gel electrophoresis using labeled size markers and autoradiography. In an extension of the procedure, *in vitro* transcription can be initiated using a dinucleotide (Janssen *et al.*, 1989). The ability of just 1 of the 16 possible dinucleotides to prime *in vitro* transcription suggests a truly unique start site, whereas the ability of two or three dinucleotides to prime could be indicative of staggered sites (which should be corroborated by high-resolution S1 nuclease mapping).

4.1.6. Reverse transcription PCR and quantitative real time RT-PCR

More recently, the presence or absence of transcripts (but not their start sites) is often determined by Reverse Transcription-PCR (RT-PCR). Although only semiquantitative, it is a rapid technique that employs synthetic oligonucleotides, reverse transcriptase, and DNA polymerase to amplify, as DNA, an internal segment of the transcript of interest. The amplified product can then be analyzed by agarose gel electrophoresis (e.g., see Stratigopoulos *et al.*, 2004). Where the transcription of a gene is not completely abolished, if the number of cycles is varied, it may be possible to gain some semiquantitative insight into the level of transcription. Preferably, quantitative real time RT-PCR (qRT-PCR) is carried out (for a review, see Kubista *et al.*, 2006). This requires fluorescently labeled oligonucleotide primers, or the use of a dye that fluoresces only in the presence of double-stranded DNA, which can be used to monitor the kinetics of formation of the amplified product and provide accurate estimates of transcript abundance over a wide dynamic range. Given the availability of different fluorescent labels, it is now possible to multiplex the qRT-PCR procedure to permit the analysis of several genes simultaneously.

The absence of a transcript, as deduced by any of the above procedures, does not prove direct regulation by a pathway-specific regulatory protein, but with the localization of the transcriptional start site in the wild-type strain, it provides the necessary information to carry out additional experiments that address this possibility in a direct manner (see Electrophoretic Mobility Shift Assay and Surface Plasmon Resonance).

4.1.7. Electrophoretic mobility shift assay (EMSA)

Once the transcriptional start site of a gene whose transcript is absent in a putative regulatory mutant has been determined, for example by S1 nuclease mapping or primer extension, it is then possible to address whether the gene is directly or indirectly regulated by the putative pathway-specific regulatory protein. The most frequently used procedure thus far is EMSA (Fried and Crothers, 1981; Garner and Revzin, 1981). In this approach, a labeled (generally by PCR) DNA fragment that contains sequences that encompass or lie upstream of the transcriptional start site is incubated with increasing amounts of the putative regulatory protein. The latter may be purified (e.g., after His-tagging), or be present in a crude cell extract. After incubation, the potential protein–DNA complexes are subjected to nondenaturing gel electrophoresis in parallel with an aliquot of the DNA fragment that has not been incubated with protein. A reduction in the mobility of the labeled DNA fragment (a "band-shift") is indicative of a protein–DNA complex. The proportion of the DNA fragment retarded should increase with increasing amounts of protein. Moreover, to ensure that a specific interaction is occurring, it should be possible to abolish the band-shift by adding excess unlabeled fragment, while an excess of a nonspecific competitor should not compete out the band-shift. The most suitable nonspecific competitor is an unrelated DNA fragment of the same size and base composition.

To locate the site of interaction more precisely, it may be possible to use shorter and/or nonoverlapping fragments derived from the DNA sequence used in the original EMSA. Beyond that, DNaseI foot-printing (Galas and Schmitz, 1978) and site-directed mutagenesis can be used to define the interaction more precisely, and to determine the specific nucleotides that play a crucial role in protein binding.

4.1.8. Surface plasmon resonance

While it may be possible to obtain some semiquantitative data about DNA–protein interactions from EMSA, Surface Plasmon Resonance (SPR, commercialized initially by Biacore, now GE Healthcare (http://www.biacore.com/lifesciences/company/presentation/introduction/index.html)) allows for much more stringent kinetic analysis. SPR enables real-time detection and monitoring of DNA–protein interactions, and provides quantitative information on rates of association and dissociation, and hence rate and equilibrium constants. In principle, either the DNA fragment containing the binding site or the regulatory protein can be attached to the Biacore chip, and the other interacting partner applied through the flow channel. SPR can also be used to quantify the effects of mutations in either the DNA-binding site or the regulatory protein on the interaction.

4.1.9. Reporter genes

Reporter genes encode protein products that can be detected and quantified readily, and include those conferring fluorescence, luminescence, antibiotic resistance, or a conveniently assayable enzyme activity (reviewed in Daunert et al., 2000). Reporters are, therefore, useful tools for quantitatively assaying the transcriptional activity of promoters of interest, and could be recruited for measuring changes in the level of transcription of different operons present in a particular antibiotic biosynthetic gene cluster that occur as a result of mutation or overexpression of a putative regulatory gene. In practice, however, few examples of this particular application exist in the literature, which perhaps reflects the fact that a convenient, reliable reporter system to rival that based on lacZ in E. coli has yet to be developed for streptomycetes (discussed in Kieser et al., 2000). Early attempts to apply the lacZ-based system to S. lividans or S. coelicolor met with only limited success due to issues with plasmid instability, suboptimal codon usage, and interference by native enzymes with β-galactosidase activity. Since then, systems based on green fluorescent protein (GFP), the kanamycin resistance gene neo, luciferase (lux), the xylE gene from the Pseudomonas TOL plasmid, and the redD gene from the S. coelicolor undecylprodigiosin antibiotic cluster have all been tried with varying degrees of success (reviewed in Kieser et al., 2000). Endogenously produced compounds that fluoresce at a similar wavelength to GFP interfere with the use of this reporter in certain Streptomyces species, while both the neo and redD systems are at best only semi-quantitative. xylE and lux show the most promise as transcriptional reporters, and in recent work the lux system has been specifically optimized for use in high-GC bacteria, including streptomycetes (Craney et al., 2007).

However, because of likely differences in transcript stability between a native mRNA and that produced from a reporter gene, regardless of the potential ease of use of reporter systems, analysis of wild-type and mutant strains at the RNA level, using the methods described earlier, will ultimately be required to confirm any quantitative changes observed using a reporter gene.

4.2. Pleiotropic regulatory genes

If the gene that has been identified is a pleiotropic regulatory gene, approaches that more globally capture patterns of gene expression will be required to help dissect the influence it exerts on the antibiotic gene clusters present in the strain. This is possible for organisms for which the genome sequence has been determined, allowing the application of DNA microarray and proteomics technologies (Hesketh et al., 2007a,b;

Huang *et al.*, 2001, 2005; Lian *et al.*, 2008; Takano *et al.*, 2005), but may also be achievable to a more limited extent in strains where genome sequence data exists only for a closely related species. The goal of any such functional genomics study in this context is to define the series of cellular events that link the identified regulatory gene with the antibiotic production phenotype observed. This may include both transcriptional and posttranscriptional effects, necessitating the use of both transcriptome and proteome analysis to view the complete picture. In reality however, costs and the local availability of equipment or expertise is likely to limit the approach chosen to only one or other of these methods, and in this chapter we confine our considerations to microarray analyzes (although the principles of experimental design and data analysis will be similar in each case).

DNA-microarrays were developed to detect the presence and abundance of labeled nucleic acids in a biological sample, and consist of a solid surface onto which many thousands of different DNA molecules have been chemically bonded (Bier *et al.* (2008) present a recent useful overview). For gene expression analysis, the labeled nucleic acids from the experimental samples are derived from mRNA and the DNA immobilized on the array has been designed so that every gene in the genome of interest is represented. Quantification of the amount of label hybridized, via Watson–Crick duplex formation, to each different DNA molecule on the array can therefore produce a simultaneous measurement of the expression levels of many thousands of genes. DNA microarrays can differ in a number of ways (the length of the DNA molecules immobilized; whether the array has been manufactured by robotic spotting or *in situ* synthesis; the number of different DNA molecules used to represent each gene; whether or not intergenic regions of the genome are also represented; cost) and will also differ in their sensitivity and accuracy of performance. Different array technologies can also require different methods for labeling of the sample to be hybridized, which can impact on how the samples are quantified. The Affymetrix platform, for instance, uses biotin-labeled cDNA or cRNA which dictates that only one sample can be hybridized to each array, whereas for many other technologies the labeled samples can be prepared using Cy3 or Cy5 fluorescent dyes which, since they emit fluorescence at significantly different wavelengths, allows two samples to be hybridized and detected per array. The former method thus yields an estimation of the absolute value of gene expression, while the latter produces a relative ratio-based measurement. Careful consideration to the choice of the array platform to be used in any transcriptomics experiment should be given, and is helped by clearly defining the desired outcomes of the study within the framework of the budget available.

Having decided on the technology to be used for the analysis, the next important consideration is the design of the experiment to produce the samples to be labeled and hybridized. Many pleiotropic regulatory genes are only required for antibiotic production under particular nutritional conditions, and the culture medium used therefore needs to be carefully selected to suit not only production of the antibiotics but also the requirement for the regulator under study. Using the optimal growth conditions, three experimental designs that can be used for investigating the function of a pleiotropic regulatory gene are described below. In practice, and if resources permit, it may be desirable to combine at least two of these to reliably identify candidate genes that play intermediary roles in the regulation of antibiotic biosynthesis. In any event, the putative involvement of candidates identified using any of these approaches will need to be confirmed experimentally.

4.2.1. The effect of gene deletion

The analysis of differences in patterns of gene expression between the wild-type strain and a mutant in which the gene of interest has been cleanly deleted (e.g., by using one of the methods described above) is usually the first step in identifying sets of genes that are likely to be controlled by the pleiotropic regulatory gene. Since antibiotic production in the wild-type strain is normally growth phase-dependent rather than constitutive, this is usually achieved by sampling (in at least triplicate) both wild-type and mutant strains at several comparable time points during growth in liquid culture, such that at least one sample from the exponential, transition, and stationary phases of growth is present. Statistical analysis is then used to identify genes that are significantly up- or down-regulated as a result of the mutation, and clustering algorithms can be applied to find groups of genes that have similar expression profiles, and thus may be coregulated (Boutros and Okey (2005), D'haeseleer (2005)). If the pleiotropic regulator in question is predicted to be a DNA-binding protein, the upstream regions of the putatively coregulated sets of genes can be searched for common promoter sequences (using tools such as MEME (http://meme.sdsc.edu/meme) and Prodoric (http://prodoric.tu-bs.de)) to provide evidence for coregulation, and any putative protein binding site identified can be assessed experimentally using the *in vitro* techniques described earlier for the analysis of pathway-specific regulatory genes. In this way, it may be possible to build up a list of the genes that are direct targets for regulation by the mutated gene, and these can be assessed bioinformatically, looking for sequence homology to known regulatory factors or enzymes that may be involved in precursor supply or cofactor biosynthesis, for those most likely to be involved in influencing production of the antibiotics. Construction of targeted mutations in the candidate genes will ultimately be required to confirm any role in antibiotic biosynthesis, and the resultant mutant strains may subsequently form the subject of additional rounds of transcriptome/proteome analysis to further characterize the links to antibiotic production.

4.2.2. The effect of gene overexpression

The enhanced expression of a regulatory gene carried on a high copy number plasmid vector, or overexpressed from a strong constitutive promoter, is expected to produce the opposite effect to gene deletion: genes found to be up-regulated as a result of gene mutation should be down-regulated in the overexpression strain (and *vice versa*). Analysis of overexpression strains along the lines outlined for the deletion mutant above will therefore provide complementary data helping to identify sets of genes that are subject to control by the pleiotropic regulator, and which may therefore also be involved in the control of antibiotic production.

4.2.3. The effect of controlled induction of gene expression

The disadvantage of using the gene deletion or overexpression approaches described above is that in both cases steady-state conditions are being analyzed in which the cells have physiologically adapted to the change in expression of the pleiotropic regulatory gene. This means that the expression of numerous genes that are not directly controlled by the gene under study will also be significantly different between the strains analyzed. These indirect effects can potentially arise via additional regulatory genes which are *bona fide* members of the regulon of the pleiotropic regulatory gene (and thus will also be of interest), or be the result of compensatory changes occurring in response to deleterious effects on cellular metabolism caused by the mutation/overexpression. One approach that helps to distinguish between the direct and indirect effects of a regulatory gene is to induce a controlled change in its expression and analyze the response dynamically over a short period of time. This is most conveniently done by expressing a copy of the gene of interest under the control of a tightly regulated promoter, and is therefore limited to strains where such systems are available (e.g., streptomycete vectors carrying the thiostrepton-inducible promoter of *tipA* (as used in Hesketh *et al.*, 2007a; Huang *et al.*, 2005), and potentially the epsilon-caprolactam-inducible promoter present in pSH19 (Herai *et al.*, 2004). Genes that are directly under the control of the induced regulatory gene should exhibit instant changes in gene expression, whereas those further down the sequence of events will take a longer time to show a response. Compensatory changes should similarly not be evident in the data until after the more direct effects have been observed.

4.2.4. Defining genome-wide DNA–protein interactions by chromatin immunoprecipitation and microarray analysis (ChIP–on-Chip)

Once genes that are likely to be controlled by a pleiotropic regulatory gene have been identified (e.g., by microarray analysis), it may be necessary to determine whether regulation occurs in a direct or indirect manner.

While in principle this can be accomplished by EMSA or SPR analyzes on individual promoter regions, a more labor-efficient and global approach is accomplished by carrying out ChIP on Chip analysis. This generally requires a genome sequence and suitable microarrays for the organism of interest, purified regulatory protein and a corresponding antibody (although it may be possible to use commercially available antibodies made to, e.g., a FLAG tag that has been engineered into the protein prior to overexpression and purification). In brief, cultures at an appropriate stage of growth are treated with formaldehyde to cross-link regulatory proteins to their cognate and occupied (*in vivo*) binding sites. DNA, with any covalently attached proteins, is isolated from the treated cultures and sheared to give fragments that are 500–1000 bp in length. The required DNA–protein complexes are then isolated using antibody to the protein or its tag. The protein is removed from the DNA fragment by heat treatment, and the DNA amplified by PCR using a fluorescent dye, such as Cy5, and used to probe a microarray that contains all of the intergenic regions present in the genome (it may also be possible to use arrays confined to protein-coding sequences, but there would need to be some confidence that sequences existed on the array that were located close enough to the $5'$-end of the coding sequences to be available for hybridization to the labeled probe). Development of the hybridized array followed by bioinformatic analysis will then reveal genes that are candidates for direct binding by the pleiotropic regulatory protein and therefore likely to be direct regulatory targets.

This in itself is not sufficient to establish direct regulation—hybridization to the array may be nonspecific or not accurately reflect the *in vivo* situation. Lack of, or altered, transcription of the newly identified genes in the pleiotropic mutant should first be confirmed by, for example, microarray or qRT-PCR analysis, and a direct interaction should ultimately be demonstrated by EMSA or SPR.

4.2.5. Systematic evolution of ligands by exponential enrichment (SELEX)

Although apparently not used thus far for streptomycetes, an alternative approach to identifying the binding sites (and hence directly controlled genes) of a pleiotropic regulatory protein is SELEX (Stoltenburg *et al.*, 2007; Tsai and Reed, 1998). In this approach, a random nucleotide sequence of ~20 bp is synthesized that is flanked by opposing PCR primer sites. This highly degenerate mixture of nucleotide sequences is then incubated with the purified regulatory protein and subjected to EMSA analysis. Although a retarded band is unlikely to be apparent upon ethidium bromide staining after the first EMSA, material is extracted from the region of the gel where a shifted band would be expected to migrate to, and subjected to PCR using primers corresponding to the flanking sites. The subsequent PCR product is then incubated with the purified regulatory protein and the

EMSA, gel extraction and PCR repeated until a band becomes visible by ethidium bromide staining. Subsequent sequencing of the PCR products should then reveal *in vitro* binding sites for the regulatory protein of interest. Bioinformatic analysis is then required to identify putative binding sites present in the genome sequence of the organism of interest. Confirmation that the protein does indeed bind to such sites *in vivo* and regulates the corresponding genes then needs to be obtained by transcriptional analysis of the regulatory mutant (e.g., by qRT-PCR). Note that while repeated rounds of SELEX should enrich for the most tightly binding nucleotide sequences *in vitro*, these are not necessarily representative of the majority of physiologically relevant binding sites present in the genome.

In a variation on this approach, Elliot *et al.* (2001), Horinouchi *et al.* (2000), and Yamazaki *et al.* (2000) used purified regulatory proteins with fragmented genomic DNA to which PCR primers had been attached as the target sequences. Multiple rounds of PCR and EMSA assays yielded a number of candidate genes for further analysis.

4.3. Concluding remarks

The regulation of antibiotic production is a highly complex process, but a plethora of technologies can now be used to gain a greater understanding of how the transcription of the often large gene clusters that encode these compounds is triggered. We hope that this article, which is not intended to be a comprehensive description of all of the approaches that can be used, will provide the reader with the initial understanding required to embark on such studies.

ACKNOWLEDGMENTS

We thank Keith Chater and Maureen Bibb for their comments on this contribution, and the UK Biotechnology and Biological Sciences Research Council for funding.

REFERENCES

Bibb, M. J. (2005). Regulation of secondary metabolism in streptomycetes. *Curr. Opin. Microbiol.* **8,** 208–215.

Bier, F. F., von Nickisch-Rosenegk, M., Ehrentreich-Forster, E., Reif, E., Henkel, J., Strehlow, R., and Andresen, D. (2008). DNA microarrays. *Adv. Biochem. Eng. Biotechnol.* **109,** 433–453.

Bierman, M., Logan, R., O'Brien, K., Seno, E. T., Rao, R. N., and Schoner, B. E. (1992). Plasmid cloning vectors for the conjugal transfer of DNA from *Escherichia coli* to *Streptomyces* spp. *Gene* **116,** 43–49.

Boutros, P. C., and Okey, A. (2005). Unsupervised pattern recognition: An introduction to the whys and wherefores of clustering microarray data. *Brief. Bioinform.* **6,** 331–343.

Chakraburtty, R., and Bibb, M. J. (1997). The ppGpp synthetase gene (*relA*) of *Streptomyces coelicolor* A3(2) plays a conditional role in antibiotic production and morphological differentiation. *J. Bacteriol.* **179,** 5854–5861.

Chater, K. F., and Bruton, C. J. (1983). Mutational cloning in *Streptomyces* and the isolation of antibiotic production genes. *Gene* **26,** 67–78.

Chater, K. F., and Chandra, G. (2008). The use of the rare UUA codon to define "expression space" for genes involved in secondary metabolism, development and environmental adaptation in *Streptomyces*. *J. Microbiol.* **46,** 1–11.

Corre, C., Song, L., O'Rourke, S., Chater, K. F., and Challis, G. L. (2008). 2-Alkyl-4-hydroxymethylfuran-3-carboxylic acids, antibiotic production inducers discovered by *Streptomyces coelicolor* genome mining. *Proc. Natl. Acad. Sci. USA* **105,** 17510–17515.

Craney, A., Hohenauer, T., Xu, Y., Navani, N. K., Li, Y., and Nodwell, J. (2007). A synthetic *luxCDABE* gene cluster optimized for expression in high-GC bacteria. *Nucleic Acids Res.* **35,** e46.

Daunert, S., Barrett, G., Feliciano, J. S., Shetty, R. S., Shrestha, S., and Smith-Spencer, W. (2000). Genetically engineered whole-cell sensing systems: Coupling biological recognition with reporter genes. *Chem. Rev.* **100,** 2705–2738.

D'haeseleer, P. (2005). How does gene expression clustering work? *Nat. Biotechnol.* **23,** 1499–1501.

Elliot, M. A., Bibb, M. J., Buttner, M. J., and Leskiw, B. K. (2001). BldD is a direct regulator of key developmental genes in *Streptomyces coelicolor* A3(2). *Mol. Microbiol.* **40,** 257–269.

Fried, M., and Crothers, D. M. (1981). Equilibria and kinetics of *lac* repressor–operator interactions by polyacrylamide gel electrophoresis. *Nucleic Acids Res.* **9,** 6505–6525.

Fujii, T., Gramajo, H. C., Takano, E., and Bibb, M. J. (1996). *redD* and *actII*-ORF4, pathway-specific regulatory genes for antibiotic production in *Streptomyces coelicolor* A3(2), are transcribed *in vitro* by an RNA polymerase holoenzyme containing sigma hrdD. *J. Bacteriol.* **178,** 3402–3405.

Galas, D. J., and Schmitz, A. (1978). DNAse footprinting: A simple method for the detection of protein–DNA binding specificity. *Nucleic Acids Res.* **5,** 3157–3170.

Garner, M. M., and Revzin, A. (1981). A gel electrophoresis method for quantifying the binding of proteins to specific DNA regions: Application to components of the *Escherichia coli* lactose operon regulatory system. *Nucleic Acids Res.* **9,** 3047–3060.

Gramajo, H. C., Takano, E., and Bibb, M. J. (1993). Stationary phase production of the antibiotic actinorhodin in *Streptomyces coelicolor* A3(2) is transcriptionally regulated. *Mol. Microbiol.* **7,** 837–845.

Guilfoyle, R. A., Leeck, C. L., Kroening, K. D., Smith, L. M., and Guo, Z. (1997). Ligation-mediated PCR amplification of specific fragments from a class-II restriction endonuclease total digest. *Nucleic Acids Res.* **25,** 1854–1858.

Gust, B., Chandra, G., Jakimowicz, D., Yuqing, T., Bruton, C. J., and Chater, K. F. (2004). Lambda red-mediated genetic manipulation of antibiotic-producing *Streptomyces*. *Adv. Appl. Microbiol.* **54,** 107–128.

Herai, S., Hashimoto, Y., Higashibata, H., Maseda, H., Ikeda, H., Omura, S., and Kobayashi, M. (2004). Hyper-inducible expression system for streptomycetes. *Proc. Natl. Acad. Sci. USA* **101,** 14031–14035.

Hesketh, A., Chen, J., Ryding, J., Chang, S., and Bibb, M. J. (2007a). The global role of ppGpp synthesis in morphological differentiation and antibiotic production in *Streptomyces coelicolor* A3(2). *Genome Biol.* **8,** R161.

Hesketh, A., Bucca, G., Laing, E., Flett, F., Hotchkiss, G., Smith, C. P., and Chater, K. F. (2007b). New pleiotropic effects of eliminating a rare tRNA from *Streptomyces coelicolor*, revealed by combined proteomic and transcriptomic analysis of liquid cultures. *BMC Genomics* **8,** 261.

Hesketh, A., Sun, J., and Bibb, M. J. (2001). Induction of ppGpp synthesis in *Streptomyces coelicolor* A3(2) grown under conditions of nutritional sufficiency elicits *actII*-ORF4 transcription and actinorhodin biosynthesis. *Mol. Microbiol.* **39**, 136–144.

Hopwood, D. A., Malpartida, F., and Chater, K. F. (1985). Gene cloning to analyse the organization and expression of antibiotic biosynthesis genes in *Streptomyces*. *In* "Regulation of Secondary Metabolic Formation" (H. Kleinkauf, H. Van Dohren, H. Dornauer, and G. Nesemann, eds.) pp. 22–33. VCH, Weinheim.

Horinouchi, S., Onaka, H., Yamazaki, H., Kameyama, S., and Ohnishi, Y. (2000). Isolation of DNA fragments bound by transcriptional factors, AdpA and ArpA, in the A-factor regulatory cascade. *Actinomycetologica* **14**, 37–42.

Huang, J., Lih, C. J., Pan, K. H., and Cohen, S. N. (2001). Global analysis of growth phase responsive gene expression and regulation of antibiotic biosynthetic pathways in *Streptomyces coelicolor* using DNA microarrays. *Genes Dev.* **15**, 3183–3192.

Huang, J., Shi, J., Molle, V., Sohlberg, B., Weaver, D., Bibb, M. J., Karoonuthaisiri, N., Lih, C. J., Kao, C. M., Buttner, M. J., *et al*. (2005). Cross-regulation among disparate antibiotic biosynthetic pathways of *Streptomyces coelicolor*. *Mol. Microbiol.* **58**, 1276–1287.

Janssen, G. R., Ward, J. M., and Bibb, M. J. (1989). Unusual transcriptional and translational features of the aminoglycoside phosphotransferase gene (*aph*) from *Streptomyces fradiae*. *Genes Dev.* **3**, 415–429.

Kieser, T., Bibb, M. J., Chater, K. F., and Hopwood, D. A. (2000). Practical *Streptomyces* Genetics Norwich: John Innes Foundation, Norwich.

Kubista, M., Andrade, J. M., Bengtsson, M., Forootan, A., Jonák, J., Lind, K., Sindelka, R., Sjöback, R., Sjögreen, B., Strömbom, L., *et al*. (2006). The real-time polymerase chain reaction. *Mol. Aspects Med.* **27**, 95–125.

Lian, W., Jayapal, K. P., Charaniya, S., Mehra, S., Glod, F., Kyung, Y. S., Sherman, D. H., and Hu, W. S. (2008). Genome-wide transcriptome analysis reveals that a pleiotropic antibiotic regulator, AfsS, modulates nutritional stress response in *Streptomyces coelicolor* A3(2). *BMC Genomics* **9**, 56.

Magnusson, L. U., Farewell, A., and Nystrom, T. (2005). ppGpp: A global regulator in *Escherichia coli*. *Trends Microbiol.* **13**, 236–242.

McKenzie, N. L., and Nodwell, J. R. (2007). Phosphorylated AbsA2 negatively regulates antibiotic production in *Streptomyces coelicolor* through interactions with pathway-specific regulatory gene promoters. *J. Bacteriol.* **189**, 5284–5292.

Ohnishi, Y., Yamazaki, H., Kato, J. Y., Tomono, A., and Horinouchi, S. (2005). AdpA, a central transcriptional regulator in the A-factor regulatory cascade that leads to morphological development and secondary metabolism in *Streptomyces griseus*. *Biosci. Biotechnol. Biochem.* **69**, 431–439.

O'Rourke, S., Wietzorrek, A., Fowler, K., Corre, C., Challis, G. L., and Chater, K. F. (2009). Extracellular signalling, translational control, two repressors and an activator all contribute to the regulation of methylenomycin production in *Streptomyces coelicolor*. *Mol. Microbiol.* **71**, 763–778.

Rigali, S., Titgemeyer, F., Barends, S., Mulder, S., Thomae, A. W., Hopwood, D. A., and van Wezel, G. P. (2008). Feast or famine: The global regulator DasR links nutrient stress to antibiotic production by *Streptomyces*. *EMBO Rep.* **9**, 670–675.

Sambrook, J., and Russell, D. W. (2001). Molecular Cloning: A Laboratory Manual 3rd ed. Cold Spring Harbor LaboratoryCold Spring Harbor, NY.

Stoltenburg, R., Reinemann, C., and Strehlitz, B. (2007). SELEX—A (r)evolutionary method to generate high-affinity nucleic acid ligands. *Biomol. Eng.* **24**, 381–403.

Stratigopoulos, G., Bate, N., and Cundliffe, E. (2004). Positive control of tylosin biosynthesis: Pivotal role of TylR. *Mol. Microbiol.* **54**, 1326–1334.

Sun, J., Hesketh, A., and Bibb, M. J. (2001). Functional analysis of *relA* and *rshA*, two relA/spoT homologues of *Streptomyces coelicolor* A3(2). *J. Bacteriol.* **183**, 3488–3498.

Takano, E. (2006). Gamma-butyrolactones: *Streptomyces* signalling molecules regulating antibiotic production and differentiation. *Curr. Opin. Microbiol.* **9,** 287–294.

Takano, E., Gramajo, H., Strauch, E., Andres, N., White, J., and Bibb, M. J. (1992). Transcriptional regulation of the *redD* transcriptional activator gene accounts for growth phase-dependent production of the antibiotic undecylprodigiosin in *Streptomyces coelicolor* A3(2). *Mol. Microbiol.* **6,** 2797–2804.

Takano, E., Kinoshita, H., Mersinias, V., Bucca, G., Hotchkiss, G., Nihira, T., Smith, C. P., Bibb, M., Wohlleben, W., and Chater, K. (2005). A bacterial hormone (the SCB1) directly controls the expression of a pathway-specific regulatory gene in the cryptic type I polyketide biosynthetic gene cluster of *Streptomyces coelicolor*. *Mol. Microbiol.* **56,** 465–479.

Tsai, R. Y. L., and Reed, R. R. (1998). Identification of DNA recognition sequences and protein interaction domains of the multiple-Zn-finger protein Roaz. *Mol. Cell. Biol.* **18,** 6447–6456.

Wright, L. F., and Hopwood, D. A. (1976). Identification of the antibiotic determined by the SCP1 plasmid of *Streptomyces coelicolor* A3(2). *J. Gen. Microbiol.* **95,** 96–106.

Yamazaki, H., Ohnishi, Y., and Horinouchi, S. (2000). An A factor-dependent extracytoplasmic function sigma factor (sAdsA) that is essential for morphological development in *Streptomyces griseus*. *J. Bacteriol.* **182,** 4596–4605.

CHAPTER FIVE

APPLYING THE GENETICS OF SECONDARY METABOLISM IN MODEL ACTINOMYCETES TO THE DISCOVERY OF NEW ANTIBIOTICS

Gilles P. van Wezel,* Nancy L. McKenzie,[†] and Justin R. Nodwell[†]

Contents

1. Introduction	118
2. Actinomycetes as Antibiotic Factories	119
3. Effects of Culture Conditions and Metabolism	120
3.1. Growth-dependent control mechanisms	120
3.2. Stringent control	121
3.3. Phosphate-mediated control	121
3.4. Interactions between metabolism and the DasR regulon	122
3.5. Morphology as determinant of productivity	123
4. Molecular Genetic Factors that Regulate Antibiotic Production	125
4.1. Pathway-specific regulation	125
4.2. Pleiotropic regulation	127
5. Applications for New Antibiotic Screening Technologies	129
5.1. Heterologous overexpression and mutant alleles	130
6. Future Prospects	133
Acknowledgments	134
References	134

Abstract

The actinomycetes, including in particular members of the filamentous genus *Streptomyces*, are the industrial source of a large number of bioactive small molecules employed as antibiotics and other drugs. They produce these molecules as part of their "secondary" or nonessential metabolism. The number and diversity of secondary metabolic pathways is enormous, with some estimates suggesting that this one genus can produce more than 100,000 distinct

* Molecular Genetics, Leiden Institute of Chemistry, Gorlaeus Laboratories, The Netherlands
[†] Michael G. DeGroote Centre for Infectious Disease Research, Department of Biochemistry and Biomedical Sciences, McMaster University, Hamilton, Ontario, Canada

molecules. However, the discovery of new antimicrobials is hampered by the fact that many wild isolates fail to express all or sometimes any of their secondary metabolites under laboratory conditions. Furthermore, the use of previously successful screening strategies frequently results in the rediscovery of known molecules: the all-important novel structures have proven to be elusive. Mounting evidence suggests that streptomycetes possess many regulatory pathways that control the biosynthetic gene clusters for these secondary metabolic pathways and that cell metabolism plays a significant role in limiting or potentiating expression as well. In this article we explore the idea that manipulating metabolic conditions and regulatory pathways can "awaken" silent gene clusters and lead to the discovery of novel antimicrobial activities.

1. Introduction

The modern antibiotic era began with the discovery and large-scale application of penicillin and streptomycin to combat diseases such as septicemia and tuberculosis (Hopwood, 2007). The spectacular success of these and other antibiotics led to the belief that infectious diseases could be conquered. Correspondingly, due in part to a perceived lack of need, the discovery of new antibiotics declined steeply after the 1960s (Hopwood, 2007). Ironically, antibiotic resistance, which was discovered soon after the clinical introduction of penicillin and streptomycin, and which has accompanied the introduction of each new drug, has increased steadily ever since to the point where it now constitutes a significant threat to human health. Meanwhile, antibiotic discovery and research in most large pharmaceutical companies has all but ceased (Katz et al., 2006; Payne et al., 2007), with the result that in the past 40 years only two new chemical classes of antibiotics have been approved for clinical use: linezolid (an oxazolidinone) in 2000 and daptomycin (a cyclic lipopeptide) in 2003 (Wright, 2007). Resistance to these new classes of antibiotics has already emerged (Bersos et al., 2004; Hayden et al., 2005).

There are several impediments to antibiotic discovery, including the low return on investment (as compared to drugs for chronic diseases) and the rigorous regulatory environment governing the clinical testing of new antibiotics. However, it is likely that technical difficulties in identifying and developing truly novel and clinically useful antibiotics is the most significant factor (Bradley et al., 2007; Katz et al., 2006; Payne et al., 2007). The success of previous screening efforts has created a situation in which conventional screening technologies tend to result in the repeated identification of known molecules. Estimates vary, but it is believed by some that if conventional technology were used it would be necessary to test millions of new isolates of environmental microbes to find a single new antimicrobial compound (Baltz, 2006). New thinking in this area is therefore a high priority (Grundmann et al., 2006; Mukherjee et al., 2004).

It is clear that novel strategies and technologies are required to avoid the rediscovery of known antibiotics and to increase the probability of finding new ones. One encouraging example is the recent discovery of platensimycin, produced by *Streptomyces platensis*, which represents a novel class of antibiotic that inhibits FabF, an enzyme involved in fatty acid biosynthesis—see Chapter 17 in Volume 459 (Wang *et al.*, 2006). Platensimycin was identified by screening 83,000 strains grown under different growth conditions against a *Staphylococcus aureus* strain in which antisense RNA against *fabF* was used to lower its gene expression (Wang *et al.*, 2006), thereby making the bacteria more sensitive to inhibitors of the targeted protein, and increasing the probability of finding a hit. Such occasional successes using novel approaches kept hopes alive to identify natural products from bacterial sources (Clardy *et al.*, 2006). New thinking on this old problem is what is required most.

2. Actinomycetes as Antibiotic Factories

Filamentous microorganisms are widely used as industrial producers of antibiotics and other medicinal agents (Bennett, 1998; Demain, 1991; Hopwood *et al.*, 1995). These organisms include the eukaryotic filamentous fungi (ascomycetes) and the prokaryotic actinomycetes (e.g., *Amycolatopsis, Nocardia, Thermobifida,* and *Streptomyces*). The market capitalization for antibiotics is greater than 30 billion US dollars per year. Approximately 30% of all antibiotics are produced by fungi (Hersbach *et al.*, 1984; van den Berg *et al.*, 2007, 2008; Wright, 1999) and 65% by filamentous actinomycetes, including mostly streptomycetes (Hopwood, 2007).

Streptomyces coelicolor is an important model system for the study of antibiotic production and its regulation in actinomycetes. Before its complete genome sequence became available, four antibiotics had been identified, namely actinorhodin (Act; Rudd and Hopwood (1979)), undecylprodigiosin (Red; Feitelson *et al.* (1985)), the calcium-dependent antibiotic (CDA; Hopwood and Wright (1983)), and the plasmid-encoded methylenomycin (Mmy; Wright and Hopwood (1976)). Later, the complete genome sequences of *S. coelicolor* (Bentley *et al.*, 2002), *Streptomyces avermitilis* (Ikeda *et al.*, 2003), and *Streptomyces griseus* (Ohnishi *et al.*, 2008) revealed that their antibiotic-producing potential had been underestimated; each genome contains more than 20 sets of putative biosynthetic genes for secondary metabolites (Challis and Hopwood, 2003). For unknown reasons, many of these are not expressed under laboratory conditions. To take advantage of this untapped source of potentially valuable natural products, we need a better understanding of the genetic and environmental parameters, particularly metabolic inputs and secondary metabolite production.

Recently, examples have been presented of dormant biosynthetic clusters that could be induced by selective growth conditions, such as the enediyne-type PKS antibiotics in several actinomycetes (Zazopoulos et al., 2003) (see Chapter 5 in Volume 459) or the *kas* (*cpk*) gene cluster in *S. coelicolor* (Rigali et al., 2008). Understanding the control of cryptic antibiotic biosynthetic clusters is a major challenge and the potential benefits are enormous (Van Lanen and Shen, 2006; Wilkinson and Micklefield, 2007). It has been estimated that the genus *Streptomyces* alone is capable of producing >100,000 distinct antimicrobial compounds of which only ~3% have been identified and investigated so far (Watve et al., 2001).

Antibiotic biosynthesis is mediated by large contiguous gene clusters ranging in size from a few to over 100 kb (Bentley et al., 2002; Ikeda et al., 2003; Ohnishi et al., 2008). The actinorhodin, undecylprodigiosin, and CDA biosynthetic genes are clustered in distinct locations on the *S. coelicolor* chromosome. The 82-kb CDA cluster is the largest secondary metabolite cluster, consisting of at least 40 genes (*Streptomyces* locus accession numbers: SCO3210-3249). The actinorhodin and undecylprodigiosin biosynthetic clusters are smaller, each consisting of 22–23 genes (actinorhodin: SCO5072-5092, 22 kb; undecylprodigiosin: SCO5877-5899, 32 kb).

In this article, we outline the metabolic and regulatory factors that are important for antibiotic production and describe how these can be manipulated for the discovery of new molecules. The regulation of antibiotic biosynthesis has been most extensively investigated in two model organisms, *S. coelicolor* and *S. griseus*, but there is a dense literature describing important regulatory phenomena in many others (Baltz, 1998; Bate et al., 2006; Cundliffe, 2006; Hopwood, 1989). We will argue that this information can be used in the search for new antibiotics: by understanding the interaction of regulatory pathways with nutrients and metabolites (including in particular carbon sources) we can create synthetic media that trigger the expression of otherwise cryptic secondary metabolic pathways. In parallel, by manipulating the regulators themselves, by overexpression, mutation, or both, we can bypass unknown road blocks for antibiotic biosynthesis and create constitutively overexpressing strains for novel small molecules.

3. Effects of Culture Conditions and Metabolism

3.1. Growth-dependent control mechanisms

Secondary metabolism, the dispensable metabolic pathways that produce antibiotics, usually occurs in a growth phase-dependent manner, peaking as cells approach or after they reach stationary phase. It is therefore influenced by a wide variety of nutritional (carbon, nitrogen, and phosphate levels), growth-related (cell density, morphology), and physiological factors (cyclic AMP, GTP, and ppGpp levels) (Bibb, 2005; Champness, 2000).

3.2. Stringent control

One factor that modulates antibiotic production is guanosine tetraphosphate or ppGpp. Under conditions of amino acid limitation, the ribosome-associated RelA protein synthesizes ppGpp in response to uncharged tRNAs occupying the ribosomal A-site (Chakraburtty and Bibb, 1997). In turn, the 50S ribosomal L11 protein encoded by *rplI* (also known as *relC*) activates RelA and thus ppGpp synthesis (Ochi, 1990). Under nitrogen-limiting conditions ppGpp causes a dramatic switch in cellular physiology, activating the expression of genes involved in stationary phase processes, such as morphogenesis and secondary metabolite production (CDA and actinorhodin), while repressing genes involved in active growth (Hesketh *et al.*, 2007). The exact mechanism by which ppGpp exerts its effect remains to be determined but it is conceivable that ppGpp affects the affinity of RNA polymerase for promoters of genes associated with stationary phase processes (Hesketh *et al.*, 2007).

3.3. Phosphate-mediated control

Whereas ppGpp is required for antibiotic biosynthesis under nitrogen starvation, a ppGpp-independent signaling mechanism operates under phosphate-limited conditions (Chakraburtty and Bibb, 1997). Inorganic phosphate (P_i) inhibits antibiotic biosynthesis at concentrations as low as a few 100 μM, whereas lower levels of P_i trigger antibiotic biosynthesis (Martin, 2004). The PhoR/PhoP two-component signal transduction system senses and responds to low P_i levels and activates genes involved in phosphate metabolism, allowing growth of the bacteria at low P_i concentrations. Deletion of *phoP* results in increased actinorhodin and undecylprodigiosin production. In a *phoP* deletion mutant, the supply of P_i is reduced and the mutant is prematurely starved of P_i. This P_i starvation is thought to be responsible for triggering antibiotic biosynthesis in a ppGpp-independent manner (Sola-Landa *et al.*, 2003).

Consistent with low levels of P_i triggering antibiotic production, disruption of another gene, *ppk*, involved in P_i metabolism also resulted in increased actinorhodin production and increased levels of transcription of the *act*II-ORF4, *redD*, and *cdaR* pathway-specific regulatory genes for actinorhodin, undecylprodigiosin, and CDA, respectively (Chouayekh and Virolle, 2002). Ppk catalyzes the reversible polymerization of the γ-phosphate of ATP into polyphosphate, a phosphate- and energy-storage polymer (Chouayekh and Virolle, 2002). *ppk* is optimally expressed under

P$_i$-limiting conditions and is positively regulated by the PhoR/PhoP two-component system (Ghorbel et al., 2006). The exact mechanism by which P$_i$ triggers antibiotic production remains unknown.

3.4. Interactions between metabolism and the DasR regulon

Antibiotics are synthesized by dedicated biosynthetic pathways, but the precursors and cofactors these pathways require are derived from primary metabolism, and high-level production of antibiotics draws heavily on the available building blocks. For example, polyketide biosynthesis requires high levels of acetyl-CoA, malonyl-CoA, and the reducing equivalents NADH and NADPH (Borodina et al., 2005; Hutchinson and Fujii, 1995), placing a burden on the pentosephosphate pathway under production conditions. Interestingly, by targeting phosphofructokinase (Pfk), a key glycolytic enzyme that catalyses the conversion of fructose-6-P to fructose-1,6-biP, the groups of Dijkhuizen and Nielsen could enhance production of actinorhodin and undecylprodigiosin in *S. coelicolor* (Borodina et al., 2008). It was demonstrated that deletion of *pfkA2* (SCO5426), encoding the major Pfk, resulted in increased carbon flux through the pentose phosphate pathway, primarily as a result of the accumulation of glucose 6-phosphate and fructose 6-phosphate. Excellent reviews on metabolic engineering of filamentous microorganisms and the impact of antibiotic production may be found elsewhere (e.g., (Heide et al., 2008; Hershberger, 1996; Martin, 1998; Stephanopoulos, 2002; Thykaer and Nielsen, 2003)).

However, one direct link between primary and secondary metabolism is of particular relevance to this article. The C/N sources glutamate and *N*-acetylglucosamine are important carbon sources for streptomycetes, and growth studies show that they are preferred over glucose by *S. coelicolor* (Nothaft et al., 2003; van Wezel et al., 2006b). Both compounds have a high energy value and are just two metabolic steps away from fructose-6-P. Furthermore, glucosamine-6-P is the major precursor for cell-wall biosynthesis. Surprisingly, higher concentrations of *N*-acetylglucosamine (>5–10 m*M*) inhibit development and antibiotic production under rich growth conditions, while they activate them under poor growth conditions. This complex phenomenon is somehow mediated via the global regulator DasR, a GntR-family regulator whose regulon includes *N*-acetylglucosamine metabolism and transport as well as antibiotic production (Rigali et al., 2006, 2008). DasR directly controls the promoter of *act*II-ORF4, the pathway-specific activator gene for actinorhodin biosynthesis, and transcription of all known chromosomally-encoded antibiotic biosynthetic clusters of *S. coelicolor* (*act, cda, red*, and the "cryptic" *kas* or *cpk* cluster) is enhanced in *dasR* mutants on minimal media (Rigali et al., 2008; van Wezel et al., 2006b). DasR repression is relieved during differentiation, when *N*-acetylglucosamine accumulates and is converted to glucosamine-6-phosphate. This is due to direct binding of glucosamine-6-phosphate to DasR, which reduces the protein's affinity for DNA (Rigali et al., 2006).

N-acetylglucosamine is the monomer of the abundant natural polymer chitin and an important constituent of the cell-wall peptidoglycan. Following chitin utilization or autolytic cell-wall degradation, N-acetylglucosamine can enter the cell via at least three different sugar transporters, namely as the monomer via the NagE2 permease (SCO2907), which is part of the PTS phosphotransferase system, or as the dimer chitobiose via DasABC (SCO5232-5234), or via the NgcEFG transporter (SCO6005-6007) (Nothaft et al., 2003; Parche et al., 2000; Saito et al., 2007; Schlösser et al., 1999). The PTS consists of several carbohydrate-specific permeases (designated Enzyme IIBC) and a global part consisting of Enzyme I (EI, encoded by *ptsI*), HPr (histidine protein, encoded by *ptsH*), and Enzyme IIA (EIIA, encoded by *crr*) (Parche et al., 2000; Postma et al., 1993; Titgemeyer et al., 1995). EI, HPr, and EIIA form the phosphate-transfer system, using phosphoenolpyruvate as the energy source, resulting in phosphorylation of the incoming sugar. Interestingly, in *Streptomyces* the general PTS components play a much more global role in antibiotic production (and sporulation) than just the transport of N-acetylglucosamine. For example, deletion of *dasR*, but also of *dasA*, *ptsH*, *ptsI*, or *crr*, results in developmental arrest and—surprisingly—this is independent of the carbon source (Colson et al., 2008; Rigali et al., 2006). The details of this pleiotropic regulatory mechanism are unknown. Application of DasR for screening purposes is discussed below.

3.5. Morphology as determinant of productivity

Streptomycetes grow as a branched multicellular network of hyphae—the mycelium—in which cell division is relatively rare and does not lead to full cytokinesis (Flardh, 2003). A reproductive cell type, which on solid medium grows up into the air, leads to regularly spaced septation events, and the production of spores. Evidence suggests that the growth of the spore-forming cells occurs at the expense of substrate hyphal lysis. In this way, a single spore will eventually develop into a large multicellular network. Antibiotic production responds as much to morphology as it does to the specific medium conditions. For example, erythromycin production by *Saccharopolyspora erythraea* requires clumps that are at least 90 µm in diameter (Bushell, 1988; Wardell et al., 2002). Thus, while fragmented growth is generally favorable in terms of growth rate and biomass accumulation, it can have a detrimental effect on antibiotic production.

This is not true for all antibiotics. Enhanced expression of the cell division activator protein SsgA induces fragmentation of *S. coelicolor* and, while this almost abolishes production of the polyketide antibiotic actinorhodin, it brings about a >20-fold increase in undecylprodigiosin production during batch fermentation (Traag and van Wezel, 2008; van Wezel *et al.*, 2000a, 2006a). Similarly, the chloramphenicol producer, *Streptomyces venezuelae*, grows in an extremely fragmented way while efficiently producing antibiotics (Bewick *et al.*, 1976; Glazebrook *et al.*, 1990). Over-expression of the cell division protein FtsZ resulted in the formation of large and dense clumps (flocks), which effected the massive overproduction of actinorhodin in *Streptomyces lividans*, which hardly produces this antibiotic under normal growth conditions (van Wezel *et al.*, 2000b). Therefore, approaches to improve antibiotic production should include analyzing the effects of changes in the morphology of liquid-grown cultures, as these can have spectacular effects on productivity. As a rule of thumb, antibiotics that are produced during late exponential growth or earlier (like undecylprodigiosin) benefit from fragmented growth (Fig. 5.1), while those produced during late transition or stationary phase (like actinorhodin) are produced much more efficiently by clumps (van Wezel *et al.*, 2006a).

Figure 5.1 Overproduction of the cell division activator SsgA leads to an ~20-fold increase in the production of the red-pigmented antibiotic undecylprodigiosin during fermentation of *S. coelicolor*. (See Color Insert.)

4. Molecular Genetic Factors that Regulate Antibiotic Production

Genetic experiments have revealed a diverse collection of regulatory proteins that control antibiotic production (Fig. 5.2); these fall naturally into three groups. The first includes pleiotropic regulators that influence both antibiotic production and sporulation. A well-understood example is *bldA*, which encodes the only tRNA that efficiently translates the UUA codon in *Streptomyces* (Lawlor *et al.*, 1987). *bldA* mutants fail to produce aerial hyphae or most of the antibiotics (Fernandez-Moreno *et al.*, 1991; White and Bibb, 1997). The second group includes pleiotropic regulators of several antibiotic biosynthetic pathways that have little or no effect on the sporulation pathway. An example is the *absA* operon, which encodes a two-component system that serves to repress antibiotic production in *S. coelicolor* and *S. griseus* (Aceti and Champness, 1998; Adamidis *et al.*, 1999; Anderson *et al.*, 1999, 2001; Brian *et al.*, 1996, 2001; Champness and Brian, 1998; Champness *et al.*, 1992; Ishizuka *et al.*, 1992; McKenzie and Nodwell, 2007; Ryding *et al.*, 2002; Sherman and Anderson, 2001; Sheeler *et al.*, 2005). The most specific regulators constitute a third group, the pathway-specific regulators.

4.1. Pathway-specific regulation

Located within the antibiotic biosynthetic clusters are genes encoding biosynthetic enzymes, resistance determinants, transporters, and pathway-specific regulators. There are several types of biosynthetic cluster-encoded,

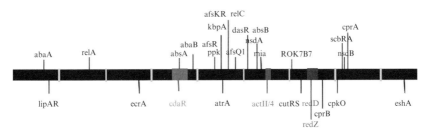

Figure 5.2 The 8,667,507-base pair linear chromosome of *S. coelicolor*. The locations of the CDA (cdaR), actinorhodin (actII/4), and undecylprodigiosin (redD/redZ) biosynthetic gene clusters and the antibiotic regulatory genes are depicted. The genes encoding the pleiotropic regulators are shown above the chromosome, and those encoding the pathway-specific regulators are below. Regulatory genes that are embedded within the biosynthetic clusters are colored. Vertical grey lines divide the chromosome into 1-Mb segments.

pathway-specific regulators, the best-characterized being the *Streptomyces* antibiotic regulatory proteins (SARPs), which are transcriptional activators.

SARPs typically bind to direct repeats that overlap the −35 regions of promoters controlling the expression of antibiotic biosynthetic genes (Wietzorrek *et al.*, 1997). Members of this family have a characteristic N-terminal OmpR-like winged helix-turn-helix DNA-binding motif (Wietzorrek *et al.*, 1997), followed by a bacterial transcriptional activation domain that may help to stabilize the DNA-bound dimer (Alderwick *et al.*, 2006). The function of the C-terminal half of these proteins remains unknown and is not required for DNA-binding specificity (Lee *et al.*, 2002; Sheldon *et al.*, 2002) or for the initiation of transcription *in vitro* (Tanaka *et al.*, 2007). SARPs activate transcription of some or all of the antibiotic biosynthetic genes in the cluster. The grouping of codirectional genes into operons controlled by a single promoter allows a SARP to activate many genes through just a few promoters. In *S. coelicolor*, *act*II-ORF4 and *redD* encode the SARPs for actinorhodin and undecylprodigiosin production, respectively, and are embedded within the biosynthetic cluster they regulate (Arias *et al.*, 1999; Takano *et al.*, 1992). These genes are essential for antibiotic biosynthesis: deletion of either of them abolished production of actinorhodin and undecylprodigiosin, respectively (Floriano and Bibb, 1996).

Transcription of *act*II-ORF4 is growth-phase-dependent and reaches a maximum during the transition from exponential to stationary phase growth (Gramajo *et al.*, 1993). Overexpression of *act*II-ORF4 in *S. coelicolor* resulted in overproduction of actinorhodin that began during exponential growth (Gramajo *et al.*, 1993). While no direct DNA-binding studies have been done with RedD, microarray analysis revealed that most of the genes involved in the biosynthesis of undecylprodigiosin are RedD-dependent and activated by this SARP (Huang *et al.*, 2001, 2005). Transcription of *redD* increases dramatically during late exponential growth and maximal levels of *redD* transcripts must be present to initiate transcription of the biosynthetic genes (Takano *et al.*, 1992). As with *act*II-ORF4, overexpression of *redD* resulted in elevated levels of undecylprodigiosin and synthesis started earlier during exponential growth (Takano *et al.*, 1992). *redD* itself is *trans*-activated by a second pathway-specific activator inside the *red* cluster, the response regulator-like RedZ (Guthrie *et al.*, 1998; White and Bibb, 1997).

Other SARPs in *S. coelicolor* include CdaR, the proposed activator of the biosynthetic genes for the calcium-dependent antibiotic, a relative of daptomycin (Huang *et al.*, 2005; Khanin *et al.*, 2007; Ryding *et al.*, 2002). Two others are encoded by *cpkO* (formally known as *kasO*) and *cpkN*, which are embedded in a cryptic type I polyketide synthase (*cpk*) gene cluster (SCO6229-6288) (Pawlik *et al.*, 2007; Takano *et al.*, 2005). (The identity of the final product of the cluster is unknown; Takano *et al.* (2005)). *cpkN* has not been characterized but it has been shown that expression of the cryptic polyketide cluster depends on *cpkO* (Takano *et al.*, 2005).

4.2. Pleiotropic regulation

There are at least fifteen pleiotropic regulators of antibiotic biosynthesis in *S. coelicolor*. Predictably, many of these are putative or demonstrated transcription factors, although others are less well understood.

AfsR exhibits sequence similarity to the SARP proteins and controls actinorhodin, undecylprodigiosin, and CDA (Floriano and Bibb, 1996). AfsR contains two extra domains: a central ATPase domain and a C-terminal tetratricopeptide repeat (TPR) domain presumed to associate with other regulatory proteins (Tanaka *et al.*, 2007). Unlike other SARPs that directly activate transcription of the antibiotic biosynthetic genes, AfsR activates *afsS*, encoding a small protein that stimulates transcription of the SARPs *actII-ORF4* and *redD* by an as yet unidentified mechanism (Floriano and Bibb, 1996; Lee *et al.*, 2002). The ATPase domain of AfsR is essential for *afsS* transcriptional activation and may serve to stimulate open complex formation with RNA polymerase (Lee *et al.*, 2002). AfsR may activate other gene(s) as well—its over-expression has been shown to stimulate antibiotic production independently of *afsS* in addition to the more familiar *afsS*–dependent mechanism (Lee *et al.*, 2002). Importantly, AfsR is subject to phosphorylation by the serine/threonine kinase AfsK, where phosphorylation stimulates its affinity for DNA and modulates its ATPase activity (Lee *et al.*, 2002). AfsK is localized to the inner side of the membrane and autophosphorylates in response to an unknown signal (Matsumoto *et al.*, 1994). In turn, AfsK is modulated by KbpA, an AfsK-binding protein that accumulates after antibiotic production has begun and is believed to prevent unlimited antibiotic production (Umeyama and Horinouchi, 2001). Thus, *afsK*, *kbpA*, *afsR*, and *afsS* genes constitute a linear signal transduction system that activates antibiotic production in *S. coelicolor*. Disruption of *afsK*, *afsR*, or *afsS* reduces antibiotic production, while disruption of *kbpA* enhances it (Floriano and Bibb, 1996; Matsumoto *et al.*, 1994; Umeyama and Horinouchi, 2001).

The *absA* locus encodes a two-component system consisting of a sensor kinase, AbsA1, and the response regulator AbsA2 (Aceti and Champness, 1998; Adamidis *et al.*, 1999; Anderson *et al.*, 1999, 2001; Brian *et al.*, 1996, 2001; Champness and Brian, 1998; Champness *et al.*, 1992; Ishizuka *et al.*, 1992; McKenzie and Nodwell, 2007; Ryding *et al.*, 2002; Sherman and Anderson, 2001; Sheeler *et al.*, 2005). Importantly, these genes appear to serve primarily as negative regulators of antibiotic production. The AbsA1 protein can both phosphorylate AbsA2 and dephosphorylate AbsA2~P (Sheeler *et al.*, 2005) and, since mutations that impair kinase activity stimulate antibiotic production while those that impair phosphatase activity inhibit it, the current model posits that AbsA2~P is a repressor of antibiotic biosynthetic gene expression. Indeed, it has been shown that AbsA2~P can interact directly with *actII-ORF4*, *redZ*, and *CDAR* (McKenzie and Nodwell, 2007).

A second two-component system, *absAQ1/2*, may also control antibiotic production in *S. coelicolor* (Ishizuka *et al.*, 1992). Unlike, the *absA* locus, the *afsQ1/2* system genes are highly conserved in other streptomycetes (McKenzie and Nodwell, unpublished).

The *mia* sequence (for multicopy inhibition of antibiotic synthesis) eliminated production of actinorhodin, undecylprodigiosin, and CDA when expressed at high copy number (Champness *et al.*, 1992). It has been suggested that the *mia* locus and at least one other newly cloned gene cluster may encode small RNAs that modulate antibiotic biosynthetic gene expression. Consistent with this, deletion of *absB*, which encodes an RNase III protein, abolishes antibiotic production (Aceti and Champness, 1998; Adamidis and Champness, 1992; Chang *et al.*, 2005; Price *et al.*, 1999; Sello and Buttner, 2008).

The *abaA* locus, another poorly characterized pleiotropic regulator, is now believed to consist of at least five open reading frames involved primarily in the control of CDA production but potentially also of other antibiotics (Fernandez-Moreno *et al.*, 1992). One of the *abaA* gene products is a putative transcription factor and a second is a protein similar in sequence to BldB, a dimeric protein of unknown function that is required both for the production of antibiotics and for the sporulation pathway (Eccleston *et al.*, 2002, 2006; Harasym *et al.*, 1990; Pope *et al.*, 1998). The other three *abaA* gene products are proteins of unknown function (Hart *et al.*, unpublished observations).

While the number of around 25 genes involved in the control of antibiotic production appears rather large for a single organism, several new regulators have just been uncovered, and the impact of others has been underestimated. As an example of the latter, the TetR-like regulator AtrA was shown to "specifically" activate the biosynthesis of actinorhodin and, apparently, no other antibiotic (Uguru *et al.*, 2005). This activation involves a direct interaction with the *actII-ORF4* promoter sequence. However, AtrA occurs in all streptomycetes, its DNA binding domain is extremely well conserved (>90% amino acid identity), and an AtrA orthologue activates streptomycin production in *S. griseus* (Hirano *et al.*, 2008), which is compelling phylogenetic evidence against a specific function in *S. coelicolor*. Indeed, evidence is accumulating that AtrA controls a more diverse regulon including metabolic genes, and recent DNA binding experiments revealed binding sites in the *cda* gene cluster, suggesting a pleiotropic (activating?) function for AtrA in the control of antibiotic production (K. McDowall, Pers. Comm.). Furthermore, induced expression of *atrA* can bypass the effects of hyper-repressive alleles of pleiotropic regulators such as *absA1* (McKenzie and Nodwell, unpublished results).

With increasing genomics efforts the landscape is expected to change further. This is exemplified by the discovery of *nsdA* and *nsdB*, which negatively control antibiotic production in *S. coelicolor*: disruption of these genes results in strongly enhanced actinorhodin production (Li *et al.*, 2006; Zhang *et al.*, 2007). Like, AfsR, the predicted gene products have a TPR domain associated with mediating protein–protein interactions, but their mode of action is unknown.

An interesting approach towards the discovery of novel antibiotic control proteins is metagenomics. In this approach, a library is constructed of genomic DNA isolated from a certain ecological niche, and introduced into a potential expression host such as *S. coelicolor*. A novel activator of antibiotic production was identified in this way from a soil-derived DNA library (Martinez *et al.*, 2005). This gene had previously been identified as *ngcR* in *S. olivaceoviridis* or *rok7B7* (SCO6008) in *S. coelicolor*. The ROK-family regulator it encodes was identified as the transcriptional activator of the *N*-acetylglucosamine and chitobiose transporter NgcEFG in *S. olivaceoviridis* (Schlösser *et al.*, 1999; Xiao *et al.*, 2002) and it also activates PTS and Act production in *S. coelicolor* (Nothaft, 2004). By affecting *N*-acetylglucosamine transport, NgcR also influences the highly pleiotropic DasR regulon, as the metabolic derivative glucosamine-6P modulates its DNA binding activity. Evidence is accumulating that ROK7B7/NgcR is the mirror image of DasR in terms of its regulon, both of which affect hundreds of genes and many antibiotic biosynthetic gene clusters (Swiatek and van Wezel, unpublished).

5. Applications for New Antibiotic Screening Technologies

At present there are many gaps in our understanding of the connectivity between the regulators, pleiotropic, or specific, that have been discovered in *S. coelicolor* and, as argued above, it is likely that many are still to be uncovered even in this single organism. How these regulators interact with signaling molecules and metabolic cues is for the most part completely unknown. One recurring theme is that regulators encoded outside the biosynthetic gene clusters often act by stimulating the expression of the activator genes in the clusters. As this field progresses, it will be interesting to learn whether there are other mechanisms for pleiotropic or pathway-specific control. An intriguing possibility, for example, is enhancing the accumulation of antibiotic precursor molecules.

Regardless of the gaps in our knowledge, it is clear that the genes and culture effects we know about can be exploited in the discovery of new antibiotics. While there are clearly big differences between the core regulatory phenomena in different streptomycetes, frequently these differences involve shuffling the order or manner in which individual proteins act (Chater and Horinouchi, 2003). It is likely, therefore, that what is known in *S. coelicolor* and other model streptomycetes can be used to manipulate many streptomycetes to increase the chances of discovering new molecules.

BLAST searches (Altschul *et al.*, 1990; Gish and States, 1993) against eight fully sequenced *Streptomyces* genomes, (*S. avermitilis* (Ikeda *et al.*, 2003), *Streptomyces clavuligerus* (Accession EDY48520), *S. coelicolor* (Bentley *et al.*, 2002), *S. griseus* (Ohnishi *et al.*, 2008), *Streptomyces pristinaespiralis* (Accession EDY64088), *S. scabies* (http://www.sanger.ac.uk/projects/S_scabies), *Streptomyces sviceus* (Accession EDY54226), and *Streptomyces* species MG1 (Accession EDX24495)) using the known *S. coelicolor* regulators as bait identified putative orthologues of AbsB, AfsQ1/2, AfsR, AtrA, DasR, ROK7B7/NgcR, and Mia in all eight species. These genes exhibit >60% sequence similarity and tend to preserve chromosomal location and organization. Putative orthologues can also be found for AbsA, AbaA, AfsK, and other regulators, though not in all streptomycetes. Orthologues for the γ-butyrolactone-binding protein ScbR were also found in all streptomycetes. The sequence conservation in these regulators is lower in the C-terminal domains than in the N-terminus, consistent with the fact that they interact with distinct signaling molecules.

5.1. Heterologous overexpression and mutant alleles

A straightforward strategy would be to simply induce the constitutive expression of a pleiotropic activator from *S. coelicolor* in a streptomycete that, through conventional screening, shows little or no antimicrobial activity. Genes could be expressed in single copy from a strong promoter such as the *ermE* promoter or the stronger *ermE** promoter (Bibb *et al.*, 1994; Schmitt-John and Engels 1992), or the thiostrepton-inducible *tipA* promoter (Takano *et al.*, 1995) on an integrating vector such as pSET152 (Bierman *et al.*, 1992). Alternatively, genes with their own promoters could be expressed at high copy number using multicopy vectors derived from pIJ101, of which pIJ486 (Ward *et al.*, 1986) is the most frequently used due to its high stability and copy number. Introduction of these cloned genes into nonproducing streptomycetes could lead to the discovery of new antibiotics. Before turning to individual genes, it is worth pointing out that metagenomics approaches, such as the one successfully used to identify ROK7B7/NgcR as an antibiotic activator (Martinez *et al.*, 2005), are very promising. As discussed, many *Streptomyces* antibiotic regulators are rather well conserved in actinomycetes, and it is quite possible that regulators from other actinomycetes may induce antibiotic production in streptomycetes.

This approach of overexpressing a single antibiotic control-related gene is most logical for activator genes such as *afsR*, but are there ways that repressors could also be employed to trigger antibiotic production in nonproducing strains? One approach is to take advantage of the more complicated genetics of some systems. For example, several alleles of the *absA1* sensor kinase have been discovered that cause overproduction of antibiotics in *S. coelicolor*, It is agreed that these act by eliminating the protein's capacity to phosphorylate AbsA2 but leaving its ability to dephosphorylate AbsA2~P (Anderson et al., 2001; McKenzie and Nodwell, 2007; Ryding et al., 2002; Sheeler et al., 2005). This is important because such alleles are dominant over wild-type alleles (McKenzie and Nodwell, unpublished), causing enhanced antibiotic production even when the normal repressing allele is present. These genetic observations have been employed to explore the possible application of *absA1* to antibiotic discovery. As shown in Fig. 5.3, introduction of two different phosphorylation-defective, dephosphorylation-competent alleles of *absA1* from *S. coelicolor* into *S. lividans* results in a dramatic stimulation of production of the blue-pigmented antibiotic actinorhodin (McKenzie and Nodwell, unpublished). This trait has been extended to screens of 10 environmental *Streptomyces* isolates and nine well-characterized producers of known antibiotics. Screens of the engineered strains against panels of model prokaryotes and pathogenic strains such as *S. aureus* demonstrated the activation of new antimicrobial activities in at least six of the engineered strains. Two of these turned out to be known antibiotics (streptomycin and blasticidin S), but at least one of them appears to be novel (McKenzie*et al.*, unpublished observations). In addition to demonstrating the potential of this approach, these observations show how important it is that we acquire a clearer understanding of how these pleiotropic regulators work.

To date perhaps the most successful application of this approach is demonstrated by efforts to enhance antibiotic biosynthesis by manipulating the DasR regulon. The DasR protein and regulatory network is highly conserved in streptomycetes, with around 75% of the DasR-binding (*dre*) sites predicted in *S. coelicolor* also found upstream of the orthologous genes in

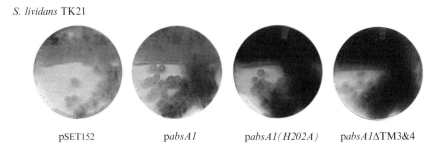

Figure 5.3 Activation of actinorhodin production by alleles of *absA1*. Mutants of the *S. coelicolor absA1* gene that encode proteins lacking AbsA2 kinase activity but having AbsA2~P phosphatase activity enhance production of the blue-pigmented antibiotic actinorhodin when expressed in *S. lividans*. The strains growing on each plate are *S. lividans* TK21 bearing a control vector (pSET152) or vectors expressing wild type *absA1*, or antibiotic activating alleles *absA1(H202A)* and *absA1(delTM3 and 4)*. (See Color Insert.)

S. *avermitilis* (Rigali, Titgemeyer and van Wezel, unpublished data and (van Wezel *et al.*, 2006b). We suggest, therefore, that the *dasR* regulon is likely to be central to the production of natural products in other actinomycetes. Indeed, a scan of available genome sequences in the databases highlights a range of putative targets, suggesting that DasR may control many important clinical drugs, such as clavulanic acid, chloramphenicol, and the glycopeptide antibiotics daptomycin and teichoplanin (Table 5.1). Whether or not these genes are indeed controlled (repressed) by DasR remains to be elucidated.

Finding a tool to manipulate the activity of DasR should therefore allow control of the expression of many industrially and medically relevant compounds (antibiotics, anti-tumor agents, agricultural compounds, and industrial enzymes) from the outside rather than by genetic engineering. As discussed above, the DNA-binding activity of DasR is inhibited by

Table 5.1 Selected antibiotic-related genes predicted or known (italics) to be controlled by DasR

Secondary metabolite	*Streptomyces*	Target gene(s)	Function
Clavulanic acid	S. *clavuligerus*	pcbR	PBP; β-lactam resistance
Actinorhodin (Act)	S. *coelicolor*	actII-ORF4	Pathway-specific activator
Undecylprodigiosin (Red)	S. *coelicolor*	redZ	Pathway-specific activator
Kas (cryptic antibiotic)	S. *coelicolor*	kasO (cpkO)	Pathway-specific activator
Calcium-dependent antibiotic (CDA)	S. *coelicolor*	cdaR	Pathway-specific activator
Valanimycin	S. *viridifaciens*	vlmM	Valanimycin transferase
Daptomycin	S. *filamentosus*	dptABC	Peptide synthetase 1, 2 and 3
Novobiocin	S. *spheroides*	novH	Peptide synthase
Actinomycin	S. *anulatus*	acmC	Peptide synthase III
Teichoplanin A47934	S. *toyocaensis*	staQ	Transcriptional regulator
Chloramphenicol	S. *venezuelae*	papABD	p-aminobenzoic acid synthase

Note that all known chromosomally-encoded antibiotics of S. *coelicolor* are controlled by DasR.

glucosamine-6-phosphate, a metabolic derivative of N-acetylglucosamine (Rigali et al., 2006). Indeed, addition of N-acetylglucosamine to growth media at 5–10 mM results in a *dasR* mutant phenocopy, namely lack of development and antibiotic production under rich growth conditions and accelerated development and enhanced antibiotic production under poor growth conditions (van Wezel et al., 2006b; Rigali et al., 2008). In some cases, this had a spectacular effect on the production of antibiotics, as illustrated by *Streptomyces hygroscopicus* and *S. clavuligerus* (Fig. 5.4).

Obvious targets for this approach are cryptic clusters, which are not expressed under normal growth conditions and therefore have not yet been identified by routine activity-based screening assays. At least for the *cpk* cluster it was established that expression of these clusters is induced by N-acetylglucosamine, and the majority of the streptomycetes tested produced more antibiotics in the presence of N-acetylglucosamine in the growth medium (Rigali et al., 2008). This is one example of novel approaches that may be employed to boost the potential of novel screening procedures.

6. Future Prospects

While there is a wealth of basic knowledge of the regulatory mechanisms governing antibiotic production, it is clear that the exploitation of this information is in its infancy. Indeed, at least one class of regulatory gene, the

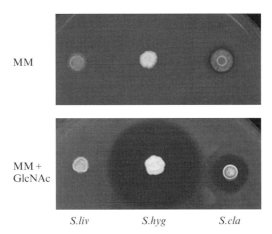

Figure 5.4 Effect of added N-acetyl glucosamine on antibiotic production by streptomycetes. *S. lividans* (*S.liv*; nonproducing control), *S. hygroscopicus* (*S.hyg*), and *S. clavuligerus* (*S. cla*) were grown on plates of minimal medium (MM) with added GlcNAc (bottom) or without (top). *Bacillus subtilis* was used as indicator strain. Halos in the indicator lawn demonstrate strong increase of antibiotic production by the streptomycetes.

small RNAs, appears to be missing from our collection. This is likely to change in the near future, however, as at least one group has evidence for control of antibiotic production via small RNA (Hindra and Elliot, McMaster University, Personal Communication). This is important because such genes might be particularly good candidates for heterologous regulation in less well-characterized strains.

Intriguing possibilities for future endeavors include manipulation of multiple regulatory pathways or exploring the effects of culture conditions on engineered strains. Indeed, one principal difference between the *Streptomyces* habitat (the soil) and the laboratory is the degree of crowding. Perhaps by mimicking the undoubtedly complex interspecific signaling events that occur in nature, it will be possible to trigger the expression of otherwise cryptic biosynthetic gene clusters.

Finally, it is clear that these approaches need to be brought to bear on more challenging target pathogenic microorganisms. Screens of engineered streptomycetes against pathogens exhibiting high levels of antibiotic resistance are clearly a high priority. If the rational engineering of streptomycete genomes yields new antimicrobial activities, screening against pathogens that lack sensitivity to known antibiotics could be a way of avoiding the rediscovery of known drugs.

ACKNOWLEDGMENTS

We thank Kenny McDowall (Leeds, UK) and Marie Elliot (McMaster University) for sharing unpublished data, Erik Vijgenboom and Ellen de Waal (Leiden University) for Fig. 5.1, Govind Chandra (Norwich, UK) for help with blast searches, and to Sébastien Rigali (Liège, Belgium) and Fritz Titgemeyer (Münster, Germany) for discussions. GvW and JRN are grateful to the Netherlands council for scientific research (NWO), the ACTS consortium for Biobased Sustainable Industrial Chemistry (B-BASIC), the Canadian Institutes for Health Research, the Natural Sciences and Engineering Research Council of Canada, and JNE Biotech Inc. for funding.

REFERENCES

Aceti, D. J., and Champness, W. C. (1998). Transcriptional regulation of *Streptomyces coelicolor* pathway-specific antibiotic regulators by the *absA* and *absB* loci. *J. Bacteriol.* **180,** 3100–3106.

Adamidis, T., and Champness, W. (1992). Genetic analysis of *absB*, a *Streptomyces coelicolor* locus involved in global antibiotic regulation. *J. Bacteriol.* **174,** 4622–4628.

Adamidis, T., Kong, R., Champness, W., and Aceti, D. J. (1999). Transcriptional regulation of *Streptomyces coelicolor* pathway-specific antibiotic regulators by the absA and absB loci. *J. Bacteriol.* **181,** 6142–6151.

Alderwick, L. J., Molle, V., Kremer, L., Cozzone, A. J., Dafforn, T. R., Besra, G. S., and Futterer, K. (2006). Molecular structure of EmbR, a response element of Ser/Thr kinase signaling in *Mycobacterium tuberculosis*. *Proc. Natl. Acad. Sci. USA* **103,** 2558–2563.

Altschul, S. F., Gish, W., Miller, W., Myers, E. W., and Lipman, D. J. (1990). Basic local alignment search tool. *J. Mol. Biol.* **215,** 403–410.

Anderson, T., Brian, P., Riggle, P., Kong, R., and Champness, W. (1999). Genetic suppression analysis of non-antibiotic-producing mutants of the *Streptomyces coelicolor* absA locus. *Microbiology* **145,** 2343–2353.

Anderson, T. B., Brian, P., and Champness, W. C. (2001). Genetic and transcriptional analysis of *absA*, an antibiotic gene cluster-linked two-component system that regulates multiple antibiotics in *Streptomyces coelicolor*. *Mol. Microbiol.* **39,** 553–566.

Arias, P., Fernandez-Moreno, M. A., and Malpartida, F. (1999). Characterization of the pathway-specific positive transcriptional regulator for actinorhodin biosynthesis in *Streptomyces coelicolor* A3(2) as a DNA-binding protein. *J. Bacteriol.* **181,** 6958–6968.

Baltz, R. H. (1998). Genetic manipulation of antibiotic-producing Streptomyces. *Trends Microbiol.* **6,** 76–83.

Baltz, R. H. (2006). Marcel Faber Roundtable: Is our antibiotic pipeline unproductive because of starvation, constipation or lack of inspiration? *J. Ind. Microbiol. Biotechnology* **33,** 507–513.

Bate, N., Bignell, D. R., and Cundliffe, E. (2006). Regulation of tylosin biosynthesis involving 'SARP-helper' activity. *Mol. Microbiol.* **62,** 148–156.

Bennett, J. W. (1998). Mycotechnology: The role of fungi in biotechnology. *J. Biotechnology* **66,** 101–117.

Bentley, S. D., Chater, K. F., Cerdeno-Tarraga, A. M., Challis, G. L., Thomson, N. R., James, K. D., Harris, D. E., Quail, M. A., Kieser, H., Harper, D., Bateman, A., Brown, S., *et al.* (2002). Complete genome sequence of the model actinomycete *Streptomyces coelicolor* A3(2). *Nature* **417,** 141–147.

Bersos, Z., Maniati, M., Kontos, F., Petinaki, E., and Maniatis, A. N. (2004). First report of a linezolid-resistant vancomycin-resistant Enterococcus faecium strain in Greece. *J. Antimicrob. Chemother.* **53,** 685–686.

Bewick, M. W., Williams, S. T., and Veltkamp, C. (1976). Growth and ultrastructure of Streptomyces venezuelae during chloramphenicol production. *Microbios* **16,** 191–199.

Bibb, M. (2005). Regulation of secondary metabolism in streptomycetes. *Curr. Opin. Microbiol.* **8,** 208–215.

Bibb, M. J., White, J., Ward, J. M., and Janssen, G. R. (1994). The mRNA for the 23S rRNA methylase encoded by the ermE gene of *Saccharopolyspora erythraea* is translated in the absence of a conventional ribosome-binding site. *Mol. Microbiol.* **14,** 533–545.

Bierman, M., Logan, R., O'Brien, K., Seno, E. T., Rao, R. N., and Schoner, B. E. (1992). Plasmid cloning vectors for the conjugal transfer of DNA from *Escherichia coli* to Streptomyces spp. *Gene* **116,** 43–49.

Borodina, I., Krabben, P., and Nielsen, J. (2005). Genome-scale analysis of *Streptomyces coelicolor* A3(2) metabolism. *Genome Res.* **15,** 820–829.

Borodina, I., Siebring, J., Zhang, J., Smith, C. P., van Keulen, G., Dijkhuizen, L., and Nielsen, J. (2008). Antibiotic overproduction in *Streptomyces coelicolor* A3 2 mediated by phosphofructokinase deletion. *J. Biol. Chem.* **283,** 25186–25199.

Bradley, J. S., Guidos, R., Baragona, S., Bartlett, J. G., Rubinstein, E., Zhanel, G. G., Tino, M. D., Pompliano, D. L., Tally, F., Tipirneni, P., Tillotson, G. S., Powers, J. H., *et al.* (2007). Anti-infective research and development--problems, challenges, and solutions. *Lancet Infect Dis.* **7,** 68–78.

Brian, P., Champness, W. C., and Anderson, T. (2001). Genetic suppression analysis of non-antibiotic-producing mutants of the *Streptomyces coelicolor* absA locus. *Mol. Microbiol.* **39,** 553–566.

Brian, P., Riggle, P. J., Santos, R. A., and Champness, W. C. (1996). Global negative regulation of *Streptomyces coelicolor* antibiotic synthesis mediated by an *absA*-encoded putative signal transduction system. *J. Bacteriol.* **178,** 3221–3231.

Bushell, M. E. (1988). Growth, product formation and fermentation technology. *In* "Actinomycetes in Biotechnology." pp. 185–217. Academic press, London, UK.

Chakraburtty, R., and Bibb, M. (1997). The ppGpp synthetase gene (*relA*) of *Streptomyces coelicolor* A3(2) plays a conditional role in antibiotic production and morphological differentiation. *J. Bacteriol.* **179,** 5854–5861.

Challis, G. L., and Hopwood, D. A. (2003). Synergy and contingency as driving forces for the evolution of multiple secondary metabolite production by Streptomyces species. *Proc. Natl. Acad. Sci. USA* **2**(100 Suppl.), 14555–14561.

Champness, W. (2000). Actinomycete development, antibiotic production, and phylogeny: Questions and challenges. *In* "Prokaryotic Development" (Y. V. Brun and L. J. Shimkets, eds.), pp. 11–31. ASM Press, Washington, DC.

Champness, W. C., and Brian, P. (1998). Global negative regulation of *Streptomyces coelicolor* antibiotic synthesis mediated by an absA-encoded putative signal transduction system. *J. Bacteriol.* **180,** 3100–3106.

Champness, W., Riggle, P., Adamidis, T., and Vandervere, P. (1992). Identification of *Streptomyces coelicolor* genes involved in regulation of antibiotic synthesis. *Gene* **115,** 55–60.

Chang, S. A., Bralley, P., and Jones, G. H. (2005). The absB gene encodes a double strand-specific endoribonuclease that cleaves the read-through transcript of the rpsO-pnp operon in *Streptomyces coelicolor*. *J. Biol. Chem.* **280,** 33213–33219.

Chater, K. F., and Horinouchi, S. (2003). Signalling early developmental events in two highly diverged Streptomyces species. *Mol. Microbiol.* **48,** 9–15.

Chouayekh, H., and Virolle, M. J. (2002). The polyphosphate kinase plays a negative role in the control of antibiotic production in *Streptomyces lividans*. *Mol. Microbiol.* **43,** 919–930.

Clardy, J., Fischbach, M. A., and Walsh, C. T. (2006). New antibiotics from bacterial natural products. *Nat. Biotechnol.* **24,** 1541–1550.

Colson, S., van Wezel, G. P., Craig, M., Noens, E. E., Nothaft, H., Mommaas, A. M., Titgemeyer, F., Joris, B., and Rigali, S. (2008). The chitobiose-binding protein, DasA, acts as a link between chitin utilization and morphogenesis in *Streptomyces coelicolor*. *Microbiology* **154,** 373–382.

Cundliffe, E. (2006). Antibiotic production by actinomycetes: The Janus faces of regulation. *J. Ind. Microbiol. Biotechnol.* **33,** 500–506.

Demain, A. L. (1991). Production of beta-lactam antibiotics and its regulation. *Proc. Natl. Sci. Counc. Repub. China B* **15,** 251–265.

Eccleston, M., Ali, R. A., Seyler, R., Westpheling, J., and Nodwell, J. (2002). Structural and genetic analysis of the BldB protein of *Streptomyces coelicolor*. *J. Bacteriol.* **184,** 4270–4276.

Eccleston, M., Willems, A., Beveridge, A., and Nodwell, J. R. (2006). Critical residues and novel effects of overexpression of the *Streptomyces coelicolor* developmental protein BldB: Evidence for a critical interacting partner. *J. Bacteriol.* **188,** 8189–8195.

Feitelson, J. S., Malpartida, F., and Hopwood, D. A. (1985). Genetic and biochemical characterization of the red gene cluster of *Streptomyces coelicolor* A3(2). *J. Gen. Microbiol.* **131,** 2431–2441.

Fernandez-Moreno, M. A., Caballero, J. L., Hopwood, D. A., and Malpartida, F. (1991). The act cluster contains regulatory and antibiotic export genes, direct targets for translational control by the bldA tRNA gene of Streptomyces. *Cell* **66,** 769–780.

Fernandez-Moreno, M. A., Martin-Triana, A. J., Martinez, E., Niemi, J., Kieser, H. M., Hopwood, D. A., and Malpartida, F. (1992). abaA, a new pleiotropic regulatory locus for antibiotic production in *Streptomyces coelicolor*. *J. Bacteriol.* **174,** 2958–2967.

Flardh, K. (2003). Growth polarity and cell division in Streptomyces. *Curr. Opin. Microbiol.* **6,** 564–571.

Floriano, B., and Bibb, M. (1996). *afsR* is a pleiotropic but conditionally required regulatory gene for antibiotic production in *Streptomyces coelicolor* A3(2). *Mol. Microbiol.* **21,** 385–396.

Ghorbel, S., Smirnov, A., Chouayekh, H., Sperandio, B., Esnault, C., Kormanec, J., and Virolle, M. J. (2006). Regulation of ppk expression and *in vivo* function of Ppk in Streptomyces lividans TK24. *J. Bacteriol.* **188,** 6269–6276.

Gish, W., and States, D. J. (1993). Identification of protein coding regions by database similarity search. *Nat. Genet.* **3,** 266–272.

Glazebrook, M. A., Doull, J. L., Stuttard, C., and Vining, L. C. (1990). Sporulation of Streptomyces venezuelae in submerged cultures. *J. Gen. Microbiol.* **136,** 581–588.

Gramajo, H. C., Takano, E., and Bibb, M. J. (1993). Stationary-phase production of the antibiotic actinorhodin in *Streptomyces coelicolor* A3(2) is transcriptionally regulated. *Mol. Microbiol.* **7,** 837–845.

Grundmann, H., Aires-de-Sousa, M., Boyce, J., and Tiemersma, E. (2006). Emergence and resurgence of meticillin-resistant Staphylococcus aureus as a public-health threat. *Lancet* **368,** 874–885.

Guthrie, E. P., Flaxman, C. S., White, J., Hodgson, D. A., Bibb, M. J., and Chater, K. F. (1998). A response-regulator-like activator of antibiotic synthesis from *Streptomyces coelicolor* A3(2) with an amino-terminal domain that lacks a phosphorylation pocket. *Microbiology* **144,** 727–738.

Harasym, M., Zhang, L. H., Chater, K., and Piret, J. (1990). The *Streptomyces coelicolor* A3(2) bldB region contains at least two genes involved in morphological development. *J. Gen. Microbiol.* **136,** 1543–1550.

Hayden, M. K., Rezai, K., Hayes, R. A., Lolans, K., Quinn, J. P., and Weinstein, R. A. (2005). Development of Daptomycin resistance *in vivo* in methicillin-resistant Staphylococcus aureus. *J. Clin. Microbiol.* **43,** 5285–5287.

Heide, L., Gust, B., Anderle, C., and Li, S. M. (2008). Combinatorial biosynthesis, metabolic engineering and mutasynthesis for the generation of new aminocoumarin antibiotics. *Curr. Top Med. Chem.* **8,** 667–679.

Hersbach, G. J. M., Van der Beek, C. P., and Van Dijck, P. W. M. (1984). ThePenicillins: Properties, Biosynthesis, and Fermentation Vol. **22,** Marcel Dekker, New York.

Hershberger, C. L. (1996). Metabolic engineering of polyketide biosynthesis. *Curr. Opin. Biotechnol.* **7,** 560–562.

Hesketh, A., Chen, W. J., Ryding, J., Chang, S., and Bibb, M. (2007). The global role of ppGpp synthesis in morphological differentiation and antibiotic production in *Streptomyces coelicolor* A3(2). *Genome Biol.* **8,** R161.

Hirano, S., Tanaka, K., Ohnishi, Y., and Horinouchi, S. (2008). Conditionally positive effect of the TetR-family transcriptional regulator AtrA on streptomycin production by *Streptomyces griseus*. *Microbiology* **154,** 905–914.

Hopwood, D. A. (1989). Antibiotics: Opportunities for genetic manipulation. *Philos. Transact. R. Soc. Lond.—Ser. B: Biol. Sci.* **324,** 549–562.

Hopwood, D. A. (2007). Streptomyces in Nature and Medicine: The Antibiotic Makers. Oxford University Press., New York.

Hopwood, D. A., Chater, K. F., and Bibb, M. J. (1995). Genetics of antibiotic production in *Streptomyces coelicolor* A3(2), a model streptomycete. *Biotechnology* **28,** 65–102.

Hopwood, D. A., and Wright, H. M. (1983). CDA is a new chromosomally-determined antibiotic from *Streptomyces coelicolor* A3(2). *J. Gen. Microbiol.* **129**(Pt 12), 3575–3579.

Huang, J., Lih, C. J., Pan, K. H., and Cohen, S. N. (2001). Global analysis of growth phase responsive gene expression and regulation of antibiotic biosynthetic pathways in *Streptomyces coelicolor* using DNA microarrays. *Genes Dev.* **15,** 3183–3192.

Huang, J., Shi, J., Molle, V., Sohlberg, B., Weaver, D., Bibb, M. J., Karoonuthaisiri, N., Lih, C. J., Kao, C. M., Buttner, M. J., and Cohen, S. N. (2005). Cross-regulation among disparate antibiotic biosynthetic pathways of *Streptomyces coelicolor*. *Mol. Microbiol.* **58,** 1276–1287.

Hutchinson, C. R., and Fujii, I. (1995). Polyketide synthase gene manipulation: A structure-function approach in engineering novel antibiotics. *Ann. Rev. Microbiol.* **49,** 201–238.

Ikeda, H., Ishikawa, J., Hanamoto, A., Shinose, M., Kikuchi, H., Shiba, T., Sakaki, Y., Hattori, M., and Omura, S. (2003). Complete genome sequence and comparative analysis of the industrial microorganism Streptomyces avermitilis. *Nat. Biotechnol.* **21,** 526–531.

Ishizuka, H., Horinouchi, S., Kieser, H. M., Hopwood, D. A., and Beppu, T. (1992). A putative two-component regulatory system involved in secondary metabolism in Streptomyces spp. *J. Bacteriol.* **174,** 7585–7594.

Katz, M. L., Mueller, L. V., Polyakov, M., and Weinstock, S. F. (2006). Where have all the antibiotic patents gone? *Nat. Biotechnol.* **24,** 1529–1531.

Khanin, R., Vinciotti, V., Mersinias, V., Smith, C. P., and Wit, E. (2007). Statistical reconstruction of transcription factor activity using Michaelis-Menten kinetics. *Biometrics* **63,** 816–823.

Lawlor, E. J., Baylis, H. A., and Chater, K. F. (1987). Pleiotropic morphological and antibiotic deficiencies result from mutations in a gene encoding a tRNA-like product in Streptomyces coelicolor A3(2). *Genes Dev.* **1,** 1305–1310.

Lee, P. C., Umeyama, T., and Horinouchi, S. (2002). afsS is a target of AfsR, a transcriptional factor with ATPase activity that globally controls secondary metabolism in Streptomyces coelicolor A3(2). *Mol. Microbiol.* **43,** 1413–1430.

Li, W., Ying, X., Guo, Y., Yu, Z., Zhou, X., Deng, Z., Kieser, H., Chater, K. F., and Tao, M. (2006). Identification of a gene negatively affecting antibiotic production and morphological differentiation in Streptomyces coelicolor A3(2). *J. Bacteriol.* **188,** 8368–8375.

Martin, J. F. (1998). New aspects of genes and enzymes for beta-lactam antibiotic biosynthesis. *Appl. Microbiol. Biotechnol.* **50,** 1–15.

Martin, J. F. (2004). Phosphate control of the biosynthesis of antibiotics and other secondary metabolites is mediated by the PhoR-PhoP system: An unfinished story. *J. Bacteriol.* **186,** 5197–5201.

Martinez, A., Kolvek, S. J., Hopke, J., Yip, C. L., and Osburne, M. S. (2005). Environmental DNA fragment conferring early and increased sporulation and antibiotic production in Streptomyces species. *Appl. Environ. Microbiol.* **71,** 1638–1641.

Matsumoto, A., Hong, S. K., Ishizuka, H., Horinouchi, S., and Beppu, T. (1994). Phosphorylation of the AfsR protein involved in secondary metabolism in Streptomyces species by a eukaryotic-type protein kinase. *Gene* **146,** 47–56.

McKenzie, N. L., and Nodwell, J. R. (2007). Phosphorylated AbsA2 negatively regulates antibiotic production in Streptomyces coelicolor through interactions with pathway-specific regulatory gene promoters. *J. Bacteriol.* **189,** 5284–5292.

Mukherjee, J. S., Rich, M. L., Socci, A. R., Joseph, J. K., Viru, F. A., Shin, S. S., Furin, J. J., Becerra, M. C., Barry, D. J., Kim, J. Y., Bayona, J., Farmer, P., et al. (2004). Programmes and principles in treatment of multidrug-resistant tuberculosis. *Lancet* **363,** 474–481.

Nothaft, H. (2004). Ph.D. thesis. Erlangen, Germany.

Nothaft, H., Dresel, D., Willimek, A., Mahr, K., Niederweis, M., and Titgemeyer, F. (2003). The phosphotransferase system of Streptomyces coelicolor is biased for N-acetylglucosamine metabolism. *J. Bacteriol.* **185,** 7019–7023.

Ochi, K. (1990). A relaxed (rel) mutant of Streptomyces coelicolor A3(2) with a missing ribosomal protein lacks the ability to accumulate ppGpp, A-factor and prodigiosin. *J. Gen. Microbiol.* **136,** 2405–2412.

Ohnishi, Y., Ishikawa, J., Hara, H., Suzuki, H., Ikenoya, M., Ikeda, H., Yamashita, A., Hattori, M., and Horinouchi, S. (2008). Genome sequence of the streptomycin-producing microorganism Streptomyces griseus IFO 13350. *J. Bacteriol.* **190,** 4050–4060.

Parche, S., Nothaft, H., Kamionka, A., and Titgemeyer, F. (2000). Sugar uptake and utilisation in *Streptomyces coelicolor*: A PTS view to the genome. *Antonie Van Leeuwenhoek* **78,** 243–251.

Pawlik, K., Kotowska, M., Chater, K. F., Kuczek, K., and Takano, E. (2007). A cryptic type I polyketide synthase (cpk) gene cluster in *Streptomyces coelicolor* A3(2). *Arch. Microbiol.* **187,** 87–99.

Payne, D. J., Gwynn, M. N., Holmes, D. J., and Pompliano, D. L. (2007). Drugs for bad bugs: Confronting the challenges of antibacterial discovery. *Nat. Rev. Drug Discov.* **6,** 29–40.

Pope, M. K., Green, B., and Westpheling, J. (1998). The bldB gene encodes a small protein required for morphogenesis, antibiotic production, and catabolite control in *Streptomyces coelicolor*. *J. Bacteriol.* **180,** 1556–1562.

Postma, P. W., Lengeler, J. W., and Jacobson, G. R. (1993). Phosphoenolpyruvate: Carbohydrate phosphotransferase systems of bacteria. *Microbiol. Rev.* **57,** 543–594.

Price, B., Adamidis, T., Kong, R., and Champness, W. (1999). A *Streptomyces coelicolor* antibiotic regulatory gene, *absB*, encodes an RNase III homolog. *J. Bacteriol.* **181,** 6142–6151.

Rigali, S., Nothaft, H., Noens, E. E., Schlicht, M., Colson, S., Muller, M., Joris, B., Koerten, H. K., Hopwood, D. A., Titgemeyer, F., and van Wezel, G. P. (2006). The sugar phosphotransferase system of *Streptomyces coelicolor* is regulated by the GntR-family regulator DasR and links N-acetylglucosamine metabolism to the control of development. *Mol. Microbiol.* **61,** 1237–1251.

Rigali, S., Titgemeyer, F., Barends, S., Mulder, S., Thomae, A. W., Hopwood, D. A., and van Wezel, G. P. (2008). Feast or famine: The global regulator DasR links nutrient stress to antibiotic production by Streptomyces. *EMBO Rep.* **9,** 670–675.

Rudd, B. A., and Hopwood, D. A. (1979). Genetics of actinorhodin biosynthesis by *Streptomyces coelicolor* A3(2). *J. Gen. Microbiol.* **114,** 35–43.

Ryding, N. J., Anderson, T. B., and Champness, W. C. (2002). Regulation of the *Streptomyces coelicolor* calcium-dependent antibiotic by *absA*, encoding a cluster-linked two-component system. *J. Bacteriol.* **184,** 794–805.

Saito, A., Shinya, T., Miyamoto, K., Yokoyama, T., Kaku, H., Minami, E., Shibuya, N., Tsujibo, H., Nagata, Y., Ando, A., Fujii, T., and Miyashita, K. (2007). The dasABC gene cluster adjacent to dasR encodes a novel ABC transporter for the uptake of N,N′-diacetylchitobiose in *Streptomyces coelicolor* A3(2). *Appl. Environ. Microbiol.* **79,** 3000–3008.

Schlösser, A., Jantos, J., Hackmann, K., and Schrempf, H. (1999). Characterization of the binding protein-dependent cellobiose and cellotriose transport system of the cellulose degrader *Streptomyces reticuli*. *Appl. Environ. Microbiol.* **65,** 2636–2643.

Schmitt-John, T., and Engels, J. W. (1992). Promoter constructions for efficient secretion expression in *Streptomyces lividans*. *Appl. Microbiol. Biotechnol.* **36,** 493–498.

Sello, J. K., and Buttner, M. J. (2008). The gene encoding RNase III in *Streptomyces coelicolor* is transcribed during exponential phase and is required for antibiotic production and for proper sporulation. *J. Bacteriol.* **190,** 4079–4083.

Sheeler, N. L., MacMillan, S. V., and Nodwell, J. R. (2005). Biochemical activities of the *absA* two-component system of *Streptomyces coelicolor*. *J. Bacteriol.* **187,** 687–696.

Sheldon, P. J., Busarow, S. B., and Hutchinson, C. R. (2002). Mapping the DNA-binding domain and target sequences of the *Streptomyces peucetius* daunorubicin biosynthesis regulatory protein, DnrI. *Mol. Microbiol.* **44,** 449–460.

Sherman, D. H., and Anderson, T. B. (2001). Genetic and transcriptional analysis of absA, an antibiotic gene cluster-linked two-component system that regulates multiple antibiotics in *Streptomyces coelicolor*. *Metab. Eng.* **3,** 15–26.

Sola-Landa, A., Moura, R. S., and Martin, J. F. (2003). The two-component PhoR-PhoP system controls both primary metabolism and secondary metabolite biosynthesis in *Streptomyces lividans*. *Proc. Natl. Acad. Sci. USA* **100,** 6133–6138.

Stephanopoulos, G. (2002). Metabolic engineering by genome shuffling. *Nat. Biotechnol.* **20,** 666–668.

Takano, E., Gramajo, H. C., Strauch, E., Andres, N., White, J., and Bibb, M. J. (1992). Transcriptional regulation of the *redD* transcriptional activator gene accounts for growth-phase-dependent production of the antibiotic undecylprodigiosin in *Streptomyces coelicolor* A3(2). *Mol. Microbiol.* **6,** 2797–2804.

Takano, E., Kinoshita, H., Mersinias, V., Bucca, G., Hotchkiss, G., Nihira, T., Smith, C. P., Bibb, M., Wohlleben, W., and Chater, K. (2005). A bacterial hormone (the SCB1) directly controls the expression of a pathway-specific regulatory gene in the cryptic type I polyketide biosynthetic gene cluster of *Streptomyces coelicolor*. *Mol. Microbiol.* **56,** 465–479.

Takano, E., White, J., Thompson, C. J., and Bibb, M. J. (1995). Construction of thiostrepton-inducible, high-copy-number expression vectors for use in Streptomyces spp. *Gene* **166,** 133–137.

Tanaka, A., Takano, Y., Ohnishi, Y., and Horinouchi, S. (2007). AfsR recruits RNA polymerase to the afsS promoter: A model for transcriptional activation by SARPs. *J. Mol. Biol.* **369,** 322–333.

Thykaer, J., and Nielsen, J. (2003). Metabolic engineering of beta-lactam production. *Metab. Eng.* **5,** 56–69.

Titgemeyer, F., Walkenhorst, J., Reizer, J., Stuiver, M. H., Cui, X., and Saier, M. H. Jr., (1995). Identification and characterization of phosphoenolpyruvate:Fructose phosphotransferase systems in three Streptomyces species. *Microbiology* **141**(Pt 1), 51–58.

Traag, B. A., and van Wezel, G. P. (2008). The SsgA-like proteins in actinomycetes: Small proteins up to a big task. *Antonie Van Leeuwenhoek* **94,** 85–97.

Uguru, G. C., Stephens, K. E., Stead, J. A., Towle, J. E., Baumberg, S., and McDowall, K. J. (2005). Transcriptional activation of the pathway-specific regulator of the actinorhodin biosynthetic genes in *Streptomyces coelicolor*. *Mol. Microbiol.* **58,** 131–150.

Umeyama, T., and Horinouchi, S. (2001). Autophosphorylation of a bacterial serine/threonine kinase, AfsK, is inhibited by KbpA, an AfsK-binding protein. *J. Bacteriol.* **183,** 5506–5512.

van den Berg, M. A., Albang, R., Albermann, K., Badger, J. H., Daran, J. M., Driessen, A. J., Garcia-Estrada, C., Fedorova, N. D., Harris, D. M., Heijne, W. H., Joardar, V., Kiel, J. A., *et al.* (2008). Genome sequencing and analysis of the filamentous fungus *Penicillium chrysogenum*. *Nat. Biotechnol.* **26,** 1161–1168.

van den Berg, M. A., Westerlaken, I., Leeflang, C., Kerkman, R., and Bovenberg, R. A. (2007). Functional characterization of the penicillin biosynthetic gene cluster of Penicillium chrysogenum Wisconsin54–1255. *Fungal Genet. Biol.* **44,** 830–844.

Van Lanen, S. G., and Shen, B. (2006). Microbial genomics for the improvement of natural product discovery. *Curr. Opin. Microbiol.* **9,** 252–260.

van Wezel, G. P., Krabben, P., Traag, B. A., Keijser, B. J., Kerste, R., Vijgenboom, E., Heijnen, J. J., and Kraal, B. (2006a). Unlocking streptomyces spp. for use as sustainable industrial production platforms by morphological engineering. *Appl. Environ. Microbiol.* **72,** 5283–5288.

van Wezel, G. P., Titgemeyer, F., and Rigali, S. (2006b). Methods and means for metabolic engineering and improved product formation by micro-organisms Patent application WO/2007/094667 patent EP20060075336.

van Wezel, G. P., van der Meulen, J., Kawamoto, S., Luiten, R. G. M., Koerten, H. K., and Kraal, B. (2000a). ssgA is essential for sporulation of *Streptomyces coelicolor* A3(2) and affects hyphal development by stimulating septum formation. *J. Bacteriol.* **182,** 5653–5662.

van Wezel, G. P., van der Meulen, J., Taal, E., Koerten, H., and Kraal, B. (2000b). Effects of increased and deregulated expression of cell division genes on the morphology and on antibiotic production of streptomycetes. *Antonie Van Leeuwenhoek* **78**, 269–276.

Wang, J., Soisson, S. M., Young, K., Shoop, W., Kodali, S., Galgoci, A., Painter, R., Parthasarathy, G., Tang, Y. S., Cummings, R., Ha, S., Dorso, K., *et al.* (2006). Platensimycin is a selective FabF inhibitor with potent antibiotic properties. *Nature* **441**, 358–361.

Ward, J. M., Janssen, G. R., Kieser, T., Bibb, M. J., Buttner, M. J., and Bibb, M. J. (1986). Construction and characterisation of a series of multi-copy promoter-probe plasmid vectors for Streptomyces using the aminoglycoside phosphotransferase gene from Tn5 as indicator. *Mol. Gen. Genet.* **203**, 468–478.

Wardell, J. N., Stocks, S. M., Thomas, C. R., and Bushell, M. E. (2002). Decreasing the hyphal branching rate of *Saccharopolyspora erythraea* NRRL 2338 leads to increased resistance to breakage and increased antibiotic production. *Biotechnol. Bioeng.* **78**, 141–146.

Watve, M. G., Tickoo, R., Jog, M. M., and Bhole, B. D. (2001). How many antibiotics are produced by the genus *Streptomyces*? *Arch. Microbiol.* **176**, 386–390.

White, J., and Bibb, M. (1997). bldA dependence of undecylprodigiosin production in *Streptomyces coelicolor* A3(2) involves a pathway-specific regulatory cascade. *J. Bacteriol.* **179**, 627–633.

Wietzorrek, A., Bibb, M., and Chakraburtty, R. (1997). A novel family of proteins that regulates antibiotic production in streptomycetes appears to contain an OmpR-like DNA-binding fold. *Mol. Microbiol.* **25**, 1181–1184.

Wilkinson, B., and Micklefield, J. (2007). Mining and engineering natural-product biosynthetic pathways. *Nat. Chem. Biol.* **3**, 379–386.

Wright, A. J. (1999). The penicillins. *Mayo Clin. Proc.* **74**, 290–307.

Wright, G. D. (2007). The antibiotic resistome: The nexus of chemical and genetic diversity. *Nat. Rev. Microbiol.* **5**, 175–186.

Wright, L. F., and Hopwood, D. A. (1976). Identification of the antibiotic determined by the SCP1 plasmid of *Streptomyces coelicolor* A3(2). *J. Gen. Microbiol.* **95**, 96–106.

Xiao, X., Wang, F., Saito, A., Majka, J., Schlosser, A., and Schrempf, H. (2002). The novel Streptomyces olivaceoviridis ABC transporter Ngc mediates uptake of N-acetylglucosamine and N,N'-diacetylchitobiose. *Mol. Genet. Genomics* **267**, 429–439.

Zazopoulos, E., Huang, K., Staffa, A., Liu, W., Bachmann, B. O., Nonaka, K., Ahlert, J., Thorson, J. S., Shen, B., and Farnet, C. M. (2003). A genomics-guided approach for discovering and expressing cryptic metabolic pathways. *Nat. Biotechnol.* **21**, 187–190.

Zhang, L., Li, W. C., Zhao, C. H., Chater, K. F., and Tao, M. F. (2007). NsdB, a TPR-like-domain-containing protein negatively affecting production of antibiotics in *Streptomyces coelicolor* A3 (2). *Wei Sheng Wu Xue Bao* **47**, 849–854.

CHAPTER SIX

REGULATION OF ANTIBIOTIC PRODUCTION BY BACTERIAL HORMONES

Nai-Hua Hsiao, Marco Gottelt, *and* Eriko Takano

Contents

1. Introduction	144
2. Rapid Small-Scale γ-Butyrolactone Purification	145
3. Antibiotic Bioassay	146
4. Kanamycin Bioassay	148
5. Identification of γ-Butyrolactone Receptors	150
6. Identification of the γ-Butyrolactone Receptor Targets	151
7. Gel Retardation Assay to Detect Target Sequences of the γ-Butyrolactone Receptors	152
7.1. Labeling of DNA fragments	153
8. Conclusions	156
Acknowledgments	156
References	156

Abstract

Antibiotic production is regulated by numerous signals, including the so-called bacterial hormones found in antibiotic producing organisms such as *Streptomyces*. These signals, the γ-butyrolactones, are produced in very small quantities, which has hindered their structural elucidation and made it difficult to assess whether they are being produced. In this chapter, we describe a rapid small-scale extraction method from either solid or liquid cultures in scales of one plate or 50 ml of medium. Also described is a bioassay to detect the γ-butyrolactones by determining either the production of pigmented antibiotic of *Streptomyces coelicolor* or kanamycin resistant growth on addition of the γ-butyrolactones. We also describe some insights into the identification of the γ-butyrolactone receptor and its targets and also the gel retardation conditions with three differently labeled probes.

Department of Microbial Physiology, GBB, University of Groningen, Haren, The Netherlands

Methods in Enzymology, Volume 458 © 2009 Elsevier Inc.
ISSN 0076-6879, DOI: 10.1016/S0076-6879(09)04806-X All rights reserved.

1. INTRODUCTION

Most secondary metabolites (including antibiotics) are the products of complex biosynthetic pathways activated in the stationary phase or during slow growth, which triggers the transition from primary to secondary metabolism (Bibb, 2005). This switch is complex and poorly understood, and involves many signals, including those by small signaling molecules called γ-butyrolactones. These signaling molecules, found mainly in *Streptomyces* species, are considered to be "bacterial hormones" because they are important in the regulation of antibiotic production and in some cases morphological differentiation (Takano, 2006). The first γ-butyrolactone, A-factor, was identified from *Streptomyces griseus* in 1967 and was isolated as a compound that could stimulate the production of streptomycin and sporulation in a mutant defective in these characteristics (Khokhlov *et al.*, 1967) (Fig. 6.1). Subsequently, the group of Sueharu Horinouchi identified the A-factor receptor (Onaka *et al.*, 1995) and the gene involved in the

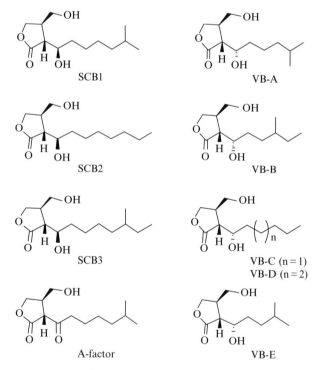

Figure 6.1 Chemical structures of the γ-butyrolactones. SCB1, 2, and 3 are produced by *S. coelicolor*; A-factor is produced by *S. griseus*; VB-A, B, C. D, and E are produced by *S. virginiae*. Figure contributed by Christian Hertweck with permission.

synthesis of A-factor (Horinouchi *et al.*, 1985) and also the pathway by which A-factor regulates streptomycin production (Ohnishi *et al.*, 1999). Other γ-butyrolactones and its receptor systems were identified in *Streptomyces virginae* (Yamada, 1999 review), *Streptomyces lavendulae* FRI-5 (Kitani *et al.*, 2008), and *Streptomyces coelicolor* (Takano *et al.*, 2005), the model organism for *Streptomyces* genetics.

The number of identified receptor homologues has more than tripled in recent years with the analysis of antibiotic biosynthetic gene cluster sequences, suggesting that the γ-butyrolactone system may be a widespread regulatory component in controlling antibiotic production in the genus *Streptomyces* and in other actinomycetes (Takano, 2006).

These signaling molecules are produced only in minute concentrations, making them challenging to purify. However, their activity in nanomolar concentration is helpful in bioactivity assays, which can be performed with small quantities of material. In all the systems that have been reported, the γ-butyrolactones have been detected by such bioassays. For example: A-factor in *S. griseus* is extracted with chloroform and detected by measuring the induction of streptomycin production on a solid culture of a non-A-factor producing mutant, *S. griseus* HH1 (Hara and Beppu, 1982); virginiae butanolides (VBs) in *Streptomyces virginae* are extracted with ethyl acetate from acidified cell-free supernatants of cultures and detected by measuring the production of virginiamycin in a liquid culture of an *S. virginae* mutant (Nihira *et al.*, 1988); IM-2 in *S. lavendulae* FRI-5 is extracted in the same way as VBs and detected by the induction of a blue pigment (Sato *et al.*, 1989).

To quickly assess whether γ-butyrolactones are a regulatory component in the antibiotics of choice, a large-scale purification or structural elucidation is not the first preference. This chapter describes the rapid small-scale extraction and detection of γ-butyrolactones in *S. coelicolor* which will aid in solving this problem. We also aim to aid identification of γ-butyrolactone receptors and their targets.

2. Rapid Small-Scale γ-Butyrolactone Purification

Intracellularly produced γ-butyrolactones are believed to diffuse passively out of the mycelium. In most protocols, liquid culture supernatants are used for production. However, the amounts of γ-butyrolactones produced in liquid cultures vary with the growth of the mycelium and extraction can be tedious. Therefore, we have developed a method using plates of solid agar medium. Here, we describe extraction protocols for both liquid and solid cultures.

1. Grow the strain either on plates up to aerial mycelium formation or early spore formation, or in liquid medium up to early stationary phase.

   Note: Both minimal and rich media can be used in both cases; however, the stability of the γ-butyrolactones in rich media is much less than in a defined medium and they must be extracted at an earlier time.

2. For plates, cut the agar into large pieces and place in a beaker. Add 30–40 ml ethyl acetate to cover the agar pieces, and shake gently. Transfer the ethyl acetate into a rotary flask.

   Note: Avoid breaking the agar into small pieces as this makes it difficult to separate the agar from the solvent. Also, it is important to test the ethyl acetate used for extraction beforehand as some ethyl acetate may already have bioactivity.

3. For liquid cultures, add a 2–3-fold volume of ethyl acetate in a separation flask. Shake 5–10 times, then separate the ethyl acetate phase from the water phase and place the ethyl acetate phase in a rotary flask.

   Note: If a small amount of second phase separates from the ethyl acetate sample, remove it before evaporation. Normally, any second phase consists of fatty acids which cannot be evaporated will affect the diffusion ability of samples in the bioassay.

4. Evaporate all the ethyl acetate in a rotary evaporator and resuspend in 100% HPLC-grade methanol.

   Note: From 3 plates or from 50 ml cultures, resuspend sample in ≤50 μl of methanol.

3. Antibiotic Bioassay

The classical approach for the bioassay of γ-butyrolactones is the observation or measurement of induced antibiotic production from an indicator strain in response to the compound of interest. Based on this concept, the antibiotic bioassay in *S. coelicolor* detects early production of the pigmented antibiotics, blue actinorhodin, and red prodiginines (Takano *et al.*, 2000).

1. Prepare spore stocks of the antibiotic bioassay indicator strain *S. coelicolor* M145 (Kieser *et al.*, 2000). Determine viable counts and store spores at −20 °C in 20% glycerol.
2. Prepare supplemented minimal media solid (SMMS) plates (20 ml agar/plate; Takano *et al.*, 2001).

   Note: The source of the agar strongly affects the timing and production of the pigmented antibiotics. Try to use only agar from the same company and the same batch.

3. For each plate, add $1 \times 10^7 – 10^8$ M145 spores (so the plate is just about confluent) in 60–100 μl of water. Using wet sterile cotton buds, spread confluently onto the plates and dry in the laminar flow cabinet.
4. Spot a sample (<2 μl) in the middle of the plate and dry. Incubate plates at 30 °C, check them every 8–12 h and record the antibiotic production by scanning the plates on a scanner (or take a photograph) (Fig. 6.2).

   Note:
   - Avoid dropping a sample before a plate is completely dry or dropping more than 2 μl of the sample. If the sample volume is more than 2 μl, spot in 2 μl aliquots several times after drying each time or concentrate the sample before spotting.
   - The extracted samples should be spotted onto the indicator plate within 8 h of spreading, the sooner the better. No induction of the pigmented antibiotics will be observed if the indicator is grown for more than 8 h.
   - The optimal amount of the positive control is 0.25 μg for the γ-butyrolactone SCB1. When using extracts made from *S. coelicolor* M145, 20–30 μl out of the 50 μl of sample either from 50 ml liquid cultures or from three plates should give bioactivity.
   - To measure the approximate concentration of γ-butyrolactones isolated: for plates start the inoculum with the exact same number of viable spore counts for each sample; for liquid cultures measure the growth phase (by OD450 nm) convert into milligrams per dry weight and also measure the volume of the supernatant; this amount then can be used for comparison between samples.
   - As seen in Fig. 6.2, the color may change depending on the amount of indicator spores used and also the length of time incubated.

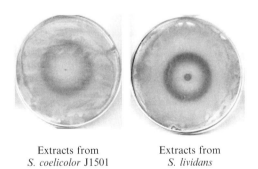

Extracts from
S. coelicolor J1501

Extracts from
S. lividans

Figure 6.2 An example of a typical bioassay. Extracts from solid plates of *S. coelicolor* J1501 and *S. lividans* were spotted onto the indicator strain M145, then incubated for 36 h at 30 °C. The pigmented halos are the result of the butyrolactones stimulating pigmented antibiotic production.

4. Kanamycin Bioassay

The kanamycin bioassay was developed as an easy and direct method to detect γ-butyrolactone. It is based on the γ-butyrolactone receptor, ScbR, and its binding to the target DNA and the γ-butyrolactones. ScbR represses transcription of its own gene and that of *cpkO*, a pathway-specific regulatory gene for the Cpk cluster, by binding to the promoter regions (Pawlik *et al.*, 2007; Takano *et al.*, 2005). This repression is abolished by SCB1, resulting in transcription of the target genes. Using these principles, a γ-butyrolactone detection plasmid, pTE134, was constructed which harbors *scbR* and its own promoter region with the ScbR binding sites, together with the *cpkO* promoter with the ScbR binding sites, fused to a promoterless-kanamycin resistance gene (Fig. 6.3). pTE134, was transferred into *S. coelicolor* LW16 (*scbA* and *scbR* double deletion mutant) to obtain the kanamycin indicator strain LW16/pTE134. LW16 was chosen as a host strain for the kanamycin bioassay because it does not produce γ-butyrolactones and it lacks *scbR*, thus avoiding competition from endogenous ScbR. The kanamycin bioassay indicator strain LW16/pTE134 is sensitive to kanamycin but when γ-butyrolactones are added the repression caused by ScbR is abolished

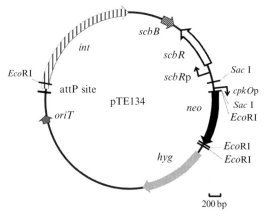

Figure 6.3 Schematic map of pTE134 used for the kanamycin bioassay. pTE134 has *scbR* (open arrow) with its own promoter region, *scbR*p (black arrow) and a *cpkO* promoter, *cpkO*p (black arrow) coupled with a promoterless kanamycin resistance gene (*neo*, solid arrow). The light gray arrow indicates *hyg* (omega hygromycin resistance gene), the deep gray arrow represents *oriT* (*RP4* origin of single-stranded DNA transfer), the hatched arrow represents *int* (phiC31 integrase), the shaded arrow indicates a partial coding region (204 bp) of *scbB* and the attP site (phage phiC31 attP site) is indicated by a vertical black line. Restriction sites used for cloning are indicated by black lines (not all *Sac*I and *Eco*RI sites are shown on the map).

and the kanamycin resistance gene is transcribed, rendering the indicator strain kanamycin resistant (Hsiao *et al.*, submitted for publication).

This assay was also tested for its ability to detect other butyrolactones than the ones produced by *S. coelicolor*. Though the sensitivity is weaker than for the native γ-butyrolactones, with enough material spotted to the plates, kanamycin resistance can be observed. Furthermore, γ-butyrolactones have been identified from those antibiotic producing *Streptomyces* strains which were not known to produce these molecules using this assay (Hsiao *et al.*, submitted for publication).

1. Prepare spore stocks of the kanamycin bioassay indicator strain LW16/pTE134 and stock spores in 50–100 μl aliquots at −20 °C in 20% glycerol.

   Note: Because this indicator strain is unstable, spores need to be collected from MS plates (Kieser *et al.*, 2000) containing 50 μg/ml hygromycin, then aliquoted. Avoid thawing and refreezing the spore stocks.

2. Prepare fresh DNAgar (Difco Nutrient Agar) plates containing kanamycin at a final concentration 3–5 μg/ml (20 ml per plate).

   Note: Because the *cpkO* promoter has very low activity (conferring resistance to <10 μg/ml kanamycin), the concentration of kanamycin should not be higher than 5 μg/ml (Fig. 6.4).

3. To each plate, add 2.6×10^6 spores of the indicator strain in about 60–100 μl of water. Using wet sterile cotton buds, spread confluently onto the plates and dry.

4. Spot the extract or sample (<2 μl) in the middle of the plate and dry. Incubate plates at 30 °C, check them after 2–3 days and record the growth by scanning the plates on a scanner. The kanamycin resistant colonies will grow around the spot where the γ-butyrolactones have diffused.

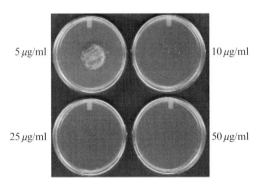

Figure 6.4 The kanamycin bioassay using different kanamycin concentrations in the indicator media. 0.1 μg of chemically synthesized SCB1 was spotted onto lawns of LW16/pTE134 (2.6×10^6 spores) on DNAgar plates containing 5, 10, 25, or 50 μg/ml kanamycin, respectively. The plates were incubated at 30 °C for 3 days.

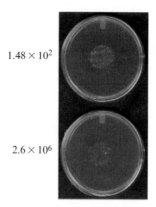

Figure 6.5 The kanamycin bioassay using different amount of indicator strains. 0.1 μg of chemically synthesized SCB1 was spotted onto lawns of LW16/pTE134 (2.6 × 10^6 or 1.48 × 10^7 spores) on DNAgar plates containing 5 μg/ml kanamycin, respectively. The plates were incubated at 30 °C for 3 days.

Note:
- Again avoid spotting a sample before a plate is completely dry or spotting more than 2 μl of the sample. If the sample volume is more than 2 μl, spot in 2 μl aliquots several times after drying each time or concentrate the sample before spotting.
- The optimal amount of the positive control is 0.1 μg for SCB1 and 0.25 μg for A-factor. To enhance the density of the growth halo, a larger amount of indicator spores and a smaller amount of kanamycin in the indicator plate can be used (Figs. 6.4 and 6.5).

5. IDENTIFICATION OF γ-BUTYROLACTONE RECEPTORS

Several groups have used different methods to identify the γ-butyrolactone receptors. The first receptor to be identified was by Horinouchi and co-workers using an A-factor affinity column and recovering the receptor from the crude extract of *S. griseus* (Onaka *et al.*, 1995). The same technique was used to identify the γ-butyrolactone receptors from *S. virginiae* (Okamoto *et al.*, 1995) and *S. lavendulae* FRI-5 (Ruengjitchatchawalya *et al.*, 1995). Another approach was via PCR using degenerate oligonucleotides to isolate the two receptor homologues in *S. coelicolor* (Onaka *et al.*, 1998). Interestingly, Onaka and co-workers failed to identify ScbR with this approach. This may be due to the limited homologous sequence information at the time.

We have identified ScbR only by cloning the γ-butyrolactone synthase gene, ScbA, which lies divergent from the ScbR gene, using PCR (Takano *et al.*, 2001). In most of the γ-butyrolactone systems identified (apart from that in *S. griseus*) synthase and receptor genes are next to or very close to each other. It is best to use this property and to identify the γ-butyrolactone synthase gene first using degenerate oligonucleotides or by an *in silico* approach. The degenerate oligonucleotides should be designed near the active sites we have identified by mutagenesis analysis, which also correspond to the AfsA repeats (Hsiao *et al.*, 2007).

Another approach is to use any sequence information available. These γ-butyrolactone receptors will resemble TetR repressors and may be found especially concentrated close to antibiotic biosynthetic gene clusters. There are usually several γ-butyrolactone receptor homologues in one strain. To determine the "real" receptor, the amino acid identity should be more than 40% compared to the functionally proven γ-butyrolactone receptors, that is, ArpA, BarA, FarA, and ScbR. These proteins also will have a pI of around 5–6 compared to the other ScbR homologues whose pIs are much higher at around 9–11.

6. IDENTIFICATION OF THE γ-BUTYROLACTONE RECEPTOR TARGETS

The first target to be identified was from the A-factor system by use of the genomic SELEX (systematic evolution of ligands by exponential enrichment) system (Ohnishi *et al.*, 1999). This method uses purified γ-butyrolactone receptor in gel retardation assays on partially digested chromosomal DNA to find the binding sites. Several groups have now proposed the target sequences for the γ-butyrolactone receptors (Folcher *et al.*, 2001; Kinoshita *et al.*, 1999; Onaka and Horinouchi, 1997) (Fig. 6.6). However, these sequences are not sufficiently conserved compared with the identified targets for ScbR (Fig. 6.6).

We have used the properties of the TetR receptor family, which the γ-butyrolactone receptors resemble. TetR proteins regulate their own genes and/or the adjacent gene. By using this property, gel retardation was performed to identify one of the ScbR targets. To identify another target of ScbR, microarray analysis was performed, which did not result in the identification of a direct target but rather the downstream genes which represent the Cpk antibiotic biosynthetic gene cluster. But doing so, the direct target, a pathway-specific activator, for the Cpk cluster was identified. Surprisingly, of the four targets of ScbR that we have identified, two, site R and site OA, have 100% consensus sequences at 6 bp at each ends. While the

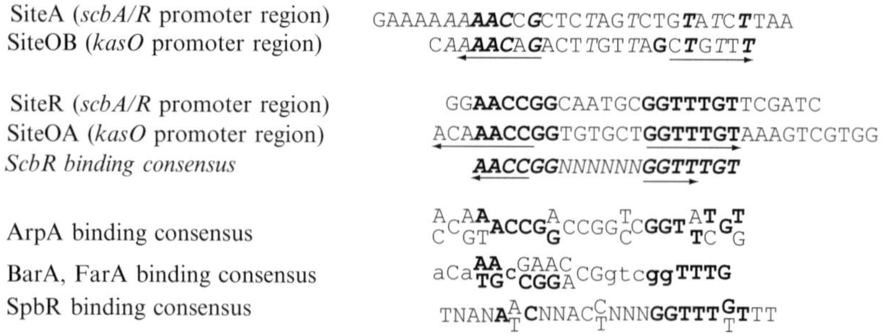

SiteA (*scbA/R* promoter region) GAAAAAA**AACC**G*CTC TAGTCTGTA TCT TAA*
SiteOB (*kasO* promoter region)   CA**AACA**G*ACT TGT TAGC T G TT T*

SiteR (*scbA/R* promoter region) GG**AACCGG**CAATGC**GGTTTGT**TCGATC
SiteOA (*kasO* promoter region)   ACA**AACCGG**TGTGCT**GGTTTGT**AAAGTCGTGG
ScbR binding consensus   **AACCGG**NNNNNNN**GGTTTGT**

ArpA binding consensus $^A_C C^{AA}_{GT}$**ACCG**^A_GCCGG^T_C**CGGT**$^{AT}_{TC}$**G**^T_G

BarA, FarA binding consensus   aCa**AA**$_{TG}$c**GAAC**$_{CGGA}$CGgtcgg**TTTG**

SpbR binding consensus   TNANA^A_T**C**NNAC^C_TNNN**GGTTT**^G_T**T**TT

Figure 6.6 Consensus sequences of different γ-butyrolactone receptors. The target sequences found so far for ScbR and its consensus sequences in *S. coelicolor* are presented on the top five lanes. The bottom three lanes are consensus sequences reported for the different receptor proteins. ArpA: Onaka and Horinouchi (1997); BarA, FarA: Kinoshita *et al.* (1999); SpbR: Folcher *et al.* (2001). Bold letters represent the conserved base pairs in ScbR binding sites. The base pairs in italic in siteA and siteOB represent the conserved base pairs between these two sites.

others, site A and site OB, only have 4 bp that are conserved with site R and site OA. In contrast, within the pair, 11 bp are conserved out of 23 bp (Fig. 6.6) (Takano *et al.*, 2005).

To identify the γ-butyrolactone receptor targets, first use the promoter region itself, and also include the promoter region of the gene adjacent to it, to perform a gel retardation assay with either a purified protein or a crude extract. The activity of the protein is stabilized when the protein is still in the crude extract and at a low concentration. As soon as the protein is purified and concentrated, it tends to aggregate and will no longer be suitable for any further experiments to analyze activity. Use the identified target sequence to identify further targets *in silico*.

7. Gel Retardation Assay to Detect Target Sequences of the γ-Butyrolactone Receptors

Gel retardation is not a new technique and can be found described in many publications (e.g., Folcher *et al.*, 2001; Kinoshita *et al.*, 1999; Onaka and Horinouchi, 1997). However, we have experience in using three differently labeled γ-butyrolactone receptor target DNA fragments, and have optimized the technique for each probe using pure protein and crude extracts from either *Escherichia coli* or *Streptomyces*.

7.1. Labeling of DNA fragments
7.1.1. Radio-labeled probes

1. The oligo was first labeled at the 5′ ends by incubation of 50 pmol of oligo, 5 μl of [γ-^{32}P]ATP (10 mCi/ml), 1 μl of ×10 kinase buffer, and 1 μl T4 nuclease kinase in a total volume of 50 μl incubated at 37 °C for 30 min^{-1} h. The reaction was stopped and cleaned by addition of ×1 volume of phenol/chloroform with vortexing, then the phenol was extracted by addition of either ×1 vol of chloroform or ether, twice. The oligo were precipitated using 100% EtOH with 3 M NaOAc and 1 μg of glycogen at −20 °C preferably O/N. The mixture was spun at 4 °C for 30 min, then the supernatant discarded, and the pellet air dried.
2. PCR was conducted using labeled- and unlabeled-oligos. 60 ng of template, 50 pmol of both unlabeled- and labeled-oligos, 200 μM dNTP, 5% DMSO, and PCR buffer (commercial) together with Taq polymerase were incubated to amplify in a total volume of 100 μl. The PCR conditions are: 95 °C for 5 min, then 30 cycles of 95 °C for 50 s, 55 °C for 40 s, 72 °C for 40 s, then extend at 72 °C for 5 min. Five microliters of the reaction was loaded on a gel to assure amplification. If extra bands are amplified, the fragment of interest can be purified using a commercial gel-extraction kit. The final volume should be eluted in 50 μl.
3. Gel retardation was conducted with 2.5 ng of labeled DNA (about 10 cpm), pure protein or 0–15 μl crude extract from *E. coli* or *Streptomyces*, 125 mM HEPES pH 7.5, 20 mM DTT, 20% glycerol, 200 mM KCl, 0.16 μg/μl calf thymus DNA in a final volume of 12.5–25 μl. The mixture was incubated at room temperature for 10 min, then 2–4 μl of loading dye (50%(v/v) glycerol, 0.25% (w/v) bromophenol blue, 10 mM Tris–HCl, pH 8, 1 mM EDTA) was added. To detect the binding of the receptor to the butyrolactones, the butyrolactones can be added either before or 10 min after the incubation. In the case of SCB1, 1 μg was added to see a release of the full shift.

> *Note:*
> - For the *E. coli* crude extract, a 10 ml LB overnight preculture of *E. coli* harboring a plasmid with the γ-butyrolactone receptor gene under the control of the *lacZ* promoter was inoculated at a 1:100 dilution in 50 ml LB and grown at 37 °C until OD$_{600}$ 0.7–0.8 when induced with 1 mM (final conc.) IPTG (Isopropyl β-D-1-thiogalactopyranoside). After another 3 h of incubation, cells were harvested by centrifugation and the cell pellet was washed twice with buffer (50 mM of Tris pH 7.0, 1 mM of EDTA, 1 mM of DTT, 100 mM of phenylmethylsulfonylfluoride (PMSF)) and resuspended in 500 μl of the same buffer. 100 μl aliquots were stored at −80 °C for later use. For crude extract

preparation, the cells were immediately disrupted by sonication. The cell lysate was clarified by centrifugation and the supernatant used as a crude extract after determination of the total protein content.
- For the *S. coelicolor* crude extract, *S. coelicolor* was grown in 50 ml of SMM then the same procedure as above was followed except the cells were resuspended in 100 μl prior to sonication and 10 μl was used for the gel retardation analysis. Due to the growth phase-dependent expression of ScbR, the time point of growth when the cells were harvested is crucial to detect any protein binding to the DNA.
- KCl was the best salt for this assay and MgCl was not suitable.

4. For detection of 100–300 bp DNA fragments, a 5% acrylamide gel (5% acrylamide: bisacrylamide = 37.5:1, 1.25 ml of ×10 TBE, 15 μl of TEMED, 87.5 μl of 10% AMPS in a total of 12.5 ml) was used in ×1 TBE (90 mM Tris–HCl, 90 mM boric acid, 2 mM EDTA, pH 8.0) running buffer. It is also possible to use a 3.5% acrylamide gel to detect a longer DNA fragment. A Bio-Rad Mini Protean kit was used for running the gel at 100 V constant for 1 h till the BPB just ran off the gel.
5. The gel was then taken off the glass and wrapped in cling film (Saran wrap) and directly placed onto an X-ray film (Super RX (Fujifilm)), and exposed for 30 min at room temperature in a cassette. The film was developed using an automated film developer (Fig. 6.7A).

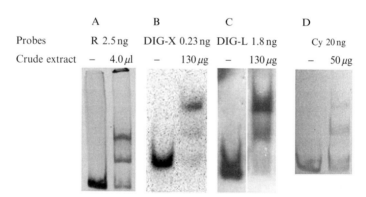

Figure 6.7 Gel retardation assay using differently labeled probes. Gel retardation using: (A) Radio-labeled probes (R); (B) DIG-labeled probes with detection using X-ray film (DIG-X); (C) DIG-labeled probes with detection using Lumi-imager (DIG-L); (D) Cy3-labeled probes (Cy). All probes were made using a 146 bp DNA fragment which includes the ScbR binding site R which is situated at -68 to -42 nt from the transcriptional start site (Takano *et al.*, 2001). The amount of each probe used is indicated next to each probe name. Crude extract (given in total protein as microl, microg) from *E. coli* harboring pIJ6120 was used and the amount is indicated on the top of the lanes.

7.1.2. DIG-labeled probes

1. DNA fragments were first amplified by PCR using both unlabeled-oligos using the PCR conditions mentioned in Point 2 of Section 7.1.1. Purified PCR product (either by a cleanup kit or by extracting from a gel), 100 ng, was used in the DIG gel shift kit 2nd generation (Roche) labeling procedure, steps 5–7 (p11 of Roche protocol). (Note: this kit labels 3′ ends.)
2. After labeling, the probes were incubated with the protein for gel retardation assays as in Point 3 of Section 7.1.1.

    Note:
    - The labeled probes can be stored at $-20\ °C$ for a very long time.
    - Depending on the sensitivity of the detection method used (see below), the amount of DIG-labeled probe may have to be varied. In our case, eightfold less probe was used for detection with X-ray films (we used 1.8 ng of labeled probe) compared to when a Lumi-imager F1 (we used 0.23 ng labeled probe) was used for detection (Fig. 6.7B and C).
    - After optimization of the amount of probe used, the protein concentration in the assay will have to be adjusted to keep the protein/DNA ratio constant.

3. After running, the gel was taken off the glass plates for direct contact-blotting (Roche protocol 3.7.2), further crosslinking (3.7.3), and chemiluminescent detection (3.8). The membrane was detected using an X-ray film exposed for 20 min, then developed in an automated X-ray film developer (Fig. 6.7B) or in a Lumi-imager F1 (Roche) for 40 min at room temperature (Fig. 6.7C).

7.1.3. Cy3-labeled probes

The DIG-labeling protocol will take up to 2 days. To improve the efficiency in time, Cy3-labeled oligos are now our preferred choice. A preliminary protocol is described below.

1. 5′-Cy3-labeled oligo was purchased from Sigma–Aldrich and used for PCR along with the unlabeled-oligo as in Point 3 of Section 7.1.1. The amplified product was purified and the concentration measured by Nanodrop (Thermo scientific).
2. The gel retardation was conducted with 20 ng of Cy3-labeled DNA fragment with 50–300 µg of crude extracts from *Streptomyces* as in Points 3 and 4 of Section 7.1.1. After running, the gel with the glass plate is exposed for 30 s and detected in a LAS 4000 (Fujifilm) (Fig. 6.7D).

    Note: A flourescence compound was detected in the *Streptomyces* crude extract which may interfere with detection.

8. Conclusions

The rapid small-scale purification of the γ-butyrolactones and the kanamycin assay have simplified the detection of the γ-butyrolactones, which are found to be one of the important factors in the regulation of antibiotic production. From genome sequences and also from the sequence of many antibiotic biosynthetic gene clusters many homologues to the γ-butyrolactone synthases and receptors have been identified, but the number of γ-butyrolactones elucidated has not increased since year 2000. This method may provide an easy solution to detect the γ-butyrolactones from those strains which have the synthase and receptor homologues and also to identify new organisms that may produce γ-butyrolactones. The binding sensitivity of the different γ-butyrolactones to ScbR has been tested and found to be lower; however, with enough material kanamycin resistance was observed (Hsiao *et al.*, submitted). Identification of the receptor target is a major job. However, with the several hints obtained from previous work and also from our experience with the gel retardation assays, it may be possible to quicken the procedure.

ACKNOWLEDGMENTS

Thanks to David Hopwood for critical reading of the manuscript. Also thanks to Christian Hertweck for providing Fig. 6.1. Hsiao's, Gottelt's, and Takano's work were funded by the Deutsche Forschungsgemeinschaft (TA428/1-1,1-2), EUFP6 ActinoGEN, Rosalind Franklin Fellowship from the University of Groningen, respectively.

REFERENCES

Bibb, M. (2005). Regulation of secondary metabolism in streptomycetes. *Curr. Opin. Microbiol.* **8,** 208–215.

Folcher, M., Gaillard, H., Nguyen, L. T., Nguyen, K. T., Lacroix, P., Bamas-Jacques, N., Rinkel, M., and Thompson, C. J. (2001). Pleiotropic functions of a *Streptomyces pristinaespiralis* autoregulator receptor in development, antibiotic biosynthesis, and expression of a superoxide dismutase. *J. Biol. Chem.* **276,** 44297–44306.

Hara, O., and Beppu, T. (1982). Mutants blocked in streptomycin production in *Streptomyces griseus*—The role of A-factor. *J. Antibiot. (Tokyo)* **35,** 349–358.

Horinouchi, S., Nishiyama, M., Suzuki, H., Kumada, Y., and Beppu, T. (1985). The cloned *Streptomyces bikiniensis* A-factor determinant. *J. Antibiot. (Tokyo)* **38,** 636–641.

Hsiao, N. H., Soeding, J., Linke, D., Lange, C., Hertweck, C., Wohlleben, W., and Takano, E. (2007). ScbA from *Streptomyces coelicolor* A3(2) has homology to fatty acid synthases and is able to synthesise γ-butyrolactones. *Microbiology* **153,** 1394–1404.

Hsiao, N. H., Nakayama, S., Merlo, M. E., Vries, M., Bunet, R., Kitani, S., Vonk, R., Nihira, T., and Takano, E. Identification of two new γ-butyrolactones and a novel detection system in *Streptomyces coelicolor* and its application. (Submitted).

Khokhlov, A. S., Tovarova, I. I., Borisova, L. N., Pliner, S. A., Shevchenko, L. N., Kornitskaia, E., Ivkina, N. S., and Rapoport, I. A. (1967). The A-factor, responsible for streptomycin biosynthesis by mutant strains of *Actinomyces streptomycini*. *Dokl. Akad. Nauk SSSR* **177,** 232–235.

Kieser, T., Bibb, M. J., Buttner, M. J., Chater, K. F., and Hopwood, D. A. (2000). *Practical Streptomyces Genetics*. John Innes Foundation, Norwich.

Kinoshita, H., Tsuji, T., Ipposhi, H., Nihira, T., and Yamada, Y. (1999). Characterization of binding sequences for butyrolactone autoregulator receptors in streptomycetes. *J. Bacteriol.* **181,** 5075–5080.

Kitani, S., Iida, A., Izumi, T. A., Maeda, A., Yamada, Y., and Nihira, T. (2008). Identification of genes involved in the butyrolactone autoregulator cascade that modulates secondary metabolism in *Streptomyces lavendulae* FRI-5. *Gene* **425,** 9–16.

Nihira, T., Shimizu, Y., Kim, H. S., and Yamada, Y. (1988). Structure-activity relationships of virginiae butanolide C, an inducer of virginiamycin production in *Streptomyces virginiae*. *J. Antibiot. (Tokyo)* **41,** 1828–1837.

Ohnishi, Y., Kameyama, S., Onaka, H., and Horinouchi, S. (1999). The A-factor regulatory cascade leading to streptomycin biosynthesis in *Streptomyces griseus*: Identification of a target gene of the A-factor receptor. *Mol. Microbiol.* **34,** 102–111.

Okamoto, S., Nakamura, K., Nihira, T., and Yamada, Y. (1995). Virginiae butanolide binding protein from *Streptomyces virginiae*. Evidence that VbrA is not the virginiae butanolide binding protein and reidentification of the true binding protein. *J. Biol. Chem.* **270,** 12319–12326.

Onaka, H., Ando, N., Nihira, T., Yamada, Y., Beppu, T., and Horinouchi, S. (1995). Cloning and characterization of the A-factor receptor gene from *Streptomyces griseus*. *J. Bacteriol.* **177,** 6083–6092.

Onaka, H., and Horinouchi, S. (1997). DNA-binding activity of the A-factor receptor protein and its recognition DNA sequences. *Mol. Microbiol.* **24,** 991–1000.

Onaka, H., Nakagawa, T., and Horinouchi, S. (1998). Involvement of two A-factor receptor homologues in *Streptomyces coelicolor* A3(2) in the regulation of secondary metabolism and morphogenesis. *Mol. Microbiol.* **28,** 743–753.

Pawlik, K., Kotowska, M., Chater, K. F., Kuczek, K., and Takano, E. (2007). A cryptic type I polyketide synthase (cpk) gene cluster in *Streptomyces coelicolor* A3(2). *Arch. Microbiol.* **187,** 87–99.

Ruengjitchatchawalya, M., Nihira, T., and Yamada, Y. (1995). Purification and characterization of the IM-2-binding protein from *Streptomyces* sp. strain FRI-5. *J. Bacteriol.* **177,** 551–557.

Sato, K., Nihira, T., Sakuda, S., Yanagimoto, M., and Yamada, Y. (1989). Isolation and structure of a new butyrolactone autoregulator from *Streptomyces* sp. FRI-5. *J. Ferment. Bioeng.* **68,** 170–173.

Takano, E. (2006). γ-Butyrolactones: *Streptomyces* signalling molecules regulating antibiotic production and differentiation. *Curr. Opin. Microbiol.* **9,** 287–294.

Takano, E., Chakraburtty, R., Nihira, T., Yamada, Y., and Bibb, M. J. (2001). A complex role for the γ-butyrolactone SCB1 in regulating antibiotic production in *Streptomyces coelicolor* A3(2). *Mol. Microbiol.* **41,** 1015–1028.

Takano, E., Kinoshita, H., Mersinias, V., Bucca, G., Hotchkiss, G., Nihira, T., Smith, C. P., Bibb, M., Wohlleben, W., and Chater, K. (2005). A bacterial hormone (the SCB1) directly controls the expression of a pathway-specific regulatory gene in the cryptic type I polyketide biosynthetic gene cluster of *Streptomyces coelicolor*. *Mol. Microbiol.* **56,** 465–479.

Takano, E., Nihira, T., Hara, Y., Jones, J. J., Gershater, C. J., Yamada, Y., and Bibb, M. (2000). Purification and structural determination of SCB1, a γ-butyrolactone that elicits antibiotic production in *Streptomyces coelicolor* A3(2). *J. Biol. Chem.* **275,** 11010–11016.

Yamada, Y (1999). Microbial Signalling and Communication, pp. 177–196. Cambridge University Press, Cambridge, UK.

CHAPTER SEVEN

Cloning and Analysis of Natural Product Pathways

Bertolt Gust

Contents

1. Introduction	160
2. Cloning and Identification of Biosynthetic Gene Clusters	161
3. Analysis of Natural Product Pathways by PCR-Targeted Gene Replacement	163
4. *In Vitro* Transposon Mutagenesis	170
5. Heterologous Expression of Biosynthetic Gene Clusters	173
6. Reassembling Entire Gene Clusters by "Stitching" Overlapping Cosmid Clones	174
7. Conclusions	177
Acknowledgments	177
References	177

Abstract

The identification of gene clusters of natural products has lead to an enormous wealth of information about their biosynthesis and its regulation, and about self-resistance mechanisms. Well-established routine techniques are now available for the cloning and sequencing of gene clusters. The subsequent functional analysis of the complex biosynthetic machinery requires efficient genetic tools for manipulation. Until recently, techniques for the introduction of defined changes into *Streptomyces* chromosomes were very time-consuming. In particular, manipulation of large DNA fragments has been challenging due to the absence of suitable restriction sites for restriction- and ligation-based techniques. The homologous recombination approach called recombineering (referred to as Red/ET-mediated recombination in this chapter) has greatly facilitated targeted genetic modifications of complex biosynthetic pathways from actinomycetes by eliminating many of the time-consuming and labor-intensive steps. This chapter describes techniques for the cloning and identification of biosynthetic gene clusters, for the generation of gene replacements within such

Pharmazeutische Biologie, Pharmazeutisches Institut, Eberhard-Karls-Universität Tübingen, Tübingen, Germany clusters, for the construction of integrative library clones and their expression in heterologous hosts, and for the assembly of entire biosynthetic gene clusters from the inserts of individual library clones. A systematic approach toward insertional mutation of a complete *Streptomyces* genome is shown by the use of an *in vitro* transposon mutagenesis procedure.

1. Introduction

Genes encoding secondary metabolite production in microorganisms are generally clustered in one contiguous region of the chromosome. Once a portion of the biosynthetic machinery is localized, it is almost certain that chromosomal walking from this locus will lead to the identification of an entire biosynthetic gene cluster. The corresponding genes that comprise the biosynthesis of natural product pathways are generally clustered in loci of about 10–200 kb (Ahlert *et al.*, 2002; Van Lanen *et al.*, 2007). The basis for the further analysis of natural product pathways is cloning and sequencing of the corresponding gene cluster and detailed analysis of the functions of genes contained therein. Besides biochemical investigations with expressed enzymes, generation of gene disruptions is an important step towards understanding the function of individual genes in the catalysis of biosynthetic reactions, in regulation mechanisms, and in self-resistance. Conventionally, gene disruptions have been constructed for several decades in *Streptomyces* by insertional inactivation via a single crossover using a DNA fragment internal to the gene of interest cloned into a vector. Another strategy involved cloning of an antibiotic resistance gene into a cloned copy of the target gene prior to allelic replacement via double crossing over. Both strategies are time-consuming, restriction- and ligation-based methods relying on the identification of suitable restriction sites (Kieser *et al.*, 2000). In addition, these technologies encounter difficulties for engineering and mutating DNA molecules greater than 15–20 kb. Therefore, a technique for DNA manipulation called recombineering (from recombinogenic engineering) or Red/ET-mediated recombination, was established which overcomes these hurdles (Datsenko and Wanner, 2000; Muyrers *et al.*, 1999, 2001; Zhang *et al.*, 1998, 2000). Recombineering is mediated through homologous recombination, which allows the exchange of DNA between two molecules in a precise and specific manner. Furthermore, recombineering requires only short regions of sequence identity (~40 bp) for efficient homologous recombination. These sequences for Red/ET-mediated recombination can easily be integrated into synthetic primers by PCR. Because most bacteria are not readily transformable with linear DNA due to the intracellular *recBCD* exonuclease (ExoV), which degrades linear DNA molecules, it has been shown that expression of the RecE and RecT proteins or the corresponding proteins Gam Bet and Exo of the bacteriophage lambda (Red proteins) can bypass DNA degradation by ExoV and greatly increase homologous recombination in *E. coli* (Court *et al.*, 2002; Muyrers *et al.*, 2000; Poteete, 2001). The use of Red/ET-mediated recombination is illustrated by the generation of gene replacements in *Streptomyces*, construction of integrative cosmid clones for expression in heterologous hosts, and the assembly of overlapping inserts to create a clone carrying an entire biosynthetic gene cluster. A second strategy for the rapid and systematic generation of gene disruptions by *in vitro* transposon mutagenesis is also described.

2. Cloning and Identification of Biosynthetic Gene Clusters

For the cloning of biosynthetic gene clusters, cosmid or BAC vectors are commonly used in order to generate genomic libraries which will then be subjected to appropriate screening procedures. For the choice of an appropriate vector different aspects have to be taken into consideration. If heterologous expression of an entire gene clusters is required, cosmid or BAC vectors with functions for site-specific integration (Thorpe and Smith, 1998) into the host genome are needed. Alternatively, an *E. coli*–*Streptomyces* shuttle vector can be used. The additional inclusion of an origin of transfer (*oriT* from RK2) allows the conjugal transfer of the cosmid or BAC clone from *E. coli* into *Streptomyces* readily, yielding exconjugants containing the desired pathway. Frequently used integrative or replicative cosmid vectors are pOJ436 or pOJ446, respectively (Bierman *et al.*, 1992). Integrative BAC vectors capable of accepting and maintaining >100 kb DNA segments have been developed for the generation of ESAC (*E. coli*–*Streptomyces* Artificial Chromosome) libraries (Alduina *et al.*, 2003; Sosio *et al.*, 2000). For cloning and sequencing of biosynthetic gene clusters, cosmid vectors can also be used: they neither autonomously replicate nor integrate into the *Streptomyces* genome. Such cosmid vectors, for example, SuperCos1 (Stratagene), have been used for the generation of an ordered cosmid library (Redenbach *et al.*, 1996), which was the starting point for the total genome sequencing of *Streptomyces coelicolor* M145 (Bentley *et al.*, 2002). A variety of possible cloning vectors are summarised in *Practical Streptomyces Genetics* (Kieser *et al.*, 2000).

Cosmid libraries are generated from the genomic DNA of the respective producer strains using standard techniques (Kretz *et al.*, 1989; Sambrook and Russell, 2001). Given an average insert size of a cosmid library clone of ~40 kb, 200 cosmid clones are required for a single coverage of an 8-Mb genome. To ensure with reasonable probability that any given DNA sequence is contained in the library of a typical *Streptomyces* genome,

1000–3000 cosmid clones are usually required (5–15-fold coverage). More precisely, the following formula (Sambrook and Russell, 2001) can be used to determine the number of clones for a complete library:

$$N = \ln(1 - P)/\ln(1 - f)$$

where P is the desired probability (expressed as a fraction), f is the proportion of the genome contained in a single clone, and N is the required number of cosmid clones.

To identify the desired library clone containing parts of or the entire natural product pathway, two different methods can be used. Colony hybridization has been shown to be an efficient and rapid procedure to identify positive clones, assuming that an appropriate probe of sufficient length and homology to the target sequence exists (Sambrook and Russell, 2001). Using PCR for the screening procedure, on the other hand, allows the direct use of degenerate primers. The CODEHOP program, for instance, designs degenerate PCR primers based on multiple sequence-alignments, even from distantly related proteins (Rose et al., 1998). For this purpose, four identical amino acids within an alignment are sufficient for the calculation of an appropriate degenerate primer. The procedure for PCR screening involves the following three steps (Fig. 7.1).

1. The library clones are stored as glycerol stocks in microtiter plates. Forty-eight clones from a microtiter plate are stamped onto two LB-plates

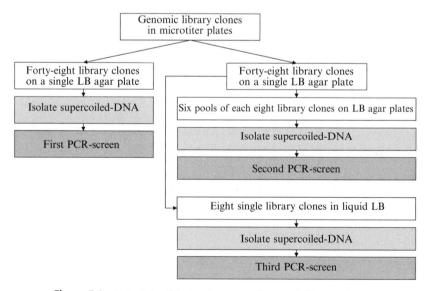

Figure 7.1 Principle of the PCR-screen of genomic library clones.

(Sambrook and Russell, 2001) containing an appropriate antibiotic. The plates are incubated overnight at 37 °C and colonies are washed from one plate using 1 ml of LB. The other plate is stored at 4 °C. Supercoiled-DNA of each pool representing 48 clones is isolated and used for the first PCR screen.
2. From positive pools of 48, 6 new pools representing each 8-library clones are streaked from the second LB plate onto fresh plates with the appropriate antibiotic and incubated overnight at 37 °C. The pools of 8 are washed from the plates again with 1 ml of LB and supercoiled-DNA is isolated and used for the second PCR screen.
3. From positive pools of eight, supercoiled-DNA from single colonies is used after cultivation in liquid LB for the third and final PCR screen.

Positive clones emerging from either colony hybridization or PCR screening are then verified to contain one continuous region of chromosomal DNA by restriction analysis. Common restriction fragments in all positive clones indicate the presence of overlapping inserts. Subsequently, end-sequencing of the inserts of positive clones will facilitate the selection of one clone for total sequencing.

3. ANALYSIS OF NATURAL PRODUCT PATHWAYS BY PCR-TARGETED GENE REPLACEMENT

The generation of gene disruptions is a principal technique for studying the functions of genes, including those contained in the biosynthetic gene clusters of natural product pathways. Red/ET-mediated recombination has been adapted for the genetic manipulation of cosmids derived from *Streptomyces* spp. (Gust *et al.*, 2003, 2004). This strategy is to replace a chromosomal sequence within a *Streptomyces* cosmid by a selectable marker that has been generated by PCR using primers with 39-nt homologous extensions. The inclusion of *oriT* in the disruption cassette allows conjugation to be used to introduce the modified cosmid DNA into *Streptomyces*. Conjugation is much more efficient than transformation of protoplasts and is readily applicable to many actinomycetes (Matsushima *et al.*, 1994). The potent methyl-specific restriction system of many streptomycetes is circumvented by passaging DNA through a methylation-deficient *E. coli* host such as ET12567 (MacNeil *et al.*, 1992). Vectors containing *oriT* are mobilisable *in trans* in *E. coli* by the self-transmissible pUB307 (Flett *et al.*, 1997) or the nontransmissible pUZ8002, which lacks a *cis*-acting function for its own transfer (Paget *et al.*, 1999). To adapt the procedure of Red/ET-mediated recombination for *Streptomyces*, cassettes for gene disruptions were constructed that can be selected both in *E. coli* and in *Streptomyces*. A list of cassettes, sequences, and a program to assist in primer design and in the analysis of the mutants generated is available at http://streptomyces.org.uk/redirect/index.html. The generation of a gene replacement involves the following steps (Fig. 7.2).

1. *Purification of the PCR template (cassette).* Using whole plasmids as templates for the PCR can result in a high proportion of antibiotic-resistant transformants without gene replacement. This is caused by traces of supercoiled DNA that compete with the linear PCR fragment and result in the occurrence of false-positive transformants. Using gel-purified disruption cassettes as templates prevents the occurrence of false positives. The purified fragment is stored in 10 mM Tris–HCl (pH 8) at a concentration of 100 ng/μl at $-20\ °$C. Absence of plasmid DNA is tested by using 100 ng of purified cassette DNA to transform highly competent *E. coli* DH5α cells ($10^8/\mu$g). If transformants appear, gel-purification of the cassette is repeated.

2. *Design of long PCR primers for recombineering.* For each gene disruption, two long PCR primers (58 nt and 59 nt) are required. Each has at its 5′ end 39 nt matching the *Streptomyces* sequence adjacent to the gene to be inactivated, and at its 3′ end a 19- or 20-nt sequence matching the right or left end of the disruption cassette (all cassettes have the same priming sites P1 and P2). The precise positioning of the 39-nt sequence

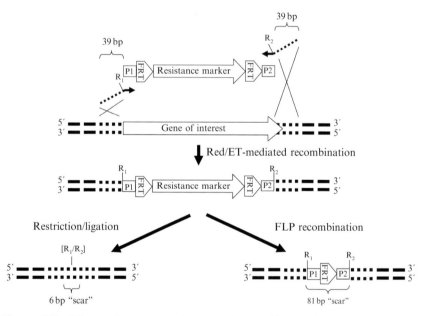

Figure 7.2 Scheme for gene replacement by Red/ET-mediated recombination and subsequent removal of the resistance marker; R_1/R_2 = compatible restriction sites; P1/P2 = priming sites for resistance marker cassette; FRT = FLP recognition target.

is important for creating in-frame deletions either by FLP recombinase-induced excision of the resistance marker or by elimination using restriction and ligation (Fig. 7.2).

3. *PCR amplification of the resistance cassette.* PCR amplification is performed in 50 μl volumes with 100 ng template, 0.2 mM dNTPs, 50 pmol of each primer, and 5% (v/v) DMSO with the Expand High Fidelity PCR system (Roche Molecular Biochemicals): denaturation at 94 °C for 2 min, then 10 cycles with denaturation at 94 °C for 45 s, annealing at 50 °C for 45 s, and elongation at 72 °C for 90 s, followed by 15 cycles with annealing at 55 °C for 45 s, and the last elongation step at 72 °C for 5 min. 5 μl of the PCR product is used for analysis by gel electrophoresis. The expected sizes are 78 bp larger than the sizes of the disruption cassettes (because of the 2 × 39 bp 5′-primer extensions). The remaining 45 μl of the PCR product is purified using the QIAGEN PCR purification kit according to the manufacturer's instructions. The PCR product is finally eluted from the columns with 12 μl of water ($\sim$200 ng/μl).

4. *Introduction of a cosmid clone into E. coli BW25113/pIJ790 (Red/ET recombination plasmid) by electroporation.* pIJ790 contains a chloramphenicol resistance marker and a temperature-sensitive origin of replication (requires 30 °C for replication).

   a. Grow *E. coli* BW25113/pIJ790 overnight at 30 °C in 10 ml LB (containing chloramphenicol (25 μg/ml).
   b. Inoculate 100 μl *E. coli* BW25113/pIJ790 from the overnight culture into 10 ml LB containing chloramphenicol (25 μg/ml). Grow for 3–4 h at 30 °C, shaking at 200 rpm to an OD$_{600}$ of $\sim$0.4.
   c. Recover the cells by centrifugation at 4000 rpm for 5 min at 4 °C in a Sorvall GS3 rotor (or equivalent). Decant the medium and resuspend the pellet by gentle mixing in 10 ml ice-cold 10% glycerol.
   d. Centrifuge as above and resuspend the pellet in 5 ml ice-cold 10% glycerol, centrifuge, and decant. Resuspend the cell pellet in the remaining $\sim$100 μl 10% glycerol.
   e. Mix 50 μl cell suspension with $\sim$100 ng of cosmid DNA. Carry out electroporation in a 0.2-cm ice-cold electroporation cuvette using a BioRad GenePulser II (or equivalent) set to 200 Ω, 25 μF, and 2.5 kV. The expected time constant should be 4.5–5.0 ms.
   f. Immediately add 1 ml ice-cold LB to shocked cells and incubate shaking for 1 h at 30 °C. Spread onto LB agar containing carbenicillin (100 μg/ml) and kanamycin (50 μg/ml) for selection of a SuperCos1-based cosmid (Stratagene), and chloramphenicol (25 μg/ml) for selection of pIJ790.
   g. Incubate overnight at 30 °C. Transfer one isolated colony into 5 ml LB containing the same antibiotics as above. Incubate with shaking

overnight at 30 °C. This culture will be used as a preculture for generating competent cells to be transformed with the PCR product (Point 3, Section 3.).

5. *PCR-targeted gene replacement.* E. coli BW25113/pIJ790 containing the desired cosmid is induced to express Red/ET genes and then electrotransformed with the PCR product. The example described uses an apramycin disruption cassette with an *oriT* for conjugal transfer from pIJ773 (Gust et al., 2003).

   a. Inoculate a 10-ml SOB culture (without $MgSO_4$)(Sambrook and Russell, 2001) containing carbenicillin (100 μg/ml), kanamycin (50 μg/ml), and chloramphenicol (25 μg/ml) with 5% of the overnight culture of *E. coli* BW25113/pIJ790 containing the SuperCos1-based cosmid.
   b. Add 100 μl 1M L arabinose stock solution for induction of the Red/ET genes (final concentration is 10 mM). Grow for 3–4 h at 30 °C shaking at 200 rpm to an OD_{600} of ~0.5.
   c. Recover the cells by centrifugation at 4000 rpm for 5 min at 4 °C in a Sorvall GS3 rotor (or equivalent). Decant the medium and resuspend the pellet by gentle mixing in 10 ml ice-cold 10% glycerol.
   d. Centrifuge as above and resuspend the pellet in 5 ml ice-cold 10% glycerol, centrifuge, and decant. Resuspend the cell pellet in the remaining ~100 μl of 10% glycerol.
   e. Mix 50 μl cell suspension with ~100 ng of PCR product. Carry out electroporation as described above. Immediately add 1 ml ice-cold SOC (Sambrook and Russell, 2001) to shocked cells, and incubate with shaking for 1 h at 37 °C. Spread onto LB agar containing carbenicillin (100 μg/ml), kanamycin (50 μg/ml), and apramycin (50 μg/ml). Incubate overnight at 37 °C to promote the loss of pIJ790.

6. *Confirmation of gene replacement by restriction analysis and/or PCR.* After 12–16 h of growth at 37 °C different colony sizes are observed. It is important to note that at this stage wild type and mutant cosmids often coexist within one cell. The transformation with a PCR product and its integration in the cosmid DNA by homologous recombination will not occur in all copies of the cosmid molecules in one cell. One copy of a cosmid containing the incoming resistance marker is sufficient for resistance to this antibiotic. Normally, the larger colonies have been shown to contain the desired modified cosmids. Supercoiled DNA is isolated, followed by phenol/chloroform extraction. Omitting the phenol/chloroform extraction step results in degradation of the cosmid DNA. Use of miniprep columns without including a phenol/chloroform extraction is not recommended. Verification of positive transformants by PCR

requires an additional pair of 18–20-nt test primers which anneal 100–200 bp upstream and downstream of the 39-bp recombination region (These primers can also be used later to verify the FLP-mediated excision of the resistance cassette). PCR amplification is performed in 50 μl volume with 50 ng template, 0.2 mM dNTPs, 20 pmol of each primer, and 5% (v/v) DMSO with the Expand High Fidelity PCR system (Roche Molecular Biochemicals): denaturation at 94 °C for 2 min, followed by 30 cycles with denaturation at 94 °C for 45 s, annealing at 55 °C for 45 s, and elongation at 72 °C for 90 s and the last elongation step at 72 °C for 5 min. 5 μl of the PCR product is used for analysis by gel electrophoresis.

7. *Transfer of the modified cosmids into Streptomyces* (Fig. 7.3). If the target *Streptomyces* for mutagenesis carries a methyl–sensitive restriction system (as is the case for *S. coelicolor* and *S. avermitilis*), it is necessary to passage the cosmid containing an apramycin resistance cassette and an *oriT* through a nonmethylating *E. coli* host. To achieve this, it is introduced by transformation into the nonmethylating *E. coli* ET12567 containing the RP4 derivative pUZ8002. The cosmid is then transferred to *Streptomyces* by intergeneric conjugation (Kieser *et al.*, 2000). If the target *Streptomyces* for mutagenesis does not carry a methyl–sensing restriction system (as is the case for *S. lividans*), common *E. coli* strains such as DH5α containing pUZ8002 can be used instead. Prepare competent cells of *E. coli* ET12567/pUZ8002 (Paget *et al.*, 1999) grown at 37 °C

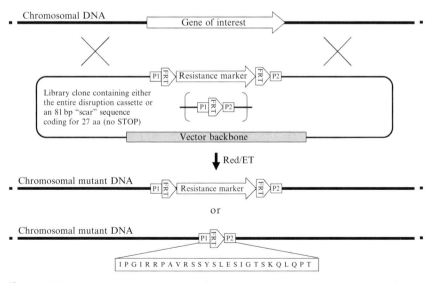

Figure 7.3 Schematic representation for the generation of a chromosomal gene replacement.

in LB containing kanamycin (25 μg/ml) and chloramphenicol (25 μg/ml) to maintain selection for pUZ8002 and the *dam* mutation, respectively.

   a. Transform competent cells with the *oriT*–containing cosmid clone and select for the incoming plasmid using only apramycin (50 μg/ml) and carbenicillin (100 μg/ml). Inoculate a colony into 10 ml LB containing apramycin (50 μg/ml), chloramphenicol (25 μg/ml), and kanamycin (50 μg/ml). Grow overnight with shaking at 37 °C.

   b. Inoculate 300 μl overnight culture into 10 ml fresh LB plus antibiotics as above and grow with shaking for ∼4 h at 37 °C to an OD_{600} of 0.4. Wash the cells twice with 10 ml of LB to remove antibiotics that might inhibit *Streptomyces*, and resuspend in 1 ml of LB.

   c. While washing the *E. coli* cells, for each conjugation add 10 μl (10^8) *Streptomyces* spores to 500 μl 2 × YT broth (Kieser *et al.*, 2000). Heat–shock at 50 °C for 10 min, then allow cooling by leaving at room temperature for 15 min.

   d. Mix 0.5 ml *E. coli* cell suspension and 0.5 ml heat–shocked spores and spin briefly. Pour off most of the supernatant and then resuspend the pellet in the 50 μl residual liquid.

   e. Make a dilution series from 10^{-1} to 10^{-4} in a total of 100 μl of water. Plate out 100 μl of each dilution on MS agar (Kieser *et al.*, 2000) +10 m*M* $MgCl_2$ (without antibiotics) and incubate at 30 °C for 16–20 h.

   f. Overlay the plate with 1 ml water containing 0.5 mg nalidixic acid and 1.25 mg apramycin (25 μl of 50 mg/ml stock). Use a spreader to lightly distribute the antibiotic solution evenly. Continue incubation at 30 °C.

   g. Replica–plate each MS agar plate with single colonies onto DNA plates containing nalidixic acid (25 μg/ml) and apramycin (50 μg/ml) with and without kanamycin (50 μg/ml). Double–crossover exconjugants are kanamycin^S and apramycin^R. Kanamycin^S clones are picked from the DNA plates and streaked for single colonies on MS agar containing nalidixic acid (25 μg/ml) and apramycin (50 μg/ml). Confirm kanamycin sensitivity by replica–plating onto DNA plates containing nalidixic acid (25 μg/ml) with and without kanamycin (50 μg/ml). Purified kanamycin–sensitive strains are then verified by PCR and Southern blot analysis.

8. *FLP-mediated excision of the disruption cassette*. The disruption cassettes are flanked by FRT sites (FLP recognition targets; Fig. 7.2). Expression of the FLP-recombinase in *E. coli* removes the central part of the disruption cassette, leaving behind a 81–bp "scar" sequence lacking stop codons (Fig. 7.3). This allows the generation of nonpolar, unmarked in–frame deletions and repeated use of the same resistance marker for making multiple knockouts in the same cosmid or in the same strain.

a. *E. coli* DH5α cells containing the temperature–sensitive FLP recombination plasmid BT340 (Datsenko and Wanner, 2000) are transformed with the mutated cosmid DNA. BT340 contains ampicillin and chloramphenicol resistance determinants and is temperature–sensitive for replication (replicates at 30 °C). FLP synthesis and loss of the plasmid are induced at 42 °C (Cherepanov and Wackernagel, 1995).
  b. Grow *E. coli* DH5α/BT340 overnight, with shaking at 30 °C in 10 ml LB containing chloramphenicol (25 μg/ml).
  c. Inoculate 100 μl *E. coli* DH5α/BT340 from the overnight culture into 10 ml LB containing chloramphenicol (25 μg/ml). Grow for 3–4 h at 30 °C shaking at 200 rpm to an OD_{600} of ~0.4.
  d. Recover the cells and wash with ice–cold 10% glycerol as described above. Mix 50 μl cell suspension with ~100 ng of mutated cosmid DNA and carry out electroporation as described above. Immediately add 1 ml ice–cold LB to shocked cells and incubate with shaking for 1 h at 30 °C. Spread onto LB agar containing apramycin (50 μg/ml) and chloramphenicol (25 μg/ml).
  e. Incubate for 2 days at 30 °C (*E. coli* DH5α/BT340 grows slowly at 30 °C). A single colony is streaked for single colonies on an LB agar plate without antibiotics and grown overnight at 42 °C to induce expression of the FLP recombinase followed by the loss of plasmid BT340.
  f. Make two master plates by streaking 20–30 single colonies with a toothpick, first on LB agar containing apramycin (50 μg/ml) and then on LB agar containing kanamycin (50 μg/ml). Grow the master plates overnight at 37 °C. Apramycin^S kanamycin^R clones indicate the successful loss of the resistance cassette and are further verified by restriction, PCR, or sequencing analysis.

9. *Replacing the resistance cassette inserts in Streptomyces with the unmarked "scar" sequence.* This is achieved by homologous recombination between the chromosome and the corresponding "scar cosmid" prepared in Point 8, Section 3. The procedure differs from Point 7, Section 3 because the cosmid now lacks an *oriT*, and the desired product is antibiotic–sensitive. Therefore, it is necessary to introduce the scar cosmid into *Streptomyces* by protoplast transformation or triparental conjugation, and then select for kanamycin resistant *Streptomyces* containing the entire scar cosmid integrated by a single crossover. Restreaking to kanamycin–free medium, followed by screening for concomitant loss of kanamycin resistance and apramycin resistance, then identifies the desired *Streptomyces* clones.
10. *An alternative to FLP recombinase.* It is possible to create unmarked in–frame deletion mutants using restriction/ligation methodology (Fig. 7.2). To do

this requires the addition of restriction sites to the PCR primers between the 39–bp section of the long primers and the universal priming sites P1 and P2. It is recommended that the restriction enzymes are used in combination. There are four rare–cutting enzymes (i.e., for high GC DNA) that, when used in combination, generate the same single–stranded overhang but, once ligated together, form a hybrid site that cannot be cut by either enzyme. The four enzymes are *Spe*I, *Xba*I, *Nhe*I, and *Avr*II (try to avoid using *Avr*II if possible because of an *Avr*II restriction site in the SuperCos1 backbone). Conventional use of the FLP recombinase is still available with such primers, but will create a 93–bp scar sequence instead of the conventional 81-bp scar. Follow the protocol described for making gene replacements in a cosmid and once the mutant cosmid has been checked by restriction analysis, the in–frame deletion can be constructed.

a. Perform a double digest with the selected rare–cutting enzymes. Run out an aliquot of the digest on an agarose gel to check complete digestion. A large band (~50 kb) and a smaller band (1.4 kb) representing the disruption cassette will be present on the gel.
b. Purify by phenol/chloroform extraction and precipitation and resuspend the pellet in 10 μl 10 mM Tris–HCl (pH 8). Set up a ligation and incubate at 16 °C overnight.
c. Transform DH5α competent cells and plate onto LB containing kanamycin (50 μg/ml) and carbenicillin (100 μg/ml). Analysis of the colonies can be performed as described in Point 8, Section 3.

4. *IN VITRO* TRANSPOSON MUTAGENESIS

A more systematic approach to insertional mutagenesis of a *Streptomyces* genome has been developed by the group of Paul Dyson at Swansea University, UK (Bishop *et al.*, 2004). This high–throughput method relies on *in vitro* transposon mutagenesis of a cosmid library and was first established for the SuperCos1–based ordered cosmid library from *S. coelicolor* M145 (Redenbach *et al.*, 1996). A Tn5–based transposon, Tn*5062* (Fig. 7.4), was engineered for this purpose. It contains the following elements: a promotorless *egfp* reporter gene containing a consensus streptomycete ribosome–binding site (RBS) to detect transcriptional fusions with a promoter of a disrupted gene; an apramycin resistance gene (*aac3(IV)*) for selection, flanked by two T4 transcriptional terminators; and an *oriT* from the broad host–range plasmid RK2 to allow mobilization of a mutagenised cosmid by intergeneric conjugation. Two Tn*5* mosaic ends (ME) flank a three–framed translational stop.

Gene disruption is achieved by obtaining insertion libraries of Tn*5062* within a library clone, for example, a cosmid clone, which is recovered

Figure 7.4 Organization of Tn*5062*; ME = Mosaic ends; stop = translational stop in all three reading frames; RBS = consensus *streptomyces* ribosome binding site; *egfp* = promoter-less copy of the green fluorescent protein gene; *aac3(IV)* = apramycin resistance gene; *oriT* = RK2 origin of transfer; T_4 = T4 transcriptional terminator. EZL2, EZR1 = position of the sequencing primers to determine the precise location of the insertion. The sequence of Tn*5062* has the accession number AJ566337.

initially in *E. coli*. The location of each insertion is then determined by sequencing using Tn*5062*–specific primers EZL2 and EZR1 (Fig. 7.4). The DNA sequence files are then processed to identify the precise site of each Tn*5062* insertion using the software package Transposon Express (Herron et al., 2004). These data are available for the complete genome of *S. coelicolor* and are incorporated into the *Streptomyces* database StrepDB at http://strepdb.streptomyces.org.uk. Individual Tn*5062* insertion are marked as a triangle on StrepDB (Fig. 7.5) and further information, such as the exact position within the genome or the corresponding cosmid and a possible fusion with *egfp*, is obtained by a mouse click on a given triangle.

Generation of an *in vitro* transposon library involves the following steps.

1. *In vitro reaction*. The 3442–bp Tn*5062* *Pvu*II fragment is isolated from an agarose gel and purified by a gel extraction and further PCR purification kit before being resuspended in distilled water. The following reaction mix is prepared according to the manufacturers' instructions (Epicentre) and incubated for 2 h at 37 °C: 1 µl of EZ::Tn 10× Reaction buffer (Epicentre); 200 ng of cosmid DNA; 17.5 ng Tn*5062* DNA; 1 µl of EZ::Tn transposase (Epicentre) and sterile distilled water to a final volume of 10 µl. After completion, the reaction is stopped by adding 1 µl of EZ::Tn stop solution (Epicentre) and incubation at 70 °C for 10 min.
2. *Transformation*. 1 µl of the transposition reaction is introduced by electroporation into competent *E. coli* JM109 (Sambrook and Russell, 2001) and plated onto LB agar plates containing apramycin (50 µg/ml) and the appropriate antibiotics to which resistance is encoded by the vector backbone.
3. *Library clone extraction*. A total of 96 colonies are picked and transferred to individual wells of a 96–square deep–well growth block (ABgene), each well containing 1 ml of L broth and apramycin (50 µg/ml), followed by overnight incubation at 37 °C at 225 rpm. 1.3 µl from each well is transferred to a second growth block containing 1.3 ml L broth and apramycin (50 mg/ml) and incubated for 18 h at 37 °C at 225 rpm.

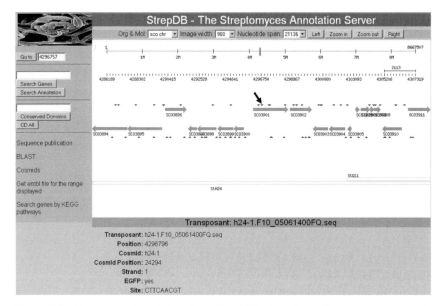

Figure 7.5 Representative window from StrepDB at http://strepdb.streptomyces.org.uk showing as triangles transposon insertions. A mouse click on a given triangle (indicated by the arrow) opens the lower window, revealing further data on the insertion.

330 μl of 60% (w/v) glycerol is added to each well of the first growth block, mixed and stored at −70 °C. Cosmid DNA is isolated from the second growth block according to the Wizard SV 96 protocol (Promega) and stored at −70 °C.

4. *Sequencing and sequence analysis.* Flanking sequences to Tn*5062* insertions are obtained by using a standard automated sequencer and the associated dideoxy–cycle sequencing chemistry using the Tn*5062*–specific primers EZL2 and EZR1 (Fig. 7.4). To identify the exact site of Tn*5062* insertions, each sequence needs to be compared with the sequence of the appropriate library clone selected for mutagenesis. In addition, the sequence up to the outside base of the inverted repeat (Mosaic end) should be identified. The aforementioned software, Transposon Express (Herron *et al.*, 2004), allows rapid location of transposon insertions and is available at http://www.swan.ac.uk/genetics/dyson/TExpress_home.htm.

5. *Construction of specific transposon insertion mutants in Streptomyces.* The mutated cosmids can be transferred from *E. coli* to *Streptomyces* by intergeneric conjugation (Mazodier *et al.*, 1989; Paget *et al.*, 1999) as described in Point 7, Section 3. Allelic replacements via double crossing over are validated by Southern hybridization analysis of independent mutants.

5. HETEROLOGOUS EXPRESSION OF BIOSYNTHETIC GENE CLUSTERS

Heterologous expression of biosynthetic gene clusters offers some principal advantages. First, the successful cloning of an entire biosynthetic gene cluster can be verified by its heterologous expression in a host that originally does not produce the desired compound. Accumulation of the desired natural product in a heterologous host gives evidence that all genes required for biosynthesis, regulation, and possible resistance are contained in one library clone. Second, rapid genetic manipulations of the clusters by Red/ET–mediated recombination can be performed in *E. coli* before integration into the genome of a *Streptomyces* strain that is more difficult to manipulate genetically. Third, working in a completely sequenced host offers the potential to influence antibiotic production by metabolic engineering. Below, the modification of a SuperCos1–based cosmid is described for heterologous expression in *Streptomyces*. For this purpose, a cassette (pIJ787) was constructed (Eustaquio *et al.*, 2005; Gust *et al.*, 2004) that makes use of the integration functions of phage ΦC31 (Thorpe and Smith, 1998), allowing for site–specific integration of the modified cosmid containing the entire gene cluster into a *Streptomyces* genome, for example, *S. coelicolor M512* (Δ*redD* Δ*actII–ORF4* SCP1⁻ SCP2⁻) (Floriano and Bibb, 1996) (Fig. 7.6). Therefore, pIJ787 contains the following elements: attP and *int* (coding for the ΦC31 integrase) for stable integration into a bacterial attB site; *oriT* for conjugal transfer of the cosmid into *Streptomyces*; and the tetracycline resistance gene *tet*. The cassette is flanked by about 100 bp

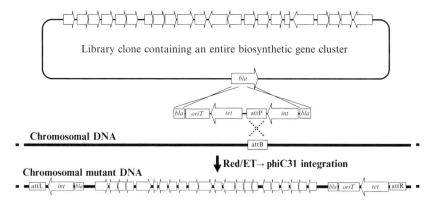

Figure 7.6 Schematic representation of the generation of an integrative library clone and site-specific integration of a gene cluster into a *Streptomyces* genome; *bla* = betalactamase; *oriT* = RK2 origin of transfer; *tet* = tetracycline resistance gene; *int* = phiC31 integrase; attP/attB = phiC31 attachment site of phage/bacteria; Red/ET = Red/ET-mediated recombination.

of β-lactamase (*bla*) sequence on one side and about 300 bp of *bla* sequence on the other side, homologous to *bla* of the SuperCos1 backbone for Red/ ET–mediated recombination. The procedure involves the following steps (Fig. 7.6).

1. The 4990–bp *Dra*I–*Bsa*I fragment of pIJ787 containing the aforementioned elements is isolated from an agarose gel and purified by a PCR purification kit as mentioned in Point 1, Section 3.
2. Red/ET–mediated recombination is performed as described in Points 4 and 5, Section 3. Notably, 200–500 ng of the integration cassette from pIJ787 is used for electroporation of competent BW25113/pIJ790 containing the cosmid with an entire gene cluster and LB plates containing only 5 μg/ml thiostrepton are used for selection of positive clones [*tet* in pIJ787 is under control of the *aac3(IV)* promoter (Eustaquio *et al.*, 2005)].
3. Positive clones are verified by restriction analysis as described in Point 6, Section 3.
4. Because of the potent methylation-restriction system of *S. coelicolor*, cosmid DNA is passed through *E. coli* ET12567 as a nonmethylating host (MacNeil *et al.*, 1992) prior to introduction into *S. coelicolor M512*.
5. The integrative cosmid, still carrying the kanamycin resistance gene *neo*, is then introduced into *S. coelicolor* M512 via polyethylene glycol–mediated protoplast transformation (Kieser *et al.*, 2000). Kanamycin-resistant clones are checked for site–specific integration into the genome, by either PCR or Southern blot analysis.

6. Reassembling Entire Gene Clusters by "Stitching" Overlapping Cosmid Clones

Many secondary product biosynthetic pathways entail dozens and sometimes up to over 100 individual biosynthetic steps (Fujimori *et al.*, 2007; Karray *et al.*, 2007). While many of the corresponding gene clusters have been cloned and sequenced, genomic library clones rarely contain an entire biosynthetic gene cluster due to the limitation of the average insert size of common cloning vectors. Since many of the producing microorganisms are difficult to handle genetically, only few manipulation techniques are established for these strains. For this reason, manipulation and heterologous expression of complete pathways in sequenced and genetic accessible strains is desirable. The application of Red/ET recombination for the functional reconstitution of large natural product biosynthetic gene clusters from overlapping inserts of separate clones was first described for myxobacterial metabolites (Perlova *et al.*, 2006; Wenzel *et al.*, 2005), more precisely for the so called "stitching" of the 57-kb myxothiazole and the 43-kb myxochromid gene clusters. Recent reports demonstrated the successful stitching of the 42-kb phenalinolactone (Binz et al., 2008), the 32.5-kb antramycin (Hu et al., 2008), the 38.5-kb coumermycin A_1 (Wolpert et al., 2008), and the 58-kb epothilone (Fu et al., 2008) biosynthetic gene clusters. A general protocol is described below based on the retrofitting of each library clone with a different selectable marker cassette that shares identical sequences of more than 500 bp between the priming site P1 and the resistance marker, providing recombination sites for Red/ET-mediated recombination. Assembling an entire gene cluster into one single cosmid involves the following steps (Fig. 7.7).

1. *PCR-targeted gene replacement in library clone 1.* A resistance marker cassette, for example, from pIJ773, is amplified by using primers with added restriction sites R_2 and R_3 between the resistance cassette and the 39-bp flanking sequences for PCR-targeted replacement (following the procedure described in Point 3, Section 3.). Natural restriction sites for R_2 and R_3 do not occur within the library clone 1 and 2 (e.g., $R_2 = Xba$I and $R_3 = Spe$I or $R_2 = Pme$I and $R_3 = Swa$I) but should produce compatible ends after restriction with the corresponding enzyme. Any sequence upstream of the first gene of the gene cluster can be deleted during Red/ET-mediated recombination (Fig. 7.7; following the procedure described in Points 4 and 5, Section 3).
2. *PCR-targeted gene replacement in library clone 2.* In the same manner as shown in Point 1, Section 6, a resistance marker cassette, for example, from pIJ778, is amplified and used for Red/ET-mediated recombination to delete the first gene of the overlapping region in the library clone 2.
3. *Recombination of the entire gene cluster on a single library clone.* The modified library clone 1 ("donor clone") is digested with restriction enzymes recognizing restriction sites R_2 and R_1. R_1 represents a unique restriction site present in the overlapping region of both library clones. The gel-purified restriction fragment from the "donor clone" is used for a Red/ET-mediated recombination with the modified library clone 2 ("recipient clone") according to the procedure described in Points 4 and 5, Section 3. Due to the identical sequences between the two resistance marker cassettes and the overlapping region of the library clone inserts, recipient clones will lose the resistance marker 2, for example, streptomycin, and will be selectable by the incoming resistance marker 1, for example, apramycin, after Red/ET recombination.
4. *Elimination of the resistance marker cassette.* Restriction with enzymes recognizing restriction sites R_2 and R_3, followed by religation and transformation of *E. coli* competent cells (according to Point 10, Section 3), will finally result in a library clone containing only a 6-bp "scar" at the region of the resistance marker cassette exchange. Such clones can be manipulated further for heterologous expression (Section 5) or used for gene inactivation experiments (Point 2, Section 3 and Point 3, Section 4).

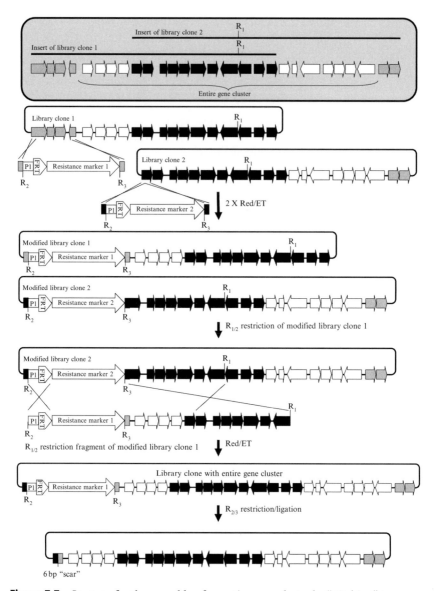

Figure 7.7 Strategy for the assembly of an entire gene cluster by "stitching" two overlapping library clones. Black arrows indicate genes on the overlapping sequence of the two library inserts and white arrows represent genes of the gene cluster within the non-overlapping region. Grey arrows are genes not involved in natural product formation; R_1 = restriction site within the overlapping region; R_2/R_3 = compatible restriction sites that do not cut within both library clones; P1 = priming site for resistance marker cassette; FRT = FLP recognition target; Red/ET = Red/ET-mediated recombination.

7. Conclusions

The cloning and sequencing of natural product pathways is now relatively routine. Most gene clusters from *Streptomyces* have been cloned on cosmid vectors. However, BAC vectors are usually more versatile because they can accommodate large secondary metabolic pathways spanning 100 kb and more. Additional functions for conjugal transfer and site-specific integration has facilitated the movement of engineered pathways from *E. coli* into *Streptomyces* expression hosts. For genetic engineering of biosynthetic gene clusters Red/ET-mediated recombination has been adapted and exploited for genetic manipulation of complex pathways, predominantly for the generation of precise gene inactivations, eliminating most of the time-consuming steps in standard genetic procedures. In addition, a high throughput *in vitro* transposon mutagenesis approach allows the systematic inactivation of nearly all genes within a library clone, or even a whole genome, in a feasible manner. However, some natural producers cannot be easily manipulated genetically and in these cases, heterologous hosts such as *S. coelicolor, S. lividans*, or *S. albus* are needed. For the production of the desired natural product, a genomic library clone containing all genes required for its biosynthesis, regulation, and self-resistance is required. If the natural product pathway is not covered by a single library clone, such a clone can be generated by assembling two (or more) clones with overlapping inserts harboring an entire gene cluster. Heterologous expression in well-studied and genetically accessible hosts can then provide the basis for increasing production rates by metabolic engineering, or to alter drug structures by genetic engineering.

ACKNOWLEDGMENTS

The author gratefully acknowledges contributions of many colleagues, coworkers and collaborators in the development of the methods described here, especially of K.F. Chater, T. Kieser, P. Dyson, A. Bishop, P. Herron, L. Heide, A. Eustáquio, M. Wolpert, and many others. The work was financially supported by the Biotechnology and Biological Sciences Research Council (BBSRC), the John Innes Foundation, the Deutsche Forschungsgemeinschaft (DFG), and from the European Community (IP 005244 ActinoGEN).

REFERENCES

Ahlert, J., Shepard, E., Lomovskaya, N., Zazopoulos, E., Staffa, A., Bachmann, B. O., Huang, K., Fonstein, L., Czisny, A., Whitwam, R. E., *et al.* (2002). The calicheamicin gene cluster and its iterative type I enediyne PKS. *Science* **297,** 1173–1176.

Alduina, R., De Grazia, S., Dolce, L., Salerno, P., Sosio, M., Donadio, S., and Puglia, A. M. (2003). Artificial chromosome libraries of *Streptomyces coelicolor* A3(2) and *Planobispora rosea. FEMS Microbiol. Lett.* **218,** 181–186.

Bentley, S. D., Chater, K. F., Cerdeno-Tarraga, A. M., Challis, G. L., Thomson, N. R., James, K. D., Harris, D. E., Quail, M. A., Kieser, H., Harper, D., et al. (2002). Complete genome sequence of the model actinomycete *Streptomyces coelicolor* A3(2). *Nature* **417**, 141–147.

Bierman, M., Logan, R., O'Brien, K., Seno, E. T., Rao, R. N., and Schoner, B. E. (1992). Plasmid cloning vectors for the conjugal transfer of DNA from *Escherichia coli* to *Streptomyces* spp. *Gene* **116**, 43–49.

Binz, T. M., Wenzel, S. C., Schnell, H. J., Bechthold, A., and Muller, R. (2008). Heterologous expression and genetic engineering of the phenalinolactone biosynthetic gene cluster by using red/ET recombineering. *Chembiochem* **9**, 447–454.

Bishop, A., Fielding, S., Dyson, P., and Herron, P. (2004). Systematic insertional mutagenesis of a streptomycete genome: A link between osmoadaptation and antibiotic production. *Genome Res.* **14**, 893–900.

Cherepanov, P. P., and Wackernagel, W. (1995). Gene disruption in *Escherichia coli*: TcR and KmR cassettes with the option of Flp-catalyzed excision of the antibiotic-resistance determinant. *Gene* **158**, 9–14.

Court, D. L., Sawitzke, J. A., and Thomason, L. C. (2002). Genetic engineering using homologous recombination. *Annu. Rev. Genet.* **36**, 361–388.

Datsenko, K. A., and Wanner, B. L. (2000). One-step inactivation of chromosomal genes in *Escherichia coli* K-12 using PCR products. *Proc. Natl. Acad. Sci. USA* **97**, 6640–6645.

Eustaquio, A. S., Gust, B., Galm, U., Li, S. M., Chater, K. F., and Heide, L. (2005). Heterologous expression of novobiocin and clorobiocin biosynthetic gene clusters. *Appl. Environ. Microbiol.* **71**, 2452–2459.

Flett, F., Mersinias, V., and Smith, C. P. (1997). High efficiency intergeneric conjugal transfer of plasmid DNA from *Escherichia coli* to methyl DNA-restricting streptomycetes. *FEMS Microbiol. Lett.* **155**, 223–229.

Floriano, B., and Bibb, M. (1996). afsR is a pleiotropic but conditionally required regulatory gene for antibiotic production in *Streptomyces coelicolor* A3(2). *Mol. Microbiol.* **21**, 385–396.

Fu, J., Wenzel, S. C., Perlova, O., Wang, J., Gross, F., Tang, Z., Yin, Y., Stewart, A. F., Muller, R., and Zhang, Y. (2008). Efficient transfer of two large secondary metabolite pathway gene clusters into heterologous hosts by transposition. *Nucleic Acids Res.* **36**, e113.

Fujimori, D. G., Hrvatin, S., Neumann, C. S., Strieker, M., Marahiel, M. A., and Walsh, C. T. (2007). Cloning and characterization of the biosynthetic gene cluster for kutznerides. *Proc. Natl. Acad. Sci. USA* **104**, 16498–16503.

Gust, B., Challis, G. L., Fowler, K., Kieser, T., and Chater, K. F. (2003). PCR-targeted *Streptomyces* gene replacement identifies a protein domain needed for biosynthesis of the sesquiterpene soil odor geosmin. *Proc. Natl. Acad. Sci. USA* **100**, 1541–1546.

Gust, B., Chandra, G., Jakimowicz, D., Yuqing, T., Bruton, C. J., and Chater, K. F. (2004). l RED-mediated genetic manipulation of antibiotic-producing *Streptomyces*. *Adv. Appl. Microbiol.* **54**, 107–128.

Herron, P. R., Hughes, G., Chandra, G., Fielding, S., and Dyson, P. J. (2004). Transposon Express, a software application to report the identity of insertions obtained by comprehensive transposon mutagenesis of sequenced genomes: Analysis of the preference for in vitro Tn5 transposition into GC-rich DNA. *Nucleic Acids Res.* **32**, e113.

Hu, Y., Phelan, V. V., Farnet, C. M., Zazopoulos, E., and Bachmann, B. O. (2008). Reassembly of anthramycin biosynthetic gene cluster by using recombinogenic cassettes. *Chembiochem* **9**, 1603–1608.

Karray, F., Darbon, E., Oestreicher, N., Dominguez, H., Tuphile, K., Gagnat, J., Blondelet-Rouault, M. H., Gerbaud, C., and Pernodet, J. L. (2007). Organization of the biosynthetic gene cluster for the macrolide antibiotic spiramycin in *Streptomyces ambofaciens*. *Microbiology* **153**, 4111–4122.

Kieser, H., Bibb, M. J., Buttner, M. J., Chater, K. F., and Hopwood, D. A. (2000). Practical Streptomyces Genetics. The John Innes Foundation.

Kretz, P. L., Reid, C. H., Greener, A., and Short, J. M. (1989). Effect of lambda packaging extract mcr restriction activity on DNA cloning. *Nucleic Acids Res.* **17**, 5409.

MacNeil, D. J., Gewain, K. M., Ruby, C. L., Dezeny, G., Gibbons, P. H., and MacNeil, T. (1992). Analysis of *Streptomyces avermitilis* genes required for avermectin biosynthesis utilizing a novel integration vector. *Gene* **111**, 61–68.

Matsushima, P., Broughton, M. C., Turner, J. R., and Baltz, R. H. (1994). Conjugal transfer of cosmid DNA from *Escherichia coli* to *Saccharopolyspora spinosa*: Effects of chromosomal insertions on macrolide A83543 production. *Gene* **146**, 39–45.

Mazodier, P., Petter, R., and Thompson, C. (1989). Intergeneric conjugation between *Escherichia coli* and *Streptomyces* species. *J. Bacteriol.* **171**, 3583–3585.

Muyrers, J. P., Zhang, Y., and Stewart, A. F. (2000). ET-cloning: Think recombination first. *Genet Eng. (N Y)* **22**, 77–98.

Muyrers, J. P., Zhang, Y., and Stewart, A. F. (2001). Techniques: Recombinogenic engineering–new options for cloning and manipulating DNA. *Trends Biochem. Sci.* **26**, 325–331.

Muyrers, J. P., Zhang, Y., Testa, G., and Stewart, A. F. (1999). Rapid modification of bacterial artificial chromosomes by ET-recombination. *Nucleic Acids Res.* **27**, 1555–1557.

Paget, M. S., Chamberlin, L., Atrih, A., Foster, S. J., and Buttner, M. J. (1999). Evidence that the extracytoplasmic function sigma factor sigmaE is required for normal cell wall structure in *Streptomyces coelicolor* A3(2). *J. Bacteriol.* **181**, 204–211.

Perlova, O., Fu, J., Kuhlmann, S., Krug, D., Stewart, A. F., Zhang, Y., and Muller, R. (2006). Reconstitution of the myxothiazol biosynthetic gene cluster by Red/ET recombination and heterologous expression in *Myxococcus xanthus*. *Appl. Environ. Microbiol.* **72**, 7485–7494.

Poteete, A. R. (2001). What makes the bacteriophage lambda Red system useful for genetic engineering: Molecular mechanism and biological function. *FEMS Microbiol. Lett.* **201**, 9–14.

Redenbach, M., Kieser, H. M., Denapaite, D., Eichner, A., Cullum, J., Kinashi, H., and Hopwood, D. A. (1996). A set of ordered cosmids and a detailed genetic and physical map for the 8 Mb *Streptomyces coelicolor* A3(2) chromosome. *Mol. Microbiol.* **21**, 77–96.

Rose, T. M., Schultz, E. R., Henikoff, J. G., Pietrokovski, S., McCallum, C. M., and Henikoff, S. (1998). Consensus-degenerate hybrid oligonucleotide primers for amplification of distantly related sequences. *Nucleic Acids Res.* **26**, 1628–1635.

Sambrook, J., and Russell, D. W. (2001). Molecular Cloning. A Laboratory Manual. Cold Spring Harbor Laboratory Press, New York.

Sosio, M., Giusino, F., Cappellano, C., Bossi, E., Puglia, A. M., and Donadio, S. (2000). Artificial chromosomes for antibiotic-producing actinomycetes. *Nat. Biotechnol.* **18**, 343–345.

Thorpe, H. M., and Smith, M. C. (1998). In vitro site-specific integration of bacteriophage DNA catalyzed by a recombinase of the resolvase/invertase family. *Proc. Natl. Acad. Sci. USA* **95**, 5505–5510.

Van Lanen, S. G., Oh, T. J., Liu, W., Wendt-Pienkowski, E., and Shen, B. (2007). Characterization of the maduropeptin biosynthetic gene cluster from *Actinomadura madurae* ATCC 39144 supporting a unifying paradigm for enediyne biosynthesis. *J. Am. Chem. Soc.* **129**, 13082–13094.

Wenzel, S. C., Gross, F., Zhang, Y., Fu, J., Stewart, A. F., and Muller, R. (2005). Heterologous expression of a myxobacterial natural products assembly line in pseudomonads via red/ET recombineering. *Chem. Biol.* **12**, 349–356.

Wolpert, M., Heide, L., Kammerer, B., and Gust, B. (2008). Assembly and heterologous expression of the coumermycin A1 gene cluster and production of new derivatives by genetic engineering. *Chembiochem* **9,** 603–612.

Zhang, Y., Buchholz, F., Muyrers, J. P., and Stewart, A. F. (1998). A new logic for DNA engineering using recombination in *Escherichia coli. Nat. Genet.* **20,** 123–128.

Zhang, Y., Muyrers, J. P., Testa, G., and Stewart, A. F. (2000). DNA cloning by homologous recombination in *Escherichia coli. Nat. Biotechnol.* **18,** 1314–1317.

CHAPTER EIGHT

Methods for *In Silico* Prediction of Microbial Polyketide and Nonribosomal Peptide Biosynthetic Pathways from DNA Sequence Data

Brian O. Bachmann* *and* Jacques Ravel[†]

Contents

1. Introduction	182
1.1. PKS and NRPS as model systems	184
1.2. Software for PKS/NRPS domain analysis	184
2. Converting Type I PKSs to Structural Elements	191
2.1. The mechanistic approach	191
2.2. Heuristic approach to PKS structure prediction	193
2.3. Conserved active-site motifs	193
2.4. Caveats	194
2.5. Methods for drawing polyketides based on PKS domain strings	196
2.6. Consensus motif analysis	200
2.7. Ketoreductase (KR) domain analysis	201
2.8. Deriving possible polyketide structures from the domain strings	202
2.9. Polyketide case studies	203
3. Converting NRPS Domain Strings to Structural Elements	205
3.1. Polypeptide chain length and amino acid side chains	206
3.2. Modifications and stereochemistry	206
3.3. Caveats	207
3.4. Method for rendering NRPs	208
3.5. NRPS Case studies	210
4. Concluding Remarks	212
Acknowledgments	214
References	214

* Department of Chemistry, Vanderbilt Institute for Chemical Biology, Vanderbilt University, Nashville, Tennessee, USA
[†] Institute for Genomic Sciences, Department of Microbiology and Immunology, University of Maryland School of Medicine, Baltimore, Maryland, USA

Methods in Enzymology, Volume 458   © 2009 Elsevier Inc.
ISSN 0076-6879, DOI: 10.1016/S0076-6879(09)04808-3   All rights reserved.

Abstract

Foreknowledge of the secondary metabolic potential of cultivated and previously uncultivated microorganisms can potentially facilitate the process of natural product discovery. By combining sequence-based knowledge with biochemical precedent, translated gene sequence data can be used to rapidly derive structural elements encoded by secondary metabolic gene clusters from microorganisms. These structural elements provide an estimate of the secondary metabolic potential of a given organism and a starting point for identification of potential lead compounds in isolation/structure elucidation campaigns. The accuracy of these predictions for a given translated gene sequence depends on the biochemistry of the metabolite class, similarity to known metabolite gene clusters, and depth of knowledge concerning its biosynthetic machinery. This chapter introduces methods for prediction of structural elements for two well-studied classes: modular polyketides and nonribosomally encoded peptides. A bioinformatics tool is presented for rapid preliminary analysis of these modular systems, and prototypical methods for converting these analyses into substructural elements are described.

1. Introduction

Microbial genome sequencing has resulted in an embarrassment of riches in terms of the sheer volume of cryptic gene clusters that apparently encode secondary metabolism. These efforts have revealed the untapped potential of microorganisms as producers of secondary metabolic diversity. The richness of this diversity underlines its importance in understanding how microorganisms adapt and thrive in various environments, be they symbiotic or solitary (Clardy, 2005; Schmidt, 2008). Recent analysis also suggests that natural product structural diversity is highly complementary to the chemical space carved out by synthetic molecules for bio-probe discovery and drug development efforts (Koch et al., 2004, 2005). Hence, there is great interest in translating genomic sequence data into accurate predictions of secondary metabolite structures, which can accelerate their subsequent isolation, investigation, and application (Challis, 2008a,b).

The chemical structure of natural products is by and large determined by the sequence of enzymatic reactions involved in their assembly. The biochemical activity of individual enzymes in these sequences (pathways) is in turn determined by their tertiary and quaternary architectures, which are largely a function of their primary amino acid sequence. While understanding of the fundamental sequence determinants of protein folding and their application to prediction of protein structure from primary sequence data are still developing research areas (Moult et al., 2007), global and local sequence similarities in proteins (homologies) can be measured with great statistical rigor and are often used as indicators of similarities in local and global structure, and therefore function, of uncharacterized proteins. For instance, variants of the BLAST algorithm, which identify and score regions of sequence similarity between nucleic acids and/or proteins, have become pre-eminent methods for analyzing local sequence similarity (McGinnis and Madden, 2004). Other techniques, such as Hidden Markov Models (HMMs), are also used to identify conserved sequence motifs for protein and conserved domain identification (Bystroff and Krogh, 2008).

In principle, if the functions of proteins in cryptic gene clusters can be accurately estimated based on sequence precedents, then hypotheses concerning the concerted action of these enzymes in a biosynthetic pathway can be formulated. By applying the extant logic of natural product biosynthesis, well-defined structural elements of secondary metabolites can be approximated from these hypotheses. In practice, the accuracy of these predictions is a function of (1) the extent of experimentally validated sequence precedents, (2) the correlation of sequence similarity to fold and function for the given protein, and (3) the depth of knowledge of the putative secondary metabolite's pathway logic.

It is therefore easier to predict structural elements of secondary metabolites from cryptic gene clusters if they are similar to those of well-studied classes of natural products. Some classes of metabolites are amenable to this sort of analysis while for others it is only possible to infer general features. For instance, the sequence hallmarks of aromatic polyketides and polysaccharides are unmistakable (iterative polyketide synthase domains and glycosyltransferases, respectively), but the extent and ordering of the biotransformations in these pathways is currently challenging to predict (Hertweck et al., 2007). In other systems, it is possible to make fairly accurate predictions. For instance, the structures of some varieties of reduced polyketides and both ribosomally and nonribosomally encoded peptide natural products are generally easier to infer, as substantially more is known about their individual biochemistries.

Structural prescience regarding the secondary metabolic potential of organisms can be leveraged to great utility in isolation/elucidation campaigns. Progress in this area has been recently reviewed and new structures of diverse structural types have been reported, including polyketides, peptides, and terpenes (Challis, 2008a,b). Commercially, this technology has been successfully employed as a platform for industrial natural product discovery at Thallion Pharmaceuticals (formerly Ecopia Biosciences) (Farnet, 2005), and products of these efforts have entered Phase II clinical trials. In addition to prioritizing natural product discovery efforts, structural foreknowledge can simplify structure elucidation. For example, in the case of ECO-2301, a 1297-Da polyketide, the two-dimensional structure was predicted prior to isolation with a high degree of accuracy (McAlpine et al., 2005).

This article provides prototypical methods for prediction of two generally amenable classes, reduced polyketides (type I PKSs) and nonribosomally encoded peptides (NRPs). Recognizing that a single comprehensive approach cannot be applied successfully to all systems, these methods represent case studies illustrating strategies for structure prediction from translated gene sequence data. Polyketides and peptides are chosen as examples as there is a comparatively large body of knowledge regarding these systems, and bioinformatics tools have been generated for their rapid analysis and parsing. Methods will be comprised of two components (1) a bioinformatics analysis of biocatalytic domains and (2) converting the deduced domain strings to putative structural elements. Case studies are provided as didactic applications of the methods described.

1.1. PKS and NRPS as model systems

Polyketides are polymers of acetate substituted at the C-2 position, with varying oxidation states at C-1 and C-2. The enzymatic systems responsible for polyketide assembly are referred to as polyketide synthases (PKSs) and can be grouped into rough classes based on their biochemical mechanisms and enzyme architecture. NRPs are relatively small peptides (2–30 amino acids) with hallmarks of nonproteinogenic amino acids and extensive "posttranslational modifications." PKSs and Nonribosomal Peptide Synthetases (NRPS) utilize complementary strategies. Substrates and intermediates are covalently bound via phosphopantetheine tethers during the assembly pathway and the order of the catalytic domains in these synth(et)ases often parallels the order of their biosynthetic pathway. Several comprehensive reviews of nonribosomal peptide and polyketide biosynthesis describe the biochemistry and architecture of the extant classes of polyketide synthase in great detail (Fischbach and Walsh, 2006; Hill, 2006; Smith and Tsai, 2007). A summary of common domains in PKSs and NRPSs, though by no means comprehensive, is found in Table 8.1. Due to the parallels between ordering of the catalytic domains in sequence space and reaction space, it is possible for some systems to convert the order of domains defined by the sequence (herein referred to as a "domain string") to a virtual biochemical reaction sequence that allows prediction, to varying degrees of accuracy, of a PKS or NRPS core scaffold.

1.2. Software for PKS/NRPS domain analysis

Relatively few bioinformatics tools are available to dissect and identify each of the domains that comprise these large multidomain enzymes. Key to performing functional assignment for these enzymes is the ability to identify each domain and to predict its catalytic function. The most commonly used tools are similarity algorithms for rapid comparison of unknown sequences

Table 8.1 Common PKS/NRPS domains

Domain	Function
AT	Acyltransferase, acylation of T domains with acyl and malonyl-CoAs
KS	Ketosynthase, C–C bond formation via Claisen condensation of acyl-S-Ts
KR	Ketoreductase, reduction of keto groups during PKS homologation
DH	Dehydratase, dehydration to $\alpha-\beta$-enoyl-T during homologation
ER	Enoylreductase, reduction of enoyl-T chains during homologation
T	Thiolation, phosphopantetheinylate acyl carrier protein shared by PKS and NRPS
Te	Thioesterase, cleavage of mature PK/NRP via macrocylization of hydrolysis
MT	Methyltransferase, methylation of PK, and N-methylation of NRP
C	Condensation, amide bond formation
A	Adenylation, amino acid activation via intermediary adenylation
E	Epimerization, epimerizing aminoacyl-S-T amino acids (also dual C/E domains)
Re	Reductase, reduction (usually terminal) of mature PK/NRP resulting in aldehyde
Cy	Cyclization, cyclase/dehydratase resulting in thiazolines and oxazolines from S,T,C
Ox	Oxidase, often converting thiazoline, oxazolines to thiazoles and oxazoles

to large sequence databases. Unfortunately, these tools do not achieve the level of resolution required for the detailed analysis of PKSs and NRPSs.

BLAST-based algorithms (Altschul et al., 1990, 1997), such as the NCBI Conserved Domain Database (CDD) and its associated search tool (CD-Search) (Marchler-Bauer and Bryant, 2004; Marchler-Bauer et al., 2009) or InterPro (Hunter et al., 2009; Mulder and Apweiler, 2008; Mulder et al., 2008), were designed for the detection of structural and functional domains in protein sequences, but are inadequate for our purpose as they often fail to identify all NRPS/PKS domains. For example, CDD analysis is unable to differentiate NRPS epimerization from condensation domains, or recognize polyketide reductive domains. To detect conserved domains in a protein sequence, the CD-Search service uses the reverse position-specific BLAST algorithm against the CDD, a collection of multiple sequence alignments and derived database search models, which represent protein domains conserved in molecular evolution. The query sequence is compared to a position-specific score matrix prepared from the underlying conserved domain alignment. While these tools are a key component of the bioinformaticist's toolkit, they can be misleading.

Dedicated tools are thus necessary to parse and assign specific functions to PKS/NRPS domains. These tools combine the use of BLAST with more precise and refined statistical models for functional assignment and substrate predictions. For NRPSs, these tools build on the independent discovery by two groups of critical residues in all known NRPS A domains that align with eight binding-pocket residues in the 3D structure of the gramicidin synthetase phenylalanine-activating A domain, GrsA (Conti *et al.*, 1997). Critical residues define sets of remarkably conserved recognition templates and allow for predictive structure-based models for the recognition of amino acid substrate by A domains to be drawn (Challis *et al.*, 2000; Stachelhaus *et al.*, 1999). These residues are identified by analyzing BLAST sequence alignments with the GrsA protein sequence. The resultant 8 or 10 residues represent a specificity-conferring code of A domains in NRPS and can be used to reliably predict the amino-acid substrate. These residues correspond to the positions 235, 236, 239, 278, 299, 301, 322, 330, 331, and 517 of the crystal structure of GrsA (1AMU, Conti *et al.*, 1997) and can be extracted by either threading or sequence alignment (Challis *et al.*, 2000; Stachelhaus *et al.*, 1999). The combination of domain identification and A domain substrate prediction forms the basis of these analytical tools. Manually curated databases of all known domains are essential to support these algorithms.

In this section, these different algorithms and databases will be described in detail.

1.2.1. Manual PKS/NRPS parsing

Manual parsing is more time-intensive than automatic parsing but may have advantages when examining systems for which automated models have not been generated. The process of manual parsing can be accomplished readily by applying pairwise sequence alignment algorithms (for instance, using the NCBI BLAST 2 sequences program) with a small sequence database of archetypal domains. Domain coordinates can be roughly estimated from aligned regions and the coordinates of domains analyzed in series recorded to produce domain strings.

The archetypal domains database can be easily generated from previously analyzed synth(et)ase sequences. Example files containing domains for PKS and NRPS systems are attached in Supplementary Material. For instance, individual FASTA format templates for KS, AT, KR, DH, ER, and T domain sequences were obtained from the erythromycin PKS for classical modular PKS systems and from the dorrigocin literature for AT-less modular PKSs. NRPS templates were generated from the tyrocidine synthetase.

1.2.2. Automated NRPS/PKS parsing

Three main automated NRPS/PKS parsing algorithms have been implemented. Each has taken a different approach to parse these large enzymes into defined functional domains. A FASTA-formatted protein sequence file of the NRPS/PKS is the input in each of these tools.

1.2.2.1. NRPS-PKS: A knowledge based resource for analyzing NRPSs and PKSs This resource is a knowledge-based tool supported by a comprehensive analysis of the sequence and structural features of experimentally characterized biosynthetic gene clusters (Ansari *et al.*, 2004; Yadav *et al.*, 2003a,b). It is composed of four integrated searchable databases for elucidating domain organization and substrate specificity of NRPSs, modular PKSs, type I iterative PKSs, and chalcone synthases (CHSs). The databases include information pertaining to domain organization, domain sequences, domain linker sequences, and specificities of domains involved in selection of starter and extender molecules. Additionally, the active-site key residue patterns of NRPS A domains and PKS AT domains are automatically extracted by the software. This information is distributed among four integrated subdatabases named NRPSDB, PKSDB, ITERDB, and CHSDB for NRPS, PKS, type I iterative PKSs, and CHSs, respectively.

The high level of manual curation of the data sets is important as domains are identified by simple BLAST analysis and by selecting the best match in these curated subdatabases. Each prediction is further characterized by searching for specific motifs, such as, for example, the DxxxxD motif for cyclization domains (Cy), which allow discrimination between C, E, and Cy domains. Domain identification is combined with both A domain and AT domain substrate prediction, implementing the specificity-conferring code previously described (Challis *et al.*, 2000; Stachelhaus *et al.*, 1999) and a novel schema for the prediction of PKS AT domain substrate selectivity, based on the structure of the *Escherichia coli* malonyl CoA acyl carrier protein transacylase (1MLA, Serre *et al.*, 1995) (Yadav *et al.*, 2003a,b).

The NRPS-PKS user-friendly web-based interface represents the front-end for two search algorithms, SEARCHNRPS and SEARCHPKS, which can be run together or as two separate programs for the analysis of hybrid PKSs/NRPSs. The graphical output combined with specific query interfaces facilitates easy identification of various domains and modules from a given polypeptide sequence. NRPS-PKS allows the user to analyze in great detail and retrieve the sequence homology of various PKS or NRPS domains extracted from the relevant databases.

One unique feature of NRPS-PKS is added functionality for modular PKS systems. By incorporating knowledge pertaining to reductive domain type and the specificity of the AT domain, automated prediction of the approximate chemical linear structure is provided. This functionality could provide valuable information in support of the manual analysis described below.

Overall, NRPS-PKS is a well-designed resource for microbial secondary metabolite gene cluster analysis. One caveat is that NRPS-PKS will only report on domains it is able to recognize and is not able at this time to provide information on potential domains of unrecognized functions.

1.2.2.2. NRPSpredictor: A domain substrate prediction using transductive support vector machines (TSVMs)

The specificity-conferring code of A domains in NRPSs as first published (Challis *et al.*, 2000; Stachelhaus *et al.*, 1999) cannot explain all known A domain substrates. It has an accuracy of about 70–80%. NRPSpredictor implements a new computational approach to predicting A domain substrate specificity (Rausch *et al.*, 2005). NRPSpredictor is freely available at http://www-ab.informatik.uni-tuebingen.de/software/NRPSpredictor.

Previously, the 8–10 specificity-conferring residues were those identified within a 5.5-Å radius around the phenylalanine bound in the A domain active site of GrsA. NRPSpredictor uses a physico-chemical fingerprint of the residues lining the active site of the enzymatic domain within an 8-Å radius around the bound substrate. Residues are encoded into feature vectors for machine learning based on a basic set of physico-chemical properties of the amino acids and utilizing an up-to-date training dataset of A domains with known specificity that was manually curated from the literature. The predictive model is based on residues extracted from 397 known and unknown specificities used to encode this physico-chemical fingerprint into normalized real-valued feature vectors based on the physico-chemical properties of the A domain substrate amino acids. Composite specificities includes large clusters, for example "apolar, aliphatic side chain" (Gly, Ala, Val, Leu, Ile, Abu, and Iva) or "aromatic side chain" (Phe, Trp, Phg, Tyr, and Bht), or "long positively charged side chain with $-NH_2$ at the end" (Orn, Lys and Arg"). It also includes small clusters such as "tiny size, hydrophobic, transition to aliphatic" (Gly and Ala) or "nonpolar aromatic ring" (Phe and Trp) or "Orn and hydroxy Orn specific" (Orn). The reliability of the models is higher than that of sequence-based analysis, and was demonstrated to correctly predict an additional 18% of the 1230 NRPS A domain currently in UniProt. None of the methods can infer specificity (with any statistical confidence) for 2.4% of the sequences, suggesting completely new types of specificity not encompassed in either model.

The input to NRPSpredictor is a FASTA- or multiFASTA-formatted protein sequence. Unfortunately, NRPSpredictor has not yet been implemented alongside NRPS domain/module identification. The output interface is not as user-friendly as other applications and currently does not provide graphical representations of the A domains and their substrates within a domain string.

1.2.2.3. PKS/NRPS analysis web site: An HMM implementation for domain parsing and a domain substrate prediction

NRPS/PKS domains share more sequence and structure similarities with each other than with other non-NRPS/PKS proteins with related functionality. In these domains, functional residues are apparently subject to different selective pressures. Multiple alignments of a sequence family (i.e., AMP binding protein family, to which NRPS A domain belong) reveal this in their pattern of conservation. Some positions are more conserved than others, and some regions of a multiple alignment seem to tolerate insertions and deletions more than other regions. Thus, it seems preferable to use position-specific information from multiple alignments when searching databases for homologous sequences. Profile methods for building position-specific scoring models from multiple alignments have been developed for this purpose (Barton, 1990; Henikoff, 1996; Taylor, 1986). Profile methods are statistical descriptions of the consensus of a multiple sequence alignment. They use *position-specific* scores for amino acids and position-specific penalties for opening and extending an insertion or deletion. Traditional pairwise alignments [e.g., those generated using BLAST (Altschul *et al.*, 1990), FASTA (Pearson and Lipman, 1988), or the Smith/Waterman algorithms (Smith and Waterman, 1981)] use position-*independent* scoring parameters. This property of profiles captures important information about the degree of conservation at various positions in the multiple alignment, and the varying degree to which gaps and insertions are permitted. Profile HMMs are statistical models of multiple protein sequence alignments, with a formal probabilistic basis. Probability theories are used to guide how the scoring parameters should be set. For an extensive review on HMMs see Eddy (1998).

In the new PKS/NRPS analysis website described herein, HMM searches are implemented using the software package HMMER (http://hmmer.janelia.org/) with a set of HMMs specifically built for NRPS/PKS domain families. HMMER is a freely distributable implementation of profile HMM software for protein sequence analysis. Multiple alignments were built that included characterized NRPS/PKS domains, which were used as seed to build domain-specific HMMs. Scoring cut-offs were established by running these HMMs against related protein families. The PKS/NRPS analysis web-site implements these HMMs for NRPS/PKS domain identification. HMMs perform better than BLAST-based searches in differentiating highly related domains, such as NRPS condensation, epimerization, and cyclization domains. Because, the set of HMMs is small, the searches are fast. HMMs can be refined when new biochemically characterized domains become available and new HMMs can be rapidly generated. A detailed description of these HMMs is located at the resource web-site (http://nrps.igs.umaryland.edu/nrps/). The implementation is simple and rapid, and the graphical output is user-friendly. Unlike the aforementioned

NRPS-PKS parsers, the domains of unknown function are identified and graphically represented. The protein sequence of these domains is easily accessible for further analysis. An example of the output is shown in Fig. 8.1.

The HMM PKS/NRPS analysis pipeline implements a BLAST-based substrate prediction of NRPS A domains, based on an extensive database of eight-amino acid specificity-conferring codes extracted from biochemically characterized NRPS A domains. A separate server dedicated to predicting A domain substrates is also available that includes a searchable database of un-characterized A domains.

1.2.2.3.1. Whole-genome PKS/NRPS analysis One advantage of using HMM-based searches is that whole genome proteomes can be rapidly parsed for NRPS and PKS gene discovery. A separate stand-alone implementation of the PKS/NRPS analysis pipeline is also available for download that inputs a multiFASTA-formatted protein sequence file. The output is a web-based graphical interface displaying all identified NRPS/PKS proteins and their detailed domain structures. This tool is freely available at http://www.microbialgenomics.org/2metdb/. With the increasing amount of genomic information being generated, genome scanning software for PKSs and NRPSs should complement genome auto-annotation tools.

1.2.2.4. NORINE: A database of nonribosomal peptides NORINE is a platform that includes a database of NRPs in which manually curated information on over 700 NRPs is fully searchable. Peptide natural product

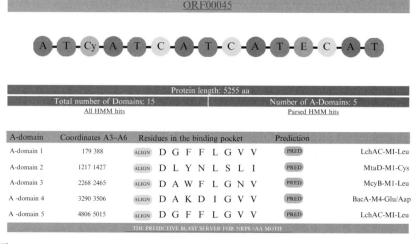

Figure 8.1 Graphical output of the PKS/NRPS analysis web site. The top panel depicts the domain organization of the NRPS; clicking on each domain allows access to the domain protein sequence. The bottom panel represents the output of the A domain substrate prediction server.

data can be searched by compound names (generic or specific), activity, structure (exact or pattern search), molecular weight, number of amino acid monomers and specific amino acids, and also by bibliographical references and producing organisms. This resource is noted due to its value to the investigator analyzing unknown NRPS genes using the methods described in this chapter. For example, NORINE could be used to gather information on all the known peptides containing a specific amino acid, or a specific sequence of amino acids. NORINE can be found at http://bioinfo.lifl.fr/norine/.

2. Converting Type I PKSs to Structural Elements

The multidomain architecture of type I PKS systems allows the inference, to varying degrees of accuracy depending upon the system, of the order of domain execution. The domains of PKSs have historically been grouped into "modules," a notation that refers to a group of collinear domains responsible for a single homologation step in polyketide synthesis. It can be argued that the growing number of exceptions to modular colinearity render the concept of the module less useful. Ultimately, it is the executed biochemical sequence (activity string) that encodes a polyketide's skeleton. However, we continue to employ the concept of the module in the following methods in order to group domains into domain strings as a first approximation to their order of execution. The process of converting domain strings to structural elements is composed of: (1) obtaining domain strings from polyketide open reading frames (ORFs); (2) analyzing consensus motifs in identified domains to test for substrate specificity, stereoselectivity, and/or defective domains; (3) assigning an order, if necessary, to modules on separate proteins; (4a) applying the predicted biochemical reactions in the colinearly determined order; or, more heuristically, (4b) render a saturated polyketide chain of the appropriate length and apply tailoring reactions retroconsecutively from the terminal carboxylate in the polyketide chain.

2.1. The mechanistic approach

The most accurate means of extrapolating structural elements from a domain string is a reaction mechanism approach (Fig. 8.2). The order of domain execution within a module is roughly (1) malonyl loading onto the T domains via the acyltransferase (AT) domain and (2) decarboxylative condensation of tethered malonyl-T precursors by ketosynthase (KS) domains. Optional accessory domains include (3) ketoreduction (KR) domains,

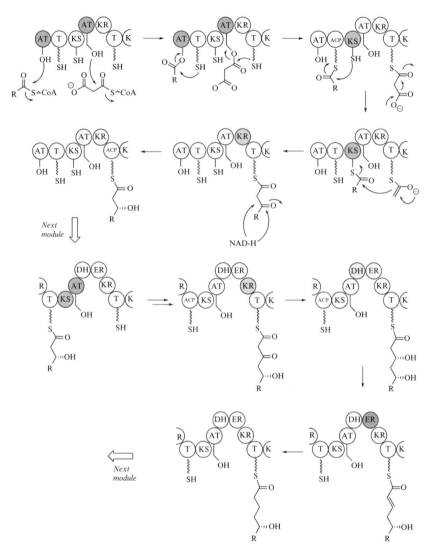

Figure 8.2 Examples of reaction mechanism-based structure prediction for a three-module hypothetical PKS, domain string AT-T-KS-AT-KR-T-KS-AT-DH-ER-KR-T.

(4) 2,3-dehydratase (DH) domains, and (5) enoylreductase (ER) domains. A minimal "elongation" module comprises the set of domains used in one homologation reaction, and is usually bracketed by KS and T domains. It should be noted that the order of the domains within a module may not correspond to their order of execution in a biosynthetic sequence (Haynes and Challis, 2007). An advantage of this approach is that it is less subject to

reductionist errors and can address newly discovered putative domains or new domain combinations for which sequence precedent exists.

Initiating or "loading" domains follow slightly different logic and may contain AT domains preceded by a decarboxylative KS domain (e.g., a KS^Q domain), which ultimately primes the first T domain with an acyl-S-CoA (e.g., acetate, propionate) unit instead of a substituted malonate. Other initiating modules lack the KS domain altogether and contain AT domains capable of loading and activating acyl-S-CoA precursors directly (Moffitt and Neilan, 2003). The terminating domain in PKSs is often the thioesterase (TE) domain that can liberate the fully extended polyketide acyl-S-T intermediate via intramolecular macrocyclization (Akey *et al.*, 2006) or hydrolysis with water (McAlpine *et al.*, 2005). Alternatively, a reductase domain can release tethered elongation intermediates as aldehydes (e.g., Hu *et al.*, 2007).

2.2. Heuristic approach to PKS structure prediction

The end result of the bichemical execution of a series of active domains is a polyketide of n-"acetates," where n is the number of AT domains, and the oxidation states are determined by the last functional domain within each module. Hence, the structure of a polyketide backbone can be alternatively deduced in a more heuristic fashion from the domain string by (a) determining the minimal length of the polyketide based on the number of AT domains, (b) applying the likely oxidation states at odd-numbered positions based on the terminal active domain within each module and stereochemistry based on established consensus motifs and (c) determining the substitution at even positions based on AT-domain specificity. This approach allows the rapid inference of the linear polyketide structure from a domain string and may save time for researchers interested in a quick approximation of the structure encoded by the sequence of genes in PKS gene clusters without resorting to extensive mechanistic analyses.

2.3. Conserved active-site motifs

PKSs may contain one or more defective domains. Inactive domains can often be identified by analyzing conserved active-site motifs (Table 8.2). The sequence and number of domains within a module varies, but chemically must be executed in the order required by fatty acid biosynthesis: AT, KS, KR, DH, and ER. Inactive domains with defective active-site residues therefore typically render domains acting subsequently nonfunctional within a module. Thus, an inactive KR will render a DH or ER within the module nonfunctional, and an inactive DH domain generally renders an ER domain superfluous. Inactive AT or KS domains may suggest module skipping.

Table 8.2 Common consensus motifs

Domain	Active site motif(s)	References
KS	Active site cysteine conserved KS^Q (Q substitutes for C) decarboxylative KS^S, KS^Y initiating	Kakavas et al. (1997) Moffitt and Neilan (2003)
AT	Malonate-activating: QQGHS [QMI]GRSHT[NS]V and "HAFH" motif Methylmalonate-activating QQGHS[LVIFAM]GR[FP]H [ANTGEDS][NHQ]V And "YASH" motif	Yadav et al. (2003a,b) Del Vecchio et al. (2003) Petkovic et al. (2008)
KR	Active consensus: GxGxxG(A)xxxA Active consensus: LXS(G)RXG (T,A) Stereochemistry: "A-type" stereochemistry W141 "B-type" stereochemistry LDD, P144, N148:	Scrutton et al. (1990) Kakavas et al. (1997) Caffrey (2003)
DH	Active consensus: LxxHxxxGxxxxP	Bevitt et al. (1992)
ER	Active consensus: GGVGxAAxQxA	Donadio and Katz (1992) Kakavas et al. (1997)

2.4. Caveats

The application of known biosynthetic "rules" to translated hypothetical ORFs has most often been applied in cases where the secondary metabolite is already known. In such cases, exceptions to the rules are handily rationalized with 20–20 hindsight. The reverse case, the prediction of natural product structure from ORF data, is more treacherous. Following is a list of some caveats to consider when attempting to derive structural elements from hypothetical ORFs and biosynthetic logic.

2.4.1. Applicable systems

The accuracy of structure predictions for polyketides is currently greatest for modular type I PKS systems, more specifically those embedding *cis*-acting acyltransferase domains (Staunton and Weissman, 2001). The domain strings in these systems are often concatenated into a small number of multimodule proteins with clear domain boundaries, facilitating the possible ordering of the domain strings ultimately encoding the sequence of reactions resulting in a polyketide core. The number of permutations increases if a large number of modules are present on multiple discrete proteins.

More difficulty is encountered when analyzing the so-called "AT-less" PKS systems, which contain *trans*-acting AT domains and residual interaction domains (IDs), where it has been proposed that AT domains dock into the megasynthase (Cheng et al., 2003). In these common systems, some bending of the rules of classical PKS logic has often been required to fit a known natural product to a sequence-derived domain string. Perhaps unsurprisingly, predictions of metabolite structures encoded by cryptic AT-less PKSs are currently significantly less accurate than their archetypal cousins from which the colinearity rules were derived. However, AT-less PKS predictions are not without utility. If a cryptic polyketide domain string is similar to a domain string of known function, the accuracy of prediction can be increased substantially. Moreover, the approximate size and general functionality of a cryptic polyketide backbone can be obtained from careful analysis of these systems.

Other PKS systems remain even more undecipherable, given the comparative paucity of structure/function data and mechanistic insight. These systems include the iterative type I systems and type II PKSs (Hertweck et al., 2007), and are beyond the scope of this article. Figure 8.3 shows a general decision tree for identifying the type of polyketides currently amenable to domain-string analysis.

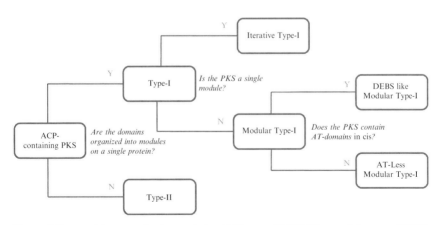

Figure 8.3 Decision tree for determining PKS type. DEBS-like modular type I PKSs are currently most amenable to analysis and are tractable with presented methods, followed by AT-less modular systems.

2.4.2. Nonlinear enzymatic logic

A growing number of divergences from colinear logic have emphasized the limitations of structural extrapolations based on current understanding (Haynes and Challis, 2007; Shen, 2003). Departures from colinear logic have been identified in the form of domain/module skipping and domain/module stuttering, among others. Skipping may be a result of the topology of the megasynthase or the presence of inactive domains, some of which are readily identifiable (see above). Stuttering (module/domain iteration within a synthase) is currently more difficult to predict and has been demonstrated for a growing number of PKS systems (e.g., Dimise *et al.*, 2008; Li *et al.*, 2008).

2.4.3. Post-PKS modifications

It should be noted that predicting the structure of a polyketide scaffold is usually an approximation of a biosynthetic intermediate. Post-PKS modifications are nearly ubiquitous in PKS systems studied to date. Common examples of post-PKS modifications include oxidation, double bond isomerization, polyketide splicing (Traitcheva *et al.*, 2007), and glycosylation, to name a few. The analysis of ORFs clustered with PKSs can sometimes reveal the scope of possible modifications, though a publically available systematic system for such analysis is yet to be described.

2.4.4. Unprecedented domains

The identification of long segments of unprecedented peptide sequence interstitially between domains of precedented function may suggest the presence of a cryptic catalytic domain. These domains can be analyzed for cofactor binding motifs or motifs of low homology to proteins of known function.

2.5. Methods for drawing polyketides based on PKS domain strings

Though no method is entirely generalizable to PKS analysis, the following prototypical outline can be employed for many modular type I systems. The method is divided into three stages: (1) domain identification/ordering (by manual parsing or automated parsing); (2) consensus motif analysis; and (3) structure derivation.

2.5.1. Domain identification and ordering of proteins/modules

Parsing of cryptic synth(et)ase sequences into domain strings is accomplished by pairwise alignment of the unknown enzyme with examples of each possible domain sequence of known function in a serial fashion. It is assumed that PKS-homologous ORFs within a given gene cluster have been identified by BLAST homology and class has been inferred (e.g., type I modular or AT-less). Domain identification can be determined either manually, using pairwise alignment software, or in an automated fashion, using parsing servers.

2.5.2. Manual parsing

1. Convert each translated ORF to FASTA format.
2. Identify domain locations in each peptide by pairwise comparison to domains of known function and record coordinates. Paste polyketide sequence into the first window of a pairwise alignment tool. For instance, on the BLAST 2 Sequences server page (http://blast.ncbi.nlm.nih.gov/bl2seq/wblast2.cgi), select "program" to align proteins and turn off filtering by unclicking the "filter" dialog box.

   a. For classical type I PKS systems, search for KS domains by pasting archetypal KS domains from DEBS (e.g., DEBS KS1) into the second sequence pane. Click "align." If significant sequence similarity is observed, record the approximate coordinates of identified domain(s). Repeat for AT, KR, DH, T, and TE. If there are discernable unassigned peptide-encoding regions between domains, search for more accessory domains by performing additional pairwise alignments against known accessory domains (e.g., cyclization Cy, oxidation Ox, N-methylation NM, etc.) and/or BLAST these regions for conserved domain analysis against GenBank or by using the NCBI CDD resource.

   b. For AT-less PKS systems, use alignment templates from known AT-less pathways (e.g., dorrigocin, leinamycin) as in 2a.

3. Based on domain coordinates, assemble the identified domains into an ordered sequence of domains (domain string). See case studies in Figs. 8.4–8.6 for example of domain strings.
4. Repeat for all PKS proteins.

2.5.3. Automated parsing

1. Convert each translated ORF sequence into FASTA format.
2. Identify domain locations in each peptide by pasting each sequence into the automated NRPS/PKS parser http://nrps.igs.umaryland.edu/nrps/ and click "analyze NPRS/PKS." A tutorial is available on the website. Record domain string from the graphical representation. It is recommended that the ("all HMM Hits") be examined to identify possible unassigned protein sequence regions (e.g., >150 amino acids) between domains and/or putative domains of low homology, which may lie beneath the threshold of the graphical user interface. Regions of low similarity can be searched against GenBank to identify conserved

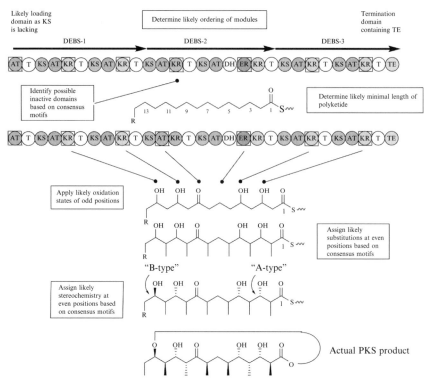

Figure 8.4 Predicted structure using the heuristic method versus actual structure of erythromycin. Boxes are drawn around the terminal activity in each domain.

sequence motifs such as cofactor-binding motifs, which may be used to infer functional elements.
3. Repeat for all PKS proteins.

The sequence of colinear domains and modules in a PKS protein generally reflects the approximate order of their application in the biosynthetic pathway. If the PKS is distributed across multiple ORFs, the ordering of domains/modules will be further determined by the ordering of the individual encoded proteins. For instance, the erythromycin PKS (Fig. 8.4) consists of three proteins, DEBS1-3, containing seven modules, and the ordering of proteins reflects the approximate order of execution of the domains contained therein. It is not uncommon for PKS systems to be contained on four or more polypeptides, and it is sometimes necessary to construct multiple permutations of protein organization. Amphotericin PKS (Fig. 8.5), for instance, is contained on six polypeptides and it is not possible to predict the ordering of modules between proteins via domain sequence alone.

In Silico Prediction of Microbial Polyketide and Nonribosomal Peptide Biosynthetic 199

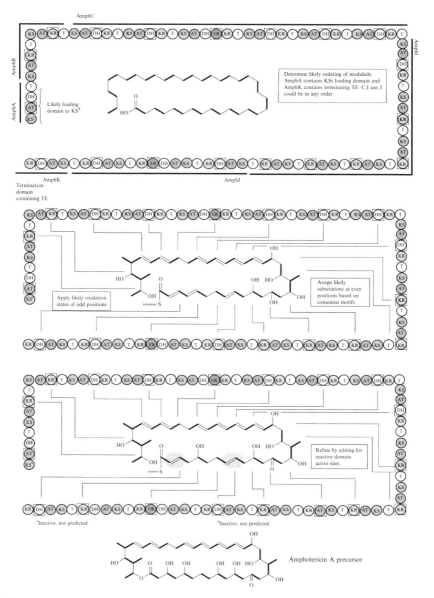

Figure 8.5 Predicted structure of amphotericin scaffold using the heuristic method versus actual structure.

2.5.4. Interprotein domain ordering

1. If possible, identify loading modules and terminating modules. These modules can sometimes be identified by:

    a. Identifying a probable initiating AT domain. For proteins beginning with a KS domain, check for replacement of the KS active-site

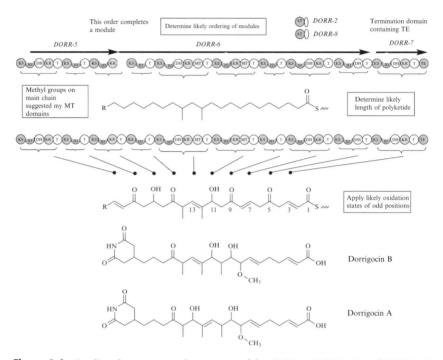

Figure 8.6 Predicted versus actual structure of dorrigicins. DORR-2 and DORR-8 are *trans* AT domains reported linked to TE and Ox domains, respectively.

conserved cysteine with glutamine (KS^Q), serine (KS^S) or tyrosine (KS^Y). These substitutions are frequently observed within initiating KS domains. KS^Q domains have been demonstrated to load their cognate T domain with acetate or propionate via malonyl, or methylmalonyl decarboxylation (Moffitt and Neilan, 2003).
  b. Identify a module that contains a termination domain, most commonly TE (thoiesterase) or RE (Reductase). This is likely to be the terminating module.

2. Arrange proteins/modules in all likely orders, conforming to colinear logic whenever possible. Ordering that preserves module integrity is favored. For instance, if one translated ORF in a gene cluster ends with a DH domain and other begins with a KR domain, colinearity will be conserved if the proteins are ordered to complete the domain.

2.6. Consensus motif analysis

Identify possible inactive domains based on apparent defects in the consensus sites. Examples of common consensus motifs used in domain activity analysis are listed in Table 8.2. Consensus motifs are readily analyzed via comparison of cryptic domain active site sequences with corresponding sequences with known activity. This is typically accomplished via multiple sequence alignment of the cryptic domain with a compiled list of domains of known function in FASTA format. Multiple sequence alignment can be performed using the Clustal algorithm, which is freely available as a freestanding program (e.g., ClustalX) or a web-based tool (e.g., ClustalW2, http://www.ebi.ac.uk/Tools/clustalw2/).

2.7. Ketoreductase (KR) domain analysis

Following is a general procedure using ClustalW2 for KR domain activity and stereochemistry:

1. Paste sequences for KR domains of known selectivity (see supplementary data file KR_domains_stereochem.txt) into the sequence pane.
2. Add the cryptic domain of interest, which can be obtained from the NRPS/PKS analysis website (clicking on domain graphics provides a FASTA sequence of an individual domain) into the sequence pane and adding the unknown to the file containing KR domains of known selectivity.
3. Perform multiple sequence alignment by clicking Run.
4. Check for consensus motifs

   a. *Activity*: The consensus motifs, GxGxxG(A)xxxA and the invariant arginine in motif LXS(G)RXG(T,A) (Kakavas *et al.*, 1997; Scrutton *et al.*, 1990) are generally found in active KR domains. For instance, in DEBS KR1 the motifs are specified by the sequences GTGGVGGQIA and LVSRSG respectively. Deviations from these motifs may indicate an inactive KR domain. For instance, DEBS KR3 is inactive and contains defects or deletions in these motifs.
   b. *Hydroxyl stereochemistry*: The consensus motifs LDD, P144, N148 suggest B-type stereochemistry (Table 8.2) and W141 suggests A-type stereochemistry (Caffrey, 2003).

Note inactive motifs in domain strings. When an inactive domain is postulated, the previous active domain (that is not contingent biochemically on the defective domain) is putatively the operative domain in a module. An inactive KR domain renders subsequent DH and ER domains superfluous. Record putative hydroxyl stereochemistry.

This procedure is repeated for AT domains (see supplementary file AT_domains_specificity). DH and ER domains can be analyzed for active site residues outlined in Table 8.2 by constructing similar multiple sequence files using the NRPS/PKS parser. Furthermore, thioesterase TE activity (macrolactonization or hydrolysis) may be postulated by multiple sequence alignments with TEs of known function.

2.8. Deriving possible polyketide structures from the domain strings

2.8.1. Mechanism-based approach

It is our opinion that a mechanism-based approach is most accurate for prediction of structural elements. In this case, the mechanism of each catalytic domain is executed in the order specified by the proposed domain string and a pathway is rendered of relayed T-bound homologation reactions, resulting in a polyketide scaffold. An example of this process for a small portion of a polyketide domain string (three modules) is shown in Fig. 8.2.

2.8.2. Heuristic approach

An alternative heuristic approach may be more rapid for researchers less familiar with the mechanistic biochemistry of PKSs.

1. Count the number of extension modules and render a polyketide backbone based on the number of extension modules:

   a. In the case of classical type I PKSs, draw the length of the polyketide chain according to the number of AT domain-containing modules or
   b. In the case of AT-less PKSs, assign the length of the polyketide according to the number of modules containing a KS and a downstream T. If there is more than one T domain adjacent, use the last T domain for counting module stops. Homologating modules may be more confidently assigned if an AT interaction domain (ID) is identified within a module.

   For example DEBS (Fig. 8.4) contains seven putative AT-containing modules, which prescribe a heptaketide (14-carbon) chain. In the foregoing discussions, numbering follows fatty acid convention and begins from the terminating carbonyl carbon.

2. Identify the last active modification domain in each module and apply the predicted oxidation state to each odd carbon (1-position, acetate numbering). For example, following are hypothetical oxidation states of C-1 positions for various domain strings (where ★ indicates an inactive domain and the underlined domain is the last domain executed):

   - KS-*AT*-T: ketone
   - KS-AT-*KR*-T: alcohol
   - KS-AT-*DH*-KR-T: alkene ($\alpha-\beta$ unsaturated from 1-position)
   - KS-AT-DH-*ER*-KR-T saturation methylene
   - KS-*AT*-DH-KR★-T: ketone

3. Append the appropriate functional group to C-2 positions by identifying AT domain specificity as malonyl, methylmalonyl (or other). Methylmalonyl-CoA-activating domains (substituting a methyl group at C-2)

may be identified from the so called YASH motif and malonyl-CoA activating domains (substituting a hydrogen atom) from the "HAFH motif," among others (Table 8.2).
4. Append C-1 stereochemistries from putative KR consensus motifs (see above: "Consensus motif analysis").
5. Apply terminating biochemistry. Multiple sequence alignments of the terminating TE domain can be sometimes aligned with groups of validated cyclizing or noncyclizing TEs to predict this biochemistry (McAlpine et al., 2005).

2.9. Polyketide case studies
2.9.1. Erythromycin
The erythromycin PKS has long been considered the archetypal modular type I PKS as it was the first to be isolated and characterized (Bevitt et al., 1992; Donadio and Katz, 1992) and is by far the most extensively studied from a biochemical perspective. The polyketide synthase protein sequences for DEBS1-3 are contained in GenBank (GI 134097327, 134097329, 134097330).

As the template for all other type I PKSs, it is unsurprising that the domains are faithfully identified. The TE domain in DEBS 3 identifies it as the terminal module (and protein) executed in the pathway. The order of domains between the first two proteins is decided by choosing the order that is maximally colinear according to polyketide logic. In this example, DEBS1 lacks a KS domain at the N-terminus. This is typical for loading modules. Were DEBS-1 to be selected as the second functional protein in the pathway, there would be a missing KS domain in the third module, violating colinear PKS logic. With the order of proteins and modules established, the heuristic method predicts a polyketide chain of seven acetate units. All AT domains (excepting the initiating domain) activate methylmalonate, as expected. The KR domain in the third module is predicted to be inactive based on divergent consensus motifs in GxGxxG(A)xxxA and LXS(G)RXG(T,A). The oxidation state of C-9 is therefore correctly determined as a ketone. The ultimate stereochemistries of ketoreductions are accurately assigned based on A-type configuration and B-type configuration sequence identifiers (Caffrey, 2003). Likely stereochemistries of C-2 substitution are not predicted by the procedures described in this method.

2.9.2. Amphotericin—A classical type I modular PKS
The amphotericin scaffold is determined by a large classical type I PKS produced by *Streptomyces nodosus* (Caffrey et al., 2001). As the PKS is contained on six separate proteins, it is difficult to determine the order of domains between proteins. The peptide sequences are available from

GenBank, accessions AmphA: AAK73512, AmphB: AAK73513, AmphC: AAK73514, AmphI: AAK73501, AmphJ: AAK73502, AmphK: AAK73503. Depicted in Fig. 8.5 is one possible ordering of the proteins and their corresponding domains. The loading and terminating domains can be identified by an initiating KS^S domain in AmphA and a TE domain in AmphK, respectively. Many other permutations of AmphB,C,I,J are possible. However, the distinctive polyene pharmacophore (a bioactivity handle for isolations, were this an unknown), is contained primarily on AmphC.

AT domains in AmphB and AmphI are predicted to activate methylmalonyl-CoA in the expected positions. The DH domain in the loading module is not functional, presumably as there is no upstream β-hydroxy group to eliminate due to its position in the initiating module. KR13 contains a putative inactive NADH binding site with an arginine disrupting the NADH binding site GxGxxG(A)xxxA). AT11 in AmphI is also methylmalonate-selective. DH17 is missing the active-site histidine in the HxxxGxxxxP motif. DH18 and DH16 are predicted to be active, but are actually inactive. ER5 is apparently partially active, as amphotericin B and A, differing only in reduction of the C18–C19 double bond, are produced by the same microorganism. KR stereochemistry was not analyzed in this example, but this analysis has been performed (Caffrey, 2003).

2.9.3. Dorrigocin/Migrastatin—An "AT-less" PKS

The dorrigicin/migrasatin gene cluster, obtained from *Streptomyces platensis* (Farnet *et al.*, 2002), GenBank locus AX598593, contains an "AT-less" PKS system. The detailed biochemistry of these systems is less understood than the classical modular type I PKS. In lieu of colinear AT domains, these systems possess putative "docking domains" (interaction domains) and *trans*-acting AT domains. These systems frequently diverge from the colinearity rules, reflecting likely additional differences in the quaternary organization of AT-less PKS systems in comparison to their classical modular cousins. These systems also frequently contain accessory methyltransferase domains (MT) that are capable of C-methylation during elongation.

The domain string for dorrigocin was determined by manual parsing using pairwise comparison of DORR-5, -6, -7 with reported dorrigocin KS, KR, DH, ER, and ID (intereaction) domains. Protein order was chosen based on the evidence for a TE domain in DORR-7 and preservation of module integrity between DORR-5 and -6. As shown in Fig. 8.6, the predicted polyketide chain diverges significantly from the reported metabolite structures when applying classical polyketide biochemical logic. It is expected that advances in understanding of structure and function of AT-less PKSs may provide new sequence-based methods for determining the ultimate order of execution of the biocatalytic domains.

3. CONVERTING NRPS DOMAIN STRINGS TO STRUCTURAL ELEMENTS

NRPS enzymes fabricate polypeptide backbones via a strategy analogous to PKS systems. NRPSs are also large enzymatic assemblages consisting of a series of catalytic domains that are used sequentially to initiate, homologate and modify a growing T-tethered polypeptide chain (Fig. 8.7).

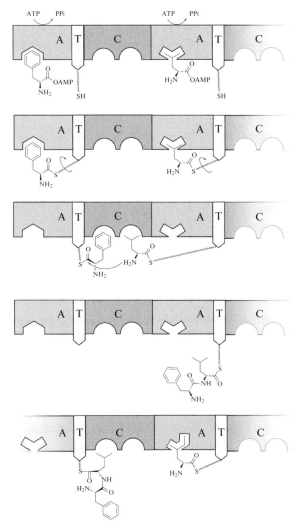

Figure 8.7 Overview of minimal NRPS chemistry: Adenylation (A), thiolation (T) and condensation (C) domains.

Amino acid monomers, thethered by phosphopantetheine linkers to peptidyl carrier proteins, are acted upon by more or less colinearly arranged domains. A minimal set of catalytic domains (module) is comprised of amino acid adenylation (A), peptide condensation (C), peptide acyl carrier protein (T or PCP), and thioesterase domains (Te) for cleavage of the ultimate peptide chain from the assemblage. Accessory catalytic domains are able to epimerize (E), heterocyclize serines, threonines and cysteines (Cy), oxidize resultant oxazoline and thiazolines (Ox), or N-methylate (NM), to name a few modifications (Fischbach and Walsh, 2006). The order of execution of the domains is determined by the quaternary organization of the multidomain enzyme complex (Tanovic et al., 2008).

In many investigated biosynthetic pathways, the organization of domains (and hence chemistry) is reflected in the order of the domains in the primary sequence of the translated ORFs. In these cases, the pathways are said to follow the "colinearity rule." As previously discussed, in recent years, a growing number of exceptions to the rule have been discovered.

3.1. Polypeptide chain length and amino acid side chains

NRPS peptide chain length is minimally defined by the number of A domains present in the megasynthetase. As the terminal carboxylate remains uncoupled, for n A domains there should in principle be $n-1$ condensation domains. Practically, an excess of C domains may indicate other condensation reactions in addition to peptide homologations, including N-acylation, coupling to polyketide biosynthesis, or coupling to other (amino) acyl-S-T species.

Initiation modules may be comprised of an A–T didomain, in the simplest case. If the first module is a tridomain, C–A–T, this may indicate ligation of the first amino acid to the donor acyl-S–T chain. A thiolation domain initiating an ORF may suggest an initiating module that accepts an activated CoA ligand (e.g., Marshall et al., 2002). If the T domain is prefaced by a CoA–ligase domain (AL), the likelihood is high that this module is initiating (e.g., Shen et al., 2002).

Terminating modules are usually terminally punctuated by a thioesterase (Te) domain, which releases the nascent peptide via macrocyclization or hydrolysis, or in some cases a reduction (Re) domain, which reduces the terminal thioester to the aldehyde oxidation state.

3.2. Modifications and stereochemistry

As mentioned, a growing number of accessory domains (Fig. 8.8) have been described, including domains involved in N-methyltransferase (NM), epimerization (E), oxidation (Ox), and cyclization (Cy) activities, to name

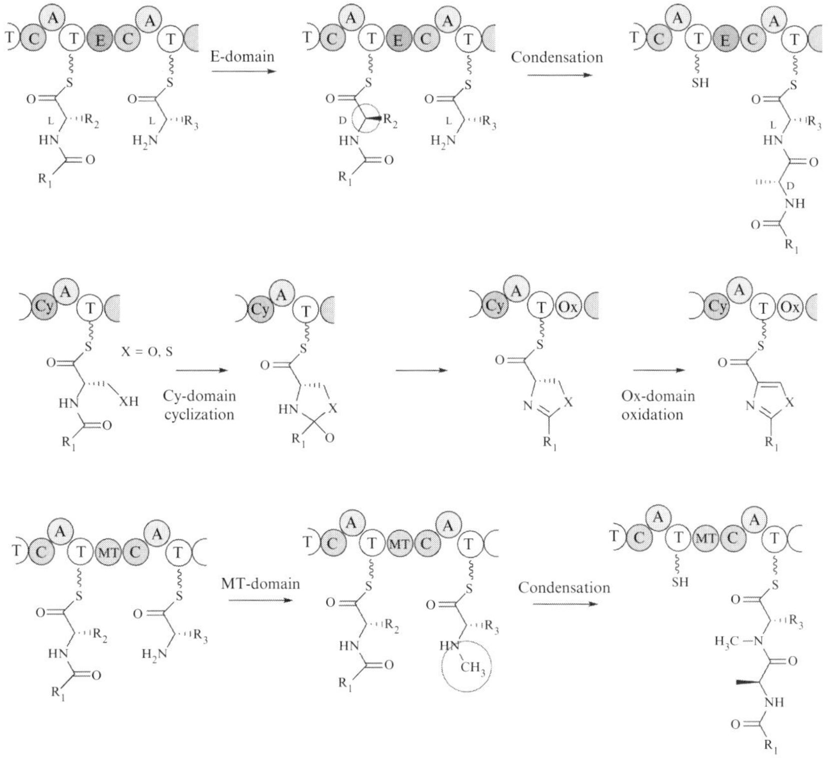

Figure 8.8 Selected accessory domains involved in NRPS biosynthesis. Epimerization (E) domains epimerize upstream tethered peptidyl-Ts prior to condensation. Cyclization domains (Cy) mediate intramolecular cyclization and dehydration of cysteines, threonines, and serines. Subsequent oxidation (Ox) domains, if present, function in a desaturative mode, aromatizing the heterocycles. Methyltransferase domains N-methylate peptidyl-S-Ts prior to condensation.

a few. These domains are frequently found in *cis*, though examples of *trans* modifications have also been described.

NRPs frequently contain amino acids with D-stereochemistry. In most cases, the D-configuration is generated by epimerization domains that racemize an upstream pendant l-peptidyl-T donor amino acid and the ensuing C domains are often selective for d-peptidyl amino acids. Additionally, dual function C/E domains are capable of both epimerizing and condensing chemistries (Rausch *et al.*, 2007).

3.3. Caveats

While many NPRS systems follow the colinear model, exceptions are coming to dominate the rule (Haynes and Challis, 2007). Modules that "stutter" contain domains that are use more than once in an assembly line and "skipping" domains/modules may be bypassed entirely. There are also notable examples of A-domains that apparently activate more than one amino acid, or accept an A domain in *trans* (see case studies, below). Regardless, prediction of A domains from module strings is a straightforward process and, while predictions may be more or less accurate, they provide a minimal approximation with which to estimate a provisional structure.

3.4. Method for rendering NRPs

Domains are readily identified by homology-based methods analogous to those described for PKS systems. Indeed, many domains are functionally shared by both systems and the preponderance of mixed PKS/NRPS systems demonstrates that the distinction between the two systems may only be artifice. Following is a general method for rendering NRPs from their colinear domain string.

3.4.1. Domain string determination

Herein we discuss only an automation-assisted domain identification protocol using the NRPS/PKS website. However, domains may also be readily identified by a manual method using archetypal domains and BLAST2Seq as in Section 2.5.

1. Identify all putative NRPS-encoding protein sequences by BLAST and convert each into FASTA format.
2. Paste sequence into PKS/NRPS parser http://nrps.igs.umaryland.edu/nrps/ and click Analyze PKS/NRPS. It is recommended that the ("all HMM Hits") be examined to identify possible unassigned protein sequence regions (e.g., >150 amino acids) between domains and/or putative domains of low homology, which may lie beneath the threshold of the graphical user interface. Regions of no similarity can be searched against GenBank to identify conserved sequence motifs such as cofactor-binding motifs, which may be used to infer functional elements.
3. Repeat for all NRPS-containing proteins.

3.4.2. Interprotein domain ordering

1. Identify terminating and initiating modules
   a. Initiating modules are often identifiable by virtue of an initiating two-domain A–T module in an ORF. A thiolation domain initiating an ORF may suggest an initiating module that accepts an activated CoA ligand (e.g., Marshall *et al.*, 2002). If the T domain is prefaced by a CoA–ligase domain (AL), the likelihood is high that this module is initiating (e.g., Shen *et al.*, 2002).

b. The identification of a putative terminal thioesterase (Te) or reductase (Re) domain is a likely indication of a terminating module (e.g., C–A–T–Te).

2. Arrange proteins in all likely orders, conforming to colinear logic wherever possible. Interprotein ordering that preserves module integrity is favored. For instance, an ORF that terminates with the domain string ...C–A– would likely be followed by an ORF that initiates with the sequence T–E–C–A..., thereby completing a module (C–A–T–E).

3.4.3. Consensus motif analysis

Active-site motifs may be scrutinized in an analogous fashion to the PKS systems described earlier. However, NRPS gene clusters with inactive domains are rarer in the literature and methods for active site consensus analysis will not be discussed in detail.

3.4.4. Heuristic structure prediction

As with PKSs, structural prediction can be attempted by a mechanism-based approach in which each domain is executed in the prescribed order. The following alternative heuristic approach may also be employed to render a rapid approximation of NRPS structure:

1. Subsequent to domain identification and interprotein domain ordering, identify homologating modules. Minimal extending modules contain C–A–T domains and possible accessory domains (e.g., C–A–NM–T, C–A–Ox–T, C–A–T–E).
2. Render a peptide chain corresponding to a polypeptide of length $n + 1$ where n is the number of A domains.

   a. Estimate A domain specificities. Draw amino acid side chain on alpha-carbons for A domains with unambiguous homology models. This can be achieved manually using the homology method (Challis *et al.*, 2000), which is automated on the NRPS/PKS website (Fig. 8.1). Alternatively, or additionally, the TSVM method can be used (Rausch *et al.*, 2005) to augment the accuracy of the prediction.
   b. Apply accessory chemistries (if necessary) for.
      - Epimerization (E) domains mostly epimerize the PPant-tethered amino acid activated by the directly preceding A domain.
      - *N*-Methylation (NM) domains most commonly *N*-methylate the amino acid activated by the A domain directly prior to the NM occurrence
      - Cyclization (Cy) domains result in the formation of thiazoles in the presence of oxidation (Ox) domains and thiazolines in their absence. As in Fig. 8.8, side-chain cys/thr/ser cyclize via nucleophilic attack on a preceding (toward the N-terminus) amide, followed by dehydration.

c. Apply terminating biochemistry.
  i. If a terminating Te is present, optionally identify if it is macrocyclizing or hydrolytic based on multiple sequence alignments with Te domains of known function. Possible locations for macrolactonization include the N-terminus and nucleophilic side chains (Roongsawang et al., 2007).
  ii. If a terminating reductase (Re) module is present, the polypeptide is terminated by an aldehyde and likely exists in solution as an intramolecularly cyclized hemiaminal and/or corresponding imine via dehydration.

3.5. NRPS Case studies

3.5.1. Fengycin

Fengycin is produced by *Bacillus amyloliquefaciens* and its biosynthetic gene sequence has been recently identified (Koumoutsi et al., 2004). Domain analysis reveals a typical organization comprised of 10 modules on five putative proteins: FenA (gi:154686279), FenB (gi:154686278), FenC (gi:154686277), (FenD gi:154686276), FenE (gi:154686275).

Aside from the terminating module, the ordering of domains/modules is difficult to ascertain from domain parsing alone and only one possible combination is shown in Fig. 8.9. The amino acid specificity is accurately predicted by A domain analysis and the stereochemistry specified by E domains is consistent with the structure of the natural product. Not evident from the domain string, the natural product is macrocyclized and acylated at its N-terminus by a long-chain β-hydroxy fatty acid.

3.5.2. Penicillin

The synthetase for penicillin was one of the first sequenced NRPSs (Martin, 2000). The tripeptide synthetase is contained on a single peptide in fungi and bacteria (*Amycolatopsis lactamdurans* GI:113316). It is unsurprising that this peptide is faithfully predicted as it was used to populate the A-domain databases. This example illustrates an unusual δ-linkage between aminoadipate and cysteine, and the degree of "posttranslational" modifications sometimes encountered in NRPs (Fig. 8.10).

3.5.3. Ramoplanin

Ramoplanin is a glycolipodepsipeptide produced by *Actinoplanes* sp. ATCC 33076. The sequence and annotation of this genome has been previously performed using proprietary software (Farnet et al., 2002) at Ecopia Biosciences (now Thallion Inc.). The ramoplanin NRPS is contained on four megasynthetases and translated ORF data has been deposited into Genback (identifiers gi:112084475, gi:112084465, gi:112084466, and gi:112084469). Our analysis, using one of many possible domain

In Silico Prediction of Microbial Polyketide and Nonribosomal Peptide Biosynthetic 211

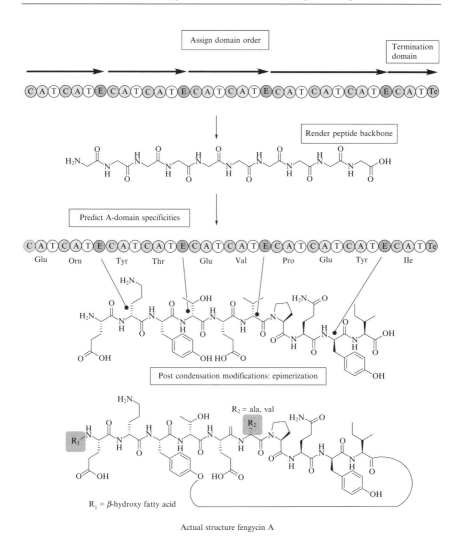

Figure 8.9 The predicted structure of fengycin for one (of several) possible interprotein domain orderings.

organizations, is shown in Fig. 8.11. A-domain specificities were estimated by using both the NRPSPredictor and the predictive NRPS/PKS servers, which resulted in the identities of three amino acids being indeterminate or incorrectly assigned: Lys-4, Thr-8, and Ala-10. Otherwise, the amino acid side-chains were predicted with overall good precision. Notable divergences from colinearty include the apparent stuttering of the first aspartate homologation module and the proposed unusual in *trans* application of the monomodule ORF-17 within module ORF-13, which lacks an A domain.

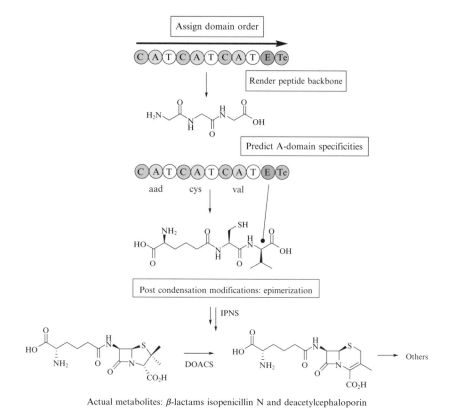

Figure 8.10 Predicted structure from ACV synthetase of penicillin biosynthesis.

In *Actinoplanes*, the ramoplanins are present as a family of peptides that are N-acylated by various fatty acids at the N-terminus and polyglycosylated.

4. Concluding Remarks

The methods presented in this chapter represent didactic examples of methodology used to deduce structural elements from translated gene sequence data, and are far from comprehensive. It is recommended that structures of interest approximated by the heuristic approach be further analyzed using a mechanisms approach. In practice, individual practitioners of the art of sequence-based metabolite structure prediction amass unique palettes of tools (additional consensus sequence models, bioinformatics tools, etc.) that aid in more refined predictions. Detailed analysis of

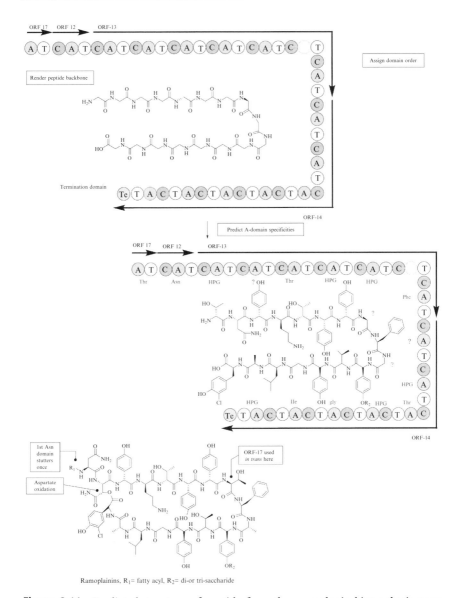

Figure 8.11 Predicted structure of peptide from the ramoplanin biosynthetic gene cluster.

individual ORFs in a cryptic gene cluster are evaluated in the context of the compendium of natural product biosynthetic transformations and principles of (bio)chemical mechanisms.

Errors in predicted structures based on PKS and NRPS ORFs are the result of equivocating the domain string with the activity string, which in this work is derived from the order of domains found in a cryptic ORF primary structure. As tertiary and quaternary architectures are further resolved, new sequence elements may be identified that aid in ascertaining an accurate *activity sequence*. From an "activity string," the accuracy of prediction of polyketide- and nonribosomally encoded scaffolds will increase substantially. With these improvements further development of the heuristic approaches will facilitate increasingly automated computer-generated substructure predictions as an aid to understanding chemical ecology and a catalyst for natural product discovery.

ACKNOWLEDGMENTS

BB acknowledges financial support from the Vanderbilt Institute of Chemical Biology and the NIH (1RO1GM077189).

REFERENCES

Akey, D. L., Kittendorf, J. D., Giraldes, J. W., Fecik, R. A., Sherman, D. H., and Smith, J. L. (2006). Structural basis for macrolactonization by the pikromycin thioesterase. *Nat. Chem. Biol.* **2,** 537–542.

Altschul, S. F., Gish, W., Miller, W., Myers, E. W., and Lipman, D. J. (1990). Basic local alignment search tool. *J. Mol. Biol.* **215,** 403–410.

Altschul, S. F., Madden, T. L., Schaffer, A. A., Zhang, J., Zhang, Z., Miller, W., and Lipman, D. J. (1997). Gapped BLAST and PSI-BLAST: A new generation of protein database search programs. *Nucleic Acids Res.* **25,** 3389–3402.

Ansari, M. Z., Yadav, G., Gokhale, R. S., and Mohanty, D. (2004). NRPS-PKS: A knowledge-based resource for analysis of NRPS/PKS megasynthases. *Nucleic Acids Res.* **32,** W405–W413.

Barton, G. J. (1990). Protein multiple sequence alignment and flexible pattern matching. *Methods Enzymol.* **183,** 403–428.

Bevitt, D. J., Cortes, J., Haydock, S. F., and Leadlay, P. F. (1992). 6-Deoxyerythronolide-B synthase 2 from *Saccharopolyspora erythraea*. Cloning of the structural gene, sequence analysis and inferred domain structure of the multifunctional enzyme. *Eur. J. Biochem./FEBS* **204,** 39–49.

Bystroff, C., and Krogh, A. (2008). Hidden Markov Models for prediction of protein features. *Methods Mol. Biol. (Clifton, NJ)* **413,** 173–198.

Caffrey, P. (2003). Conserved amino acid residues correlating with ketoreductase stereospecificity in modular polyketide synthases. *Chembiochem* **4,** 654–657.

Caffrey, P., Lynch, S., Flood, E., Finnan, S., and Oliynyk, M. (2001). Amphotericin biosynthesis in *Streptomyces nodosus*: Deductions from analysis of polyketide synthase and late genes. *Chem. Biol.* **8,** 713–723.

Challis, G. L. (2008a). Genome mining for novel natural product discovery. *J. Med. Chem.* **51,** 2618–2628.

Challis, G. L. (2008b). Mining microbial genomes for new natural products and biosynthetic pathways. *Microbiology (Reading, England)* **154,** 1555–1569.

Challis, G. L., Ravel, J., and Townsend, C. A. (2000). Predictive, structure-based model of amino acid recognition by nonribosomal peptide synthetase adenylation domains. *Chem. Biol.* **7,** 211–224.

Cheng, Y. Q., Tang, G. L., and Shen, B. (2003). Type I polyketide synthase requiring a discrete acyltransferase for polyketide biosynthesis. *Proc. Natl. Acad. Sci. USA* **100,** 3149–3154.

Clardy, J. (2005). Using genomics to deliver natural products from symbiotic bacteria. *Genome Biol.* **6,** 232.

Conti, E., Stachelhaus, T., Marahiel, M. A., and Brick, P. (1997). Structural basis for the activation of phenylalanine in the non-ribosomal biosynthesis of gramicidin S. *EMBO J.* **16,** 4174–4183.

Del Vecchio, F., Petkovic, H., Kendrew, S. G., Low, L., Wilkinson, B., Lill, R., Cortes, J., Rudd, B. A., Staunton, J., and Leadlay, P. F. (2003). Active-site residue, domain and module swaps in modular polyketide synthases. *J. Ind. Microbiol. Biotechnol.* **30,** 489–494.

Dimise, E. J., Widboom, P. F., and Bruner, S. D. (2008). Structure elucidation and biosynthesis of fuscachelins, peptide siderophores from the moderate thermophile Thermobifida fusca. *Proc. Nat. Acad. Sci. USA* **105,** 15311–15316.

Donadio, S., and Katz, L. (1992). Organization of the enzymatic domains in the multifunctional polyketide synthase involved in erythromycin formation in Saccharopolyspora erythraea. *Gene* **111,** 51–60.

Eddy, S. R. (1998). Profile hidden Markov models. *Bioinformatics* **14,** 755–763.

Farnet, C. M. (2005). *In* "Natural Products Drug Discovery and Therapeutic Medicine: Drug Discovery and Therapeutic Medicine" (L. Zhang and Arnold L. Demain, eds.), pp. 95–106. Springer, New York.

Farnet, C. M., Zazopoulos, E., and Staffa, A. (2002). *In* (P. WIPO, ed.). This is an international patent application: WO/2002/031155.

Farnet, C. M., Zazopoulos, E., Staffa, A., and Xianshu, Y. (2002). *In* (P. WIPO, ed.). International patent application: WO 02/088176 A2.

Fischbach, M. A., and Walsh, C. T. (2006). Assembly-line enzymology for polyketide and nonribosomal Peptide antibiotics: Logic, machinery, and mechanisms. *Chem. Rev.* **106,** 3468–3496.

Haynes, S. W., and Challis, G. L. (2007). Non-linear enzymatic logic in natural product modular mega-synthases and -synthetases. *Curr. Opin. Drug Discov. Dev.* **10,** 203–218.

Henikoff, S. (1996). Scores for sequence searches and alignments. *Curr. Opin. Struct. Biol.* **6,** 353–360.

Hertweck, C., Luzhetskyy, A., Rebets, Y., and Bechthold, A. (2007). Type II polyketide synthases: Gaining a deeper insight into enzymatic teamwork. *Nat. Prod. Rep.* **24,** 162–190.

Hill, A. M. (2006). The biosynthesis, molecular genetics and enzymology of the polyketide-derived metabolites. *Nat. Prod. Rep.* **23,** 256–320.

Hu, Y., Phelan, V., Ntai, I., Farnet, C. M., Zazopoulos, E., and Bachmann, B. O. (2007). Benzodiazepine biosynthesis in Streptomyces refuineus. *Chem. Biol.* **14,** 691–701.

Hunter, S., Apweiler, R., Attwood, T. K., Bairoch, A., Bateman, A., Binns, D., Bork, P., Das, U., Daugherty, L., Duquenne, L., Finn, R. D., Gough, J., *et al.* (2009). Interpro: The integrative protein signature database. *Nucleic Acids Res.* **37,** D211–D215.

Kakavas, S. J., Katz, L., and Stassi, D. (1997). Identification and characterization of the niddamycin polyketide synthase genes from Streptomyces caelestis. *J. Bacteriol.* **179,** 7515–7522.

Koch, M. A., Schuffenhauer, A., Scheck, M., Wetzel, S., Casaulta, M., Odermatt, A., Ertl, P., and Waldmann, H. (2005). Charting biologically relevant chemical space: A structural classification of natural products (SCONP). *Proc. Natl. Acad. Sci. USA* **102,** 17272–17277.

Koch, M. A., Wittenberg, L. O., Basu, S., Jeyaraj, D. A., Gourzoulidou, E., Reinecke, K., Odermatt, A., and Waldmann, H. (2004). Compound library development guided by protein structure similarity clustering and natural product structure. *Proc. Natl. Acad. Sci. USA* **101,** 16721–16726.

Koumoutsi, A., Chen, X. H., Henne, A., Liesegang, H., Hitzeroth, G., Franke, P., Vater, J., and Borriss, R. (2004). Structural and functional characterization of gene clusters directing nonribosomal synthesis of bioactive cyclic lipopeptides in *Bacillus amyloliquefaciens* strain FZB42. *J. Bacteriol.* **186,** 1084–1096.

Li, L., Deng, W., Song, J., Ding, W., Zhao, Q. F., Peng, C., Song, W. W., Tang, G. L., and Liu, W. (2008). Characterization of the saframycin A gene cluster from *Streptomyces lavendulae* NRRL 11002 revealing a nonribosomal peptide synthetase system for assembling the unusual tetrapeptidyl skeleton in an iterative manner. *J. Bacteriol.* **190,** 251–263.

Marchler-Bauer, A., Anderson, J. B., Chitsaz, F., Derbyshire, M. K., Deweese-Scott, C., Fong, J. H., Geer, L. Y., Geer, R. C., Gonzales, N. R., Gwadz, M., He, S., Hurwitz, D. I., *et al.* (2009). CDD: Specific functional annotation with the Conserved Domain Database. *Nucleic Acids Res.* **37,** D205–D210.

Marchler-Bauer, A., and Bryant, S. H. (2004). CD-Search: Protein domain annotations on the fly. *Nucleic Acids Res.* **32,** W327–W331.

Marshall, C. G., Burkart, M. D., Meray, R. K., and Walsh, C. T. (2002). Carrier protein recognition in siderophore-producing nonribosomal peptide synthetases. *Biochemistry* **41,** 8429–8437.

Martin, J. F. (2000). Alpha-aminoadipyl-cysteinyl-valine synthetases in beta-lactam producing organisms. From Abraham's discoveries to novel concepts of non-ribosomal peptide synthesis. *J. Antibiot.* **53,** 1008–1021.

McAlpine, J. B., Bachmann, B. O., Piraee, M., Tremblay, S., Alarco, A. M., Zazopoulos, E., and Farnet, C. M. (2005). Microbial genomics as a guide to drug discovery and structural elucidation: ECO-02301, a novel antifungal agent, as an example. *J. Nat. Prod.* **68,** 493–496.

McGinnis, S., and Madden, T. L. (2004). BLAST: At the core of a powerful and diverse set of sequence analysis tools. *Nucleic Acids Res.* **32,** W20–W25.

Moffitt, M. C., and Neilan, B. A. (2003). Evolutionary affiliations within the superfamily of ketosynthases reflect complex pathway associations. *J. Mol. Evol.* **56,** 446–457.

Moult, J., Fidelis, K., Kryshtafovych, A., Rost, B., Hubbard, T., and Tramontano, A. (2007). Critical assessment of methods of protein structure prediction-Round VII. *Proteins* **69**(Suppl. 8), 3–9.

Mulder, N. J., and Apweiler, R. (2008). The InterPro database and tools for protein domain analysis *Current protocols in bioinformatics/editoral board, Andreas D. Baxevanis... et al.* Chapter 2, Unit 2 7.

Mulder, N. J., Kersey, P., Pruess, M., and Apweiler, R. (2008). *In silico* characterization of proteins: Uniprot, interpro and Integr8. *Mol. Biotechnol.* **38,** 165–177.

Pearson, W. R., and Lipman, D. J. (1988). Improved tools for biological sequence analysis. *Proc. Natl. Acad. Sci. USA* **85,** 2444–2448.

Petkovic, H., Sandmann, A., Challis, I. R., Hecht, H. J., Silakowski, B., Low, L., Beeston, N., Kuscer, E., Garcia-Bernardo, J., Leadlay, P. F., Kendrew, S. G., Wilkinson, B., *et al.* (2008). Substrate specificity of the acyl transferase domains of epoc from the epothilone polyketide synthase. *Org. Biomol. Chem.* **6,** 500–506.

Rausch, C., Hoof, I., Weber, T., Wohlleben, W., and Huson, D. H. (2007). Phylogenetic analysis of condensation domains in NRPS sheds light on their functional evolution. *BMC Evol. Biol.* **7,** 78.

Rausch, C., Weber, T., Kohlbacher, O., Wohlleben, W., and Huson, D. H. (2005). Specificity prediction of adenylation domains in nonribosomal peptide synthetases

(NRPS) using transductive support vector machines (tsvms). *Nucleic Acids Res.* **33,** 5799–5808.

Roongsawang, N., Washio, K., and Morikawa, M. (2007). In vivo characterization of tandem C-terminal thioesterase domains in arthrofactin synthetase. *Chembiochem* **8,** 501–512.

Schmidt, E. W. (2008). Trading molecules and tracking targets in symbiotic interactions. *Nat. Chem. Biol.* **4,** 466–473.

Scrutton, N. S., Berry, A., and Perham, R. N. (1990). Redesign of the coenzyme specificity of a dehydrogenase by protein engineering. *Nature* **343,** 38–43.

Serre, L., Verbree, E. C., Dauter, Z., Stuitje, A. R., and Derewenda, Z. S. (1995). The *Escherichia coli* malonyl-coa:acyl carrier protein transacylase at 1.5-A resolution. Crystal structure of a fatty acid synthase component. *J. Biol. Chem.* **270,** 12961–12964.

Shen, B. (2003). Polyketide biosynthesis beyond the type I, II and III polyketide synthase paradigms. *Curr. Opin. Chem. Biol.* **7,** 285–295.

Shen, B., Du, L., Sanchez, C., Edwards, D. J., Chen, M., and Murrell, J. M. (2002). Cloning and characterization of the bleomycin biosynthetic gene cluster from *Streptomyces verticillus* ATCC15003. *J. Nat. Prod.* **65,** 422–431.

Smith, S., and Tsai, S. C. (2007). The type I fatty acid and polyketide synthases: A tale of two megasynthases. *Nat. Prod. Rep.* **24,** 1041–1072.

Smith, T. F., and Waterman, M. S. (1981). Identification of common molecular subsequences. *J. Mol. Biol.* **147,** 195–197.

Stachelhaus, T., Mootz, H. D., and Marahiel, M. A. (1999). The specificity-conferring code of adenylation domains in nonribosomal peptide synthetases. *Chem. Biol.* **6,** 493–505.

Staunton, J., and Weissman, K. J. (2001). Polyketide biosynthesis: A millennium review. *Nat. Prod. Rep.* **18,** 380–416.

Tanovic, A., Samel, S. A., Essen, L. O., and Marahiel, M. A. (2008). Crystal structure of the termination module of a nonribosomal peptide synthetase. *Science (New York, NY)* **321,** 659–663.

Taylor, W. R. (1986). Identification of protein sequence homology by consensus template alignment. *J. Mol. Biol.* **188,** 233–258.

Traitcheva, N., Jenke-Kodama, H., He, J., Dittmann, E., and Hertweck, C. (2007). Non-colinear polyketide biosynthesis in the aureothin and neoaureothin pathways: An evolutionary perspective. *Chembiochem* **8,** 1841–1849.

Yadav, G., Gokhale, R. S., and Mohanty, D. (2003a). Computational approach for prediction of domain organization and substrate specificity of modular polyketide synthases. *J. Mol. Biol.* **328,** 335–363.

Yadav, G., Gokhale, R. S., and Mohanty, D. (2003b). SEARCHPKS: A program for detection and analysis of polyketide synthase domains. *Nucleic Acids Res.* **31,** 3654–3658.

CHAPTER NINE

Synthetic Probes for Polyketide and Nonribosomal Peptide Biosynthetic Enzymes

Jordan L. Meier *and* Michael D. Burkart

Contents

1. Introduction	220
2. Synthetic Probes of PKS and NRPS Mechanism	221
2.1. Carrier protein posttranslational modification and PPTase promiscuity	221
2.2. Utility of CoA analogues to study PKS and NRPS mechanism	223
2.3. One-pot chemoenzymatic synthesis of CoA and carrier protein analogues	225
3. Synthetic Probes of PKS and NRPS Structure	233
3.1. Challenges in the structural analysis of modular biosynthetic enzymes	233
3.2. Synthetic probes of PKS and NRPS protein–substrate interactions	234
3.3. Synthetic probes of PKS and NRPS protein–protein interactions	236
4. Synthetic Probes for Proteomic Identification of PKS and NRPS Enzymes	241
4.1. Proteomic study and identification of PKS and NRPS biosynthetic enzymes	241
4.2. Chemoenzymatic labeling of *apo*-carrier proteins in cell lysates	241
4.3. Complementary labeling of PKS and NRPS domains by activity-based probes	244
4.4. *In vivo* labeling of carrier proteins	247
5. Conclusions	250
References	250

Department of Chemistry and Biochemistry, University of California, San Diego, La Jolla, California, USA

Abstract

Polyketides and nonribosomal peptides constitute two classes of small molecule natural products that are well-known for their ability to impact important biological processes in a multitude of ways. The modular biosynthetic enzymes responsible for production of these compounds (PKS and NRPS enzymes) have been the subject of extensive genetic, biochemical, and structural characterization, in part due to the potential utility their successful reengineering may have for the production of new therapeutics. In this chapter, we provide background as well as specific techniques in which synthetically produced small molecule probes have been applied to help better understand the mechanism and structure of PKS and NRPS biosynthetic pathways, as well as to help streamline their discovery process. The continued development and application of these methods has the potential to greatly complement our current approaches to the study of natural product biosynthesis.

1. Introduction

Polyketide and nonribosomal peptide natural products have demonstrated utility in a wide range of therapeutic applications, including as antibiotics (erythromycin, daptomycin, vancomycin), antiparasitics (avermectin), anticancer agents (epothilone, C-1027), cholesterol lowering agents (statins), and immunosuppressants (FK506, cyclosporin) (Fischbach and Walsh, 2006). Perhaps surprisingly, the diverse chemical structures which these compounds employ to interact with their medicinally relevant targets arise through a common biosynthetic paradigm, in which individual monomer units are activated, condensed, and functionalized by large, multienzymatic catalysts. These modular biosynthetic enzymes, known as polyketide synthases (PKS) and nonribosomal peptide synthetases (NRPS), have been the subject of intense genetic, mechanistic, and structural study, topics which have been thoroughly reviewed both in this series and elsewhere (Sieber and Marahiel, 2005; Smith and Tsai, 2007; Staunton and Weissman, 2001). The discovery and characterization of PKS and NRPS enzymes remains an important goal in the study of natural product biosynthesis, due to the often unprecedented enzymatic chemistry which these modular biosynthetic systems utilize, as well as for the potential biotechnological applications their successful reengineering may have for so-called "combinatorial biosynthetic" approaches to drug discovery.

This chapter aims to describe some of the ways in which these twin goals of discovery and characterization of PKS and NRPS biosynthetic enzymes have been facilitated by synthetically constructed probes. We have split the topic into three parts, covering the use of synthetically constructed small molecules to probe the (i) mechanism and (ii) structure of modular biosynthetic enzymes, as well as their use in (iii) discovery and labeling of these enzymes in proteomic settings. As this is a relatively young field of study, we have chosen to introduce each section by outlining some notable examples from the literature in which synthetic probes have been applied towards the study of these topics, followed by the description of detailed protocols from our own laboratory for the chemoenzymatic synthesis and carrier protein (CP) loading of coenzyme A (CoA) analogues, site-specific crosslinking of acyl CP and partner domains, and labeling of PKS and NRPS CPs in proteomic preparations. Where applicable we have provided improved synthetic methods and notes to assist in the production of referenced compounds. This information should aid those interested in the use of these techniques to study modular biosynthetic enzymes, as well as help provide inspiration for the design of next-generation probes of PKS and NRPS biochemistry.

2. Synthetic Probes of PKS and NRPS Mechanism

2.1. Carrier protein posttranslational modification and PPTase promiscuity

One common characteristic shared by PKS and NRPS multienzymes is their use of CP domains to covalently tether intermediates throughout biosynthesis (Mercer and Burkart, 2007). The site of tethering is the terminal thiol of a posttranslationally appended $4'$-phosophopantetheine ($4'$-PPant) arm. Covalent tethering to CP domains facilitates biosynthesis by aiding in substrate channeling and isolation of unstable intermediates from the cellular milieu. The terminal thiol of the $4'$-PPant moiety is ideally suited for this purpose, as the thermodynamically activated thioester bond facilitates the acyl transfer and condensation reactions which are responsible for production of polyketides and nonribosomal peptides from their constituent monomer units. The $4'$-PPant posttranslational modification itself is introduced into PKS and NRPS CP domains through the action of dedicated phosphopantetheinyltransferase (PPTase) enzymes, which transfer the $4'$-PPant group from CoA to a conserved serine residue of the acyl (ACP) or peptidyl carrier protein (PCP) domain (Lambalot et al., 1996; Fig. 9.1).

This mechanism of CP posttranslational modification by PPTases is especially notable from the perspective of the design and utilization of synthetic probes for PKS and NRPS biochemistry. This is due to the phenomenon, observed early in the study of PKS and NRPS PPTases, that a number of these enzymes show a remarkable lack of specificity for the identify of the CoA analogue, and can be used to transfer a variety of unnatural $4'$-PPant arms to their cognate CPs (Gehring et al., 1997; Fig. 9.2A). The biological rationale for this substrate permissivity is thought to be due to the fact that a large portion of the intracellular CoA

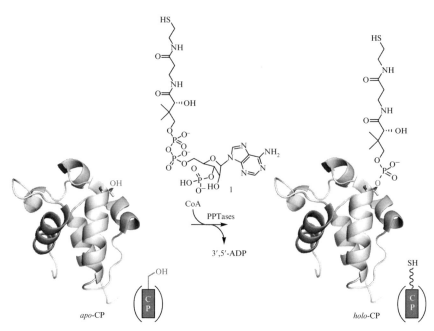

Figure 9.1 Posttranslational modification of carrier protein domains in PKS and NRPS biosynthesis. PPTase enzymes transfer 4′-phosphopantetheine from CoA to the conserved serine of carrier protein domains. The terminal thiol of this prosthetic group functions to covalently tether growing biosynthetic intermediates to the carrier protein during both polyketide and nonribosomal peptide biosynthesis. Pictured: *Escherichia coli* ACP, PDB ID 1T86. (See Color Insert.)

pool at any given time is made up of acetyl CoA, and in order to efficiently posttranslationally modify and activate PKS and NRPS megasynthases (produced at a colossal energy cost to the microbial organism), their PPTases must be able to utilize this substrate. Along with CoA substrate promiscuity some PPTases, most notably Sfp from the surfactin biosynthetic pathway, have also shown the ability to modify noncognate CPs (Quadri *et al.*, 1998). The doubly promiscuous nature of PPTase enzymes such as Sfp has been extremely useful both in providing a method for the expression of *holo*-PKS and NRPS enzymes in heterologous hosts, as well as by allowing the loading of recombinantly produced *apo*-PKS and -NRPS CPs with CoA analogues prepared by combined synthetic and chemoenzymatic routes. Depending on the structure of the CoA analogue, the resulting CP-appended 4′-PPant prosthetic group can be used in the probing of PKS and NRPS biosynthetic mechanisms, substrate specificity, protein–protein interactions, as well as for visualization and affinity purification of their CP active sites, a strategy which figures prominently in some of the examples and protocols described in this chapter.

Synthetic Probes for Polyketide and Nonribosomal Peptide Biosynthetic Enzymes 223

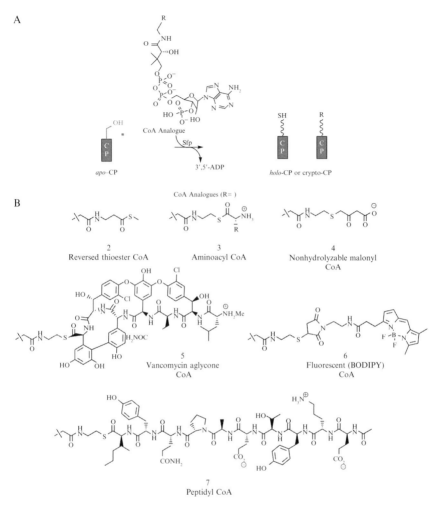

Figure 9.2 PPTase promiscuity in the CP posttranslational modification reaction. (A) Many PPTases such as Sfp (from the *B. subtilis* surfactin biosynthetic pathway) are capable of utilizing analogues of CoA and modifying noncognate CP domains during the PPTase reaction. (B) This can be used to prime the carrier protein with probes of NRPS condensation (3), PKS condensation (4), NRPS oxidative cross-coupling (5), PKS/NRPS *in vivo* posttranslational modification (6), NRPS macrocyclization (7), and a host of other processes.

2.2. Utility of CoA analogues to study PKS and NRPS mechanism

One of the first uses of the PPTase Sfp in combination with synthetically prepared CoA analogues was to study the mechanism of the initial condensation reaction in the tyrocidine NRPS (Belshaw *et al.*, 1999). The NRPS modules TycA (A/PCP/E) and TycB1 (C/A/PCP) catalyze loading, epimerization, and condensation of L-Phe and L-Pro to form the PCP-bound dipeptide intermediate D-Phe-L-Pro-S-TycB1 during the biosynthesis of tyrocidine. In order to gain insight into the timing of the epimerization and condensation reactions in nonribosomal peptide biosynthesis, Belshaw and coworkers (1999) used Sfp in combination with aminoacyl CoA analogues (Fig. 9.2, analogue 3) to bypass the A domain substrate selectivity of these enzymes and load *apo*-GrsA (structurally and functionally analogous to TycA) and *apo*-TycB1 with a number of different aminoacyl $4'$-PPant arms. Notable findings of this study included the observation that upon loading of the donor site, GrsA, with L-Phe-S-CoA, TycB1 loaded with the natural substrate L-Pro formed the D-Phe-L-Pro dipeptide almost exclusively, providing evidence that epimerization of L-Phe preceded the condensation reaction. Also interesting was the finding that, while TycB1 loaded with the natural substrate L-Pro was able to efficiently catalyze dipeptide formation with several different TycA-loaded amino acids, a similar substrate promiscuity was not observed upon loading of TycB1 with the unnatural amino acids and TycA with its natural L-Phe substrate, indicating the condensation reaction has a greater substrate specificity for the downstream acceptor than the upstream donor (Belshaw *et al.*, 1999). This study showcased the power of synthetic probes and PPTase modification by providing some of the first insights into the timing and fidelity-conferring determinants of NRPS biosynthesis. Since this original study of NRPS initiation C-domains, peptidyl CoAs have found use in probing the timing of the condensation and epimerization reactions of the internal C-domain housed on TycC1, finding that epimerization preceded condensation in this setting as well (Clugston *et al.*, 2003), and in the study of the dual C/E domains used in biosynthesis of the cyclic lipopeptide arthrofactin (Balibar *et al.*, 2005). Peptidyl CoAs have also been used to probe the cyclization mechanism of thioesterase (TE) domains which do not efficiently utilize N-acetyl cysteamine (NAC) analogues as CP surrogates and to probe the substrate selectivity of E domains (Sieber *et al.*, 2003, 2005).

An example in which a synthetically prepared CoA analogue has been used to probe the condensation reaction in a very different manner lies in the use of nonhydrolyzable malonyl CoA analogue 4 to capture intermediates formed in type III polyketide biosynthesis (Spiteller *et al.*, 2005). Stilbene synthase, a canonical type III PKS enzyme, catalyzes formation of phenylpropanoids such as resveratrol by condensation of a *p*-coumaroyl CoA and three malonyl CoA units, followed by cyclization (Austin and Noel, 2003). Recalling the normal PKS condensation reaction, in which a malonate (in type III PKSs, malonyl CoA; in type I and II PKSs, malonyl-S-ACP) undergoes decarboxylation and the resulting enolate attacks the enzyme-bound polyketide to affect *trans*-thioesterification and condensation, Spiteller and colleagues (2005) developed a nonhydrolyzable malonyl

CoA analogue (**4**) in which the thioester bond was replaced by a nonhydrolyzable thioether linkage. This substitution ensured that upon decarboxylation and capture of the polyketide intermediate, the nonhydrolyzable polyketide CoA would be incapable of further *trans*-acylation reactions, effectively trapping the polyketide intermediate at that step. By incubating stilbene synthase with an aromatic CoA starter unit and varying the ratios of malonyl CoA to nonhydrolyzable malonyl CoA **4**, di- and triketide CoA intermediates were observable by LC-MS (Spiteller *et al.*, 2005). While this approach was first tested in a type III PKS system, it should be compatible also with CP loading by Sfp. The use of this approach in combination with the top-down observation of CP-bound intermediates by Fourier Transform mass spectrometry (FT-MS) (Dorrestein and Kelleher, 2006) could be a powerful tool for mechanistic analysis of types I and II PKS systems in the future.

Another way in which CoA analogues can be used to study PKS and NRPS systems is through fluorescent and/or affinity labeling of CP active sites. This approach was first pursued by our group in 2004, when we described a simple synthesis of reporter-labeled CoA analogues which could be used to label *apo*-CPs using the broad substrate tolerance of Sfp (La Clair *et al.*, 2004). These modified CPs, in which the terminal 4′-PPant thiol is "hidden" by a nonhydrolyzable linkage, are referred to as *crypto*-CPs. In this study fluorescent labeling of CPs was used to quantify the posttranslational modification of heterologous PKS and NRPS CPs by endogenous CoA during cooverexpression with Sfp. Similar approaches have since been applied towards identification of CP active site fragments for observation of biosynthetic intermediates in reconstituted PKS and NRPS systems by FT-MS (McLoughlin *et al.*, 2005), as well as in a number of site-specific protein labeling strategies (Foley and Burkart, 2007). This approach also has modest utility in labeling and identification of PKS and NRPS CPs from natural product producer proteomes, which we will discuss later in this chapter (Section 4.2).

2.3. One-pot chemoenzymatic synthesis of CoA and carrier protein analogues

One major challenge in the use of CoA analogues and PPTase loading to probe the structure, mechanism, and selectivity of PKS and NRPS biosynthetic systems is the difficulty involved in synthetic manipulation and purification of CoA analogues. CoA itself is expensive (0.5–3.8 dollars/milligram, http://sial.com, 9/20/08) and difficult to work with synthetically due to its highly polar nature, which often requires the use of aqueous-compatible coupling conditions followed by reverse phase HPLC purification of the thioester product. Furthermore, some acyl CoA analogues, particularly aminoacyl CoAs and succinyl CoA, are inherently unstable and undergo rapid hydrolysis of the thioester bond, complicating their use in CP labeling strategies (Belshaw et al., 1999). Nonhydrolyzable CoA analogues, which can be useful in fluorescent labeling and chemoenzymatic crosslinking (Section 3.3) of CPs, can be prepared directly from CoA through its reaction with suitably functionalized maleimides, haloacetamides, or halomethylketones. However, incorporation of the resulting maleimide or methylene unit into the resulting $4'$-PPant arm of the crypto-CP may hinder applications in which substrate recognition and presentation is essential. For example, when applying their nonhydrolyzable malonyl CoA analogue to capture intermediates in type III PKS biosynthesis, Spencer and coworkers found they were unable to observe capture of the tetraketide intermediate under a variety of conditions, presumably due to the inability of the large CoA analogue (one methylene unit longer than the natural substrate) to access the enzyme active site (Spiteller et al., 2005).

In order to overcome these limitations, we commonly utilize a one-pot chemoenzymatic method to prepare CPs with modified $4'$-Ppant arms directly from CoA precursors, referred to herein as pantetheine analogues (Worthington and Burkart, 2006). Pantetheine itself consists of three components, pantoic acid, β-alanine, and a terminal cysteamine, which are coupled through amide linkages. This material can be converted into CoA by the combined enzymatic activities of recombinant Escherichia coli pantothenate kinase (PanK), $4'$-phosphopantetheine adenyltransferase (PPAT), and dephospho-CoA kinase (DPCK) (Martin and Drueckhammer, 1992). Perhaps more importantly, these enzymes will also accept a wide variety of usefully functionalized pantetheine analogues, and can be applied in combination with Sfp to produce crypto-CP domains with modified $4'$-PPant arms. For this reason we have devised a number of synthetic routes toward derivatives of this small molecule, which is considerably more amenable to traditional synthetic manipulations than CoA (Fig. 9.3). Pantethine (the disulfide of pantetheine) is commercially available and a number of protocols exist for reduction and acylation of the resulting free thiol (Davis et al., 1987; Palmer et al., 1991). The majority of our work has focused on the synthesis of acyl-pantetheine analogues in which the thioester bond is replaced by a nonhydrolyzable amide linkage, some examples of which are given in Fig. 9.3 (**11, 13–15**). Routes to these molecules proceed through a set of common protected precursors, a number of which we provide optimized protocols for below. These nonhydrolyzable pantetheine analogues can be used with the one-pot chemoenzymatic method (Fig. 9.4A, Section 2.3.2) for PKS and NRPS inhibition by nonhydrolyzable substrate analogues, capture of enzymatic intermediates, site-specific crosslinking, and fluorescent visualization of CPs.

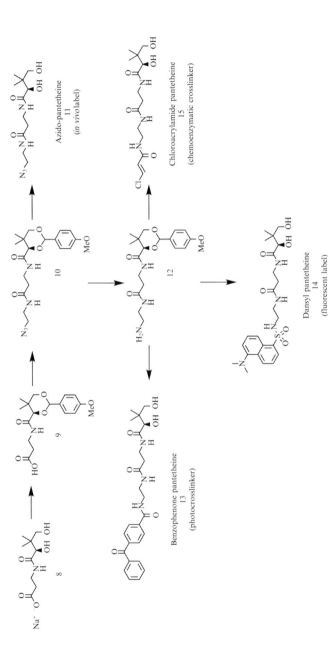

Figure 9.3 Synthetic routes to CoA precursors (pantetheine analogues). These analogues can be applied in combination with the CoA biosynthetic enzymes and Sfp in order to prime CP domains with 4′-PPant arms useful in *in vivo* labeling (**11**), photocrosslinking (**13**), fluorescent visualization (**14**), chemoenzymatic crosslinking (**15**) of CP domains.

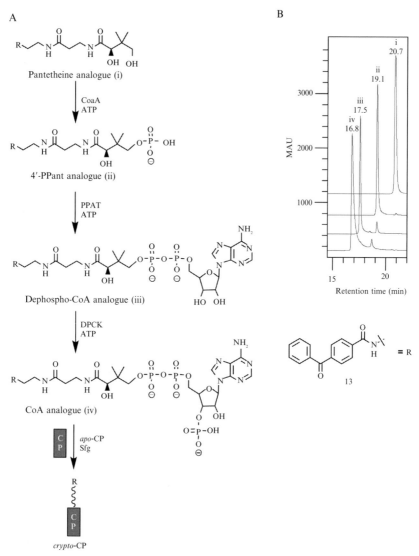

Figure 9.4 One-pot chemoenzymatic synthesis of modified CPs. (A) Pantetheine analogue (top) is processed by the sequential action of the CoA biosynthetic enzymes PanK, PPAT, and DPCK to form a CoA analogue, which can then be transferred to the CP domain by Sfp. (B) The compatibility of the pantetheine analogue with each step of the recombinant CoA biosynthetic pathway can be followed by HPLC, which demonstrates a characteristic shift to lower retention times the pantetheine analogue (13 eluting at 20.7 min) is converted to the 4′-phosphopantetheine analogue (19.1 min), dephospho-CoA analogue (17.5 min), and CoA analogue (16.8 min). Compatibility with the PPTase reaction can then be monitored by previously described gel-based or MALDI methods (Meier *et al.*, 2006).

2.3.1. Synthetic methods

2.3.1.1. General synthetic methods
All chemical reagents were obtained from commercial suppliers and used without further purification unless otherwise noted. Rhodamine alkyne, biotin alkyne, and FP-DMACA (applied in Sections 4.3 and 4.4) are prepared according to previously reported procedures (Meier *et al.*, 2006, 2008; Speers *et al.*, 2003). The following protocols describe improved synthetic routes and notes for compounds **6, 10–12**, and **16** developed since their initial publication.

2.3.1.2. Synthesis of BODIPY-Coenzyme A (6)
To a 1.6-ml aqueous solution of MES acetate (75 mM) and Mg(OAc)$_2$ (100 mM) at pH 6.0 was added 300 μl of CoA trilithium salt (1 mg/ml, 0.382 μmol, 1 eq) and 300 μl of DMSO. The reaction was vortexed and 202 μl of BODIPY FL N-(2-aminoethyl)maleimide (Molecular Probes, http://probes.invitrogen.com) dissolved in DMSO (1.11 mg/ml, 0.541 μmol, 1.4 eq) was added. The resulting solution was vortexed briefly and allowed to react at 4 °C for 3 h. Note that the reaction should be protected from light during this period to avoid photobleaching of the BODIPY fluorophore. Formation of the fluorescent CoA analogue could be followed by thin layer chromatography (butanol/acetic acid/water, 5:2:4) or HPLC. Extraction of the completed reaction with ethyl acetate (3 × 4 ml) provided BODIPY-Coenzyme A as a stock solution of ∼100–160 μM. The exact concentration of the CoA analogue can be determined by comparative fluorescence analysis to a standard solution of BODIPY FL (λ_{Ex} ∼504 nm, λ_{Em} ∼510 nm) using a PTI-Alphascan fluorimeter.

2.3.1.3. Synthesis of PMB-pantothenic acid (9)
A 2-l round-bottom flask equipped with a magnetic stir bar and drying tube was charged with 750 ml of dichloromethane, followed by sequential addition of sodium pantothenate (50g, 207 mmol) and p-anisaldehyde dimethyl acetal (45g, 42 ml, 247 mmol). To this suspension was added an anhydrous solution of 4 M hydrochloric acid in dioxanes (∼50 ml) over 30 min. The best yields were found to occur when HCl addition was performed slowly enough to avoid formation of a pink color, associated with *p*-methoxybenzylidene protonation and deprotection. The reaction was stirred for 2 h at room temperature and vacuum-filtered to remove precipitated NaCl. This precipitate was washed with cold dichloromethane and the resulting filtrate loaded directly onto a short plug of silica gel and eluted with dichloromethane into 1 l Erlenmeyer flasks. Fractions containing a significant amount of the protected acid (visualized by TLC staining with 2,4-dinitrophenyl hydrazine) were recrystallized by addition of hexanes and overnight incubation at 4 °C to give the pure PMB-protected pantothenic acid **9** as a white crystalline powder (38g, 55%). Yields typically range from 30–55%. This material can alternately be purified through acid-base extraction by addition of saturated NaHCO$_3$, removal of organic impurities by

washing with dichloromethane, followed by careful acidification with citric acid to pH ~4.0 and extraction with dichlormethane, although this procedure causes significant deprotection during the reacidification step resulting in lower yields. ^{1}H NMR (400 MHz, (CD$_3$)$_2$SO) 7.41 (d, 2H, J = 8.8 Hz), 6.91 (d, 2H, J = 8.8 Hz), 5.50 (s, 1H), 4.07 (s, 1H), 3.74 (s, 3 H), 3.62 (d, H, J = 10.8 Hz), 3.59 (d, 1H, J = 10.8 Hz), 3.34 (m, 1H), 3.25 (m, 1H), 2.38 (t, 2H, J = 6.8 Hz), 0.99 (s, 3 H), 0.93 (s, 3 H). ^{13}C NMR (400 MHz, (CD$_3$)2SO) 173.8, 168.9, 160.3, 131.1, 128.4, 114.0, 101.1, 83.8, 78.0, 55.8, 34.9, 34.4, 33.2, 22.2, 19.7. HRMS (FAB) (m/z): [M + H]$^+$ calcd for C$_{17}$H$_{23}$O$_6$N, 338.1598, found 338.1594.

2.3.1.4. Synthesis of PMB-pantetheine azide (10)
Synthesis of PMB-protected pantetheine azide proceeds through coupling of PMB-pantothenic acid and 2-azidoethylamine. 2-azidoethylamine is most easily prepared as its hydrochloride salt starting from the commercially available 2-bromoethylamine hydrobromide according to the method of Ritter and König (Ritter and Konig, 2006). In our hands this procedure has provided 2-azidoethylamine hydrochloride on 5–20 g scales in high yields (90–95%). This material can then be coupled to PMB-pantothenic acid **9** according to our originally reported protocol using EDC, or in a more timely fashion through the mixed-anhydride method using isobutylchloroformate. Briefly, PMB-pantothenic acid **9** (16 g, 48.0 mmol) and N-methylmorpholine (10.7 ml, 100.5 mmol) are dissolved in dry THF (450 ml) with stirring. Isobutylchloroformate (5.5 ml, 42 mmol) is added dropwise, resulting in formation of a white precipitate, and stirred at room temperature for 10 min. A solution of 2-azidoethylamine hydrochloride (3.7 g, 52.5 mmol) dissolved in the minimal amount of DMF is added to the reaction dropwise, and the solution is allowed to stir for 30 min. The reaction is quenched by addition of methanol (~15 ml), followed by removal of the solvent under reduced pressure. The resulting oil is redissolved in ethyl acetate (~500 ml) and extracted with water (200 ml × 1), saturated sodium bicarbonate (200 ml × 3) and brine (200 ml × 1). Removal of the solvent under reduced pressure gives the product **10** as a clear oil (15 g, 88%). ^{1}H NMR (500 MHz, (CD$_3$)$_2$SO) δ 8.12 (1H, bt), 7.44 (1H, bt), 7.41 (d, 2H, J = 8.5 Hz), 6.90 (d, 2H, J = 8.5Hz), 5.50 (s, 1H), 4.06 (s, 1H), 3.74 (s, 3H), 3.62 (d, 1H, J = 11.5 Hz), 3.58 (d, 1H, J = 11.5 Hz), 3.35–3.18 (m, 6H), 2.27 (t, 2H, J = 7.0 Hz), 0.98 (s, 3H), 0.93 (s, 3H). ^{13}C- NMR (75.5 MHz, (CD$_3$)2CO) δ 171.7, 169.7, 160.4, 130.3, 127.8, 113.9, 101.5, 84.0, 78.5, 50.7, 39.1, 35.8, 35.1, 33.2, 22.0, 19.3. HRMS (EI) (m/z): [M + H]$^+$ calcd for C$_{19}$H$_{27}$O$_5$N$_5$, 405.2007, found 405.2005.

2.3.1.5. Synthesis of pantetheine azide (11)
PMB-pantetheine azide **10** (14.6 g, 36 mmol) is dissolved in THF (800 ml) with stirring. An aqueous solution of 1 M HCl (100 ml) is added dropwise via an addition funnel, and the mixture is allowed to stir at room temperature. The deprotection can be

hastened by running the reaction at higher concentrations (i.e., 0.1 M) with a 1:1 ratio of THF/1M HCl, but in our experience this also results in greater amounts of product degradation, mainly due to acid-catalyzed formation of pantolactone and concomitant release of the β-alanine-2-azidoethylamine conjugate. The reaction is followed by TLC and usually judged to be complete after ~3–6 h. Upon completion the reaction is quenched by addition of AG1X8 Strong Basic anion exchange resin until neutral, followed by vacuum filtration and removal of the solvent under reduced pressure. The crude product is redissolved in dichlormethane with the minimal amount of methanol and purified by flash chromatography (CH$_2$Cl$_2$ to 7% methanol:CH$_2$Cl$_2$) to yield pantetheine azide **11** (9.4 g, 91%) as a clear oil. ^{1}H NMR (500 MHz, (CD$_3$)$_2$SO) δ 8.10 (bt, 1H), 7.67 (bt, 1H), 5.34 (d, J = 5.5 Hz, 1H), 4.44 (t, J = 5.5 Hz, 1H), 4.07 (dd, J = 10.5, 5.5 Hz, 1H), 3.67 (d, J = 5.5 Hz, 1H), 3.33–3.14 (m, 6H), 2.25 (t, J = 7.0 Hz, 2H), 0.77 (s, 3H), 0.75 (s, 3H). ^{13}C NMR (75.5 MHz, CDCl$_3$) δ 174.2, 172.2, 77.6, 70.9, 50.8, 39.5, 39.1, 35.9, 35.4, 21.5, 20.7. HRMS (EI) (m/z) [M]$^+$ calcd for C$_{11}$H$_{21}$O$_4$N$_5$, 287.1592, found 287.1588.

2.3.1.6. Synthesis of pantetheine chloroacrylamide (15) PMP-pantetheine amine **12** is prepared from PMP-pantetheine azide by Staudinger reduction according to our previously reported procedure. The free amine is then coupled to the commercially available *trans*-chloroacrylic acid (Sigma Aldrich, http://sial.com) according to the method of Worthington (2008). In our experience PyBOP has proven to be the coupling reagent of choice for this transformation due to its compatibility with α,β-unsaturated acids, while carbodiimides such as EDC and DCC give extremely low yields. Acid-catalyzed deprotection is performed using THF/1M HCl as above, with purification by flash chromatography (EtOAc to 5% MeOH:EtOAc) yielding the deprotected pantetheine chloroacrylamide **15** as white crystals. ^{1}H NMR (400 MHz, CD$_3$OD): δ7.13 (d, J = 13.2 Hz, 1H), 6.25 (d, J = 13.2 Hz, 1H), 3.83 (s, 1H), 3.35 (m, 8H), 2.31 (m, 2H), 0.85 (s, 3H), 0.82 (s, 3H). ^{13}C NMR (100.5 MHz, CD$_3$OD): δ 175.0, 173.2, 173.1, 133.2, 127.2, 76.1, 69.1, 39.2, 38.9, 38.8, 35.5, 35.3, 20.3, 19.7. HRMS (ESI) (m/z): [M + H]$^+$ calcd for C$_{14}$H$_{24}$ClN$_3$O$_5$ 350.1476, found 350.1484.

2.3.2. Chemoenzymatic reaction protocol
2.3.2.1. General materials and protocols PanK, PPAT, and DPCK were cloned from *E. coli* genomic DNA into pET24b vectors (Novagen) for overexpression as previously described (Clarke *et al.*, 2005). The overexpression vector encoding *E. coli* KASI (pCA24N) was obtained from the ASKA library of Andrei Osterman (Burnham Institute for Medical Research). Vectors encoding Fren ACP (pET22b) and Sfp (pREP4) were gifts from Christopher Walsh (Harvard Medical School). All proteins were

overexpressed and purified according to standard protocols and stored as 15–40% glycerol stocks at −80 °C. *E. coli apo*-ACP was prepared as previously described (Keating *et al.*, 1995). For chemoenzymatic reaction protocols, the enzyme amount added to the reaction is given in μg according to quantitation of protein concentration by absorbance at 280 nm using the extinction coefficient of the specified protein. For proteomic identification and in gel digests (used in Sections 3.3.1, 4.2, and 4.3) bands are excised from SDS-PAGE gels and prepared for identification by MS/MS according to the method of Worthington (2006).

2.3.2.2. HPLC analysis of pantetheine analogue compatibility with PanK/ PPAT/DPCK In order for a pantetheine analogue to be useful in our one-pot chemoenzymatic loading of CPs, it must be compatible not only with Sfp (whose permissivity is well-known) but also with each of the CoA biosynthetic enzymes, whose substrate selectivity has been less well studied. While a spectrophotometric-coupled enzyme assay can be used to determine the kinetics of pantetheine analogues with PanK, a quick qualitative analysis of this variable can be performed using HPLC. Conversion of a pantetheine analogue by each step in the *in vitro* reconstituted CoA biosynthetic pathway introduces an additional negative charge to the transformed substrate, resulting in a characteristic shift to shorter retention time. A standard assay involves sequential addition of the enzymes PanK, PPAT, and DPCK (2 μg each) to a 30-μl reaction containing the pantetheine analogue (4 mM) in a CoA reaction buffer (300 mM Tris–Cl, pH 7.5, 16 mM KCl, 12 mM MgCl$_2$, 40 mM ATP). The excess of ATP (10:1 relative to pantetheine analogue) used in this protocol serves the purpose of pushing the reaction to completion, obviating the need to use an ATP-recycling system such as pyruvate kinase, inorganic pyrophosphatase, and phosphoenolpyruvate. After 30 min, 10 μl of the reaction is analyzed by reverse-phase HPLC using a Burdick and Jackson OD5 column (4.6 mm by 25 cm) with monitoring at 220 and 254 nm. Solvents used are 0.05% TFA in H$_2$O (Solvent A) and 0.05% TFA in ACN (solvent B). Compounds are eluted at a flow rate of 1 ml/min. Our method utilizes an isocratic step from 0 to 5 min with 100% A, followed by a linear gradient to 45% B over 15 min, followed by an increasing gradient with solution B until at 25 min the solvent composition was 100% solution B. In the case of analogue **13** this results in elution at $\sim$20.7 min for the pantetheine analogue, 19.1 min for the 4$'$-phosphopantetheine analogue, 17.5 min for the dephospho-CoA analogue, and 16.8 min for the fully formed CoA analogue (Fig. 9.4A).

2.3.2.3. One-pot chemoenzymatic loading of CPs by pantetheine analogues Pantetheine analogues that are substrates for the recombinant *E. coli* CoA biosynthetic pathway can be loaded onto recombinant CP *in vitro* using Sfp. To date there has not been a CoA analogue reported which Sfp is incapable of processing, with perhaps the greatest example of this remarkable substrate permissivity being the loading of a CoA analogue incorporating the entire vancomycin aglycone (**5**, MW of 4′-PPant arm = 1.4 kDa) onto a CP by Vitali and coworkers (2003). Our protocol for one-pot chemoenzymatic synthesis and CP loading of CoA analogues involves addition of PanK (2 μg), PPAT (2 μg), DPCK (3 μg), and Sfp (0.5 μg) to a reaction mixture consisting of the pantetheine analogue (0.4 mM) and CP (5 μg) in a one-pot reaction buffer (100 mM potassium phosphate, pH 7.0, 40 mM KCl, 20 mM MgCl$_2$, 10 mM ATP), followed by incubation for 1 h at 37 °C. We have observed that it is important to have MgCl$_2$ in excess of ATP, as ATP is known to chelate Mg^{2+} which can prevent Sfp from accessing this cofactor and cause the reaction to fail. Using this procedure a pantetheine: CP ratio of ~8:1 is usually sufficient for full loading of the CP, and the lower salt concentrations used in this protocol facilitate analysis of the reaction products by MALDI-MS. Loading of fluorescent pantetheine analogues can be analyzed directly by SDS-PAGE, while complete loading of nonfluorescent pantetheine analogues can be followed by quenching the one-pot reaction with BODIPY CoA (10 μM) and an additional aliquot of Sfp, followed by analysis of the resulting fluorescence by SDS-PAGE. CPs which have been successfully loaded with a nonfluorescent CoA precursor by the one-pot chemoenzymatic method should not show labeling under these conditions.

3. Synthetic Probes of PKS and NRPS Structure

3.1. Challenges in the structural analysis of modular biosynthetic enzymes

For years, the spatial arrangement and quaternary architecture of multienzymatic biosynthetic catalysts have represented one of the most mysterious and impenetrable topics in PKS and NRPS biochemistry. Contributing to these difficulties are the extremely large size of PKS and NRPS enzymes (often >200 kDa), as well as their incorporation of one or more CP domains. These CP domains must retain an inherent flexibility in order to shuttle reactive biosynthetic intermediates between multiple enzymatic partners, a property which ultimately hinders their crystallographic resolution. Despite these obstacles, several researchers have thrown themselves at this seemingly Sisyphean task with admirable drive and persistence, recognizing that the considerable challenges involved in PKS and NRPS structure elucidation are balanced by the rewards which may result from their successful redesign. Underscoring the potential of structure to guide PKS reengineering is the example of the type III PKS chalcone synthase, the first polyketide biosynthetic catalyst to be crystallized (Ferrer *et al.*, 1999), for which the existence of high quality structural data has been used to guide rational reprogramming of starter unit specificity, chain length, and cyclization mechanism (Ferrer *et al.*, 2008; Watanabe *et al.*, 2007). The application of a similar approach to the reengineering of type I PKS and NRPS proteins would be extremely attractive in terms of the broad spectrum of small molecules it could potentially access, and has lead to several recent successes in the field, including the generation of high-resolution crystallographic data for multidomain PKS fragments from the DEBS system (Khosla *et al.*, 2007; Tang *et al.*, 2007), multiple refinements of the PKS-related fungal (Jenni *et al.*, 2007; Leibundgut *et al.*, 2007) and animal (Maier *et al.*, 2006, 2008) fatty acid synthase (FAS) crystal structures, and most recently both crystallization and NMR studies of the terminal module of the surfactin NRPS (Frueh *et al.*, 2008; Tanovic *et al.*, 2008). While these studies have answered several questions concerning the quaternary architecture of PKS and NRPS enzymes many more remain, oftentimes related to the precise nature of the protein–protein interactions which allow the CP domain to successfully interact with its many partner domains in polyketide and nonribosomal peptide biosynthesis (Weissman and Muller, 2008). Here we provide a brief background into the ways synthetic probes have been used to facilitate the structural study of PKS and NRPS biosynthetic processes, as well as protocols for the chemoenzymatic synthesis and application of electrophile-conjugated *crypto*-CP domains, which can be used to immobilize these enzymes in well-defined conformations for structural analysis of the protein–protein interactions underlying polyketide and nonribosomal peptide biosynthesis.

3.2. Synthetic probes of PKS and NRPS protein–substrate interactions

Small molecules have been used to provide insight into the molecular interactions necessary for PKS and NRPS function since the advent of high quality structural data for these enzymes. This approach has been most notably applied to enhance the data provided by crystallographic study of these enzymes, some examples being the structures of chalcone synthase in complex with CoA as well as its product naringenin (Ferrer *et al.*, 1999), the crystallization of the excised gramicidin adenylation domain with its hydrolyzed aminoacyl-AMP (a landmark study which provided the molecular basis for the now commonly used "nonribosomal code" of NRPS substrate utilization) (Challis *et al.*, 2000; Conti *et al.*, 1997), and the recent crystallization of the actinorhodin ketoreductase with the product analogue emodin by Tsai and coworkers (Korman *et al.*, 2008).

In contrast to these studies, which have for the most part utilized naturally occurring or commercially available compounds, efforts to specifically design and apply synthetic ligands to the study of PKS and NRPS biosynthetic enzymes are a more recent innovation. The TE domain of PKS and NRPS biosynthetic enzymes has proven an especially appealing target, due to its possible utility in chemoenzymatic approaches to peptide and polyketide macrocyclization, as well as the wealth of knowledge which exists concerning inhibition of serine hydrolase-type enzymes (Powers *et al.*, 2002). Synthetic probes were first used to study peptide macrocyclization by Tseng and coworkers (2002), who designed peptidyl-boronic acid inhibitors such as **16** to probe protein–substrate interactions of the surfactin TE. While clear electron density was observed only for the two C-terminal residues of the inhibitor in complex with surfactin TE, the structural data gleaned was consistent with biochemical observations suggesting that the fully-formed peptidyl chain assumes a product-like conformation prior to macrocycle formation (Tseng *et al.*, 2002).

A similar approach was applied to PKS TE domains by Fecik and coworkers, who utilized linear polyketide substrate analogues incorporating irreversible inhibitors to study TE-catalyzed macrolactonization in the pikromycin (Pik) biosynthetic pathway (Akey *et al.*, 2006; Giraldes *et al.*, 2006). While synthesis of synthetically challenging polyketide-diphenylphosphonates **17–19** constituted a major feat in and of itself soaking of Pik TE crystals with these substrate analogues yielded unexpected findings. These included the observations that the enzyme did not form any specific hydrogen bonds with the inhibitors, nor did it undergo any substantial conformational change upon inhibitor binding. These analyses, along with molecular modeling studies of the full length polyketide bound in the Pik TE active site, were used to argue against an induced-fit model of PKS macrocyclization, and instead lead the authors to propose a model of Pik macrocyclization in which conformational restrictions inherent in the polyketide substrate itself were responsible for constraining the linear polyketide into a product-like conformation for formation of the intramolecular ester bond (Akey *et al.*, 2006). This is consistent with previous biochemical studies of linear Pik substrate analogues (He *et al.*, 2006).

Synthetic probes have also been successfully applied toward study of the accessory enzymes involved in formation of the nonproteinogenic amino acids used in nonribosomal peptide biosynthesis. This was first demonstrated in the study of DpgC, an enzyme which catalyzes a key dioxygenation step in the biosynthesis of 3,5-dihydroxyphenylglycine, an important building block used in biosynthesis of the nonribosomal peptide antibiotic vancomycin. By designing a nonhydrolyzable mimic of the 3,5-dihydroxyphenylacetyl CoA substrate (**20**), Widboom *et al.* (2007) were able to trap an intermediate in the enzyme-catalyzed reaction and visualize both the CoA analogue and electron density interpretable as molecular oxygen in the active site of this cofactor-free oxygenase. This provided some of the first mechanistic information regarding cofactor-free enzymatic oxygen activation. Irreversible inhibitors also proved useful in the study of SgTAM, a tyrosine aminomutase which converts L-tyrosine to (S)-β-tyrosine in the pathway responsible for the biosynthesis of the anticancer enediyne C-1027. By synthesizing a product analogue (α-difluoro-β-tyrosine, **21**) Christianson *et al.* (2007) were able to observe a stable complex between the inhibitor and the electrophilic 4-methylideneimidazole-5-one (MIO) cofactor of the enzyme, providing evidence for a covalent interaction between the L-tyrosine amine and MIO prosthetic group of SgTAM. This mechanism may be general to MIO-utilizing aminomutases and ammonia lyases. Each of these studies demonstrates the ability of small molecule probes to yield fundamental advances in our understanding of the molecular basis for natural product biosynthesis, knowledge which will likely be crucial to future efforts at rational reprogramming of these enzymes for combinatorial biosynthetic endeavors (Fig. 9.5).

3.3. Synthetic probes of PKS and NRPS protein–protein interactions

3.3.1. Chemoenzymatic Crosslinking of ACP and Partner Enzymes

In addition to studies of discrete PKS and NRPS domains, recent years have also seen the noteworthy publication of several crystal structures of multienzymatic fragments from PKS, NRPS, and the related FAS biosynthetic systems. However, one element common to each of these studies has been a lack of electron density corresponding to the CP domain. This is presumably due the inherent conformational flexibility of the CP, which must interact with several enzymatic partners during PKS or NRPS biosynthesis. For this reason, synthetic probes have been developed to aid in the study of the multienzymatic protein–protein interactions which lie at the heart of polyketide and nonribosomal peptide biosynthesis, specifically through covalent immobilization of the transient interactions of the CP and partner domains (Fig. 9.6). This crosslinking approach has now been applied to study the interactions of PKS and NRPS CPs with ketosynthase (KS) (Worthington *et al.*, 2006), adenylation (A) (Qiao *et al.*, 2007), and TE (Liu and Bruner, 2007) domains with varying degrees of success. Here we focus on the first of these methods, site-specific crosslinking of the ACP-KS pair.

Early in the study of the mammalian FAS it was shown that bifunctional electrophiles such as 1,3-dibromopropanone are capable of crosslinking ACP and KS domains (Stoops and Wakil, 1981). While this approach was useful for the study of FAS multimerization and domain architecture, the low yields of crosslinked product and lack of site-selectivity with which these strong electrophiles act preclude its use in the preparation of homogeneous ACP-KS complexes for crystallographic analysis (Kapur *et al.*, 2008). To circumvent this problem Worthington *et al.* (2006) examined

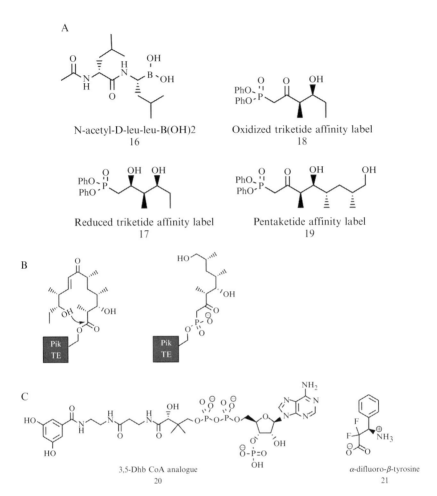

Figure 9.5 Structures of small-molecule probes of PKS and NRPS protein–substrate interactions. (A) Probes of NRPS (16) and PKS (17–19) macrocyclization. (B) Comparison of structures of natural hexaketides cyclized by Pik TE and pentaketide affinity label 19 bound to Pik TE. (C) Mechanistic probes of enzymes involved in non-proteinogenic amino acid biosynthesis in the vancomycin (20) and C-1027 (21) biosynthetic pathways.

the ability of Sfp, whose substrate promiscuity for substituted CoA analogues had at this point been well-established, to transfer electrophilic CoA analogues to the CP active site. Using the one-pot chemoenzymatic method (Section 2.3.2), chloroacrylamide- and epoxide-containing pantetheine analogues (**15** and **22**) were transformed into CoA analogues and loaded onto the *apo*-ACP of the *E. coli* FAS using Sfp. Upon addition of the KS domain to this electrophilic *crypto*-ACP a higher molecular weight species was observed to form, presumably due to reaction of the catalytic

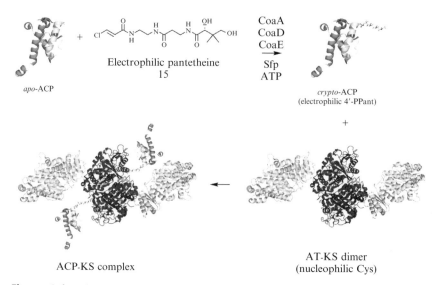

Figure 9.6 Chemoenzymatic crosslinking of ACP and KS domains. *Apo*-ACP domains (depicted here as the ACP from module 2 of the 6-deoxyerythronolide synthase) can be posttranslationally modified with electrophilic 4′-PPant arms using the one-pot chemoenzymatic synthesis to form *crypto*-ACPs. Upon addition of a nucleophilic KS domain (depicted here as the KS-AT didomain from module 3 of the 6-deoxyerythronolide synthase), which contains complementary protein–protein interactions to the ACP, covalent crosslinking will occur between the electrophilic 4′-PPant arm and the nucleophilic cysteine of the KS, resulting in a crosslinked complex. PDB IDs: DEBS2 ACP, 2JU2; DEBS3 KS-AT, 2QO3. (See Color Insert.)

cysteine of the KS domain with the ACP-loaded 4′-PPant electrophile (Fig. 9.7A). The identity of the ACP-KS crosslinked species was further verified by tandem MS analysis of this gel-shifted band.

Since this initial report, site-specific crosslinking of ACP and KS domains has also been applied to a number of PKS systems, including the ACP and KS-CLF of the type II enterocin PKS (Worthington *et al.*, 2008), and the discrete ACPs and AT-KS didomains from modules 3 and 5 of the modular type I DEBS PKS (Kapur *et al.*, 2008). In addition to the potential discoveries which may arise from structural determination of well-ordered ACP-KS pairs by X-ray crystallography, this method can also provide insight into the suitability of protein–protein interactions between unnatural ACP-KS pairs, as it has been shown that KS domains exhibit distinct preferences for crosslinking with their cognate ACP domain. For example, study of the DEBS PKS by this method found that ACP and KS-AT didomains from the same module of the DEBS synthase showed preferential crosslinking (Fig. 9.6) (Kapur *et al.*, 2008). Also notable during these studies was the finding that the crosslinking efficiency of *crypto*-ACPs incorporating fatty-acyl or aromatic groups

Synthetic Probes for Polyketide and Nonribosomal Peptide Biosynthetic Enzymes 239

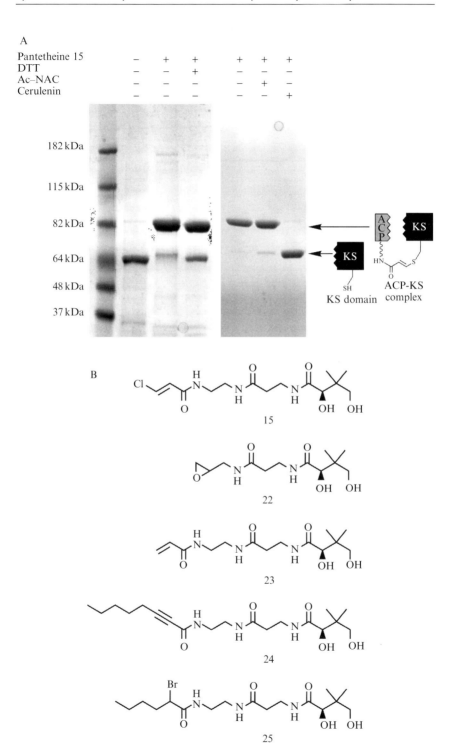

into the electrophilic 4′-PPant arm showed a good correspondence with substrate specificity of the KS in the study of ACP-KS crosslinking of type II PKS and FAS systems. This is an indication that through careful design of electrophilic CoA analogues it may be possible to use ACP-KS crosslinking to measure the relative contributions of the substrate specificity conferred by the group located on the 4′-PPant chain (i.e., fatty acyl-pantetheine), and compatibility of protein–protein interactions for ACP-KS association (Worthington et al., 2008). The use of such methods to probe the complementary nature of CP-partner protein interactions will be useful to future combinatorial biosynthetic efforts.

3.3.2. Chemoenzymatic crosslinking of ACP and KS domains

3.3.2.1. Reaction protocol
This protocol describes chemoenzymatic crosslinking of the ACP and KS domains from the type II fatty acid biosynthetic pathway of E. coli, as this reaction has been the most extensively characterized by our group. Identical procedures have also been shown to be applicable to crosslinking of type I and II PKS domains. The first step of the ACP-KS crosslinking protocol involves loading of the apo-ACP with an electrophilic 4′-PPant arm. This is accomplished through a one-pot chemoenzymatic synthesis (2.3.2) of modified CPs in which an electrophilic pantetheine analogue such as pantetheine chloroacrylamide 15 (0.15 mM) is added to a reaction mixture of PanK (0.5 μg), PPAT (0.5 μg), DPCK (1 μg), Sfp (0.5 μg), and apo-ACP (10 μg) in reaction buffer (50 mM potassium phosphate, pH 7.0, 50 mM $MgCl_2$, 25 mM ATP) and incubated for 30 min at 37 °C. The KS domain can also be added at this time, although our studies have shown that higher yields of ACP-KS complex result from preloading of the ACP before addition of the KS domain when crosslinking type II systems. After this time, KASI (2.3 μg) is added and the reaction is allowed to incubate at room temperature for 1 h, followed by SDS-PAGE and staining by Coomassie. Successful crosslinking causes a shift in molecular weight of KASII of ~+20 kDa (Fig. 9.7A). Evidence for the site-selectivity of this process can be obtained by predenaturation of the KS by boiling or treatment with a known inhibitor of the catalytic cysteine such as cerulenin. Factors which can cause the reaction to fail include the use of ATP at greater concentrations than Mg^{2+} (causes Sfp to fail), the use of high concentrations (>0.5 mM) of electrophilic pantetheine (inhibits KS in *trans*, blocking

Figure 9.7 Representative data from ACP-KS crosslinking experiments. (A) E. coli ACP-KS crosslinking. Addition of KASI (KS) to *crypto*-ACP posttranslationally modified with pantetheine analogue 15 results in covalent crosslinking, visualized as a gel shift of the KS from ~64 kDa to ~82 kDa. Gel-shift is not observed if 15 is omitted from the reaction mixture (so that *crypto*-ACP is not formed), or if KS domain is preincubated with cerulenin, a known KS active site cysteine affinity label. (B) Pantetheine analogues (15, 22–25) which have been usefully applied to the study of ACP-KS crosslinking in PKS, NRPS, and FAS systems.

crosslinking), and degradative hydrolysis of electrophilic pantetheine analogues during the ACP preloading step (only a problem if the ACP-pantetheine preincubation is performed for ≫30 min). This last factor can be remedied by the use of nonhalogen-containing Michael acceptors, as we found in a recent study that acrylamide **23** and 2-alkynyl substituted pantetheine analogue **24** were valid substitutes for **15** in the crosslinking reaction (Worthington *et al.*, 2008). In general the conditions above are amenable to scale-up, and have lead to high yields of ACP-KS product even when performed on scales of up to $10^6 \, \mu l$. Orthogonally tagged CoA biosynthetic enzymes and native Sfp can be used to facilitate isolation of the ACP-KS complex for structural studies (Haushalter *et al.*, 2008).

4. Synthetic Probes for Proteomic Identification of PKS and NRPS Enzymes

4.1. Proteomic study and identification of PKS and NRPS biosynthetic enzymes

In addition to the continued development of techniques for the study of recombinant PKS and NRPS enzymes *in vitro*, recent years have seen the development of synthetic probes designed to enable study of these systems *in vivo* as well as in crude proteomic preparations. Methods to specifically label PKS and NRPS proteins in these settings have the potential to complement the way we currently study these enzymes, allowing us to observe the dynamics and trafficking of CP-mediated biosynthetic processes, streamline the identification of PKS and NRPS gene clusters from unsequenced organisms, better understand the signaling pathways underlying natural product biosynthesis, and even possibly determine key mechanistic steps in polyketide and nonribosomal peptide biosynthesis using enzymes isolated directly from the natural product producer itself. While some of these approaches are still far from maturity, to facilitate their continued development and optimization we provide here a thorough description of currently used methods for specific labeling of PKS and NRPS enzymes in proteomic settings.

4.2. Chemoenzymatic labeling of *apo*-carrier proteins in cell lysates

4.2.1. Background
The well-known tolerance of the PPTase Sfp for noncognate CPs as well as analogues of CoA provides an avenue for selective labeling of PKS and NRPS biosynthetic enzymes in proteomic preparations, namely through labeling of their CP domains with affinity-tagged or fluorescent CoA analogues (Fig. 9.8A). In our initial report on the use of fluorescent CoA

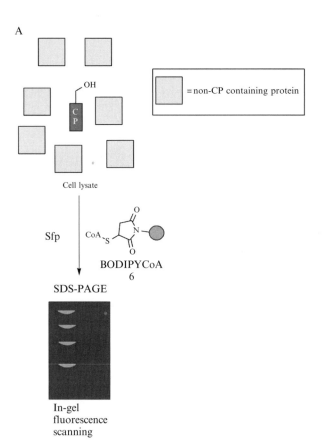
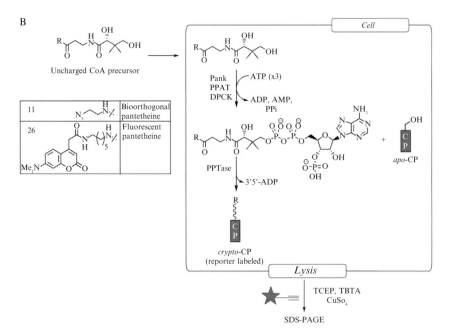

Figure 9.8 Strategies for proteomic labeling of PKS and NRPS enzymes. (A) *Apo*-CP domains in crude cell lysate can be directly labeled by use of reporter-labeled CoA analogues (6) and Sfp. This strategy is most useful in organisms in which the secondary PPTase has been inactivated, such as *B. subtilis* 168. (B) *In vivo* labeling of CP domains by bioorthogonal (11) or fluorescent (26) CoA precursors. Cellular uptake and processing of 11 by the endogenous CoA biosynthetic pathway results in formation of reporter-labeled CoA *in vivo*, which can be utilized by endogenous PPTases to form labeled CP domains. Labeled carrier proteins are visualized after cell lysis by chemoselective ligation to a bioorthogonal reporter molecule followed by SDS-PAGE. (C) Multienzymatic labeling of PKS and NRPS megasynthases. The treatment of natural product producer proteomes with reporter-labeled electrophiles can be used to distinguish PKS and NRPS multienzymes from monofunctional hydrolases, provided the inhibitors target orthogonal PKS active sites (pictured are inhibitors targeting KS and TE domains). (See Color Insert.)

analogues to visualize CPs, we observed labeling of the native DEBS PKS from proteomic preparations of *Saccharopolyspora erythrea*, albeit at extremely low levels (La Clair *et al.*, 2004). We attributed the low labeling efficiency to the fact that the majority of CP domains (>95%) from native producer organisms exist in their *holo*-form, posttranslationally modified by their endogenous PPTase and acetyl CoA. This modification blocks chemoenzymatic labeling by reporter-labeled CoAs and Sfp, and to date no reliable methods for cleavage of the 4′-PPant prosthetic group have been reported in PKS and NRPS systems. This limits the utility of this approach to organisms in which the secondary PPTase has been inactivated and *apo*-CP domains are abundant. One such organism is *Bacillus subtilis* strain 168 (ATCC 23857), which contains an in-frame deletion in the gene coding for its secondary PPTase, Sfp (Lambalot *et al.*, 1996). This results not only in production of its secondary metabolism CPs in *apo*- form (making them accessible to labeling by fluorescent CoA analogues/Sfp) but also in an apparently modest up-regulation in production of these enzymes, as qualitatively observed by comparison of Coomassie-stained SDS-PAGE gels of unfractionated proteomes of *B. subtilis* 168 and the wild type (6051) organism (Meier *et al.*, 2008).

4.2.2. Proteomic preparation

Proteomes are prepared by overnight growth of *B. subtilis* in LB medium to stationary phase (OD_{600} ~1.5), followed by harvesting of 1-l cultures by centrifugation. Interestingly, we have seen higher NRPS levels resulting from growth of *B. subtilis* 168 in LB in contrast to richer media such as

YEME7, although we have not explored this phenomenon in detail. After resuspension in 10 ml lysis buffer (25 mM potassium phosphate, pH 7.0, 100 mM NaCl), cell lysis is performed by two passes through a French pressure cell, followed by treatment with DNase I for 30 min at 0 °C and clearing of cell debris by centrifugation. This typically results in isolation of unfractionated proteomes of ~10–20 mg/ml as determined by Bradford assay. For long-term use, glycerol is added to the proteomic preparations to a final concentration of 10% before storage at −80 °C.

4.2.3. Labeling and visualization

For chemoenzymatic labeling of *apo*-CPs, we typically dilute B. *subtilis* 168 proteomes to a final concentration of 5 mg/ml with 100 mM potassium phosphate, pH 7.0, 10 mM DTT. Labeling is achieved by sequential addition of BODIPY CoA **6** (10 μM), MgCl$_2$ (40 mM, pH 7.0), and Sfp (0.5 μg). Note that the order of addition is important, as addition of Sfp prior to BODIPY CoA **6** results in greatly decreased fluorescent labeling, most likely due to the presence of endogenous B. *subtilis* CoA in the reaction mixture (Fig. 9.9A). Negative controls can be performed by omission of Sfp and MgCl$_2$, or addition of EDTA. After 1 h the reaction is quenched by addition of an equal volume of 2× loading buffer and subjected to SDS-PAGE. We have found the utilization of low percentage gradient gels (3–8% Bis-Tris NuPAGE gels, Invitrogen) enhances separation of the high molecular weight proteins commonly labeled by this method. Fluorescent visualization of *crypto*-CP domains is then performed using a Typhoon Laser Flatbed Scanner (GE Healthcare). The labeled bands can be excised and prepared for tandem MS identification according to standard protocols.

4.3. Complementary labeling of PKS and NRPS domains by activity-based probes

4.3.1. Background

Besides their use of the 4'-PPant postranslational modification and CP domains, another distinguishing feature of type I PKS and NRPS biosynthetic enzymes is the presence of multiple active sites within a single protein. Provided these active sites can be targeted by orthogonal protein labels, this provides another method for selective elabeling and identification of these enzymes (Fig. 9.8B). This approach was recently explored in a preliminary study which used fluorescent analogues of well-known KS and TE active-site affinity labels in combination with chemoenzymatic CP labeling to label NRPS and type I FAS enzymes in unfractionated proteomes (Meier et al., 2008). Reporter-labeled fluorophosphonates (FPs) (Liu et al., 1999) were first

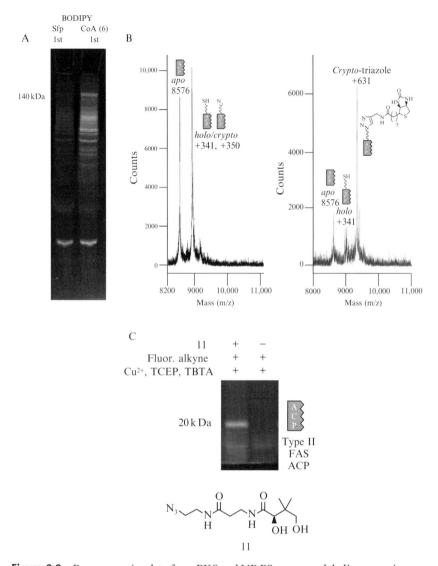

Figure 9.9 Representative data from PKS and NRPS proteome labeling experiments. (A) Labeling of *B. subtililis* 168 proteome with BODIPY CoA (6) and Sfp. Addition of Sfp before BODIPY-CoA (left) greatly decreases labeling, presumably due to the presence of endogenous CoA in unfractionated proteomic samples. (B) MALDI-MS analysis following *in vivo* labeling of overexpressed frenolicin ACP by 11. Following lysis the PKS ACP Fren can be observed in cell lysate. While *holo*- and *crypto*-ACP (mass difference ~9 Da) cannot be distinguished on this low resolution instrument, following Cu-catalyzed [3 + 2] cycloaddition with biotin alkyne (Meier *et al.*, 2006) a clear mass shift is observed. (C) Gel-analysis (0–30 kDa region) following *in vivo* labeling of native *E. coli* K12 by azide 11. Labeled FAS ACP (~20 kDa—verified by LC-MS/MS) is observed only in lysate from *E. coli* grown with 11.

tested against PKS enzymes *in vitro*, where they showed excellent labeling of TE- but not AT-type serine hydrolases. Similarly, reporter-labeled haloacetamides showed active-site directed labeling of KS domains, although the labeling and site-selectivity were modest compared to FP-probes.

The ability of these reagents to facilitate identification of NRPS enzymes in proteomic extracts was demonstrated by their application to the unfractionated proteome of *B. subtilis* strain 168. Application of the FP-DMACA and chemoenzymatic CP probe **6** (2.3.2) showed colabeling of a single high molecular weight band, suggesting that this protein was a CP/TE-containing multienzyme. Inspection of the sequenced *B. subtilis* genome indicated the most likely PKS/NRPS candidate in the observed molecular weight range to be SrfAC, the terminal module of the surfactin NRPS. The identity of the labeled protein as SrfAC was confirmed by tandem MS identification of the excised band. This initial study was simplified by both the ease of CP labeling in this particular strain, as well the availability of genomic information. Still, this method is promising in its ability to simplify complex protein labeling patterns generated by orthogonal protein labels into a subset of candidate PKS and NRPS proteins. In the future this approach will likely benefit from coupling to more sensitive gel-free mass spectrometry-based detection methods, similar to the activity-based protein profiling/multidimensional protein identification technology (ABPP-MUDPIT) technology pioneered by Cravatt (Jessani *et al.*, 2005), in order to facilitate identification of low-abundance type I PKS and NRPS enzymes.

4.3.2. Proteomic preparation
Lysates from *B. subtilis* 168 were prepared as specified above.

4.3.3. Labeling and visualization
We constrain our focus here to application of reporter-labeled FPs, as these compounds have been shown to be extremely site-selective and robust active-site labeling reagents (Liu *et al.*, 1999). *B. subtilis* 168 proteome is diluted to a final concentration of 5 mg/ml with 100 mM potassium phosphate, pH 7.0, and 10 mM DTT. Labeling is initiated by addition of FP-DMACA (25 μM and allowed to proceed for 1 h at room temperature, followed by addition of an equal volume of 2× loading buffer and SDS-PAGE (3–8% Bis-Tris NuPAGE gels, Invitrogen). Negative controls are performed by predenaturation of the lysate by heating or addition of 2% SDS. SDS is the milder of the two methods, but can also cause poor staining of total protein content when using Coomassie-based stains following fluorescent visualization of FP-labeling (Meier and Burkart, personal communication). Fluorescent visualization of FP-labeled proteins is then performed using BioRad FluorS Gel Doc equipped with a 460 nm emission filter. The FP labeling reaction is remarkably tolerant of small-molecule

nucleophiles such as DTT, and shows very little nonspecific background labeling of *B. subtilis* proteins at concentrations up to 200 μM, although previous reports using more sensitive detection methods have recommended an FP concentration of $\sim$5 μM for optimal labeling (Kidd *et al.*, 2001). For identification of candidate PKS and NRPS enzymes labeled by this method, an orthogonal protein labeling method such as chemoenzymatic CP labeling (4.3) can be performed on a separate proteomic preparation and run on the same gel—bands labeled in both are good candidates for containing terminal CP-TE modules of PKS or NRPS enzymes and can be excised and prepared for tandem MS identification according to standard protocols. In addition to the fluorophosphonate, chemoenzymatic CP labels and haloacetamide KS labels explored in our study, recently reported affinity-tagged inhibitors of alternate PKS and NRPS active-sites, including the CoA-acetyltransferase (AT) (Hwang *et al.*, 2007) and A domain (Finking *et al.*, 2003), should also be compatible with this approach.

4.4. *In vivo* labeling of carrier proteins

4.4.1. Background

As mentioned earlier, the fact that the CP domains of PKS and NRPS enzymes from most organisms bear an endogenous 4'-PPant posttranslational modification greatly reduces the utility of chemoenzymatic CP labeling approaches for identification of PKS and NRPS enzymes in crude cell lysate. One method to circumvent this limitation would be to label the CP domains with reporter-labeled CoA analogues such as BODIPY CoA **6** *in vivo*, as the *apo*-CP domain comes off the ribosomal assembly line. However, because the highly charged nature of CoA renders it membrane impermeable, the only way to incorporate reporter-labeled CoA analogues into the intracellular CoA pool is through the use of reporter-labeled CoA precursors, specifically pantetheine analogues such as **11** and **26**. These compounds can cross the cell membrane and be transformed to CoA analogues by the endogenous CoA biosynthetic pathway of the organism (Fig. 9.8B). We refer here to the use of reporter-labeled pantetheine analogues for metabolic labeling of CPs as *in vivo* CP labeling. In addition to its use in the proteomic identification of CP-containing PKS and NRPS enzymes after cell lysis, this method also has the potential to allow us to study the dynamics and trafficking of these massive multienzymatic catalysts *in vivo*.

Much of our work to date has focused on determining the ideal structural characteristics possessed by reporter-labeled pantetheine analogues capable of coopting the CoA biosynthetic pathway *in vivo*. To analyze the ability of a pantetheine analogue to be processed by the endogenous *E. coli* CoA biosynthetic pathway (PanK, PPAT, and DPCK), we use an *in vivo* assay which overexpresses an *apo*-CP and PPTase. Observation of CP labeling after cell lysis by MALDI-MS mass shift (Fig. 9.9B) provides a quick assessment of the ability of our synthetic CoA precursor to compete with naturally occurring CoA precursors, while negating extraneous variables such as PPTase substrate selectivity and CP expression that may vary widely between natural product producing organisms. In addition, through metabolic incorporation of electrophilic pantetheine analogues, this overexpression method could prove useful in itself as a method for the crosslinking of CP-containing enzymes that cannot be readily expressed in their *apo*-form to partner enzymes *in vivo*, although this application has not been explored in detail.

4.4.2. *In vivo* labeling of overexpressed CPs

To begin the assay, *E. coli* (BL-21) is cotransformed by electroporation with the plasmids pET22b-Fren and pREP4-Sfp, which encode the CP Fren (type II PKS ACP from the frenolicin biosynthetic pathway) and Sfp, and streaked onto LB-Amp/Kam agar plates. After growth a single colony is picked and used to inoculate 10 ml of LB-Amp/Kan liquid medium, followed by growth overnight at 37 °C. This starter culture is then diluted 1:50 (20 μl/ml) in 10 ml of fresh LB-Amp/Kan media and split into five 2-ml aliquots, each of which can be used for an *in vivo* labeling experiment. We typically run at least one negative control to which no analogue is added, one positive control using an analogue which is known to work in these systems (such as **11** or **26**), and two experimentals to study pantetheine analogues of unknown activity. Cultures are grown at 37 °C to an OD$_{600}$ of 0.6 before induction of VibB/Sfp expression with 1 mM IPTG. At the same time 10 μl of the pantetheine analogue (200 mM in DMSO) being assayed is added to the culture, bringing its final concentration to 1 mM. After growth for 4 h, cells are pelleted the supernatant is removed, and the pellet resuspended in 10 ml lysis buffer (50 mM sodium phosphate, pH 7.0, 300 mM NaCl). After two more 10 ml washes to remove extracellular pantetheine analogue, the pellet is resuspended in a volume of 0.30 ml lysis buffer and lysed by incubation with lysozyme (3 mg/ml) for 1 h on ice, followed by sonication (3 × 30 s pulses, low power), treatment with DNase I for 30 min on ice, and centrifugation. The protein supernatant can be stored at −80 °C for upwards of 2 months with no noticeable signs of degradation. Samples are prepared for MALDI-MS by 1:5 dilution of crude cell lysate with saturated cinnapinic acid matrix solution. 1-μl spots are then analyzed using an Applied Biosystems Voyager DE-STR MALDI-TOF mass spectrometer (Fig. 9.9B).

4.4.3. *In vivo* labeling of endogenous CPs

In addition to the labeling of PKS and NRPS CP domains in the above described overexpression system, very recently we have also developed methods for the labeling of endogenous CPs in native (genetically unmodified) bacterial organisms using pantetheine azide **11** (Mercer *et al.*, 2008).

While to date we have only observed labeling of FAS ACPs by this method (Fig. 9.9C), this method should also be applicable to labeling of PKS and NRPS biosynthetic enzymes when applied under natural product-inducing growth conditions, facilitating their proteomic identification and potentially their *in vivo* visualization.

Briefly, for *in vivo* labeling of endogenous CPs by **11**, native natural product-producing organisms are first grown under conditions specified in the literature or on simple LB medium unless otherwise specified. For labeling of the fatty acid ACP in wild type *B. subtilis* strain 6051 (ATCC 6051), bacteria are first streaked on LB-agar plates and grown overnight. A single colony is picked and used to inoculate 10 ml of LB medium, followed by growth overnight at 37 °C. This starter culture is then diluted 1:100 (10 μl/ml) in 10–1000 ml of fresh LB medium supplemented with 1 mM of pantetheine azide **11**. We typically run a negative control as well, in which the starter culture is used to inoculate LB media supplemented with vehicle DMSO. Bacteria are then grown for 12–17 h with orbital shaking at 37 °C. Cells are harvested by centrifugation and resuspension in lysis buffer (25 mM potassium phosphate, pH 7.0, 100 mM NaCl). For small cultures (<10 ml), cells can be lysed by addition of 3-mg/ml lysozyme (Worthington), incubation for 1 h on ice, and sonication (3 × 30 s pulses, low power). For larger cultures, cells are resuspended in 10 ml of lysis buffer per liter of culture with 0.1 mg/ml lysozyme, incubated for 1 h on ice, and lysed by two passes through a French pressure cell, followed by treatment with DNase I for 30 min at 0 °C and clearing of cell debris by centrifugation.

4.4.4. Visualization of azide-labeled CPs via Cu-catalyzed [3 + 2] cycloaddition with fluorophore or biotin alkyne

Following centrifugation, lysate protein concentrations are quantified according to the method of Bradford and diluted to a concentration of 1 mg/ml with lysis buffer (25 mM potassium phosphate, pH 7.0, 100 mM NaCl). Proteins labeled by pantetheine azide are detected by Cu-catalyzed [3 + 2] cycloaddition with a fluorophore or biotin alkyne (Fig. 9.9C). Our reaction conditions are identical to those optimized by Speers and Cravatt (2004), involving incubation of labeled lysate with 100-μM fluorophore alkyne, 1 mM TCEP, 100 μM tris-(benzylriazolylmethyl)amine ligand (TBTA), and CuSO$_4$ (1 mM) for 1 h at room temperature, followed by SDS-PAGE or Western blot analysis. Storage of the TBTA ligand as a stock solution in 1:4 DMSO: t-butanol allows for the addition of 5% t-butanol to the reaction mixture, which has been reported to enhance the yield of the cycloaddition reaction (Speers and Cravatt, 2004). The order of addition of the reagents is not critical, although we typically add CuSO$_4$ as the last step. In our hands the most common cause of cycloaddition reaction failure is a high protein concentration (>1 mg/ml), which is thought to inhibit the cycloaddition by Cu^{2+} chelation, and oxidation of the TCEP solution, which can be remedied by its fresh preparation. The choice of the alkyne detection agent is also key, as in the past we have observed differential detection of labeled CP domains by biotin- and fluorophore-alkynes, most notably in labeling of type II bacterial FAS ACP domains. This is thought to be due to the fact that failure to completely denature these small, heat stable proteins can lead to sequestration of the 4′-PPant-linked biotin moiety in their hydrophobic core, resulting in false negatives when attempting to visualize CP labeling by Western blot (Mercer and Burkart, personal communication). For this reason we typically perform fluorescence and Western blot visualization in parallel. After cycloaddition and SDS-PAGE, fluorescently labeled bands can be excised and prepared for tandem MS identification according to standard protocols.

5. Conclusions

The methods and examples highlighted in this article have demonstrated utility in facilitating the study of PKS and NRPS mechanism and structure, and with further development will likely be useful in future approaches to the proteomic identification of these enzymes. One major challenge to the future use of synthetic probes for the study of PKS and NRPS biochemistry will be the continued development of chemical tools powerful enough to provide unique insights into these systems, yet simple enough to be generally accessible to a wide range of researchers. A cause for optimism in this respect is the radical success of the promiscuous PPTase Sfp, which in the decade since its discovery has became an everyday tool in the study of PKS and NRPS biosynthesis and is being applied in increasingly innovative ways. It will be extremely interesting to see which, if any, of the more recent approaches described herein will be seen as having had a similar impact on the field of natural product biosynthesis 10 years from now.

REFERENCES

Akey, D. L., Kittendorf, J. D., Giraldes, J. W., Fecik, R. A., Sherman, D. H., and Smith, J. L. (2006). Structural basis for macrolactonization by the pikromycin thioesterase. *Nat. Chem. Biol.* **2,** 537–542.

Austin, M. B., and Noel, J. P. (2003). The chalcone synthase superfamily of type III polyketide synthases. *Nat. Prod. Rep.* **20,** 79–110.

Balibar, C. J., Vaillancourt, F. H., and Walsh, C. T. (2005). Generation of D amino acid residues in assembly of arthrofactin by dual condensation/epimerization domains. *Chem. Biol.* **12,** 1189–1200.

Belshaw, P. J., Walsh, C. T., and Stachelhaus, T. (1999). Aminoacyl-CoAs as probes of condensation domain selectivity in nonribosomal peptide synthesis. *Science* **284,** 486–489.

Challis, G. L., Ravel, J., and Townsend, C. A. (2000). Predictive, structure-based model of amino acid recognition by nonribosomal peptide synthetase adenylation domains. *Chem. Biol.* **7,** 211–224.

Christianson, C. V., Montavon, T. J., Festin, G. M., Cooke, H. A., Shen, B., and Bruner, S. D. (2007). The mechanism of MIO-based aminomutases in beta-amino acid biosynthesis. *J. Am. Chem. Soc.* **129,** 15744–15745.

Clarke, K. M., Mercer, A. C., La Clair, J. J., and Burkart, M. D. (2005). In vivo reporter labeling of proteins via metabolic delivery of coenzyme A analogues. *J. Am. Chem. Soc.* **127,** 11234–11235.

Clugston, S. L., Sieber, S. A., Marahiel, M. A., and Walsh, C. T. (2003). Chirality of peptide bond-forming condensation domains in nonribosomal peptide synthetases: The C5 domain of tyrocidine synthetase is a (D)C(L) catalyst. *Biochemistry* **42,** 12095–12104.

Conti, E., Stachelhaus, T., Marahiel, M. A., and Brick, P. (1997). Structural basis for the activation of phenylalanine in the non-ribosomal biosynthesis of gramicidin S. *EMBO J.* **16,** 4174–4183.

Davis, J. T., Chen, H. H., Moore, R., Nishitani, Y., Masamune, S., Sinskey, A. J., and Walsh, C. T. (1987). Biosynthetic thiolase from Zoogloea ramigera. II. Inactivation with haloacetyl CoA analogs. *J. Biol. Chem.* **262,** 90–96.

Dorrestein, P. C., and Kelleher, N. L. (2006). Dissecting non-ribosomal and polyketide biosynthetic machineries using electrospray ionization Fourier-Transform mass spectrometry. *Nat. Prod. Rep.* **23,** 893–918.

Ferrer, J. L., Austin, M. B., Stewart, C. Jr., and Noel, J. P. (2008). Structure and function of enzymes involved in the biosynthesis of phenylpropanoids. *Plant Physiol. Biochem.* **46,** 356–370.

Ferrer, J. L., Jez, J. M., Bowman, M. E., Dixon, R. A., and Noel, J. P. (1999). Structure of chalcone synthase and the molecular basis of plant polyketide biosynthesis. *Nat. Struct. Biol.* **6,** 775–784.

Finking, R., Neumuller, A., Solsbacher, J., Konz, D., Kretzschmar, G., Schweitzer, M., Krumm, T., and Marahiel, M. A. (2003). Aminoacyl adenylate substrate analogues for the inhibition of adenylation domains of nonribosomal peptide synthetases. *Chembiochem* **4,** 903–906.

Fischbach, M. A., and Walsh, C. T. (2006). Assembly-line enzymology for polyketide and nonribosomal Peptide antibiotics: Logic, machinery, and mechanisms. *Chem. Rev.* **106,** 3468–3496.

Foley, T. L., and Burkart, M. D. (2007). Site-specific protein modification: Advances and applications. *Curr. Opin. Chem. Biol.* **11,** 12–19.

Frueh, D. P., Arthanari, H., Koglin, A., Vosburg, D. A., Bennett, A. E., Walsh, C. T., and Wagner, G. (2008). Dynamic thiolation-thioesterase structure of a non-ribosomal peptide synthetase. *Nature* **454,** 903–906.

Gehring, A. M., Lambalot, R. H., Vogel, K. W., Drueckhammer, D. G., and Walsh, C. T. (1997). Ability of *Streptomyces* spp. acyl carrier proteins and coenzyme A analogs to serve as substrates *in vitro* for *E. coli* holo-ACP synthase. *Chem. Biol.* **4,** 17–24.

Giraldes, J. W., Akey, D. L., Kittendorf, J. D., Sherman, D. H., Smith, J. L., and Fecik, R. A. (2006). Structural and mechanistic insights into polyketide macrolactonization from polyketide-based affinity labels. *Nat. Chem. Biol.* **2,** 531–536.

Haushalter, R. W., Worthington, A. S., Hur, G. H., and Burkart, M. D. (2008). An orthogonal purification strategy for isolating crosslinked domains of modular synthases. *Bioorg. Med. Chem. Lett.* **18,** 3039–3042.

He, W., Wu, J., Khosla, C., and Cane, D. E. (2006). Macrolactonization to 10-deoxymethynolide catalyzed by the recombinant thioesterase of the picromycin/methymycin polyketide synthase. *Bioorg. Med. Chem. Lett.* **16,** 391–394.

Hwang, Y., Thompson, P. R., Wang, L., Jiang, L., Kelleher, N. L., and Cole, P. A. (2007). A selective chemical probe for coenzyme A-requiring enzymes. *Angew. Chem. Int. Ed. Engl.* **46,** 7621–7624.

Jenni, S., Leibundgut, M., Boehringer, D., Frick, C., Mikolasek, B., and Ban, N. (2007). Structure of fungal fatty acid synthase and implications for iterative substrate shuttling. *Science* **316,** 254–261.

Jessani, N., Niessen, S., Wei, B. Q., Nicolau, M., Humphrey, M., Ji, Y., Han, W., Noh, D. Y., Yates, J. R. 3rd, Jeffrey, S. S., and Cravatt, B. F. (2005). A streamlined platform for high-content functional proteomics of primary human specimens. *Nat. Methods* **2,** 691–697.

Kapur, S., Worthington, A., Tang, Y., Cane, D. E., Burkart, M. D., and Khosla, C. (2008). Mechanism based protein crosslinking of domains from the 6-deoxyerythronolide B synthase. *Bioorg. Med. Chem. Lett.* **18,** 3034–3038.

Keating, D. H., Carey, M. R., and Cronan, J. E. Jr. (1995). The unmodified (apo) form of *Escherichia coli* acyl carrier protein is a potent inhibitor of cell growth. *J. Biol. Chem.* **270,** 22229–22235.

Khosla, C., Tang, Y., Chen, A. Y., Schnarr, N. A., and Cane, D. E. (2007). Structure and mechanism of the 6-deoxyerythronolide B synthase. *Annu. Rev. Biochem.* **76,** 195–221.

Kidd, D., Liu, Y., and Cravatt, B. F. (2001). Profiling serine hydrolase activities in complex proteomes. *Biochemistry* **40,** 4005–4015.

Korman, T. P., Tan, Y. H., Wong, J., Luo, R., and Tsai, S. C. (2008). Inhibition kinetics and emodin cocrystal structure of a type II polyketide ketoreductase. *Biochemistry* **47,** 1837–1847.

La Clair, J. J., Foley, T. L., Schegg, T. R., Regan, C. M., and Burkart, M. D. (2004). Manipulation of carrier proteins in antibiotic biosynthesis. *Chem. Biol.* **11,** 195–201.

Lambalot, R. H., Gehring, A. M., Flugel, R. S., Zuber, P., LaCelle, M., Marahiel, M. A., Reid, R., Khosla, C., and Walsh, C. T. (1996). A new enzyme superfamily—the phosphopantetheinyl transferases. *Chem. Biol.* **3,** 923–936.

Leibundgut, M., Jenni, S., Frick, C., and Ban, N. (2007). Structural basis for substrate delivery by acyl carrier protein in the yeast fatty acid synthase. *Science* **316,** 288–290.

Liu, Y., and Bruner, S. D. (2007). Rational manipulation of carrier-domain geometry in nonribosomal peptide synthetases. *Chembiochem* **8,** 617–621.

Liu, Y., Patricelli, M. P., and Cravatt, B. F. (1999). Activity-based protein profiling: The serine hydrolases. *Proc. Natl. Acad. Sci. USA* **96,** 14694–14699.

Maier, T., Jenni, S., and Ban, N. (2006). Architecture of mammalian fatty acid synthase at 4.5 A resolution. *Science* **311,** 1258–1262.

Maier, T., Leibundgut, M., and Ban, N. (2008). The crystal structure of a mammalian fatty acid synthase. *Science* **321,** 1315–1322.

Martin, D. P., and Drueckhammer, D. G. (1992). Combined chemical and enzymic synthesis of coenzyme A analogues. *J. Am. Chem. Soc.* **114,** 7287–7288.

McLoughlin, S. M., Mazur, M. T., Miller, L. M., Yin, J., Liu, F., Walsh, C. T., and Kelleher, N. L. (2005). Chemoenzymatic approaches for streamlined detection of active site modifications on thiotemplate assembly lines using mass spectrometry. *Biochemistry* **44,** 14159–14169.

Meier, J. L., Mercer, A. C., and Burkart, M. D. (2006). Synthesis and evaluation of bioorthogonal pantetheine analogues for *in vivo* protein modification. *J. Am. Chem. Soc.* **128,** 12174–12184.

Meier, J. L., Mercer, A. C., Rivera, H. Jr., and Burkart, M. D. (2008). Fluorescent profiling of modular biosynthetic enzymes by complementary metabolic and activity based probes. *J. Am. Chem. Soc.* **130,** 5443–5445.

Mercer, A. C., and Burkart, M. D. (2007). The ubiquitous carrier protein–a window to metabolite biosynthesis. *Nat. Prod. Rep.* **24,** 750–773.

Mercer, A. C., Meier, J. L., Torpey, J. W., and Burkart, M. D. (2008). Metabolic labeling of endogenous acyl carrier proteins Submitted for publication.

Palmer, M. A., Differding, E., Gamboni, R., Williams, S. F., Peoples, O. P., Walsh, C. T., Sinskey, A. J., and Masamune, S. (1991). Biosynthetic thiolase from Zoogloea ramigera. Evidence for a mechanism involving Cys-378 as the active site base. *J. Biol. Chem.* **266,** 8369–8375.

Powers, J. C., Asgian, J. L., Ekici, O. D., and James, K. E. (2002). Irreversible inhibitors of serine, cysteine, and threonine proteases. *Chem. Rev.* **102,** 4639–4750.

Qiao, C., Wilson, D. J., Bennett, E. M., and Aldrich, C. C. (2007). A mechanism-based aryl carrier protein/thiolation domain affinity probe. *J. Am. Chem. Soc.* **129,** 6350–6351.

Quadri, L. E., Weinreb, P. H., Lei, M., Nakano, M. M., Zuber, P., and Walsh, C. T. (1998). Characterization of Sfp, a *Bacillus subtilis* phosphopantetheinyl transferase for peptidyl carrier protein domains in peptide synthetases. *Biochemistry* **37,** 1585–1595.

Ritter, S. C., and Konig, B. (2006). Signal amplification and transduction by photo-activated catalysis. *Chem. Commun. (Camb)* 4694–4696.

Sieber, S. A., and Marahiel, M. A. (2005). Molecular mechanisms underlying nonribosomal peptide synthesis: Approaches to new antibiotics. *Chem. Rev.* **105,** 715–738.

Sieber, S. A., Walsh, C. T., and Marahiel, M. A. (2003). Loading peptidyl-coenzyme A onto peptidyl carrier proteins: A novel approach in characterizing macrocyclization by thioesterase domains. *J. Am. Chem. Soc.* **125,** 10862–10866.

Smith, S., and Tsai, S. C. (2007). The type I fatty acid and polyketide synthases: A tale of two megasynthases. *Nat. Prod. Rep.* **24,** 1041–1072.

Speers, A. E., Adam, G. C., and Cravatt, B. F. (2003). Activity-based protein profiling *in vivo* using a copper(i)-catalyzed azide-alkyne [3 + 2] cycloaddition. *J. Am. Chem. Soc.* **125,** 4686–4687.

Speers, A. E., and Cravatt, B. F. (2004). Profiling enzyme activities *in vivo* using click chemistry methods. *Chem. Biol.* **11,** 535–546.

Spiteller, D., Waterman, C. L., and Spencer, J. B. (2005). A method for trapping intermediates of polyketide biosynthesis with a nonhydrolyzable malonyl-coenzyme A analogue. *Angew Chem. Int. Ed. Engl.* **44,** 7079–7082.

Staunton, J., and Weissman, K. J. (2001). Polyketide biosynthesis: A millennium review. *Nat. Prod. Rep.* **18,** 380–416.

Stein, D. B., Linne, U., and Marahiel, M. A. (2005). Utility of epimerization domains for the redesign of nonribosomal peptide synthetases. *FEBS J.* **272,** 4506–4520.

Stoops, J. K., and Wakil, S. J. (1981). Animal fatty acid synthetase. A novel arrangement of the beta-ketoacyl synthetase sites comprising domains of the two subunits. *J. Biol. Chem.* **256,** 5128–5133.

Tang, Y., Chen, A. Y., Kim, C. Y., Cane, D. E., and Khosla, C. (2007). Structural and mechanistic analysis of protein interactions in module 3 of the 6-deoxyerythronolide B synthase. *Chem. Biol.* **14,** 931–943.

Tanovic, A., Samel, S. A., Essen, L. O., and Marahiel, M. A. (2008). Crystal structure of the termination module of a nonribosomal peptide synthetase. *Science* **321,** 659–663.

Tseng, C. C., Bruner, S. D., Kohli, R. M., Marahiel, M. A., Walsh, C. T., and Sieber, S. A. (2002). Characterization of the surfactin synthetase C-terminal thioesterase domain as a cyclic depsipeptide synthase. *Biochemistry* **41,** 13350–13359.

Vitali, F., Zerbe, K., and Robinson, J. A. (2003). Production of vancomycin aglycone conjugated to a peptide carrier domain derived from a biosynthetic non-ribosomal peptide synthetase. *Chem. Commun. (Camb)* 2718–2719.

Watanabe, K., Praseuth, A. P., and Wang, C. C. (2007). A comprehensive and engaging overview of the type III family of polyketide synthases. *Curr. Opin. Chem. Biol.* **11,** 279–286.

Weissman, K. J., and Muller, R. (2008). Protein-protein interactions in multienzyme megasynthetases. *Chembiochem* **9,** 826–848.

Widboom, P. F., Fielding, E. N., Liu, Y., and Bruner, S. D. (2007). Structural basis for cofactor-independent dioxygenation in vancomycin biosynthesis. *Nature* **447,** 342–345.

Worthington, A. S., and Burkart, M. D. (2006). One-pot chemo-enzymatic synthesis of reporter-modified proteins. *Org. Biomol. Chem.* **4,** 44–46.

Worthington, A. S., Hur, G. H., Meier, J. L., Cheng, Q., Moore, B. S., and Burkart, M. D. (2008). Probing the compatibility of type II ketosynthase-carrier protein partners. *Chembiochem.* **9,** 2096–2103.

Worthington, A. S., Rivera, H., Torpey, J. W., Alexander, M. D., and Burkart, M. D. (2006). Mechanism-based protein cross-linking probes to investigate carrier protein-mediated biosynthesis. *ACS Chem. Biol.* **1,** 687–691.

CHAPTER TEN

Using Phosphopantetheinyl Transferases for Enzyme Posttranslational Activation, Site Specific Protein Labeling and Identification of Natural Product Biosynthetic Gene Clusters from Bacterial Genomes

Murat Sunbul, Keya Zhang, *and* Jun Yin

Contents

1. Introduction	256
2. Experimental Procedures	263
2.1. Expression of AcpS from *E. coli*	263
2.2. Expression of Sfp from *E. coli*	263
2.3. Coexpression of NRPS or PKS modules with Sfp in *E. coli*	264
2.4. *In vitro* activation of the PKS or NRPS modules by Sfp	265
2.5. Preparation of small molecule–CoA conjugates	265
2.6. Preparation of Qdot–CoA conjugates	266
2.7. Sfp catalyzed protein labeling with small molecule–CoA conjugates	267
2.8. Sfp-catalyzed protein labeling on the cell surface	268
2.9. Construction of the genomic library for phage selection of NRPS and PKS fragments	269
2.10. Phage selection by Sfp catalyzed carrier protein modification with biotin–SS–CoA 2	270
3. Conclusion	271
References	271

Department of Chemistry, The University of Chicago, Chicago, Illinois, USA

Abstract

Phosphopantetheinyl transferases (PPTases) covalently attach the phosphopantetheinyl group derived from coenzyme A (CoA) to acyl carrier proteins or peptidyl carrier proteins as part of the enzymatic assembly lines of fatty acid synthases (FAS), polyketide synthases (PKS), and nonribosomal peptide synthetases (NRPS). PPTases have demonstrated broad substrate specificities for cross-species modification of carrier proteins embedded in PKS or NRPS modules. PPTase Sfp from *Bacillus subtilis* and AcpS from *Escherichia coli* also transfer small molecules of diverse structures from their CoA conjugates to the carrier proteins. Short peptide tags have thus been developed as efficient substrates of Sfp and AcpS for site-specific labeling of the peptide-tagged fusion proteins with biotin or organic fluorophores. This chapter discusses the use of PPTases for *in vivo* and *in vitro* modification of PKS and NRPS enzymes and for site-specific protein labeling. We also describe a phage selection method based on PPTase-catalyzed carrier protein modification for the identification of PKS or NRPS genes from bacterial genomes.

1. Introduction

Phosphopantetheinyl transferases (PPTase) represent an important class of protein posttranslational modification enzymes that activate the biosynthesis of primary and secondary metabolites, including fatty acids, polyketides, and nonribosomal peptides (Lambalot *et al.*, 1996; Walsh *et al.*, 1997). PPTases are widely distributed in organisms ranging from microbes to plants and mammals (Copp and Neilan, 2006; Crawford *et al.*, 2008; Joshi *et al.*, 2003; Praphanphoj *et al.*, 2001). They catalyze the covalent transfer of the 4′-phosphopantetheinyl (Ppant) group derived from coenzyme A (CoA) to a specific Ser residue in acyl carrier proteins (ACP) and peptidyl carrier proteins (PCP) (Lai *et al.*, 2006; Mercer and Burkart, 2007) as integral parts of fatty acid synthases (FAS) (White *et al.*, 2005), polyketide synthases (PKS) (Khosla *et al.*, 1999; Staunton and Weissman, 2001), and nonribosomal peptide synthetases (NRPS) (Fig. 10.1) (Fischbach and Walsh, 2006; Sieber and Marahiel, 2005). The free thiols at the end of the Ppant prosthetic groups on the modified ACP or PCP provide the

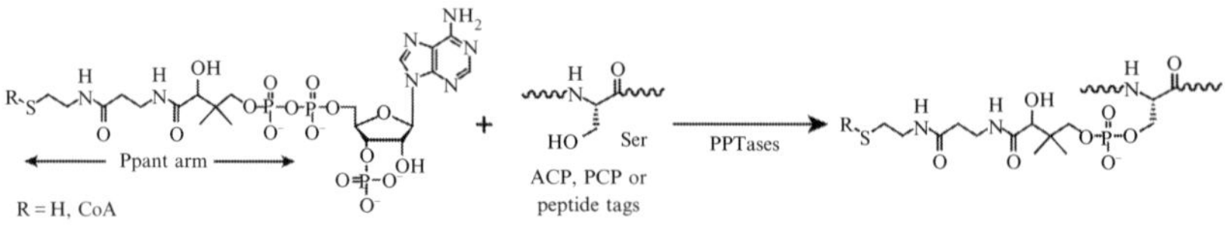

Figure 10.1 PPTase-catalyzed carrier protein modification with CoA or small-molecule CoA conjugates.

anchoring points to attach substrates and biosynthetic intermediates to FAS, PKS, and NRPS enzymes. In this way, the carrier proteins act as "swinging domains" and the Ppant prosthetic groups act as "swinging arms" spanning 20 Å in length to channel the intermediates through the various catalytic sites on the enzymatic assembly lines (Perham, 2000). A variety of reactions can be carried out on the biosynthetic intermediates loaded on the carrier proteins, including decarboxylative condensation or amide bond formation for chain elongation and additional tailoring reactions such as oxidation, reduction, dehydration, hydrogenation, epimerization, and cyclization for the assembly of the complex structures of natural products (Cane and Walsh, 1999; Walsh, 2004).

Based on the sequence homology, structural assembly and the modification targets of PPTases, Finking and Marahiel *et al.* divided the enzyme superfamily into three subgroups (Finking *et al.*, 2002). The first group, AcpS-type PPTases, is represented by *Escherichia coli* AcpS (Flugel *et al.*, 2000; Lambalot and Walsh, 1995), which is about 120 residues in length with substrate specificity for ACPs in the FAS and PKS enzymes. Crystal structures of AcpS from *Bacillus subtilis* and *Streptococcus pneumoniae* have been solved (Chirgadze *et al.*, 2000; Parris *et al.*, 2000), showing a trimeric assembly of the enzyme with the substrates CoA and ACP bound at the interface between two AcpS monomers. The second group of PPTases, the Sfp type, is named after the *B. subtilis* enzyme Sfp, which is associated with the NRPS gene cluster of surfactin synthetase (Mootz *et al.*, 2001; Quadri *et al.*, 1998). Members of this group of PPTases are about twice the size of AcpS with broader substrate specificities, modifying both PCPs and ACPs of NRPS, PKS and FAS origins (Mofid *et al.*, 2002, 2004). The crystal structure of Sfp reveals that the protein consists of two nearly identical domains arranged in a "pseudo twofold symmetry" with respect to each other and each of the domains closely resembles the α/β fold of an AcpS monomer (Reuter *et al.*, 1999). CoA is bound to Sfp at the interface between the two domains. It has thus been postulated that Sfp-type PPTases evolved by gene duplication of the AcpS enzymes (Finking *et al.*, 2002, 2004; Joshi *et al.*, 2003). Human PPTase also belongs to this group and the crystal structure of the enzyme shows the same homodimeric fold as Sfp (Bunkoczi *et al.*, 2007). The third group of PPTases functions as an integral domain at the C-terminus of FASs for *cis* modification of ACP within the same polypeptide (Fichtlscherer *et al.*, 2000). These enzymes are structurally similar to group I AcpS. Recently the C-terminal domain of PKS responsible for synthesizing enediyne in *Streptomyces globisporus* was found to be an integral PPTase. However, it shares significant homology with Sfp, and catalyzes Ppant modification of the ACP domain in the same PKS module (Zhang *et al.*, 2008).

Many organisms are equipped with multiple PPTases. For examples, AcpS, and Sfp have been identified in *B. subtilis* (Mootz *et al.*, 2001) and AcpS, EntD, and AcpT in *E. coli* (De Lay and Cronan, 2006). AcpS in those two organisms catalyzes phosphopantetheinylation of ACP to activate fatty acid biosynthesis (Lambalot and Walsh, 1995; Polacco and Cronan, 1981). Sfp in *B. subtilis* and Sfp-type enzyme EntD in *E. coli* are responsible for the modification of PCPs in NRPS modules synthesizing the peptide antibiotic surfactin and the siderophore enterobactin, respectively (Lambalot *et al.*, 1996). AcpT has recently been found to modify two carrier proteins encoded in a pathogenic *E. coli* strain (De Lay and Cronan, 2006). The yeast *Saccharomyces cerevisiae* has three PPTases with PPT2 responsible for ACP modification in mitochondrial FAS, Lys5 for PCP modification in α-aminoadipate reductase involved in lysine biosynthesis, and another PPTase as a constituent domain in the cytosolic FAS for *cis* ACP modification (Ehmann *et al.*, 1999; Fichtlscherer *et al.*, 2000; Mootz *et al.*, 2002; Stuible *et al.*, 1998). In contrast, only one PPTase has been identified in mammals (Joshi *et al.*, 2003). For example, the human genome encodes a single PPTase that is responsible for ACP modification in both cytosolic and mitochondrial FAS systems (Joshi *et al.*, 2003). *Pseudomonas aeruginosa* also has only one PPTase to activate the synthesis of both fatty acids and siderophores (Finking *et al.*, 2002).

A number of PPTases characterized so far have shown broad substrate specificities, with respect to both carrier proteins and CoA derivatives functionalized with a variety of chemical substitutions at the end of the free thiol group (Fig. 10.2). This feature has been most extensively demonstrated with the *B. subtilis* enzyme Sfp (La Clair *et al.*, 2004; Quadri *et al.*, 1998). Sfp has been shown to modify ACPs and PCPs of different origins, including *E. coli, Streptomyces*, and yeast (Chen and Walsh, 2001; Quadri *et al.*, 1998). In fact, Sfp has been most commonly used to transform apo

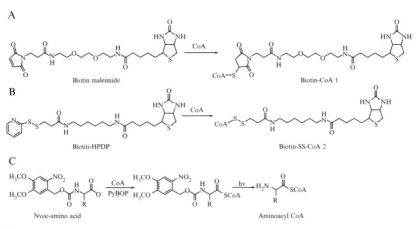

Figure 10.2 Synthesis of small-molecule CoA conjugates by formation of a thioether bond (A), a disulfide bond (B), or a thioester bond (C).

PKS and NRPS modules expressed from *E. coli* or other bacterial strains to their holo counterparts with catalytic activities (Admiraal *et al.*, 2002; Gokhale *et al.*, 1999; Keating *et al.*, 2000; Suo *et al.*, 2000). The gene for Sfp has also been coexpressed with the genes for PKS or NRPS modules to achieve *in vivo* holo enzyme formation in expression hosts such as *E. coli* and *S. cerevisiae* (Admiraal *et al.*, 2002; Kealey *et al.*, 1998; Wattanachaisaereekul *et al.*, 2007). Pfeifer, Khosla, and colleagues engineered *E. coli* to incorporate the gene for Sfp for the expression of active deoxyerythronolide B (6dEB) synthase (DEBS) and produced a panel of 6dEB analogues from *E. coli* cultures (Pfeifer *et al.*, 2001). Thus heterologous expression of PKS or NRPS genes with appropriate PPTases would produce catalytically active enzymes and allow combinatorial manipulation of the biosynthetic gene clusters in order to produce structurally diversified natural products (Cane *et al.*, 1998). To serve this purpose, other PPTases have been cloned and tested for cross-reactivity with carrier proteins from different species. Among them, PcpS from *P. aeruginosa* (Finking *et al.*, 2002), PPT_{NS} from *Nodularia spumigena* (Copp *et al.*, 2007), Svp from *Streptomyces verticillus* (Sanchez *et al.*, 2001), Gsp from *Bacillus brevis* (Ku *et al.*, 1997) have been found to be capable of modifying ACP or PCP domains non-native to the host strains of the PPTases.

The broad substrate specificity of PPTases with small molecules conjugated to CoA also attracts significant interest to develop site-specific protein labeling methods based on PPTase-catalyzed carrier protein modification (Johnsson *et al.*, 2005; Yin *et al.*, 2006). The crystal structure of Sfp in complex with CoA shows that the $3'$-phospho-$5'$-ADP moiety of CoA is bound to the enzyme at the cleft formed between the two homologous domains (Reuter *et al.*, 1999). The Ppant part of CoA is not visible in the crystal structure except for the pyrophosphate that is coordinated with a magnesium ion at the enzyme active site. This indicates that the Ppant group of CoA could be conformationally flexible due to its lack of interaction with the enzyme active-site residues and explains the tolerance of the enzyme with diverse structures of the substitutes at the thiol end of CoA (Mofid *et al.*, 2004). Belshaw, Walsh, and colleagues first took advantage of the broad substrate specificity of Sfp to load non-native aminoacyl substrates onto the PCP within a NRPS module (Belshaw *et al.*, 1999). In this way, the strict substrate specificity imposed by the adenylation domain embedded in the NRPS module could be bypassed for the enzymatic synthesis of new chemical structures. Later, CoA-conjugated peptides or crosslinked peptides such as the vancomycin aglycone have been shown to be the substrates for Sfp for PCP modification (Sieber *et al.*, 2003; Vitali *et al.*, 2003). Burkart and colleagues further expanded the potential substrate pool of Sfp and demonstrated that biotin, carbohydrates, and fluorophores of diverse structures can be site-specifically attached to PCP through the Ppant arm by Sfp-catalyzed transfer from their CoA derivatives (Clarke et al., 2005; La Clair et al., 2004; Meier et al., 2006).

Yin, Walsh, and colleagues used PCP as a tag for protein labeling by expressing the target proteins as fusions to an 80-residue PCP from an NRPS module, GrsA (Yin et al., 2004). Biotin labeling of the PCP tag was accomplished by Sfp-catalyzed PCP modification using biotin-CoA 1 as the substrate (Fig. 10.2) (Yin et al., 2004). Phage-displayed PCP can also be loaded with small molecules such as biotin, glutathione, porphyrin, or carbohydrates by Sfp-catalyzed transfer from the corresponding CoA conjugates (Yin et al., 2004). Short nucleotide sequences inserted in the phagemid can then be used to encode the small molecules loaded on the phage particles. After selection for small molecule binding with a protein receptor, the DNA coding region of the selected phage clones can be sequenced or hybridized with a decoding array to deconvolute the identity of the small molecules that show high binding affinity with the designated receptor. Sfp was also used for site-specific labeling of the target proteins on live cell surfaces site-specific labeling of the target proteins on live cell surfaces to image the distribution and trafficking of the cell surface receptors and their interaction with the protein ligands (Yin et al., 2005, 2006). Because of the broad substrate specificity of Sfp with CoA conjugates, various fluorophores can be attached to the cell surface receptors and their interactions with the cognate ligands can be monitored by fluorescence energy transfer (Yin et al., 2005). Sfp catalyzed protein labeling has also been adopted by Wong, Micklefield, and colleagues for protein immobilization on synthetic polymers functionalized with CoA (Wong et al., 2008). Besides Sfp, other PPTases have also been used for site-specific protein labeling. George, Johnsson, and colleagues used AcpS from E. coli to attach small molecule fluorophores to the ACP tag fused with target proteins on the surface of yeast cells (George et al., 2004). They also demonstrated the use of AcpS to label yeast cell surface proteins with different fluorophores to visualize the spatial and temporal distribution of the target proteins at various stages of cell development (Vivero-Pol et al., 2005).

There is also an interest in developing short peptide tags as surrogate substrates for PPTase-catalyzed protein modification reactions. Commonly used peptide tags in biochemical assays such as flag tag, myc tag or $6 \times$ His tags are less than 10 residues in length. In contrast, full-length PCP or ACP domains are typically 75–80 residues long (George et al., 2004; Yin et al., 2004). The large size of the carrier proteins may affect the expression of the tagged fusion proteins and their physiological functions. By phage selection with a B. subtilis genomic library and randomized peptide libraries, short peptides of 11–12 residues long, namely ybbR and S6, were identified as efficient substrates of Sfp, and A1 peptide as an efficient substrate of AcpS (Yin et al., 2005; Zhou et al., 2007). Furthermore, the S6 tag and the A1 tag showed substantial orthogonality with their reactivity for Sfp- and AcpS-catalyzed peptide modification. Sfp catalyzes S6 modification with small-molecule CoA conjugates with a k_{cat}/K_m 400-fold higher than AcpS, while AcpS catalyzed A1 modification with a k_{cat}/K_m more than 30-fold higher than Sfp (Zhou et al., 2007). Recently Zhou, Walsh, and colleagues identified a shorter version of the A1 peptide that is only eight residues long and can serve as an efficient substrate of AcpS for the labeling of the fusion proteins (Zhou et al., 2008). A1 and S6 peptide tags were attached to the C-terminus of the transferrin receptor and the N-terminus of the epidermal growth factor receptor (EGFR), respectively. The tagged receptors were labeled with distinctive fluorophores by Sfp and AcpS on the surface of the same cell (Zhou et al., 2007). Sunbul, Yin and colleagues further conjugated CoA to quantum dots (Qdots) that are nano-sized inorganic crystals with bright fluorescence and demonstrated that Sfp and AcpS can site-specifically attach Qdots to target protein receptors on the cell surface (Fig. 10.3) (Sunbul et al., 2008).

A high-throughput method has been developed to identify PKS and NRPS genes by phage selection (Fig. 10.4) (Yin et al., 2007). This method employs Sfp from B. subtilis and AcpS from E. coli to covalently modify phage-displayed PCP or ACP domains of NRPS or PKS origin with a biotin–CoA conjugate (Yin et al., 2007). To use this method for cloning NRPS and PKS genes from a bacterial genome, a shotgun library of genomic DNA is constructed in the pComb3H phagemid for M13 phage display. Sfp and biotin–SS–CoA 2 are then used to covalently label phage-displayed apo-PCP or apo-ACP domains with biotin. Biotin-labeled phage particles are then selectively bound to streptavidin-coated plates and iterative rounds of phage selection are carried out until the phage library is converged toward clones displaying fragments of NRPS or PKS enzymes

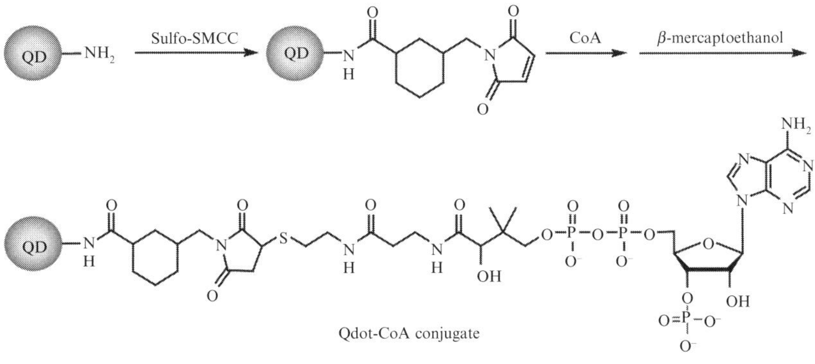

Figure 10.3 Synthesis of CoA-conjugated Qdots. Sulfo-SMCC, sulfosuccinimidyl-4-(N-maleimidomethyl) cyclohexane-1-carboxylate).

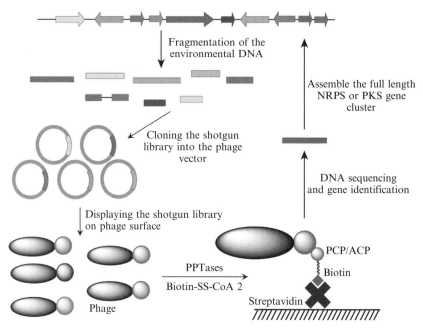

Figure 10.4 Phage selection for NRPS or PKS gene fragments encoded in a bacterial genome. (See Color Insert.)

containing PCP or ACP domains. DNA sequencing of the selected phage clones would provide the gene sequences of the enriched NRPS or PKS fragments displayed on the phage surface. Hybridization probes can then be designed based on the DNA sequence of the identified PKS or NRPS fragments to screen a cosmid library of the bacterial genomic DNA in order to clone the full-length biosynthetic gene clusters. Alternatively, the gene clusters can be assembled by polymerase chain reactions (PCR) based on the sequences of the selected NRPS and PKS fragments. It has been shown that phage selection identified almost half of the carrier proteins in *B. subtilis* and 20% of the carrier proteins in *Myxococcus xanthus* from their shotgun genomic libraries, validating phage display as a high-throughput platform for cloning PKS and NRPS genes from the bacterial genomes (Yin et al., 2007). Sfp-catalyzed phage selection thus offers a simple and efficient method for genome-wide cloning of NRPS and PKS gene clusters. This method can potentially be applied to clone PKS and NRPS clusters directly from environmental DNA in order to discover new biosynthetic pathways from the metagenome. Heterologous expression of the cloned clusters may lead to the identification of natural product molecules with novel structures and bioactivities that are previously inaccessible to drug development programs due to the difficulties in culturing their native host strains.

2. Experimental Procedures

2.1. Expression of AcpS from *E. coli*

The AcpS gene from the *E. coli* genome was cloned in the pET28b vector (Novagen) between the *Nde*I and *Xho*I restriction sites to give the expression plasmid pET-AcpS (Zhou *et al.*, 2007). The AcpS enzyme can then be expressed in *E. coli* BL21 (DE3) pLysS chemically competent cells (Invitrogen). BL21 DE3 competent cells are transformed with pET-AcpS and single colonies are grown on an LB agar plate supplemented with 50 μg/mL kanamycin. An overnight culture is then prepared by inoculating 10 mL of LB supplemented with 50 μg/mL kanamycin with a single colony of BL21 DE3 cells transformed with pET-AcpS. The next day, 10 mL of the overnight culture is added to 1 L freshly prepared LB medium supplemented with 50 μg/mL kanamycin and the culture is grown at 37 °C to an OD of 0.5 at 600 nm (OD_{600}). The temperature of the shaker is then reduced to 30 °C and 1 mM isopropyl-d-thiogalactopyranoside (IPTG) is added to the cell culture. The culture is allowed to grow for another 6 h before the cells are harvested by centrifugation (4000g, 15 min). The cell pellets are resuspended in 20 mL lysis buffer (50 mM Tris HCl, 0.5 M NaCl, 5 mM imidazole, pH 8.0) with 1 unit/mL DNase I and disrupted by French Press with two passes at 16,000 psi. Cell debris is removed by ultracentrifugation (95,000g, 30 min). The clarified cell extract is then incubated with Ni-NTA resin (Qiagen) with the addition of 1 mL 50% resin suspension to the cell lysate prepared from 1 L of cell culture. The resin and lysate are allowed to incubate for 3–4 h at 4 °C in a batch-binding format with gentle mixing. The suspension is then loaded on a gravity column and washed with 10 mL lysis buffer. The Ni-NTA resin is further washed with 20 mL wash buffer (50 mM Tris HCl, 0.5 M NaCl, 20 mM imidazole, pH 8.0). Protein bound to the column is eluted with 6 mL of 250 mM imidazole in lysis buffer. The purity of the fractions containing purified AcpS enzyme is checked by SDS-PAGE with Coomassie blue staining. Fractions with the desired purity are pooled and dialyzed twice against 1 L of 100 mM Bis-Tris propane (pH 6.0), 500 mM NaCl, and 10% glycerol. Protein solutions are then concentrated to 5 mg/mL with a Centriprep YM-10 concentrator and aliquots are stored at −80 °C. A typical yield of AcpS protein from this expression protocol is about 8 mg/L cell culture.

2.2. Expression of Sfp from *E. coli*

Sfp was cloned as a C-terminal 6 × His tagged protein in the pET29 vector carrying a kanamycin resistance gene (Yin *et al.*, 2006). For Sfp expression, pET29-Sfp is transformed into *E. coli* BL21(DE3) pLysS chemically competent cells (Invitrogen) following the vendor's protocol. The cells from the transformation reaction are streaked on an LB agar plate containing 50 μg/mL kanamycin. The plate is then incubated at 37 °C overnight and a single colony is picked to inoculate a 10 mL LB starting culture containing 50 μg/mL kanamycin. The starting culture is shaken at 200 rounds per minute (rpm) at 37 °C overnight. The next day, the starting culture is used to inoculate 1 L LB containing 50 μg/mL kanamycin. The culture is shaken at 200 rpm at 37 °C until OD_{600} reaches 0.6. IPTG is added to the LB culture to a final concentration of 1 mM. After IPTG addition, shaking is continued at 200 rpm at room temperature for 6 h. The cells are then harvested by centrifugation in two 500 mL centrifuge bottles (4000g, 15 min), and the cell pellets are resuspended in 20 mL lysis buffer containing 1 unit/mL DNase I. The cell lysate is prepared by passing the cell suspension twice through a French pressure cell at 16,000 psi. The cellular debris is removed from the lysate by ultracentrifugation (95,000g, 30 min). The Sfp protein is purified by affinity chromatography using Ni-NTA resin with the same procedure as for AcpS purification. The fractions of the eluant from the Ni-NTA column are analyzed by SDS-PAGE, and fractions containing Sfp (26 kD) with greater than 90% purity are pooled. The combined fractions are dialyzed twice against 1 L of 10 mM 4-(2-hydroxyethyl)-1-piperazineethanesulfonic acid (HEPES) (pH 8.0), 120 mM NaCl, 5 mM dithiothreitol (DTT), and 10% glycerol. After dialysis, the protein solution is concentrated to more than 5 mg/mL with a Centriprep YM-10 concentrator. The Sfp stock solution is aliquoted into Eppendorf tubes, flash-frozen in liquid nitrogen, and stored at −80 °C. This purification procedure typically yields 25 mg Sfp/L cell culture.

2.3. Coexpression of NRPS or PKS modules with Sfp in *E. coli*

NRPS or PKS modules can be coexpressed with Sfp to achieve *in vivo* modification of the embedded carrier protein domains with the Ppant group (Admiraal *et al.*, 2002; Gokhale *et al.*, 1999; Keating *et al.*, 2000; Suo *et al.*, 2000). Plasmids carrying the genes for NRPS or PKS modules can be combined with the Sfp expression plasmid pRSG56 (Gokhale *et al.*, 1999) to cotransform *E. coli* expression strains such as BL21(DE3). Since pRSG56 carries a kanamycin resistance gene, the expression plasmid of the NRPS or PKS genes should harbor a different type of antibiotic resistance gene, typically, a β-lactamase gene for ampicillin resistance, in order to maintain both plasmids in the cotransformed *E. coli* cells. The cell culture of the transformation reaction is plated on LB agar plates supplemented with 100 μg/mL ampicillin and 50 μg/mL kanamycin. Single colonies growing on the plate should carry plasmids for both Sfp and apo enzyme expression and are used to inoculate overnight cultures supplemented with both ampicillin and kanamycin. Protein expression and purification follow the same protocols described above. Since the Sfp gene is not fused to the 6 × His tag in pRSG56, the coexpressed Sfp would not be retained on Ni-NTA resin and only NRPS or PKS fragments fused to the 6 × His tag is purified by Ni-NTA affinity chromatography. Typically, more than 90% of the NRPS or PKS modules are purified as holo enzymes that are catalytically active for substrate loading.

2.4. *In vitro* activation of the PKS or NRPS modules by Sfp

Sfp can be used to prime apo PKS or NRPS modules by phosphopantetheinylation of the carrier protein domains using CoA as the substrate. The priming reaction can be carried out in a reaction mixture containing 10 mM $MgCl_2$, 5 mM DTT, and 75 mM HEPES, pH 7.0. Apo protein is added to a concentration of 100–200 μM with 10–20% excess in the concentration of CoA. The reaction is initiated by adding 0.1 μM Sfp, followed by incubation at 37 °C for 30 min. Subsequent assays of the holo NRPS or PKS modules for substrate loading or chain elongation can be directly performed on the priming reaction mixture without further purification.

A trichloroacetic acid (TCA) assay has been developed to measure the kinetics and the stoichiometry of apo-to-holo enzyme conversion catalyzed by Sfp using tritium labeled CoA ([^{3}H]CoA) (Lambalot *et al.*, 1996). In a typical assay, [^{3}H]CoA at a concentration of 50 μM is incubated with 5 μM apo protein in a 200 μL reaction mixture as described above containing 0.1 μM Sfp. 20 μL aliquots of the reaction mixture are withdrawn at various time points and the reaction is quenched by adding the aliquot to 0.5 mL 10% TCA. Precipitated protein is pelleted by centrifugation and the pellets are washed three times with 0.8 mL 10% TCA before being dissolved in 0.5 mL 88% formic acid and counting for radioactivity on a liquid scintillation counter.

2.5. Preparation of small molecule–CoA conjugates

Small-molecule CoA conjugates can serve as efficient substrates of Sfp and AcpS for site-specific attachment of small-molecule probes to the carrier protein domains or short peptide tags through the Ppant arm (Johnsson *et al.*, 2005; Yin *et al.*, 2006). Synthesis of small-molecule CoA conjugates can be accomplished by one-step Michael condensation of the maleimide-functionalized small molecules with the free thiol group at the end of the Ppant arm of CoA. Maleimide-linked biotin and fluorophores are commercially available from Pierce and Molecular Probes. The synthesis of biotin-CoA conjugate 1 (La Clair *et al.*, 2004; Yin *et al.*, 2004) is used as an example here (Fig. 10.2A).

To a solution of biotin maleimide (Pierce) (10 mg, 0.019 mmol) in 300 μL DMSO, CoA lithium salt (Sigma) (18.2 mg, 0.023 mmol) in 2 mL MES acetate 50 mM, pH 6.0 is added and the reaction mixture is stirred at room temperature overnight. The reaction mixture is then purified by preparative high-performance liquid chromatography (HPLC) on a reversed-phase C18 column with a 0–60% gradient of acetonitrile in 0.1% TFA/water over 35 min. The purified compound is lyophilized and its identity can be confirmed by liquid chromatography-mass spectrometry (LC-MS) operating in the positive-ion mode.

Small-molecule probes can be also conjugated to CoA through a disulfide linkage so that the probe can be selectively released from the labeled protein by disulfide cleavage with DTT. Biotin was conjugated to CoA by disulfide formation to afford biotin–SS–CoA 2, which played an important role for the phage selection of NRPS or PKS fragments from bacterial genomes (Yin et al., 2007). During the selection process, Sfp catalyzed the labeling of phage-displayed carrier proteins with biotin–SS–CoA 2. Subsequently, biotin-attached phage particles were bound to immobilized streptavidin and eluted by DTT cleavage of the disulfide bond between biotin and the Ppant arm on PCP. Here we provide an example for the synthesis of biotin–SS–CoA 2 (Fig. 10.2B). To a solution of N-(6-(biotinamido)hexyl)-3'-(2'-pyridyldithio)-propionamide (biotin HPDP) (Pierce) (10.8 mg, 0.020 mmol) in 500 μL DMSO, CoA lithium salt (Sigma) (16.5 mg, 0.022 mmol) in 2 mL sodium phosphate 100 mM, pH 7.0 is added and the reaction mixture is stirred at room temperature for 1 h. The reaction mixture is then purified by preparative HPLC on a reversed-phase C18 column with a 0–60% gradient of acetonitrile in 0.1% TFA/water over 35 min. The purified compound is lyophilized and its identity is confirmed by LC–MS.

Amino acids, carboxylic acids or the elongation intermediates of non-ribosomal peptides or polyketides can also be conjugated to CoA through the formation of thioester bonds and the corresponding CoA conjugates can be recognized by Sfp for specific attachment to the carrier proteins within PKS or NRPS modules (Belshaw et al., 1999; Clugston et al., 2003; Sieber et al., 2003). A general procedure for the synthesis of aminacyl CoA has been reported (Fig. 10.2C) (Belshaw et al., 1999). Briefly, 1 equivalent of Nvoc protected amino acid is incubated with 1 equivalent of CoA lithium salt, 4 equivalent of potassium carbonate and 1.5 equivalent of PyBOP in 1:1 mixture of THF/H$_2$O. The reaction mixture is stirred at room temperature for 2 h, followed by photolysis to remove the Nvoc protecting group. The aminoacyl CoA product is then purified by preparative HPLC. A similar procedure has also been reported for the synthesis of peptide CoA conjugate with a thioester linkage (Clugston et al., 2003; Sieber et al., 2003).

2.6. Preparation of Qdot–CoA conjugates

We recently showed that CoA conjugated Qdots can be recognized by Sfp or AcpS for site-specific attachment of Qdots to target proteins fused with carrier proteins or small peptide tags (Sunbul et al., 2008). Qdot–CoA conjugates are prepared with Qdot ITK Amino (PEG) (Invitrogen) which is supplied at 8 μM in 50 mM borate buffer (Fig. 10.3). Buffer exchange is performed by suspending 75 μL of Qdot into ~4 mL of 25 mM HEPES (pH 7.4) and transferring the solution into a 100-kDa Amicon Ultra centrifugal filter device (Millipore). The filter unit is centrifuged at low speed (1250g) until the retentate volume is ~100 μL. Subsequently, retentate is transferred into a 2 mL polypropylene microfuge tube and 500 equivalents of sulfosuccinimidyl 4-[N-maleimidomethyl]cyclohexane-1-carboxylate (sulfo-SMCC) (Pierce) are added (15 μL of 20 mM solution, in water). The mixture is allowed to react for 2 h at 28 °C under gentle shaking (200 rpm). Excess sulfo-SMCC is removed by exchanging the reaction mixture twice with fresh HEPES buffer, as described above. Following the second round of centrifugation, the retentate (~100 μL total volume) is transferred to a 2-mL microfuge tube and 10 equivalents of CoA (0.6 μL of 10 mM solution in water) are added to the mixture and allowed to react under the same condition as above. After 3 h, 500 equivalents of β-mecaptoethanol (3 μL of a 150 mM solution, in water) are added to cap the unreacted maleimide groups on the Qdot surface under identical conditions. Following a 30-min reaction, excess CoA and β-mercaptoethanol are removed by a gel filtration column (Superdex 200, GE Healthcare) equilibrated with 50 mM borate buffer (pH 8.3). Approximately five to six column fractions are collected at various intervals and analyzed by a UV spectrophotometer (800–200 nm). Fractions lacking a free CoA peak at 260 nm are pooled together, filtered through a 0.2 μm Spin-X microfuge tube filter (Costar), and stored at 4 °C before use. The Qdot–CoA conjugates are stable for more than 3 months when stored at 4 °C.

2.7. Sfp catalyzed protein labeling with small molecule–CoA conjugates

The labeling reaction can be carried out under similar conditions to prime PKS or NRPS modules by carrier protein modification with CoA. Here, we use Sfp-catalyzed PCP labeling with biotin CoA 1 as an example (Yin et al., 2006). To a total volume of 100 μL containing 10 mM MgCl$_2$ and 50 mM HEPES pH 7.5, is added 0.1 μM Sfp, 5 μM biotin CoA 1 and 5 μM target protein fused with PCP. The reaction mixture is incubated at room temperature for 30 min. Sfp-catalyzed labeling of ybbR- or S6-tagged proteins, or AcpS-catalyzed labeling of ACP or A1-tagged proteins, can be performed under the same conditions (Yin et al., 2005; Zhou et al., 2007). Target proteins fused with carrier protein domains or small peptide tags can also be directly labeled by Sfp or AcpS in cell lysates following the same protocol (Yin et al., 2005; Zhou et al., 2007). Construction of PCP or ybbR fusions with target proteins has been reported elsewhere (Yin et al., 2006).

The formation of biotinylated protein in the labeling reaction mixture can be verified by Western blotting probed with streptavidin–horseradish peroxidase (HRP) conjugates (Yin et al., 2006). To detect biotin labeling, a labeling reaction mixture containing 0.5 μg protein is loaded on a 4–15% SDS-PAGE gel. After electrophoresis, the protein bands are electroblotted onto a piece of polyvinylidene fluoride (PVDF) membrane. The membrane is then blocked with 3% BSA in Tris-buffered saline (TBS) for 2 h, followed by incubation with 0.1 μg/mL streptavidin–HRP conjugate in 1% BSA for 1 h. The membrane is washed five times with 0.05% Tween 20 and 0.05% Triton X-100 in TBS followed by five washes in TBS alone. Streptavidin binding can then be detected with an ECL^{TM} luminescence detection kit following the manufacturer's instructions.

HPLC assay has also been developed to monitor the kinetics for Sfp- or AcpS-catalyzed modification of carrier proteins or short peptide substrates (ybbR, S6, or A1) by CoA and their conjugates with small-molecule probes (Yin et al., 2005; Zhou et al., 2007). After incubating the labeling reaction mixture for 10–30 min at 37 °C, reactions are quenched by adding 30 μL 4% trifluoroacetic acid (TFA) to 100 μL of the reaction mixture. Product formation is monitored at 220 nm on an analytical HPLC with a reverse-phase C_{18} column, using a 0–60% gradient of acetonitrile in 0.1% TFA/H_2O over 30 min. PCP loaded with Ppant or Ppant small-molecule conjugates would have different retention times from the apo PCP protein. Formation of the labeled PCP can be quantified based on the area of the product peak on the HPLC chromatogram. Similarly, ybbR, S6, and A1 peptide modification catalyzed by Sfp or AcpS can also be analyzed by HPLC. For peptide modification, 200 μM biotin–CoA 1 and 1 μM Sfp are incubated with 100 μM peptide in a 100 μL solution of 10 mM $MgCl_2$ and 50 mM HEPES, pH 7.5. After 30 min incubation at 37 °C, reactions are quenched by adding 30 μL 4% TFA and analyzed at 220 nm by analytical HPLC with a reverse-phase C_{18} column, using a 0–60% gradient of acetonitrile in 0.1% TFA/H_2O over 30 min (Yin et al., 2005; Zhou et al., 2007).

2.8. Sfp-catalyzed protein labeling on the cell surface

Live-cell labeling of transferrin receptor 1 (TfR1) is used as an example (Yin et al., 2005). PCP is fused to TfR1 at the C-terminus of the receptor, which is exposed on the cell surface. Construction of TfR1–PCP fusion in the pcDNA3.1(+) plasmid has been reported (Yin et al., 2005). TRVb cells are grown in Hams F12 medium containing 50 U/mL penicillin, 50 μg/mL streptomycin, and 10% fetal bovine serum (FBS). For each 35-mm well in a 6-well plate, 1 vial of transfection medium is prepared, containing 3 μl of FuGENE 6 transfection reagent (Roche Diagnostics Corporation) and 1 μg of the relevant plasmid diluted in 100 μl of a 1:1 mixture of Dulbecco minimal essential medium and Hams F12. Transfection medium is mixed well and allowed to rest for 20 min to reach equilibrium. TRVb cells are grown on sterilized cover slips to 50–60% confluency, incubated with transfection medium for 5 h, and then incubated in regular cell medium (with serum) for 24 h to allow time for protein expression.

Cells transfected with TfR1–PCP are incubated in serum-free media for 2 h prior to labeling. To label TfR1–PCP, cells are incubated with 0.5 μM Sfp and 1 μM CoA-Alexa Fluor 488 in serum-free medium for 30 min at 37 °C (Yin et al., 2005). Labeled cells are washed three times with PBS and reimmersed in medium. If needed, cells are then incubated with 10 $\mu g/mL$ Alexa Fluor 568 conjugated transferrin (Molecular Probes Inc.) for various time periods and washed three times with PBS. Cells are then fixed using a 3.7% formaldehyde solution in PBS, and mounted with SlowFade® Antifade Kit (Molecular Probes Inc.) for optical microscopy studies.

Details on cell imaging studies and labeling of cell surface receptors fused with A1 or S6 peptide tags with AcpS and Sfp have been previously reported (Yin et al., 2006; Zhou et al., 2007). Protocols for Sfp- or AcpS-catalyzed Qdot labeling of cell surface receptors and subsequent cell imaging studies have also been reported (Sunbul et al., 2008).

2.9. Construction of the genomic library for phage selection of NRPS and PKS fragments

A genomic library of *B. subtilis* was constructed and displayed on the phage surface (Fig. 10.4) (Yin et al., 2007). Sfp was used to label carrier proteins displayed on the phage surface with biotin–SS–CoA 2. Biotinylated phage particles were then selected by streptavidin for the enrichment of phages displaying fragment of PKS pr NRPS enzymes (Yin et al., 2007). The construction of a genomic DNA library for phage selection is as follows.

Genomic DNA is extracted from 50 mL of a log-phase culture of *B. subtilis* strain 3610 (Harwood and Cutting, 1990). One hundred micrograms of RNaseA-treated genomic DNA is randomly sheared with a sonicator (Misonix, Inc., Farmingdale, NY). Sheared DNA is size-selected on an agarose prep gel for fragments ranging from 500 bp to 1.5 kb. Ten micrograms of sheared DNA fragments are treated with 15U of T4 polymerase to create blunt-ended fragments. The pComb phagemid vector, which was modified to contain *Eco*RI and *Hin*dIII restriction sites (pComb-HE), is digested with *Sac*I and *Spe*I and treated with T4 polymerase to create blunt ends and with calf intestinal phosphatase to prevent vector self-ligation. The blunt fragments are ligated into the pComb-HE phagemid vector using 5 μg *B. subtilis* genomic DNA fragments and 17 μg pComb-HE vector. Ligated plasmids are cloned into electrocompetent XL1 Blue *E. coli* (Stratagene, La Jolla, CA). *E. coli* clones containing the vector are screened by colony PCR for insertion of *B. subtilis* genomic DNA fragments. The efficiency of insertion should be ~90% of the total clones recovered.

2.10. Phage selection by Sfp catalyzed carrier protein modification with biotin–SS–CoA 2

Phage particles are produced with the following procedures (Yin et al., 2004). Briefly, E. coli XL1-Blue cells are transformed with pComb3H vectors, and shaken at 37 °C in 2 × yeast/tryptone (2 × YT) broth and 100 μg/mL of ampicillin. At an OD_{600} of 0.5, helper phage VCSM13 (Stratagene) is added to a final concentration of 1.5×10^8 cfu/mL, and the infection mixture is incubated at 37 °C for 1 h without shaking. The cells are centrifuged and re-suspended in 2 × YT, supplemented with 100 μM IPTG, 100 μg/mL ampicillin, and 50 μg/mL kanamycin, and shaken for 14 h at room temperature. Cells are centrifuged and phage particles in the supernatant are precipitated by polyethylene glycol and resuspended in TBS (25 mM Tris HCl, pH 7.4, 140 mM NaCl, 2.5 mM KCl). Phage titrations are performed with E. coli XL1-blue cells using standard procedures (Kay et al., 1996).

Phage-displayed B. subtilis genomic libraries are then labeled with biotin by Sfp using biotin–SS–CoA 2 as the substrate. For the first round of selection, the labeling reactions are carried out with 10^{12} phage particles in 1 mL 10 mM $MgCl_2$ and 50 mM HEPES pH 7.5 with 5 μM biotin–SS–CoA 2, 1 μM Sfp and incubated at 30 °C for 30 min. For the subsequent rounds of selections, the number of input phage particles, concentrations of enzymes and biotin–SS–CoA 2, and the reaction time are decreased step by step and eventually for the fifth round of selection, only 10^{10} phage particles are incubated with 0.08 μM enzyme and 1 μM biotin–SS–CoA 2 for 7 min at 30 °C. Control reactions are also run in parallel without the addition of enzymes or biotin–SS–CoA 2.

After the labeling reaction, the reaction mixtures are added to 250 μL 20% (w/v) polyethylene glycol 8000 with 2.5 M NaCl followed by 10-min incubation on ice. The phage particles in the reaction mixture are precipitated by centrifugation at 4 °C (10,000g, 30 min). The phage pellet is resuspended in 1 mL TBS supplemented with 1% (w/v) BSA and distributed in 100 μL aliquots to the wells of streptavidin-coated 96-well plates (Pierce). The plates are incubated at room temperature for 1 h before the supernatant is discarded. Each well is washed 30 times with 0.05% (v/v) Tween 20, 0.05% (v/v) Triton X-100 in TBS and 30 times with TBS, each time with 200 μL of solution. After washing, phages bound to the streptavidin surface are eluted by adding 100 μL 20 mM DTT in TBS to each well. The DTT solution induces the cleavage of the disulfide bond that links the biotin group with Ppant on phage displayed carrier proteins. Eluted phage particles are combined, added to 10 mL of log phase XL1-Blue cells and shaken at 37 °C for 1 h to infect the cells. The cells are then plated on LB agar plates supplemented with 2% (w/v) glucose and 100 μg/mL ampicillin. After incubation at 37 °C overnight, colonies on the plates are scraped off and the phagemid DNA is extracted by a Qiagen Plasmid Maxi kit. The phagemid DNA is then used for the next round of phage production and selection. Also phage particles eluted from the wells loaded with either selection or the control reactions are titered in order to count the number of phage particles selected by each round. After the fifth round of selection, phage clones are sequenced using the primer Jun13 (5′- ACT TTA TGC TTC CGG CTC GTA TGT).

3. Conclusion

PPTases catalyze phosphopantetheinylation of the carrier proteins for posttranslational activation of the FAS, PKS, and NRPS enzymes that are responsible for the biosynthesis of fatty acid, polyketide, and nonribosomal peptide natural products. PPTases can be used for *in vitro* modification of the carrier protein domains with CoA or coexpressed with PKS or NRPS modules for the production of holo enzymes upon protein expression. Recent discovery of the broad substrate specificity of PPTases with CoA-conjugated small molecules prompts new applications of this class of enzymes in site-specific protein labeling and cell imaging. Sfp from *B. subtilis* and AcpS from *E. coli* have been used for the covalent attachment of biotin, fluorophore, and nano-sized Qdots to target protein receptors on the cell surface. Further protein engineering work has identified short peptide tags as surrogate substrates of Sfp and AcpS and facilitates the construction of fusion proteins for PPTase-catalyzed protein labeling. Sfp and AcpS have also been used for high throughput selection of PKS and NRPS fragments from bacterial genomic DNA libraries for the identification of new biosynthetic gene clusters.

REFERENCES

Admiraal, S. J., Khosla, C., and Walsh, C. T. (2002). The loading and initial elongation modules of rifamycin synthetase collaborate to produce mixed aryl ketide products. *Biochemistry* **41,** 5313–5324.

Belshaw, P. J., Walsh, C. T., and Stachelhaus, T. (1999). Aminoacyl-CoAs as probes of condensation domain selectivity in nonribosomal peptide synthesis. *Science* **284,** 486–489.

Bunkoczi, G., Pasta, S., Joshi, A., Wu, X., Kavanagh, K. L., Smith, S., and Oppermann, U. (2007). Mechanism and substrate recognition of human holo ACP synthase. *Chem. Biol.* **14,** 1243–1253.

Cane, D. E., and Walsh, C. T. (1999). The parallel and convergent universes of polyketide synthases and nonribosomal peptide synthetases. *Chem. Biol.* **6,** R319–R325.

Cane, D. E., Walsh, C. T., and Khosla, C. (1998). Harnessing the biosynthetic code: Combinations, permutations, and mutations. *Science* **282,** 63–68.

Chen, H., and Walsh, C. T. (2001). Coumarin formation in novobiocin biosynthesis: Beta-hydroxylation of the aminoacyl enzyme tyrosyl-S-NovH by a cytochrome P450 NovI. *Chem. Biol.* **8,** 301–312.

Chirgadze, N. Y., Briggs, S. L., McAllister, K. A., Fischl, A. S., and Zhao, G. (2000). Crystal structure of Streptococcus pneumoniae acyl carrier protein synthase: An essential enzyme in bacterial fatty acid biosynthesis. *EMBO J.* **19,** 5281–5287.

Clarke, K. M., Mercer, A. C., La Clair, J. J., and Burkart, M. D. (2005). In vivo reporter labeling of proteins via metabolic delivery of coenzyme A analogues. *J. Am. Chem. Soc.* **127,** 11234–11235.

Clugston, S. L., Sieber, S. A., Marahiel, M. A., and Walsh, C. T. (2003). Chirality of peptide bond-forming condensation domains in nonribosomal peptide synthetases: The C5 domain of tyrocidine synthetase is a (D)C(L) catalyst. *Biochemistry* **42,** 12095–12104.

Copp, J. N., and Neilan, B. A. (2006). The phosphopantetheinyl transferase superfamily: Phylogenetic analysis and functional implications in cyanobacteria. *Appl. Environ. Microbiol.* **72,** 2298–2305.

Copp, J. N., Roberts, A. A., Marahiel, M. A., and Neilan, B. A. (2007). Characterization of PPTNs, a cyanobacterial phosphopantetheinyl transferase from Nodularia spumigena NSOR10. *J. Bacteriol.* **189,** 3133–3139.

Crawford, J. M., Vagstad, A. L., Ehrlich, K. C., Udwary, D. W., and Townsend, C. A. (2008). Acyl-carrier protein-phosphopantetheinyltransferase partnerships in fungal fatty acid synthases. *Chem. Bio. Chem.* **9,** 1559–1563.

De Lay, N. R., and Cronan, J. E. (2006). A genome rearrangement has orphaned the *Escherichia coli* K-12 AcpT phosphopantetheinyl transferase from its cognate *Escherichia coli* O157:H7 substrates. *Mol. Microbiol.* **61,** 232–242.

Ehmann, D. E., Gehring, A. M., and Walsh, C. T. (1999). Lysine biosynthesis in *Saccharomyces cerevisiae*: Mechanism of alpha-aminoadipate reductase (Lys2) involves posttranslational phosphopantetheinylation by Lys5. *Biochemistry* **38,** 6171–6177.

Fichtlscherer, F., Wellein, C., Mittag, M., and Schweizer, E. (2000). A novel function of yeast fatty acid synthase. Subunit alpha is capable of self-pantetheinylation. *Eur. J. Biochem.* **267,** 2666–2671.

Finking, R., Mofid, M. R., and Marahiel, M. A. (2004). Mutational analysis of peptidyl carrier protein and acyl carrier protein synthase unveils residues involved in protein-protein recognition. *Biochemistry* **43,** 8946–8956.

Finking, R., Solsbacher, J., Konz, D., Schobert, M., Schafer, A., Jahn, D., and Marahiel, M. A. (2002). Characterization of a new type of phosphopantetheinyl transferase for fatty acid and siderophore synthesis in *Pseudomonas aeruginosa*. *J. Biol. Chem.* **277,** 50293–50302.

Fischbach, M. A., and Walsh, C. T. (2006). Assembly-line enzymology for polyketide and nonribosomal Peptide antibiotics: Logic, machinery, and mechanisms. *Chem. Rev.* **106,** 3468–3496.

Flugel, R. S., Hwangbo, Y., Lambalot, R. H., Cronan, J. E., Jr., and Walsh, C. T. (2000). Holo-(acyl carrier protein) synthase and phosphopantetheinyl transfer in *Escherichia coli*. *J. Biol. Chem.* **275,** 959–968.

George, N., Pick, H., Vogel, H., Johnsson, N., and Johnsson, K. (2004). Specific labeling of cell surface proteins with chemically diverse compounds. *J. Am. Chem. Soc.* **126,** 8896–8897.

Gokhale, R. S., Tsuji, S. Y., Cane, D. E., and Khosla, C. (1999). Dissecting and exploiting intermodular communication in polyketide synthases. *Science* **284,** 482–485.

Harwood, C. R., and Cutting, S. M. (eds.), (1990). Molecular Biological Methods for *Bacillus*, John Wiley & Sons, Chichester, UK.

Johnsson, N., George, N., and Johnsson, K. (2005). Protein chemistry on the surface of living cells. *Chem. Bio. Chem.* **6,** 47–52.

Joshi, A. K., Zhang, L., Rangan, V. S., and Smith, S. (2003). Cloning, expression, and characterization of a human 4′-phosphopantetheinyl transferase with broad substrate specificity. *J. Biol. Chem.* **278,** 33142–33149.

Kay, B. K., Winter, J., and McCafferty, J. (1996). Phage Display of Peptides and Proteins, Academic Press, Inc., Boston.

Kealey, J. T., Liu, L., Santi, D. V., Betlach, M. C., and Barr, P. J. (1998). Production of a polyketide natural product in nonpolyketide-producing prokaryotic and eukaryotic hosts. *Proc. Natl. Acad. Sci. USA* **95,** 505–509.

Keating, T. A., Suo, Z., Ehmann, D. E., and Walsh, C. T. (2000). Selectivity of the yersiniabactin synthetase adenylation domain in the two-step process of amino acid activation and transfer to a holo-carrier protein domain. *Biochemistry* **39,** 2297–2306.

Khosla, C., Gokhale, R. S., Jacobsen, J. R., and Cane, D. E. (1999). Tolerance and specificity of polyketide synthases. *Annu. Rev. Biochem.* **68,** 219–253.

Ku, J., Mirmira, R. G., Liu, L., and Santi, D. V. (1997). Expression of a functional non-ribosomal peptide synthetase module in *Escherichia coli* by coexpression with a phosphopantetheinyl transferase. *Chem. Biol.* **4,** 203–207.

La Clair, J. J., Foley, T. L., Schegg, T. R., Regan, C. M., and Burkart, M. D. (2004). Manipulation of carrier proteins in antibiotic biosynthesis. *Chem. Biol.* **11,** 195–201.

Lai, J. R., Koglin, A., and Walsh, C. T. (2006). Carrier protein structure and recognition in polyketide and nonribasomal peptide biosynthesis. *Biochemistry* **45,** 14869–14879.

Lambalot, R. H., Gehring, A. M., Flugel, R. S., Zuber, P., LaCelle, M., Marahiel, M. A., Reid, R., Khosla, C., and Walsh, C. T. (1996). A new enzyme superfamily – the phosphopantetheinyl transferases. *Chem. Biol.* **3,** 923–936.

Lambalot, R. H., and Walsh, C. T. (1995). Cloning, overproduction, and characterization of the *Escherichia coli* holo-acyl carrier protein synthase. *J. Biol. Chem.* **270,** 24658–24661.

Meier, J. L., Mercer, A. C., Rivera, H., Jr., and Burkart, M. D. (2006). Synthesis and evaluation of bioorthogonal pantetheine analogues for *in vivo* protein modification. *J. Am. Chem. Soc.* **128,** 12174–12184.

Mercer, A. C., and Burkart, M. D. (2007). The ubiquitous carrier protein–a window to metabolite biosynthesis. *Nat. Prod. Rep.* **24,** 750–773.

Mofid, M. R., Finking, R., Essen, L. O., and Marahiel, M. A. (2004). Structure-based mutational analysis of the 4′-phosphopantetheinyl transferases Sfp from *Bacillus subtilis*: Carrier protein recognition and reaction mechanism. *Biochemistry* **43,** 4128–4136.

Mofid, M. R., Finking, R., and Marahiel, M. A. (2002). Recognition of hybrid peptidyl carrier proteins/acyl carrier proteins in nonribosomal peptide synthetase modules by the 4′-phosphopantetheinyl transferases AcpS and Sfp. *J. Biol. Chem.* **277,** 17023–17031.

Mootz, H. D., Finking, R., and Marahiel, M. A. (2001). 4′-phosphopantetheine transfer in primary and secondary metabolism of *Bacillus subtilis*. *J. Biol. Chem.* **276,** 37289–37298.

Mootz, H. D., Schorgendorfer, K., and Marahiel, M. A. (2002). Functional characterization of 4′-phosphopantetheinyl transferase genes of bacterial and fungal origin by complementation of Saccharomyces cerevisiae lys5. *FEMS Microbiol. Lett.* **213,** 51–57.

Parris, K. D., Lin, L., Tam, A., Mathew, R., Hixon, J., Stahl, M., Fritz, C. C., Seehra, J., and Somers, W. S. (2000). Crystal structures of substrate binding to *Bacillus subtilis* holo-(acyl carrier protein) synthase reveal a novel trimeric arrangement of molecules resulting in three active sites. *Structure* **8,** 883–895.

Perham, R. N. (2000). Swinging arms and swinging domains in multifunctional enzymes: Catalytic machines for multistep reactions. *Annu. Rev. Biochem.* **69,** 961–1004.

Pfeifer, B. A., Admiraal, S. J., Gramajo, H., Cane, D. E., and Khosla, C. (2001). Biosynthesis of complex polyketides in a metabolically engineered strain of *E. coli*. *Science* **291,** 1790–1792.

Polacco, M. L., and Cronan, J. E., Jr. (1981). A mutant of *Escherichia coli* conditionally defective in the synthesis of holo-[acyl carrier protein]. *J. Biol. Chem.* **256,** 5750–5754.

Praphanphoj, V., Sacksteder, K. A., Gould, S. J., Thomas, G. H., and Geraghty, M. T. (2001). Identification of the alpha-aminoadipic semialdehyde dehydrogenase-phosphopantetheinyl transferase gene, the human ortholog of the yeast LYS5 gene. *Mol. Genet. Metab.* **72,** 336–342.

Quadri, L. E., Weinreb, P. H., Lei, M., Nakano, M. M., Zuber, P., and Walsh, C. T. (1998). Characterization of Sfp, a *Bacillus subtilis* phosphopantetheinyl transferase for peptidyl carrier protein domains in peptide synthetases. *Biochemistry* **37,** 1585–1595.

Reuter, K., Mofid, M. R., Marahiel, M. A., and Ficner, R. (1999). Crystal structure of the surfactin synthetase-activating enzyme sfp: A prototype of the 4'-phosphopantetheinyl transferase superfamily. *EMBO J.* **18,** 6823–6831.

Sanchez, C., Du, L., Edwards, D. J., Toney, M. D., and Shen, B. (2001). Cloning and characterization of a phosphopantetheinyl transferase from *Streptomyces verticillus* ATCC15003, the producer of the hybrid peptide-polyketide antitumor drug bleomycin. *Chem. Biol.* **8,** 725–738.

Sieber, S. A., and Marahiel, M. A. (2005). Molecular mechanisms underlying nonribosomal peptide synthesis: Approaches to new antibiotics. *Chem. Rev.* **105,** 715–738.

Sieber, S. A., Walsh, C. T., and Marahiel, M. A. (2003). Loading peptidyl-coenzyme A onto peptidyl carrier proteins: A novel approach in characterizing macrocyclization by thioesterase domains. *J. Am. Chem. Soc.* **125,** 10862–10866.

Staunton, J., and Weissman, K. J. (2001). Polyketide biosynthesis: A millennium review. *Nat. Prod. Rep.* **18,** 380–416.

Stuible, H. P., Meier, S., Wagner, C., Hannappel, E., and Schweizer, E. (1998). A novel phosphopantetheine: Protein transferase activating yeast mitochondrial acyl carrier protein. *J. Biol. Chem.* **273,** 22334–22339.

Sunbul, M., Yen, M., Zou, Y., and Yin, J. (2008). Enzyme catalyzed site-specific protein labeling and cell imaging with quantum dots. *Chem. Commun. (Camb)* 5927–5929.

Suo, Z., Chen, H., and Walsh, C. T. (2000). Acyl-CoA hydrolysis by the high molecular weight protein 1 subunit of yersiniabactin synthetase: Mutational evidence for a cascade of four acyl-enzyme intermediates during hydrolytic editing. *Proc. Natl. Acad. Sci. USA* **97,** 14188–14193.

Vitali, F., Zerbe, K., and Robinson, J. A. (2003). Production of vancomycin aglycone conjugated to a peptide carrier domain derived from a biosynthetic non-ribosomal peptide synthetase. *Chem. Commun. (Camb)* 2718–2719.

Vivero-Pol, L., George, N., Krumm, H., Johnsson, K., and Johnsson, N. (2005). Multicolor Imaging of Cell Surface Proteins. *J. Am. Chem. Soc.* **127,** 12770–12771.

Walsh, C. T. (2004). Polyketide and nonribosomal peptide antibiotics: Modularity and versatility. *Science* **303,** 1805–1810.

Walsh, C. T., Gehring, A. M., Weinreb, P. H., Quadri, L. E., and Flugel, R. S. (1997). Post-translational modification of polyketide and nonribosomal peptide synthases. *Curr. Opin. Chem. Biol.* **1,** 309–315.

Wattanachaisaereekul, S., Lantz, A. E., Nielsen, M. L., Andresson, O. S., and Nielsen, J. (2007). Optimization of heterologous production of the polyketide 6-MSA in *Saccharomyces cerevisiae*. *Biotechnol. Bioeng.* **97,** 893–900.

White, S. W., Zheng, J., Zhang, Y. M., and Rock, Y. M. (2005). The structural biology of type II fatty acid biosynthesis. *Annu. Rev. Biochem.* **74,** 791–831.

Wong, L. S., Thirlway, J., and Micklefield, J. (2008). Direct site-selective covalent protein immobilization catalyzed by a phosphopantetheinyl transferase. *J. Am. Chem. Soc.* **130,** 12456–12464.

Yin, J., Lin, A. J., Buckett, P. D., Wessling-Resnick, M., Golan, D. E., and Walsh, C. T. (2005). Single-cell FRET imaging of transferrin receptor trafficking dynamics by Sfp-catalyzed, site-specific protein labeling. *Chem. Biol.* **12,** 999–1006.

Yin, J., Lin, A. J., Golan, D. E., and Walsh, C. T. (2006). Site-specific protein labeling by Sfp phosphopantetheinyl transferase. *Nat. Protoc.* **1,** 280–285.

Yin, J., Liu, F., Li, X., and Walsh, C. T. (2004). Labeling proteins with small molecules by site-specific posttranslational modification. *J. Am. Chem. Soc.* **126,** 7754–7755.

Yin, J., Liu, F., Schinke, M., Daly, C., and Walsh, C. T. (2004). Phagemid encoded small molecules for high throughput screening of chemical libraries. *J. Am. Chem. Soc.* **126,** 13570–13571.

Yin, J., Mills, J. H., and Schultz, P. G. (2004). A catalysis-based selection for peroxidase antibodies with increased activity. *J. Am. Chem. Soc.* **126,** 3006–3007.

Yin, J., Straight, P. D., Hrvatin, S., Dorrestein, P. C., Bumpus, S. B., Jao, C., Kelleher, N. L., Kolter, R., and Walsh, C. T. (2007). Genome-wide high-throughput mining of natural-product biosynthetic gene clusters by phage display. *Chem. Biol.* **14,** 303–312.

Yin, J., Straight, P. D., McLoughlin, S. M., Zhou, Z., Lin, A. J., Golan, D. E., Kelleher, N. L., Kolter, R., and Walsh, C. T. (2005). Genetically encoded short peptide tag for versatile protein labeling by Sfp phosphopantetheinyl transferase. *Proc. Natl. Acad. Sci. USA* **102,** 15815–15820.

Zhang, J., Van Lanen, S. G., Ju, J., Liu, W., Dorrestein, P. C., Li, W., Kelleher, N. L., and Shen, B. (2008). A phosphopantetheinylating polyketide synthase producing a linear polyene to initiate enediyne antitumor antibiotic biosynthesis. *Proc. Natl. Acad. Sci. USA* **105,** 1460–1465.

Zhou, Z., Cironi, P., Lin, A. J., Xu, Y., Hrvatin, S., Golan, D. E., Silver, P. A., Walsh, C. T., and Yin, J. (2007). Genetically encoded short peptide tags for orthogonal protein labeling by Sfp and AcpS phosphopantetheinyl transferases. *ACS Chem. Biol.* **2,** 337–346.

Zhou, Z., Koglin, A., Wang, Y., McMahon, A. P., and Walsh, C. T. (2008). An eight residue fragment of an acyl carrier protein suffices for post-translational introduction of fluorescent pantetheinyl arms in protein modification *in vitro* and *in vivo*. *J. Am. Chem. Soc.* **130,** 9925–9930.

CHAPTER ELEVEN

SUGAR BIOSYNTHESIS AND MODIFICATION

Felipe Lombó, Carlos Olano, José A. Salas, *and* Carmen Méndez

Contents

1. Introduction	278
2. Deoxysugar Biosynthesis	279
3. Deoxysugar Transfer	283
4. Modification of the Glycosylation Pattern through Gene Inactivation	284
4.1. Generation of a mithramycin derivative by inactivating the *mtmU* gene	285
5. Modification of the Glycosylation Pattern through Heterologous Gene Expression	288
5.1. Selection of suitable expression vectors	289
6. Modification of the Glycosylation Pattern through Combinatorial Biosynthesis	290
7. Gene Cassette Plasmids for Deoxysugar Biosynthesis	292
7.1. Construction of pLN2	296
8. Generation of Glycosylated Compounds	299
8.1. Generation of *S. albus* GB16	300
8.2. Feeding experiments with 8-demethyl-tetracenomycin C	301
9. Tailoring Modifications of the Attached Deoxysugars	301
10. Detection of Glycosylated Compounds	303
Acknowledgments	303
References	303

Abstract

Many bioactive compounds contain as part of their molecules one or more deoxysugar units. Their presence in the final compound is generally necessary for biological activity. These sugars derive from common monosaccharides, like D-glucose, which have lost one or more hydroxyl groups (monodeoxysugars, dideoxysugars, trideoxysugars) during their biosynthesis. These deoxysugars are transferred to the final molecule by the action of a glycosyltransferase. Here, we first summarize the different biosynthetic steps required for the

Departamento de Biología Funcional and Instituto Universitario de Oncología del Principado de Asturias (I.U.O.P.A), Universidad de Oviedo, Oviedo, Spain generation of the different families of deoxysugars, including those containing extra methyl or amino groups, or tailoring modifications of the glycosylated compounds. We then give examples of several strategies for modification of the glycosylation pattern of a given bioactive compound: inactivation of genes involved in the biosynthesis of deoxysugars; heterologous expression of genes for the biosynthesis or transfer of a specific deoxysugar; and combinatorial biosynthesis (including the use of gene cassette plasmids). Finally, we report techniques for the isolation and detection of the new glycosylated derivatives generated using these strategies.

1. INTRODUCTION

Deoxysugars are a large family of sugars synthesized by replacing one or more hydroxyl groups by hydrogen in the initial monosaccharide, usually D-glucose or D-mannose. Deoxysugars lack the energy storage and supply roles of other common sugars such as glucose and fructose, or polymers like glycogen, and they usually play a more structural role when attached to compounds or cellular structures, modifying the chemical properties of the corresponding aglycone or the immunological aspects of the cell. They form part of nucleic acids, lipopolysaccharides, secondary metabolites, etc. (He and Liu, 2002).

Lipopolysaccharides are constituents of the Gram-negative bacterial cell wall, where they participate in aspects of microbial immunogenicity and endotoxicity. These lipopolysaccharides contain several different deoxysugars as part of their structure, including the external region or O-chain, whose variability between species and strains determines the serological specificity of the host bacteria (Schnaitman and Klena, 1993).

Many secondary metabolites from microorganisms and plants, such as antibiotics, antiparasitic compounds, antifungals, cardioglycosides, and antitumor compounds, contain deoxysugars as part of the molecule, and their presence is usually necessary for the corresponding biological activity. Changes in the deoxysugar portion of an active molecule modify pharmacokinetic parameters such as solubility, hydrophobicity, activity, toxicity, etc. This is the case for the anthracyclines doxorubicin and epirubicin, which differ only in the stereochemistry of the 4′-hydroxyl group of the deoxysugar attached to the aglycone, L-daunosamine or 4′-epi-L-daunosamine, respectively (O'Kunewick *et al.*, 1990). Another example is the macrolide antibiotic erythromycin, produced by the actinomycete *Saccharopolyspora erythraea*, which binds to the 70S ribosomal subunit, inhibiting bacterial protein synthesis by blocking the tunnel used by the nascent peptide on its way from the peptidyl transferase centre (Vázquez-Laslop *et al.*, 2008). In this macrolide, the presence of the deoxysugars, L-cladinose and D-desosamine, is very important since intermediates with only one deoxysugar are less active (Gaisser *et al.*, 2000). A third example is the antitumor compound mithramycin, produced by *Streptomyces argillaceus*, which contains disaccharide and trisaccharide chains formed by D-deoxysugars. Dimers of this compound associated with a Mg^{2+} ion bind noncovalently to the DNA minor groove in GC-rich regions. The presence of at least three of these deoxysugars is necessary to maintain the biological activity, as they contribute with hydrogen bonds to the stability of the mithramycin–DNA complex (Pérez *et al.*, 2008; Sastry and Patel, 1993). Depending on the number of hydroxyl groups which are replaced by hydrogen, as well as their position, deoxysugars can be classified into several categories. The most abundant monodeoxysugar is 2-deoxy-D-ribose, a key constituent of DNA. 2-deoxy-D-glucose is another monodeoxysugar, present in cardiac glycosides. Cordycepose is a 3-deoxysugar from the nucleoside antibiotic cordycepin. 4-deoxy-D-arabino-hexose is present in the lipopolysaccharide of *Citrobacter braakii*. Among monodeoxysugars present in bioactive compounds, 6-deoxysugars are the most abundant, as in the case of D-mycinose found in tylosin or L-rhamnose present in spinosyn.

Dideoxysugars are common in many secondary metabolites, especially 2,6-dideoxyhexoses, as in the case of D-olivose (in mithramycin and urdamycin), D-digitoxose (in cardiac glycosides), and L-oleandrose (an O-methylated derivative of L-olivose found in avermectin and oleandomycin). Less abundant are 4,6-dideoxyhexoses such as D-lankavose found in the macrolide lankamycin. A few secondary metabolites contain 2,3,6-trideoxysugars, such as L-rhodinose (in urdamycin, landomycin, rhodomycin, and granaticin).

Aminodeoxysugars contain an amino group replacing one of the hydrogens in the initial monosaccharide. They are common in lipopolysaccharides and in some antibiotics, especially macrolides (D-desosamine in erythromycin), polyenes (D-mycosamine in amphotericin B), and anthracyclines (L-daunosamine in doxorubicin).

Finally, branched-chain deoxysugars are usually synthesized by attaching an alkyl side chain (from *S*-adenosylmethionine (SAM) in the case of methyl groups) to one of the carbons in the monosaccharide, for example D-mycarose in mithramycin.

2. Deoxysugar Biosynthesis

The most common form of glucose inside the cell is D-glucose-6-phosphate, generated from D-glucose via the action of a hexokinase. Glucose-6-phosphate plays a central role in energy metabolism as a starting point for glycolysis, as well as being the substrate for the initial step in the pentose phosphate pathway. However, most deoxysugars derive from

D-glucose-1-phosphate (Fig. 11.1). This molecule is generated from D-glucose-6-phosphate by the reversible enzyme phosphoglucomutase (Lu and Kleckner, 1994).

The first step during biosynthesis of any deoxysugar derived from glucose is the activation of D-glucose-1-phosphate to NDP-D-glucose by the action of NDP-hexose synthetase (also known as hexose-1-phosphate nucleotidyltransferase or NDP-hexose pyrophosphorylase). Depending on the specific deoxysugar pathway, this reaction can yield dTDP-D-glucose (as in the case of MtmD for generation of D-olivose, D-oliose, and D-mycarose in the antitumor mithramycin pathway, Lombó et al., 1997) or CDP-D-glucose (RfbF for D-paratose in lipopolysaccharides biosynthesis, Lindqvist et al., 1994). Detailed mechanistic aspects of these reactions and other biosynthetic steps in deoxysugar biosynthesis are covered in Chapter 21 of Volume 459.

Most deoxysugars that are present in bioactive compounds are 6-deoxy derivatives and, in these cases, the second biosynthetic step is the irreversible deoxygenation at C6, which is catalyzed by an NAD^+-dependent NDP-hexose 4,6-dehydratase, producing the common key intermediate NDP-4-keto-6-deoxy-D-hexose (Fig. 11.1). This intermediate serves as branching point for the biosynthesis of all 6-deoxysugars (He and Liu, 2002).

In the case of 2,6-dideoxysugar biosynthesis, formation of NDP-4-keto-6-deoxy-D-hexose is followed by the loss of the hydroxyl group at C2 by replacing it by hydrogen. This reaction takes place in two steps carried out by two different enzymes, a 2,3-dehydratase and a 3-ketoreductase (Fig. 11.1). The first enzyme of the pair, such as GraOrf27 involved in D-olivose biosynthesis in the granaticin pathway, eliminates the hydroxyl group at C2 and a hydrogen from C3, generating a 3-keto intermediate (Draeger et al., 1999). The second enzyme is a 3-ketoreductase, which generates a hydroxyl group at C3 by using NADPH. This C3 hydroxyl group can retain the original equatorial configuration, for example, resulting from the action of GraOrf 26-like enzyme as in the case of granaticin (Draeger et al., 1999), or it can possess an axial configuration, as the result of the action of a TylC1-like 3-ketoreductase from the tylosin pathway, involved in L-mycarose biosynthesis (Chen et al., 1999).

In the case of 3,6-dideoxysugars biosynthesis, the 4-keto group of NDP-4-keto-6-deoxy-D-hexose is attacked by a pyridoxamine 5-phosphate (PMP)-containing dehydrase (like AscC for L-ascarylose biosynthesis during lipopolysaccharide formation), which eliminates the C3 hydroxyl group, generating an PMP-bound glucoseen intermediate, which contains a C3, C4 double bond (Lei et al., 1995). Finally, a 3,4-glucoseen reductase flavoprotein (such as AscD) reduces this intermediate, generating NDP-4-keto-3,6-dideoxy-hexose (Han et al., 1990).

With respect to 4,6-dideoxysugar biosynthesis, the intermediate NDP-4-keto-6-deoxy-D-hexose is transformed first by a transaminase into an intermediate with an amino group attached at C4, as in the case of DesI during D-desosamine formation in pikromycin biosynthesis (Zhao et al., 2001).

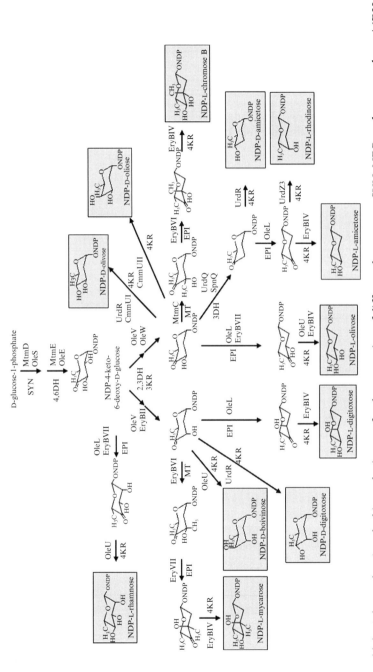

Figure 11.1 Scheme showing the biosynthetic steps for the generation of different deoxysugars. SYN, NDP-D-glucose-synthase; 4,6DH, 4,6-dehydratase; 2,3DH, 2,3-dehydratase; 3DH, 3-dehydratase; 3KR, 3-ketoreductase; EPI, epimerase; MT, methyltransferase; 4KR, 4-ketoreductase.

Then, a SAM-dependent flavoprotein like DesII acts as a deaminase, generating NDP-3-keto-4,6-dideoxy-D-hexose (Szu et al., 2005).

Formation of L-deoxysugars requires the involvement of epimerases. These can be 5-epimerases, as in the case of EryBVII from the erythromycin pathway involved in L-mycarose biosynthesis (Lombó et al., 2004; Summers et al., 1997), or 3,5-epimerases like OleL (Fig. 11.1) during L-olivose biosynthesis in the oleandomycin pathway (Aguirrezabalaga et al., 2000).

In the case of aminodeoxysugars like D-desosamine, L-daunosamine (3-aminosugars), or D-forosamine (4-aminosugar), a pyridoxal 5-phosphate (PLP)-containing aminotransferase is involved in the pathway. These enzymes act on a keto group of the corresponding intermediate, incorporating an amino group at that position (Nedal and Zotchev, 2004).

Branched-chain deoxysugars contain extra methyl groups attached mainly to C3, as in the case of D-mycarose (mithramycin), L-mycarose (erythromycin), L-nogalose (anthracyclines), and L-chromose B (chromomycin), or to C5 as in the case of L-noviose (novobiocin). These methyltransferases use SAM as cofactor for these reactions, as in the case of TylCIII (tylosin pathway) (Chen et al., 2001).

The last step during the biosynthesis of the activated deoxysugar is the reduction of 4-keto groups that have been maintained during the biosynthetic pathway. These 4-ketoreductases are highly stereospecific, generating axial or equatorial hydroxyl moieties depending on the enzyme. The cofactor for these reactions is usually NADPH (Chen et al., 2000).

In the biosynthesis of glycosylated compounds, once the final activated NDP-deoxysugar has been generated, the next step is its transfer by a specific glycosyltransferase to the corresponding aglycone. The aglycone and the deoxysugar can be linked by O-, C–C (urdamycin), or C–N (rebeccamycin) glycosidic bonds. A certain degree of substrate flexibility has been described for these enzymes, which has allowed the generation of new glycosylated products by combinatorial biosynthesis (Luzhetskyy et al., 2008; Méndez et al., 2008; Salas and Méndez, 2005). Although most glycosyltransferases described so far act as independent enzymes, a few cases have been described in which the canonical glycosyltransferase requires an auxiliary protein for in vivo and in vitro activity, as in the case of DesVII and DesVIII (pikromycin-methymycin, Borisova et al., 2004), AknS and AknT (aclacinomycin, Lu et al., 2005), TylM2 and TylM3 (tylosin), and MycB and MydC (mycinamycin) (Melançon et al., 2004). For a more complete review of glycosyltransferases, see Chapter 12, in this volume.

Once the deoxysugar has been attached to the aglycone by the action of the glycosyltransferase, in some cases a final enzymatic reaction carried out by a specific methyltransferase or acyltransferse introduces methoxy, acetoxy, or N-methyl groups at corresponding hydroxyl or amino residues on the attached deoxysugar (Salas and Méndez, 2007).

3. Deoxysugar Transfer

In order to generate new glycosylated derivatives, it is necessary to produce the desired activated deoxysugar to be transferred to the chosen aglycone. Once the activated deoxysugar is available, aspects of substrate flexibility of glycosyltransferases must be taken into account.

Flexibility with respect to the sugar donor has been described for several glycosyltransferases. This is the case for the elloramycin ElmGT (Blanco *et al.*, 2001), which naturally transfers L-rhamnose to its aglycone, 8-demethyl-tetracenomycin C, but is also able to transfer other L-sugars such as L-olivose, L-rhodinose, L-digitoxose, or L-amicetose; D-sugars like D-olivose, D-digitoxose, D-amicetose, or D-boivinose; branched-chain sugars such as D-mycarose, L-mycarose, or L-chromose B; or even a D-olivose disaccharide (Fischer *et al.*, 2002; Lombó *et al.*, 2004; Pérez *et al.*, 2005; Rodríguez *et al.*, 2000, 2002). Another example of a flexible glycosyltransferase is StaG, the staurosporine glycosyltransferase, which naturally transfers L-ristosamine to the indolocarbazole K252c aglycone through a C–N bond. In a second step, the monooxygenase StaN establishes a second C–N bond between the aglycone and this bound sugar. StaG is able to transfer other deoxysugars such as L-rhamnose, L-olivose, L-digitoxose, or D-olivose. StaN is also able to generate the second C–N link as long as the bound moiety is an L-sugar (Salas *et al.*, 2005). The glycosyltransferases GtfC and GtfD transfer 4-epi-L-vancosamine (during chloroeremomycin biosynthesis) and L-vancosamine (during vancomycin biosynthesis), respectively, but they can also transfer an array of different structurally related deoxysugars (Losey *et al.*, 2002; Oberthür *et al.*, 2005). Similarly, the mycaminosyltransferase TylM2 from the tylosin pathway can transfer to its tylactone aglycone D-glucose and D-desosamine in addition to its natural substrate, D-mycaminose (Gaisser *et al.*, 2000). Also, the glycosyltransferase StfG involved in the biosynthesis of the anthracycline steffimycin has been found to transfer, as well as its natural substrate L-rhamnose, other deoxysugars such as D-boivinose, D-olivose, D-digitoxose, L-olivose, L-digitoxose, and L-amicetose (Olano *et al.*, 2008).

The oleandomycin OleG2 glycosyltransferase transfers L-olivose to its natural aglycone, oleandolide. This enzyme is an example of flexibility for both the sugar donor and the aglycone acceptor, as it is able also to transfer L-rhamnose to an unnatural aglycone, erythronolide B (Doumith *et al.*, 1999). In *Streptomyces venezuelae*, two types of macrolides are produced by the same biosynthetic pathway, methymycin (12-membered) and pikromycin (14-membered). The glycosyltransferase involved, DesVII, transfers D-desosamine to both kinds of aglycone, and it can also transfer D-quinovose, D-mycaminose, D-olivose, and L-rhamnose (Borisova *et al.*,

1999; Hong et al., 2004; Yamase et al., 2000). An interesting glycosyltransferase is UrdGT2, involved in urdamycin biosynthesis. This enzyme transfers D-olivose to its aglycone through a C–C bond, and it is also able to transfer both rhodinose forms, D and L (Hoffmeister et al., 2003). Interestingly, this enzyme can also transfer other deoxysugars such as D-olivose or D-mycarose, but to a completely different aglycone, premithramycinone (a biosynthetic intermediate from the mithramycin pathway), which is linear, instead of being angular like its natural aglycone (Trefzer et al., 2002).

AraGT, the glycosyltransferase which transfers L-rhamnose during aranciamycin biosynthesis in *Streptomyces echinatus*, is able to transfer other deoxysugars to its natural aglycone. A cosmid containing the necessary genes for aranciamycin aglycone generation and the *araGT* glycosyltransferase gene was introduced into *Streptomyces fradiae* A0, the producer of urdamycin A, and *Streptomyces diastatochromogenes* Tü6028, the producer of polyketomycin. AraGT was able to transfer to its aglycone the deoxysugars L-rhodinose in *S. fradiae*, and L-rhamnose, L-axenose, and D-amicetose in *S. diastatochromogenes* (Luzhetskyy et al., 2007).

4. MODIFICATION OF THE GLYCOSYLATION PATTERN THROUGH GENE INACTIVATION

The easiest way to obtain natural products with an altered glycosylation pattern is by inactivating genes involved in deoxysugar biosynthesis of a known gene cluster. These gene inactivation experiments shed light on the biosynthetic functions of the gene under study, and eventually may cause the accumulation of a new glycosylated intermediate or shunt product. In order to accomplish this objective the relevant glycosyltransferase must be able to recognize and transfer the modified deoxysugar that is now synthesized in this mutant. An example of this kind of approach is the inactivation of the 4-ketoreductase gene *urdR*, which is responsible for the last step during bisynthesis of NDP-D-olivose, one of the deoxysugars present in urdamycin A in *S. fradiae*. This mutant strain is unable to generate NDP-D-olivose, but the accumulated 4-keto intermediate is recognized by the 3-deoxygenase UrdQ and reduced by the 4-ketoreductase UrdZ3 (both involved in L-rhodinose biosynthesis), generating NDP-D-rhodinose, a new sugar which is transferred to the aglycone (Hoffmeister et al., 2003).

In *Sac. erythraea*, the erythromycin gene cluster contains several *eryB* genes for L-mycarose biosynthesis and several *eryC* genes for D-desosamine generation. Inactivation of several of these genes leads to the accumulation of small amounts of erythromycin intermediates and shunt products which were useful for establishing some of the biosynthetic steps for both deoxysugars (Salah-Bey et al., 1998).

S. *venezuelae* is the producer of methymycin and pikromycin, two macrolides which contain D-desosamine. Inactivation of the gene coding for dehydratase DesI abolishes the elimination of the 4-hydroxyl group during D-desosamine biosynthesis. In this mutant, the accumulated deoxysugar intermediate is reduced at C4 by an unknown host 4-ketoreductase, generating D-quinovose, a new sugar that is attached by the glycosyltransferase to the aglycone 10-deoxymethynolide (Borisova *et al.*, 1999).

Aclacinomycin is an anthracycline produced by *Streptomyces galilaeus*, which contains a trisaccharide of L-rhodosamine, 2-deoxy-L-fucose, and L-rhodinose. Several mutants in deoxysugar biosynthetic genes from the gene cluster were generated in this strain, leading to an array of different glycosylated aclacinomycin intermediates, whose structures helped in establishing the corresponding gene functions (Räty *et al.*, 2002).

In *S. argillaceus*, the mithramycin producer, inactivation of the deoxysugar biosynthesis genes *mtmV*, *mtmU*, *mtmC*, and *mtmTIII* was carried out in order to establish the functions of the encoded proteins. MtmV codes for a 2,3-dehydratase necessary for the biosynthesis of all deoxysugars in mithramycin (D-olivose, D-oliose, and D-mycarose), and its inactivation abolishes the formation of any transferable deoxysugar, and therefore this mutant only accumulates the aglycone. MtmU is the 4-ketoreductase for D-oliose generation, and its inactivation causes the accumulation of an intermediate containing only the first attached D-olivose moiety. MtmC is the C-methyltransferase needed for D-mycarose biosynthesis, but this mutant accumulates an intermediate with D-olivose containing a 4-keto group, indicating that this enzyme could code as well for a D-olivose 4-ketoreductase. MtmTIII is the 4-ketoreductase for D-mycarose biosynthesis since the mutant strain defective in this enzyme accumulates a compound with an attached 4-keto-D-mycarosyl moiety (González *et al.*, 2001; Remsing *et al.*, 2002).

We describe below a detailed procedure for the generation of a mutant in a deoxysugar biosynthetic gene.

4.1. Generation of a mithramycin derivative by inactivating the *mtmU* gene

For the construction of a strain with a mutated copy of *mtmU* (Fig. 11.2), the following steps are necessary: construction of a plasmid containing a mutated copy of the gene; transformation of *S. argillaceus* protoplasts to introduce the mutated gene; and characterization of mutant *S. argillaceus* M7U1.

1. Isolation of a 2.7-kb *Eco*RI-*Sph*I fragment from cosmid cosAR3, obtained from the *S. argillaceus* gene library (Lombó *et al.*, 1996). This fragment contains the 5′ region of the *mtmU* gene, and the *mtmV* and *mtmW* genes.

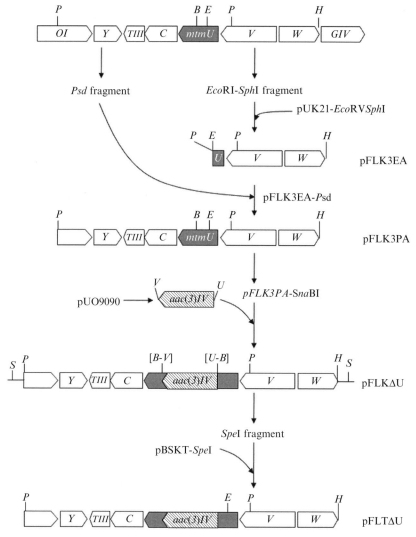

Figure 11.2 Construction of plasmid pFLTΔU for the generation of the *S. argillaceus* mutant M7U1. B, SnaBI; E, EcoRI; H, SphI; P, PstI; S, SpeI; U, StuI; V, EcoRV.

The procedure for fragment isolation is described in "Illustra GFX PCR DNA and Gel Band Purification Kit," GE Healthcare.

2. Digestion of vector pUK21 (Vieira and Messing, 1991) with *Eco*RI-*Sph*I.
3. Subcloning of the 2.7-kb fragment into digested pUK21, generating pFLK3EA.

4. Isolation of a 5-kb *Pst*I fragment from cosAR3. This fragment contains the 3′ region of the genes *mtmV, mtmU, mtmC, mtmTIII,* and *mtmY* and the 3′ end of *mtmOI*.
5. Digestion of pFLK3EA with *Pst*I.
6. Subcloning of the 5-kb *Pst*I fragment into *Pst*I-digested pFLK3EA, generating pFLK3PA, which contains the *mtmU* gene and flanking DNA regions.
7. Isolation and purification of the apramycin resistance cassette *aac(3)IV* from pOU9090 (Prado *et al.*, 1999) as a 1.7-kb *Stu*I-*Eco*RV fragment.
8. Digestion of vector pFLK3PA with *Sna*BI. There is a single *Sna*BI site in the middle of *mtmU* gene.
9. Subcloning of the 1.7-kb apramycin resistance gene into *Sna*BI-digested pFLK3PA, generating pFLKΔU.
10. Isolation and purification of the insert from pFLKΔU as a 9.4-kb *Spe*I fragment, using flanking polylinker sites.
11. Digestion of vector pBSKT (Prado *et al.*, 1999) with *Spe*I. pBSKT is an *Escherichia coli* vector derived from pBSK (Stratagene), which contains a thiostrepton resistance gene. This vector cannot replicate in *Streptomyces* unless it is integrated into the chromosome through homologous recombination. Integrants can be selected with thiostrepton (50 μg/ml).
12. Subcloning of the 9.4-kb *Spe*I fragment into digested pBSKT. This generates pFLTΔU (Fig. 11.2), which contains the disrupted *mtmU* gene flanked by 2- and 4-kb DNA regions. Once pFLTΔU is generated, it must be introduced into *S. argillaceus* protoplasts, via the next steps.
13. Inoculation of 25 ml of YEME medium (17% sucrose, 0.5% glycine) (Kieser *et al.*, 2000) with *S. argillaceus* spores, in 250-ml baffled flask. Incubate at 30 °C for 36 h.
14. Washing of mycelium (twice) with 10.3% sucrose and resuspension in 10 ml supplemented P-buffer (Kieser *et al.*, 2000) including 1 mg/ml lysozyme. Incubate at 30 °C for 1 h.
15. Filtering of formed protoplasts through a cotton syringe. Centrifuge at 3000 rpm room temperature and wash pellet twice with 5 ml supplemented P-buffer (Kieser *et al.*, 2000).
16. Resuspension of protoplasts pellet in 1 ml supplemented P-buffer and divide into 200-μl volumes.
17. Addition of 20 μl plasmid pFLTΔU to a protoplast aliquot together with 500 μl of 25% PEG 6000 in supplemented P-buffer and spread on four R5 medium agar plates (previously 10% dehydrated) (Kieser *et al.*, 2000). Incubate at 30 °C for 24 h.
18. Addition of an overlay to plates: 1.5 ml sterile distilled water containing enough thiostrepton achieves a final concentration of 50 μg/ml. Thiostrepton must be kept in DMSO at 50 mg/ml and stored at 4 °C.

19. Incubation at 30 °C for 7 days and then pick up the resulting transformant colonies on A-medium agar plates (González et al., 2001) with apramycin (25 μg/ml) or with apramycin and thiostrepton (50 μg/ml). In this way, it is possible to distinguish between colonies in which a single crossover event or the desired double crossover event has taken place. Only apramycin-resistant and thiostrepton-sensitive colonies which have lost the plasmid pBSKT after a double crossover event have the wild type *mtmU* gene replaced by the mutated copy.
20. The last step consists of the genetic characterization of mutant *S. argillaceus* M7U1, which can be carried out by determining, with appropiate PCR primers, the mutant genotype of the selected mutant strain.

5. MODIFICATION OF THE GLYCOSYLATION PATTERN THROUGH HETEROLOGOUS GENE EXPRESSION

Wild type or mutant strains, which are affected in one or more sugar biosynthetic genes, can be used as hosts for the heterologous expression of deoxysugar biosynthetic genes from other pathways. If the new deoxysugar generated by the action of the host genes and the heterologously expressed ones is recognized by the corresponding glycosyltransferase, a new derivative may be produced. A good example is the generation of 4′-epidoxorubicin, a medically important anthracycline, which differs from doxorubicin in the stereochemistry at C4 of the deoxysugar moiety L-daunosamine. This compound was usually generated by an expensive semisynthetic process after the fermentation of doxorubicin by *Streptomyces peucetius*. Engineering the sugar biosynthetic pathway in *S. peucetius* has allowed the production of 4′-epidoxorubicin in a more specific way. First, a mutant in the 4-ketoreductase *dnmV* responsible for this step during L-daunosamine was generated in *S. peucetius*, which caused the accumulation of a 4-keto sugar biosynthetic intermediate. Expression in this mutant of genes coding for 4-ketoreductases generating a 4-hydroxyl group with the opposite stereochemistry, like *eryBIV* (for L-mycarose biosynthesis in the erythromycin pathway) or *avrE* (for L-oleandrose biosynthesis in avermectin), caused the formation of L-epidaunosamine, which was transferred to the aglycone, generating 4′-epidoxorubicin (Madduri et al., 1998). Another example is the use of a mutant of *S. peucetius* which accumulates the doxorubicin aglycone, which was transformed with a plasmid containing several glycosylation genes from the aclacinomycin gene cluster in *S. galilaeus*, leading to the accumulation of L-rhamnosyl derivatives (Räty et al., 2000).

S. fradiae is the producer of the macrolide tylosin, which contains three deoxysugars, D-mycinose, D-mycaminose, and L-mycarose. D-Mycaminose is an aminodeoxysugar with several common biosynthetic steps with D-desosamine, the sugar moiety of the macrolide narbomycin. The only difference between the two aminosugars is that D-desosamine lacks the 4-hydroxyl group present in D-mycaminose. Two genes from the narbomycin pathway, *nbmK* and *nbmJ* (which are responsible for the elimination of this 4-hydroxyl group), were introduced into the tylosin producer, causing the accumulation of tylosin derivatives containing D-desosamine instead of D-mycaminose (Butler *et al.*, 2002).

The D-forosamine glycosyltransferase SpnP from the spinosyn gene cluster was expressed in a triple mutant of the erythromycin producer affected in the biosynthesis of the aglycone as well as in both erythromycin glycosyltransferases, EryBV (for L-mycarose) and EryCIII (for D-desosamine). Feeding the spinosyn aglycone to this mutant led to the production of new compounds in which an L-mycarose had been transferred to the aglycone (Gaisser *et al.*, 2002).

Streptomyces roseochromogenes produces the aminocoumarin clorobiocin. The gene cluster has been expressed using *Streptomyces coelicolor* as heterologous host. Inactivation of the gene coding for the C5-methyltransferase CloU, involved in the biosynthesis of the sugar moiety L-noviose, led to the production in very small amounts of a new aminocoumarin with an L-rhamnosyl moiety instead of L-noviose. The cause of the low production was that the L-noviose 4-ketoreductase CloS did not process at a good rate the accumulated NDP-4-keto-L-rhamnose intermediate. In order to overcome this bottleneck, *oleU*, which codes for a 4-ketoreductase able to use efficiently this deoxysugar intermediate, was incorporated into the strain, greatly increasing the production levels of the new aminocoumarin (Freitag *et al.*, 2006).

5.1. Selection of suitable expression vectors

In order to express heterologously a desired deoxysugar biosynthetic gene in a selected host, several parameters must be taken into account.

1. The use of multicopy replicative vectors such as pWHM3 (Vara *et al.*, 1989) or pEM4 (Quirós *et al.*, 1998) greatly enhances the probability of getting a high enough level of gene expression. One disadvantage is the need for maintaining antibiotic selection during the whole procedure.
2. The use of integrative (and therefore single-copy) vectors such as pKC796 (Kuhstoss *et al.*, 1991), pSET152 (Bierman *et al.*, 1992), or pHM8a (Motamedi *et al.*, 1995) makes the recombinant strain more stable even without antibiotic selection, but expression levels of the desired gene (and therefore production levels of the hybrid compound) may not be as high as in the previous case.
3. Selection of the antibiotic marker for the expression vector is crucial, because it must be compatible with any selection marker already present

in the host to be used in the experiments. Particularly, this must be considered when using a mutant generated by genetic engineering. Different versions of multicopy plasmids with thiostrepton such as pWHM3, pIAGO (Aguirrezabalaga et al., 2000), and pEM4, or apramycin (pEM4A) (Blanco et al., 2001) or erythromycin (pEM4E, N. Allende, unpublished data) markers can be used. Several integrative vectors exist with apramycin (pKC796, pSET152), hygromycin (pHM8a, pHM11a) (Motamedi et al., 1995), or erythromycin (pKCE, Aguirrezabalaga et al., 2000) markers.

4. Although some genes could be expressed from their own promoter regions, better results are usually obtained by using a strong promoter preceding the cloned gene, in such a way that its high level transcription is ensured. We routinely use the constitutive promoter from the *ermE* gene (Bibb et al., 1985), P*ermE* (pIAGO), or P*ermE** (pEM4, pEM4A, pEM4E).
5. When dealing with host strains in which protoplast transformation can be difficult, it is possible to transfer from *E. coli* conjugative versions of the expression vectors, carrying the *oriT* region, as is the case for integrative vectors pSET152 and pHM11a, and also replicative vectors pEM4T and pEM4AT (Menéndez et al., 2004).

6. Modification of the Glycosylation Pattern through Combinatorial Biosynthesis

Creating plasmids containing the necessary genes for the biosynthesis of a deoxysugar is a good tool to study specific gene functions and enzymatic steps involved in deoxysugar biosynthesis. Also, these plasmids can be used to endow a specific strain with the ability to synthesize new deoxysugars. These deoxysugar plasmids can be expressed in a producer or a nonproducer host, in this case together with the corresponding glycosyltransferase (cloned in the same plasmid or integrated in the host chromosome). The relevant aglycone to be glycosylated once the activated deoxysugar has been generated can be produced by the strain itself or it can be provided to the recombinant strain by feeding or by expressing the necessary aglycone genes in the heterologous host using a compatible plasmid/cosmid vector.

There are several examples of plasmids containing native DNA fragments from the chromosome of producing strains, which are able to direct the biosynthesis of a desired activated deoxysugar. During doxorubicin biosynthesis, the aminodeoxysugar L-daunosamine is attached to the aglycone ε-rhodomycinone. All the necessary *dnm* genes for L-daunosamine biosynthesis are distributed in four different regions in the doxorubicin gene cluster. Several variations of these four regions were cloned together in a multicopy plasmid vector containing all *dnm* genes or versions with individual *dnm* genes deleted. Simultaneously, the doxorubicin resistance ABC transporter genes, *drrAB*, were also cloned in another plasmid together with the two genes necessary for D-glucose-1-phosphate activation and 4,6-dehydration from the mithramycin pathway, *mtmDE*. Transcription of all these genes was ensured by using native promoters from the pathway, which are controlled by the positive regulator DnrI, whose gene was cloned as well in one of the plasmids. These plasmids were introduced into *Streptomyces lividans*, and the resultant recombinant strains fed with the aglycone, ε-rhodomycinone. These bioconversion experiments allowed definition of the minimum set of *dnm* genes for the biosynthesis of this aminodeoxysugar (Olano et al., 1999).

Tang and McDaniel (2001) generated a recombinant strain of *S. lividans* containing all *pik* genes for biosynthesis of L-desosamine (the aminodeoxysugar decorating the macrolides pikromycin/methymycin produced by *Streptomyces venezuelae*) integrated in the chromosome, as well as the *desVII* glycosyltransferase and *desVIII* auxiliary protein genes. Another gene, *desR*, was also included, which codes for a β-glucosidase that removes a glucose residue attached to the C2′ hydroxyl of desosamine. *S. lividans* contains a general macrolide resistance mechanism, which consists of the inactivation of macrolides by glycosylating D-desosamine residues (Cundliffe, 1992). Therefore, inclusion of *desR* in the plasmid is necessary for generation of active pikromycins. All these genes naturally form an operon in the pikromycin producer, thus facilitating their cloning and plasmid construction. Expression of all these genes was under the control of the *actI* promoter and the actinorhodin activator coded by *actII-orf4*. This recombinant strain was then transformed with a collection of replicative vectors harboring different polyketide synthase genes for the biosynthesis of several macrolide aglycones. All these strains produced desosaminyl derivatives, which showed some bioactivity against *Bacillus subtilis*.

The macrolide oleandomycin contains two deoxysugars, D-desosamine, and L-oleandrose, which is an O-methyl derivative of L-olivose. The genes responsible for L-olivose biosynthesis include *oleS* and *oleE* (for D-glucose-1-phosphate activation and 4,6-dehydration), *oleV* and *oleW* (for 2-deoxygenation), *oleL* (for 3,5-epimerization), and *oleU* (for 4-ketoreduction) (Fig. 11.1). All these genes were cloned in a multicopy plasmid as a single DNA fragment from the chromosome of the producer strain, *Streptomyces antibioticus*, maintaining the natural transcriptional promoters, and generating pOLV. This construct was introduced by transformation into an *Streptomyces albus* NAG2 strain which contained in its chromosome a copy of *oleG2* under the control of the constitutive P*ermE*★ promoter. OleG2 is the glycosyltransferase responsible for L-olivose transfer to the oleandomycin aglycone (Olano et al., 1998). *S. albus* NAG2 also contains an integrated copy of the erythromycin resistance gene *ermE*, so it can survive the production of active macrolides. By feeding erythromycin aglycone to *S. albus* NAG2 transformed with pOLV, about 50% of this aglycone was glycosylated with L-olivose (Aguirrezabalaga et al., 2000).

pRHAM is a simplified version of pOLV in which genes responsible for 2-deoxygenation (*oleVW*) were deleted and P*ermE* now controls expression of *oleLSEU*. This plasmid directs the biosynthesis of the 6-deoxysugar L-rhamnose (Fig. 11.1, Rodríguez *et al.*, 2000). *S. albus* carrying cos16F4 contains all the gene functions for generating the tetracyclic aromatic aglycone 8-demethyl-tetracenomycin C from the elloramycin pathway, as well as *elmGT*, coding for its glycosyltransferase (Blanco *et al.*, 2001; Ramos *et al.*, 2008). Transformation of *S. albus* cos16F4 with pOLV or pRHAM allowed the production of L-olivosyl-8-demethyl-tetracenomycin C and L-rhamnosyl-8-demethyl-tetracenomycin C, respectively (Rodríguez *et al.*, 2000).

The antitumor compound staurosporine, produced by *Streptomyces longisporoflavus*, is glycosylated by the aminodeoxysugar L-ristosamine, which is attached to the aglycone through two C–N linkages. Genes for the biosynthesis of L-ristosamine are located together on the chromosome. All these genes were cloned together as a single DNA fragment, in the multicopy vector pEM4, under the control of the P*ermE** promoter. The resultant plasmid pETAS was used in *S. albus* for reconstitution of the staurosporine biosynthetic pathway and the generation of new glycosylated derivatives (Salas *et al.*, 2005). For this, two versions of a vector including all the gene functions necessary for staurosporine aglycone generation and sugar transfer were constructed. These included the genes *rebO*, *rebD*, *staC*, and *rebP*, responsible for the generation of the staurosporine aglycone, K-252c (Sánchez *et al.*, 2005), the *staG* glycosyltransferase, and the *staN* P_{450} oxygenase. All these genes were cloned in an integrative vector under the control of P*ermE*, generating pKC39GNT. A version containing only the K-252c genes and *staG* was also constructed (pKC39GT). Gene *rebT* was also present in these plasmids in order to confer resistance to the indolocarbazole derivatives. *S. albus* containing pKC39GNT and pETAS was able to produce L-ristosaminylated K252c, with this deoxysugar bound to the aglycone through two C–N bonds. The enzyme in charge of the generation of the second C–N linkage was demonstrated to be the cytochrome P_{450} StaN, since in *S. albus* containing pKC39GT and pETAS, L-ristosamine was bound just by a single *N*-glycosidic linkage.

7. GENE CASSETTE PLASMIDS FOR DEOXYSUGAR BIOSYNTHESIS

An alternative to the use of deoxysugar biosynthesis plasmids which contain native DNA fragments from the producer chromosome are the "plug and play" gene cassette plasmids. These plasmids have been specifically designed for directing the biosynthesis of a certain deoxysugar in such a way that the final product can be easily modified by just exchanging some of the genes contained in the plasmid. These plasmids are bifunctional for *E. coli* and *Streptomyces*, derived from the multicopy vector pWHM3, and contain one or two copies of P*erm*E* to control the expression of deoxysugar biosynthetic genes. The first of these plasmids was pLN2 (Fig. 11.3) (Rodríguez *et al.*, 2002), which contained seven genes from the oleandomycin pathway from *S. antibioticus* involved in the biosynthesis of activated L-oleandrose. These seven genes (Table 11.1) were amplified as PCR cassettes flanked by unique restriction sites for enzymes which cut only once in the final plasmid. Genes were

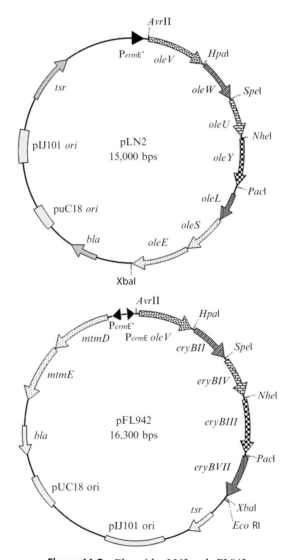

Figure 11.3 Plasmids pLN2 and pFL942.

Table 11.1 Chemical structure of the corresponding deoxysugars generated by gene cassette plasmids derived from pLN2 and FL942

Plasmid	Gene cassettes	Deoxysugar
pLN2	*oleVWUYoleLSE*	NDP-L-olivose
pLN2Δ	*oleUYoleLSE*	NDP-L-rhamnose
pLNR	*oleVWurdRoleYoleLSE*	NDP-D-olivose
pLNBIV	*oleVWeryBIVoleYoleLSE*	NDP-L-digitoxose
pLNRHO	*oleVWurdZ3oleYoleLSEurdQ*	NDP-L-rhodinose
pFL844	*oleVWeryBIVoleYoleLSEurdQ*	NDP-L-amicetose
pFL845	*oleVWurdRoleYoleLSEurdQ*	NDP-D-amicetose

pFL942	*mtmDEoleVeryBIIeryBIVBIIIBVII*	NDP-L-mycarose	(structure)
pFL947	*mtmDEoleVWeryBIVmtmCeryBVII*	NDP-L-chromose B	(structure)
pMP1★UII	*mtmDEoleVWcmmUIIoleY*	NDP-D-oliose	(structure)
pMP3★BII	*mtmDEoleVeryBIIurdRoleY*	NDP-D-digitoxose	(structure)
pMP1★BII	*mtmDEoleVeryBIIoleUoleY*	NDP-D-boivinose	(structure)

arranged in such a way that these unique restriction sites were compatible in a certain order and this facilitated the easy removal, addition, or exchange of each gene cassette at a time (Fig. 11.3).

7.1. Construction of pLN2

1. Amplification of genes *oleV*, *oleW*, *oleU*, and *oleY* from *S. antibioticus* chromosomal DNA (Fig. 11.4). Primer pairs used for these amplifications contain specific recognition sequences for unique restriction enzymes, and thus each amplicon is flanked by its corresponding pair of restriction sites, which are also compatible with those from adjacent genes in the final plasmid. All amplicons also contain a *Hin*dIII and an *Xba*I restriction sites at the 5′ and the 3′ end, respectively, for facilitating cloning steps. *oleV* contains a *Spe*I site instead of *Hin*dIII.
2. Ligation of each gene (except *oleV*) into pUC18 digested with *Hin*dIII and *Xba*I and transformation of *E. coli* DH10B competent cells using 100 µg/ml ampicillin as selection antibiotic.
3. In the case of *oleV*, subcloning of the gene as a *Spe*I-*Xba*I band into pEM4 digested with *Xba*I (Fig. 11.4). Introduce this pEM4V construct into *E. coli* DH10B by transformation.
4. Obtaining plasmid DNA from the pUC clone containing *oleW* and purification of the gene cassette after *Hpa*I-*Xba*I digestion. The *Hpa*I site is at the 5′-end of the *oleW* gene cassette.
5. Cloning of the *Hpa*I-*Xba*I *oleW* band into pEM4V digested with *Hpa*I-*Xba*I. The *Hpa*I site is at the 3′-end of the *oleV* gene cassette. Introduce this pEM4VW construct into *E. coli* DH10B by transformation.
6. Obtaining of plasmid DNA from the pUC clone containing *oleU* and purification of the gene cassette after *Spe*I-*Xba*I digestion.
7. Cloning of the *Spe*I-*Xba*I *oleU* band into pEM4VW digested with *Spe*I-*Xba*I. Introduce this pEM4VWU construct into *E. coli* DH10B by transformation.
8. Obtaining of plasmid DNA from the pUC clone containing *oleY* gene and purification of the gene cassette after *Nhe*I-*Xba*I digestion.
9. Cloning of the *Nhe*I-*Xba*I *oleY* band into pEM4VWU digested with *Nhe*I-*Xba*I. Transformation of this pEM4VWUY construct into *E. coli* DH10B. pEM4VWUY contains these four genes under the control of P*ermE**★ (Fig. 11.4).
10. Obtaining of genes *oleLSE* from cosmid cosAB61 (from an *S. antibioticus* gene library) as a 2.9-kb *Sph*I-*Sma*I fragment from an agarose gel.
11. Cloning of *oleLSE* genes into the *Sph*I-*Eco*RV sites of pUK21 and transformation of *E. coli* DH10B, using 50 µg/ml kanamycin as selection marker. This generates pLR2347Δ7 (Fig. 11.4).
12. Obtaining a 2.9-kb *Spe*I-*Xba*I DNA fragment containing the *oleLSE* genes from pLR234Δ7 from an agarose gel.

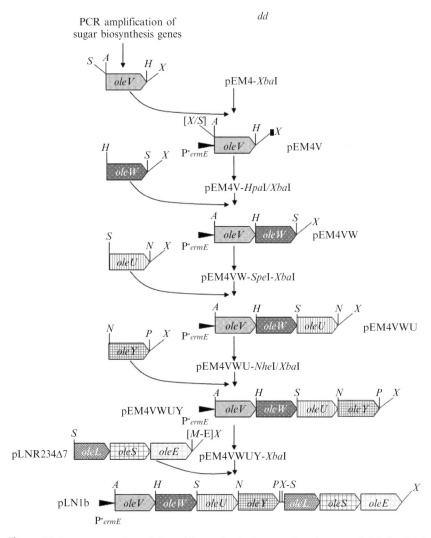

Figure 11.4 Construction of plasmid pLN1b. A, AvrII; E, EcoRV; H, HpaI; M, SmaI; N, NheI; P, PacI; S, SpeI; X, XbaI.

13. Ligation of the 2.9-kb DNA fragment into pEM4VWUY digested with XbaI. Transformation of E. coli DH10B and selection with 100 µg/ml ampicillin. This generates pLN1b (Fig. 11.4).
14. Isolation of the HindIII-XbaI DNA fragment from pLN1b which contains PermE★ and all the deoxysugar genes and ligation into pWHM3 digested with same enzymes. This generates pLN2.

pLN2 (Fig. 11.3) contains all the necessary genes for the generation of L-olivose and its derivative, L-olcandrose (Table 11.1) (Rodríguez et al., 2002). These are the *oleSE* genes coding for enzymes involved in glucose-1-phosphate activation and 4,6-dehydration; the *oleVW* genes, which encode the enzymes in charge of 2-deoxygenation (with equatorial 3-hydroxyl group generation); *oleL*, whose translated product, OleL, is a 3,5-epimerase; and *oleU*, which encodes the 4-ketoreductase, which generates the equatorial 4-hydroxyl group present in L-olivose (Fig. 11.1) and L-oleandrose. pLN2 also contains the gene encoding the O-methyltransferase OleY, which incorporates the methyl group at the L-olivose 3-hydroxyl group once this deoxysugar has been attached to its natural aglycone, oleandolide, generating L-oleandrosyl-oleandolide (Rodríguez *et al.*, 2001).

Using pLN2 as starting plasmid, it is easy to generate other plasmids able to direct the biosynthesis of activated dideoxysugars. For example, in order to generate D-olivose, the gene cassette *oleU* (Fig. 11.3), which codes for a 4-ketoreductase in charge of generating an equatorial 4-hydroxyl group during L-olivose biosynthesis is exchanged for the gene cassette *urdR*, a gene from the urdamycin gene cluster which codes for a 4-ketoreductase involved in D-olivose biosynthesis (Hoffmeister *et al.*, 2000), generating pLNR (Table 11.1). Both deoxysugar pathways share a common intermediate, dTDP-4-keto-6-deoxy-D-glucose (Fig. 11.1), previous to the 3,5-epimerization event during L-olivose biosynthesis. The ketoreductase UrdR is able to reduce this intermediate, producing dTDP-D-olivose. Another possibility is to add and exchange gene functions in order to generate a different deoxysugar. L-rhodinose is a 2,3,6-trideoxysugar with an axial 4-hydroxyl group. In order to synthesize this deoxysugar, it is first necessary to exchange the gene cassette *oleU* with *urdZ3*, which codes for a 4-ketoreductase from the urdamycin gene cluster (Hoffmeister *et al.*, 2000) able to act on a trideoxysugar, yielding a hydroxyl group with an axial stereochemistry at C4. In a second step, it is necessary to endow the new plasmid with the gene *urdQ* (Fig. 11.1), which codes for an enzyme responsible for the 3-deoxygenation step, generating pLNRHO (Rodríguez *et al.*, 2002). Substitution of the 4-ketoreduction gene cassette *urdZ3* by *urdR* in pLNRHO generates pFL845, which directs the biosynthesis of the 2,3,6-trideoxysugar D-amicetose (Fig. 11.1, Table 11.1) (Pérez *et al.*, 2005).

Two of the gene cassettes included in pLN2, *oleVW*, are responsible for deoxygenation at C2 during L-olivosyl biosynthesis (Fig. 11.1). By deleting these two gene cassettes in pLN2 a new plasmid is created, pLN2Δ, which directs the biosynthesis of L-rhamnose (Rodríguez *et al.*, 2002). L-rhamnose is a 6-deoxysugar whose only difference with respect to L-olivose is the presence of a 2-hydroxyl group (Fig. 11.1, Table 11.1).

A slight modification of the pLN2 plasmid system is the pFL942 family of plasmids (Fig. 11.3). In pLN2, the gene functions responsible for glucose-1-phosphate activation and 4,6-dehydration (*oleS* and *oleE* respectively) and for epimerization (*oleL*) are present as a native DNA fragment from the *S. antibioticus* chromosome (Fig. 11.3). This means that in pLN2 it is difficult to exchange the epimerization function or to delete it. pLN2 also contains a gene cassette for *oleY* (Fig. 11.3), an *O*-methyltransferase which only recognizes an L-olivose already attached to its macrolide aglycone, oleandolide; and this function is useless when another aglycone is used. To overcome these problems a new plasmid, pFL942, was created containing two divergent P*ermE* promoters (Fig. 11.4). One of them controls expression of *mtmDE*, cloned as a native DNA fragment from the mithramycin pathway, which codes for a glucose-1-phosphate synthase and 4,6-dehydratase, respectively (Lombó *et al.*, 1997). The other promoter controls expression of several gene cassettes: those for the 2-deoxygenation gene functions, *oleV* and *eryBII*; one for a C3-methyltransferase gene, *eryBIII;* another for a 5-epimerase gene, *eryBVII*; and another for a 4-ketoreductase gene, *eryBIV* (Fig. 11.3). OleV and EryBII deoxygenate the 2-hydroxyl group, generating an axial hydroxyl group at C3, which is recognized by EryBIII, which then introduces a methyl group at C3 (Fig. 11.1). This intermediate is epimerized at C5 by EryBVII, and finally EryBIV generates an equatorial 4-hydroxyl group, producing dTDP-L-mycarose (Fig. 11.1), a branched-chain deoxysugar (Lombó *et al.*, 2004). Substitution in pFL942 of the gene cassette *eryBII* by *oleW* causes the accumulation of an intermediate in which the generated 3-hydroxyl group after the 2-deoxygenation event is equatorial instead of axial. The simultaneous substitution of the L-mycarose methyltransferase gene, *eryBIII*, by the D-mycarose methyltransferase gene *mtmC* (from the mithramycin pathway) (Fig. 11.1) yields pFL947 (Table 11.1), which codes for L-chromose B (Lombó *et al.*, 2004). From these plasmids, a series of plasmids were generated lacking the epimerase gene function, which direct the biosynthesis of several D-sugars (Fig. 11.1, Table 11.1) (Pérez *et al.*, 2006).

8. Generation of Glycosylated Compounds

Once a plasmid for deoxysugar biosynthesis has been created, it is necessary to introduce it into a cell system where the activated deoxysugar generated by that plasmid is transferred to an appropriate aglycone by a glycosyltransferase.

The host for the plasmid can be a strain which naturally produces glycosylated compounds, or a mutant generated in such a way that the intracellular pathways for specific host deoxysugar generation have been eliminated (completely or partially). In the absence of the natural activated deoxysugars, the host glycosyltransferase can now recognize the new deoxysugars and transfer them to the aglycone, generating hybrid compounds by the joint action of host genes and plasmid-borne genes, as has been illustrated above with several examples.

Mithramycin is an aromatic antitumor aureolic acid polyketide containing a disaccharide of D-olivose and a trisaccharide of D-olivose–D-oliose–D-mycarose. Transformation of *S. argillaceus*, the producer of mithramycin, with several gene cassette plasmids yielded several new glycosylated derivatives of mithramycin, some of which with improved antitumor activities. These experiments showed how flexible mithramycin glycosyltransferases are with respect to other available activated deoxysugars inside the cell (Baig et al., 2008; Pérez et al., 2008).

Another possibility is to express the glycosyltransferase and the plasmid for deoxysugar biosynthesis in a heterologous host that does not produce glycosylated compounds. Then, a suitable aglycone for that glycosyltransferase can be either fed to this strain, or synthesized within the host through the expression of all genes for aglycone biosynthesis on a separate plasmid.

S. albus has been used as heterologous host to generate glycosylated derivatives of staurosporine (Salas et al., 2005). Two strains of *S. albus* carrying plasmids directing the biosynthesis of the staurosporine aglycone and the glycosyltransferase StaG gene with and without the StaN P_{450}-oxygenase were created and used as hosts for expressing several plasmids for deoxysugar biosynthesis: pRHAM for L-rhamnose biosynthesis; pLN2 for L-olivose; pLNR for D-olivose, and pLNBIV for L-digitoxose. This resulted in production of the corresponding L-rhamnosyl-, L-olivosyl-, L-digitoxosyl-, and D-olivosyl-derivatives, demonstrating the flexibility of the glycosyltransferase StaG. Formation of the second C–N linkage by StaN was only produced with L-deoxysugars (Salas et al., 2005).

The *S. albus* GB16 strain (Blanco et al., 2001) was created in order to generate glycosylated derivatives of 8-demethyl-tetracenomycin C by bioconversion. This strain contains integrated into the chromosome a stable copy of *elmGT* controlled by the P*ermE* promoter. *elmGT* codes for the elloramycin glycosyltransferase, and its introduction into this species offered an easy way to glycosylate its aglycone, 8-demethyl-tetracenomycin C, with several deoxysugar moieties. In order to test the flexibility of ElmGT, several "deoxysugar plasmids" were introduced into *S. albus* GB16 and the resultant recombinants were used to generate new glycosylated tetracenomycins by feeding them with the aglycone. In this way, L-olivosyl-, L-digitoxosyl-, L-mycarosyl-, L-chromosyl-, L-rhamnosyl-, L-rhodinosyl-, L-amicetosyl-, D-olivosyl-, D-boivinosyl-, D-digitoxosyl-, and D-amicetosyl-tetracenomycins were generated (Lombó et al., 2004; Pérez et al., 2005, 2006; Rodríguez et al., 2000, 2002).

8.1. Generation of *S. albus* GB16

1. Amplification of *elmGT* gene from cos16F4 (from a *Streptomyces olivaceus* gene library). Primer pairs used for this amplification contain recognition sequences for *Xba*I and *Hin*dIII, at the 5′ and 3′-ends, respectively.
2. Digestion of the *elmGT* DNA band with *Xba*I and *Hin*dIII, using the restriction sites present in the PCR primers.

3. Ligation of this DNA band to vector pIAGO digested with the same enzymes. Transformation of *E. coli* and selection with ampicillin (100 μl/ml). This generates pGB15, where *elmGT* is under the control of P*ermE*.
4. Digestion of pGB15 with *Eco*RI and *Hin*dIII, in order to isolate the P*ermE*-elmGT 1.6-kb DNA fragment.
5. Ligation of this DNA fragment into pIJ2925 (Janssen and Bibb, 1993) digested with same restriction enzymes. pIJ2925 is a pUC18 derivative whose polylinker contains two flanking *Bgl*II sites.
6. Isolation from the former construct of the 1.5-kb *Bgl*II DNA fragment which contains the P*ermE* promoter and the *elmGT* gene.
7. Ligation of this DNA fragment into the unique *Bam*HI site of vector pKC796, generating pGB16. pKC796 is an *E. coli–Streptomyces* vector which contains the apramycin resistance gene and the *attP* attachment site from phage ΦC31. This attachment site allows integration of the vector into the *Streptomyces* chromosome through site-specific recombination.
8. Transformation of pGB16 into *S. albus* protoplasts and selection with apramycin (25 μg/ml).

8.2. Feeding experiments with 8-demethyl-tetracenomycin C

1. Inoculation of 5 ml TSB medium (containing 25 μg/ml apramycin) in a 100-ml flask with 10 μl of a spore suspension of *S. albus* GB16. Incubation at 30 °C, 250 rpm, 24 h.
2. Use of 100 μl of this culture to inoculate 5 ml R5A medium in 100-ml flask.
3. Incubation at 30 °C, 250 rpm, 24 h.
4. Addition of 8-demethyl-tetracenomycin C at 100 μg/ml final concentration.
5. Incubation at 30 °C, 250 rpm, 24 h.
6. Extraction of a 0.9-ml sample from the culture with 0.3 ml ethyl acetate.
7. Centrifugation for 2 min at 12,000 rpm to separate the organic and aqueous phases.
8. Removal of the upper organic phase and evaporation under vacuum.
9. Resuspension of the residue in 10 μl methanol for TLC or HPLC analysis.

9. Tailoring Modifications of the Attached Deoxysugars

Once the corresponding deoxysugars have been transferred to the aglycone by the action of glycosyltransferases, in some cases, there are still one or a few more biosynthetic steps in order to obtain the final active compound. These tailoring modifications, in the case of deoxysugars, can be methylations or acylations, at specific hydroxyl or amino groups of the transferred deoxysugar.

The last step during erythromycin biosynthesis consists of O-methylation at C3 of L-mycarose, which decorates this macrolide, giving rise to the L-cladinose residue present in the final erythromycin A compound. This methylation is carried out by the SAM-dependent enzyme EryG (Haydock et al., 1991; Weber et al., 1989). During the biosynthesis of the macrolide oleandomycin, the methyltransferase OleY converts L-olivose to L-oleandrose (its 3-O-methylated derivative), once this deoxysugar has been transfered to its aglycone (Rodríguez et al., 2001). In addition, OleY has been shown to introduce O-methylations in deoxysugars carrying a C3 hydroxyl group, like D-boivinose, L-digitoxose, and 2-O-methyl-L-rhamnose, when these deoxysugars are attached to the anthracycline steffimycin aglycone (Olano et al., 2008). The methyltransferase CmmMIII catalyzes the penultimate enzymatic step during chromomycin A_3 biosynthesis: conversion of the attached D-olivose of the disaccharide chain into 4-O-methyl-D-olivose (Menéndez et al., 2004). Three O-methyltransferases are responsible for the generation of a permethylated L-rhamnosyl moiety once this deoxysugar has been transferred to the elloramycin aglycone. ElmMI, ElmMII, and ElmMIII modify consecutively positions at C2, C3, and C4 in the deoxysugar, respectively (Patallo et al., 2001). A D-glucose moiety decorates the rebeccamycin molecule. This sugar residue suffers a C4 O-methylation, carried out by the SAM-dependent methyltransferase RebM (Sánchez et al., 2002; Singh et al., 2008). One of the last steps during staurosporine biosynthesis is the methylation of the amino group, carried out by the methyltransferase StaMA (Salas et al., 2005).

The last enzyme acting during chromomycin A_3 biosynthesis is the acetyltransferase CmmA. This membrane-bound enzyme transfers first an acetyl residue to the 4-hydroxyl group of the last deoxysugar from the chromomycin trisaccharide, L-chromose B, and then another acetyl moiety to the 4-hydroxyl group of the first D-oliose of the disaccharide, giving rise to a final product with high biological activity (Menéndez et al., 2004; B. García, unpublished data). The macrolide carbomycin contains two deoxysugars, D-mycaminose and a derivative of L-mycarose, which contains an isovaleryl group decorating the 4-hydroxyl group. CarE is the isovaleryl transferase responsible for this tailoring modification. This enzyme is also able to modify L-mycarose from another macrolide, spiramycin, generating the hybrid compound isovaleryl-spiramycin (Epp et al., 1989). An acyltransferase has been described in the teicoplanin gene cluster, which is responsible for the transfer of a decanoyl moiety (or similar acyl chain) to the 2-amino group of the 2-aminoglucose present in this lipoglycopeptide. This acyltransferase was also able to modify glucose and 6-aminoglucose, showing sugar substrate flexibility (Kruger et al., 2005).

10. Detection of Glycosylated Compounds

Once the respective sugar has been transferred by a suitable glycosyltransferase to an appropiate aglycone, this glycosylated molecule can be extracted from the culture and analyzed in order to detect the expected compound, and eventually to purify it. Detection and purification of glycosylated compounds is usually carried out by HPLC–MS analysis. In the case of those compounds derived from 8-demethyl-tetracenomycin C, this was done using an Alliance Waters 2695 separations HPLC module coupled to a Waters 2996 photodiode array detector. MS detection was carried out with a coupled Waters-Micromass ZQ4000 mass spectrometer. The solvent for extraction of the culture was ethyl acetate, and this organic phase was dessicated using a speed vac.

Conditions for HPLC:

1. μBondapak C_{18} column (4.6 × 250 mm), Waters.
2. Mobile phase: acetonitrile and 0.1% trifluoroacetic acid in water.
3. Linear gradient from 10% to 100% acetonitrile in 30 min.
4. Flow rate of 1 ml/min.
5. Detection of peaks with the photodiode array at 280 nm using Millennium software (Waters).

Conditions for MS:

1. Electrospray ionization in positive mode.
2. Capillary voltage of 3 kV and cone voltage of 20 kV.

ACKNOWLEDGMENTS

Research in the authors' laboratory has been supported by grants from the Spanish Ministry of Science and Innovation (BFU2006-00404 to J.A.S. and BIO2005-04115 to C.M), the Red Temática de Investigación Cooperativa de Centros de Cáncer to J.A.S. (Ministry of Health, Spain; ISCIII-RETIC RD06/0020/0026), and from the UE FP6 (Integrated project no 005224). We thank Obra Social Cajastur for financial support to Carlos Olano and Felipe Lombó.

REFERENCES

Aguirrezabalaga, I., Olano, C., Allende, N., Rodríguez, L., Braña, A. F., Méndez, C., and Salas, J. A. (2000). Identification and expression of genes involved in biosynthesis of L-oleandrose and its intermediate L-olivose in the oleandomycin producer *Streptomyces antibioticus*. *Antimicrob. Agents Chemother.* **44,** 1266–1275.

Baig, I., Pérez, M., Braña, A. F., Gomathinayagam, R., Damodaran, C., Salas, J. A., Méndez, C., and Rohr, J. (2008). Mithramycin analogues generated by combinatorial biosynthesis show improved bioactivity. *J. Nat. Prod.* **71,** 199–207.

Bibb, M. J., Janssen, G. R., and Ward, J. M. (1985). Cloning and analysis of the promoter region of the erythromycin resistance gene (*ermE*) of *Streptomyces erythraeus*. *Gene* **38,** 215–226.

Bierman, M., Logan, R., O'Brien, K., Seno, E. T., Rao, R. N., and Schoner, B. E. (1992). Plasmid cloning vectors for the conjugal transfer of DNA from *Escherichia coli* to *Streptomyces spp*. *Gene* **116,** 43–49.

Blanco, G., Patallo, E. P., Braña, A. F., Trefzer, A., Bechthold, A., Rohr, J., Méndez, C., and Salas, J. A. (2001). Identification of a sugar flexible glycosyltransferase from *Streptomyces olivaceus*, the producer of the antitumor polyketide elloramycin. *Chem. Biol.* **3,** 253–263.

Borisova, S. A., Zhao, L., Melançon, C. E., Kao, C. L., and Liu, H. W. (2004). Characterization of the glycosyltransferase activity of *desVII*: Analysis of and implications for the biosynthesis of macrolide antibiotics. *J. Am. Chem. Soc.* **126,** 6534–6535.

Borisova, S. A., Zhao, L., Sherman, D. H., and Liu, H. W. (1999). Biosynthesis of desosamine: Construction of a new macrolide carrying a genetically designed sugar moiety. *Org. Lett.* **1,** 133–136.

Butler, A. R., Bate, N., Kiehl, D. E., Kirst, H. A., and Cundliffe, E. (2002). Genetic engineering of aminodeoxyhexose biosynthesis in *Streptomyces fradiae*. *Nat. Biotechnol.* **20,** 713–716.

Chen, H., Agnihotri, G., Guo, Z., Que, N. L. S., Chen, X. H., and Liu, H. W. (1999). Biosynthesis of mycarose: Isolation and characterization of enzymes involved in the C-2 deoxygenation. *J. Am. Chem. Soc.* **121,** 8124–8125.

Chen, H., Thomas, M. G., Hubbard, B. K., Losey, H. C., Walsh, C. T., and Burkart, M. D. (2000). Deoxysugars in glycopeptide antibiotics: Enzymatic synthesis of TDP-L-epivancosamine in chloroeremomycin biosynthesis. *Proc. Natl. Acad. Sci. USA* **97,** 11942–11947.

Chen, H., Zhao, Z., Hallis, T. M., Guo, Z., and Liu, H. W. (2001). Insights into the branched-chain formation of mycarose: Methylation catalyzed by an (S)-adenosylmethionine-dependent methyltransferase. *Angew. Chem. Int. Ed.* **40,** 607.

Cundliffe, E. (1992). Glycosylation of macrolide antibiotics in extracts of *Streptomyces lividans*. *Antimicrob. Agents Chemother.* **36,** 348–352.

Doumith, M., Legrand, R., Lang, C., Salas, J. A., and Raynal, M. C. (1999). Interspecies complementation in *Saccharopolyspora erythraea*: Elucidation of the function of *oleP1, oleG1* and *oleG2* from the oleandomycin biosynthetic gene cluster of *Streptomyces antibioticus* and generation of new erythromycin derivatives. *Mol. Microbiol.* **34,** 1039–1048.

Draeger, G., Park, S. H., and Floss, H. G. (1999). Mechanism of the 2-deoxygenation step in the biosynthesis of the deoxyhexose moieties of the antibiotics granaticin and oleandomycin. *J. Am. Chem. Soc.* **121,** 2611–2612.

Epp, J. K., Huber, M. L., Turner, J. R., Goodson, T., and Schoner, B. E. (1989). Production of a hybrid macrolide antibiotic in *Streptomyces ambofaciens* and *Streptomyces lividans* by introduction of a cloned carbomycin biosynthetic gene from *Streptomyces thermotolerans*. *Gene* **85,** 293–301.

Fischer, C., Rodríguez, L., Patallo, E. P., Lipata, F., Braña, A. F., Méndez, C., Salas, J. A., and Rohr, J. (2002). Digitoxosyltetracenomycin C and glucosyltetracenomycin C, two novel elloramycin analogues obtained by exploring the sugar donor substrate specificity of glycosyltransferase ElmGT. *J. Nat. Prod.* **65,** 1685–1689.

Freitag, A., Méndez, C., Salas, J. A., Kammerer, B., Li, S. M., and Heide, L. (2006). Metabolic engineering of the heterologous production of clorobiocin derivatives and elloramycin in *Streptomyces coelicolor* M512. *Metab. Eng.* **8,** 653–661.

Gaisser, S., Martin, C. J., Wilkinson, B., Sheridan, R. M., Lill, R. E., Weston, A. J., Ready, S. J., Waldron, C., Crouse, G. D., Leadlay, P. F., and Staunton, J. (2002).

Engineered biosynthesis of novel spinosyns bearing altered deoxyhexose substituents. *Chem. Commun.* **6,** 618–619.

Gaisser, S., Reather, J., Wirtz, G., Kellenberger, L., Staunton, J., and Leadlay, P. F. (2000). A defined system for hybrid macrolide biosynthesis in *Saccharopolyspora erythraea*. *Mol. Microbiol.* **36,** 391–401.

González, A., Remsing, L. L., Lombó, F., Fernández, M. J., Prado, L., Braña, A. F., Künzel, E., Rohr, J., Méndez, C., and Salas, J. A. (2001). The *mtmVUC* genes of the mithramycin gene cluster in *Streptomyces argillaceus* are involved in the biosynthesis of the sugar moieties. *Mol. Gen. Genet.* **264,** 827–835.

Han, O., Miller, V. P., and Liu, H. W. (1990). Mechanistic studies of the biosynthesis of 3,6-dideoxyhexoses in *Yersinia pseudotuberculosis*. Purification and characterization of CDP-6-deoxy-delta 3,4-glucoseen reductase based on its NADH:Dichlorophenolindolphenol oxidoreductase activity. *J. Biol. Chem.* **265,** 8033–8041.

Haydock, S. F., Dowson, J. A., Dhillon, N., Roberts, G. A., Cortés, J., and Leadlay, P. F. (1991). Cloning and sequence analysis of genes involved in erythromycin biosynthesis in *Saccharopolyspora erythraea*: Sequence similarities between EryG and a family of S-adenosylmethionine-dependent methyltransferases. *Mol. Gen. Genet.* **230,** 120–128.

He, X. M., and Liu, H. W. (2002). Formation of unusual sugars: Mechanistic studies and biosynthetic applications. *Annu. Rev. Biochem.* **71,** 701–754.

Hoffmeister, D., Ichinose, K., Domann, S., Faust, B., Trefzer, A., Dräger, G., Kirschning, A., Fischer, C., Künzel, E., Bearden, D., Rohr, J., and Bechthold, A. (2000). The NDP-sugar co-substrate concentration and the enzyme expression level influence the substrate specificity of glycosyltransferases: Cloning and characterization of deoxysugar biosynthetic genes of the urdamycin biosynthetic gene cluster. *Chem. Biol.* **7,** 821–831.

Hoffmeister, D., Dräger, G., Ichinose, K., Rohr, J., and Bechthold, A. (2003). The C-Glycosyltransferase UrdGT2 is unselective toward D- and L-configured nucleotide-bound rhodinoses. *J. Am. Chem. Soc.* **125,** 4678–4679.

Hong, J. S., Park, S. H., Choi, C. Y., Sohng, J. K., and Yoon, Y. J. (2004). New olivosyl derivatives of methymycin/pikromycin from an engineered strain of *Streptomyces venezuelae*. *FEMS Microbiol. Lett.* **238,** 391–399.

Janssen, G. R., and Bibb, M. J. (1993). Derivatives of pUC18 that have *Bgl*II sites flanking a modified multiple cloning site and that retain the ability to identify recombinant clones by visual screening of *Escherichia coli* colonies. *Gene* **124,** 133–134.

Kieser, T., Bibb, M. J., Buttner, M. J., Chater, K. F., and Hopwood, D. A. (2000). Practical *Streptomyces* Genetics. The John Innes Foundation, Norwich.

Kruger, R. G., Lu, W., Oberthür, M., Tao, J., Kahne, D., and Walsh, C. T. (2005). Tailoring of glycopeptide scaffolds by the acyltransferases from the teicoplanin and A-40,926 biosynthetic operons. *Chem. Biol.* **12,** 131–140.

Kuhstoss, S., Richardson, M. A., and Rao, N. R. (1991). Plasmid cloning vectors that integrate site-specifically in *Streptomyces spp*. *Gene* **97,** 143–146.

Lei, Y., Ploux, O., and Liu, H. W. (1995). Mechanistic studies on CDP-6-deoxy-L-threo-D-glycero-4-hexulose 3-dehydrase identification of His-220 as the active-site base by chemical modification and site-directed mutagenesis. *Biochemistry* **34,** 4643–4654.

Lindqvist, L., Kaiser, R., Reeves, P. R., and Lindberg, A. A. (1994). Purification, characterization, and high performance liquid chromatography assay of *Salmonella* glucose-1-phosphate cytidylyltransferase from the cloned *rfbF* gene. *J. Biol. Chem.* **269,** 122–126.

Lombó, F., Blanco, G., Fernández, E., Méndez, C., and Salas, J. A. (1996). Characterization of *Streptomyces argillaceus* genes encoding a polyketide synthase involved in the biosynthesis of the antitumor mithramycin. *Gene* **172,** 87–91.

Lombó, F., Gibson, M., Greenwell, L., Braña, A. F., Rohr, J., Salas, J. A., and Méndez, C. (2004). Engineering biosynthetic pathways for deoxysugars: Branched-chain sugar pathways and derivatives from the antitumor tetracenomycin. *Chem. Biol.* **11,** 1709–1718.

Lombó, F., Siems, K., Braña, A. F., Méndez, C., Bindseil, K., and Salas, J. A. (1997). Cloning and insertional inactivation of *Streptomyces argillaceus* genes involved in the earliest steps of biosynthesis of the sugar moieties of the antitumor polyketide mithramycin. *J. Bacteriol.* **179**, 3354–3357.

Losey, H. C., Jiang, J., Biggins, J. B., Oberthür, M., Ye, X. Y., Dong, S. D., Kahne, D., Thorson, J. S., and Walsh, C. T. (2002). Incorporation of glucose analogs by GtfE and GtfD from the vancomycin biosynthetic pathway to generate variant glycopeptides. *Chem. Biol.* **9**, 1305–1314.

Lu, M., and Kleckner, N. (1994). Molecular cloning and characterization of the pgm gene encoding phosphoglucomutase of *Escherichia coli. J. Bacteriol.* **176**, 5847–5851.

Lu, W., Leimkuhler, C., Gatto, G. J., Kruger, R. G., Oberthür, M., Kahne, D., and Walsh, C. T. (2005). AknT is an activating protein for the glycosyltransferase AknS in L-aminodeoxysugar transfer to the aglycone of aclacinomycin A. *Chem. Biol.* **12**, 527–534.

Luzhetskyy, A., Mayer, A., Hoffmann, J., Pelzer, S., Holzenkämper, M., Schmitt, B., Wohlert, S. E., Vente, A., and Bechthold, A. (2007). Cloning and heterologous expression of the aranciamycin biosynthetic gene cluster revealed a new flexible glycosyltransferase. *Chembiochem* **8**, 599–602.

Luzhetskyy, A., Méndez, C., Salas, J. A., and Bechthold, A. (2008). Glycosyltransferases, important tools for drug design. *Curr. Top. Med. Chem.* **8**, 680–709.

Madduri, K., Kennedy, J., Rivola, G., Inventi-Solari, A., Filippini, S., Zanuso, G., Colombo, A. L., Gewain, K. M., Occi, J. L., MacNeil, D. J., and Hutchinson, C. R. (1998). Production of the antitumor drug epirubicin (4'-epidoxorubicin) and its precursor by a genetically engineered strain of *Streptomyces peucetius. Nat. Biotechnol.* **16**, 69–74.

Melançon, C. E., Takahashi, H., and Liu, H. W. (2004). Characterization of tylM3/tylM2 and mydC/mycB pairs required for efficient glycosyltransfer in macrolide antibiotic biosynthesis. *J. Am. Chem. Soc.* **126**, 16726–16727.

Méndez, C., Luzhetskyy, A., Bechthold, A., and Salas, J. A. (2008). Deoxysugars in bioactive natural products: Development of novel derivatives by altering the sugar pattern. *Curr. Top. Med. Chem.* **8**, 710–724.

Menéndez, N., Nur-E-Alam, M., Braña, A. F., Rohr, J., Salas, J. A., and Méndez, C. (2004). Tailoring modification of deoxysugars during biosynthesis of the antitumour drug chromomycin A_3 by *Streptomyces griseus ssp. griseus. Mol. Microbiol.* **53**, 903–915.

Motamedi, H., Shafiee, A., and Cai, S. J. (1995). Integrative vectors for heterologous gene expression in *Streptomyces spp. Gene* **160**, 25–31.

Nedal, A., and Zotchev, S. B. (2004). Biosynthesis of deoxyaminosugars in antibiotic-producing bacteria. *Appl. Microbiol. Biotechnol.* **64**, 7–15.

Oberthür, M., Leimkuhler, C., Kruger, R. G., Lu, W., Walsh, C. T., and Kahne, D. (2005). A systematic investigation of the synthetic utility of glycopeptide glycosyltransferases. *J. Am. Chem. Soc.* **127**, 10747–10752.

O'Kunewick, J. P., Kociban, D. L., and Buffo, M. J. (1990). Comparative hematopoietic toxicity of doxorubicin and 4'-epirubicin. *Proc. Soc. Exp. Biol. Med.* **195**, 95–99.

Olano, C., Abdelfattah, M. S., Gullón, S., Braña, A. F., Rohr, J., Méndez, C., and Salas, J. A. (2008). Glycosylated derivatives of steffimycin: Insights into the role of the sugar moieties for the biological activity. *Chembiochem* **9**, 624–633.

Olano, C., Lomovskaya, N., Fonstein, L., Roll, J. T., and Hutchinson, C. R. (1999). A two-plasmid system for the glycosylation of polyketide antibiotics: Bioconversion of epsilon-rhodomycinone to rhodomycin D. *Chem. Biol.* **6**, 845–855.

Olano, C., Rodríguez, A. M., Michel, J. M., Méndez, C., Raynal, M. C., and Salas, J. A. (1998). Analysis of a *Streptomyces antibioticus* chromosomal region involved in oleandomycin biosynthesis, which encodes two glycosyltransferases responsible for glycosylation of the macrolactone ring. *Mol. Gen. Genet.* **259**, 299–308.

Patallo, E. P., Blanco, G., Fischer, C., Brana, A. F., Rohr, J., Méndez, C., and Salas, J. A. (2001). Deoxysugar methylation during biosynthesis of the antitumor polyketide elloramycin by *Streptomyces olivaceus*. Characterization of three methyltransferase genes. *J. Biol. Chem.* **276,** 18765–18774.

Pérez, M., Baig, I., Braña, A. F., Salas, J. A., Rohr, J., and Méndez, C. (2008). Generation of new derivatives of the antitumor antibiotic mithramycin by altering the glycosylation pattern through combinatorial biosynthesis. *ChemBioChem* **9,** 2295–2304.

Pérez, M., Lombó, F., Baig, I., Braña, A. F., Rohr, J., Salas, J. A., and Méndez, C. (2006). Combinatorial biosynthesis of antitumor deoxysugar pathways in *Streptomyces griseus*: Reconstitution of "unnatural natural gene clusters" for the biosynthesis of four 2,6-D-dideoxyhexoses. *Appl. Environ. Microbiol.* **72,** 6644–6652.

Pérez, M., Lombó, F., Zhu, L., Gibson, M., Braña, A. F., Rohr, J., Salas, J. A., and Méndez, C. (2005). Combining sugar biosynthesis genes for the generation of L- and D-amicetose and formation of two novel antitumor tetracenomycins. *Chem. Commun.* **12,** 1604–1606.

Prado, L., Lombó, F., Braña, A. F., Méndez, C., Rohr, J., and Salas, J. A. (1999). Analysis of two chromosomal regions adjacent to genes for a type II polyketide synthase involved in the biosynthesis of the antitumor polyketide mithramycin in *Streptomyces argillaceus*. *Mol. Gen. Genet.* **261,** 216–225.

Quirós, L. M., Aguirrezabalaga, I., Olano, C., Méndez, C., and Salas, J. A. (1998). Two glycosyltransferases and a glycosidase are involved in oleandomycin modification during its biosynthesis by *Streptomyces antibioticus*. *Mol. Microbiol.* **28,** 1177–1185.

Ramos, A., Lombó, F., Braña, A. F., Rohr, J., Méndez, C., and Salas, J. A. (2008). Biosynthesis of elloramycin in *Streptomyces olivaceus* requires glycosylation by enzymes encoded outside the aglycon cluster. *Microbiology* **154,** 781–788.

Räty, K., Hautala, A., Torkkell, S., Kantola, J., Mäntsälä, P., Hakala, J., and Ylihonko, K. (2002). Characterization of mutations in aclacinomycin A-non-producing *Streptomyces galilaeus* strains with altered glycosylation patterns. *Microbiology* **148,** 3375–3384.

Räty, K., Kunnari, T., Hakala, J., Mäntsälä, P., and Ylihonko, K. (2000). A gene cluster from *Streptomyces galilaeus* involved in glycosylation of aclarubicin. *Mol. Gen. Genet.* **264,** 164–172.

Remsing, L. L., García-Bernardo, J., González, A., Künzel, E., Rix, U., Braña, A. F., Bearden, D. W., Méndez, C., Salas, J. A., and Rohr, J. (2002). Ketopremithramycins and ketomithramycins, four new aureolic acid-type compounds obtained upon inactivation of two genes involved in the biosynthesis of the deoxysugar moieties of the antitumor drug mithramycin by *Streptomyces argillaceus*, reveal novel insights into post-PKS tailoring steps of the mithramycin biosynthetic pathway. *J. Am. Chem. Soc.* **124,** 1606–1614.

Rodríguez, L., Aguirrezabalaga, I., Allende, N., Braña, A. F., Méndez, C., and Salas, J. A. (2002). Engineering deoxysugar biosynthetic pathways from antibiotic-producing microorganisms. A tool to produce novel glycosylated bioactive compounds. *Chem. Biol.* **9,** 721–729.

Rodríguez, L., Oelkers, C., Aguirrezabalaga, I., Braña, A. F., Rohr, J., Méndez, C., and Salas, J. A. (2000). Generation of hybrid elloramycin analogs by combinatorial biosynthesis using genes from anthracycline-type and macrolide biosynthetic pathways. *J. Mol. Microbiol. Biotechnol.* **3,** 271–276.

Rodríguez, L., Rodríguez, D., Olano, C., Braña, A. F., Méndez, C., and Salas, J. A. (2001). Functional analysis of OleY L-oleandrosyl 3-O-methyltransferase of the oleandomycin biosynthetic pathway in *Streptomyces antibioticus*. *J. Bacteriol.* **183,** 5358–5363.

Salah-Bey, K., Doumith, M., Michel, J. M., Haydock, S., Cortés, J., Leadlay, P. F., and Raynal, M. C. (1998). Targeted gene inactivation for the elucidation of deoxysugar biosynthesis in the erythromycin producer *Saccharopolyspora erythraea*. *Mol. Gen. Genet.* **257,** 542–553.

Salas, J. A., and Méndez, C. (2005). Biosynthesis pathways for deoxysugars in antibiotic-producing actinomycetes: Isolation, characterization and generation of novel glycosylated derivatives. *J. Mol. Microbiol. Biotechnol.* **9,** 77–85.

Salas, J. A., and Méndez, C. (2007). Engineering the glycosylation of natural products in actinomycetes. *Trends Microbiol.* **15,** 219–232.

Salas, A. P., Zhu, L., Sánchez, C., Braña, A. F., Rohr, J., Méndez, C., and Salas, J. A. (2005). Deciphering the late steps in the biosynthesis of the anti-tumour indolocarbazole staurosporine: Sugar donor substrate flexibility of the StaG glycosyltransferase. *Mol. Microbiol.* **58,** 17–27.

Sánchez, C., Butovich, I. A., Braña, A. F., Rohr, J., Méndez, C., and Salas, J. A. (2002). The biosynthetic gene cluster for the antitumor rebeccamycin: Characterization and generation of indolocarbazole derivatives. *Chem. Biol.* **9,** 519–531.

Sánchez, C., Zhu, L., Braña, A. F., Salas, A. P., Rohr, J., Méndez, C., and Salas, J. A. (2005). Combinatorial biosynthesis of antitumor indolocarbazole compounds. *Proc. Natl. Acad. Sci. USA* **102,** 461–466.

Sastry, M., and Patel, D. J. (1993). Solution structure of the mithramycin dimer–DNA complex. *Biochemistry* **32,** 6588–6604.

Schnaitman, C. A., and Klena, J. D. (1993). Genetics of lipopolysaccharide biosynthesis in enteric bacteria. *Microbiol. Rev.* **57,** 655–682.

Singh, S., McCoy, J. G., Zhang, C., Bingman, C. A., Phillips, G. N., and Thorson, J. S. (2008). Structure and mechanism of the rebeccamycin sugar $4'$-O-methyltransferase RebM. *J. Biol. Chem.* **283,** 22628–22636.

Summers, R. G., Donadio, S., Staver, M. J., Wendt-Pienkowski, E., Hutchinson, C. R., and Katz, L. (1997). Sequencing and mutagenesis of genes from the erythromycin biosynthetic gene cluster of *Saccharopolyspora erythraea* that are involved in L-mycarose and D-desosamine production. *Microbiology* **143,** 3251–3262.

Szu, P. H., He, X., Zhao, L., and Liu, H. W. (2005). Biosynthesis of TDP-D-desosamine: Identification of a strategy for C4 deoxygenation. *Angew. Chem. Int. Ed. Engl.* **44,** 6742–6746.

Tang, L., and McDaniel, R. (2001). Construction of desosamine containing polyketide libraries using a glycosyltransferase with broad substrate specificity. *Chem. Biol.* **8,** 547–555.

Trefzer, A., Blanco, G., Remsing, L., Künzel, E., Rix, U., Lipata, F., Braña, A. F., Méndez, C., Rohr, J., Bechthold, A., and Salas, J. A. (2002). Rationally designed glycosylated premithramycins: Hybrid aromatic polyketides using genes from three different biosynthetic pathways. *J. Am. Chem. Soc.* **124,** 6056–6062.

Vara, J. A., Lewandowska-Skarbek, M., Wang, Y. G., Donadio, S., and Hutchinson, C. R. (1989). Cloning of genes governing the deoxysugar portion of the erythromycin biosynthetic pathway in *Saccharopolyspora erythraea* (*Streptomyces erythreus*). *J. Bacteriol.* **171,** 5872–5881.

Vazquez-Laslop, N., Thum, C., and Mankin, A. S. (2008). Molecular mechanism of drug-dependent ribosome stalling. *Mol. Cell* **30,** 190–202.

Vieira, J., and Messing, J. (1991). New pUC-derived cloning vectors with different selectable markers and DNA replication origins. *Gene* **100,** 189–194.

Weber, J. M., Schoner, B., and Losick, R. (1989). Identification of a gene required for the terminal step in erythromycin A biosynthesis in *Saccharopolyspora erythraea* (*Streptomyces erythreus*). *Gene* **75,** 235–241.

Yamase, H., Zhao, L., and Liu, H. W. (2000). Engineering a hybrid sugar biosynthetic pathway: Production of L-rhamnose and its implication on dihydrostreptose biosynthesis. *J. Am. Chem. Soc.* **122,** 12397–12398.

Zhao, L., Borisova, S., Yeung, S. M., and Liu, H. (2001). Study of C-4 deoxygenation in the biosynthesis of desosamine: Evidence implicating a novel mechanism. *J. Am. Chem. Soc.* **123,** 7909–7910.

CHAPTER TWELVE

The Power of Glycosyltransferases to Generate Bioactive Natural Compounds

Johannes Härle *and* Andreas Bechthold

Contents

1. Introduction 310
   1.1. The role of naturally occurring sugars as components of natural products 311
   1.2. Mechanism of glycosylation 312
   1.3. GT classification 312
   1.4. Reactions catalyzed by GTs 313
   1.5. Flexible GTs 315
   1.6. Bi-functional GTs 321
   1.7. GTs dependent on auxiliary proteins 322
2. Application of GTs in Producing Unnatural Bioactive Molecules 322
   2.1. Detection of new GTs 322
   2.2. Strategies for altering glycosylation to generate novel bioactive natural products 324
   2.3. Conclusion and prospects 328
References 328

Abstract

Glycosyltransferases (GTs), which catalyze the attachment of a sugar moiety to an aglycone are key enzymes for the biosynthesis of many valuable natural products. Their use in pharmaceutical biotechnology is becoming more and more visible.

The promiscuity of GTs has prompted efforts to modify sugar structures and alter the glycosylation patterns of natural products. Here, we present the state of the art in this field. After describing the importance of GTs in determining the functions of natural products, a general survey of glycosyltransferase-catalyzed reactions is documented. This is followed by an overview of crystallized GT-B superfamily members and a discussion of the amino acids of these GTs involved

Institut für Pharmazeutische Wissenschaften, Lehrstuhl für Pharmazeutische Biologie und Biotechnologie, Albert-Ludwigs-Universität Freiburg, Freiburg, Germany

Methods in Enzymology, Volume 458 © 2009 Elsevier Inc.
ISSN 0076-6879, DOI: 10.1016/S0076-6879(09)04812-5 All rights reserved.

in substrate binding. The main chapter is concerned with emphasizing the application of GTs in metabolic pathway engineering leading to novel unnatural bioactive compounds. A strategy to explore new GTs is presented as well as strategies to generate artificial GTs either randomly or in a rational design.

1. INTRODUCTION

Microorganisms, primarily gram-positive Actinomycetes, produce an array of significant bioactive molecules that include antibiotics, antitumor drugs, and immunosuppressant. Many of the natural products are regio- and stereospecifically modified via glycosylation. Often these glycosylated patterns are important for biological activity. Components of natural products are very often deoxysugars. Different enzymes, such as dehydratases, epimerases, and amino- and methyltransferases are responsible for the biosynthesis of these sugars which derive mostly from D-glucose. The attachment of these carbohydrates (mono- di- or oligo- saccharides) is catalyzed by glycosyltransferases (GTs), one of the most diverse group of enzymes in nature. They are involved in the biosynthesis of glycolipids, glycoproteins, polysaccharides, and secondary metabolites, including pharmaceutically important bioactive natural products.

The length of a sugar side-chain might be relevant for bioactivity as exemplified by the landomycins. Landomycin A, containing a hexasaccharide sugar chain, is the largest and the most active compound of the landomycins (Zhu et al., 2007). The vast majority of GTs catalyze transfer of nucleotide-activated hexoses to an OH-group. Other GTs are known to target an NH_2-group [as in rebeccamycins, (Sánchez et al., 2002) or a CH-group (as in urdamycins) (Hoffmeister et al., 2000; Mendez and Salas, 2001)] (Fig. 12.1).

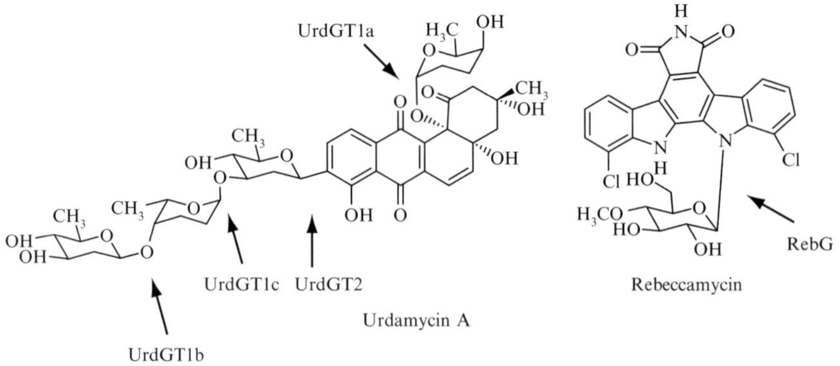

Figure 12.1 GTs forming O-, N- and C-glycosidic linkages involved in urdamycin and rebeccamycin biosynthesis.

An overview of GTs involved in natural product biosynthesis is given by Luzhetskyy *et al.* (2008). As carbohydrates can be connected by a wide range of regio- and stereochemical linkages the application of traditional synthetic chemistry for the synthesis of desired glucan structures requires multiple protection and deprotection schemes, making chemical strategies extremely difficult. A highly attractive alternative is the use of GTs for the following reasons.

1. GTs, belonging to the so called "tailoring enzymes", are often involved in late biosynthetic steps. This makes these enzymes ideal for reprogramming and/or combinatorial approaches (see 2.2).
2. GTs sometimes show broad substrate specificity, allowing the glycosylation of a library of compounds (see 1.5).
3. Although the high specificity of other GTs provides a limiting factor in natural product diversification, using structural information, the specificity of these GTs can be broadened by genetic engineering (see 2.2.2).
4. GTs are relatively stable, making their use in biochemical approaches possible.

In conclusion, GTs have an enormous potential to be used in pharmaceutical biotechnology to generate novel products. Useful candidates can either be found in nature or can be generated by genetic engineering. GTs used in smartly developed applications can exhibit endless opportunities in creating chemical diversity.

1.1. The role of naturally occurring sugars as components of natural products

1.1.1. Biological relevance

Although there is a tremendous diversity in both, types and numbers of sugar units attached to naturally occurring aglycons, the biological relevance of sugars in natural products is not well understood. In some cases the sugar serves to protect the producer cell. As an example, the oleandomycin producer glycosylates its own antibiotic as long it remains in the cell (Vilches *et al.*, 1992). Once the antibiotic is secreted an extracellular β-glycosidase reactivates the antibiotic activity by removal of the sugar moiety.

1.1.2. Pharmaceutical relevance

The functional contribution of carbohydrates to pharmaceutically active natural compounds has been reviewed in a variety of publications (Luzhetskyy *et al.*, 2008; Newman *et al.*, 2003; Weymouth-Wilson, 1997) and it has been reported that the attachment of sugars to natural products or changing a sugar moiety can improve the parent compound's pharmacological properties and specificity at multiple levels (Ahmed *et al.*, 2006).

1.2. Mechanism of glycosylation

GTs can be categorized into two main subgroups, inverting and retaining enzymes, since they form glycosidic bonds by either inverting or retaining the configuration at the anomeric centre. According to the proposed mechanism, in inverting GTs general a base initiates catalysis by abstracting a proton from the reactive aglycone nucleophile. This is followed by a direct nucleophilic attack, which occurs at the donor sugar C1 carbon centre (Unligil and Rini, 2000). In principle, the mechanism of retaining GTs is proposed to proceed via a double displacement mechanism involving the formation of a glycosyl–enzyme complex.

1.3. GT classification

Given the specificity and functional diversity of the products of glycosyl transfer, combined with the often distant evolutionary relationships between GTs, it is not surprising that sequence homologies between GTs are low. What is surprising is that the majority of GTs belong to only two structural superfamilies, implying that nature has come up with only a few solutions to the ubiquitous problem of how to catalyze glycosyl transfer. The two main superfamilies are termed GT-A and GT-B, which each have an N-terminal and a C-terminal domain. The enzymes of the GT-A fold have two dissimilar domains. The N-terminal domain, consisting of several β-strands each flanked by α-helices (Rossmann folds), recognizes the sugar-nucleotide donor. The C-terminal domain, which consists largely of mixed β-sheets, contains the acceptor-binding site. Enzymes of the GT-B fold contain two similar Rossmann-like folds. The N-terminal domain provides the acceptor-binding site, whereas the C-terminal domain is responsible for binding the donor sugar. In both types of fold, domains are connected by a linker region and the active site is located between the two domains.

Two additional superfamilies have been described. The GT-C superfamily mainly represents integral membrane GTs and the GT-D superfamily includes GTs which do not fit into the other families. (Hu and Walker, 2002; Unligil and Rini, 2000). A second classification system named Carbohydrate-active enzymes (CAZy) is mainly based on sequence similarities. (Campbell et al., 1997; Coutinho et al., 2003). Currently it contains 91 different families (see http://afmb.cnrs-mrs.fr/CAZY). GTs described in this review belong to the GT-1 family.

1.3.1. Characteristics of the GT-B superfamily

The GT-B superfamily includes around 200 genes putatively encoding GTs involved in natural product biosynthesis. The GT-B fold was originally found in phage T4 DNA glycosyltransferase (Morera et al., 1999). The two N- and C-terminal domains are connected via a linker and the reaction site is located between the two domains (Mulichak *et al.*, 2001, 2003, 2004). Inter-domain salt bridges (Mulichak *et al.*, 2001, 2003) stabilize the catalytically active conformation (Hu *et al.*, 2003). Experimental evidence for the involvement of the N-terminal part in determining sugar specificity comes from studies on the GTs UrdGT1b and UrdGT1c. Changes in the N-terminal part led to GTs with novel specificities (Hoffmeister *et al.*, 2001, 2002). A hydrophobic patch on the surface of the N-terminal domain is proposed to be the binding site of the acceptor substrate.

1.3.2. GT-B X-ray determined protein structures
The structures of several GTs belonging to the GT-B superfamily, MurG (Hu *et al.*, 2003), GtfA (Mulichak *et al.*, 2003), GtfB (Mulichak *et al.*, 2001), GtfD (Mulichak *et al.*, 2004), OleD and OleI (Bolam *et al.*, 2007), UrdGT2 (Mittler *et al.*, 2007), VinC (Nango *et al.*, 2008), and CalG3 (Zhang *et al.*, 2008), have been recently resolved. Of particular interest are the X-ray crystal structures of GtfA GtfD, OleD, and OleI as a complex with ligands as they provide direct information about the interactions between enzyme and substrate (Fig. 12.2).

1.4. Reactions catalyzed by GTs

In the so-called "classical sugar transfer" GTs catalyze the attachment of a sugar from a NDP-activated sugar donor to an acceptor molecule (Blanchard and Thorson, 2006; Rupprath *et al.*, 2005). Alternatively GTs catalyze a "reverse transfer", the deglycosylation and coupled formation of a NDP sugar from a glycosylated compound in the presence of NDP (Minami *et al.*, 2005; Zhang *et al.*, 2006b, 2007).

Several scientific groups are involved in glycodiversification of bioactive natural products, as described in several outstanding publications (Fu *et al.*, 2003, 2005; Griffith *et al.*, 2005; Langenhan *et al.*, 2005).

1.4.1. Classical sugar transfer
A standard protocol for performing the "classical sugar transfer", in which a sugar is transferred from an NDP-activated donor to an aglycone acceptor forming a glycosidic bond is given below.

1. Subclone the PCR-amplified GT-encoding sequence into a suitable expression vector; express the gene either in *E. coli* or in another suitable host (*Streptomyces lividans, Streptomyces albus*) as a His-tagged protein.
2. Purify the protein by nickel-affinity chromatography. Perform SDS-PAGE to monitor the expression rate (a yield of 30 mg l^{-1} enzyme is convenient).
3. Perform MALDI-MS analysis for verification.

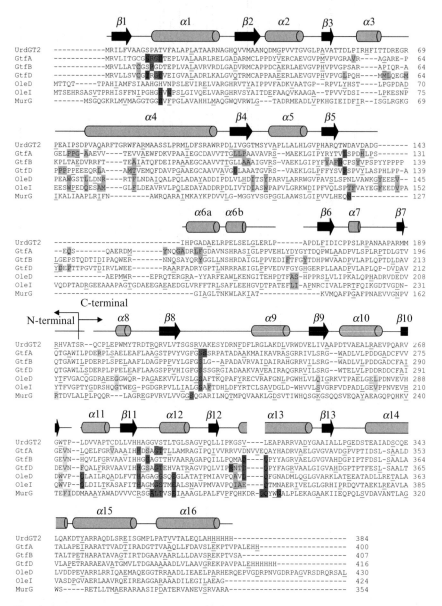

Figure 12.2 Alignment of the amino acid sequences of GT-B superfamily members. The predicted secondary structure of UrdGT2 (Prediction by PSIPRED Protein Structure Prediction Server http://bioinf.cs.ucl.ac.uk/psipred/psiform.html) is shown. The amino acids involved in substrate binding based on X-ray structure determination are highlighted in colors. Red: hexose-binding-site; pink: phosphat-binding-site; yellow: base-binding-site; green: aglycone acceptor-binding-site; turquoise: both hexose- and aglycone acceptor-binding-site; blue: proposed catalytic base. UrdGT2 is from *S. fradiae* Tü2717; GtfA, GtfB, and GtfD are from *Amycolatopsis orientalis*; OleD and OleI are from *S. antibioticus*; MurG is from *E. coli*. (See Color Insert.)

4. Use a donor sugar which can be synthesized chemically, semi- or fully-enzymatically, as described in several publications (Chen *et al.*, 2000; Rupprath *et al.*, 2005; Zhang *et al.*, 2008). Use an acceptor molecule, usually prepared by hydrolysis of available glycosides or isolated from a mutant unable to perform the glycosylation step.
5. Combine 50 μM aglycone, 300 μM NDP sugar, and 10 μM purified enzyme in a total volume of 100 μl Tris–HCl buffer (10 mM, pH 7.5) containing 1 mM MgCl$_2$.
6. Perform the incubation (30 °C or 37 °C and 60 min to 3 h are recommended).
7. Perform chemical analysis of your sample (e.g., thin-layer-chromatography (TLC), high pressure liquid chromatography (HPLC), or HPLC coupled to a mass spectrometry (HPLC-MS).
8. Perform NMR analysis to confirm the structure of the glycosylated compound.

1.4.2. Sugar transfer including the reverse reaction

While "classical sugar transfer" has been known for many years, "sugar transfer including the reverse reaction" was described in detail for the first time in 2005. In the last 3 years several GTs have been shown to perform this type of sugar transfer (Table 12.1). A protocol for performing the "reverse sugar transfer" is given below. Applied in an appropriate way the reversibility of GTs can be used for NDP sugar synthesis through sugar and/or aglycone exchange strategies in a one-pot reaction (Minami *et al.*, 2005; Zhang *et al.*, 2006a, 2007).

1. Handle recombinant expression and purification of the reversible GT as described in 1.4.1.
2. Combine 10 μM purified enzyme, 100 μM glycoside, and 2 mM NDP in a total volume of 100 μl 10 mM Tris–HCl buffer containing 1 mM MgCl$_2$ (pH = 7.5).
3. Perform the incubation (30 °C or 37 °C and 3 h are recommended).
4. Perform analysis as described in 1.4.1.

1.5. Flexible GTs

In order to create libraries of natural products, GTs flexible towards the substrates are required. Studies on the function and specificity of GTs have been performed by several groups. A surprisingly high number of GTs shows flexibility towards the sugar donor and/or the acceptor molecule (Table 12.2, Fig. 12.3).

Table 12.1 GTs shown to catalyze a reverse reaction

Name of reversible glycosyltransferase	Function	Biosynthetic pathway (Strain)	Reference
VinC	L-vicenisaminyl-transferase	Vicenistatin (*Streptomyces halstedii* HC-34)	Minami et al. (2005)
CalG1	L-rhamnosyltransferase	Calicheamycin (*Micromonospora echinospora*)	Zhang et al. (2006b, 2008)
CalG3	D-glucosyltransferase		
CalG4	D-aminopentosyl-transferase		
GtfD	L-vancosaminyl-transferase	Vancomycin (*Amycolatopsis orientalis*)	Zhang et al. (2006b)
GtfE	D-glycosyl-transferase		
AveBI	L-oleandrosyl-transferase	Avermectin (*Streptomyces avermitilis* MA-4680)	Zhang et al. (2006a)
EryBV	L-mycarosyl-transferase	Erythromycin (*Saccharopolyspora erythraea*)	Zhang et al. (2007)

Table 12.2 Flexible GTs from actinomycete strains producing bioactive natural compounds

Glycosyltransferase (Reference)	Function	Substrate flexibility
AknN Lu et al. (2004)	L-2-desoxyfucosyltransferase	**dNDP activated:** 52, 54 **Acceptor:** rhodosaminyl aklavinone daunomycin, adriamycin, idarubicin
AraGT Luzhetskyy et al. (2007)	L-rhamnosyltransferase	**dNDP activated:** 4, 50, 61, 62 **Acceptor:** aranciamycin aglycone
AveBI Zhang et al. (2006a,b)	L-oleandrosyltransferase	**dTDP activated:** 13, 14, 16, 22, 23, 30, 32, 34, 44, 48, 59 **Acceptor:** avermectin and ivermectin derivatives
CalG1 Zhang et al. (2006a,b)	L-rhamnosyltransferase	**dTDP activated:** 11, 13, 17, 22, 23, 24, 30, 32, 44, 61, 62 **Acceptor:** calcheamycin derivatives
CalG3 Zhang et al. (2008)	4,6-dideoxy-4-hydroxylamino-α-D-glycosyltransferase	**all five dNDP activated:** 11 **dTDP activated:** 13, 14, 15, 16, 18, 23, 30, 32, 48 **Acceptor:** calcheamycin derivatives
CalG4 Zhang et al. (2006a,b)	D-aminopentosyltransferase	**dTDP activated:** aminopentosyl derivatives **Acceptor:** calcheamycin derivatives
DesVII Borisova et al. (2004)	D-desosaminyltransferase	**dTDP activated:** 6, 44 **Acceptor:** 12-, 14- and 16 membered macrolactone derivatives
ElmGT Blanco et al.(2001)	L-rhamnosyltransferase	**dTDP activated:** 4, 5, 7, 11, 35, 36, 42, 43, 50, 52, 54, 58, 59, 60, 62, 63, D-diolivosyl, **Acceptor:** elloramycin aglycone

(continued)

Table 12.2 (continued)

Glycosyltransferase (Reference)	Function	Substrate flexibility
EryBV Zhang et al. (2007)	L-mycarosyltransferase	**dTDP activated:** 57, 58, 62, 59, 11, 48, 44, 30, 22, 23, 24, 13, 14, 18 **Acceptor:** erythromycin derivatives
GtfA Oberthur et al., (2005)	4-*epi*-vancosaminyltransferase	**dTDP activated:** 66, 49, 56 **Acceptor:** vancomycin aglycone
GtfC Oberthur et al. (2005)	4-*epi*-vancosaminyltransferase	**dTDP activated:** 65, 66, 53, 49, 60, 64, 56 **Acceptor:** chloroeremomycin derivatives
GtfD Zhang et al. (2006a,b)	L-vancosaminyltransferase	**dTDP activated:** 65, 19, 66, 53, 49, 60, 64, 56 **Acceptor:** monoglycosylated vancomycin and teicoplanin derivatives
GtfE Zhang et al. (2006a,b)	D-glucosyltransferase	**dTDP activated:** 11, 46, 40, 48, 3, 10, 37, 8, 1, 9, 44, 28, 29, 30, 31, 27, 22, 33, 23, 34, 24, 16, 21, 13, 14, 15, 19, 18, 17, 32, 20, 26, 25 **Acceptor:** vancomycin and teicoplanin derivatives
MycB Melancon III et al. (2004)	D-desosaminyltransferase	**dTDP activated:** 6 **Acceptor:** diverse 14- and 16-membered ring aglycone
OleD Yang et al. (2005)	D-glycosyltransferase	**dUDP activated:** 11, 48, 8, 9, 10, 44, 30, 27, 23, 16, 13, 14, 37, 18, 17, 32 **Acceptor:** oleandomycin, erythromycin, tylosin, carbomycin, baicalein, umbelliferone, 4-methyl-umbelliferone, 3,4-dichloroaniline, luteolin, fisetin, kaempferol, scopoletin, esculetin, 7-hydroxycoumerin-3-carboxylic acid, novobiocic acid, daidzein, genistein

OleG2 Doumith et al. (1999)	L-oleandrosyltransferase	**dTDP activated:** 59, 62, 58 **Acceptor:** oleandomycin aglycon erythronolide B, 6-deoxyerythronolide B
OleI Yang et al. (2005)	D-glycosyltransferase	**dUDP activated:** 11, 8, 37 **Acceptor:** oleandomycin, baicalein, umbelliferone, 4-methyl-umbelliferone, 3,4-dichloroaniline, luteolin, fisetin, kaempferol, scopoletin
RebG (N-GT) Mendez et al. (2008)	D-glucoslytransferase	**dNDP activated:** 11, 62, 60, 54, 43 **Acceptor:** staurosporine aglycon, carbazole, indol-[2,3-a]-carbazole, EJG-III-108 A
StaG (N-GT) Salas et al. (2005)	L-ristosaminyltransferase	**dNDP activated:** 64, 62, 60, 54, 43 **Acceptor:** staurosporine aglycone
TylM2 Melancon III et al. (2004)	D-mycaminosyltransferase	**dTDP activated:** 41, 6 **Acceptor:** tylosin aglycone
UrdGT2 (C-GT) Durr et al. (2004)	D-olivosyltransferase	**dNDP activated:** 43, 42, 45, 63 **Acceptor:** alizarin, 1,2-dihydroxyanthraquinon, premithramycinone
VinC Minami and Eguchi (2007)	D-vicenisaminyltransferase	**dTDP activated:** 47, 42, 7, 43, 12, 58 **UDP and ADP activated:** 47 **Acceptor:** brefeldin A, α and β-zeararenol, β-estradiol, pregnenolone, neovicenilactam

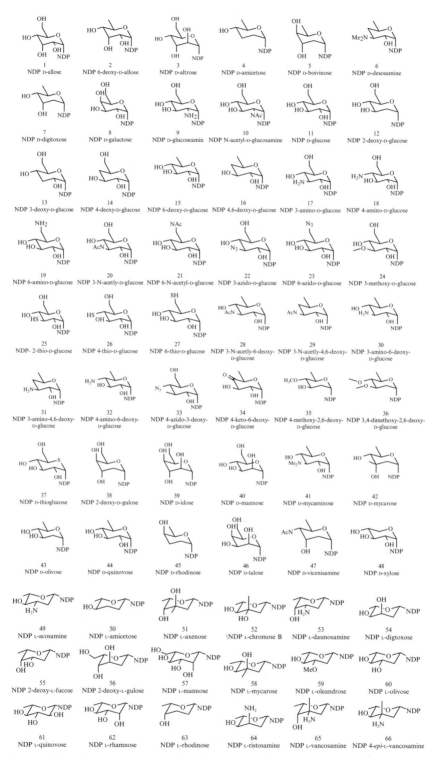

Figure 12.3 Structures of the NDP-activated deoxysugars shown in Table 12.2.

1.6. Bi-functional GTs

A few GTs—(LanGT1, LanGT4 (Luzhetskyy *et al.*, 2005b), AknK (Lu *et al.*, 2004), CmmGIV (Menéndez *et al.*, 2004), and MtmGIV (Nur-e-Alam *et al.*, 2005)—involved in natural product biosynthesis act twice (iterative function), and so belong to a special type of GT (bi-functional GT) (Fig. 12.4).

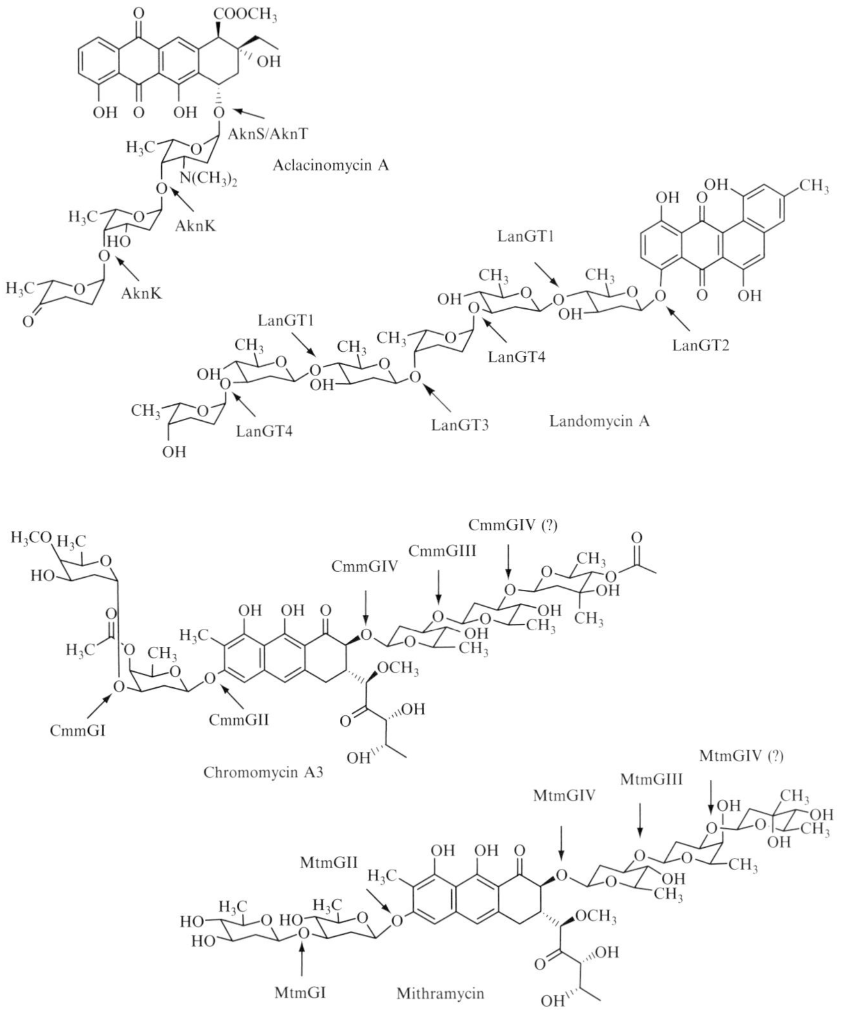

Figure 12.4 Bifunctional GTs. LanGT1 and LanGT4 are involved in landomycin biosynthesis, AknK is involved in aclacinomycin A biosynthesis, CmmGIV is involved in chromomycin A3 biosynthesis, and MtmGIV is involved in mithramycin biosynthesis.

1.7. GTs dependent on auxiliary proteins

A few GTs require an auxiliary activator protein for full activity. Examples are listed in Table 12.3. The exact function of these additional proteins remains unknown.

2. APPLICATION OF GTs IN PRODUCING UNNATURAL BIOACTIVE MOLECULES

This section summarizes methods for cloning GT genes and protocols used to generate novel unnatural natural products.

2.1. Detection of new GTs

The cloning of GT genes presents special problems as GTs do not share strongly conserved amino acid regions. To overcome this, Luzhetskyy and coworkers used the CODEHOP (COnsensus-DEgenerate Hybrid Oligo-nucleotide Primer) strategy (Rose et al., 2003) and succeeded in cloning novel so far unidentified GTs (Luzhetskyy et al., 2007). A protocol for cloning GT genes is given below.

1. Align several GTs to determine two conserved regions with ~3–4 amino acids (Fig. 12.5A). Use these amino acid blocks to design oligonucleotide primers according to the CODEHOP computer programme (http//www.bioinformatics.weizmann.ac.il/blocks/codehop.html). The 3′ degenerated core should contain all codon possibilities for the targeted motif. An identical 5′ consensus clamp (Fig. 12.5B) represents the conserved amino acids flanking the target region.
2. Use as template 500 ng of cosmid DNA preselected by hybridization with polyketide or sugar biosynthesis genes (the use of chromosomal DNA is possible but less reliable).
3. Perform PCR under the following conditions: 35 cycles of denaturing at 94 °C, annealing at 45 °C for 45 s and extending at 72 °C for 1 min. Once the primer is incorporated it becomes the template for subsequent amplifications (Fig. 12.5C and D).
4. Perform preparative agarose gel electrophoresis to separate amplified GT DNA from unspecific PCR products. Use a standard gel elution kit to elute the GT fragment with the expected size (Fig. 12.5E).
5. Ligate the restricted GT fragment into a suitable vector (e.g., clone HindIII and XbaI restricted insert into similarly treated pGEMT.easy vector) and sequence the fragment (Fig. 12.5F).
6. Design primers of the ends of the GT fragment and identify the full-length GT gene by primer walking. As template use the original cosmid DNA (Fig. 12.5G).

Table 12.3 GTs dependent on auxiliary proteins

Name of glycosyltransferase and auxiliary protein	Function	Biosynthetic pathway (Strain)	Reference
DesVII and DesVIII	D-desosaminyl transferase	Pikromycin/Methymycin (*Streptomyces venezuelae*)	Borisova et al. (2004)
EryCIII and GroEl/ES chaperone complex	D-desosaminyl transferase	Erythromycin (*Saccharopolyspora erythrea*)	Lee et al. (2004)
AknS and AknT	L-2-deoxyfucosyl transferase	Aclacinomycin A3 (*Streptomyces galilaeus*)	Lu et al. (2005)
TylM2 and TylM3	D-mycaminosyl transferase	Tylosin (*S. fradiae*)	Melancon III et al. (2004)
MycB and MydC	D-desosaminyl transferase	Mycinamicin (*Micromonospora griseorubida*)	Melancon III et al. (2004)

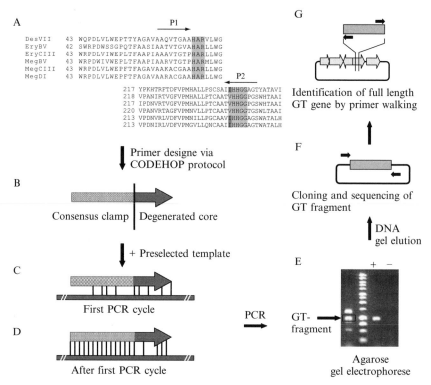

Figure 12.5 Strategy for PCR-based GT screening. (A) Alignment of several GT amino acid sequences to determine the conserved region necessary for the CODHOP primers. Blocks in grey represent the region for the degenerate parts of the CODEHOP primers. (B) Schematic presentation of the 2-part CODEHOP primers. (C) Annealing of the primer to the template. (D) Annealing of the primer to the product. (E) Preparative agarose gel electrophoresis. (F) Cloning and sequencing of the gel eluted GT DNA. (G) Identification of the full length GT gene by primer walking.

2.2. Strategies for altering glycosylation to generate novel bioactive natural products

2.2.1. Metabolic engineering and combinatorial biosynthesis

GTs have been used for the generation of novel compounds either in *in vitro* or in *in vivo* experiments. Several different technologies have been developed to engineer natural product glycosylation (Blanchard and Thorson, 2006; Luzhetskyy and Bechthold, 2005a) (Fig. 12.6).

2.2.1.1. Gene inactivation The generation of mutants with deletions in glycosyltransferase genes is a very efficient way to generate new drug candidates. This has been shown in several examples (Hofmann *et al.*,

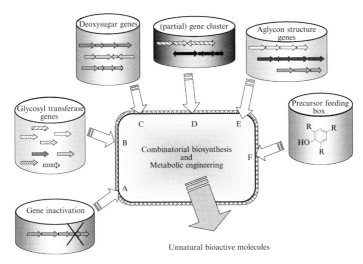

Figure 12.6 Different strategies for the generation of novel drugs (see text for explanation).

2005; Luzhetskyy et al., 2005c; Trefzer et al., 2000) and has been reviewed recently (Mendez et al., 2008) (Fig. 12.6A).

2.2.1.2. Heterologous expression of a GT gene Expression of glycosyltransferase genes is also an interesting way to create novel compounds. One example describes the expression of *lanGT4* in *Streptomyces fradiae* Tü2717. The recombinant strain elongates a natural product saccharide chain beyond the wild-type length (Hoffmeister et al., 2004) (Fig. 12.6B).

2.2.1.3. Over-expression of a GT gene Gene over-expression is a powerful method for unbalancing biosynthetic pathways in order to generate new metabolites. This is exemplified by the over-expression of the D-olivosyltransferase *lanGT3* in a mutant of the landomycin producer *Streptomyces cyanogenus* S136 and the generation of two novel compounds landomycin I and J (Zhu et al., 2007) (Fig. 12.6B).

2.2.1.4. Expression of parts of an entire gene cluster Expression of gene clusters (or parts of a gene cluster (Fig. 12.6D and E) is also a very common way to generate new drug candidates. One of the most famous example was presented with cosmid 16F4 containing all genes necessary for the biosynthesis of the polyketide moiety of elloramycin. When this cosmid was expressed together with different plasmids encoding sugar biosynthetic genes, due to the flexibility of ElmGT, novel glycosylated compounds were detected. (Blanco *et al.*, 2001; Fischer *et al.*, 2002; Lombo *et al.*, 2004; Perez *et al.*, 2005, 2006; Rodriguez *et al.*, 2002).

2.2.1.5. Expression of a GT gene in a GT mutant The C-GT UrdGT2, responsible for the attachment of the first D-olivose moiety during urdamycin A biosynthesis, shows substrate flexibility (Durr *et al.*, 2004). Heterologous expression of *urdGT2* from *S. fradiae* Tü2717 in *S. cyanogenus* Δ*lanGT2* resulted in the production of a novel C-glycosylated angucycline (Luzhetskyy *et al.*, 2005c).

One of the most encouraging examples of combinatorial biosynthesis was achieved by coexpressing *urdGT2* from the urdamycin producer *S. fradiae* Tü2717 and *lanGT1* from the landomycin producer *S. cyanogenus* S136 in a *mtmGIV* mutant of *Steptomyces argilaceus*. This resulted in the production of a new C-glycosylated di-olivosyl-premithramycinone (Trefzer *et al.*, 2002) (Fig. 12.6A and B).

2.2.1.6. Biotransformation This procedure is a very convenient method to glycosylate natural compounds. One example has been described by Zhan and Gunatilaka, who used *Beauveria bassiana* ATCC 7159, which was able to glycosylate commercially available hydroxy- and aminoanthraquinones (Zhan and Gunatilaka, 2006) (Fig. 12.6F).

2.2.1.7. Biotransformation using genetically engineered strains Strains harbouring the glycosyltransferase genes *elmGT* or *rbmA* were extensively used for biotransformation studies (Blanco *et al.*, 2001; Ohuchi *et al.*, 2000; Sánchez *et al.*, 2002). More details are given in Chapter 11 of this volume (Fig. 12.6B, C, and F).

2.2.1.8. Coexpression of deoxysugar-biosynthetic genes, a glycosyltransferase gene and genes responsible for the aglycone structure in one host organism Expression of two different plasmids in one strain can also result in the formation of novel compounds (Hutchinson, 1998; Tang and McDaniel, 2001). Salas and coworkers generated series of plasmids containing genes responsible for the formation of several deoxysugar. In combination with an additional plasmid containing genes responsible for aglycone formation and a flexible GT several compounds were successfully generated (Salas *et al.*, 2005). More details are given in Chapter 11 of this volume (Fig. 12.6B, C, and E).

2.2.2. Altering the substrate specificity of a GT

The specificity of many GTs provides a limiting factor in natural product diversification (Albermann *et al.*, 2003; Zhang *et al.*, 2007). Genetic engineering of the enzyme regions responsible for selecting sugar and acceptor substrates can provide GTs with novel specificity. Despite the wealth of GT structural and biochemical information (Hu and Walker, 2002), attempts to alter GT-donor/acceptor specificity via rational engineering is not trivial and is till yet mainly limited to sequence-guided single-site mutagenesis (Hancock *et al.*, 2006). In 2006, Aharoni *et al.* presented the application of a high throughput screening method for the directed evolution of the sialyltransferase CstII (GT-A superfamily member). With this approach he invented the *in vivo* selection of GTs (Aharoni *et al.*, 2006). Analogously Thorson *et al.* achieved this for the GT-B member OleD (Williams *et al.*, 2007). Hot spots which represent the protein region that governs enzyme's functionality were identified. Secondary mutagenesis was performed to provide GTs with improved novel activity (Williams *et al.*, 2008).

2.2.2.1. Determination of a GT's protein region responsible for substrate specificity and enzymatic activity Hoffmeister and coworkers presented a possibility to determine sequence elements responsible for substrate specificity by generating chimeric GTs from the GTs of interest.

Ten chimeric urdGT1b and urdGT1c genes were constructed and functionally investigated in a suitable host strain. With this strategy a 31-amino acid region located close to the N-terminus of these enzymes was deduced to control both sugar donor and acceptor substrate specificity (Hoffmeister *et al.*, 2001).

2.2.2.2. Mutagenesis methods for altering GT's substrate specificity Methods to mutate genes are available. Different approaches have been performed to alter the function of GT genes in Actinomycetes.

2.2.2.2.1. Mutagenesis via PCR-site-directed mutagenesis Hoffmeister *et al.* genetically engineered protein regions of UrdGT1b and UrdGT1c which are responsible for nucleotide sugar and acceptor substrate specificity. One modified GT produced a branched saccharide side-chain attached to the urdamycin aglycone (Hoffmeister *et al.*, 2002).

2.2.2.2.2. Mutagenesis via error prone PCR Williams *et al.* generated an OleD library by error-prone PCR using the wild-type OleD gene as template (Williams *et al.*, 2007). Activity of the generated OleD variants was screened using a fluorescence-based GT assay. Some artificial GTs showed improved activity in comparison to the wild-type OleD. Functional mutations were then combined via site-directed mutagenesis creating one triple mutant P67T/S132F/A242V which exhibited a 30-fold higher specific activity than wild type OleD. Additionally the evolved triple mutant had a broader substrate flexibility towards sugar nucleotide donors.

2.2.2.2.3. Mutagenesis via domain swapping LanGT1 and LndGT1 are two GTs involved in the biosynthesis of landomycins produced by different strains. The two enzymes show differences in substrate specificity. Bechthold and coworkers produced two chimere genes, hybrids h1 and h2, consisting of about half the gene of *lanGT1* connected to the other half of *lndGT1* and vice versa. Expression of these genes indicated that both enzymes were active and that the specificity determining region is located in the N-terminal domain (Krauth *et al.*, 2008 submitted for publication).

2.2.2.2.4. Mutagenesis via saturation mutagenesis As indicated above, an OleD sequence variant has been generated with improved activity in comparison to the wild type OleD (Williams *et al.*, 2007). This mutant also displayed an improvement in glycosylation of the novobiocin aglycone. To further optimize the catalytic efficiency the selected amino acids of the generated triple mutant were altered by single-site saturation mutagenesis resulting in a 150-fold higher activity in comparison to the wild-type OleD (Williams *et al.*, 2008).

2.3. Conclusion and prospects

GTs have enormous potential for the synthesis of pharmaceutically relevant carbohydrate structures. "Glyco-technologies" have been used for the generation of novel glycosylated compounds either in *in vitro* or *in vivo* experiments. Several GTs are sufficiently suitable for altering glycosylation patterns, but strict substrate specificity remains a limiting factor in natural-product diversification. Therefore engineering of GTs is the most promising way to find and develop GTs with clearly defined specificities. The requirement of widely applicable rapid selection or screening strategies has to be covered. Suitable high-throughput screening systems will support the glycodiversification technology. Further structure elucidation of GTs will help understanding the mode of action of these enzymes. Altering the specificity of GTs and improving the new "glyco-technologies" towards higher product yields will be important issues for scientists in the next decade.

REFERENCES

Aharoni, A., Thieme, K., Chiu, C. P., Buchini, S., Lairson, L. L., Chen, H., Strynadka, N. C., Wakarchuk, W. W., and Withers, S. G. (2006). High-throughput screening methodology for the directed evolution of glycosyltransferases. *Nat. Methods* **3**, 609–614.

Ahmed, A., Peters, N. R., Fitzgerald, M. K., Watson, J. A., Jr., Hoffmann, F. M., and Thorson, J. S. (2006). Colchicine glycorandomization influences cytotoxicity and mechanism of action. *J. Am. Chem. Soc.* **128**, 14224–14225.

Albermann, C., Soriano, A., Jiang, J., Vollmer, H., Biggins, J. B., Barton, W. A., Lesniak, J., Nikolov, D. B., and Thorson, J. S. (2003). Substrate specificity of novm: Implications for novobiocin biosynthesis and glycorandomization. *Org. Lett.* **5,** 933–936.

Blanchard, S., and Thorson, J. S. (2006). Enzymatic tools for engineering natural product glycosylation. *Curr. Opin. Chem. Biol.* **10,** 263–271.

Blanco, G., Patallo, E. P., Brana, A. F., Trefzer, A., Bechthold, A., Rohr, J., Mendez, C., and Salas, J. A. (2001). Identification of a sugar flexible glycosyltransferase from *Streptomyces Olivaceus*, the producer of the antitumor polyketide elloramycin. *Chem. Biol.* **8,** 253–263.

Bolam, D. N., Roberts, S., Proctor, M. R., Turkenburg, J. P., Dodson, E. J., Martinez-Fleites, C., Yang, M., Davis, B. G., Davies, G. J., and Gilbert, H. J. (2007). The crystal structure of two macrolide glycosyltransferases provides a blueprint for host cell antibiotic immunity. *Proc. Natl. Acad. Sci. USA* **104,** 5336–5341.

Borisova, S. A., Zhao, L., Melancon, C. E., III, Kao, C. L., and Liu, H. W. (2004). Characterization of the glycosyltransferase activity of DesVII: Analysis of and implications for the biosynthesis of macrolide antibiotics. *J. Am. Chem. Soc.* **126,** 6534–6535.

Campbell, J. A., Davies, G. J., Bulone, V., and Henrissat, B. (1997). A classification of nucleotide-diphospho-sugar glycosyltransferases based on amino acid sequence similarities. *Biochem. J.* **326**(Pt 3), 929–939.

Chen, H., Thomas, M. G., Hubbard, B. K., Losey, H. C., Walsh, C. T., and Burkart, M. D. (2000). Deoxysugars in glycopeptide antibiotics: Enzymatic synthesis of TDP-L-epivancosamine in chloroeremomycin biosynthesis. *Proc. Natl. Acad. Sci. USA* **97,** 11942–11947.

Coutinho, P. M., Deleury, E., Davies, G. J., and Henrissat, B. (2003). An evolving hierarchical family classification for glycosyltransferases. *J. Mol. Biol.* **328,** 307–317.

Doumith, M., Legrand, R., Lang, C., Salas, J. A., and Raynal, M. C. (1999). Interspecies complementation in *saccharopolyspora erythraea*: Elucidation of the function of OleP1, OleG1 and OleG2 from the oleandomycin biosynthetic gene cluster of *streptomyces antibioticus* and generation of new erythromycin derivatives. *Mol. Microbiol.* **34,** 1039–1048.

Durr, C., Hoffmeister, D., Wohlert, S. E., Ichinose, K., Weber, M., von Mulert, U., Thorson, J. S., and Bechthold, A. (2004). The glycosyltransferase UrdGT2 catalyzes both C- and O-Glycosidic sugar transfers. *Angew. Chem. Int. Ed. Engl.* **43,** 2962–2965.

Fischer, C., Rodriguez, L., Patallo, E. P., Lipata, F., Brana, A. F., Mendez, C., Salas, J. A., and Rohr, J. (2002). Digitoxosyltetracenomycin C and glucosyltetracenomycin C, two novel elloramycin analogues obtained by exploring the sugar donor substrate specificity of glycosyltransferase ElmGT. *J. Nat. Prod.* **65,** 1685–1689.

Fu, X., Albermann, C., Jiang, J., Liao, J., Zhang, C., and Thorson, J. S. (2003). Antibiotic optimization via in vitro glycorandomization. *Nat. Biotechnol.* **21,** 1467–1469.

Fu, X., Albermann, C., Zhang, C., and Thorson, J. S. (2005). Diversifying vancomycin via chemoenzymatic strategies. *Org. Lett.* **7,** 1513–1515.

Griffith, B. R., Langenhan, J. M., and Thorson, J. S. (2005). 'Sweetening' natural products via glycorandomization. *Curr. Opin. Biotechnol.* **16,** 622–630.

Hancock, S. M., Vaughan, M. D., and Withers, S. G. (2006). Engineering of glycosidases and glycosyltransferases. *Curr. Opin. Chem. Biol.* **10,** 509–519.

Hoffmeister, D., Ichinose, K., and Bechthold, A. (2001). Two sequence elements of glycosyltransferases involved in urdamycin biosynthesis are responsible for substrate specificity and enzymatic activity. *Chem. Biol.* **8,** 557–567.

Hoffmeister, D., Ichinose, K., Domann, S., Faust, B., Trefzer, A., Drager, G., Kirschning, A., Fischer, C., Künzel, E., Bearden, D., *et al.* (2000). The NDP-sugar co-substrate concentration and the enzyme expression level influence the substrate specificity of glycosyltransferases: Cloning and characterization of deoxysugar biosynthetic genes of the urdamycin biosynthetic gene cluster. *Chem. Biol.* **7,** 821–831.

Hoffmeister, D., Weber, M., Drager, G., Ichinose, K., Durr, C., and Bechthold, A. (2004). Rational saccharide extension by using the natural product glycosyltransferase LanGT4. *Chembiochem* **5,** 369–371.

Hoffmeister, D., Wilkinson, B., Foster, G., Sidebottom, P. J., Ichinose, K., and Bechthold, A. (2002). Engineered urdamycin glycosyltransferases are broadened and altered in substrate specificity. *Chem. Biol.* **9,** 287–295.

Hofmann, C., Boll, R., Heitmann, B., Hauser, G., Durr, C., Frerich, A., Weitnauer, G., Glaser, S. J., and Bechthold, A. (2005). Genes encoding enzymes responsible for biosynthesis of L-lyxose and attachment of eurekanate during avilamycin biosynthesis. *Chem. Biol.* **12,** 1137–1143.

Hu, Y., Chen, L., Ha, S., Gross, B., Falcone, B., Walker, D., Mokhtarzadeh, M., and Walker, S. (2003). Crystal structure of the MurG:UDP-GlcNAc complex reveals common structural principles of a superfamily of glycosyltransferases. *Proc. Natl. Acad. Sci. USA* **100,** 845–849.

Hu, Y., and Walker, S. (2002). Remarkable structural similarities between diverse glycosyltransferases. *Chem. Biol.* **9,** 1287–1296.

Hutchinson, C. R. (1998). Combinatorial biosynthesis for new drug discovery. *Curr. Opin. Microbiol.* **1,** 319–329.

Krauth, C., Fedoryshyn, M., Schleberger, C., Luzhetskyy, A., and Bechthold, A. (2009). Engineering a function into a glycosyltransferase. *Chem. Biol.* **16,** 28–35.

Langenhan, J. M., Griffith, B. R., and Thorson, J. S. (2005). Neoglycorandomization and chemoenzymatic glycorandomization: Two complementary tools for natural product diversification. *J. Nat. Prod.* **68,** 1696–1711.

Lee, H. Y., Chung, H. S., Hang, C., Khosla, C., Walsh, C. T., Kahne, D., and Walker, S. (2004). Reconstitution and characterization of a new desosaminyl transferase, EryCIII, from the erythromycin biosynthetic pathway. *J. Am. Chem. Soc.* **126,** 9924–9925.

Liu, T., Kharel, M. K., Fischer, C., McCormick, A., and Rohr, J. (2006). Inactivation of GilGT, encoding a C-glycosyltransferase, and GilOIII, encoding a P450 enzyme, allows the details of the late biosynthetic pathway to gilvocarcin V to be delineated. *Chembiochem* **7,** 1070–1077.

Lombo, F., Gibson, M., Greenwell, L., Braña, A. F., Rohr, J., Salas, J. A., and Méndez, C. (2004). Engineering biosynthetic pathways for deoxysugars: Branched-chain sugar pathways and derivatives from the antitumor tetracenomycin. *Chem. Biol.* **11,** 1709–1718.

Lu, W., Leimkuhler, C., Oberthur, M., Kahne, D., and Walsh, C. T. (2004). AknK is an L-2-deoxyfucosyltransferase in the biosynthesis of the anthracycline aclacinomycin A. *Biochemistry* **43,** 4548–4558.

Lu, W., Leimkuhler, C., Gatto, G. J., Jr., Kruger, R. G., Oberthur, M., Kahne, D., and Walsh, C. T. (2005). AknT is an activating protein for the glycosyltransferase AknS in L-Aminodeoxysugar transfer to the aglycone of Aclacinomycin A. *Chem. Biol.* **12,** 527–534.

Luzhetskyy, A., and Bechthold, A. (2005a). It works: Combinatorial biosynthesis for generating novel glycosylated compounds. *Mol. Microbiol.* **58,** 3–5.

Luzhetskyy, A., Fedoryshyn, M., Durr, C., Taguchi, T., Novikov, V., and Bechthold, A. (2005b). Iteratively acting glycosyltransferases involved in the hexasaccharide biosynthesis of landomycin A. *Chem. Biol.* **12,** 725–729.

Luzhetskyy, A., Mendez, C., Salas, J. A., and Bechthold, A. (2008). Glycosyltransferases, important tools for drug design. *Curr. Top Med. Chem.* **8,** 680–709.

Luzhetskyy, A., Taguchi, T., Fedoryshyn, M., Durr, C., Wohlert, S. E., Novikov, V., and Bechthold, A. (2005c). LanGT2 catalyzes the first glycosylation step during landomycin a biosynthesis. *Chembiochem* **6,** 1406–1410.

Luzhetskyy, A., Weiss, H., Charge, A., Welle, E., Linnenbrink, A., Vente, A., and Bechthold, A. (2007). A strategy for cloning glycosyltransferase genes involved in natural product biosynthesis. *Appl. Microbiol. Biotechnol.* **75,** 1367–1375.

Melancon, C. E., III, Takahashi, H., and Liu, H. W. (2004). Characterization of TylM3/TylM2 and MydC/MycB pairs required for efficient glycosyltransfer in macrolide antibiotic biosynthesis. *J. Am. Chem. Soc.* **126,** 16726–16727.

Mendez, C., Luzhetskyy, A., Bechthold, A., and Salas, J. A. (2008). Deoxysugars in bioactive natural products: Development of novel derivatives by altering the sugar pattern. *Curr. Top Med. Chem.* **8,** 710–724.

Mendez, C., and Salas, J. A. (2001). Altering the glycosylation pattern of bioactive compounds. *Trends Biotechnol.* **19,** 449–456.

Menéndez, N., Nur-e-Alam, A. F., Braña, A. F., Rohr, J., Salas, A. F., and Méndez, C. (2004). Biosynthesis of the antitumor chromomycin A(3) in *Streptomyces Griseus*: analysis of the gene cluster and rational design of novel chromomycin analogs. *Chem. Biol.* **11,** 21–32.

Minami, A., Uchida, R., Eguchi, T., and Kakinuma, K. (2005). Enzymatic approach to unnatural glycosides with diverse aglycon scaffolds using glycosyltransferase VinC. *J. Am. Chem. Soc.* **127,** 6148–6149.

Minami, A., and Eguchi, T. (2007). Substrate flexibility of vicenisaminyltransferase VinC involved in the biosynthesis of vicenistatin. *J. Am. Chem. Soc.* **129,** 5102–5107.

Mittler, M., Bechthold, A., and Schulz, G. E. (2007). Structure and action of the C-C bond-forming glycosyltransferase UrdGT2 involved in the biosynthesis of the antibiotic urdamycin. *J. Mol. Biol.* **372,** 67–76.

Morera, S., Imberty, A., Aschke-Sonnenborn, U., Ruger, W., and Freemont, P. S. (1999). T4 phage beta-glucosyltransferase: Substrate binding and proposed catalytic mechanism. *J. Mol. Biol.* **292,** 717–730.

Mulichak, A. M., Losey, H. C., Lu, W., Wawrzak, Z., Walsh, C. T., and Garavito, R. M. (2003). Structure of the TDP-Epi-Vancosaminyltransferase GtfA from the chloroeremomycin biosynthetic pathway. *Proc. Natl. Acad. Sci. USA* **100,** 9238–9243.

Mulichak, A. M., Losey, H. C., Walsh, C. T., and Garavito, R. M. (2001). Structure of the UDP-glucosyltransferase GtfB that modifies the heptapeptide aglycone in the biosynthesis of vancomycin group antibiotics. *Structure* **9,** 547–557.

Mulichak, A. M., Lu, W., Losey, H. C., Walsh, C. T., and Garavito, R. M. (2004). Crystal structure of vancosaminyltransferase GtfD from the vancomycin biosynthetic pathway: Interactions with acceptor and nucleotide ligands. *Biochemistry* **43,** 5170–5180.

Nango, E., Minami, A., Kumasaka, T., and Eguchi, T. (2008). Crystallization and preliminary X-ray analysis of vicenisaminyltransferase VinC. *Acta Crystallogr. Sect F Struct. Biol. Cryst. Commun.* **64,** 558–560.

Newman, D. J., Cragg, G. M., and Snader, K. M. (2003). Natural products as sources of new drugs over the period 1981–2002. *J. Nat. Prod.* **66,** 1022–1037.

Nur-e-Alam, C., Mendez, C., Salas, J. A., and Rohr, J. (2005). Elucidation of the glycosylation sequence of mithramycin biosynthesis: Isolation of 3A-deolivosylpremithramycin B and its conversion to premithramycin B by glycosyltransferase MtmGII. *Chembiochem* **6,** 632–636.

Oberthur, M., Leimkuhler, C., Kruger, R. G., Lu, W., Walsh, C. T., and Kahne, D. (2005). A systematic investigation of the synthetic utility of glycopeptide glycosyltransferases. *J. Am. Chem. Soc.* **127,** 10747–10752.

Ohuchi, T., Ikeda-Araki, A., Watanabe-Sakamoto, A., Kojiri, K., Nagashima, M., Okanishi, M., and Suda, H. (2000). Cloning and expression of a gene encoding N-glycosyltransferase (Ngt) from Saccarothrix Aerocolonigenes ATCC39243. *J. Antibiot. (Tokyo)* **53,** 393–403.

Perez, M., Lombo, F., Baig, I., Brana, A. F., Rohr, J., Salas, J. A., and Mendez, C. (2006). Combinatorial biosynthesis of antitumor deoxysugar pathways in Streptomyces Griseus: Reconstitution of "unnatural natural gene clusters" for the biosynthesis of four 2,6-D-dideoxyhexoses. *Appl. Environ. Microbiol.* **72,** 6644–6652.

Perez, M., Lombo, F., Zhu, L., Gibson, M., Brana, A. F., Rohr, J., Salas, J. A., and Mendez, C. (2005). Combining sugar biosynthesis genes for the generation of L- and D-amicetose and formation of two novel antitumor tetracenomycins. *Chem. Commun. (Camb)* 1604–1606.

Rodriguez, L., Aguirrezabalaga, I., Allende, N., Brana, A. F., Mendez, C., and Salas, J. A. (2002). Engineering deoxysugar biosynthetic pathways from antibiotic-producing microorganisms. A tool to produce novel glycosylated bioactive compounds. *Chem. Biol.* **9,** 721–729.

Rose, T. M., Henikoff, J. G., and Henikoff, S. (2003). CODEHOP (COnsensus-DEgenerate Hybrid Oligonucleotide Primer) PCR primer design. *Nucleic Acids Res.* **31,** 3763–3766.

Rupprath, C., Schumacher, T., and Elling, L. (2005). Nucleotide deoxysugars: Essential tools for the glycosylation engineering of novel bioactive compounds. *Curr. Med. Chem.* **12,** 1637–1675.

Salas, A. P., Zhu, L., Sanchez, C., Brana, A. F., Rohr, J., Mendez, C., and Salas, J. A. (2005). Deciphering the late steps in the biosynthesis of the anti-tumour indolocarbazole staurosporine: Sugar donor substrate flexibility of the StaG glycosyltransferase. *Mol. Microbiol.* **58,** 17–27.

Sánchez, C., Butovich, I. A., Braña, A. F., Rohr, J., Méndez, C., and Salas, J. A. (2002). The biosynthetic gene cluster for the antitumor rebeccamycin. characterization and generation of indolocarbazole derivatives. *Chem. Biol.* **9,** 519–531.

Tang, L., and McDaniel, R. (2001). Construction of desosamine containing polyketide libraries using a glycosyltransferase with broad substrate specificity. *Chem. Biol.* **8,** 547–555.

Trefzer, A., Blanco, G., Remsing, L., Künzel, E., Rix, U., Lipata, F., Braña, A. F., Méndez, C., Rohr, J., Bechthold, A., and Salas, J. A. (2002). Rationally designed glycosylated premithramycins: Hybrid aromatic polyketides using genes from three different biosynthetic pathways. *J. Am. Chem. Soc.* **124,** 6056–6062.

Trefzer, A., Hoffmeister, D., Kunzel, E., Stockert, S., Weitnauer, G., Westrich, L., Rix, U., Fuchser, J., Bindseil, K. U., Rohr, J., and Bechthold, A. (2000). Function of glycosyltransferase genes involved in urdamycin A biosynthesis. *Chem. Biol.* **7,** 133–142.

Unligil, U. M., and Rini, J. M. (2000). Glycosyltransferase structure and mechanism. *Curr. Opin. Struct. Biol.* **10,** 510–517.

Vilches, C., Hernandez, C., Mendez, C., and Salas, J. A. (1992). Role of glycosylation and deglycosylation in biosynthesis of and resistance to oleandomycin in the producer organism, *streptomyces antibioticus*. *J. Bacteriol.* **174,** 161–165.

Weymouth-Wilson, A. C. (1997). The role of carbohydrates in biologically active natural products. *Nat. Prod. Rep.* **14,** 99–110.

Williams, G. J., Goff, R. D., Zhang, C., and Thorson, J. S. (2008). Optimizing glycosyltransferase specificity via "hot spot" saturation mutagenesis presents a catalyst for novobiocin glycorandomization. *Chem. Biol.* **15,** 393–401.

Williams, G. J., Zhang, C., and Thorson, J. S. (2007). Expanding the promiscuity of a natural-product glycosyltransferase by directed evolution. *Nat. Chem. Biol.* **3,** 657–662.

Yang, M., Brazier, M., Edwards, R., and Davis, B. G. (2005). High-throughput mass-spectrometry monitoring for multisubstrate enzymes: Determining the kinetic parameters and catalytic activities of glycosyltransferases. *Chembiochem* **6,** 346–357.

Zhan, J., and Gunatilaka, A. A. (2006). Microbial transformation of amino- and hydroxyanthraquinones by *Beauveria bassiana* ATCC 7159. *J. Nat. Prod.* **69,** 1525–1527.

Zhang, C., Albermann, C., Fu, X., and Thorson, J. S. (2006a). The *in vitro* characterization of the iterative avermectin glycosyltransferase AveBI reveals reaction reversibility and sugar nucleotide flexibility. *J. Am. Chem. Soc.* **128**, 16420–16421.

Zhang, C., Bitto, E., Goff, R. D., Singh, S., Bingman, C. A., Griffith, B. R., Albermann, C., Phillips, G. N., Jr., and Thorson, J. S. (2008). Biochemical and structural insights of the early glycosylation steps in calicheamicin biosynthesis. *Chem. Biol.* **15**, 842–853.

Zhang, C., Fu, Q., Albermann, C., Li, L., and Thorson, J. S. (2007). The *in vitro* characterization of the erythronolide mycarosyltransferase EryBV and its utility in macrolide diversification. *Chembiochem* **8**, 385–390.

Zhang, C., Griffith, B. R., Fu, Q., Albermann, C., Fu, X., Lee, I. K., Li, L., and Thorson, J. S. (2006b). Exploiting the reversibility of natural product glycosyltransferase-catalyzed reactions. *Science* **313**, 1291–1294.

Zhu, L., Luzhetskyy, A., Luzhetska, M., Mattingly, C., Adams, V., Bechthold, A., and Rohr, J. (2007). Generation of new landomycins with altered saccharide patterns through over-expression of the glycosyltransferase gene LanGT3 in the biosynthetic gene cluster of landomycin A in *Streptomyces cyanogenus* S-136. *Chembiochem* **8**, 83–88.

SECTION TWO

PEPTIDES

CHAPTER THIRTEEN

Nonribosomal Peptide Synthetases: Mechanistic and Structural Aspects of Essential Domains

M. A. Marahiel *and* L.-O. Essen

Contents

1. Introduction	338
2. Mechanistic and Structural Aspects of Essential NRPS Domains	339
3. Structural Insights into an Entire Termination Module	347
References	349

Abstract

A widespread class of therapeutically important natural products is of peptidic origin. They are produced nonribosomally by large "assembly line"-like multienzyme complexes, the nonribosomal peptide synthetases (NRPS). In contrast to ribosomal peptide synthesis, nonribosomally assembled peptides contain unique structural features such as D-amino acids, N-terminally attached fatty acid chains, *N*- and *C*-methylated amino acids, *N*-formylated residues, heterocyclic elements, glycosylated amino acids, and phosphorylated residues. In recent research using genetic, biochemical, and structural methods, experiments have revealed profound insights into the molecular mechanism of nonribosomal peptide synthesis. Based on this, it was possible to alter existing nonribosomally produced peptides either by changing their biosynthetic templates or by the combined action of chemical peptide synthesis and subsequent enzyme catalysis. An overview of the structural aspects of the NRPS machinery with a focus on mechanistic and structural aspects of essential domains is presented.

Biochemistry-Department of Chemistry, Philipps-University Marburg, Marburg, Germany

1. Introduction

Nonribosomal peptide synthetases (NRPS) are multimodular biocatalysts that mediate the synthesis of complex bioactive macrocyclic peptides independently of a nucleic acid-derived template (Finking and Marahiel, 2004; Grünewald and Marahiel, 2006). In addition to the 20 proteinogenic amino acids, nonribosomal peptides are built from a huge number of nonproteinogenic amino acids, which are often essential for bioactivity. To date more than 422 monomers are known to be assembled by NRPS enzymes (Caboche et al., 2008). Research in recent years has focused on understanding the structure–function relationships of some prototypal multimodular NRPSs as a prerequisite to alter their basic activities, for example, for the generation of new therapeutic molecules (Baltz, 2008; Baltz et al., 2005; Felnagle et al., 2008; Fischbach and Walsh, 2006). NRPS modules are the repetitive building blocks of these megasynthetases (Fig. 13.1), with each being responsible for the incorporation of a monomeric precursor into the peptidic product. Accordingly, the order and specificity of NRPS modules within a given biosynthetic cluster reflects the sequence of the nonribosomal peptide product. The NRPS modules can be further broken down into individual catalytic domains (Schwarzer et al., 2003; Sieber and Marahiel, 2005). Three domains are ubiquitous in each NRPS module catalyzing elongation of peptidic intermediates. One domain is for substrate recognition and activation as acyladenylate (A-domain). A second, the peptidyl-carrier-protein (PCP) domain, is the site of ppan (4′-phosphopantetheine) cofactor binding, to which all substrates and intermediates of the NRPS assembly line are covalently bound. A third, the condensation (C)

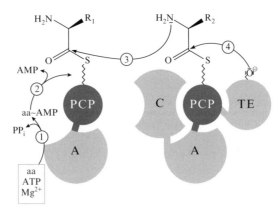

Figure 13.1 Overall mechanism of nonribosomal peptide elongation and release. The diagram depicts the sequential reactions catalyzed by individual NRPS domains during the elongation and termination reactions.

domain, finally catalyzes peptide bond formation between two adjacent PCP-bound intermediates. Therefore, an elongation module consists of three domains in the order C-A-PCP, in most cases on a single polypeptide chain. A fourth essential NRPS catalytic unit associated with product release is the thioesterase (TE) domain. In contrast to the other, repetitively occurring NRPS domains, TE domains are only present once within an NRPS assembly line, being attached to a termination module in the order C-A-PCP-TE. In addition, NRPS modules can comprise auxiliary domains to catalyze substrate modifications, such as epimerization at C_α atoms (E), *N*-methylation (M), *N*-formylation (F), or heterocyclization (Cy) of the peptidic backbone, utilizing serine and threonine side chains (Walsh, 2008). A detailed article describing all reactions catalyzed by mature and recombinant NRPS was recently published in *Methods in Enzymology* Vol. 388 (Linne and Marahiel, 2004). This article therefore focuses mainly on recent structural aspects of these multimodular enzymes.

2. Mechanistic and Structural Aspects of Essential NRPS Domains

In each NRPS module, the A-domains (~550 amino acids) act as gatekeepers for incoming monomeric building blocks. They catalyze substrate selection and activation as aminoacyl-*O*-AMP with ATP consumption. The crystal structures of three different A-domains activating 2,3-dihydroxy benzoate (DhbE), L-phenylalanine (PheA), and D-alanine (DltA) were determined, showing a highly conserved fold for all members of this adenylate-forming enzyme family (Fig. 13.2), albeit with low sequence identity (Conti *et al.*, 1997; May *et al.*, 2002; Yonus *et al.*, 2008). The fold is also highly conserved in other adenylate-forming enzymes, such as firefly luciferase from *Photinus pyralis* and acetyl–CoA synthetase (ACS) from *Salmonella enterica*. The structures show two distinct subdomains, a large and compact N-terminal subdomain (A_{core}, ~450 residues) and a small C-terminal subdomain (A_{sub}, ~100 residues), connected by a hinge region of 5–10 residues. The small C-terminal subdomain can adopt different orientations relative to the N-terminal large domain during the catalytic cycle: an open state in which the active site between the two subdomains endorses substrate (amino acid, ATP, Mg^{2+}) entry, followed by a closed state that promotes the two-step reaction involving substrate adenylation and the transfer of the aminoacyl moiety onto the ppan-arm of the downstream PCP domain. Combined biochemical, structural, and phylogenetic studies on a large number of A-domains revealed the so called "specificity-conferring code" of A-domains (Stachelhaus *et al.*, 1999). This code consists of 10 amino acid residues in the active site of A-domains that coordinate the aminoacyl adenylate substrate. In all A-domains activating α-amino acids, 2 of the 10

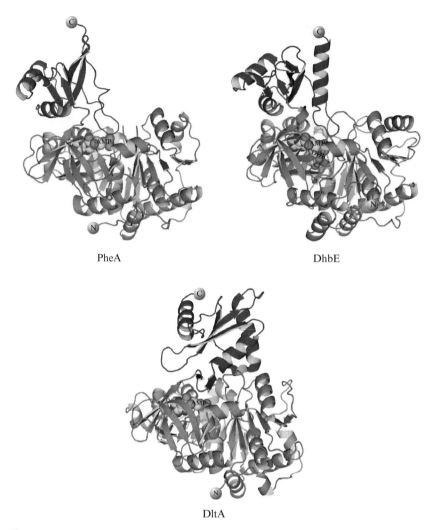

Figure 13.2 Domain organization observed for the adenylate forming enzymes activating L-phenylalanine (PheA), 2,3- dihydroxybenzoate (DhbE), and D-alanine (DltA), depicted in equivalent orientations for the large N-terminal (red) subdomain. The small C-terminal subdomains (blue) show different orientations during the catalytic cycle. The active sites are tightly clamped between the two subdomains and accommodate either ATP (PheA and DltA) or 2,3-dihydroxybenzoic-AMP (DhbE). (See Color Insert.)

residues are invariant, a lysine and an aspartate, whose side chains form salt bridges with the α-carboxyl and α-amino groups, respectively. Three other residues of the active site are in most cases hydrophobic, whereas the other five active site residues are not conserved and therefore control substrate specificity. In many cases, this discovery was helpful in predicting the substrate specificity of A-domains directly from the amino acid sequence without biochemical analysis (Rausch et al., 2005) and also led in several cases to the rational switch of their substrate specificity *in vitro* and *in vivo* (Eppelmann et al., 2001; Mootz et al., 2000, 2002; Stachelhaus et al., 1995). The evolution of the NRPS A-domain fold is independent of that of aminoacyl-tRNA-synthetases (class I and II) of the ribosomal system, although both synthetases utilize the same sequence of chemical reactions. However, A-domains lack a mechanism for proof-reading and some of them even have relaxed substrate specificities which direct in several systems the synthesis of different variants of a nonribosomal peptide by the same NRPS machinery.

The small peptidyl carrier protein (PCP) domain located downstream of the A-domain is the site of cofactor binding. Within each PCP domain ($\sim$80 residues), a highly conserved serine residue is posttranslationally modified with the cofactor 4′-phosphopantetheine (ppan) by the action of a cognate coenzyme A (CoA)-ppan-transferase (Reuter et al., 1999; Weber et al., 2000). To achieve functional NRPS assembly lines each PCP domain has to be posttranslationally modified with the ppan-arm, to which all substrates and peptidyl intermediates are attached as thioesters during synthesis (Finking et al., 2002; Mootz et al., 2001). All acyl-S-PCP intermediates must be delivered to the catalytic sites of the adjacent NRPS domains, for modification (e.g., epimerization), peptide bond formation, or product release. How such a simple four-helix bundle of the PCP domain recognizes in space and time the right partner domains is unknown. However, conformational changes within the PCP carrier domain, depending on its chemical modification state (apo, holo, or acylated), seem to be crucial for specific partner recognition. Recent NMR studies on the TycC3-PCP domain, a carrier domain dissected from module 7 of the tyrocidine biosynthetic cluster, in its apo- and holo-forms revealed distinct chemical shifts, suggesting the existence of three different and slowly interconverting conformations, the A (apo), H (holo), and A/H states (Koglin et al., 2006). For apo-PCP, the A and A/H conformations coexist and for the holo-PCP the presence of H and A/H conformations was identified (Fig. 13.3). The common conformer for both the apo- and the holo-forms is hence the A/H state. The conformational transitions can be frozen in the A-state by changing the active site serine in apo-PCP (cofactor binding site) to alanine or in the H-state by acylation of the ppan-cofactor in the holo-PCP. The most compact PCP conformation that resembles a classical four-α-helix bundle is present in the common A/H form. The apo-form (A-state) is less compact and has a large loop region between helix I and helix II, whereas these helices change their relative orientation in the H-state. By following the interactions of the amide protons and nitrogens of the cofactor in the H- and A/H-states, it was found that the cofactor relocates its position on the PCP surface by some $100°$ and as a consequence

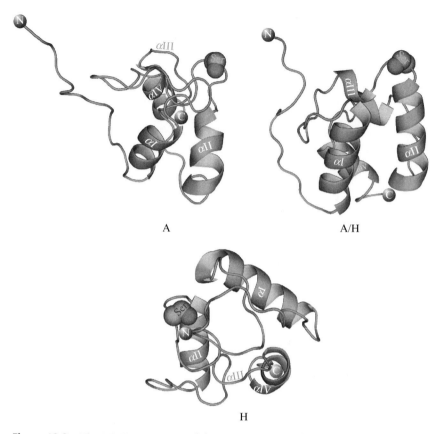

Figure 13.3 The NMR structures of the TycC$_3$-PCP conformers for apo- (A, A/H) and holo- (H, A/H) PCP. The location of the active site serine, to which the ppan cofactor in the holo-PCP form is attached, is shown in red. (See Color Insert.)

moves its terminal SH-group by 16 Å. This was the first evidence for a possible directed "swinging" movement of the loaded cofactor during NRPS peptide assembly. The biological significance of the directed movement was tested by following the chemical shifts during NMR titration of the A- and the H-states of PCP with two specific enzymes that were shown to interact with the apo- (A) and the holo (H)-forms. The CoA-ppan-transferase, Sfp, in the presence of CoA and Mg^{2+}, shows the strongest interaction with the A-state PCP, whereas the type II thioesterase (TEII), which regenerates the PCP domain by acting on mischarged ppan cofactors, selected the H-state (Koglin et al., 2008). These studies provide evidence that protein dynamics during NRPS assembly plays an important role in domain interaction. However, it remains to be established whether this multiple-state model for the TycC3-PCP domain can be generalized to other PCP domains. Structural work on PCP domains, which are part of either a PCP-C didomain (Samel et al., 2007) or a PCP-TE didomain (Frueh et al., 2008) and harboring Ser → Ala mutations at the ppan attachment site, were found to exist exclusively in the A/H state. Further, NMR titration studies between PCP and other *cis*-acting domains (such as A, C, TE, E, Cy) will hence be necessary to understand how such a simple fold can differentiate between different domains during nonribosomal peptide synthesis.

The condensation (C) domains (~450 residues) within elongation modules (C-A-PCP) are the sites of peptide bond formation and usually their number within an NRPS coincides with the number of peptide bonds in the final linear product. Biochemical and structural studies of C-domains revealed an acceptor and a donor site that accommodate the aminoacyl-S-PCP and peptidyl-S-PCP substrates, respectively (Belshaw et al., 1999). Peptide bond formation within the active site of the C-domain is believed to be initiated by nucleophilic attack of the α-amino group of the acceptor substrate onto the thioester group of the donor substrate, resulting in amide bond formation and the transfer of the entire peptide from the upstream donor site to the downstream acceptor site. The elongated peptide then serves as a donor substrate in a subsequent condensation step catalyzed by the C-domain of the next downstream module. All condensation reactions are strictly unidirectional, transporting the growing peptide chain towards the C-terminal module. Probing the substrate specificity of C-domains *in vitro* is difficult because the upstream donor and downstream acceptor substrates are defined by the restrictive A-domains. For PCP misloading studies, various synthetic aminoacyl–CoA substrates were attached to apo-PCPs of minimal dimodular NRPSs using the promiscuous ppan-transferase Sfp (Bruner et al., 2002). These mischarged carrier molecules were then tested in condensation assays, which revealed for the C-domains a high substrate specificity at the acceptor side and a relaxed one at the donor site. The substrate specificity at the acceptor site was shown to discriminate against the noncognate D-enantiomer as well as differences in the side chain of the substrate (Linne and Marahiel, 2000; Lusong et al., 2002). At the donor site, low side chain selectivity was observed for the PCP-bound electrophile substrates. These biochemical studies were supported at the molecular level by solving the crystal structure of the stand-alone C-domain VibH (from the vibriobactin biosynthetic gene cluster) and the structure of a PCP-C bidomain derived from modules 5 and 6 of the tyrocidine synthetase C (Keating et al., 2002; Samel et al., 2007). The overall architecture of both C-domains (Fig. 13.4) clearly shows that there is no classical deep cavity for substrate binding but rather a canyon-like structure in which the two PCP-bound substrates are positioned from both sides (acceptor and donor sites) of the canyon towards the highly conserved catalytic histidine of the HHxxxDG motif. Both condensation domains consist of two mainly

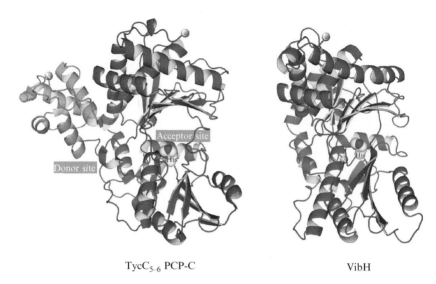

Figure 13.4 Structures of the C-domains in PCP-C and for the stand-alone VibH C-domain are shown in grey, whereas the PCP carrier domain located on the donor site of the C-domain in PCP-C is shown in green. PCP adopts an A/H form and its active-site serine (red) is located some 50 Å from the active-site histidine, located at the bottom of the C-domain canyon. (See Color Insert.)

separated and structurally similar subdomains, an N-terminal and a C-terminal subdomain, which are arranged in a V-shaped fashion. The subdomains belong to the well-known chloramphenicol acetyltransferase (CAT) fold. The two CAT-folds of the C-domains show only two major contact sites, thus building a canyon-like active site with the His-motif located on the middle of its floor. A model supported by mutational analysis suggests that the second histidine (His 224 in PCP-C) of this motive may act as a catalytic base by deprotonating the α-ammonium group of the acceptor substrate to restore its nucleophilicity to attack the carboxyl-thioester group of the donor substrate. However, recent mutational studies and pK value analysis of active site residues in PCP-C suggest that peptide bond formation may depend mainly on electrostatic interactions rather than on general acid/base catalysis (Samel et al., 2007). In this PCP-C didomain structure, the PCP domain attached to the donor site of the C-domain shows a compact four-helix bundle like that of the A/H-state and is connected by a linker to the N-terminus of the C-domain. The relative arrangement of the two domains places their active sites some 50 Å apart, suggesting a conformational state prior to peptide transfer from the donor PCP to the acceptor PCP domain and the existence of additional conformations during catalysis.

The thioesterase domain (TE) is the fourth catalytic unit of NRPSs. In contrast to the other three domains, which are repeated in each elongation module, TE-domains catalyze product release and are therefore located only in the C-terminal modules of most peptide synthetases. TE-domains are about 230–270 residues in length and very often catalyze product release by a two-step reaction. This involves formation of an acyl-O-TE intermediate that is subsequently cleaved by a regio- and stereoselective intramolecular macrocyclization, using an internal nucleophile to generate either cyclic or cyclic-branched products (Kopp and Marahiel, 2007a,b; Trauger et al., 2000). Some TE domains also catalyze product release by hydrolysis of the thioester-bound linear peptide. Alternatively, the release of linear peptides from NRPSs can be achieved by reduction of the thioester within the peptidyl-S-PCP intermediate that is attached to the termination module. This reaction, which results in the release of linear aldehydes or alcohols, is catalyzed by a C-terminal, NADPH-dependent reductase domain (R). However, macrocyclic release seems to be the favored mechanism. Among all essential domains, TE domains must have undergone a tremendous diversification to catalyze various reactions such as lactone and lactam formation as well as oligomerization of identical peptide units during iterative peptide synthesis. The crystal structures of two TE domains associated with the synthesis of the lipopeptide lactones surfactin (Srf TE) and fengycin (FenTE) have been determined, confirming the affiliation of these catalytic units with the group of α/β-hydrolyases (Fig. 13.5), such as serine esterases and lipases (Bruner et al., 2002; Samel et al., 2006). A feature that is common to the condensation domains is the shape of the substrate binding site, which forms a canyon to accommodate the linear peptide precursor. This canyon is preferably lined by hydrophobic and aromatic residues and harbors the catalytic triad (for Srf TE: Ser80, His207, Asp107) centrally at its bottom. The serine within the catalytic triad is the site of tetrahedral intermediate formation, which is stabilized by an oxyanion hole on the way to the acyl-O-TE intermediate. The acyl-O-enzyme intermediate breaks down by the nucleophilic attack of an internal nucleophile derived from the linear peptide precursor. Due to a lack of structural information about substrate or product complexes of TE domains there is still no common view as to how the active site is sufficiently sealed from bulk water to catalyze preferentially peptide cyclization rather than hydrolysis. In one model (Bruner et al., 2002), the substrate binding site resembles a bowl whose walls provide specific interactions with the peptidic substrate. A lid region consisting of a long helix protruding over the globular domain and a variably long connection to the domain's core hover over the active site canyon. This lid can then adopt either an open conformation for substrate entry or a closed conformation for excluding water from the active site. An alternative model (Samel et al., 2006) assumes that the peptide intermediate binds in a preformed conformation "edge-on" into the active site canyon and mostly seals the active site serine from bulk water by itself. The latter model was consistent with a variety of biochemical data and explained why TE

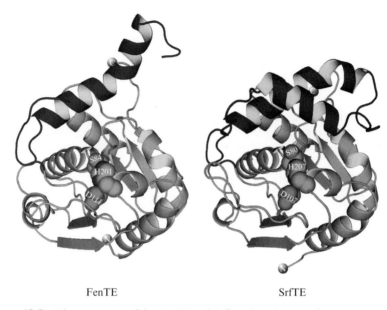

Figure 13.5 The structures of the FenTE and SrfTE domains reveal a core comprising an α/β-hydrolyase fold (gold) connected to an α-helical lid-region (blue). The catalytic triad residues (Ser, His, Asp) are located at the bottom of the canyon-like substrate-binding site. (See Color Insert.)

domains have diminished specificity for the amino acids located between the C-terminus and the attacking nucleophile. Clearly, the issue of substrate binding and efficient water exclusion can only be resolved by crystallographic or NMR elucidation of suitable TE domain complexes.

To study the substrate specificity of TE domains without the use of entire NRPSs, a chemoenzymatic approach was established that utilized excised TE domains and chemically synthesized peptidyl-S-N-acetylcysteamines (SNACs) as substrate surrogates. Initial studies with the excised tyrocidine TE domain have demonstrated that the recombinant protein is capable of cyclizing the chemically derived peptidyl-thioester substrates that were C-terminally activated with SNAC (Kopp and Marahiel, 2007a,b). Further studies with other excised TE domains have shown that these macrocyclization catalysts can utilize peptidyl-thiophenols and peptidyl-S-PCPs substrates (Grünewald and Marahiel, 2006). Most of the studied TEs were shown to be very promiscuous for most of the substrate residues except those associated directly with the cyclization reaction. In conclusion, only substitution of the nucleophile or the electrophile, and drastic changes in the substrate length, were not tolerated by the excised TEs.

3. STRUCTURAL INSIGHTS INTO AN ENTIRE TERMINATION MODULE

The crystal and NMR structures of the individual NRPS domains associated with substrate selection and activation (A-domain), substrate shuttling among the active sites (PCP-domain), and building block condensation (C-domain) into the growing peptide chain provided important insights on the catalytic mechanisms of these domains. The substrate specificity code (NRPS code) (Stachelhaus *et al.*, 1999), which is now extensively used to predict the products of orphan NRPSs, was discovered and the molecular mechanism and rules of product release by the TE domain are now better understood. NMR spectroscopy studies also revealed alternative conformational states for the substrate carrier domains (PCP) and suggested a critical role for these structural switches in specific domain–domain interaction and partner protein recognition during assembly line synthesis. However, this structural knowledge about individual catalytic units cannot provide information on how these catalytic domains are oriented and connected in relationship to each other in the three-dimensional structure of an intact NRPS module. This information is crucial for our understanding of how the covalently bound substrates are correctly shuttled between the catalytic centers and how domain movement is coordinated. The recent elucidation of the crystal structure of four domains comprising an entire termination module, SrfA-C (C-A-PCP-TE), from the surfactin synthetase (Fig. 13.6) is an important contribution towards understanding the molecular aspects that govern such complex enzymes (Tanovic *et al.*, 2008). SrfA-C (1274 residues) was crystallized as a mutant protein in which the active site serine of PCP was mutated to alanine. This mutation in PCP was shown to trap the PCP domain in one of its three conformations and may therefore contribute to the reduction of motion within the SrfA-C protein during crystallization. The structure solved at 2.6 Å resolution covered the entire SrfA-C molecule, with the four catalytic domains and all linker regions in-between (Tanovic *et al.*, 2008). The individual C, A, PCP, and TE domains within the SrfA-C structure show the same overall fold, as previously found for the dissected domains. However, the structural core of the SrfA-C module is a compact rectangular catalytic platform built by the C-domain and a major part of the A-domain (A_{core}) with both active sites arrayed on the same side of the platform. The complex is "coalesced" together by an extensive interface between C and A_{core} subdomains and a close association of the catalytic regions with the well-defined 32-residue intervening linker. The C-terminal subdomain of the A-domain (A_{sub}) and the PCP domain, which are tethered together through a flexible 15-residue linker, are located on top of the catalytic platform. In contrast to the static

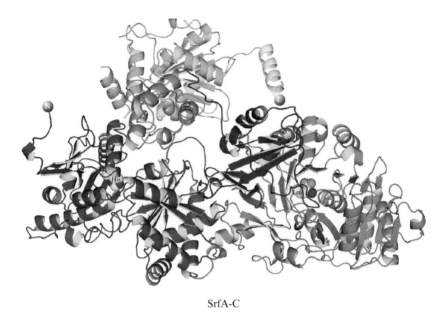

SrfA-C

Figure 13.6 Overall structure of the SrfA-C termination module (C-A-PCP-TE) showing the relative orientation of the four essential domains. Shown are a leucine residue in the active site of the A-domain and the active site histidine in the C-domain. Linkers connecting the domains are shown in blue and all domains have the same color code as in other figures. The peptide stretch arising from parts of an affinity tag is shown in yellow. (See Color Insert.)

$C-A_{core}$ platform, the PCP and $A_{sub\ domain}$ can move, as they have little interaction with the platform. This movement is required during the catalytic cycle, as the active sites of A and C are separated by over 60 Å, a distance that cannot be bridged by the 20-Å 4′-phosphopantetheine cofactor attached to PCP. In the SrfA-C structure, the PCP domain is stalled into the acceptor site of the C-domain, hence the catalytic center of PCP is only 16 Å from the active-site histidine of the C-domain. The TE-domain, which is connected to the C-terminus of SrfA-C through a short linker (9 residues) to PCP, forms a distinct region within the module and shows the same fold as the isolated TE domain.

The SrfA-C structure also provides some insight into how successive NRPS modules may act in *trans*. It was found that parts of a myc-His$_6$ affinity tag (~15 residues) fused to the C-terminus of SrfA-C form a helical segment that perfectly fits into a hand-shaped motif formed by the N-terminal part of the C-domain in SrfA-C. These two structural motifs resemble in function the so-called communication-mediating (COM) domains at the C- and N-terminal ends of NRPSs, which may contribute to selective partner recognition, as was shown by biochemical studies (Hahn and Stachelhaus, 2004). In the biochemical model of NRPS–NRPS interaction, it was postulated that a C-terminal COM helix communicates with an N-terminal COM-helix derived from the C-domain of the following NRPS. In contrast, the SrfA-C structure suggests that the C-domain does not provide a COM helix, but rather has a more complex motif, the "COM hand," which includes the N-terminal COM region and three additional β-strands.

On the basis of the SrfA-C structures a few lessons concerning future attempts to rationally manipulate NRPSs can be learned. The intimate structural association of C- and A-domains provides evidence that they should be shuffled together in any module-swapping experiment. This fact is also underlined by the importance of the high substrate specificity of the C-domain at its acceptor site, as shown by several biochemical studies (Lusong *et al.*, 2002). In such an approach, the heterologous PCP domain has to be optimized for productive C–PCP and C–A interactions. Also, rational engineering of COM-helix donor and COM-hand acceptor motifs, guided by the SrfA-C structure, would help to improve the in *trans* association of engineered NRPSs.

REFERENCES

Baltz, R. H. (2008). Biosynthesis and genetic engineering of lipopeptide antibiotics related to daptomycin. *Curr. Top. Med. Chem.* **8,** 618–638.

Baltz, R. H., Miao, V., and Wrigley, S. K. (2005). Natural products to drugs: Daptomycin and related lipopeptide antibiotics. *Nat. Prod. Rep.* **22,** 717–741.

Belshaw, P. J., Walsh, C. T., and Stachelhaus, T. (1999). Aminoacyl-CoAs as probes of condensation domain selectivity in nonribosomal peptide synthesis. *Science* **284,** 486–489.

Bruner, S. D., Weber, T., Kohli, R. M., Schwarzer, D., Marahiel, M. A., Walsh, C. T., and Stubbs, M. T. (2002). Structural basis for the cyclization of the lipopeptide antibiotic surfactin by the thioesterase domain Srf TE. *Structure* **10,** 301–310.

Caboche, S., Pupin, M., Leclère, V., Fontaine, A., Jacques, P., and Kucherov, G. (2008). NORINE: A database of nonribosomal peptides. *Nucleic Acids Res.* **36,** 326–331.

Conti, E., Stachelhaus, T., Marahiel, M. A., and Brick, P. (1997). Structural basis for the activation of phenylalanine in the non-ribosomal biosynthesis of gramicidin S. *EMBO J.* **16,** 4174–4183.

Eppelmann, K., Doekel, S., and Marahiel, M. A. (2001). Engineered biosynthesis of the peptide antibiotic bacitracin in the surrogate host *Bacillus subtilis*. *J. Biol. Chem.* **276,** 34824–34831.

Felnagle, E. A., Jackson, E. E., Chan, Y. A., Podevels, A. M., Berti, A. D., McMahon, M. D., and Thomas, M. G. (2008). Nonribosomal peptide synthetases involved in the production of medically relevant natural products. *Mol. Pharm.* **5,** 191–211.

Finking, R., and Marahiel, M. A. (2004). Biosynthesis of nonribosomal peptides. *Annu. Rev. Microbiol.* **58,** 453–488.

Finking, R., Solsbacher, J., Konz, D., Schobert, M., Schäfer, A., Jahn, D., and Marahiel, M. A. (2002). Characterization of a new type of phosphopantetheinyl transferase for fatty acid and siderophore synthesis in *Pseudomonas aeruginosa*. *J. Biol. Chem.* **277,** 50293–50302.

Fischbach, M. A., and Walsh, C. T. (2006). Assembly-line enzymology for polyketide and nonribosomal peptide antibiotics: Logic, machinery, and mechanisms. *Chem. Rev.* **106,** 3468–3496.

Frueh, D. P., Arthanari, H., Koglin, A., Vosburg, D. A., Bennett, A. E., Walsh, C. T., and Wagner, G. (2008). Dynamic thiolation-thioesterase structure of a non-ribosomal peptide synthetase. *Nature* **454,** 903–906.

Grünewald, J., and Marahiel, M. A. (2006). Chemoenzymatic and template-directed synthesis of bioactive macrocyclic peptides. *Microbiol. Mol. Biol. Rev.* **70,** 121–146.

Hahn, M., and Stachelhaus, T. (2004). Selective interaction between nonribosomal peptide synthetases is facilitated by short communication-mediating domains. *Proc. Natl. Acad. Sci. USA* **101,** 15585–15590.

Keating, T. A., Marshall, C. G., Walsh, C. T., and Keating, A. E. (2002). The structure of VibH represents nonribosomal peptide synthetase condensation, cyclization and epimerization domains. *Nat. Struct. Biol.* **9,** 522–526.

Koglin, A., Löhr, F., Bernhard, F., Rogov, V. V., Frueh, D. P., Strieter, E. R., Mofid, M. R., Güntert, P., Wagner, G., Walsh, C. T., Marahiel, M. A., and Dötsch, V. (2008). Structural basis for the selectivity of the external thioesterase of the surfactin synthetase. *Nature* **454,** 907–911.

Koglin, A., Mofid, M. R., Löhr, F., Schäfer, B., Rogov, V. V., Blum, M.-M., Mittag, T., Marahiel, M. A., Bernhard, F., and Dötsch, V. (2006). Conformational switches modulate protein interactions in peptide antibiotic synthetases. *Science* **312,** 273–276.

Kopp, F., and Marahiel, M. A. (2007a). Where chemistry meets biology: The chemoenzymatic synthesis of nonribosomal peptides and polyketides. *Curr. Opin. Biotechnol.* **18,** 513–520.

Kopp, F., and Marahiel, M. A. (2007b). Macrocyclization strategies in polyketide and nonribosomal peptide biosynthesis. *Nat. Prod. Rep.* **24,** 735–749.

Linne, U., and Marahiel, M. A. (2000). Control of directionality in nonribosomal peptide synthesis: Role of the condensation domain in preventing misinitiation and timing of epimerization. *Biochemistry* **39,** 10439–10447.

Linne, U., and Marahiel, M. A. (2004). Reactions catalyzed by mature and recombinant nonribosomal peptide synthetases. *Methods Enzymol.* **388,** 293–315.

Lusong, L., Kohli, R. M., Onishi, M., Linne, U., Marahiel, M. A., and Walsh, C. T. (2002). Timing of epimerization and condensation reactions in nonribosomal peptide assembly lines: Kinetic analysis of phenylalanine activating elongation modules of tyrocidine synthetase B. *Biochemistry* **41,** 9184–9196.

May, J. J., Kessler, N., Marahiel, M. A., and Stubbs, M. T. (2002). Crystal structure of DhbE, an archetype for aryl acid activating domains of modular nonribosomal peptide synthetases. *Proc. Natl. Acad. Sci. USA* **99,** 12120–12125.

Mootz, H. D., Finking, R., and Marahiel, M. A. (2001). 4'-Phosphopantetheine transfer in primary and secondary metabolism of *Bacillus subtilis*. *J. Biol. Chem.* **276,** 37289–37298.

Mootz, H. D., Kessler, N., Linne, U., Eppelmann, K., Schwarzer, D., and Marahiel, M. A. (2002). Decreasing the ring size of a cyclic nonribosomal peptide antibiotic by in-frame module deletion in the biosynthetic genes. *J. Am. Chem. Soc.* **124,** 10980–10981.

Mootz, H. D., Schwarzer, D., and Marahiel, M. A. (2000). Construction of hybrid peptide synthetases by module and domain fusions. *Proc. Natl. Acad. Sci. USA* **97,** 5848–5853.

Rausch, C., Weber, T., Kohlbacher, O., Wohlleben, W., and Huson, U. (2005). Specificity prediction of adenylation domains in nonribosomal peptide synthetases (NRPS) using transductive support vector machines (TSVMs). *Nucleic Acid Res.* **18,** 5799–5808.

Reuter, K., Mofid, M. R., Marahiel, M. A., and Ficner, R. (1999). Crystal structure of the surfactin synthetase-activating enzyme Sfp: A prototype of the 4'-phosphopantetheinyl transferase superfamily. *EMBO J.* **18,** 6823–6831.

Samel, S. A., Schönafinger, G., Knappe, T. A., Marahiel, M. A., and Essen, L.-O. (2007). Structural and functional insights into a peptide bond-forming bidomain from a nonribosomal peptide synthetase. *Structure* **15,** 781–792.

Samel, S. A., Wagner, B., Marahiel, M. A., and Essen, L.-O. (2006). The thioesterase domain of the fengycin biosynthesis cluster: A structural base for the macrocyclization of a non-ribosomal lipopeptide. *J. Mol. Biol.* **359,** 876–889.

Schwarzer, D., Finking, R., and Marahiel, M. A. (2003). Nonribosomal peptides: From genes to products. *Nat. Prod. Rep.* **20,** 275–287.

Sieber, S. A., and Marahiel, M. A. (2005). Molecular mechanisms underlying nonribosomal peptide synthesis: Approaches to new antibiotics. *Chem. Rev.* **105,** 715–738.

Stachelhaus, T., Mootz, H. D., and Marahiel, M. A. (1999). The specificity-conferring code of adenylation domains in nonribosomal peptide synthetases. *Chem. Biol.* **6,** 493–505.

Stachelhaus, T., Schneider, A., and Marahiel, M. A. (1995). Rational design of peptide antibiotics by targeted replacement of bacterial and fungal domains. *Science* **269,** 69–72.

Tanovic, A., Samel, S. A., Essen, L.-O., and Marahiel, M. A. (2008). Crystal structure of the termination module of a nonribosomal peptide synthetase. *Science* **322,** 659–663.

Trauger, J. W., Kohli, R. M., Mootz, H. D., Marahiel, M. A., and Walsh, C. T. (2000). Peptide cyclization catalysed by the thioesterase domain of tyrocidine synthetase. *Nature* **407,** 215–218.

Walsh, C. T. (2008). The chemical versatility of natural-product assembly lines. *Acc. Chem. Res.* **41,** 4–10.

Weber, T., Baumgartner, R., Renner, C., Marahiel, M. A., and Holak, T. A. (2000). Solution structure of PCP, a prototype for the peptidyl carrier domains of modular peptide synthetases. *Structure* **8,** 407–418.

Yonus, H., Neumann, P., Zimmermann, S., May, J. J., Marahiel, M. A., and Stubbs, M. T. (2008). Crystal structure of DltA. Implications for the reaction mechanism of nonribosomal peptide synthetase adenylation domains. *J. Biol. Chem.* **283,** 32484–32491.

CHAPTER FOURTEEN

Biosynthesis of Nonribosomal Peptide Precursors

Barrie Wilkinson* and Jason Micklefield[†]

Contents

1. Introduction	354
2. Precursors from Amino Acid Metabolism	355
2.1. Biosynthesis of (2S,3R)-methylglutamic acid: *In vivo* studies	359
2.2. Biosynthesis of (2S,3R)-methylglutamic acid: *In vitro* studies	360
3. Fatty Acid Precursor Biosynthesis	360
3.1. Biosynthesis of the 2,3-epoxyhexanoyl side chain of CDA	363
4. Polyketide Precursors	364
4.1. 3,5-Dihydroxyphenylglycine biosynthesis	365
4.2. *In vivo* studies of DHPG biosynthesis	366
4.3. *In vitro* studies of DHPG biosynthesis	367
4.4. Variation of the balhimycin DHPG moiety by mutasynthesis	368
4.5. Bmt biosynthesis (cyclosporin precursor)	369
4.6. 3-Methoxy-5-methylnapthoic acid biosynthesis	370
5. Glycosyl Building Blocks	371
6. Conclusion	372
References	373

Abstract

Nonribosomal peptides are natural products typically of bacterial and fungal origin. These highly complex molecules display a broad spectrum of biological activities, and have been exploited for the development of immunosuppressant, antibiotic, anticancer, and other therapeutic agents. The nonribosomal peptides are assembled by nonribosomal peptide synthetase (NRPS) enzymes comprising repeating modules that are responsible for the sequential selection, activation, and condensation of precursor amino acids. In addition to this, fatty acids, α-keto acids and α-hydroxy acids, as well as polyketide derived units, can also be utilized by NRPS assembly lines. Final tailoring-steps, including glycosylation and prenylation, serve to further decorate the nonribosomal peptides

---

* Biotica, Chesterford Research Park, Little Chesterford, Essex, United Kingdom
[†] School of Chemistry and Manchester Interdisciplinary Biocentre, The University of Manchester, Manchester, United Kingdom

produced. The wide range of experimental methods that are employed in the elucidation of nonribosomal peptide precursor biosynthesis will be discussed, with particularly emphasis on genomics based approaches which have become wide spread over the last 5 years.

1. Introduction

Nonribosomal peptides are amongst the most structurally diverse secondary metabolites in nature. Along with nonproteinogenic amino acids (Marahiel et al., 1997; Sieber and Marahiel, 2005) many of these peptides also comprise polyketide (Pfeifer et al., 2001; Zhao et al., 2008) or fatty acid-derived moieties (Powell et al., 2007b; Wittmann et al., 2008). Glycosylation (Chen et al., 2000; Losey et al., 2002) and to a lesser extent prenylation (Edwards and Gerwick, 2004; Schultz et al., 2008) can also further add to the structural diversification of these highly complex natural products. In this paper, the methods used to elucidate the biosynthetic pathways leading to nonribosomal peptide precursors derived from amino acid, fatty acid, polyketide, and carbohydrate metabolism will be described. The biosynthetic steps that occur on the nonribosomal peptide synthetase (NRPS) or related hybrid NRPS–polyketide synthase (PKS) assembly lines, and during post-NRPS tailoring are covered elsewhere in this volume and will not be discussed in detail here.

Recent studies on the biosynthesis of nonribosomal peptides have been guided by genomics data. With the sequence of a biosynthetic gene cluster in hand, *in vivo* approaches can be explored. This would typically involve deletion of genes within the biosynthetic gene cluster, followed by structural analysis of the products formed (Neary et al., 2007; Powell et al., 2007a,b). Supplementing deletion mutants with putative synthetic intermediates (Hojati et al., 2002; Milne et al., 2006) or genes encoding similar enzymes (Miao et al., 2005) can also be very informative. Hypotheses generated from the *in vivo* experiments can then be tested by further *in vitro* studies, which involve overproduction and characterization of the individual biosynthetic enzymes (Kopp et al., 2008; Mahlert et al., 2007; Strieker et al., 2007; Widboom et al., 2007). In the absence of genomic sequence information, it is necessary to resort to classical techniques, including isotopic labeling experiments (Hammond et al., 1982, 1983; McGahren et al., 1980; Zmijewski et al., 1987) which provide useful information about the origins of the biosynthetic precursors. Moreover, stereospecific labeling experiments can also illuminate the stereochemical course of individual steps in a pathway, which help elucidate the mechanisms of the enzymes involved in the these steps (Heidari, 2007a,b).

2. Precursors from Amino Acid Metabolism

Nonribosomal peptides contain many L- and D-configured proteinogenic as well as unusual α-amino acids. Typically, L-amino acids are first activated by adenylation (A) domains and then epimerized during peptide assembly by an epimerization (E) domain within the NRPS assembly line (Marahiel *et al.*, 1997; Sieber and Marahiel, 2005). Condensation (C) domains with dual epimerization and condensation activity have also been discovered (Balibar *et al.*, 2005; Yin and Zabriskie, 2006). Occasionally, an external racemase enzyme is responsible for the production of D-amino acids which are then activated by D-specific A-domains within the NRPS (Cheng and Walton, 2000; Li and Jensen, 2008). In addition to α-amino acids, α-keto and α-hydroxy acids derived from amino acid metabolism are utilized as precursors in nonribosomal peptide biosynthesis. For example, cyclodepsipeptides exist that contain alternating ester and amide bonds which arise through the condensation of α-amino and α-hydroxy acids. Typically, in fungal depsipeptides such as enniatin B (Feifel *et al.*, 2007) (Fig. 14.1) and the cyclooctadepsipeptides PF1022A (Yanai *et al.*, 2004), α-keto acids are reduced to α-hydroxy acids prior to activation by the NRPS A-domain. However, in the biosynthesis of the bacterial cyclododecadepsipeptides cereulide and valinomycin, α-keto acids are activated by the A-domain of the NRPS. In this case, an α-ketoreductase (α-KR) domain within the NRPS assembly line is responsible for chiral reduction to give peptidyl carrier protein (PCP)-tethered α-hydroxy acids (Magarvey *et al.*, 2006). Furthermore, some peptides possess both β- and α-amino acids. For example, 4-methylideneimidazole-5-one (MIO)-dependent aminomutases (CmdF and AdmH) are responsible for generating the β-Tyr and β-Phe precursors of the hybrid nonribosomal peptide–polyketides chondramide (Rachid *et al.*, 2007) and andrimid (Fortin *et al.*, 2007), respectively (Fig. 14.1). MIO participates in covalent catalysis, enabling the elimination followed by conjugate addition of ammonia onto an α,β-unsaturated acid intermediate (Christianson *et al.*, 2007). Alternatively, for example, a lysine 2,3-aminomutase (VioP) is implicated in the biosynthesis of the β-lysine precursor of viomycin (Thomas *et al.*, 2003) (Fig. 14.1). The lysine 2,3-aminomutases belong to the radical *S*-adenosyl methionine (SAM) superfamily utilizing pyridoxal-5′-phosphate, [4Fe–4S]$^{2+}$ and SAM cofactors to transfer the amino group via a radical mechanism (Lepore *et al.*, 2005).

β-Hydroxylation of amino acids is also frequently encountered in the biosynthesis of nonribosomal peptides (Chen *et al.*, 2001). For example, in the biosynthesis of viomycin, mannopeptimycins (Fig. 14.1) and the calcium dependent antibiotics (CDAs), the free amino acids L-Arg, L-enduracididine, and L-Asn are hydroxylated by nonhaem Fe(II)/2-oxoglutarate-dependent

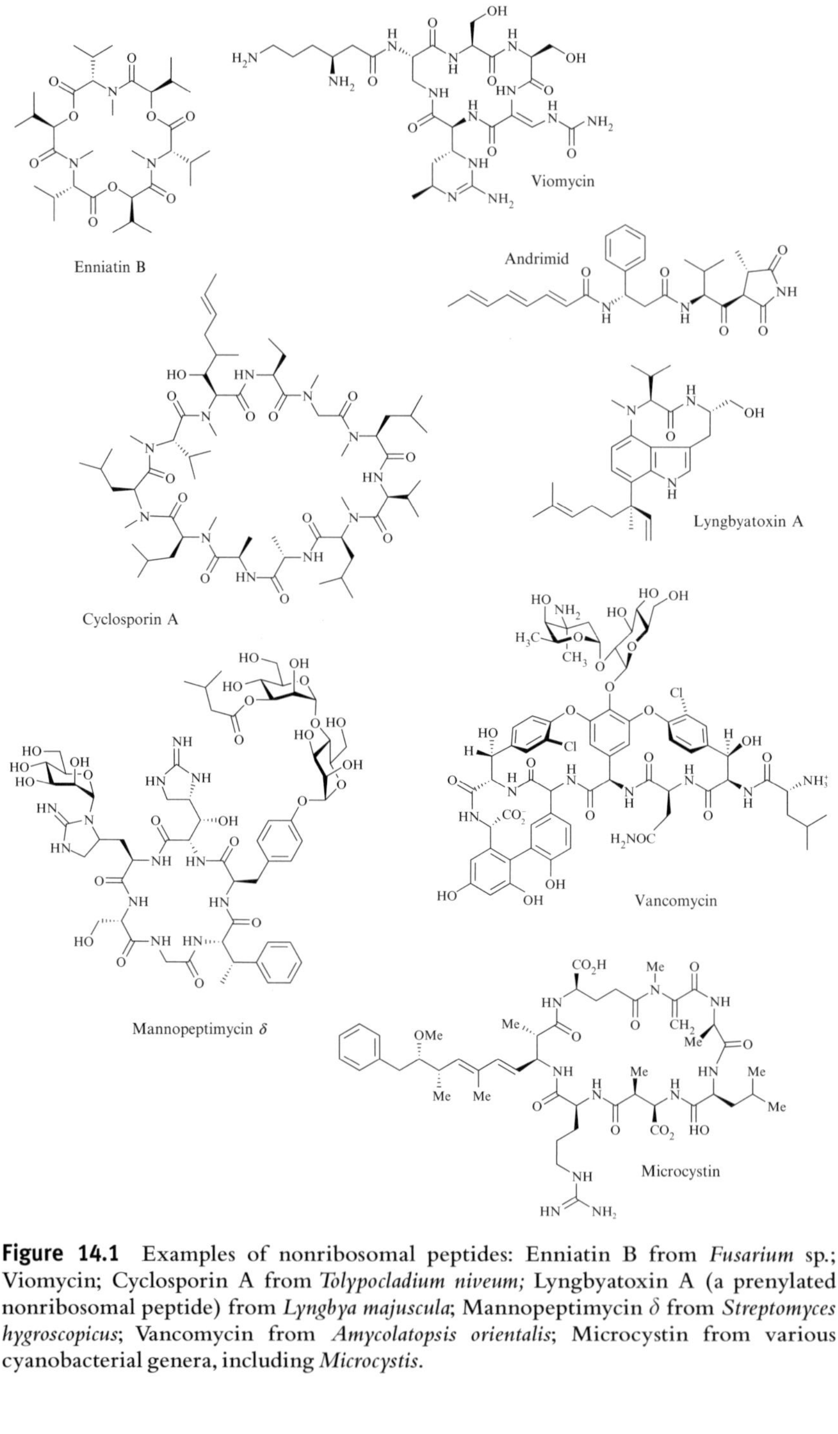

Figure 14.1 Examples of nonribosomal peptides: Enniatin B from *Fusarium* sp.; Viomycin; Cyclosporin A from *Tolypocladium niveum;* Lyngbyatoxin A (a prenylated nonribosomal peptide) from *Lyngbya majuscula;* Mannopeptimycin δ from *Streptomyces hygroscopicus;* Vancomycin from *Amycolatopsis orientalis;* Microcystin from various cyanobacterial genera, including *Microcystis*.

oxygenases, VioC (Ju et al., 2004; Yin and Zabriskie, 2004), MppO (Haltli et al., 2005), and AsnO (Neary et al., 2007; Strieker et al., 2007), respectively (Fig. 14.2A). In addition, there is a family of haem protein hydroxylases (e.g., ORF20, NovI, and NikQ) that catalyze the hydroxylation of PCP-tethered L-Tyr and L-His residues during the biosynthesis of chloroeremomycin, novobiocin (Chen et al., 2001), and nikkomycin (Chen et al., 2002). The subsequent processing of β-hydroxylated amino acids by further oxidation (Chen et al., 2002), glycosylation (Lu et al., 2004), macrolactonization (McCafferty et al., 2002), methylation (Miao et al., 2006), or phosphorylation (Neary et al., 2007; Strieker et al., 2007) further serves to increase the structural diversity of nonribosomal peptides.

A number of nonribosomal peptide contain α,β-unsaturated amino acids. Due to the instability of enamines, these residues are most likely generated during or after peptide assembly. For example, the dehydrothreonine and N-methyldehydroalanine residues found in syringomycin (Singh et al., 2007) and microcystin (Tillett et al., 2000) (Fig. 14.1) are predicted to arise by dehydration of threonine and N-methylalanine residues of PCP-peptidyl intermediates. In the biosynthesis of the β-ureidodehydroalanine residue of viomycin (Fig. 14.1), a putative FAD-dependent oxidase/dehydrogenase (VioJ) is predicted to catalyze α,β-desaturation of a 2,3-diaminopropionyl-S-PCP intermediate prior to carbamylation (Thomas et al., 2003). Finally, in the CDAs there exists a C-terminal Z-dehydrotryptophan residue. In this case, Trp dehydrogenation was shown to occur through loss of the 2′ and pro-3′S hydrogen atoms with overall *syn* stereochemistry, consistent with FAD-dependent dehydrogenation (Heidari, 2007a,b) (Fig. 14.2B).

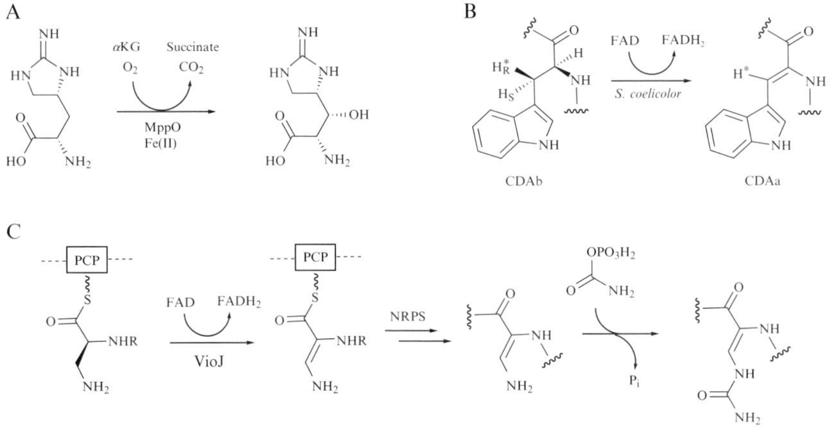

Figure 14.2 (A) β-hydroxylation of L-enduracididine. (B) Trp dehydrogenation during CDA biosynthesis. (C) Proposed biosynthesis of the β-ureidodehydroalanine residue of viomycin.

Of the many halogenated amino acid residues found in nonribosomal peptides, chlorinated derivatives are most prevalent. In the case of electron-rich substrates such as aromatic groups, chlorine–carbon bond formation is typically catalyzed by flavin-dependent halogenases (Dong et al., 2005; Yeh et al., 2007). For example, chlorination of β-hydroxytyrosine in vancomycin (Puk et al., 2004) (Fig. 14.1) and other aromatic residues in related antibiotics is catalyzed by flavin-dependent enzymes (Puk et al., 2002). On the other hand, chlorination of inactivated carbon centers requires the more powerful nonhaem Fe(II)/2-oxoglutarate-dependent halogenases, such as SyrB2 (Vaillancourt et al., 2005), which catalyzes the C4-chlorination of a L-Thr-S-PCP intermediate during syringomycin biosynthesis (Fig. 14.3A). Similarly, the enzyme KtzD is predicted to catalyze C5-chlorination of an (allo)Ile-S-PCP intermediate during the biosynthesis of a 2-(1-methylcyclopropyl)-D-glycine precursor of the kutzneride nonribosomal peptides (Fujimori et al., 2007).

Methylation is another general theme in the structural diversification of nonribosomal peptides and precursors. In addition to the N-methyl transferases (N-MT) domains within NRPSs responsible for the N-methyl amino acids in a wide range of products including cyclosporin and PF1022A (Walsh et al., 2001), there are a number of O-methyltransferases. For example, a SAM-dependent catechol $4'$-O-methyltransferase (SafC) delivers a $4'$-O-methyl-L-dopa precursor in saframycin biosynthesis (Nelson et al., 2007). In addition, an O-methyltransferase (LptI) is implicated in the biosynthesis of the 3-O-methyl-aspartic acid precursor of the A54145 lipopeptide antibiotics (Miao et al., 2006). There are also a number of C-methyl transferases, including MppJ and CymG, which are suggested to catalyze methylation reactions in the biosynthesis of the β-methylphenylalanine and

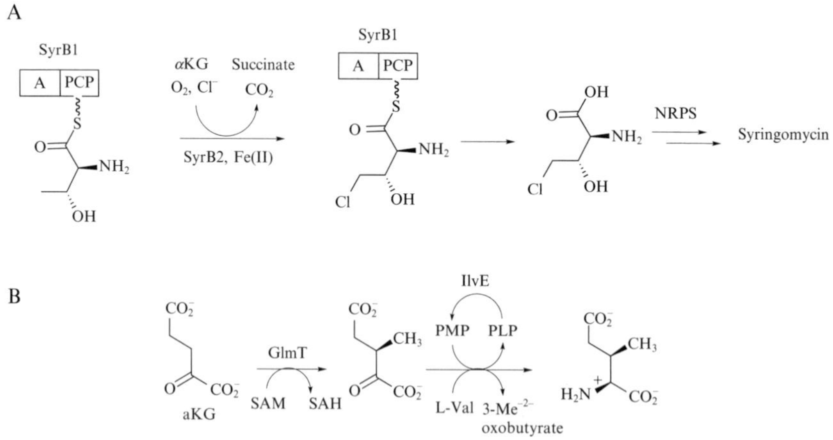

Figure 14.3 (A) Chlorination of a L-Thr-S-PCP intermediate during syringomycin biosynthesis. (B) Biosynthesis of (2S,3R)-methylglutamic acid precursor of for CDA and daptomycin.

2-amino-3,5-dimethylhex-4-enoic acid (ADH), precursors of the mannopeptimycins (Magarvey et al., 2006a,b) and cyclomarins (Schultz et al., 2008), respectively.

2.1. Biosynthesis of (2S,3R)-methylglutamic acid: In vivo studies

3-Methylglutamic acid is found at the same relative positions in the decapeptide lactone core of CDA, daptomycin and A54145, and is known to be important for antimicrobial activity of these lipopeptides (Milne et al., 2006; Nguyen et al., 2006). A gene, glmT (SCO3215), in the cda biosynthetic gene cluster from *Streptomyces coelicolor* was identified which resembles a number of C-methyl transferases (Hojati et al., 2002). A standard double crossover gene replacement strategy was used to delete glmT from the chromosome of S. coelicolor. Briefly, two gene fragments which flank the chromosomal sequence to be deleted were cloned into the pMAH vector (Bucca et al., 2003). *Escherichia coli* strain ET12567 was then transformed with the plasmid containing the deletion construct. The resulting nonmethylated plasmid was then was used to transform protoplasts of S. coelicolor. Homologous recombination and subsequent screening of second crossover recombination events resulted in the mutant strain ΔglmT. This deletant strain was grown in liquid medium (Milne et al., 2006) and the resulting culture supernatant was acidified to pH 2.0 and extracted using C-18 Bond Elute cartridges. The crude CDA was eluted with water/methanol and analyzed by LC–MS, which indicated production of the Glu-containing variant CDA3a, with no 3-MeGlu containing CDAs evident. Under identical growth conditions the parental strain produces predominantly CDA4a, which contains 3-MeGlu. Following this a gene, dptI, identified in the daptomycin biosynthetic gene cluster (Miao et al., 2005) was similarly deleted from the daptomycin producer, *Streptomyces roseosporus*. The resulting mutant also produced a lipopeptide variant possessing Glu instead of 3-MeGlu. Moreover, complementation of the mutant with glmT was shown to re-establish production of daptomycin possessing a 3-MeGlu residue.

The stereochemistry of the 3-MeGlu residues in both CDA and daptomycin was determined by comparing the 3-MeGlu derived from hydrolysis of the lipopeptides with synthetic standards. This showed that 3-MeGlu from both CDA and daptomycin possesses the 2S,3R absolute configuration. Synthetic 2S,3R-MeGlu was also used to "complement" the S. coelicolor deletant strain ΔglmT. In addition to (2S,3R)-MeGlu, synthetic 3-methyl-2-oxogluterate was shown to re-establish production of MeGlu-containing CDA4a. This led to the suggestion that GlmT was a SAM-dependent methyltransferase that methylates 2-oxoglutarate to give (3R)-methyl-2-oxoglutarate, which is then transaminated to (2S,3R)-3-MeGlu (Fig. 14.3B).

2.2. Biosynthesis of (2S,3R)-methylglutamic acid: In vitro studies

To investigate the biosynthesis of (2S,3R)-3-MeGlu, GlmT and DptI were overproduced in E. coli (Mahlert et al., 2007), and the recombinant enzymes tested with a range of putative substrates. Free glutamic acid, PCP-tethered glutamate, and Glu-containing lipopeptide precursors were not methylated in vitro. However, under identical conditions, 2-oxoglutarate undergoes efficient SAM-dependent methylation catalyzed by both GlmT and DptI. Presumably, the SAM-dependent GlmT-methylation of α-KG by the standard 2-electron transfer mechanism is electronically more feasible, via a stabilized enol or enolate, than the direct methylation at the β-position of Glu (Fig. 14.3B).

BLAST searches failed to identify any aminotransferases in the cda or dpt biosynthetic gene clusters. In light of this a variety of commercially available bacterial aminotransferases were screened with synthetic racemic 3-methyl-2-oxoglutarate, (PLP), and appropriate amino donors (Mahlert et al., 2007). From this, it was shown that a bacterial branched-chain amino acid aminotransferase, IlvE, was able to catalyze the formation of 3-MeGlu with L-Val as the amino donor. Based on this the gene SCO5523 was identified in S. coelicolor which encodes a protein with sequence similarity to IlvE. Accordingly, the SCO5523 gene product was overproduced in E. coli. The resulting transaminase was incubated with the amino donor L-Val and putative substrate 3-methyl-2-oxoglutarate (prepared from GlmT, SAM-dependant methylation of 2-oxoglutarate). The product of the reaction was derivatized with dabsyl chloride and subjected to LC–MS analysis (Mahlert et al., 2007). Comparisons with synthetic dabsylated (2S,3R)- and-(2S,3R)-3-MeGlu verified the enzymatic product as (2S,3R)-3-MeGlu. This confirmed the pathway and enzymology to 3-MeGlu in S. coelicolor (Fig. 14.3B). Using this knowledge, a mutasynthesis strategy (see Chapter 18 of Volume 459) was also developed which resulted in lipopeptide variants possessing modified glutamate residues (Powell et al., 2007a). This was achieved by feeding the $\Delta glmT$ mutant strain synthetic 3-ethyl and 3-trifluoromethyl glutamic acid derivatives.

3. Fatty Acid Precursor Biosynthesis

A number of nonribosomal lipopeptides, including daptomycin (Fig. 14.4), A54145, ramoplanin, and surfactin, possess N-terminal fatty acid side chains. In the case of daptomycin, it has been shown that endogenous fatty acids are activated by an adenylating enzyme (DptE), and transferred to a specific ACP (DptF) (Wittmann et al., 2008) (Fig. 14.5A). It is suggested that the fatty acid moiety is then transferred from the ACP (DptF) to the first amino

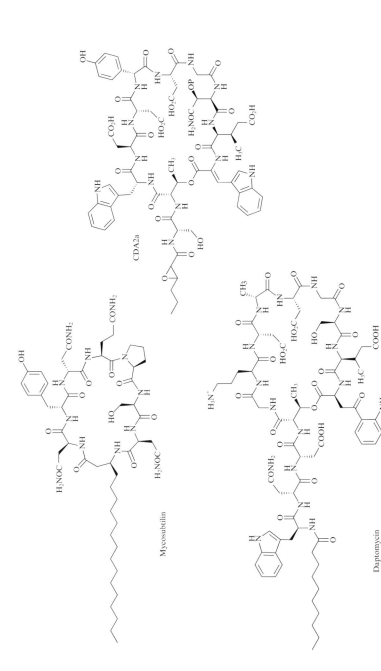

Figure 14.4 Nonribosomal lipopeptides: Mycosubtilin from *Bacillus subtilis*, CDA from *Streptomyces coelicolor* and daptomycin from *Streptomyces roseosporus*.

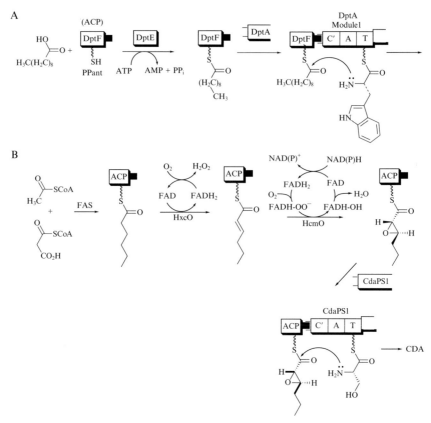

Figure 14.5 (A) Transfer of the daptomycin fatty acid moeity. (B) Biosynthesis of the 2,3-*trans*-epoxyhexanoyl fatty acid moiety and initiation of CDA–lipopeptide assembly.

acid (Trp) tethered to module 1 of the daptomycin NRPS (DptA) (Powell et al., 2007b; Wittmann et al., 2008). It is likely that the unusual N-terminal condensation domain (C′) of the NRPS catalyzes transfer of the fatty acid, initiating lipopeptide assembly (Fig. 14.5A). Kinetic studies revealed that DptE is promiscuous, allowing activation and transfer of a broad range of fatty acids. This is consistent with the fact that the A21978C family of lipopeptides possess a variety of different C10–13 fatty acid side chains (presumably derived from the primary fatty acid pool), and that simply by feeding *n*-decanoic acid to the producer *S. roeseosporus* during fermentation it is possible to produce daptomycin as the major product (Huber et al., 1988).

The iturin lipopeptide family, which includes mycosubtilin and bacillomycin, are cyclic lipopeptides incorporating β-amino fatty acids (Fig. 14.4) (Duitman et al., 1999). For this class of compounds, an N-terminal acyl-AMP ligase (AL) domain of a hybrid PKS–NRPS assembly line is responsible for the activation of a precursor fatty acid and its transfer to a downstream ACP domain (Hansen *et al.*, 2007). Subsequent PKS extension, utilizing a free-standing acyltransferase (FenF) which delivers a malonyl precursor in *trans*, results in a β-keto acyl-ACP intermediate (Aron *et al.*, 2007). This is then transaminated in *cis*, by an assembly line amino transferase domain (AMT) to give the β-amino fatty acid precursor. In addition to this, lipoglycopeptides of the vancomycin group, including teicoplanin (Kruger *et al.*, 2005; Li *et al.*, 2004) and A40926 (Sosio *et al.*, 2003), possess N-acyl aminosugars. In this case, fatty acid moieties are transferred from acyl–CoA donors to aminosugars of glycopeptide intermediates by acyltransferase tailoring enzymes (Kruger *et al.*, 2005; Li *et al.*, 2004; Sosio *et al.*, 2003).

3.1. Biosynthesis of the 2,3-epoxyhexanoyl side chain of CDA

The CDA family of molecules are unusual amongst lipopeptides in that they possess the same short (C6) 2,3-epoxyhexanoyl fatty acid moiety (Fig. 14.4). Most other related lipopetides with N-terminal fatty acid groups belong to a family of related products possessing longer (>C10) fatty acid moieties, which differ significantly in their substitution patterns, degree of saturation, and chain length. In addition, there are no genes within the *cda* biosynthetic gene cluster that encode proteins which resemble DptE and DptF, indicating that CDA fatty acid biosynthesis and loading is distinct from that of other lipopeptides (Powell *et al.*, 2007b). To explore the *in vivo* biosynthesis of the CDA fatty acid side chain the gene *hxcO* within the *cda fab* operon (including genes: *acp, fabF3, hxcO, fabH4, hcmO*) was targeted for deletion. Accordingly, a deletion construct containing gene fragments flanking the *hxcO* chromosomal sequence to be deleted was cloned into the pMAH vector (Bucca *et al.*, 2003). This plasmid was then isolated in the demethyl form and used to effect a double crossover gene replacement of *hxcO*. The Δ*hxcO* lipopeptide extracts were purified by semipreparative reversed phase HPLC and subjected to MS–MS analysis on a 9.4T Bruker Daltonics Apex III FT-ICR mass spectrometer using direct infusion electrospray. The gas-phase precursor ions were then fragmented by infrared multiphoton dissociation (IRMPD). Accurate mass measurements for the lipopeptide tail and core fragments revealed several CDA-lipopeptides with hexanoyl fatty acid side chains (hCDA4a and hCDA 4b) (Powell *et al.*, 2007b). This was consistent with HxcO, functioning as an oxidase in the desaturation of a hexanoyl fatty acid precursor, prior to epoxidation (Fig. 14.5B).

To further probe the biosynthesis of the unusual CDA fatty acid moiety HxcO was overproduced in *E. coli*. The resulting his$_6$-fusion proteins were assayed with the putative hexanoyl-CoA substrate under conditions established for similar FAD-dependent acyl-CoA dehydrogenases and oxidases (Powell *et al.*, 2007b). However, the predicted *trans*-hexenoyl-CoA product was not observed. This led to the conclusion that hexanoyl-*S*-ACP was the most likely substrate for HxcO (Powell et al., 2007b). Accordingly, the ACP encoded within the *fab* operon was overproduced and incubated with hexanoyl-CoA and the *Bacillus subtilis* 4′-phosphopantetheinetransferase (PPTase) Sfp (Kopp et al., 2008). The resulting hexanoyl-ACP was then incubated with HxcO and FAD and the ACP-tethered product was digested with trypsin and analyzed by high-resolution ESI–FTICR–MS. This showed evidence of both hexenoyl-ACP and 2,3-epoxyhexanoyl-ACP products. To confirm this result the ACP-tethered products were incubated with a 1000-fold excess of D-Phe-OMe at 60 °C for 2 h. The resulting N-acylated D-Phe methyl esters were then analyzed by LC–MS and chiral GC. From comparison with synthetic standards, it was concluded that the products of the HxcO-catalyzed oxidation of hexanoyl-ACP was a mixture of *trans*-2,3-hexenoyl-ACP and (2R,3S)-epoxyhexanoyl-S-ACP (Kopp et al., 2008). This was unexpected given that the *fab* operon of the *cda* cluster also possesses a gene, *hcmO*, which encodes a protein similar to a number of FAD-dependent monooxygenases, including a zeaxanthin-epoxidase. Indeed, incubation of *trans*-2,3-hexenoyl-S-ACP with HcmO, FAD, and NADPH resulted in the 2,3-epoxyhexanoyl-S-ACP product. However, acyl-exchange with D-Phe-OMe resulted in a product which, by comparison with synthetic standards, was assigned the opposite (2S,3R)-configuration (Kopp et al., 2008). This led to the suggestion (Kopp et al., 2008), that both (2S,3R)- and (2R,3S)-2,3-epoxyhexanoyl fatty acid moieties are incorporated into CDAs. However, earlier detailed NMR analysis of CDAs failed to identify CDA epoxide diastereoisomers (Hojati et al., 2002; Kempter et al., 1997). It, therefore, remains to be seen whether the CDA 2,3-epoxyhexanoyl fatty acid moiety arises from HxcO- or HcmO-catalyzed epoxidation. Despite this, the combined *in vitro* and *in vivo* studies (Kopp et al., 2008; Powell et al., 2007b) suggest that a mixed fatty acid synthase, comprised of enzymes from primary metabolism along with dedicated enzymes encoded by the *cda fab* operon (ACP, FabF3, FabH4) delivers the hexanoyl-ACP precursor, which is then desaturated and epoxidized by the flavin-dependent oxidase (HxcO) and monooxygenase (HcmO) enzymes (Powell et al., 2007b). Following this, it is suggested that the 2,3-epoxyhexanoyl moiety is transferred from the ACP to the NRPS, by the C′-domain of module 1 (Fig. 14.5B), in a similar fashion to the initiation step in the assembly of other related lipopeptides.

4. Polyketide Precursors

Some nonproteinogeneic amino acid residues which serve as NRPS substrates are synthesized by dedicated pathways involving the action of PKSs. These freely diffusible compounds are distinct from the products of hybrid PKS–NRPS synthetases, such as that required for microcystin biosynthesis, for example. Analysis of the microcystin structure indicates the apparent incorporation of an unusual β-amino acid substrate (Fig. 14.1). However, the microcystin biosynthetic gene cluster (Tillet *et al.*, 2000) clearly reveals a hybrid system in which an NRPS-like initiation module (Hicks *et al.*, 2006) and four PKS modules interface with an NRPS in McyE. This hybrid protein, McyE, is interesting as it contains an embedded aminotransferase domain at the PKS–NRPS interface, which most likely converts the β-keto function of the processing chain into an amino moiety. This amino group then acts as the nucleophilic residue for cyclization and release of the NRPS product. Such hybrid biosynthetic pathways are relatively commonplace, but are not the subject of this chapter and will be described in greater detail elsewhere in this volume. The investigation of PKS-derived NRPS substrates is discussed below.

4.1. 3,5-Dihydroxyphenylglycine biosynthesis

3,5-Dihydroxyphenylglycine (DHPG) is a precursor of the therapeutically important glycopeptide antibiotics. All vancomycin (Fig. 14.1) and teicoplanin family glycopeptide antibiotics contain DHPG-derived residues, and they arise from the polyketide pathway as was initially determined by the use of stable isotope labeling studies (Hammond *et al.*, 1982,1983; McGahren *et al.*, 1980; Zmijewski *et al.*, 1987). Feeding ^{13}C-labeled sodium acetates clearly identified the residue as a tetraketide, with the most likely folding pattern being C8–C3 to give 3,5-dihydroxyphenylacetate (DHPA) as its coenzyme-A thioester after dehydration and tautomerization (Fig. 14.6). It was also shown that ^{13}C-labeled DHPA fed to the chloroeromomycin-producing organism, *Amycolatopsis orientalis*, is specifically incorporated into the DHPG residue of this antibiotic (Sandercock *et al.*, 2001). Sequencing of the chloroeromomycin biosynthetic gene cluster identified a type-III, chalcone synthase-like PKS-encoding gene, *dpgA* (originally called *orf*27) (van Wageningen *et al.*, 1998).

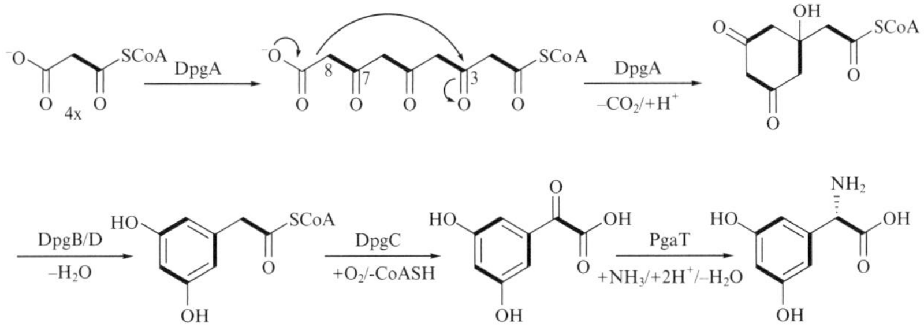

Figure 14.6 Biosynthesis of the glycopeptide precursor DHPG: DpgA, PKS; DpgB/D, putative enolhydratases; DpgC, C2-oxygenase; PgaT, transaminase.

This gene is part of a subcluster present in all of the vancomycin (Pelzer et al., 1999; van Wageningen et al., 1998) and teichoplanin (Li et al., 2004; Pootoolal et al., 2002; Sosio et al., 2003) family biosynthetic gene clusters sequenced to date, which includes three further genes (*dpgB-D*), all members of the crotonase super-family. The gene products DpgB and DpgD are most similar to enoyl-CoA hydratases (and with each other), whilst DpgC shares most similarity with characterized dehydrogenases.

4.2. *In vivo* studies of DHPG biosynthesis

Early efforts to investigate glycopeptide biosynthesis *in vivo* were frustrated by difficulty in transforming the producing organisms. A major breakthrough arose with the development of a method for transforming vegetative cell mass of the balhimycin producer *Amycolatopsis balhimycina* (formally *Amycolatopsis mediterranei*) (Pelzer et al., 1997, 1999). Mutation of *dpgA* was achieved by in-frame chromosomal deletion of 954 of the 1119 bp of coding DNA (Pfeifer et al., 2001). Unlike the parent strain, the resulting mutant, *A. balhimycina* VP1-2, was unable to produce any antibacterial substances as determined by bioassay versus glycopeptide-sensitive *B. subtilis*. Supplementation of the mutant VP1-2 with synthetic DHPA restored antibiotic formation, indicating the intermediacy of this compound in balhimycin, and thus DHPG biosynthesis. Further evidence for the action of DpgA was obtained after its heterologous expression in *Streptomyces lividans* T7 (the resulting mutant was termed VP2). Culture filtrates of *S. lividans* T7/VP2 were extracted and shown to contain DHPA using LC–MS analysis versus authentic standards, whereas the untransformed strain did not yield this compound. Expression of *dpgA-D* in this same strain led to the accumulation of 3,5-dihydroxyphenylglyoxylate, the penultimate compound in the biosynthesis of DHPG. Thus, the action of at least one of these gene products is sufficient for the 4-electron oxidation of the DHPA(-CoA) to 3,5-dihydroxyphenylglyoxylate.

The final step of DHPG biosynthesis requires transamination of the α-keto group of 3,5-hihydroxyphenylglyoxylate. Analysis of the balhimycin gene cluster identified two aminotransferases, one of which is involved in deoxysugar biosynthesis. The second gene, *pgaT*, significantly resembles the characterized 4-hydroxyphenylglycine aminotransferase *hpgT* from the chloroeromomycin cluster, leading to the assumption that the same enzyme may be used to make both amino acid precursors. To investigate this possibility *pgaT* was mutated through in-frame deletion of 657 bp, and the resulting mutant was unable to produce any antibiotic as determined by bio- and HPLC-assay (Pfeifer et al., 2001). Analysis of culture filtrates indicated the accumulation of both DHPA, which is not detected in the parent strain, and 3,5-dihydroxyphenylglyoxylate at around 10-fold greater concentration than in the parent strain, indicating a block in the subsequent conversion of this intermediate. Simultaneous addition of both (S)-DHPG and (S)-4-hydroxyphenylglycine resulted in the production of balhimycin, whereas in contrast, when either compound was fed to the mutant independently, no balhimycin was produced. These combined data clearly identified the dual function of HpgT.

4.3. *In vitro* studies of DHPG biosynthesis

Biosynthesis of a tetraketide by DpgA requires the condensation of a starter acid and three malonyl-CoA extender units. Given the β-polyketone folding pattern shown in Fig. 14.6, cyclization requires either the deprotonation of a methylketone rather than the more usual (more acidic) methylene position. An alternative possibility is the use of malonyl-CoA as starter acid, with the more favorable decarboxylation of this terminal carboxylate group generating the required anion. To study this, DpgA from the chloroeromomycin cluster was expressed in and isolated from *E. coli* BL21 using pET28 (Li *et al.*, 2001). The N-terminal hexahistidine-tagged protein was purified by nickel affinity chromatography and studied using $[1,2,3\text{-}^{13}C_3]$ malonyl-CoA, which was generated *in situ* by transfer of coenzyme-A from acetoacetyl-CoA to $[1,2,3\text{-}^{13}C_3]$malonic acid using succinyl-CoA transferase. The mass spectrum of the resulting DHPA gave only an $[M + 8]^+$ ion indicating that all the carbon atoms came from malonyl-CoA. This was consistent with data using cell-free extracts of the *dpgA*-expressing mutant *S. lividans* T7/VP2 (described above), which converted ^{14}C-malonyl-CoA to ^{14}C-labeled DHPA (Pfeifer *et al.*, 2001). These authors further showed that no radioactivity from ^{14}C-labeled acetyl-CoA was incorporated into DHPA.

In subsequent work, DpgA-D from the chloroeromomycin cluster were examined in greater detail (Chen *et al.*, 2001). Surprisingly, almost no turnover was observed for DpgA until DpgB was added, giving a net K_{cat} of 1–2 min^{-1} at a 3:1 ratio of DpgB:DpgA. This corresponded to a 17-fold rate increase over DpgA alone (calculated as the rate of appearance of CoASH). DpgD had no effect on DpgA alone, but when added as part of a three-component mixture it produced a further twofold increase in rate. It is unclear exactly what role these two proteins play in the production of DHPA-CoA, but it can be envisaged that they act to aid in performing the dehydration reactions necessary to produce DHPA-CoA. When assayed independently they were shown to exhibit (albeit weak) enoyl-CoA hydratase activity.

The next, extremely interesting, step in the pathway involves a 4-electron oxidation of the α-carbon of DHPA-CoA and subsequent hydrolysis to yield 3,5-hihydroxyphenylglyoxylate. This reaction is catalyzed by DpgC which displays a K_{cat} of 10 min^{-1} (Chen *et al.*, 2001). Phenylacetyl-CoA is also a substrate but with a 500-fold lower K_{cat}/K_m. DpgC does not require any cofactors and is oxygen dependent, indicating a new catalytic role for this member of the crotonase super-family (Tseng *et al.*, 2004). Incubation of DpgC in D_2O under anaerobic conditions demonstrated an enzyme-dependent exchange of the α-protons of DHPA-CoA, implicating a C2-anion as an early pathway intermediate. Further experiments using $^{18}O_2$ established that both oxygen atoms of dioxygen are transferred to DHPA.

The biochemistry of this pathway was further clarified with the crystal structure of DpgC in complex with a synthetic substrate analogue (Fig. 14.7) (Widboom *et al.*, 2007). Thioester enolate formation is facilitated by the presence of an oxyanion hole, and an ordered water molecule is perfectly located to deprotonate the pro*R* hydrogen atom at C2. Molecular oxygen sits in a hydrophobic pocket adjacent to the substrate and is correctly orientated through a backbone NH (Ile324) to react with the substrate. The reaction is then believed to occur as a two-step process with transfer of an electron from the electron-rich bound substrate anion to triplet oxygen, giving a conjugated radial cation/superoxide pair. These then collapse to a peroxide intermediate in a spin-allowed process, followed by reaction with the correctly orientated thioester to eliminate CoASH. The α-proton of the resulting 1,2-dioxetanone is then correctly located for abstraction by the ordered water molecule used for enolate formation to give 3,5-hihydroxyphenylglyoxylate.

4.4. Variation of the balhimycin DHPG moiety by mutasynthesis

As described above, the mutant *A. balhimycina* VP1-2 disrupted in *dpgA* does not produce balhimycin unless fed with DHPA or a later biosynthetic intermediate (Pfeifer *et al.*, 2001). This strain was used for the mutasynthesis

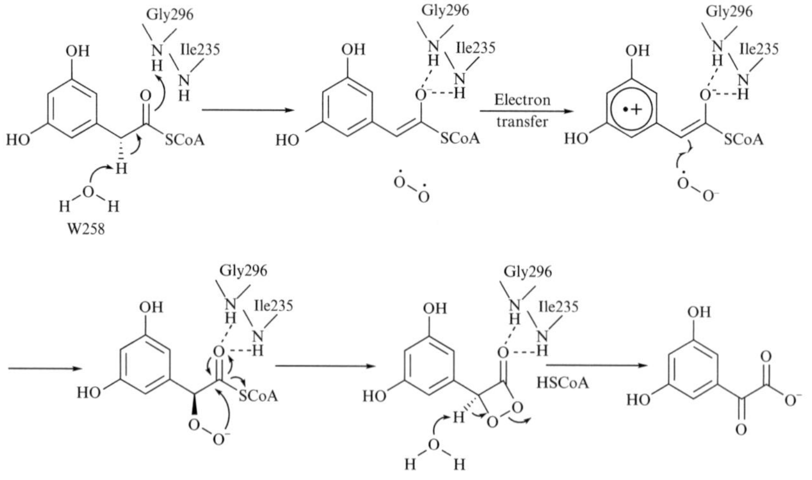

Figure 14.7 Catalytic mechanism of DHPA oxidation by DpgC.

of novel balhimycins through the exogenous addition of DHPG analogues (Weist et al., 2004). Four differentially substituted analogues of DHPG were accepted and processed to the mature tricyclic scaffold as detected through bioassay against *B. subtilis*, and confirmed by high-resolution electrospray mass spectroscopy. For 3-hydroxyPG and 3-methoxyPG, significant quantities of bicyclic compounds lacking the AB-biaryl linkage were additionally produced, indicating a narrow tolerance for the processing oxygenases. As is the case for the parent balhimycins, a spectrum of post-NRPS glycosylation was observed; all products contained the D-glucose moiety attached at the 4-hydroxyPG residue, but only those derived from 3-hydroxy-5-methoxyPG and 3,5-dimethoxyPG displayed a second glycosylation with D-4-oxovancosamine.

4.5. Bmt biosynthesis (cyclosporin precursor)

Cyclosporin (Fig. 14.1) is an NRPS-derived natural product of fungal origin (*Tolypocladium niveum*) which exhibits potent immunosuppressive activity and is used clinically to prevent allograft rejection. Two of the undecapeptide residues are nonproteinogenic, including the unique (4R)-[(E)-2-butenyl)]-4-methyl-L-threonine (Bmt) (Fig. 14.8). Bmt is a freely diffusible intermediate with biosynthetic origins in polyketide metabolism, as determined after feeding isotope-labeled sodium acetate, methionine, and glucose (Kobel et al., 1983; Senn et al., 1991) (Fig. 14.8). Subsequent *in vivo* study with doubly labeled [1-^{13}C$_1$,^{18}O$_2$]sodium acetate clearly identified the origin of the oxygen atom at C3 as acetate (from a malonyl-CoA extender unit) by virtue of an appropriate upfield shift in the ^{13}C-NMR spectrum of the isolated cyclosporin (Offenzeller et al., 1993). This ruled out introduction of the α-amino group via attack of ammonia upon a

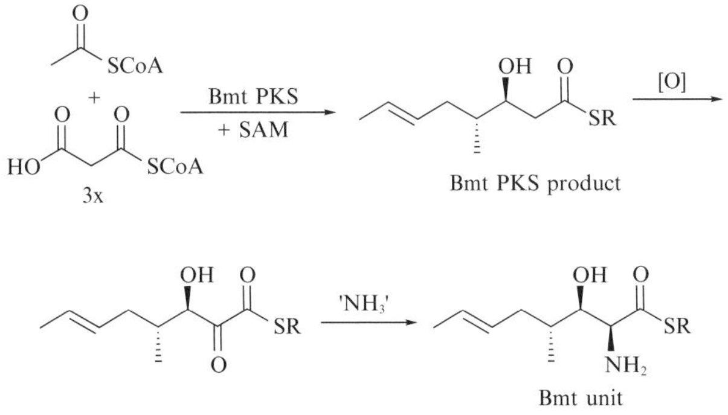

Figure 14.8 Biosynthesis of the cyclosporine Bmt unit: the Bmt PKS-derived intermediate, probably as the CoA thioester (R = SCoA), is subjected to 4-electron oxidation at C2 and subsequent transamination. Presumably, the final Bmt unit is utilized by the NRPS as the free acid (R = OH). SAM = S-adenosylmethionine.

2,3-epoxide intermediate (which would arise from post-PKS modification of 4(R)-methyl-(E,E)-2,6-octadienoic acid). Further *in vitro* experiments with partially purified PKS and authentic (synthetic) samples of the polyketide product and its analogues then identified 3(R)-hydroxy-4(R)-methyl-6 (E)-octenoic acid as the PKS product (Offenzeller et al., 1993). Subsequent work with purified PKS showed that methylation occurs during the second chain-extension cycle, and that the final PKS product was offloaded as the CoA-thioseter (Offenzeller et al., 1996). These steps are presumably followed by an (overall) 4-electron oxidation at C2 to yield an α-keto product. Given the DpgC-catalyzed oxidation in DHPG biosynthesis discussed above, it is tempting to speculate that the CoA-thioester produced by the Bmt PKS may be similarly oxidized and cleaved to the free α-ketoacid product for subsequent transamination to yield free Bmt for use by the cyclosporin NRPS.

4.6. 3-Methoxy-5-methylnapthoic acid biosynthesis

Azinomycin B is a natural product from *Streptomyces sahachiroi*, which uses an NRPS template mechanism to assemble precursors of complex biosynthetic origin (Zhao et al., 2008) (Fig. 14.9). Among these precursors is 3-methoxy-5-methylnapthoic acid (NPA), a hexaketide whose origins were confirmed by stable-isotope labeling studies (Corre et al., 2004a,b), and whose biosynthesis is catalyzed by an iterative type I PKS, AziB (Fig. 14.9). AziB displays significant similarity to a number of PKSs that generate component moieties for numerous natural products, in particular to NcsB of the neocarzostatin pathway responsible for biosynthesis of

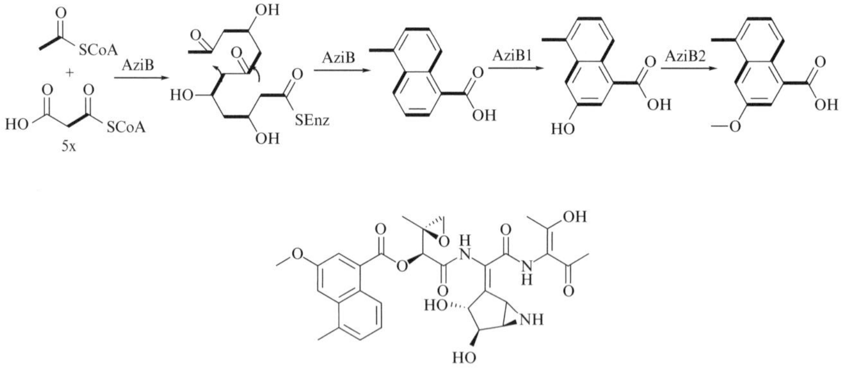

Figure 14.9 Biosynthesis of 3-methoxy-5-methyl-NPA by *S. sahachiroi*: AziA catalyzes five iterative rounds of chain extension with β-keto reduction occurring during rounds 2, 5, and 5. After cyclization/aromatization, the hexaketide product is subject to regiospecific oxidation and O-methylation. The structure of azinomycin B is shown. The incorporation of labeled [1,2-$^{13}C_2$]acetate units is shown as bold bonds.

5-methylnapthoic acid, which is most likely linked to neocarzostatin through a CoA-ligase-mediated step (Liu et al., 2005).

The initial product of AziB was shown to be 5-methylnapthoic acid by heterologous expression in *Streptomyces albus* (Zhao et al., 2008). Biosynthesis of this hexaketide requires three cycle-specific β-keto reduction events during the second, fourth, and fifth rounds of chain extension. The PKS product is converted to NPA by the action of a cytochrome P450 monooxygenase (AziB1) and O-methyltransferase (AziB2) whose functions were also confirmed by heterologous coexpression with *aziB* leading to NPA biosynthesis in *S. albus*.

5. Glycosyl Building Blocks

Although the biosynthesis of deoxysugars and the action of glycosyltransferases are the topic of other chapters in this volume, it is worth noting that numerous NRPS products are further modified by glycosylation to yield final biologically active molecules. To exemplify this, we use the theme of glycopeptide antibiotics already introduced in this chapter. The peptide core of molecules such as vancomycin, chloroemomycin, and balhimycin are all glycosylated during the final steps of biosynthesis, specifically with L-vancosamine, L-4-epivancosamine, and L-4-ketovancosamine, respectively (these are also transferred with differing regiochemistry). To investigate the biosynthesis of the chloroemomycin constituent L-4-epivancosamine, five of the proteins involved in its biosynthesis were isolated and biochemically characterized after heterologous expression in *E. coli* (Chen et al., 2000) (Fig. 14.10). This allowed their function to be verified and the order of biosynthetic steps to be elucidated. In a powerful example of *in vitro*

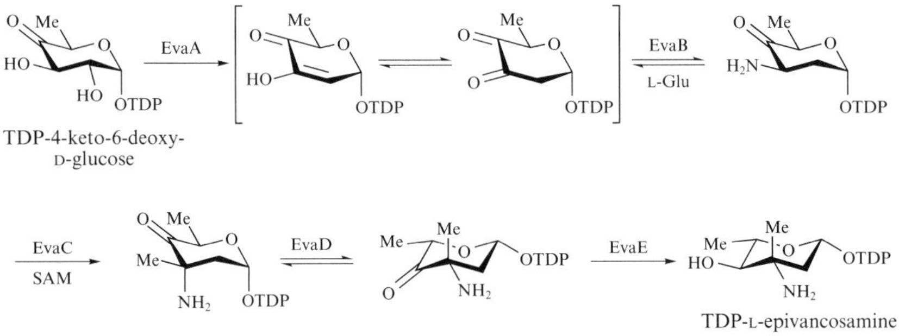

Figure 14.10 Biosynthesis of TDP-L-epivancosamine from TDP-4-keto-6-deoxy-D-glucose: EvaA, 2,3-dehydratase; EvaB, 3-transaminase (L-Glu = L-glutamate); EvaC, 3-methyltransferase (SAM = S-adenosylmethionine); EvaD, 5-epimerase; EvaE, 4-ketoreductase.

synthesis, the entire pathway from the common biosynthetic precursor TDP-6-deoxy-4-keto-D-glucose to L-4-epivancosamine was reconstructed. The details of L-4-epivancosamine biosynthesis are shown in Fig. 14.10.

The sugar moieties of glycopeptides have been targeted for alteration via engineering methods, as their nature and subsequent modification is crucial in determining biological activity. For example, the second L-4-epivancosamine moiety of chloroemomycin is thought to be responsible of its enhanced activity against vancomycin-resistant enterococci, as is the N-acylation of teichoplanin. Through the predominately *in vitro* application of glycosyltransferases from the vancomycin and chloroemomycin pathways, a range of novel compounds have been prepared. Of particular note was a study in which GtfE from the vancomycin pathway was used to attach 2-, 3-, 4-, and 6-deoxy and -aminoglucose sugars onto the vancomycin and teichoplanin aglycones (Losey *et al.*, 2002). GtfD from the same pathway was then able to further modify these compounds with L-4-epivancosamine (apart from those with a 2-deoxyglucose) to generate disaccharides. In other reports, synthetic sugars containing azide functions were incorporated attached to the vancomycin aglycone (Fu *et al.*, 2003). Subsequent use of click-type chemistry using substituted alkynes extended the structural variation of these monoglycosylated analogues, and allowed identification of a compound rivaling vancomycin for potency and with an altered antibacterial spectrum.

6. Conclusion

Nonribosomal peptide building blocks are derived from a wide range of sources including amino acid, fatty acid, polyketide, carbohydrate, and terpene metabolism. Assembly of these building blocks by NRPS and NRPS–PKS multienzyme systems, followed by further tailoring steps gives rise to some of the most complex and structurally diverse secondary metabolites in nature. It is hardly surprising that nonribosomal peptide precursor biosynthesis has been an area of intense study. Indeed, the massive increase in genomics data over the last 5 years has led to the discovery of many new pathways and novel enzymes associated with the production of nonribosomal peptides in bacteria and fungi. Here, using some important selected examples, the main experimental methodologies that have been brought to bear in the elucidation of precursor supply pathways are described. This draws on a wide variety of techniques and expertise from a range of scientific disciplines including molecular genetics, enzymology, and chemistry. The fundamental knowledge generated through these studies underpins the development of new strategies for biosynthetic engineering of nonribosomal peptides, which can lead to new products for therapeutic and other applications.

REFERENCES

Aron, Z. D., Fortin, P. D., Calderone, C. T., and Walsh, C. T. (2007). FenF: Servicing the mycosubtilin synthetase assembly line *in trans*. *ChemBioChem* **8,** 613–616.

Balibar, C. J., Vaillancourt, F. H., and Walsh, C. T. (2005). Generation of D-amino acid residues in assembly of arthrofactin by dual condensation/epimerization domains. *Chem. Biol.* **12,** 1189–1200.

Bucca, G., Brassington, A. M. E., Hotchkiss, G., Mersinias, V., and Smith, C. P. (2003). Negative feedback regulation of *dnaK, clpB* and *lon* expression by the DnaK chaperone machine in *Streptomyces coelicolor* identified by transcriptome and *in vivo* DnaK-depletion analysis. *Mol. Microbiol.* **50,** 153–166.

Chen, H., Hubbard, B. K., O'Connor, S. E., and Walsh, C. T. (2002). Formation of β-hydroxyhistidine in the biosynthesis of nikkomycin antibiotics. *Chem. Biol.* **9,** 103–112.

Chen, H., Thomas, M. G., Hubbard, B. K., Losey, H. C., Walsh, C. T., and Burkart, M. D. (2000). Deoxysugars in glycopeptide antibiotics: Enzymatic synthesis of TDP-L-epivancosamine in chloroeremomycin biosynthesis. *Proc. Natl. Acad. Sci. USA* **97,** 11942–11947.

Chen, H., Thomas, M. G., O'Connor, S. E., Hubbard, B. K., Burkart, M. D., and Walsh, C. T. (2001). Aminoacyl-S-enzyme intermediates in β-hydroxylation and α,β-desaturation of amino acids in peptide antibiotics. *Biochemistry* **40,** 11651–11659.

Chen, H., Tseng, C. C., Hubbard, B. K., and Walsh, C. T. (2001). Glycopeptide antibiotic biosynthesis: Enzymatic assembly of the dedicated amino acid monomer (S)-3,5-dihydroxyphenylglycine. *Proc. Natl. Acad. Sci. USA* **98,** 14901–14906.

Cheng, Y.-Q., and Walton, J. D. (2000). A eukaryotic alanine racemase gene involved in cyclic peptide biosynthesis. *J. Biol. Chem.* **275,** 4906–4911.

Christianson, C. V., Montavon, T. J., Festin, G. M., Cooke, H. A., Shen, B., and Bruner, S. D. (2007). The mechanism of MIO-based aminomutases in β-amino acid biosynthesis. *J. Am. Chem. Soc.* **129,** 15744–15745.

Corre, C., Landreau, C. A., Shipman, M., and Lowden, P. A. (2004a). Biosynthetic studies on the azinomycins: The pathway to the naphthoate fragment. *Chem. Commun.* 2600–2601.

Corre, C., and Lowden, P. A. (2004b). The first biosynthetic studies of the azinomycins: Acetate incorporation into azinomycin B. *Chem. Commun.* 990–991.

Dong, C., Flecks, S., Unversucht, S., Haupt, C., van Pée, K.-H., and Naismith, J. H. (2005). Tryptophan 7-halogenase (PrnA) structure suggests a mechanism for regioselective chlorination. *Science* **309,** 2216–2218.

Duitman, E. H., Hamoen, L. W., Rembold, M., Venema, G., Seitz, H., Saenger, W., Bernhard, F., Reinhardt, R., Schmidt, M., Ullrich, C., Stein, T., Leenders, F., *et al.* (1999). The mycosubtilin synthetase of *Bacillus subtilis* ATCC6633: A multifunctional hybrid between a peptide synthetase, an amino transferase, and a fatty acid synthase. *Proc. Natl. Acad. Sci. USA* **96,** 13294–13299.

Edwards, D. J., and Gerwick, W. H. (2004). Lyngbyatoxin biosynthesis: Sequence of biosynthetic gene cluster and identification of a novel aromatic prenyltransferase. *J. Am. Chem. Soc.* **126,** 11432–11433.

Feifel, S. C., Schmiederer, T., Hornbogen, T., Berg, H., Süssmuth, R. D., and Zocher, R. (2007). *In vitro* synthesis of new enniatins: Probing the α-D-hydroxy carboxylic acid binding pocket of the multienzyme enniatin synthetase. *ChemBioChem* **8,** 1767–1770.

Fortin, P. D., Walsh, C. T., and Magarvey, N. A. (2007). A transglutaminase homologue as a condensation catalyst in antibiotic assembly lines. *Nature* **448,** 824–828.

Fu, X., Albermann, C., Jiang, J., Liao, J., Zhang, C., and Thorson, J. S. (2003). Antibiotic optimization via *in vitro* glycorandomization. *Nat. Biotech.* **21,** 1467–1469.

Fujimori, D. G., Hrvatin, S., Neumann, C. S., Strieker, M., Marahiel, M. A., and Walsh, C. T. (2007). Cloning and characterization of the biosynthetic gene cluster for kutznerides. *Proc. Natl. Acad. Sci. USA* **104**, 16498–16503.

Haltli, B., Tan, Y., Magarvey, N. A., Wagenaar, M., Yin, X., Greenstein, M., Hucul, J. A., and Zabriskie, T. M. (2005). Investigating β-hydroxyenduracididine formation in the biosynthesis of the mannopeptimycins. *Chem. Biol.* **12**, 1163–1168.

Hammond, S. J., Williams, D. H., and Neilsen, R. V. (1983). The biosynthesis of ristocetin. *J. Chem. Soc. Chem. Commun.* 116–117.

Hammond, S. J., Williamson, M. P., Williams, D. H., Boeck, L. D., and Marconi, G. G. (1982). On the biosynthesis of the antibiotic vancomycin. *J. Chem. Soc. Chem. Commun.* 344–345.

Hansen, D. B., Bumpus, S. B., Aron, Z. D., Kelleher, N. L., and Walsh, C. T. (2007). The loading module of mycosubtilin: An adenylation domain with fatty acid selectivity. *J. Am. Chem. Soc.* **129**, 6366–6367.

Heidari, B. A., and Micklefield, J. (2007a). NMR confirmation that tryptophan dehydrogenation occurs with syn-stereochemistry during the biosynthesis of cda in *Streptomyces coelicolor*. *J. Org. Chem.* **72**, 8950–8953.

Heidari, B. A., Thirlway, J., and Micklefield, J. (2007b). Stereochemical course of tryptophan dehydrogenation during calcium-dependent lipopeptide antibiotic biosynthesis. *Org. Lett.* **9**, 1513–1516.

Hicks, L. M., Moffit, M. C., Beer, L. L., Moore, B. S., and Kelleher, N. L. (2006). Structural characterization of *in vitro* and *in vivo* intermediates on the loading module of microcystin synthetase. *ACS Chem. Biol.* **1**, 93–102.

Hojati, Z., Milne, C., Harvey, B., Gordon, L., Borg, M., Flett, F., Wilkinson, B., Rudd, B. A. M., Hayes, M. A., Smith, C. P., and Micklefield, J. (2002). Structure, biosynthetic origin, and engineered biosynthesis of calcium-dependent antibiotics from *Streptomyces coelicolor*. *Chem. Biol.* **9**, 1175–1187.

Huber, F. M., Pieper, R. L., and Tietz, A. J. (1988). The formation of daptomycin by supplying decanoic acid to *Streptomyces roseosporus* cultures producing the antibiotic complex A21978C. *J. Biotechnol.* **7**, 283–292.

Ju, J., Ozanick, S. G., Shen, B., and Thomas, M. G. (2004). Conversion of (2S)-arginine to (2S,3R)-capreomycidine by VioC and VioD from the viomycin biosynthetic pathway of *Streptomyces* sp. strain ATCC11861. *ChemBioChem* **5**, 1281–1285.

Kempter, C., Kaiser, D., Haag, S., Nicholson, G., Gnau, V., Walk, T., Gierling, G. H., Decker, H., Zahner, H., Jung, G., and Metzger, J. W. (1997). CDA: Calcium-dependent peptide antibiotics from *Streptomyces coelicolor* A3(2) containing unusual residues. *Angew. Chem. Int. Ed. Engl.* **36**, 498–501.

Kobel, H., Loosli, H. R., and Voges, R. (1983). Contribution to knowledge of the biosynthesis of cyclosporin A. *Experientia (Basel)* **39**, 873–876.

Kopp, F., Linne, U., Oberthür, M., and Marahiel, M. A. (2008). Harnessing the chemical activation inherent to carrier protein-bound thioesters for the characterization of lipopeptide fatty acid tailoring enzymes. *J. Am. Chem. Soc.* **130**, 2656–2666.

Kruger, R. G., Oberthür, W. L., Tao, J., Kahne, D., and Walsh, C. T. (2005). Tailoring of glycopeptide scaffolds by the acyltransferases from the teicoplanin and A-40,926 biosynthetic operons. *Chem. Biol.* **12**, 131–140.

Lepore, B. W., Ruzicka, F. J., Frey, P. A., and Ringe, D. (2005). The x-ray crystal structure of lysine-2,3-aminomutase from *Clostridium subterminale*. *Proc. Natl. Acad. Sci. USA* **102**, 13819–13824.

Li, J., and Jensen, S. E. (2008). Nonribosomal biosynthesis of fusaricidins by paenibacillus polymyxa PKB1 involves direct activation of a D-amino acid. *Chem. Biol.* **15**, 118–127.

Li, T.-L., Choroba, O. W., Hong, H., Williams, D. H., and Spencer, J. B. (2001). Biosynthesis of the vancomycin group of antibiotics: Characterisation of a type III polyketide synthase in the pathway to (S)-3,5-dihydroxyphenylglycine. *Chem. Commun.* 2156–2157.

Li, T.-L., Huang, F., Haydock, S. F., Mironenko, T., Leadlay, P. F., and Spencer, J. B. (2004). Biosynthetic gene cluster of the glycopeptide antibiotic teicoplanin: Characterization of two glycosyltransferases and the key acyltransferase. *Chem. Biol.* **11,** 107–119.

Liu, W., Nonaka, K., Nie, L., Zhang, J., Christenson, S. D., Bae, J., Van Lanen, S. G., Zazopoulos, E., Farnet, C. M., Yang, C. F., and Shen, B. (2005). The neocarzinostatin biosynthetic gene cluster from *Streptomyces carzinostaticus* ATCC 15944 involving two iterative type I polyketide synthases. *Chem. Biol.* **12,** 293–302.

Losey, H. C., Jiang, J., Biggins, J. B., Oberthür, M., Ye, Y.-Y., Dong, S. D., Kahne, D., Thorson, J. S., and Walsh, C. T. (2002). Incorporation of glucose analogs by GtfE and GtfD from the vancomycin biosynthetic pathway to generate variant glycopeptides. *Chem. Biol.* **9,** 1305.

Lu, W., Oberthur, M., Leimkuhler, C., Tao, J., Kahne, D., and Walsh, C. T. (2004). Characterization of a regiospecific epivancosaminyl transferase GtfA and enzymatic reconstitution of the antibiotic chloroeremomycin. *Proc. Natl. Acad. Sci. USA* **101,** 4390–4395.

Magarvey, N. A., Ehling-Schulz, M., and Walsh, C. T. (2006). Characterization of the cereulide NRPS α-hydroxy acid specifying modules: Activation of α-keto acids and chiral reduction on the assembly line. *J. Am. Chem. Soc.* **128,** 10698–10699.

Magarvey, N. A., Haltli, B., He, M., Greenstein, M., and Hucul, J. A. (2006). Biosynthetic pathway for mannopeptimycins, lipoglycopeptide antibiotics active against drug-resistant gram-positive pathogens. *Antimicrob. Agents Chemother.* **50,** 2167–2177.

Mahlert, C., Kopp, F., Thirlway, J., Micklefield, J., and Marahiel, M. A. (2007). Stereospecific enzymatic transformation of α−ketoglutarate to (2S,3R)-3-methyl glutamate during acidic lipopeptide biosynthesis. *J. Am. Chem. Soc.* **129,** 12011–12018.

Marahiel, M. A., Stachelhaus, T., and Mootz, H. D. (1997). Modular peptide synthetases involved in nonribosomal peptide synthesis. *Chem. Rev.* **97,** 2651–2673.

McCafferty, D. G., Cudic, P., Frankel, B. A., Barkallah, S., Kruger, R. G., and Li, W. (2002). Chemistry and biology of the ramoplanin family of peptide antibiotics. *Pept. Sci.* **66,** 261–284.

McGahren, W. J., Martin, J. H., Morton, G. O., Hargreaves, R. T., Leese, R. A., Lovell, F. M., Ellestad, G. A., O'Brien, E., and Holker, J. S. E. (1980). Structure of avoparcin components. *J. Am. Chem. Soc.* **102,** 1671–1684.

Miao, V., Brost, R., Chapple, J., She, K., Coëffet-Le Gal, M.-F., and Baltz, R. H. (2006). The lipopeptide antibiotic A54145 biosynthetic gene cluster from *Streptomyces fradiae*. *J. Ind. Microbiol. Biotechnol.* **33,** 129–140.

Miao, V., Coëffet-Le Gal, M.-F., Brian, P., Brost, R., Penn, J., Whiting, A., Martin, S., Ford, R., Parr, I., Bouchard, M., Silva, C. J., Wrigley, S. K., *et al.* (2005). Daptomycin biosynthesis in *Streptomyces roseosporus*: Cloning and analysis of the gene cluster and revision of peptide stereochemistry. *Microbiology* **151,** 1507–1523.

Milne, C., Powell, A., Jim, J., Al Nakeeb, M., Smith, C. P., and Micklefield, J. (2006). Biosynthesis of the (2S,3R)-3-methyl glutamate residue of nonribosomal lipopeptides. *J. Am. Chem. Soc.* **128,** 11250–11259.

Neary, J., Powell, A., Gordon, L., Milne, C., Flett, F., Wilkinson, B., Smith, C. P., and Micklefield, J. (2007). An asparagine oxygenase (AsnO) and a 3-hydroxyasparaginyl phosphotransferase (HasP) are involved in the biosynthesis of calcium dependent lipopeptide antibiotics. *Microbiology* **153,** 768–776.

Nelson, J. T., Lee, J., Sims, J. W., and Schmidt, E. W. (2007). Characterization of SafC, a catechol 4-O-methyltransferase involved in saframycin biosynthesis. *Appl. Environ. Microbiol.* **73,** 3575–3580.

Nguyen, K. T., Kau, D., Gu, J.-Q., Brian, P., Wrigley, S. K., Baltz, R. H., and Miao, V. (2006). A glutamic acid 3-methyltransferase encoded by an accessory gene locus important for daptomycin biosynthesis in *Streptomyces roseosporus*. *Mol. Microbiol.* **61,** 1294–1307.

Offenzeller, M., Santer, G., Totschnig, K., Su, Z., Moser, H., Traber, R., and Schneider-Scherzer, E. (1996). Biosynthesis of the unusual amino acid (4R)-4-[(E)-2-butenyl]-4-methyl-L-threonine of cyclosporin A: Enzymatic analysis of the reaction sequence including identification of the methylation precursor in a polyketide pathway. *Biochemistry* **35,** 8401–8412.

Offenzeller, M., Su, Z., Santer, G., Moser, H., Traber, R., Memmert, K., and Schneider-Scherzer, E. (1993). Biosynthesis of the unusual amino acid (4R)-4-[(E)-2-butenyl]-4-methyl-L-threonine of cyclosporin A. Identification of 3(R)-hydroxy-4(R)-methyl-6(E)-octenoic acid as a key intermediate by enzymatic *in vitro* synthesis and by *in vivo* labeling techniques. *J. Biol. Chem.* **268,** 26127–26134.

Pelzer, S., Huppert, M., Reichert, W., Heckmann, D., and Wohlleben, W. (1997). Cloning and analysis of a peptide synthetase gene of the balhimycin producer *Amycolatopsis mediterranei* DSM5908 and development of a gene disruption/replacement system. *J. Biotechnol.* **56,** 115–128.

Pelzer, S., Süssmuth, R. D., Heckmann, D., Recktenwald, J., Huber, P., Jung, G., and Wohlleben, W. (1999). Identification and analysis of the balhimycin biosynthetic gene cluster and its use for manipulating glycopeptide biosynthesis in *Amycolatopsis mediterranei* DSM5908. *Antimicrob. Agents Chemother.* **43,** 1565–1573.

Pfeifer, V., Nicholson, G. J., Ries, J., Recktenwald, J., Schefer, A. B., Shawky, R. M., Schröder, J., Wohlleben, W., and Pelzer, S. (2001). A polyketide synthase in glycopeptide biosynthesis: The biosynthesis of the non-proteinogenic amino acid (S)-3,5-dihydroxyphenylglycine. *J. Biol. Chem.* **276,** 38370–38377.

Pootoolal, J., Thoma, M. G., Marshall, C. G., Neu, J. M., Hubbard, B. K., Walsh, C. T., and Wright, G. D. (2002). Assembling the glycopeptide antibiotic scaffold: The biosynthesis of A47934 from *Streptomyces toyocaensis* NRRL15009. *Proc. Natl. Acad. Sci. USA* **99,** 8962–8967.

Powell, A., Al Nakeeb, M., Wilkinson, B., and Micklefield, J. (2007a). Precursor-directed biosynthesis of nonribosomal lipopeptides with modified glutamate residues. *Chem. Commun.* 2683–2685.

Powell, A., Borg, M., Amir-Heidari, B., Neary, J. M., Thirlway, J., Wilkinson, B., Smith, C. P., and Micklefield, J. (2007b). Engineered biosynthesis of nonribosomal lipopeptide antibiotics with modified fatty acid side chains. *J. Am. Chem. Soc.* **129,** 15182–15191.

Puk, O., Bischoff, D., Kittel, C., Pelzer, S., Weist, S., Stegmann, E., Süssmuth, R. D., and Wohlleben, W. (2004). Biosynthesis of chloro-β-hydroxytyrosine, a nonproteinogenic aminoacid of the peptidic backbone of glycopeptide antibiotics. *J. Bacteriol.* **186,** 6093–6100.

Puk, O., Huber, P., Bischoff, D., Recktenwald, J., Jung, G., Süssmuth, R. D., van Pée, K.-H., Wohlleben, W., and Pelzer, S. (2002). Glycopeptide biosynthesis in *Amycolatopsis mediterranei* DSM5908: Function of a halogenase and a haloperoxidase/perhydrolase. *Chem. Biol.* **9,** 225–235.

Rachid, S., Krug, D., Weissman, K. J., and Müller, R. (2007). Biosynthesis of (R)-β-tyrosine and its incorporation into the highly cytotoxic chondramides produced by *Chondromyces crocatus*. *J. Biol. Chem.* **282,** 21810–21817.

Sandercock, A. M., Charles, E. H., Scaife, W., Kirkpatrick, P. N., O'Brien, S. W., Papageorgiou, E. A., Spencer, J. B., and Williams, D. H. (2001). Biosynthesis of the di-meta-hydroxyphenylglycine constituent of the vancomycin-group antibiotic chloroeremomycin. *Chem. Commun.* 1252–1253.

Schultz, A. W., Oh, D.-C., Carney, J. R., Williamson, R. T., Udwary, D. W., Jensen, P. R., Gould, S. J., Fenical, W., and Moore, B. A. (2008). Biosynthesis and structures of cyclomarins and cyclomarazines, prenylated cyclic peptides of marine actinobacterial origin. *J. Am. Chem. Soc.* **130,** 4507–4516.

Senn, H., Weber, C., Kobel, H., and Traber, R. (1991). Selective ^{13}C-labelling of cyclosporin A. *Eur. J. Biochem.* **199,** 653–658.

Sieber, S. A., and Marahiel, M. A. (2005). Molecular mechanisms underlying nonribosomal peptide synthesis: Approaches to new antibiotics. *Chem. Rev.* **105,** 715–738.

Singh, G. M., Vaillancourt, F. H., Yin, J., and Walsh, C. T. (2007). Characterization of SyrC, an aminoacyltransferase shuttling threonyl and chlorothreonyl residues in the syringomycin biosynthetic assembly line. *Chem. Biol.* **14,** 31–40.

Sosio, M., Stinchi, S., Beltrametti, F., Lazzarini, A., and Donadio, S. (2003). The gene cluster for the biosynthesis of the glycopeptide antibiotic A40926 by Nonomuraea species. *Chem. Biol.* **10,** 541–549.

Strieker, M., Kopp, F., Mahlert, C., Essen, L.-O., and Marahiel, M. A. (2007). *ACS Chem. Biol.* **2,** 187.

Thomas, M. G., Chan, Y. A., and Ozanick, S. G. (2003). Deciphering tuberactinomycin biosynthesis: Isolation, sequencing, and annotation of the viomycin biosynthetic gene cluster. *Antimicrob. Agents Chemother.* **47,** 2823–2830.

Tillet, D., Dittmann, E., Erhard, M., von Döhren, H., Börner, T., and Neilan, B. A. (2000). Structural organization of microcystin biosynthesis in *Microcystis aeruginosa* PCC7806: An integrated peptide-polyketide synthetase system. *Chem. Biol.* **7,** 753–764.

Tseng, C. C., Vaillancourt, F. H., Bruner, S. D., and Walsh, C. T. (2004). DpgC is a metal- and cofactor-free 3,5-dihydroxyphenylacetyl-CoA 1,2-dioxygenase in the vancomycin biosynthetic pathway. *Chem. Biol.* **11,** 1195–1203.

Vaillancourt, F. H., Yin, J., and Walsh, C. T. (2005). SyrB2 in syringomycin E biosynthesis is a nonheme FeII α-ketoglutarate- and O$_2$-dependent halogenase. *Proc. Natl. Acad. Sci. USA* **102,** 10111–10116.

van Wageningen, A. M. A., Kirkpatrick, P. N., Williams, D. H., Harris, B. R., Kershaw, J. K., Lennard, N. J., Jones, M., Jones, S. J. M., and Solenberg, P. J. (1998). Sequencing and analysis of genes involved in the biosynthesis of a vancomycin group antibiotic. *Chem. Biol.* **5,** 155–162.

Walsh, C. T., Chen, H., Keating, T. A., Hubbard, B. K., Losey, H. C., Luo, L., Marshall, C. G., Miller, D. A., and Patel, H. M. (2001). Tailoring enzymes that modify nonribosomal peptides during and after chain elongation on NRPS assembly lines. *Curr. Opin. Chem. Biol.* **5,** 525–534.

Weist, S., Kittel, C., Bischoff, D., Pfeifer, V., Nicholson, G. J., Wohlleben, W., and Süssmuth, R. D. (2004). Mutasynthesis of glycopeptide antibiotics: Variations of vancomycin's AB-ring amino acid 3,5-dihydroxyphenylglycine. *J. Am. Chem. Soc.* **126,** 5942–5943.

Widboom, P. F., Fielding, E. N., Liu, Y., and Bruner, S. D. (2007). Structural basis for cofactor-independent dioxygenation in vancomycin biosynthesis. *Nature* **447,** 342–345.

Wittmann, M., Linne, U., Pohlmann, V., and Marahiel, M. A. (2008). Role of DptE and DptF in the lipidation reaction of daptomycin. *FEBS J.* **275,** 5343–5354.

Yanai, K., Sumida, N., Okakura, K., Moriya, T., Watanabe, M., and Murakami, T. (2004). *Nat. Biotech.* **22,** 848–855.

Yeh, E., Blasiak, L. C., Koglin, A., Drennan, C. L., and Walsh, C. T. (2007). Chlorination by a long-lived intermediate in the mechanism of flavin-dependent halogenases. *Biochemistry* **46,** 1284–1292.

Yin, X., and Zabriskie, T. M. (2004). VioC is a non-heme iron, α-ketoglutarate-dependent oxygenase that catalyzes the formation of 3S-hydroxy-l-arginine during viomycin biosynthesis. *ChemBioChem* **5,** 1274–1277.

Yin, X., and Zabriskie, T. M. (2006). The enduracidin biosynthetic gene cluster from *Streptomyces fungicidicus*. *Microbiology* **152,** 2969–2983.

Zhao, Q., He, Q., Ding, W., Tang, M., Kang, Q., Yu, Y., Deng, W., Zheng, Q., Fang, J., Tang, G., and Liu, W. (2008). Characterization of the azinomycin B biosynthetic gene cluster revealing a different iterative type I polyketide synthase for naphthoate biosynthesis. *Chem. Biol.* **15,** 693–705.

Zmijewski, M. J. Jr., Briggs, B., Logan, R., and Boeck, L. D. (1987). Biosynthetic studies on antibiotic A47934. *Antimicrob. Agents Chemother.* **31,** 1497–1501.

CHAPTER FIFTEEN

Plasmid-Borne Gene Cluster Assemblage and Heterologous Biosynthesis of Nonribosomal Peptides in *Escherichia coli*

Kenji Watanabe,* Alex P. Praseuth,[†] Mike B. Praseuth,[†] *and* Kinya Hotta[‡]

Contents

1. Introduction	380
2. Biosynthetic Pathway of Nonribosomal Peptides	382
3. Echinomycin Biosynthetic Pathway	383
4. Construction of A Multigene Assembly on Expression Vectors	387
5. Heterologous Gene Expression and NRP Biosynthesis in *E. coli*	390
6. Self-Resistance Mechanism	391
6.1. Echinomycin resistance assay	392
7. Stability of Transformants Carrying Multiple Very Large Plasmids	392
7.1. Examining the stability of transformant cells	393
8. Engineering of Heterologous NRP Biosynthetic Pathways in *E. coli*	393
9. Conclusion	395
References	396

Abstract

Nonribosomal peptides (NRPs) are synthesized by modular mega-enzymes called NRP synthetases (NRPSs) that catalyze a peptide bond-forming reaction using natural amino acids as substrates. Most members of this class of natural products exhibit remarkable biological activities, but many of these valuable compounds are often difficult to obtain in sufficient quantities from their natural sources due to low production levels in the producing organisms or difficulty in culturing them. Harnessing recent progress in our genetic and biochemical understanding of the biosynthesis of these nonprimary metabolites, our

---

* Research Core for Interdisciplinary Sciences, Okayama University, Okayama, Japan
[†] Department of Pharmacology and Pharmaceutical Sciences, University of Southern California, Los Angeles, California, USA
[‡] Department of Biological Sciences, National University of Singapore, Singapore

laboratory has successfully developed an alternative, straightforward approach for obtaining desired natural products by placing the entire biosynthetic gene cluster in our heterologous host of choice, *Escherichia coli*. This effort led to the first successful *de novo* production of heterologous bioactive complex NRPs in *E. coli*. Through developing our heterologous biosynthetic system, we were able to construct a novel platform suitable for generating an NRP library through rational engineering of the natural modular assembly-line array composed of NRPSs and the auxiliary enzymes. This chapter describes the basic concept in establishing an *E. coli*-based plasmid-borne heterologous NRP biosynthetic system, and gives selected protocols that have been used successfully for engineering NRP biosynthesis.

1. Introduction

Numerous secondary metabolites have been isolated and characterized from a variety of natural sources. Typically, these compounds possess complex structural features that endow them with diverse biological activities. Because these secondary metabolites are frequently used by the producing organisms as effective chemical weapons against other organisms for defending against assaults or preventing invasion of the habitat, the production of these compounds has been subjected to constant environmental selection for developing a wide range of bioactivities. As such, these compounds carry high potential as drugs or lead compounds for developing promising therapeutics that can address conditions, such as infectious diseases, that are running low on available remedies.

Through the impressive recent progress in the genetics, biochemistry, and structural biology of natural product biosynthesis, we have learned that the source of the diverse chemical architecture of the secondary metabolites comes from the broad range of activities exhibited by distinct yet rather closely related biosynthetic enzymes that most likely arose through the process of natural evolution via gene duplication and spontaneous mutagenesis under selection pressure. So, can we simulate the nature's approach and "synthesize" complex natural products by artificially evolving the natural product biosynthetic pathways rationally or randomly, without relying on the use of chemical catalysts or traditional synthetic methods? If so, then such a technology would permit us to perform a comprehensive one-pot (bio)synthesis of a desired molecule, exclusive of intermediate purification steps or disposal of expensive and environmentally harmful by-products, not to mention the cost of maintaining a staff of highly knowledgeable and skilled synthetic chemists. To date, some successes have already been obtained in realizing this goal (Donia *et al.*, 2008; Kealey *et al.*, 1998; Mutka *et al.*, 2006; Peiru *et al.*, 2005; Watanabe *et al.*, 2003, 2006).

However, to accomplish this artificial evolution of secondary metabolite biosynthetic pathways in a timely manner, a simple, robust yet flexible heterologous production system that allows quick modification and assemblage of biosynthetic genes from a wide range of sources is essential. Moreover, such a system ideally should parallel, if not surpass, the production level of the target natural product in its original producing host. Although several different organisms have been used for heterologous production of various natural products (Zhang *et al.*, 2008), we believe that *Escherichia coli* has certain advantages over others. First, genetic and molecular biological methods of manipulating *E. coli* and its vectors are well-established and reasonably straightforward to perform. For instance, introduction of plasmids that allow expression of various different genes in *E. coli* can be accomplished in a very fast, straightforward, and efficient manner. Established molecular biological techniques make insertion, deletion, or exchange of the genes in these plasmids fast and simple. A wealth of tools is already available for controlling the expression of, and modifying foreign genes in *E. coli*. For example, numerous plasmids with various useful properties are available commercially and noncommercially. These features include combinations of replication origins and selectable antibiotic resistance markers for proper maintenance of multiple plasmids within a single strain, as well as various promoters with different strength that can be induced by different inducer molecules. All of these technical advantages provide *E. coli* system flexibility and plasticity essential for the engineering of the heterologous biosynthetic system. In addition, the knowledge of the sequence of the entire *E. coli* genome helps greatly in designing and implementing modifications to the organism. Furthermore, *E. coli* is highly tolerant to expressing exogenous genes, and grows readily with a very short doubling time in a simple, inexpensive growth medium without requiring particularly difficult-to-establish culture conditions. Also, *E. coli* is naturally capable of producing siderophore NRPs (Valdebenito *et al.*, 2006) and a PK–NRP hybrid molecule colibactin (Nougayrède *et al.*, 2006), albeit at a limited level, suggesting that it should be metabolically feasible for *E. coli* to biosynthesize foreign natural products.

Thus far, a number of groups have exploited *E. coli* successfully as a heterologous host for the production of secondary metabolites (Gruenewald *et al.*, 2004; Kealey *et al.*, 1998; Mutka *et al.*, 2006; Peiru *et al.*, 2005; Pfeifer *et al.*, 2001, 2003; Watanabe *et al.*, 2003). We have also demonstrated recently *de novo* production of an antitumor NRP antibiotic, echinomycin (1), and its precursor triostin A (2) (Fig. 15.1) in *E. coli* (Watanabe *et al.*, 2006). Here, using our *E. coli* production of (1) and (2) as an example, we will describe the general method and discuss points to consider in establishing and engineering *E. coli*-based plasmid-borne systems for the heterologous biosynthesis of NRPs.

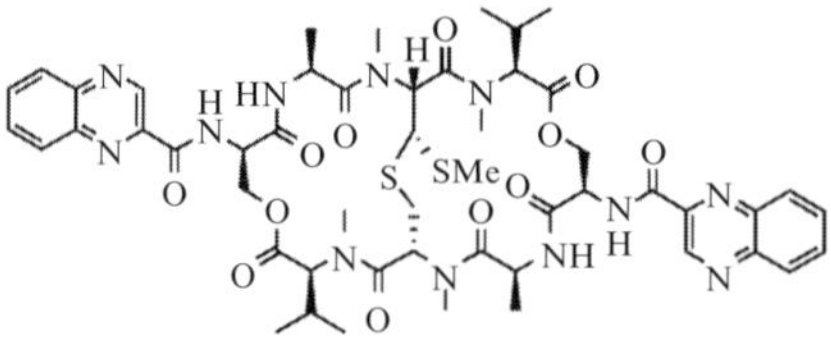

Echinomycin (1)

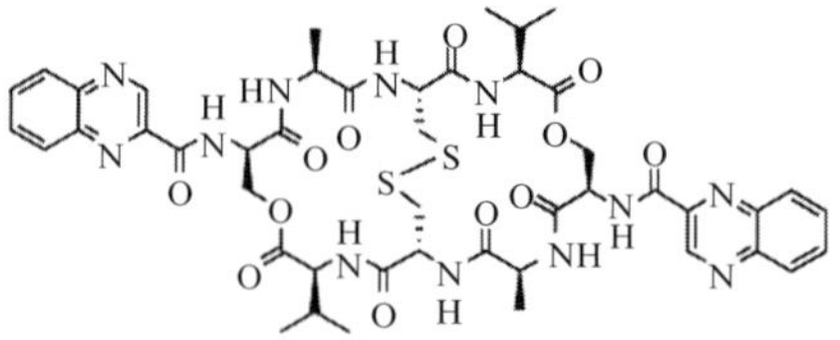

Tandem (8)

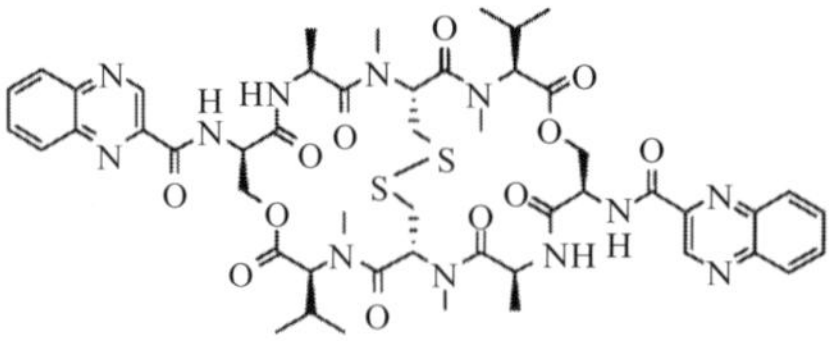

Triostin A (2)

Figure 15.1 Chemical structures of quinomycin antibiotics and a synthetic analog, TANDEM.

2. BIOSYNTHETIC PATHWAY OF NONRIBOSOMAL PEPTIDES

Genes responsible for the biosynthesis of nonribosomal peptides (NRPs), including modularly organized large enzyme NRP synthetases (NRPSs), are generally arranged in clusters on the chromosome or plasmid of the NRP-producing host organism (Cane et al., 1998). A single module of NRPS can range in size from 120 to 180 kDa and possesses the ability to independently extend a peptide chain using an amino acid as a building unit (Belshaw et al., 1999; Trauger and Walsh, 2000). Each module is comprised of a core complex containing three domains commonly referred to as condensation (C), adenylation (A), and thiolation (T) domain (Marahiel et al., 1997) (Fig. 15.2). The adenylation domain recognizes a specific amino acid as its substrate (Chiu et al., 2001), and the condensation domain catalyzes the peptide bond formation reaction to add the amino acid to the C-terminal of the growing peptide chain to biosynthesize the core peptide structure. Throughout the peptide elongation process, the nascent peptide chain is tethered to the phosphopantetheinyl moiety of the thiolation domain. These phosphopantetheinyl groups are derived from coenzyme A and added posttranslationally to a conserved serine residue of the thiolation domains by a $4'$-phosphopantetheinyl transferase (Lambalot et al., 1996). The phosphopantetheinyl moiety acts as a swinging arm for delivery of the peptide chain from the upstream module to the downstream module of the NRPS, making this complex mega-enzyme a preprogrammed biosynthetic array. Further modifications to the core peptide structure are performed by auxiliary domains, including N-methyltransferase (M),

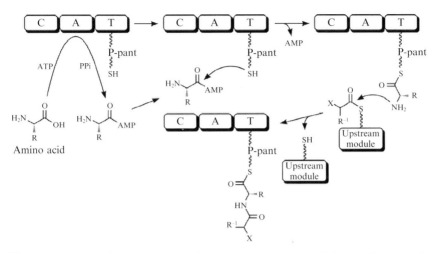

Figure 15.2 Modular organization of an NRPS. An NRPS module contains catalytic domains responsible for peptide chain elongation. A hypothetical NRPS core module is shown with its three key domains: C, condensation; A, adenylation; and T, thiolation domains. X represents the rest of the nascent peptide chain. Other abbreviations are: AMP, adenosine 5′-monophosphate; ATP, adenosine 5′-triphosphate; P-pant, 4′-phosphopantetheine; and PPi, pyrophosphate.

epimerization (E), and cyclization (Cy) domains. In the past, engineering of NRPS for production of altered products has been accomplished by various means, including site-directed mutagenesis of A domains (Eppelmann et al., 2002), domain swapping (Duerfahrt et al., 2004; Mootz et al., 2000), and inactivation of M domains (Patel and Walsh, 2001). These observations suggest that, despite the rather strict substrate amino acid specificity of the adenylation domains, the seemingly more relaxed substrate tolerance of the condensation domain (Ehmann et al., 2000) will likely permit the generation of a diverse set of peptide scaffolds through rational engineering or directed evolution of the biosynthetic machinery.

3. Echinomycin Biosynthetic Pathway

Carrying the distinctive quinoxaline-2-carboxylic acid (QXC, **3**) chromophore (Glund et al., 1990; Reid et al., 1984; Schmoock et al., 2005), compound **1** (Fig. 15.1) is a complex peptide natural product isolated from streptomycete as a secondary metabolite (Fig. 15.3). This compound is a member of the quinomycin antibiotics that share a set of common structural features characterized by a pair of quinoxaline or quinoline chromophores (Boger et al., 2001; Ciardelli et al., 1978) that is attached to

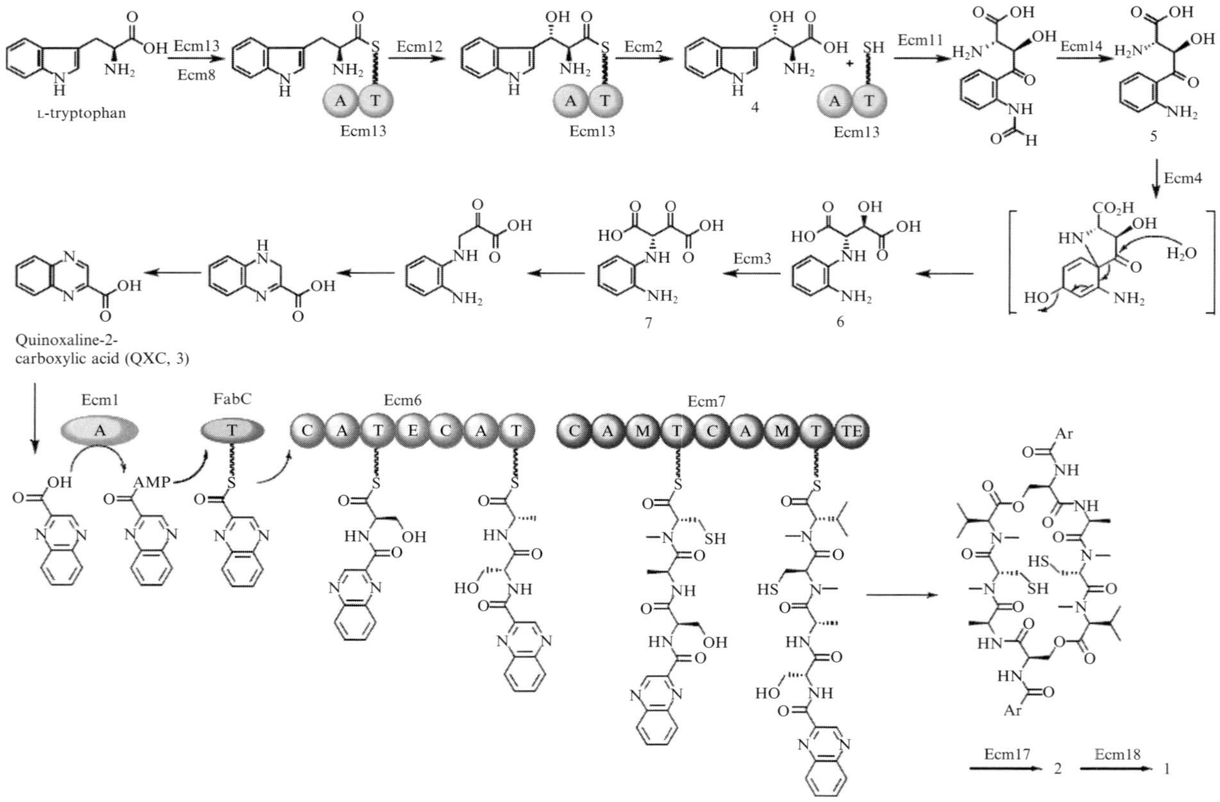

Figure 15.3 Proposed biosynthetic pathway and modular organization of the NRPS for echinomycin biosynthesis. The NRPS catalytic domains are represented by the same abbreviation as in Fig. 15.2 except for E, epimerization; M, methylation; and TE, thioesterase.

the C_2-symmetric depsipeptide core backbone structure synthesized by NRPSs. This class of natural products is an excellent DNA bis-intercalating agent. These molecules achieve high-affinity, sequence-selective DNA binding by inserting their highly conjugated chromophore substituents in between the base pairs of a double helical DNA and forming hydrogen bonds to the bases in the minor groove with the peptide backbone (Quigley et al., 1986; Wang et al., 1984). Great interest in the biosynthetic mechanism of **1** and other members of the quinomycin antibiotic class led us to isolate the echinomycin biosynthetic gene cluster from a cosmid library prepared from total DNA (Kieser et al., 2000; Kinashi et al., 1987) of *Streptomyces lasaliensis* that produces **1** (Steinerová et al., 1987). The construction and screening of the library were performed according to the procedure described elsewhere (Watanabe et al., 2006) using manufacturer's recommended protocols. Cosmid pCD4 carrying an open reading frame (ORF) for the previously identified QXC-activating enzyme was isolated. DNA sequence analysis of the 36-kb cluster found in pCD4 using FramePlot (Ishikawa and Hotta, 1999) identified 18 ORFs. A BLAST (McGinnis and Madden, 2004) sequence homology search on these ORFs was able to assign eight genes to the biosynthesis of **3** (*ecm2–4, 8, 11–14*), five genes to peptide backbone formation and modifications (*ecm1, 6, 7, 17, 18*) and one gene to self-resistance (*ecm16*) (Table 15.1).

Based on the BLAST results and past reports by others (Chen et al., 2002), we hypothesized that L-tryptophan is the precursor for **3** and Ecm2–4, Ecm8, and Ecm11–14 are responsible for biosynthesizing **3** in a process that parallels the first stage of nikkomycin biosynthesis (Fig. 15.3). Ecm12 initiates the biosynthesis of **3** by hydroxylating L-tryptophan bound to the thiolation domain of Ecm13. The product, (2S,3S)-β-hydroxytryptophan **4**, determined as an intermediate by substrate feeding experiments (Koketsu et al., 2006), is released from Ecm13 by the thioesterase activity of Ecm2. Then, much like the first two steps of kynurenine biosynthesis (Kurnasov et al., 2003), an oxidative ring opening of **4** by Ecm11, followed by hydrolysis by Ecm14 yields β-hydroxykynurenine **5**. Subsequently, oxidative cyclization and hydrolysis of **5** by Ecm4 can form N-(2′-aminophenyl)-β-hydroxyaspartic acid **6**, and oxidization of **6** by Ecm3 will give N-(2′-aminophenyl)-β-ketoaspartic acid **7**. Finally, **7** can undergo a spontaneous decarboxylation, cyclic imine formation, and oxidative aromatization to give **3**.

Curiously, the aryl carrier protein required for incorporating **3** into **1** was absent from the isolated biosynthetic gene cluster. However, as in the proposed biosynthetic scheme for triostin A (**2**, Fig. 15.1) (Schmoock et al., 2005), we speculated and later confirmed (Watanabe et al., 2006) that the A domain-containing Ecm1 activates and transfers **3** to the phosphopantetheine arm of FabC, a fatty acid biosynthesis acyl carrier protein. The first module of the bimodular NRPS Ecm6 then accepts QXC–S–FabC as the starter unit, while Ecm6 and the second NRPS Ecm7 can catalyze the

Table 15.1 Deduced functions of the ORFs from the *S. lasaliensis* echinomycin biosynthetic gene cluster and fatty acid synthase acyl carrier protein

ORF	Amino acids	Sequence homolog (conserved domain code or protein accession number)	Putative function
ecm1	527	Peptide arylation enzyme *entE* (COG1021)	QXC activation
ecm2	248	Type II thioesterase *grsT* (COG3208)	QXC biosynthesis
ecm3	362	Isopropylmalate dehydrogenase *leuB* (COG0473)	QXC biosynthesis
ecm4	472	FAD-dependent oxidoreductase *ubiH* (COG0654)	QXC biosynthesis
ecm5	n.a.	Transposase (inactive)	Unknown
ecm6	2608	Nonribosomal peptide synthetase *teiC* (CAG 15011)	Peptide synthesis modules 1–2
ecm7	3135	Nonribosomal peptide synthetase *acmC* (AAF42473)	Peptide synthesis modules 3–4
ecm8	70	MbtH-like protein *mbtH* (pfam03621)	Unknown
ecm9	181	DNA-binding response regulator *ompR* (COG0745)	Regulation
ecm10	252	TetR family transcriptional regulator *pip* (AAG31690)	Regulation
ecm11	220	Tryptophan 2,3-dioxygenase *tdo2* (COG3483)	QXC biosynthesis
ecm12	395	Cytochrome P450 oxidase *cypX* (COG2124)	QXC biosynthesis
ecm13	598	Mannopeptimycin peptide synthetase *mppB* (AAU34203)	QXC biosynthesis
ecm14	402	Erythromycin esterase *ereB* (pfam05139)	QXC biosynthesis
ecm15	285	Helix-turn-helix transcription regulator *marR* (Cd00592)	Regulation
ecm16	806	Excinuclease ATPase *uvrA* (COG0178)	Self resistance
ecm17	313	Thioredoxin reductase *trxB* (COG0492)	Disulfide formation
ecm18	224	SAM-dependent methyltransferase *smtA* (COG0500)	Thioacetal formation
fabC	82	Fatty acid synthase acyl carrier protein *acpP* (COG0236)	QXC carrier protein

n.a., not applicable; QXC, quinoxaline-2-carboxylic acid.

remaining 17 chemical reactions to assemble the peptide core. Ecm7 contains a terminal thioesterase domain that appears to homodimerize and cyclorelease the peptide chain. The cyclized product can then become the substrate for an oxidoreductase Ecm17, which can catalyze the oxidation reaction within a highly reducing cytoplasmic environment to generate the disulfide bond in **2**.

The last echinomycin biosynthetic step involves an unusual transformation of the disulfide bond into a thioacetal bridge in **2** and other related quinomycin natural products (Barrett *et al.*, 1997; Cornish *et al.*, 1983; Rance *et al.*, 1989; Takahashi *et al.*, 2001). Ecm18, which is highly homologous to other *S*-adenosyl-L-methionine (SAM)-dependent methyltransferases, is thought to be responsible for this transformation (Fig. 15.4). To verify this unprecedented proposal for an enzymatic transformation, we demonstrated *in vitro* that purified Ecm18 is able to catalyze the transformation of **2** to **1** in the presence of SAM (Watanabe *et al.*, 2006).

4. Construction of A Multigene Assembly on Expression Vectors

To express the requisite biosynthetic genes for the production of exogenous secondary metabolites in *E. coli*, it is necessary to introduce the genes into *E. coli* in such a way that they are all maintained stably and expressed functionally in a controllable manner. We have chosen to introduce the genes on a set of plasmids for the reconstitution of the heterologous biosynthetic pathway in *E. coli* for the reasons given above. Below, we will describe how to assemble the entire heterologous NRP biosynthetic pathway in *E. coli*, using the echinomycin biosynthetic system as an example.

Echinomycin biosynthesis requires 16 genes: two genes from outside the echinomycin biosynthetic gene cluster, *Bacillus subtilis* 4′-phosphopantetheinyl transferase *sfp* (0.5 kb) and *S. lasaliensis fabC* (0.2 kb); two 10 kb-long genes encoding four NRPS modules; and 12 genes ranging in size from 0.2 to 3 kb with an average size of 1 kb. Three expression plasmids are needed in order to assemble the entire biosynthetic pathway due to its large

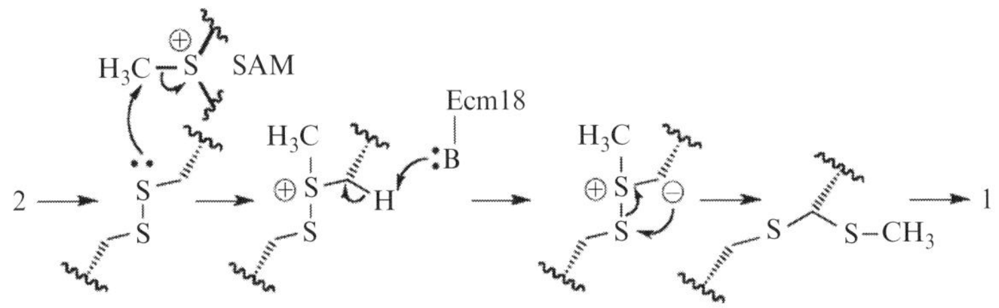

Figure 15.4 Proposed enzymatic reaction mechanism for biotransformation of a disulfide bond in 2 to a thioacetal bridge in 1.

size and number of genes required for the production of **1**. To transform and assure that all three plasmids are stably retained in *E. coli*, matched pairs of replication origins and selectable markers for *E. coli* are required. Because of these restrictions, it was essential for us to prepare three parent vectors, pKW407, pKW408, and pKW423, which contain the following essential features:

1. pKW407 carries a p*RSF*1030 origin of replication and a kanamycin resistance gene as a selectable marker;
2. pKW408 carries a CloDF13 origin of replication and a spectinomycin resistance gene for selection;
3. pKW423 carries a pBR322 origin of replication and an ampicillin resistance gene for selection.

In addition, the unique *Xba*I site in these plasmids was moved to the 5′ side of the T7 promoter, and a *Spe*I site was created on the 3′ side of the T7 terminator. This creates a "promoter–gene–terminator" cassette, containing a T7 promoter, ribosome binding site, *Nde*I site, *Eco*RI site and T7 terminator sandwiched between *Xba*I and *Spe*I sites, in each of the three plasmids (Figs. 15.5 and 15.6). All modifications were made to suitable pET plasmids obtained from Novagen. Because *Streptomyces* genomes are exceptionally GC-rich, most of the AT-rich restriction endonuclease recognition sites are absent from NRP biosynthetic genes despite their large size. Commonly used restriction endonuclease recognition sites, such as *Nde*I (CA↓TATG), *Eco*RI (G↓AATTC), *Xba*I (T↓CTAGA), and *Spe*I (A↓CTAGT), are rarely found in NRP biosynthetic genes from organisms with GC-rich genomes. Thus, using these sites facilitates cloning of the genes. Each of the biosynthetic genes was subcloned from the BAC or cosmid carrying the target gene cluster into one of the three plasmids (pKW407, pKW408, or pKW423) by amplifying the gene while creating an *Nde*I restriction site at the 5′ and an *Eco* RI site at the 3′ side of the gene by PCR. Accuracy of the DNA sequence of the subcloned gene was confirmed by DNA sequencing. A small-scale test using the *E. coli* strain BL21 (DE3) in Luria–Bertani (LB) medium with an isopropylthio-β-D-galactoside (IPTG) concentration of 0.05–1 mM was performed to determine how well each of the genes was expressed as a soluble protein in *E. coli*. The optimal temperature at which the culture is to be incubated after induction with IPTG was also determined during this test run. Frequently, an IPTG concentration of 200 μM and an incubation temperature of 15 °C have worked well in our hands (Watanabe *et al.*, 2006).

Once the optimal condition has been established, the necessary genes are assembled together on the plasmids. To minimize plasmid instability due to their large size, gene assemblies were kept to a minimal length while maintaining as many genes within a contiguous metabolic pathway on the same plasmid as possible. Specifically, we placed the genes for the

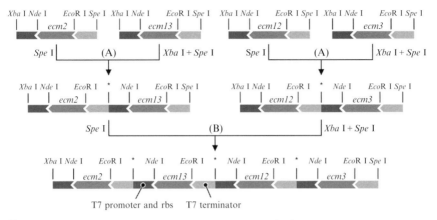

Figure 15.5 Strategy employed for constructing a multiple monocistronic expression system. (A) The first round of ligation joins the cohesive ends of an *Xba*I–*Spe*I restricted promoter–gene–terminator cassette with the compatible ends created by either *Spe*I- or *Xba*I-digestion of an acceptor plasmid (*Spe*I-digestion is shown in the diagram). This ligation creates a DNA sequence resistant to both *Xba*I and *Spe*I digestion (as indicated by ★), leading to the formation of a dual promoter–gene–terminator cassette that can be isolated as an intact single fragment upon *Xba*I–*Spe*I double digestion. (B) A second round of ligation fuses two dual promoter–gene–terminator cassettes to give a single fragment consisting of a quadruple promoter–gene–terminator cassette. T7 promoter and ribosome binding site are shown in blue, while T7 terminator is in green. Abbreviations are: rbs, ribosome binding site; and ★, *Xba*I- and *Spe*I-resistant ligation product between cohesive overhangs generated by *Xba*I and *Spe*I. (See Color Insert.)

biosynthesis of **3** on one plasmid (pKW532), the two NRPS genes on another (pKW541), and the rest of the genes on the third plasmid (pKW538) (Fig. 15.6). Also, to reduce premature transcriptional termination and mRNA degradation (Sørensen and Mortensen, 2005) in transcribing an excessively long polycistronic gene assembly, the genes were put together in a multimonocistronic manner. To generate a multimonocistronic gene assembly on a plasmid, the donor plasmid carrying a single gene is digested with *Xba*I and *Spe*I endonucleases simultaneously. This generates a "promoter–gene–terminator" cassette flanked by *Xba*I and *Spe*I overhangs (Fig. 15.5, top row). However, because these enzymes generate compatible cohesive ends, this "promoter–gene–terminator" cassette can be ligated into either an *Xba*I or a *Spe*I site of the acceptor plasmid carrying another "promoter–gene–terminator" cassette. Once ligated, however, one of the two ligation sites becomes resistant to *Xba*I or *Spe*I, effectively creating a new *Xba*I–*Spe*I cassette comprised of two "promoter–gene–terminator" cassettes (Fig. 15.5, middle row). This process can be performed iteratively to prepare plasmids carrying multiple monocistronic genes (Fig. 15.5, bottom row). Because each gene is under the control of its own promoter, this approach simplifies plasmid construction and reconfiguration efforts by

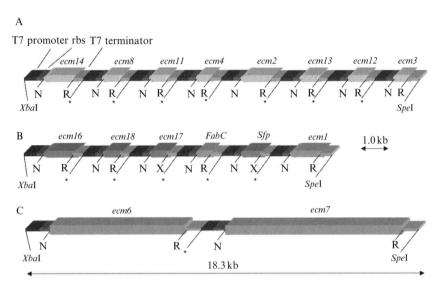

Figure 15.6 Organization of the plasmid-borne echinomycin biosynthetic gene cluster on three plasmids. T7 promoters, T7 terminators, and ribosome binding sites are shown in dark blue, light blue, and red, respectively. Abbreviation: rbs, ribosome binding site. Restriction endonuclease recognition sites are: N, *Nde*I; E, *Eco*RI; X, *Xho* I; and *, *Xba*I- and *Spe*I-resistant ligation product between cohesive overhangs generated by *Xba*I and *Spe*I. (A) Map of plasmid pKW532 for production of the quinoxaline-2-carboxylic acid biosynthetic proteins, carrying *ecm2–4*, *ecm8*, and *ecm11–14* (yellow), a carbenicillin resistance gene (not shown) and a pBR322 origin of replication (not shown). (B) Map of pKW538 for production of the peptide-forming proteins coded by *ecm1* and *fabC* (green), peptide-modifying proteins coded by *ecm17* and *ecm18* (pink), and auxiliary proteins coded by *ecm16* and *sfp* (orange). This plasmid also carries a streptomycin resistance gene (not shown) and a CDF origin of replication (not shown). (C) Map of pKW541 for the production of the NRPSs, carrying ecm6 and ecm7 (green), a kanamycin resistance gene (not shown), and a RSF origin of replication (not shown). (See Color Insert.)

eliminating the need to clone every gene unidirectionally. Furthermore, our approach of transcribing each gene independently has actually helped us achieve comparable protein production levels among the many exogenous genes we have incorporated into *E. coli*.

5. Heterologous Gene Expression and NRP Biosynthesis in *E. coli*

For gene expression and production of **1** in the *E. coli* cytosol, BL21 (DE3) was transformed by the three plasmids prepared as described above. The standard electroporation method for transformation was used on electrocompetent BL21 (DE3) cells (Sambrook and Russel, 2001) with the three plasmids simultaneously. The cells were then grown at 37 °C overnight on a LB agar plate supplemented with 50 μg/ml kanamycin, 100 μg/ml carbenicillin, and 50 μg/ml spectinomycin to select for cells capable of maintaining all three plasmids. Then a well-isolated colony was picked to inoculate 2 ml LB medium and cultured at 37 °C overnight. The culture was subsequently transferred to 100 ml M9 minimal medium for another overnight incubation at 37 °C. The entire culture was used to inoculate 1.5 l of M9 minimal medium maintained at 37 °C and pH 7.0 by BioFlo110 bioreactor system (New Brunswick Scientific). Once the initial supply of glucose was exhausted from the medium as indicated by a sudden increase of the dissolved oxygen level, the temperature was reduced to 15 °C. IPTG was then added to a final concentration of 200 μM, and subsequently feeding of the feed medium (Pfeifer *et al.*, 2003) was initiated. Following 8 days of incubation, the culture was harvested. Then the cells and the supernatant were separated by centrifugation. The supernatant and cell pellet were extracted with ethyl acetate and acetone, respectively. The extracts were combined and concentrated *in vacuo* to give an oily residue which was fractionated by silica gel flash column chromatography followed by preparative thin layer chromatography (Watanabe *et al.*, 2006) to afford purified **1** (Cheung *et al.*, 1978) at a yield of 300 μg/l of culture.

6. Self-Resistance Mechanism

Heterologous production of secondary metabolites with significant biological activities, such as the antibiotic activities of **1** and **2**, can have serious consequences for the viability of the surrogate host that may lack a resistance mechanism or effective efflux pump normally found in the original production hosts. Introducing an adequate functional resistance system that is nondestructive to the bioactive compound should sustain the viability of the heterologous host and ensure continued production of the target compound, a feature essential for achieving a higher titer of the biologically active molecules during fermentation.

For echinomycin biosynthesis, the homology between Ecm16 and the daunorubicin resistance-conferring factor DrrC (Lomovskaya *et al.*, 1996), and the similarity in the mode of action between **1** and daunorubicin (Neidle *et al.*, 1987) suggested that Ecm16 would likely provide *S. lasaliensis* with nondestructive resistance against **1**. Furthermore, expression of *drrC* in *E. coli* provided significant resistance against daunorubicin (Lomovskaya *et al.*, 1996). Combining these observations, we speculated that *ecm16* would be able to provide the necessary nondestructive resistance against **1** in *E. coli* and this was confirmed (Watanabe *et al.*, 2006). We used the following assay to test for the ability of Ecm16 to confer resistance to **1** on *E. coli*.

6.1. Echinomycin resistance assay

1. *E. coli* strain BL21 (DE3) was transformed with pKW409 carrying *ecm16*.
2. The transformant was grown in 3 ml of LB medium at 37 °C for 5 h.
3. The culture was then spread on LB agar plates containing two different concentrations of echinomycin (10 and 100 µg/ml) and supplemented either with or without 300 µM IPTG.
4. The plates were incubated at 37 °C overnight to allow the expression of *ecm16*.
5. On the following day, the plates were inspected for colony formation.

For the plates with IPTG, 200 and 70 colonies were observed at 10 and 100 µg/ml of echinomycin, respectively, while no colonies formed on either plate when IPTG was not present. In addition, we noted that the growth of the host was hampered in the absence of *ecm16*, suggesting that sufficient amounts of **1** or **2** would have been unattainable had there not been a nondestructive self-resistance mechanism in place.

7. STABILITY OF TRANSFORMANTS CARRYING MULTIPLE VERY LARGE PLASMIDS

Many procedures have been reported to date that describe fermentation of bacteria, including *E. coli*, that successfully produced desired compounds or improved the yield of target compound through metabolic engineering. More relevant to our cases, there have been reports of the use of such fermentation technology for the heterologous production of complex molecules in *E. coli* using plasmid-based biosynthetic systems, including the production of erythromycin C and its macrocyclic aglycone, 6-deoxyerythronolide B (Lau *et al.*, 2004; Peiru *et al.*, 2005). Despite using a multiplasmid expression system in *E. coli*, these systems allow the cell density of the culture to reach very high levels (optical density at 600 nm of 11.5 after 45 h of incubation). Also, the maximum titer of the compound was attained on Day 12 of fermentation. Thus, *E. coli* must be able to maintain and transmit the plasmids in intact form for an extended period of time over numerous generations to be able to achieve effective production of the target compound. Moreover, our *E. coli* system uses three plasmids carrying 16 genes flanked by a T7 promoter and a T7 terminator with identical sequence, and plasmids containing regions with homologous DNA sequences are known to be prone to undesired recombination (Lovett *et al.*, 1994). In addition, our plasmids encode proteins capable of producing antibiotic NRPs in their bioactive form in the cytosol. Although a self-resistance mechanism is installed to protect the cells from the potentially toxic effects of the NRPs being produced, there was a concern that *E. coli* may eliminate the plasmids spontaneously during the fermentation process. Thus, we monitored the stability of the three plasmids to assure the integrity and retention of all three plasmids in *E. coli* during the fermentation.

7.1. Examining the stability of transformant cells

1. *E. coli* strains DH5α and BL21 (DE3) were transformed with pKW532, pKW538, and pKW541 and cultured for 10 days under the same fermentation condition used for the production of **1**.
2. Retention of the plasmids in the BL21 (DE3) and DH5α transformants was assessed by resistance to the relevant antibiotics. An aliquot from the DH5α and BL21 (DE3) cultures were separately plated on nonselective LB agar medium and incubated overnight at 37 °C. Fifty well-separated colonies were transferred onto selective LB agar plates containing the three antibiotics to which the three plasmids confer resistance, 100 µg/ml carbenicillin, 50 µg/ml kanamycin, and 50 µg/ml spectinomycin. The plates were then incubated overnight at 37 °C, and the number of remaining colonies was recorded.
3. The intactness of the plasmids was examined using the DH5α transformant. DH5α was used in this analysis due to the inability of BL21 (DE3) to generate well-resolved plasmid preparations. Plasmid DNA was prepared from the culture of DH5α carrying pKW532, pKW538, and pKW541. Then the plasmids were digested with *Xba*I, *Xba*I and *Spe*I, or *Eco*RI, and the digests were analyzed by agarose gel electrophoresis.

For the plasmid stability test, 49 out of 50 DH5α transformants and 46 out of 50 BL21 (DE3) transformants retained resistance against all three antibiotics. Thus, the majority of the transformants retained all three plasmids throughout the fermentation. As to plasmid integrity, the restriction pattern of all the plasmids was as expected (Fig. 15.7), suggesting that the plasmids did not undergo significant alteration during fermentation. These results showed that the plasmids bearing large biosynthetic gene clusters in multimonocistronic fashion are stable and do not suffer significant elimination or mutation in *E. coli* throughout the duration of the fermentation.

8. Engineering of Heterologous NRP Biosynthetic Pathways in *E. coli*

To explore the potential of using our *E. coli* expression system for the heterologous production of various complex natural products, and to demonstrate the ease and speediness of engineering the biosynthetic pathway established on a plasmid-based system for future rational drug design, we set out to engineer the *E. coli* echinomycin biosynthetic pathway for the production of related natural and unnatural analogs of echinomycin.

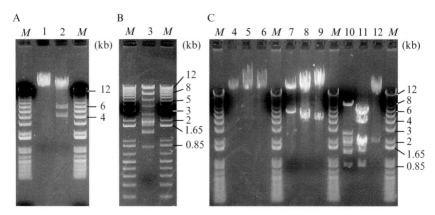

Figure 15.7 Agarose gel electrophoretic analysis of restriction digested plasmids for echinomycin production prepared from *E. coli* DH5α. Below, restriction enzymes used are given in brackets, and the expected DNA fragmentation pattern for each sample is listed. Asterisks indicate the presence of two fragments of similar size in the sample. The molecular weight marker (Invitrogen) lanes are labeled "*M*," and the band sizes are given in kb.

(A) Digests of a mixture of pKW532, pKW538, and pKW541 resolved on a 0.6% gel

lane 1 [*Xba*I]: 22, 17.2, and 15.1 kb
lane 2 [*Xba*I + *Spe*I]: 18.2, 11.9, 11.3, 5.3, and 3.8★ kb

(B) Digests of total plasmids resolved on a 1.0% gel

lane 3 [*Eco*RI]: 10.1, 9.8, 6.6, 4.8, 3.4, 2.3, 2.1, 1.9★, 1.7, 1.6★, 1.5, 1.3, 1.2, 1.1, and 0.7★ kb

(C) Digests of individual plasmid resolved on a 0.6% gel

pKW532	lane 4 [*Xba*I]:	17.2 kb
	lane 7 [*Xba*I + *Spe*I]:	11.9 and 5.3 kb
	lane 10 [*Eco*RI]:	6.6, 2.3, 1.9, 1.7, 1.6, 1.3, 1.1, and 0.7 kb
pKW538	lane 5 [*Xba*I]:	15.1 kb
	lane 8 [*Xba*I + *Spe*I]:	11.3 and 3.8 kb
	lane 11 [*Eco*RI]:	4.8, 3.4, 1.9, 1.6, 1.5, 1.2, and 0.7 kb
pKW541	lane 6 [*Xba*I]:	22.0 kb
	lane 9 [*Xba*I + *Spe*I]:	18.2 and 3.8 kb
	lane 12 [*Eco*RI]:	10.1, 9.8, and 2.1 kb

First, we chose to convert the echinomycin biosynthetic pathway into a biosynthetic pathway for the production of another member of the quinomycin antibiotics, triostin A **2**, by simply removing *ecm18*. The resulting strain produced the expected compound **2** (Blake *et al.*, 1977) at a level comparable to the observed yield of compound **1** (Watanabe *et al.*, 2006). Next, we further engineered the triostin A biosynthetic pathway for the production of a synthetic analog of **2**, TANDEM **8** (Fig. 15.1). This conversion was accomplished by introducing six point mutations into one of the NRPS genes using a standard site-directed mutagenesis method to abolish the *N*-methylation activity of the enzyme (Watanabe *et al.*, to be published elsewhere). Compound **3** (200 μg/l of culture) was obtained at slightly lower but similar yield as **1** (300 μg/l of culture) or **2** (600 μg/l of culture). These results suggest that NRPSs and their associated auxiliary enzymes are susceptible to mutagenesis to produce novel proteins and substrates, and that *E. coli* can tolerate such modifications.

Another important goal in developing an *E. coli* system for heterologous biosynthesis of natural products and their analogs is increasing the yield of desired compounds (Murli *et al.*, 2003). It is imperative that we improve on this aspect of the system to make it useful for future drug discovery and downstream commercial applications. For the echinomycin and triostin A biosynthetic pathways, compound **3** is a product of a multistep transformation of L-tryptophan as shown in Fig. 15.3. Furthermore, **3** is required to initiate the condensation of four amino acids by Ecm6 and Ecm7 to advance the assembly of the peptide backbone of **1**. Therefore, we suspected that an insufficient level of **3** *in vivo* was a likely bottleneck for efficient production of **1** and **2** in *E. coli*. This may have been associated with the poor expression level of *ecm4* and *ecm12*, but *in vivo* degradation of **3** in *E. coli* could not be ruled out. Thus, following the establishment of the biosynthetic system for the production of **2**, we attempted to improve the titer of **2** by supplementing **3** in the culture medium and testing various culture conditions using small-scale shake flask cultures. From this study, we found that simply supplying 5 mg of the primer unit per day over 7 days to the shake flask cultures increased the productivity of **2**. This result corroborated our initial concern that the biosynthesis of **3** could become the rate-limiting step in the production of **2** in *E. coli*. Subsequent optimization of the condition afforded 13 mg/l of culture of **2**, an increase of more than 130-fold relative to the yield of 100 μg/l of culture from shake flask cultures (Praseuth *et al.*, 2008).

9. Conclusion

Microbes are notable source of many natural products that are valuable to humans. According to the NCBI ENTREZ Genome Project database (http://www.ncbi.nlm.nih.gov/genomes/lproks.cgi), the genomes of nearly 750 microbes have been sequenced completely, and there are currently over 1200 microbial genome sequencing projects in progress. In addition, metagenomes, or DNA libraries prepared from environmental samples, have further expanded the available genomic information of different organisms (Venter et al., 2004). The use of metagenomics can be extremely powerful, considering that the vast majority (estimated to be more than 99%) of microbes are not culturable in the laboratory (Pace, 1997). Moreover, because many of the biosynthetic gene clusters will not be expressed by the host organisms under standard culture conditions (Firn and Jones, 2003), it is frequently not possible to isolate the compounds from culture samples to identify the products of different gene clusters. Nonetheless, computational techniques have been developed that allow successful identification of secondary metabolite biosynthetic gene clusters, especially for polyketides and NRPs (Ansari et al., 2004; Yadav et al., 2003), and prediction of the metabolites produced by such clusters from the available sequence information (Lautru et al., 2005). Development of such bioinformatic methods is crucial, especially for rapidly expanding the search for previously unknown natural products that may possess new and useful activities. Here, having an *E. coli*-based system that can readily accommodate an identified but uncharacterized gene cluster and heterologously produce the corresponding secondary metabolite should facilitate greatly the discovery of next-generation novel bioactive compounds. Moreover, such an *E. coli*-based system can serve as a platform for fast and straightforward rational engineering of the newly identified biosynthetic pathway for the preparation of analogs useful for structure–activity relationship studies or expanding the properties of the parent compound. These discovery and engineering processes will in turn allow accumulation of knowledge on how nature manages to accomplish complex natural product biosynthesis, and this accumulated knowledge can be applied to improving further the discovery and engineering processes. Ultimately, the process can be extended toward establishing combinatorial biosynthesis for efficient generation of libraries of unnatural bioactive compounds. We envision that progress in bioinformatics, genetics, biochemistry, structural biology, and metabolic engineering will eventually lead to streamlining of the method for establishing *E. coli*-based plasmid-borne heterologous secondary metabolite biosynthetic system. Generation of libraries of compounds will no longer be performed using the conventional reaction vessel, but rather in a flexible biological reaction vessel, *E. coli*.

REFERENCES

Ansari, M. Z., Yadav, G., Gokhale, R. S., and Mohanty, D. (2004). NRPS-PKS: A knowledge-based resource for analysis of NRPS/PKS megasynthases. *Nucleic Acids Res.* **32**, W405–W413.

Barrett, A. G. M., Hamprecht, D., White, A. J. P., and Williams, D. J. (1997). Iterative cyclopropanation: A concise strategy for the total synthesis of the hexacyclopropane cholesteryl ester transfer protein inhibitor U-106305. *J. Am. Chem. Soc.* **119**, 8608–8615.

Belshaw, P. J., Walsh, C. T., and Stachelhaus, T. (1999). Aminoacyl-CoAs as probes of condensation domain selectivity in nonribosomal peptide synthesis. *Science* **284**, 486–489.

Blake, T. J., Kalman, J. R., and Williams, D. H. (1977). Two symmetrical conformations of the triostin antibiotics in solution. *Tetrahedron Lett.* **18**, 2621–2624.

Boger, D. L., Ichikawa, S., Tse, W. C., Hedrick, M. P., and Jin, Q. (2001). Total syntheses of thiocoraline and BE-22179 and assessment of their DNA binding and biological properties. *J. Am. Chem. Soc.* **123**, 561–568.

Cane, D. E., Walsh, C. T., and Khosla, C. (1998). Harnessing the biosynthetic code: Combinations, permutations, and mutations. *Science* **282**, 63–68.

Chen, H., Hubbard, B. K., O'Connor, S. E., and Walsh, C. T. (2002). Formation of beta-hydroxy histidine in the biosynthesis of nikkomycin antibiotics. *Chem. Biol.* **9**, 103–112.

Cheung, H. T., Feeney, J., Roberts, G. C. K., Williams, D. H., Ughetto, G., and Waring, M. J. (1978). The conformation of echinomycin in solution. *J. Am. Chem. Soc.* **100**, 46–54.

Chiu, H. T., Hubbard, B. K., Shah, A. N., Eide, J., Fredenburg, R. A., Walsh, C. T., and Khosla, C. (2001). Molecular cloning and sequence analysis of the complestatin biosynthetic gene cluster. *Proc. Natl. Acad. Sci. USA* **98**, 8548–8553.

Ciardelli, T. L., Chakravarty, P. K., and Olsen, R. K. (1978). Des-*N*-tetramethyltriostin A and bis-L-seryldes-*N*-tetramethyltriostin A, synthetic analogues of the quinoxaline antibiotics. *J. Am. Chem. Soc.* **100**, 7684–7690.

Cornish, A., Waring, M. J., and Nolan, R. D. (1983). Conversion of triostins to quinomycins by protoplasts of *Streptomyces echinatus*. *J. Antibiot. (Tokyo)* **36**, 1664–1670.

Donia, M. S., Ravel, J., and Schmidt, E. W. (2008). A global assembly line for cyanobactins. *Nat. Chem. Biol.* **4**, 341–343.

Duerfahrt, T., Eppelmann, K., Muller, R., and Marahiel, M. A. (2004). Rational design of a bimodular model system for the investigation of heterocyclization in nonribosomal peptide biosynthesis. *Chem. Biol.* **11**, 261–271.

Ehmann, D. E., Trauger, J. W., Stachelhaus, T., and Walsh, C. T. (2000). Aminoacyl-SNACs as small-molecule substrates for the condensation domains of nonribosomal peptide synthetases. *Chem. Biol.* **7**, 765–772.

Eppelmann, K., Stachelhaus, T., and Marahiel, M. A. (2002). Exploitation of the selectivity-conferring code of nonribosomal peptide synthetases for the rational design of novel peptide antibiotics. *Biochemistry* **41**, 9718–9726.

Firn, R. D., and Jones, C. G. (2003). Natural products—A simple model to explain chemical diversity. *Nat. Prod. Rep.* **20**, 382–391.

Glund, K., Schlumbohm, W., Bapat, M., and Keller, U. (1990). Biosynthesis of quinoxaline antibiotics: Purification and characterization of the quinoxaline-2-carboxylic acid activating enzyme from *Streptomyces triostinicus*. *Biochemistry* **29**, 3522–3527.

Gruenewald, S., Mootz, H. D., Stehmeier, P., and Stachelhaus, T. (2004). *In vivo* production of artificial nonribosomal peptide products in the heterologous host *Escherichia coli*. *Appl. Environ. Microbiol.* **70**, 3282–3291.

Ishikawa, J., and Hotta, K. (1999). FramePlot: A new implementation of the frame analysis for predicting protein-coding regions in bacterial DNA with a high G + C content. *FEMS Microbiol. Lett.* **174**, 251–253.

Kealey, J. T., Liu, L., Santi, D. V., Betlach, M. C., and Barr, P. J. (1998). Production of a polyketide natural product in nonpolyketide-producing prokaryotic and eukaryotic hosts. *Proc. Natl. Acad. Sci. USA* **95**, 505–509.

Kieser, T., Bibb, M. J., Buttner, M. J., Chater, K. F., and Hopwood, D. A. (2000). Practical Streptomyces Genetics. The John Innes Foundation, Norwitch, UK.

Kinashi, H., Shimaji, M., and Sakai, A. (1987). Giant linear plasmids in *Streptomyces* which code for antibiotic biosynthesis genes. *Nature* **328**, 454–456.

Koketsu, K., Oguri, H., Watanabe, K., and Oikawa, H. (2006). Identification and stereochemical assignment of the beta-hydroxytryptophan intermediate in the echinomycin biosynthetic pathway. *Org. Lett.* **8,** 4719–4722.

Kurnasov, O., Jablonski, L., Polanuyer, B., Dorrestein, P., Begley, T., and Osterman, A. (2003). Aerobic tryptophan degradation pathway in bacteria: Novel kynurenine formamidase. *FEMS Microbiol. Lett.* **227,** 219–227.

Lambalot, R. H., Gehring, A. M., Flugel, R. S., Zuber, P., LaCelle, M., Marahiel, M. A., Reid, R., Khosla, C., and Walsh, C. T. (1996). A new enzyme superfamily—The phosphopantetheinyl transferases. *Chem. Biol.* **3,** 923–936.

Lau, J., Tran, C., Licari, P., and Galazzo, J. (2004). Development of a high cell-density fed-batch bioprocess for the heterologous production of 6-deoxyerythronolide B in *Escherichia coli*. *J. Biotechnol.* **110,** 95–103.

Lautru, S., Deeth, R. J., Bailey, L. M., and Challis, G. L. (2005). Discovery of a new peptide natural product by *Streptomyces coelicolor* genome. *Nat. Chem. Biol.* **1,** 265–269.

Lomovskaya, N., Hong, S. K., Kim, S. U., Fonstein, L., Furuya, K., and Hutchinson, R. C. (1996). The *Streptomyces peucetius drrC* gene encodes a UvrA-like protein involved in daunorubicin resistance and production. *J. Bacteriol.* **178,** 3238–3245.

Lovett, S. T., Gluckman, T. J., Simon, P. J., Sutera, V. A. Jr., and Drapkin, P. T. (1994). Recombination between repeats in *Escherichia coli* by a recA-independent, proximity-sensitive mechanism. *Mol. Gen. Genet.* **245,** 294–300.

Marahiel, M. A., Stachelhaus, T., and Mootz, H. D. (1997). Modular peptide synthetases involved in nonribosomal peptide synthesis. *Chem. Rev.* **97,** 2651–2674.

McGinnis, S., and Madden, T. L. (2004). BLAST: At the core of a powerful and diverse set of sequence analysis tools. *Nucleic Acids Res.* **32,** W20–W25.

Mootz, H. D., Schwarzer, D., and Marahiel, M. A. (2000). Construction of hybrid peptide synthetases by module and domain fusions. *Proc. Natl. Acad. Sci. USA* **97,** 5848–5853.

Murli, S., Kennedy, J., Dayem, L. C., Carney, J. R., and Kealey, J. T. (2003). Metabolic engineering of *Escherichia coli* for improved 6-deoxyerythronolide B production. *J. Ind. Microbiol. Biotechnol.* **30,** 500–509.

Mutka, S. C., Carney, J. R., Liu, Y., and Kennedy, J. (2006). Heterologous production of epothilone C and D in *Escherichia coli*. *Biochemistry* **45,** 1321–1330.

Neidle, S., Pearl, L. H., and Skelly, J. V. (1987). DNA structure and perturbation by drug binding. *Biochem. J.* **243,** 1–13.

Nougayrède, J. P., Homburg, S., Taieb, F., Boury, M., Brzuszkiewicz, E., Gottschalk, G., Buchrieser, C., Hacker, J., Dobrindt, U., and Oswald, E. (2006). *Escherichia coli* induces DNA double-strand breaks in eukaryotic cells. *Science* **313,** 848–851.

Pace, N. R. (1997). A molecular view of microbial diversity and the biosphere. *Science* **276,** 734–740.

Patel, H. M., and Walsh, C. T. (2001). *In vitro* reconstitution of the *Pseudomonas aeruginosa* nonribosomal peptide synthesis of pyochelin: Characterization of backbone tailoring thiazoline reductase and N-methyltransferase activities. *Biochemistry* **40,** 9023–9031.

Peiru, S., Menzella, H. G., Rodriguez, E., Carney, J., and Gramajo, H. (2005). Production of the potent antibacterial polyketide erythromycin C in *Escherichia coli*. *Appl. Environ. Microbiol.* **71,** 2539–2547.

Pfeifer, B. A., Admiraal, S. J., Gramajo, H., Cane, D. E., and Khosla, C. (2001). Biosynthesis of complex polyketides in a metabolically engineered strain of *E. coli*. *Science* **291,** 1790–1792.

Pfeifer, B. A., Wang, C. C., Walsh, C. T., and Khosla, C. (2003). Biosynthesis of yersiniabactin, a complex polyketide-nonribosomal peptide, using *Escherichia coli* as a heterologous host. *Appl. Environ. Microbiol.* **69,** 6698–6702.

Praseuth, A. P., Praseuth, M. B., Oguri, H., Oikawa, H., Watanabe, K., and Wang, C. C. (2008). Improved production of triostin A in engineered *Escherichia coli* with furnished

quinoxaline chromophore by design of experiments in small-scale culture. *Biotechnol. Prog.* **24,** 134–139.
Quigley, G. J., Ughetto, G., van der Marel, G. A., van Boom, J. H., Wang, A. H., and Rich, A. (1986). Non-Watson–Crick G.C and A.T base pairs in a DNA-antibiotic complex. *Science* **232,** 1255–1258.
Rance, M. J., Ruddock, J. C., Pacey, M. S., Cullen, W. P., Huang, L. H., Jefferson, M. T., Whipple, E. B., Maeda, H., and Tone, J. (1989). UK-63,052 Complex, new quinomycin antibiotics from *Streptomyces braegensis* subsp. *japonicus*; taxonomy, fermentation, isolation, characterisation and antimicrobial activity. *J. Antibiot. (Tokyo)* **42,** 206–217.
Reid, D. G., Doddrell, D. M., Williams, D. H., and Fox, K. R. (1984). A ^{15}N nuclear magnetic resonance study of the biosynthesis of quinoxaline antibiotics. *Biochim. Biophys. Acta* **798,** 111–114.
Sambrook, J., and Russel, D. W. (2001). Molecular Cloning: A Laboratory Manual, 3rd ed. Cold Spring Harbor Laboratory Press, Cold Spring Harbor, NY.
Schmoock, G., Pfennig, F., Jewiarz, J., Schlumbohm, W., Laubinger, W., Schauwecker, F., and Keller, U. (2005). Functional cross-talk between fatty acid synthesis and nonribosomal peptide synthesis in quinoxaline antibiotic-producing streptomycetes. *J. Biol. Chem.* **280,** 4339–4349.
Sørensen, H. P., and Mortensen, K. K. (2005). Advanced genetic strategies for recombinant protein expression in *Escherichia coli*. *J. Biotechnol.* **115,** 113–128.
Steinerová, N., Lipavská, H., Stajner, K., Cáslavská, J., Blumauerová, M., Cudlín, J., Van, Z., and k, Z. (1987). Production of quinomycin A in *Streptomyces lasaliensis*. *Folia Microbiol. (Praha)* **32,** 1–5.
Takahashi, K., Koshino, H., Esumi, Y., Tsuda, E., and Kurosawa, K. (2001). SW-163C and E, novel antitumor depsipeptides produced by *Streptomyces sp.* II. Structure elucidation. *J. Antibiot. (Tokyo)* **54,** 622–627.
Trauger, J. W., and Walsh, C. T. (2000). Heterologous expression in *Escherichia coli* of the first module of the nonribosomal peptide synthetase for chloroeremomycin, a vancomycin-type glycopeptide antibiotic. *Proc. Natl. Acad. Sci. USA* **97,** 3112–3117.
Valdebenito, M., Crumbliss, A. L., Winkelmann, G., and Hantke, K. (2006). Environmental factors influence the production of enterobactin, salmochelin, aerobactin, and yersiniabactin in *Escherichia coli* strain Nissle 1917. *Int. J. Med. Microbiol.* **296,** 513–520.
Venter, J. C., Remington, K., Heidelberg, J. F., Halpern, A. L., Rusch, D., Eisen, J. A., Wu, D., Paulsen, I., Nelson, K. E., Nelson, W., Fouts, D. E., Levy, S., *et al.* (2004). Environmental genome shotgun sequencing of the Sargasso Sea. *Science* **304,** 66–74.
Wang, A. H., Ughetto, G., Quigley, G. J., Hakoshima, T., van der Marel, G. A., van Boom, J. H., and Rich, A. (1984). The molecular structure of a DNA-triostin A complex. *Science* **225,** 1115–1121.
Watanabe, K., Hotta, K., Praseuth, A. P., Koketsu, K., Migita, A., Boddy, C. N., Wang, C. C., Oguri, H., and Oikawa, H. (2006). Total biosynthesis of antitumor nonribosomal peptides in *Escherichia coli*. *Nat. Chem. Biol.* **2,** 423–428.
Watanabe, K., Rude, M. A., Walsh, C. T., and Khosla, C. (2003). Engineered biosynthesis of an ansamycin polyketide precursor in *Escherichia coli*. *Proc. Natl. Acad. Sci. USA* **100,** 9774–9778.
Yadav, G., Gokhale, R. S., and Mohanty, D. (2003). SEARCHPKS: A program for detection and analysis of polyketide synthase domains. *Nucleic Acids Res.* **31,** 3654–3658.
Zhang, H., Wang, Y., and Pfeifer, B. A. (2008). Bacterial hosts for natural product production. *Mol. Pharm.* **5,** 212–225.

CHAPTER SIXTEEN

Enzymology of β-Lactam Compounds with Cephem Structure Produced by Actinomycete

Paloma Liras[*,†] and Arnold L. Demain[‡]

Contents

1. Introduction	402
2. Biosynthesis of Cephamycins: Enzymes and Genes	405
3. Early Steps Specific for Cephamycin Biosynthesis	405
4. Common Steps in Cephamycin-Producing Actinomycetes and Penicillin- or Cephalosporin-Producing Filamentous Fungi	409
4.1. Isopenicillin N synthase	412
4.2. Isopenicillin N epimerase	415
4.3. Deacetoxycephalosporin C synthase	417
4.4. Deacetoxycephalosporin C hydroxylase	419
5. Specific Steps for Tailoring the Cephem Nucleus in Actinomycetes	420
6. Regulation of Cephamycin C Production	422
Acknowledgments	423
References	423

Abstract

Cephamycins are β-lactam antibiotics with a cephem structure produced by actinomycetes. They are synthesized by a pathway similar to that of cephalosporin C in filamentous fungi but the actinomycetes pathway contains additional enzymes for the formation of the α-aminoadipic acid (AAA) precursor and for the final steps specific to cephemycins. Most of the biochemical and genetic studies on cephemycins have been made on cephamycin C biosynthesis in the producer strains *Streptomyces clavuligerus* ATCC27064 and *Amycolatopsis lactamdurans* NRRL3802. Genes encoding cephamycin C biosynthetic enzymes are clustered in both actinomycetes. Ten enzymatic steps are involved in the formation of cephamycin C. The precursor α-AAA is formed by the sequential action of

---

[*] Área de Microbiología, Facultad de Ciencias Biológicas y Ambientales, Universidad de León, León, Spain
[†] Biotechnological Institute INBIOTEC, Scientific Park of León, Spain
[‡] Research Institute for Scientists Emeriti (R.I.S.E.), Drew University, Madison, New Jersey, USA

lysine-6-aminotransferase and piperideine-6-carboxylate dehydrogenase. Steps common to cephalosporin C biosynthesis include the formation of the tripeptide L-δ—α-aminoadipyl-L-cysteinyl-D-valine (ACV) by ACV synthetase, the cyclization of ACV to form isopenicillin N (IPN) by IPN synthase, the epimerization of IPN to penicillin N by isopenicillin N epimerase, the ring expansion of penicillin N to a six member cephem ring by deacetoxycephalosporin C synthase (DAOCS) and the hydroxylation at C-3' by deacetylcephalosporin C hydroxylase. However, in actinomycetes, the epimerization step is different from that in cephalosporin-producing fungi, and the expansion of the ring and its hydroxylation are performed by separate enzymes. Specific steps in cephamycin biosynthesis include the carbamoylation at C-3' by cephem carbamoyl transferase and the introduction of a methoxyl group at C-7 by the joint action of a C-7 cephem-hydroxylase and a methyltransferase. All the enzymes of the pathway have been purified almost to homogeneity and the DAOC synthase and 7-hydroxycephem-methyltransferase (CmcI) of *S. clavuligerus* have been crystallized giving insights into the mode of action of these enzymes. The *cefE* gene of *S. clavuligerus*, encoding DAOCS, has been extensively used to expand the penicillin ring in filamentous fungi *in vivo* using DNA recombinant technology.

1. INTRODUCTION

The actinomycetes are able to produce a whole array of compounds with the β-lactam structure, that is, compounds containing a four-membered β-lactam ring. This ring is closed by an amide bond, which is the target for enzymes with β-lactamase activity. Most β-lactam compounds produced by actinomycetes, such as cephamycins, carbapenems, and nocardicins, are antibiotics inhibiting peptidoglycan cross-linking and therefore affecting cell wall biosynthesis. This is the same mechanism as for the penicillins and cephalosporins produced by filamentous fungi. However, others such as clavulanic acid have β-lactamase inhibitory properties and compete irreversibly with the β-lactams for the active site of these enzymes. Discovery of most of these compounds was via new screening methods employed since 1970. The first group of β-lactams discovered from actinomycetes were the cephamycins (Nagarajan *et al.*, 1971; Stapley *et al.*, 1972). These are modified cephalosporins produced by many *Streptomyces* species, for example, *Streptomyces clavuligerus* and *Streptomyces cattleya,* and by other actinomycetes such as *Amycolatopsis lactamdurans*. The cephamycins contain a cephem nucleus with a β-lactam ring and a five-membered dihidrothiazine ring containing a sulphur atom (Fig. 16.1A). In addition, cephamycins possess an α-aminoadipyl side chain and a methoxyl group at C-7, as well as different side chains at C-3'. The methoxyl group makes the cephamycins relatively insensitive to hydrolysis by most β-lactamases.

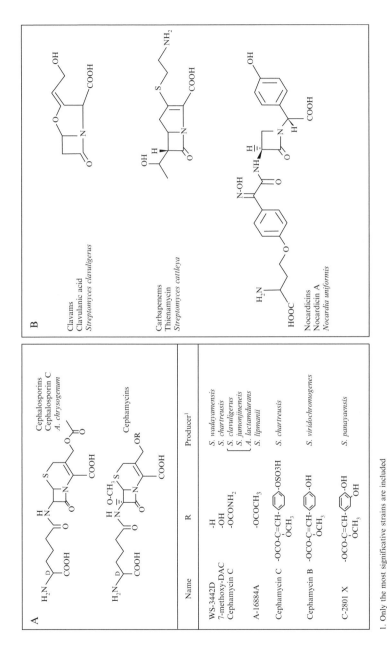

Figure 16.1 (A) Chemical structure of cephamycins produced by actinomycetes. For comparison, the structure of cephalosporin C, produced by *A. chrysogenum*, is included. The structures of different cephamycins and the most representative producer strains are indicated below. (B) Structures of different types of unconventional β-lactam compounds produced by actinomycetes.

Different cephamycins are modified only at the substituent on the methylene group at C-3', which might be an H atom (WS-3442D), a hydroxyl group (7 methoxy-deacetylcephalosporin C, produced by *Streptomyces chartreusis*), an acetyl group (in A-16884A, produced by *Streptomyces lipmanii*), or the complex structures present in cephamycins A, B, and C-2801X. The latter are very widely distributed, and contain phenyl or phenyl-sulfonyl groups. Cephamycin C, containing a carbamoyl group at C-3', is produced by more than eight different *Streptomyces* species and by *A. lactamdurans*. It is the only cephamycin well studied from the biochemical and genetic points of view.

A different group of compounds produced by *Streptomyces* species have a clavam nucleus (Brown *et al.*, 1976). The bicyclic structure of this nucleus possesses a β-lactam four-membered ring and a five-membered oxazolidine ring containing an oxygen atom (Fig. 16.1B). Two types of compounds with the clavam nucleus are known. Clavulanic acid, which is produced by *S. clavuligerus, Streptomyces jumonjinesis*, and *Streptomyces katsuharamanus*, has a 5R configuration and a carboxyl group at C-3 and is a potent β-lactamase inhibitor. Many other compounds with a clavam nucleus have been isolated that lack the carboxyl group at C-3 and have 5S stereochemistry. These compounds, globally named clavams, lack β-lactamase inhibitory ability but have been found to be active as antifungal or antibacterial agents. Clavams, such as clavam-2-carboxylate, 2-formyloxymethylclavam, 2-hydroxymethylclavam, and alanylclavam, were discovered in *S. clavuligerus* culture broths. Valclavams are produced by *Streptomyces antibioticus* TU1718 and clavamycins by *Streptomyces hygroscopicus* (Baggaley *et al.*, 1997; Liras and Rodríguez-García, 2000).

Other groups of unconventional β-lactams produced by actinomycetes include the carbapenems and the olivanic acid family (Kahan *et al.*, 1979; Okamura *et al.*, 1979). They contain the β-lactam ring fused to a five-membered ring containing a carbon atom instead of the oxygen present in the clavams or the sulphur atom characteristic of the penicillins produced by filamentous fungi (Fig. 16.1B). These compounds are cell-wall inhibitory antibiotics and include thienamycins, carpetimycins, asparenomycins, PS antibiotics, and olivanic acids.

Finally, there are monobactams with a monocyclic structure, that is, they only contain the β-lactam ring and different side chains (Aoki *et al.*, 1976). Many monobactams are produced by nonactinomycete bacteria, but only the nocardicins are produced, as a family of seven members, by *Nocardia uniformis* var *tsuyamanensis* and by *Actinosynnema mirum, Nocardiopsis atra* and *Microtetraspora caesia*.

Cephamycin C has been produced industrially from *A. lactamdurans* as the base molecule to produce the commercial semisynthetic cefoxitin. Many studies have been performed on cephamycins due to their similarity to the cephalosporins produced by the fungus *Acremonium chrysogenum* and to the interest of using cephamycin biosynthetic enzymes and genes to genetically obtain β-lactam-producing recombinant filamentous fungi. Thienamycin, produced by *S. cattleya*, although a potent broad-spectrum antibacterial compound resistant to β-lactamases, is not produced industrially due to its chemical instability. However, chemically produced modified derivatives, such as imipenem or meropenem, have had great medical success.

Clavulanic acid, produced industrially by fermentation of *S. clavuligerus*, is an interesting compound that is used synergistically with β-lactams in several clinical preparations. It protects the β-lactams against hydrolysis by β-lactamases.

Each group of compounds (cephamycins, clavams, carbapenems, or monobactams) is produced from different precursors by specific enzymatic pathways. In this article, we refer only to the cephamycin biosynthetic enzymes and the genes encoding them. These are related to similar enzymes or genes present in other β-lactam-producing actinomycetes.

2. Biosynthesis of Cephamycins: Enzymes and Genes

The biosynthetic pathway for cephamycins has several steps in common with those of the penicillins and cephalosporins produced by filamentous fungi. There has been a continuous exchange of knowledge between scientists dealing with prokaryotic- and eukaryotic-produced β-lactams. Many comprehensive reviews have been published on β-lactams produced by filamentous fungi and actinomycetes (Brakhage *et al.*, 2005; Demain and Elander, 1999; Martín and Liras, 1989). In this article, we shall center specifically on what is known about the enzymes and genes in actinomycetes. The biosynthesis of cephamycins contains specific enzymatic steps for the formation of the precursor L-α—aminoadipic acid (AAA) which are only present in actinomycetes. Five steps are common with filamentous fungi, starting with the synthesis of the tripeptide L-α-AAA-L-cys-D-val up to the formation of deacetylcephalosporin C (DAC). Finally, there are several late steps for the tailoring of the DAC molecule which are again specific for the actinomycetes (for a review see Liras, 1999) (Fig. 16.2A). The cluster of genes for cephamycin C biosynthesis is larger than that for cephalosporin biosynthesis in filamentous fungi and contains genes for structural enzymes, regulatory genes, and genes encoding enzymes for cephamycin resistance (Fig. 16.2B).

3. Early Steps Specific for Cephamycin Biosynthesis

Cephamycins, penicillins, and cephalosporins are formed by the condensation of three amino acids, α-AAA, L-cysteine, and L-valine to make the tripeptide L-α-AAA-L-cys-D-val. However, since prokaryotic actinomycetes

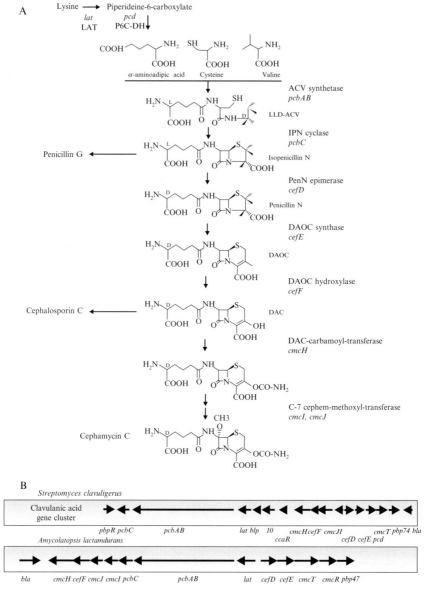

Figure 16.2 (A) Biosynthetic pathway of cephamycin C in actinomycetes. (B) Gene clusters for cephamycin C biosynthesis in *S. clavuligerus* and *A. lactamdurans*. Notice the presence of the clavulanic acid gene cluster lying next to the cephamycin cluster in the *S. clavuligerus* genome.

form lysine by the diaminopimelic acid pathway in which α-AAA is not an intermediate, they require specific enzymatic steps to produce α-AAA. The standard pathway for lysine utilization in streptomycetes is the cadaverine pathway, but cephamycin-producing actinomycetes possess, in addition, a specific pathway to form α-AAA from lysine using lysine-6-aminotransferase (LAT).

LAT activity was first described in *A. lactamdurans* by Kern *et al.* (1980). Piperideine-6-carboxylate (P6C) formed from lysine was detected colorimetrically by reaction with *o*-aminobenzaldehyde. The LAT is encoded by the *lat* gene, present in the cephamycin C gene clusters of *A. lactamdurans* (Coque *et al.*, 1991b) and *S. clavuligerus* (Madduri *et al.*, 1991) but not in cephamycin nonproducing actinomycetes. This was shown by hybridization studies using the *lat* gene. LAT activity peaks earlier than cephamycin formation in producer strains and an increase in *lat* copy number results in a two- to fivefold increase of cephamycin formation. This suggests that formation of α-AAA is a limiting step for cephamycin formation. LAT of *A. lactamdurans* was partially purified from a high-producing strain of *A. lactamdurans* (Kern *et al.*, 1980) and later, using *Streptomyces lividans* transformants carrying the *lat* gene, was further purified by ammonium sulphate precipitation and gel filtration (Coque *et al.*, 1991a,b). LAT is a 450-amino acid protein (Mr 48811) that requires 2-ketoglutarate as amino acceptor, and pyridoxal phosphate, which has a binding site at Lys^{300} of the *A. lactamdurans* LAT protein, as cofactor.

It was assumed initially that the unstable P6C could be converted spontaneously to α-AAA; however, this was never found *in vitro* using LAT extracts of *S. clavuligerus* or *A. lactamdurans* (Fuente *et al.*, 1997). This led to the discovery of a new enzymatic activity, piperideine-6-carboxylate dehydrogenase (P6C-DH), that was purified to homogeneity by Fuente *et al.* (1997) from *S. clavuligerus* cultures (see protocol below). Again, this enzymatic activity is only detectable in cephamycin-producing actinomycetes. P6C-dehydrogenase is a monomer of 56 kDa and converts piperideine-6-carboxylate into α-AAA with high efficiency (K_m 14 μM) in the presence of NAD (K_m 115 μM). P6C-DH activity can be measured by two methods, a spectrophotometric assay based on the formation of NADH when P6C is used as substrate, and a radiometric assay in which labeled P6C is converted to α-AAA and detected by TLC and autoradiography. P6C-DH does not use pyrroline-5-carboxylate, a similar five-carbon intermediate in proline biosynthesis, as substrate. N-terminal sequencing of the P6C-DH protein revealed the location of the *pcd* gene, which is present downstream of *cefE* in the cephamycin C gene cluster of *S. clavuligerus*, but has not been located yet in *A. lactamdurans*. The gene encodes a protein of 496 amino acids and a deduced Mr of 52499 Da. The function of *pcd* was confirmed by the presence of P6C-DH activity in cephamycin nonproducing *S. lividans* transformants carrying the *pcd* gene. As occurs with the *lat* gene, *pcd* has not been detected in cephamycin nonproducing strains.

Recently, it was found that *S. clavuligerus pcd*-null mutants are still able to form 30–50% of the cephamycin produced by the wild type strain (Alexander et al., 2007). This suggests that an additional enzyme is able to convert P6C into α-AAA, or alternatively, that *in vivo*, some P6C, which is chemically unstable in aqueous solution, might be converted spontaneously into α-AAA, since the next enzyme in the pathway is unable to use P6C directly instead of α-AAA (Coque et al., 1996b). The possibility exists that the *pcd* gene has been recruited by the organism to assure a more efficient formation of α-AAA, an apparent bottleneck in cephamycin biosynthesis.

Protocol for the purification of P6C-DH from *S. clavuligerus* (Fuente et al., 1997)

1. Inoculate spores of *S. clavuligerus* into 500-ml tripled baffled flasks containing 100 ml of trypticase soy broth (TSB) medium.
2. Incubate the cultures for 48 h at 28 °C at 250 rpm. Use 5 ml of this seed culture to inoculate identical flasks containing TSB medium. Grow under the same conditions for 48 h.
3. Collect the cells by centrifugation, wash them with 0.9% NaCl, and suspend them in 1/25 of the initial volume of buffer A (20 mM Tris/HCl pH 8.0, 0.1 mM dithiothreitol, 0.4 mM EDTA, 3 mM Mg Cl$_2$, 5% v/v glycerol).
4. Treat the cell suspension at 4 °C with six cycles of 10 s sonication (with 1 min intervals between cycles) using a Branson B-12 sonifier.
5. Centrifuge at 10,000g for 30 min and desalt the extract by filtration through a PD-10 column (Pharmacia).
6. Apply the cell extract (5 mg protein/ml) to a Blue-Sepharose CL6B column (7 × 1.6 cm) equilibrated with buffer A, and wash the column with a flow of 0.34 ml/min of the same buffer. Once the 280 nm absorbance reaches zero, apply a 120-ml volume linear gradient of 0–6 mM NAD$^+$.
7. Apply the active fractions of the previous step to a 3-ml bed DEAE-Sepharose fast-flow column (2.5 × 1.6 cm) equilibrated with buffer A and elute with a 0–400 mM NaCl gradient.
8. Apply the active fractions (released at about 190 mM NaCl) to a Sephadex G-75 column (68 × 2.6 cm) equilibrated with buffer A supplemented with 0.4 mM NAD$^+$ and elute the enzyme with the same buffer. The enzyme elutes with a $K_{av} = 0.066$, which corresponds to a molecular weight of 56.2 kDa.

4. COMMON STEPS IN CEPHAMYCIN-PRODUCING ACTINOMYCETES AND PENICILLIN- OR CEPHALOSPORIN-PRODUCING FILAMENTOUS FUNGI

The next five steps of the pathway are similar but not identical to the biosynthetic pathways of cephalosporins in *A. chrysogenum*. First is the condensation of L-δ-α-AAA, L-cysteine, and L-valine to form the tripeptide L-δ-α-AAA-L-cysteinyl-D-valine (LLD-ACV) by the L-δ-α-AAA-L-cysteinyl-D-valine synthetase (ACVS). This enzyme is a nonribosomal peptide synthetase (NRPS) able to form ribosome-independent peptide bonds (for general reviews on NRPS, see Linne and Marahiel, 2004; Grünewald and Marahiel, 2006, as well as Chapter 13, in this volume). Most studies on ACVSs have been performed with *Penicillium chrysogenum* or *A. chrysogenum* ACVSs involved in the formation of penicillins and cephalosporins. ACVSs are multifunctional enzymes that activate amino acids as aminoacyladenylates, forming a mixed anhydride with the α-phosphate group of ATP and releasing pyrophosphate. Then, they form peptide bonds between the activated amino acids, epimerize L-valine to the D-configuration and release the peptides formed through a thioesterase activity (Martín, 2000). The ACVS assay is based on the general ATP/PPi exchange reaction used for nonribosomal peptide synthetases (Theilgaard *et al.*, 1997). ACVS of *S. clavuligerus* was purified 12-fold by Jensen *et al.* (1990) using a combination of two successive chromatography steps on MonoQ columns separated by ultrafiltration through XM300. The enzyme was reported to be heterodimeric, containing two subunits of 283 kDa and a subunit of 32 kDa. However, additional purifications by other authors and the sequence of the *pcbAB* genes revealed that the ACVS is a high-molecular-weight multifunctional monomer. *S. clavuligerus* ACVS was later purified 2700-fold by Schwecke *et al.* (1992) by gel filtration, ion exchange, and phenyl sepharose and superose TM-6 chromatography. The protein, although apparently sensitive to ultrafiltration, was found to be a 560-kDa monomer that appears as a 500-kDa band in SDS-PAGE. It activates *in vitro* the carboxyl groups of L-cysteine and L-valine but this reaction has not been experimentally demonstrated for L-α-AAA. Assuming independent activation sites, by using different amino acid concentrations or different ATP concentrations at a fixed amino acid concentration, the binding constants have been calculated to be 1.25 and 1.5 mM for L-cysteine and ATP, respectively; and 2.4 and 0.25 mM for valine and ATP, respectively. These constants are relatively high, especially when compared to the dissociation constants of aminoacyl-tRNA synthetases.

Zhang *et al.* (1992) characterized the partially purified ACVS of *S. clavuligerus*. The enzyme was shown to have an optimal pH of 8–8.5

and the activity decreased 30% in the presence of 100 mM phosphate, being also affected by AMP and pyrophosphate, products of ATP hydrolysis. However, the major effect on ACVS activity was exerted by thiol-blocking agents such as N-ethylmaleimide, 5–5′dithiobis-2-nitrobenzoate and iodoacetamide, which almost completely inhibited the activity at 1 mM concentration, confirming the importance of thiol groups in ACVS activity.

The *S. clavuligerus* ACVS is rather specific, being unable to produce glycyl-ACV or glutathione but able to substitute, with low efficiency, L-*S*-carboxyethylcysteine for α-AAA, DL-homocysteine for L-cysteine and L-*allo*-isoleucine, L-α-aminobutyrate, norvaline, L-allylglycine, or L-leucine at increasing lower specificity for L-valine. D-valine was not a substrate for the reaction (Zhang *et al.*, 1992).

ACVS of *A. lactamdurans* was purified as a recombinant protein from a *S. lividans* transformant, carrying the *pcbAB* gene. The recombinant protein was purified 2785-fold to near homogeneity by a combination of gel filtration, ultrafiltration, and ion-exchange chromatography (See protocol for purification below). The enzyme activity was followed by the ATP-PPi interchange assay using ^{14}C-valine. The presence of ACVS protein was confirmed by SDS-PAGE and immunoblotting of the purified fractions using anti-*Aspergillus nidulans* ACVS polyclonal antibodies.

Pure *A. lactamdurans* ACVS is able to activate α-AAA or its lactam 6-oxopiperideine-2-carboxylic acid, a compound that is easily converted to α-AAA, but is unable to use piperideine-6-carboxylate or pipecolic acid as substrates. This might explain the requirement for a previous oxidation by a dehydrogenase, as indicated above. The enzyme is also able to use L-cystathionine with the same efficiency as L-cysteine. This indicates the existence of an additional function for the ACVS, permitting the hydrolysis of L-cystathionine by a pyridoxal-phosphate, propargylglycine-independent mechanism different from that of cystathionine-γ-lyase (Coque *et al.*, 1996b).

The *pcbAB* gene from *A. lactamdurans* is 10.9 kb long and encodes a 3649 amino acid ACVS of Mr 404134. Only 2424 nucleotides from the 5′end of the homologous gene of *S. clavuligerus* have been sequenced (Yu *et al.*, 1994) and 675 nucleotides of the 3′end have been reported (Doran *et al.*, 1990). In both species, as well as in *Streptomyces griseus* (García-Domínguez, unpublished results), *pcbAB* is upstream of *pcbC* (encoding the next step in the pathway) in a head-to-tail organization. This is different from filamentous fungi where the two genes lie side by side but in opposite orientation in the genome.

The amino acid sequence of ACVSs, deduced from the sequence of the *pcbAB* genes, indicates that they possess three domains: (A) adenylate-containing modules for ATP binding and specific activation of either α-AAA, L-cysteine, or L-valine as aminoacyladenylates; (C) modules for the condensation of adjacent activated amino acids; and (T) modules encoding a thioesterase activity for the release of the final peptide. They are present as CAT units.

The thiolation module contains a serine residue that binds a thiol phosphopantetheine arm required to form a thioester with the activated amino acids. It has been proposed that at the end of the third module there is an epimerase domain (E) for the conversion of L- to D-valine. Seven consensus sequences (E1–E7) found in the ACVS epimerase domain of *P. chrysogenum* are fully conserved in the *A. lactamdurans* enzyme (Martín, 2000). A final region bearing the thioesterase activity for the hydrolysis of the thioester bond between ACV and the enzyme might correspond in *A. lactamdurans* ACVS to the conserved motif GWSFGGVL.

Protocol for the purification of ACVS from *A. lactamdurans* (Coque *et al.*, 1996b)

S. lividans (pIJ699-ABC) carrying a 17.8-kb DNA insert containing the *pcbAB* gene of *A. lactamdurans* was used as source of ACVS.

1. Inoculate *S. lividans* (pIJ699-ABC) spores in YEME medium (g/l: yeast extract 3, malt extract 3, peptone 5, glucose 10, and sucrose 340) supplemented with thiostrepton (10 µg/ml) at 30 °C for 48 h with 250 rpm shaking.
2. Inoculate 10 ml of this culture into 100 ml of minimal medium LAT (containing 15 mM L-lysine, 15 mM asparagine, KH_2PO_4 10.5 g/l, $MgSO_4 \cdot 7H_2O$ 0.2 g/l, NaCl 0.09 g/l, 1 ml of 1000x salt solution (g/l: $ZnSO_4 \cdot 7H_2O$ 4, $FeSO_4 \cdot 7H_2O$ 9, $CuSO_4 \cdot 7H_2O$ 0.18, H_3BO_3 0.026, $[NH_4]_3Mo_4O_7 \cdot 4H_2O$ 0.017, $MnSO_4 \cdot 4H_2O$ 0.027, $CaCl_2$ 9; pH 7.0)) supplemented with thiostrepton (10 µg/ml), in 500-ml baffled flasks and incubate at 30 °C for 48 h at 250 rpm.
3. Centrifuge the cells at 10,000g for 30 min at 4 °C. Wash them with saline solution and resuspend the mycelium in disruption buffer A (10 mM $MgCl_2$, 30 mM dithiothreitol, 20 mM EDTA, 50 mM KCl, 50% w/v glycerol, 10 µg/ml DNAse, 50 mM MOPS pH 7.8). Keep the enzyme at 4 °C during all purification steps.
4. Disrupt the cells, kept in ice, by sonication using 20-s pulses with 5-min intervals. Centrifuge the cell-extract at 30,000g for 30 min at 4 °C and desalt the cell-extract in a PD-10 column (Pharmacia) equilibrated with buffer B (1 mM dithiothreitol, 10% glycerol, 0.1 mM EDTA, 50 mM MOPS, pH 7.8).
5. Dilute the protein suspension fourfold to avoid glycerol interference and precipitate the proteins with 30–50% ammonium sulphate. Centrifuge at 17,000g for 20 min.
6. Resuspend the proteins in buffer B and apply to a Sephacryl S-300 (Pharmacia) column (2.5 × 40 cm). Elute the proteins with a flow of 0.225 ml/min of B buffer. The ACVS activity is detected by

(continued)

> enzymatic assay and confirmed by the presence of a band of protein of high molecular weight in SDS-PAGE and by immunoblotting with anti-ACVS antibodies.
> 7. Pool the active fractions and filter them through an Amicon XM300 ultrafilter (43 cm diameter) under pressure of 276 kPa.
> 8. Apply the retained proteins to a DEAE-Sepharose (Fast-Flow, Pharmacia) column (10 ml) and wash with 10 volumes of buffer B. Elute the proteins with a linear gradient 0–0.6 M of NaCl. Active fractions (at about 0.37 M NaCl) possess nearly homogeneous ACVS.

The next four steps of the pathway are: (i) cyclization of ACV to form isopenicillin N by the isopenicillin N synthase (IPNS or cyclase); (ii) conversion of the L-configuration of the α-AAA lateral chain in the isopenicillin N to the D-isomer of penicillin N by IPN epimerase; (iii) expansion of the five-membered thiazolidine ring in penicillin N to a six-membered dihydrothiazine ring in deacetoxycephalosporin C (DAOC) by the DAOC synthase (DAOCS or expandase); and (iv) hydroxylation at the C-3 carbon of DAOC to form deacetylcephalosporin C (DAC) by DAC synthase (DACS), also named DAOC hydroxylase.

4.1. Isopenicillin N synthase

The IPNSs of actinomycetes have been purified in parallel, using the same assay as for fungal cyclases (Ramos *et al.*, 1985; Hollander *et al.*, 1989). These include IPNSs from *S. clavuligerus* (Jensen *et al.*, 1986a) and *A. lactamdurans* (Castro *et al.*, 1988), and also recombinant proteins in *Escherichia coli* originating from *S. jumonjinensis*, *S. clavuligerus*, and *S. lipmanii* (Doran *et al.*, 1990; Landman *et al.*, 1991). The enzyme of *S. clavuligerus* was purified from the producer organism by ammonium sulfate precipitation, ion-exchange and gel filtration, yielding a purification of 130-fold (Jensen *et al.*, 1986a).

IPNSs are nonheme iron-dependent oxygenases that catalyze cyclization of the peptide L-δ-α-AAA-L-cys-D-val, added to the reaction as bis-ACV. They transfer four hydrogen atoms from LLD-ACV to molecular oxygen forming, in a single reaction, the bicyclic nucleus of the penam molecule. Therefore, they have a requirement for oxygen and Fe^{2+} ions. The reducing agents DTT and ascorbate stimulate IPNS activity, which is very sensitive to thiol-specific inhibitors such as N-ethylmaleimide. The IPNS assay is based on the formation of isopenicillin N, a compound with antibiotic properties, active against *Micrococcus luteus*. Differently from the following oxygenases of the pathway (DAOCS, DACS), the IPNS does not use 2-ketoglutarate as hydrogen acceptor.

The molecular size of S. clavuligerus cyclase is 33 kDa which agrees well with the Mr deduced from the pcbC gene. The K_m values for ACV exhibited for different cyclases are on the order of 0.2–0.3 mM. Metal ions (Co^{2+}, Zn^{2+}, Mn^{2+}) inhibit the IPNS activity of the A. lactamdurans protein, probably by competing with Fe^{2+} (Castro et al., 1988).

4.1.1. Other substrates

Replacement of natural substrates of IPNS (as well as other enzymes in the pathway) was encouraged by the possibility of obtaining by biotransformation natural β-lactams or synthetic β-lactams with new structures. Cyclases of fungal origin (Luengo et al., 1986) and those produced by S. clavuligerus (Jensen et al., 1986b) or A. lactamdurans (Castro et al., 1986) are able to cyclize the peptide bis-phenylacetyl-L-cysteinyl-D-valine (PCV), directly producing penicillin G, which is detected by its sensitivity to β-lactamases, as well as by HPLC. The K_m values for PCV and ACV of the A. lactamdurans enzyme are 3.6 and 0.18 mM, respectively.

Highly purified cyclase from S. clavuligerus converts DLD-ACV into penicillin N but the conversion of the natural substrate LLD-ACV to isopenicillin N is 15-fold faster. The slow utilization of DLD-ACV is partially due to differences in the enzyme activity but also to inhibition of the cyclization reaction by penicillin N (Demain et al., 1986). Many other substrate analogs have been used as substrates for IPNSs of different origins (see in Martín and Liras, 1989). Substrate analogs in which the L-α-AAA side of the natural peptide was substituted by L-glutamyl-L-, L-aspartyl-, N-acetyl-L-α-AAA-, or glycyl-L-α-AAA were inactive substrates for S. clavuligerus IPNS but adipyl-L-cysteine-D-valine is cyclized to carboxybutylpenicillin (Wolfe et al., 1984).

4.1.2. Genes

The gene encoding IPNS is called pcbC. Using, as probe, mixed oligonucleotides or internal fragments of pcbC genes of fungal or prokaryotic origin, the genes of many actinomycete species have been isolated and characterized, that is, those of S. clavuligerus (Leskiw et al., 1988), S. jumonjinensis (Shiffman et al., 1988), A. lactamdurans (Coque et al., 1991a,b), S. griseus (García-Domínguez et al., 1991), and S. lipmanii (Loke et al., 2000). In all cases, the pcbC gene lies downstream of the pcbAB gene, encoding ACVS, in a head to tail organization. This is different from that occurring in fungal pcbC genes, which are divergent from pcbAB with a bidirectional promoter. A promoter region has been located upstream of the pcbC gene of S. clavuligerus using promoter-probe plasmids. This has been confirmed by the presence of a 1.2-kb transcript hybridizing to pcbC internal probes (Petrich et al., 1992) and by the different regulation patterns of pcbAB and pcbC (A. Kurt, personal communication) as measured by RT-PCR. The pcbC gene of S. clavuligerus encodes a 329-amino acid protein with an Mr of

36917. These data are very similar to those of all the other *pcbC* genes cloned, which encode proteins with an identity of about 70% in amino acids.

4.1.3. Cyclization of β-lactam compounds in other actinomycetes

Formation of the bicyclic cephem ring in cephamycin is mediated by a single enzyme, IPNS; however, this is not the case in the clavams and carbapenem. Formation of the clavam structure proceeds in two steps. The first is mediated by the β-lactam synthetase, β-LS, a protein similar to asparagine synthetases which, in the presence of Mg^{2+} and ATP, forms directly the four-membered β-lactam ring by an intramolecular amide bond in the carboxyethylarginine precursor molecule (Bachmann et al., 1998). The second ring is formed by clavaminate synthetase (CAS), a multifunctional nonheme iron dioxygenase, similar to IPNS, that requires 2-ketoglutarate and Fe^{2+}. In *S. cattleya*, the producer of thienamycin, the cluster of genes for thienamycin biosynthesis was detected by hybridization with a probe internal to the *bls* gene of *S. clavuligerus*. A gene in the cluster, *thnM*, encodes a protein with 29% identity to the β-lactam synthetase of *S. clavuligerus*, suggesting that the mechanism for cyclization of the carbapenem ring is similar in this respect to that of clavam biosynthesis. However, no purification or biochemical studies on this enzyme have been reported (Núñez et al., 2003). The presence of this mechanism in carbapenems is supported by cyclization of the (5S-carboxymethyl)-S-proline precursor of the carbapenem nucleus, in *Erwinia carotovora*. This precursor is cyclized by the CarA protein, an ATP/Mg^{2+}-dependent enzyme, similar to the β-LS, that has been crystallized (Miller et al., 2003).

4.1.4. Mutagenesis and crystallization

Spectroscopic studies of the IPNS proteins support a physical model for the coordination of the ferrous atom to three histidines and an aspartic acid in the IPNS holoenzyme (Cooper, 1993). Biochemical analysis of seven histidine and five aspartic acid residues changed to alanine by *in vitro* mutagenesis in the *S. jumonjinesis* IPNS indicate that only two histidine residues (H^{212}, H^{268}) and one aspartic acid (D^{214}) are essential for activity. Multiple alignment of representative nonheme Fe^{2+}-containing dioxygenases established that these three amino acids are entirely conserved in all of them. Mutation of the other conserved residues resulted in mutants retaining 5–68% activity (Borovok et al., 1996). Although the crystal structure of the manganese form of *A. nidulans* IPNS (Roach et al., 1997) suggests that the fourth ligand for the ferrous atom is glutamine Q^{329}, mutagenesis studies in *S. jumonjinensis* IPNS indicate that glutamine Q^{230}, which is highly conserved among dioxygenases and proximal to the active site, is the fourth ligand for the Fe^{2+} atom (Landman et al., 1997). However, a $Q^{230}L$ mutant of the *S. clavuligerus* IPNS only showed diminished enzyme activity (Loke and Sim, 1999). Phenylalanine P^{238} has also been found to be important for activity (Wong et al., 2001). Several other highly conserved residues have been proposed to be responsible for the binding of the valine in ACV (Roach et al., 1997). The crystal structure of the enzyme revealed that the active site is buried in a characteristic "jelly-roll" motif that has been found in other oxygenases. Schemes for the mechanisms of ACV cyclization by the IPNSs can be found in Kreisberg-Zakarin et al. (2000) and Blackburn et al. (1995).

4.2. Isopenicillin N epimerase

Isopenicillin N epimerase epimerizes the side chain of α-AAA in isopenicillin N from the L- to the D-configuration, forming penicillin N. This is an important step since the next enzyme in the pathway, DAOCS, is unable to use isopenicillin N as substrate, in spite of the apparent minor structural difference between the two molecules. The epimerases of *A. lactamdurans* (Láiz et al., 1990) and *S. clavuligerus* (Jensen et al., 1983; Usui and Yu, 1989; see purification protocol) were purified from their own producer organisms. The enzyme has been quantified by bioassay based on the different sensitivity of *M. luteus* and *E. coli* Ess22-31 to isopenicillin N in relation to penicillin N; also by HPLC analysis and quantification of *o*-phthalaldehyde (OPA) (Usui and Yu, 1989) or 2,3,4,6 tetra-O-glucosyl isothiocyanate (GITC)-derivatized-penicillin N and -isopenicillin N (Ullan et al., 2002). Both enzymes had, by gel filtration, molecular weights of 59–63 kDa, although electrophoresis by SDS-PAGE gave consistently a protein band of 50 kDa. The molecular weight deduced from the cloned *cefD* genes was even smaller, that is, 43,622 for the 398-amino acid *A. lactamdurans* epimerase. The bacterial epimerases are very unstable but can be partially stabilized with pyridoxal phosphate. This cofactor is essential for activity and is supposed to bind the conserved amino acid sequence –SXHKXL- deduced from the *cefD* genes. Purified IPN epimerase is, in fact, able to act in both orientations, leading

---

Protocol for the purification of the isopenicillin N epimerase from *S. clavuligerus* (Usui and Yu, 1989)

1. Inoculate 250-ml Erlenmeyer flasks containing 50 ml of medium (g/l: glycerol 10, sucrose 20, Nutrisoy grits 15, Amber BYF300 5, Tryptone 5, K_2HPO_4 0.2, pH 6.5) with spores from a *S. clavuligerus* ATCC27064 slant. Grow for 48 h at 30 °C and 250 rpm and use to inoculate at 5% (v/v) 250-ml Erlenmeyer flasks containing 50 ml of fermentation medium (g/l: NZ Amine typeA 3, glycerol 7.5, cornstarch 45, Nutrisoy Flour 20, Nadrisol 1, $FeSO_4 \cdot 7H_2O$ 0.1). Incubate for 48 h at 30 °C with 250-rpm shaking.

(*continued*)

2. Harvest by centrifugation and wash the cells with 15 mM Tris–HCl (pH 7.5) buffer containing 1.0 M KCl and 20% (w/w) ethanol and then with 15 mM Tris–HCl pH 7.5. Keep the cells (250 g) frozen at $-80\,°C$ until used.
3. All the steps of the purification are carried out at 0–4 °C. Thaw and homogenize the cells in 1 l buffer A (10 mM pyrophosphate-HCl, pH 8.0; 2.2 M glycerol, 0.1 mM dithiothreitol). Aliquots of the cell suspension are sonicated four times for 20 s with 30 s intervals.
4. Centrifuge the cell-extract at 30,000g for 30 min and submit to ammonium sulphate precipitation. Dissolve the 35–70% ammnonium sulphate precipitate in buffer B (10 mM pyrophosphate buffer at pH 8.0 containing 0.1 mM dithiothreitol and 0.15 mM NaCl). Dialyze overnight against buffer B.
5. Apply to a DE-52 column (3.7 × 25 cm) equilibrated with buffer B and elute with a linear gradient formed by buffer B containing NaCl from 0.15 to 0.3 M. Pool the active fractions and concentrate by ultrafiltration with an Amicon PM-10 membrane. Dialyze overnight against buffer C (10 mM potassium phosphate at pH 7.0 containing 0.1 mM dithiothreitol and 0.1 M NaCl).
6. Load onto a DEAE Affi-Gel blue column (1.6 × 15 cm) equilibrated with buffer C. The enzyme, retained by the column, is eluted with a linear gradient of buffer C containing from 0.1 to 0.3 M NaCl.
7. Concentrate the active fractions by ultrafiltration and apply them to a Sephadex G-200 (2.5 × 42 cm) column equilibrated with buffer D (10 mM potassium phosphate, pH 6.0; 0.1 mM dithiothreitol, 10 μM pyridoxal phosphate).
8. Pool, concentrate by ultrafiltration, and apply the active fractions to a calcium phosphate-cellulose (1.6 × 10 cm) column equilibrated with buffer D. Wash the column with the same buffer and elute the enzyme with buffer D containing 10–100 mM potassium phosphate.
9. Combine the active fractions, concentrate them by ultrafiltration and dialyze them overnight against buffer E (10 mM pyrophosphate, pH 8.0, 0.1 mM dithiothreitol, 10 μM pyridoxal phosphate).
10. Apply the enzyme to a FPLC MonoQ column (0.5 × 0.5 cm) equilibrated with buffer E. Elute the enzyme with a linear gradient of 0–0.4 M NaCl in buffer E.

The protocol results in a 650-fold purification to electrophoretic homogeneity of the Isopenicillin N epimerase, with a yield of 18%.

(continued)

> **Assay for Isopenicillin N Epimerase**
> The standard assay mixture contains 1.4 mM isopenicillin N or penicillin N, 0.2 mM dithiothreitol, 0.1 mM pyridoxal phosphate, and buffer (50 mM pyrophosphate pH 8.3) in a final volume of 0.5 ml. The mixture is incubated for 20 min at 37 °C and the reaction is stopped by placing the assay mixture in boiling water for 10 min.
>   Penicillin N and isopenicillin N in the samples (20 µl) are derivatized with 5 µl of OPA reagent (containing 4 mg OPA dissolved in 300 ml methanol, 250 µl of sodium borate at 0.4 M (pH 9.4), 390 µl H_2O, 60 µl N-acetylcysteine at 1 M and NaOH to adjust the pH to 5–6). After 2 min of reaction, the samples are diluted with 200 µl of 50 mM sodium acetate at pH 5.0 and aliquots (50 µl) are analyzed by HPLC.
>   The analysis is done in a Varian HPLC system including a Vista 5560 fluorescent detector. Separation of derivatized isopenicillin N and penicillin N is done using a Cyclobond I column (0.46 × 25 cm) with a mobile phase formed by 59% (v/v) 50 mM sodium phosphate, 1% (v/v) 1 M sodium acetate, 40% (v/v) methanol, pH 6.0, and a flow of 1 ml/min.

in vitro to an equimolecular mixture of isopenicillin N and penicillin N. The K_m is 0.3 mM for isopenicillin N and 0.78 mM for penicillin N.

The *cefD* gene, encoding IPN epimerase, is located in *A. lactamdurans* (Coque *et al.*, 1993) and *S. clavuligerus* (Kovacevic *et al.*, 1990) next to *cefE* which encodes DAOCS, responsible for the next step in the pathway. The two genes are cotranscribed, assuring the presence of equimolecular amounts of both enzymes. This is important since DAOCS will capture penicillin N shifting the IPN-penicillin N equilibrium toward the formation of more penicillin N. IPN-epimerase is sensitive to divalent ions, to –SH-modifying reagents such as N-ethyl maleimide and p-chlorobenzoate, as well as to fluorescamine, a compound binding primary amines such as lysine located in the SXHKXL sequence required for pyridoxal phosphate binding.

Differently from other enzymes involved in either penicillin or cephalosporin biosyntheses, which were first purified and characterized from filamentous fungi, the IPN-epimerase was first purified and characterized from actinomycetes. The homologous enzyme from *A. chrysogenum* was very unstable and further studies demonstrated that the epimerization of IPN in fungi is carried out by a specific pathway requiring three enzymatic steps (Ullan *et al.*, 2002).

4.3. Deacetoxycephalosporin C synthase

DAOCS was discovered by Kohsaka and Demain in 1976. Purification of the DAOCS synthases of *A. lactamdurans* (Cortés *et al.*, 1987) and *S. clavuligerus* (Dotzlaf and Yeh, 1989; Rollins *et al.*, 1988) to near homogeneity showed that the enzyme was a monomer with relatively unstable activity. It requires Fe^{2+}, penicillin N, and 2-ketoglutarate and is stimulated and stabilized by ascorbate and dithiothreitol. Fe^{2+} could not be substituted by other divalent metal ions nor penicillin N by isopenicillin N or 6-aminopenicillanic acid (6-APA). In addition, the natural enzyme is only able to use 2-ketoglutarate but not 2-ketobutyrate, 3-ketoadipate, 2-ketohexanoic acid, or 2-keto-4-methylpentanoic acid as cosubstrates. The apparent K_m values for penicillin N, 2-ketoglutarate, and Fe^{2+} were calculated to be 52, 3, and 71 μM, respectively (Cortés et al., 1987).

The *cefE* gene, encoding DAOCS in actinomycetes has been cloned from *S. clavuligerus* (Kovacevic et al., 1990) and *A. lactamdurans* (Coque et al., 1993). DAOCSs, of actinomycetes possess some DAC hydroxylating activity (see below) while the DAOCSs of filamentous fungi are fully bifunctional enzymes with both DACS and DAOCS activity. Most of the studies on this enzyme have been performed using r-DAOCS of *S. clavuligerus* expressed in *E. coli* as inclusion bodies where it is easily extracted and activated. The enzyme of *S. clavuligerus*, differently from those of filamentous fungi, has very little DAOC hydroxylase activity (see below) and is more active and stable than that from *A. lactamdurans*. Therefore, recombinant DAOCS of *S. clavuligerus* has been crystallized (Valegärd et al., 1998) and used in further studies.

Recombinant DAOCS of *S. clavuligerus* is a monomer (Mr 28.9 kDa) but is in equilibrium with a trimeric form (Mr 92.9 kDa), which is the form that crystallizes. The C-terminal end of the protein is responsible for the oligomerization process since it projects a fold toward the active site of the adjacent monomer in a cyclic manner, forming a trimer. This organization precludes the formation of crystals of monomeric DAOCS bound to penicillin N. Addition of Fe^{2+} or 2-ketoglutarate shifts the equilibrium toward the monomeric form that appears to be the active form in solution. A sequential binding of 2-ketoglutarate, penicillin N, and oxygen has been proposed and X-ray absorption spectroscopic studies as well X-ray absorption fine structure techniques have been used to propose a mechanistic model of DAOCS activity (Lloyd et al., 1999; Valegärd et al., 1998). From the crystal structure of DAOCS, different amino acids have been proposed to bind the substrate and the Fe^{2+} atom. A substantial amount of work on site-directed mutagenesis of the *cefE* gene has been done to evaluate the role of different amino acid residues on ACVS activity (Goo et al., 2008a). The importance for substrate binding of eight arginine residues (Lee et al., 2001a; Lipscomb et al., 2002) and of different cysteine residues (Lee et al., 2000) has been assessed using *in vitro* mutagenesis and crystalization of mutant DAOCSs. Programs of random mutagenesis and screening have also been performed and DAOCS mutants with 13-fold more activity on penicillin G have been obtained (Wei et al., 2003). Small deletions in the C-terminal end also result in significant differences in the way that the enzyme catalyzes penicillin N oxidation as compared to penicillin G oxidation (Lee et al., 2001b). Mutations altering the specificity for 2-ketoglutarate arise by mutagenesis of arginine R[258] (Lee et al., 2003) and 2-ketoacids with up to six carbons appear to be efficient cosubstrates with an efficient coupling to penicillin oxidation.

Mutations to obtain a wider substrate specificity for DAOCS (Chin and Sim, 2002; Chin et al., 2004; Goo et al., 2008 a,b) are very important since modification of the DAOCS activity to accept hydrophobic penicillins is of great industrial interest. 7-aminodeacetylcephalosporanic acid (7-ADCA) is the molecule of importance for production of semisynthetic cephalosporins and is currently produced by chemical expansion of penicillin G into phenylacetyl-7-ADCA, followed by removal of the phenylacetyl side chain. An efficient enzymatic expansion of penicillin G using DAOCS is ecologically and economically desirable.

Another approach for the bioconversion of penicillin G has been the use of a hybrid expandase gene originating from *S. clavuligerus* and *A. lactamdurans* (Gao et al., 2003) or the use of *in vitro* DNA recombination (family shuffling) of *S. clavuligerus cefE* genes isolated from several actinomycetes (Hsu et al., 2004).

In vivo methods of metabolic engineering have also used the *cefE* gene of *S. clavuligerus*. This gene was incorporated into *P. chrysogenum* to carry out, in fermentations fed with adipic acid as a side-chain precursor, the conversion of adipyl-6-aminopenicillanic acid to adipyl-7-aminodeacetoxycephalosporanic acid, which can be hydrolyzed by a glutaryl acylase to 7-aminodesacetyl-cephalosporanic acid (7-ADCA) (Bovenberg et al., 1998). The *cefF* gene has also been introduced by transformation into a *A. chrysogenum cefEF*-negative mutant to produce DAOC that can be purified and processed *in vitro* by a two-step enzymatic conversion, to 7-ADCA (Velasco et al., 2000). Both the *cefD* epimerase of *S. lipmanii* and a hybrid *pcbC-cefE* gene have been introduced by transformation into *P. chrysogenum* to directly produce DAOC (Cantwell et al., 1992; Queener et al., 1994).

4.4. Deacetoxycephalosporin C hydroxylase

DAOC hydroxylase catalyzes hydroxylation of the deacetoxycephalosporin molecule at C-3′. DAOC hydroxylase of *S. clavuligerus* was purified to near homogeneity by anion-exchange chromatography, ammonium sulphate precipitation, and gel filtration. The enzyme is a nonheme Fe^{2+} oxygenase of the same type as IPNS or DAOCS. It requires Fe^{2+} (K_m 20 μM), molecular oxygen, 2-ketoglutarate (K_m 10 μM) as acceptor of two hydrogen atoms and DAOC (K_m 59 mM) as substrate. In addition, it catalyzes hydroxylation of 3-exomethylenecephalosporin to DAOC (Baker et al., 1991).

Cloning of *cefF* genes, encoding DAOC hydroxylase, has been reported in *S. clavuligerus* (Kovacevic and Miller, 1991) and *A. lactamdurans* (Coque et al., 1996a). The amino acid sequences of DAOCS hydroxylase and DACS of *S. clavuligerus* are 59% identical. The gene *cefF* from *A. lactamdurans* has been expressed in *E. coli* and found to lack C-7 hydroxylating activity. However, both the protein purified from the producer organisms and the recombinant protein expressed in *E. coli* show a residual DAC synthase activity. Since a single gene, *cefEF*, has been found in *A. chrysogenum* encoding both DAC synthase and DAOC hydroxylase activities (Samson et al., 1987), it has been proposed that in actinomycetes, an ancestral *cefEF* gene was duplicated, one of the duplicated genes evolving to improve its expandase activity while the other specializing in hydroxylase activity (Kovacevic and Miller, 1991).

5. Specific Steps for Tailoring the Cephem Nucleus in Actinomycetes

The order of the last steps for cephamycin C biosynthesis is still unclear. They consist of the introduction of a methoxyl group at C-7 and a carbamoyl group at the exocyclic CH_2OH of the C-3' carbon in the cephamycin molecule. However, recent crystallization studies on CmcI suggest that the carbamoylation step might occur first.

Information on the carbamoylation at C-3' was obtained using cell-extracts of *S. lividans* transformants carrying the *cmcH* gene of *A. lactamdurans* (Coque et al., 1995a). The assay requires $[^{14}C]$-carbamoylphosphate, Mn^{2+} ions, ATP, and the unnatural substrate decarbamoyl-cefuroxime (Glaxo Pharmaceuticals). The labeled carbamoyl-cefuroxime formed can be easily extracted and the radioactivity in the samples reflects the enzyme activity. The genes encoding the cephamycin carbamoyl transferase (*cmcH*) and the enzymes for the methoxylation steps (*cmcI*, *cmcJ*) lie side by side in *A. lactamdurans* and *S. clavuligerus* (Coque et al., 1995a,b; Alexander and Jensen, 1998). The deduced CmcH protein of *A. lactamdurans* has 520 amino acids (Mr 57149) and low overall homology with ornithine- or aspartyl-carbamoyltransferase (OTCase and ATCase), which also use carbamoylphosphate (CP) as substrate, but the three types of enzymes contain a common region for CP binding (aa 29–38 and 119–123 in the CmcH protein of *A. lactamdurans*).

Initial biochemical studies on the C-7 α-hydroxylating system were made in *S. clavuligerus*. The C-7 hydroxylating activity was partially purified by standard techniques of ion-chromatography, ammonium sulphate precipitation, gel filtration, and affigel-blue chromatography. The almost pure preparation showed a major protein band of Mr 32 kDa (Xiao et al., 1991), a size that fits well with the molecular weight that was later deduced from the gene. The C-7 hydroxylation assay was based on the HPLC detection of 7-α-hydroxylcephalosporin C formed using cephalosporin C as substrate in the presence of Fe^{2+} ions, ascorbic acid, and α-ketoglutarate. The enzyme was shown to have a K_m value of 0.72 mM for cephalosporin C and no DAOC synthase or DAC hydroxylase activity was detected in the preparation. However, crystallization studies of CmcI (see below) indicate that the enzyme does not require iron or 2-ketoglutarate.

In *A. lactamdurans,* the *cmcI* and *cmcJ* genes appear to be expressed in a coordinate manner by translational coupling (Coque *et al.*, 1995b) and encode proteins of Mr 27364 and 32090, respectively. The *cmcI*-encoded protein resembles *O*-methyltransferases of different origins and has consensus sequences in the primary structure to domains of catechol-*O*-methyl transferase and SAM-dependent DNA methylases. However, the presence of domains similar to 7-α-monooxygenases, a weak 7-α-cephalosporin C hydroxylase activity and affinity to NADH in the purified recombinant CmcI monomer preclude a clear relationship between the hydroxylase or methyltransferase activities and the CmcI or CmcJ proteins. Expression of both genes, *cmcI* and *cmcJ*, in *S. lividans* resulted in 7-cephem hydroxylase and 7-hydroxycephem methyltransferase activities, as shown by HPLC detection of the corresponding products and by the formation of labeled 7-methoxycephalosporin C when *S*-adenosyl-L-[methyl-^{14}C] methionine was used in the assays, However, transformants carrying the individual genes did not possess either of the activities (Coque *et al.*, 1995b).

Using recombinant proteins purified from *E. coli,* Enguita *et al.* (1996) obtained antibodies against CmcI and CmcJ. Both antibodies, independently, purified the CmcI–CmcJ complex using either *A. lactamdurans* or *S. clavuligerus* cell-free extracts. Codenaturation and glutaraldehyde cross-linking indicated a 1:1 heterodimer with an apparent dissociation constant of 47 μM, as shown by fluorescence spectroscopy. The complex, but not the separate proteins, showed a strong interaction with *S*-adenosylmethionine and cephalosporin C and initially it was suggested that the methoxylation activity was on CmcI and that CmcJ acts as a helper protein for efficient methoxylation. However, recent crystallization studies of CmcI indicate that this protein has methyltransferase activity (Öster *et al.*, 2006).

The CmcI protein of *S. clavuligerus* was crystallized as a selenomethionine-substituted protein (Lester *et al.*, 2004), and was later crystallized again, after correction of the initially published sequence of the *S. clavuligerus* gene (Öster *et al.*, 2006). CmcI is a hexamer formed by a trimer of dimers and occurs also as a hexamer in solution, with the N-terminal domain being responsible for the oligomerization. The polypeptide forms a C-terminal Rossmann domain which is the binding site of *S*-adenosylmethionine (through Tyr^{91} and Asp^{187}). It also binds Mg^{2+} (through Asp^{160}, Glu^{186}, Asp^{187}) which agrees with the magnesium requirement of catechol-*O*-methyltransferases, although this requirement was not found in the different enzyme assays. The structure of CmcI does not contain the jelly-roll fold common to 2-ketoglutarate-dependent dioxygenases or the histidine-ferrous iron binding motif characteristic of IPNS or DAOCS.

NADH is positioned in the cavity between molecules and only three NADH molecules bind per hexamer. The binding of NADH to CmcI is not the expected mode of binding of NAD/NADH to nucleotide-binding motifs. A model for the CmcI–SAM–Mag^{2+} complex has been proposed (Öster et al., 2006).

The protein encoded by *cmcJ* shows weak similarity to a gibberellin A desaturase with hydroxylase activity but no homology to the 2-ketoglutarate dioxygenases required for early steps in cephamycin biosynthesis. A clearer picture of the methoxylation system will require the crystallization of the protein encoded by *cmcJ*.

6. Regulation of Cephamycin C Production

Production of antibiotics depends on the presence of precursors, on general mechanisms of carbon, nitrogen, or phosphate regulation, and on ppGpp intracellular content. It is also controlled by general cascade mechanisms involving butyrolactones, regulatory proteins such as AdpA or butyrolactone receptor proteins (Brp), and finally by antibiotic, cluster specific, regulatory proteins of the SARP type (reviewed by Liras et al., 2008; see also Chapters 4, 5, and 6 in this volume). Many of these systems have been reported by several authors to control cephamycin production, although the exact control mechanisms have not been fully elucidated.

The major influence on cephamycin production in defined medium is due to lysine, a precursor of α-AAA, which acts both as the substrate of LAT, providing α-AAA, and by increasing LAT cellular activity (Rius et al., 1996). It has not been reported whether this occurs at the transcriptional or enzymatic level. Addition of 150 m*M* lysine to the cultures results in a global increase of cephalosporin production of four- to fivefold. Regulatory mechanisms of cephamycin C control by carbon and nitrogen sources have been reported but, again, no molecular studies have been made on this subject.

The cluster-specific regulatory protein in *S. clavuligerus* is named CcaR, an autoregulatory protein that controls the formation of cephamycin C by binding to the *cefD-cmcI* bidirectional promoter that also controls formation of clavulanic acid by a still unclarified mechanism (Santamarta et al., 2002). Several biosynthetic enzymes such as LAT, IPNS, or DAOCS depend on the presence of CcaR. The transcripts encoding these enzymes and that encoding ACVS are not present in CcaR-defective mutants.

CcaR formation itself is sensitive to the modulatory effect of other regulators such as the AreB protein, binding an ARE sequence present upstream of the *ccaR* promoter, or the Brp protein for butyrolactone-binding

(Santamarta et al., 2005). However, the presence of butyrolactones in S. clavuligerus has not been studied in detail. The mutant strain S.clavuligerus bldG, which is deficient in a putative anti-anti-sigma factor that regulates morphological differentiation and formation of antibiotics (Bignell et al., 2005), is unable to produce CcaR. The stringent response has a dual effect on cephamycin C production. While a ppGpp-deficient relC mutant of S. clavuligerus does not produce cephamycin C, the ppGpp-null relA mutant produces 2.5-fold more cephamycin C (Gomez-Escribano et al., 2008), indicating that ppGpp itself is not the controlling molecule.

A regulatory gene, similar to S. clavuligerus ccaR, has not been located in the cephamycin producer A. lactamdurans but a protein with high similarity to CcaR is encoded by the tnhU gene, located in the thienamycin gene cluster of S. cattleya, a microorganism that also produces cephamycin C (Núñez et al., 2003). The possibility exists that this protein controls the formation of both thienamycin and cephamycin C in this microorganism as occurs with clavulanic acid and cephamycin C in S. clavuligerus.

The presence of many steps regulated either at the transcriptional or translational level in relation to cephamycin C production opens the way to improve the yield of this compound using molecular biological techniques.

ACKNOWLEDGMENTS

This research was supported by Grants from the Spanish CICYT (Bio 2006-14853) Junta de Castilla y León GR117 and by the European Project LSHM-CT-2004-005224.

REFERENCES

Alexander, D. C., Anders, C. L., Lee, L., and Jensen, S. E. (2007). pcd mutants of *Streptomyces clavuligerus* still produce cephamycin C. *J. Bacteriol.* **189,** 5867–5874.

Alexander, D. C., and Jensen, S. E. (1998). Investigation of the *Streptomyces clavuligerus* cephamycin C gene cluster and its regulation by the CcaR protein. *J. Bacteriol.* **180,** 4068–4079.

Aoki, H., Sakai, H., Kohsaka, M., Konomi, T., and Hosoda, J. (1976). Nocardicin A, a new monocyclic beta-lactam antibiotic. I. Discovery, isolation and characterization. *J. Antibiot.* **29,** 492–500.

Bachmann, B. O., Li, R., and Townsend, C. A. (1998). Beta-lactam synthetase: A new biosynthetic enzyme. *Proc. Natl. Acad. Sci. USA* **95,** 9082–9086.

Baggaley, K. H., Brown, A. G., and Schofield, C. J. (1997). Chemistry and biosynthesis of clavulanic acid and other clavams. *Nat. Prod. Rep.* **14,** 309–333.

Baker, B. J., Dotzlaf, J. E., and Yeh, W. u. K. (1991). Deacetoxycephalosporin C hydroxylase of *Streptomyces clavuligerus*: Purification, characterization and evolutionary implication. *J. Biol. Chem.* **266,** 5087–5093.

Bignell, D. R., Tahlan, K., Colvin, K. R., Jensen, S. E., and Leskiw, B. K. (2005). Expression of ccaR, encoding the positive activator of cephamycin C and clavulanic acid production in *Streptomyces clavuligerus*, is dependent on *bldG*. *Antimicrob. Agents Chemother.* **49,** 1529–1541.

Blackburn, J. M., Sutherland, J. D., and Baldwin, J. E. (1995). A heuristic approach to the analysis of enzymic catalysis: Reaction of δ-(L-α-Aminoadipoyl)-L-cysteinyl-D-α-aminobutyrate and δ-(L-α-aminoadipoyl)-L-cysteinyl-D-allylglycine catalized by isopenicillin N synthase isozymes. *Biochemistry* **34,** 7548–7562.

Borovok, I., Landman, O., Kreisberg-Zakarin, R., Aharonowitz, Y., and Cohen, G. (1996). Ferrous active site of isopenicillin N synthase: Genetic and sequence analysis of the endogenous ligands. *Biochemistry* **35,** 1981–1987.

Bovenberg, R. A. L., Vollebregt, A. W. H., van der Berg, M. A., Kerkman, R., Nieboer, M., and Schipper, D. (1998). Applications of Biosynthetic Genes to Modify β-Lactam Formation by *P. chrysogenum* 8th GIM Abstracts Program, pp. 21–22. Jerusalem.

Brakhage, A. A., Al-Abdallah, Q., Tüncher, A., and Spröter, P. (2005). Evolution of beta-lactam biosynthesis genes and recruitment of trans-acting factors. *Phytochemistry* **66,** 1200–1210.

Brown, A. G., Butterworth, D., Cole, M., Hanscomb, G., Hood, J. D., Reading, C., and Rolinson, G. N. (1976). Naturally-occuring beta-lactamase inhibitors with antibacterial activity. *J. Antibiot.* **29,** 668–669.

Cantwell, C., Beckmann, R., Whiteman, P., Queener, S. W., and Abraham, E. P. (1992). Isolation of deacetoxycephalosporin C from fermentation broths of *Penicillium chrysogenum* transformants: Construction of a new fungal biosynthetic pathway. *Proc. Biol. Sci.* **248,** 283–289.

Castro, J. M., Liras, P., Cortés, J., and Martín, J. F. (1986). Conversion of phenylacetyl-cysteinyl-valine *in vitro* into penicillin G by isopenicillin N synthase of *S. lactamdurans*. *FEMS Microbiol. Lett.* **34,** 349–353.

Castro, J. M., Liras, P., Laíz, L., Cortés, J., and Martín, J. F. (1988). Purification and characterization of the isopenicillin N synthase of *Streptomyces lactamdurans*. *J. Gen. Microbiol.* **134,** 133–141.

Chin, H. S., Goo, K. S., and Sim, T. S. (2004). A complete library of amino acid alterations at N304 in *Streptomyces clavuligerus* deacetoxycephalosporin C synthase elucidates the basis for enhanced penicillin analogue conversion. *Appl. Environ. Microbiol.* **70,** 607–609.

Chin, H. S., and Sim, T. S. (2002). C-terminus modification of *Streptomyces clavuligerus* deacetoxycephalosporin C synthase improves catalysis with an expanded substrate specificity. *Biochem. Biophys. Res. Commun.* **295,** 55–61.

Cooper, R. D. (1993). The enzymes involved in biosynthesis of penicillin and cephalosporin; their structure and function. *Bioorg. Med. Chem.* **1,** 1–17.

Coque, J. J. R., de la Fuente, J. L., Liras, P., and Martín, J. F. (1996a). Overexpression of the *Nocardia lactamdurans* α-aminoadipyl-L-cysteinyl-D-valine synthetase in *Streptomyces lividans*. *Eur. J. Biochem.* **242,** 264–270.

Coque, J. J. R., Enguita, F. J., Cardoza, R. E., Martín, J. F., and Liras, P. (1996b). Characterization of the *cefF* gene of *Nocardia lactamdurans* encoding a 3′-methylcephem hydroxylase different from the 7-cephem hydroxylase. *Appl. Microbiol. Biotechnol.* **44,** 605–609.

Coque, J. J. R., Enguita, F. J., Martin, J. F., and Liras, P. (1995b). A two protein component 7α-cephem methoxylase encoded by two genes of the cephamycin C cluster converts cephalosporin C to 7-methoxycephalosporin C. *J. Bacteriol.* **177,** 2230–2235.

Coque, J. J. R., Liras, P., Laiz, L., and Martín, J. F. (1991b). A gene encoding lysine 6-aminotransferase, which forms the beta-lactam precursor alpha-aminoadipic acid, is located in the cluster of cephamycin biosynthetic genes in *Nocardia lactamdurans*. *J. Bacteriol.* **73,** 6258–6264.

Coque, J. J. R., Martín, J. F., Calzada, J. G., and Liras, P. (1991a). The cephamycin biosynthetic genes *pcbAB*, encoding a large multidomain peptide synthetase, and *pcbC* of *Nocardia lactamdurans* are clustered together in an organization different from the same genes in *Acremonium chrysogenum* and *Penicillium chrysogenum*. *Mol. Microbiol.* **5,** 1125–1133.

Coque, J. J. R., Martín, J. F., and Liras, P. (1993). Characterization and expression in *Streptomyces lividans* of *cefD* and *cefE* genes from *Nocardia lactamdurans*: The organization of the cephamycin gene cluster differs from that in *Streptomyces clavuligerus*. *Mol. Gen. Genet.* **236,** 453–458.

Coque, J. J. R., Pérez-Llarena, F. J., Enguita, F. J., Fuente, J. L., Martín, J. F., and Liras, P. (1995a). Characterization of the *cmcH* genes of *Nocardia lactamdurans* and *Streptomyces clavuligerus*, encoding a functional 3′-hydroxymethylcephem O-carbamoyltransferase for cephamycin biosynthesis. *Gene* **162,** 21–27.

Cortés, J., Martín, J. F., Castro, J. M., Láiz, L., and Liras, P. (1987). Purification and characterization of a 2-oxoglutarate-linked ATP-independent deacetoxycephalosporin C synthase of *Streptomyces lactamdurans*. *J. Gen. Microbiol.* **133,** 3165–3174.

Demain, A. L., and Elander, R. P. (1999). The β-lactam antibiotics: Past, present and future. *Antonie van Leeuwenhoek* **75,** 5–19.

Demain, A. L., Shen, Y. Q., Jensen, S. E., Westlake, D. W., and Wolfe, S. (1986). Further studies on the cyclization of the unnatural tripeptide δ-(L-α-aminoadipyl)-L-cysteinyl-D-valine to penicillin N. *J. Antibiot.* **39,** 1007–1010.

Doran, J. L., Leskiw, B. K., Petrich, A. K., Westlake, D. W., and Jensen, S. E. (1990). Production of *Streptomyces clavuligerus* isopenicillin N synthase in *Escherichia coli* using two-cistron expression systems. *J. Ind. Microbiol.* **5,** 197–206.

Dotzlaf, J. E., and Yeh, W. K. (1989). Purification and properties of deacetoxycephalosporin C synthase from recombinant *Escherichia coli* and its comparison with the native enzyme purified from *Streptomyces clavuligerus*. *J. Biol. Chem.* **264,** 10219–10227.

Enguita, F. J., Liras, P., Leitao, A. L., and Martín, J. F. (1996). Interaction of the two proteins of the methoxylation system involved in cephamycin C biosynthesis. *J. Biol. Chem.* **271,** 33225–33230.

Fuente, J. L., Rumbero, A., Martín, J. F., and Liras, P. (1997). Delta-1-piperideine-6-carboxylate dehydrogenase, a new enzyme that forms alpha-aminoadipate in *Streptomyces clavuligerus* and other cephamycin C-producing actinomycetes. *Biochem. J.* **327,** 59–64.

Gao, Q., Piret, J. M., Adrio, J. L., and Demain, A. L. (2003). Performance of a recombinant strain of *Streptomyces lividans* for bioconversion of penicillin G to deacetoxycephalosporin G. *J. Ind. Microbiol. Biotechnol.* **30,** 190–194.

García-Domínguez, M., Liras, P., and Martín, J. F. (1991). Cloning and characterization of the isopenicillin N synthase gene of *Streptomyces griseus* NRRL 3851 and studies of expression and complementation of the cephamycin pathway in *Streptomyces clavuligerus*. *Antimicrob. Agents Chemother.* **35,** 44–52.

Gomez-Escribano, J. P., Martín, J. F., Hesketh, A., Bibb, M. J., and Liras, P. (2008). *Streptomyces clavuligerus* relA-null mutants overproduce clavulanic acid and cephamycin C: Negative regulation of secondary metabolism by (p)ppGpp. *Microbiology* **154,** 744–755.

Goo, K. S., Chua, S. C., and Sim, T. S. (2008a). A complete library of amino acid alterations at R306 in *Streptomyces clavuligerus* deacetoxycephalosporin C synthase demonstrates its structural role in the ring-expansion activity. *Proteins* **70,** 739–747.

Goo, K. S., Chua, C. S., and Sim, T. S. (2008b). Relevant double mutations in bioengineered *Streptomyces clavuligerus* deacetoxycephalosporin C synthase result in higher binding specificities which improve penicillin bioconversion. *Appl. Environ. Microbiol.* **74,** 1167–1175.

Grünewald, J., and Marahiel, M. A. (2006). Chemoenzymatic and template-directed synthesis of bioactive macrocyclic peptides. *Microbiol. Mol. Biol. Rev.* **70,** 121–146.

Hollander, I. J., Shen, Y. Q., Heim, J., Demain, A. L., and Wolfe, S. (1984). A pure enzyme catalyzing penicillin biosynthesis. *Science* **224**, 610–612.

Hsu, J. S., Yang, Y. B., Deng, C. H., Wei, C. L., Liaw, S. H., and Tsai, Y. C. (2004). Family shuffling of expandase genes to enhance substrate specificity for penicillin G. *Appl. Environ. Microbiol.* **70**, 6257–6263.

Jensen, S. E., Leskiw, B. K., Vining, L. C., Aharonowitz, Y., Westlake, D. W., and Wolfe, S. (1986a). Purification of isopenicillin N synthetase from *Streptomyces clavuligerus*. *Can. J. Microbiol.* **32**, 953–958.

Jensen, S. E., Westlake, D. W., Bowers, R. J., Lyubechansky, L., and Wolfe, S. (1986b). Synthesis of benzylpenicillin by cell-free extracts from *Streptomyces clavuligerus*. *J. Antibiot.* **39**, 822–826.

Jensen, S. E., Westlake, D. W., and Wolfe, S. (1983). Partial purification and characterization of isopenicillin N epimerase activity from *Streptomyces clavuligerus*. *Can. J. Microbiol.* **29**, 1526–1531.

Jensen, S. E., Wong, A., Rollins, M. J., and Westlake, D. S. W. (1990). Purification and partial characterization of the δ-(L-α-aminoadipyl)-L-cysteinyl-D-valine synthetase from *Streptomyces clavuligerus*. *J. Bacteriol.* **172**, 7269–7271.

Kahan, J. S., Kahan, F. M., Goegelman, R., Currie, S. A., Jackson, M., Stapley, E. O., Miller, T. W., Miller, A. K., Hendlin, D., Mochales, S., Hernández, S., Woodruff, H. B., et al. (1979). Thienamycin, a new β-lactam antibiotic. I. Dicovery, taxonomy, isolation and physical properties. *J. Antibiot.* **32**, 1–12.

Kern, B. A., Hendlin, D., and Inamine, E. (1980). l-lysine epsilon-aminotransferase involved in cephamycin C synthesis in *Streptomyces lactamdurans*. *Antimicrob. Agents Chemother.* **17**, 679–685.

Kohsaka, M., and Demain, A. L. (1976). Conversion of penicillin N to cephalosporin(s) by cell-free extracts of *Cephalosporium acremonium*. *Biochem. Biophys. Res. Commun.* **70**, 465–473.

Kovacevic, S., and Miller, J. R. (1991). Cloning and sequencing of the β-lactam hydroxylase gene (*cefF*) from *Streptomyces clavuligerus*: Gene duplication may have led to separate hydroxylase and expandase. *J. Bacteriol.* **173**, 398–400.

Kovacevic, S., Tobin, M. B., and Miller, J. R. (1990). The beta-lactam biosynthesis genes for isopenicillin N epimerase and deacetoxycephalosporin C synthetase are expressed from a single transcript in *Streptomyces clavuligerus*. *J. Bacteriol.* **172**, 3952–3958.

Kreisberg-Zakarin, R., Borovok, I., Yanko, M., Frolow, F., Aharonowitz, Y., and Cohen, G. (2000). Structure-function studies of the non-heme iron active site of isopenicillin N synthase: Some implications for catalysis. *Biophys. Chem.* **86**, 109–118.

Láiz, L., Liras, P., Castro, J. M., and Martín, J. F. (1990). Purification and characterization of the isopenicillin N epimerase from *Nocardia lactamdurans*. *J. Gen. Microbiol.* **136**, 663–671.

Landman, O., Borovok, I., Aharonowitz, Y., and Cohen, G. (1997). The glutamine ligand in the ferrous iron active site of isopenicillin N synthase of *Streptomyces jumonjinensis* is not essential for catalysis. *FEBS Lett.* **405**, 172–174.

Landman, O., Shiffman, D., Av-Gay, Y., Aharonowitz, Y., and Cohen, G. (1991). High level expression in *Escherichia coli* of isopenicillin N synthase genes from *Flavobacterium* and *Streptomyces*, and recovery of active enzyme from inclusion bodies. *FEMS Microbiol. Lett.* **68**, 239–244.

Lee, H. J., Dai, Y. F., Shiau, C. Y., Schofield, C. J., and Lloyd, M. D. (2003). The kinetic properties of various R258 mutants of deacetoxycephalosporin C synthase. *Eur. J. Biochem.* **270**, 1301–1307.

Lee, H. J., Lloyd, M. D., Clifton, I. J., Harlos, K., Dubus, A., Baldwin, J. E., Frere, J. M., and Schofield, C. J. (2001a). Alteration of the co-substrate selectivity of deacetoxycephalosporin C synthase. The role of arginine 258. *J. Biol. Chem.* **276**, 18290–18295.

Lee, H. J., Lloyd, M. D., Harlos, K., Clifton, I. J., Baldwin, J. E., and Schofield, C. J. (2001b). Kinetic and crystallographic studies on deacetoxycephalosporin C synthase (DAOCS). *J. Mol. Biol.* **308,** 937–948.

Lee, H. J., Lloyd, M. D., Harlos, K., and Schofield, C. J. (2000). The effect of cysteine mutations on recombinant deacetoxycephalosporin C synthase from *S. clavuligerus*. *Biochem. Biophys. Res. Commun.* **267,** 445–448.

Leskiw, B. K., Aharonowitz, Y., Mevarech, M., Wolfe, S., Vining, L. C., Westlake, D. W., and Jensen, S. E. (1988). Cloning and nucleotide sequence determination of the isopenicillin N synthetase gene from *Streptomyces clavuligerus*. *Gene* **62,** 187–196.

Lester, D. R., Öster, L. M., Svenda, M., and Anderson, I. (2004). Expression, purification, crystallization and preliminary X-ray diffraction studies of the *cmcI* component of *S. clavuligerus* 7α-cephem-methoxylase. *Acta Crystallogr.* **D60,** 1618–1621.

Linne, U., and Marahiel, M. A. (2004). Reactions catalyzed by mature and recombinant nonribosomal peptide synthetases. *Methods Enzymol.* **388,** 293–315.

Lipscomb, S. J., Lee, H. J., Mukherji, M., Baldwin, J. E., Schofield, C. J., and Lloyd, M. D. (2002). The role of arginine residues in substrate binding and catalysis by deacetoxycephalosporin C synthase. *Eur. J. Biochem.* **269,** 2735–2739.

Liras, P. (1999). Biosynthesis and molecular genetics of cephamycins. Cephamycins produced by actinomycetes. *Antonie Van Leeuwenhoek* **75,** 109–124.

Liras, P., Gomez-Escribano, J. P., and Santamarta, I. (2008). Regulatory mechanisms controlling antibiotic production in *Streptomyces clavuligerus*. *J. Ind. Microbiol. Biotechnol.* **35,** 667–676.

Liras, P., and Rodríguez-García, A. (2000). Clavulanic acid, a beta-lactamase inhibitor: Biosynthesis and molecular genetics. *Appl. Microbiol. Biotechnol.* **54,** 467–475.

Lloyd, M. D., Lee, H. J., Harlos, K., Zhang, Z. H., Baldwin, J. E., Schofield, C. J., Charnock, J. M., Garner, C. D., Hara, T., Terwisscha van Scheltinga, A. C., Valegard, K., Viklund, J. A., *et al.* (1999). Studies on the active site of deacetoxycephalosporin C synthase. *J. Mol. Biol.* **287,** 943–960.

Loke, P., Ng, C. P., and Sim, T. (2000). PCR cloning, heterologous expression, and characterization of isopenicillin N synthase from *Streptomyces lipmanii* NRRL 3584. *Can. J. Microbiol.* **46,** 166–170.

Loke, P., and Sim, T. (1999). Site-directed mutagenesis of arginine-89 supports the role of its guanidino side-chain in substrate binding by *Cephalosporium acremonium* isopenicillin N synthase. *FEMS Microbiol. Lett.* **179,** 423–429.

Luengo, J. M., Alemany, M. T., Salto, F., Ramos, F., López-Nieto, M. J., and Martín, J. F. (1986). Direct enzymatic synthesis of penicillin G using cyclases of *Penicillium chrysogenum* and *Acremonium chrysogenum*. *Biotechnology* **4,** 44–47.

Madduri, K., Stuttard, C., and Vining, L. C. (1991). Cloning and location of a gene governing lysine epsilon-aminotransferase, an enzyme initiating beta-lactam biosynthesis in *Streptomyces* spp. *J. Bacteriol.* **173,** 985–988.

Martín, J. F. (2000). Alpha-aminoadipyl-cysteinyl-valine synthetases in beta-lactam producing organisms. From Abraham's discoveries to novel concepts of non-ribosomal peptide synthesis. *J. Antibiot.* **53,** 1008–1021.

Martín, J. F., and Liras, P. (1989). Enzymes involved in penicillin, cephalosporin and cephamycin biosynthesis. *In* "Advances in Biochemical Engineering/Biotechnology", (A. Fiechter, ed.), Vol. 39, pp. 153–187. Springer-Verlag, Berlin, Heidelberg.

Miller, M. T., Gerratana, B., Stapon, A., Townsend, C. A., and Rosenzweig, A. C. (2003). Crystal structure of carbapenam synthetase (CarA). *J. Biol. Chem.* **28,** 40996–41002.

Nagarajan, R., Boeck, L. D., Gorman, M., Hamill, R. L., Higgens, C. E., Hoehn, M. M., Stark, W. M., and Whitney, J. G. (1971). Beta-lactam antibiotics from *Streptomyces*. *J. Am. Chem. Soc.* **93,** 2308–2310.

Núñez, L. E., Méndez, C., Braña, A. F., Blanco, G., and Salas, J. A. (2003). The biosynthetic gene cluster for the beta-lactam carbapenem thienamycin in *Streptomyces cattleya*. *Chem. Biol.* **10**, 301–311.

Okamura, K., Hirata, S., Koki, A., Hori, K., Shibamoto, N., Okumura, Y., Okabe, M., Okamoto, R., Kouno, K., Fukagawa, Y., Shimauchi, Y., Ishikura, T., *et al.* (1979). PS-5, a new beta-lactam antibiotic. I. Taxonomy of the producing organism, isolation and physico-chemical properties. *J. Antibiot.* **32**, 262–271.

Öster, L. M., Lester, D. R., Terwisscha van Scheltinga, A., Svenda, M., van Lun, M., Genereux, C., and Anderson, I. (2006). Insights into cephamycin biosynthesis: The crystal structure of CmcI from *Streptomyces clavuligerus*. *J. Mol. Biol.* **358**, 546–558.

Petrich, A. K., Wu, X., Roy, K. L., and Jensen, S. E. (1992). Transcriptional analysis of the isopenicillin N synthase-encoding gene of *Streptomyces clavuligerus*. *Gene* **111**, 77–84.

Queener, S. W., Beckmann, R. J., Cantwell, C. A., Hodges, R. L., Fisher, D. L., Dotzlaf, J. E., Yeh, W. K., McGilvray, D., Greaney, M., and Rosteck, P. (1994). Improved expression of a hybrid *Streptomyces clavuligerus cefE* gene in *Penicillium chrysogenum*. *Ann. N. Y. Acad. Sci.* **721**, 178–193.

Ramos, F. R., López-Nieto, M. J., and Martín, J. F. (1985). Isopenicillin N synthetase of *Penicillium chrysogenum*, an enzyme that converts δ-(L-α-aminoadipyl)-L-cysteinyl-D-valine to isopenicillin N. *Antimicrob. Agents Chemother.* **27**, 380–387.

Rius, N., Maeda, K., and Demain, A. L. (1996). Induction of L-lysine e-aminotransferase by L-lysine in *Streptomyces clavuligerus*, producer of cephalosporins. *FEMS Microbiol. Lett.* **144**, 207–211.

Roach, P. L., Clifton, I. J., Hensgens, C. M., Shibata, N., Schofield, C. J., Hajdu, J., and Baldwin, J. E. (1997). Structure of isopenicillin N synthase complexed with substrate and the mechanism of penicillin formation. *Nature* **387**, 827–830.

Rollins, M. J., Westlake, D. W., Wolfe, S., and Jensen, S. E. (1988). Purification and initial characterization of deacetoxycephalosporin C synthase from *Streptomyces clavuligerus*. *Can. J. Microbiol.* **34**, 1196–1202.

Samson, S. M., Dotzlaf, J. E., Slisz, M. L., Becker, G. W., van Frank, R. M., Veal, L. E., Yeh, W. K., Miller, J. R., Queener, S. W., and Ingolia, T. D. (1987). Cloning and expression of the fungal expandase/hydroxylase gene involved in cephalosporin biosynthesis. *Biotechnology (N.Y.)* **5**, 1207–1214.

Santamarta, I., Pérez-Redondo, R., Lorenzana, L. M., Martín, J. F., and Liras, P. (2005). Different proteins bind to the butyrolactone receptor protein ARE sequence located upstream of the regulatory *ccaR* gene of *Streptomyces clavuligerus*. *Mol. Microbiol.* **56**, 824–835.

Santamarta, I., Rodríguez-García, A., Pérez-Redondo, R., Martín, J. F., and Liras, P. (2002). CcaR is an autoregulatory protein that binds to the *ccaR* and *cefD-cmcI* promoters of the cephamycin C-clavulanic acid cluster in *Streptomyces clavuligerus*. *J. Bacteriol.* **184**, 3106–3113.

Schwecke, T., Aharonowitz, Y., Palissa, H., van Döhren, H., Kleinkauf, H., and van Liempt, H. (1992). Enzymatic characterization of the multifunctional enzyme δ-(L-α-aminoadipyl)-L-cysteinyl-D-valine synthetase from *Streptomyces clavuligerus*. *Eur. J. Biochem.* **205**, 687–694.

Shiffman, D., Mevarech, M., Jensen, S. E., Cohen, G., and Aharonowitz, Y. (1988). Cloning and comparative sequence analysis of the gene coding for isopenicillin N synthase in *Streptomyces*. *Mol. Gen. Genet.* **214**, 562–569.

Stapley, E. O., Jackson, M., Hernandez, S., Zimmerman, S. B., Currie, S. A., Mochales, S., Mata, J. M., Woodruff, H. B., and Hendlin, D. (1972). Cephamycins, a new family of beta-lactam antibiotics. I. Production by actinomycetes, including *Streptomyces lactamdurans* sp. n. *Antimicrob. Agents Chemother.* **2**, 122–131.

Theilgaard, H. B., Kristiansen, K. N., Henriksen, C. M., and Nielsen, J. (1997). Purification and characterization of delta-(L-alpha-aminoadipyl)-L-cysteinyl-D-valine synthetase from *Penicillium chrysogenum*. *Biochem. J.* **327,** 185–191.

Ullan, R. V., Casqueiro, J., Bañuelos, O., Fernández, F. J., Gutiérrez, S., and Martín, J. F. (2002). A novel epimerization system in fungal secondary metabolism involved in the conversion of isopenicillin N into penicillin N in *Acremonium chrysogenum*. *J. Biol. Chem.* **277,** 46216–46225.

Usui, S., and Yu, C. A. (1989). Purification and properties of isopenicillin N epimerase from *Streptomyces clavuligerus*. *Biochim. Biophys. Acta* **999,** 78–85.

Valegard, K., van Scheltinga, A. C., Lloyd, M. D., Hara, T., Ramaswamy, S., Perrakis, A., Thompson, A., Lee, H. J., Baldwin, J. E., Schofield, C. J., Hajdu, J., and Andersson, I. (1998). Structure of a cephalosporin synthase. *Nature* **394,** 805–809.

Velasco, J., Adrio, J., Moreno, M. A., Díez, B., Soler, G., and Barredo, J. L. (2000). Environmentally safe production of 7-aminodeacetoxycephalosporanic acid (7-ADCA) using recombinant strains of *Acremonium chrysogenum*. *Nat. Biotechnol.* **18,** 857–861.

Wei, C. L., Yang, Y. B., Wang, W. C., Liu, W. C., Hsu, J. S., and Tsai, Y. C. (2003). Engineering *Streptomyces clavuligerus* deacetoxycephalosporin C synthase for optimal ring expansion activity toward penicillin G. *Appl. Environ. Microbiol.* **69,** 2306–2312.

Wolfe, S., Demain, A. L., Jensen, S. E., and Westlake, D. W. (1984). Enzymatic approach to syntheses of unnatural beta-lactams. *Science* **226,** 1386–1392.

Wong, E., Sim, J., and Sim, T. S. (2001). The invariant F283 and its strategic position in the hydrophobic cleft of *Streptomyces jumonjinensis* isopenicillin N synthase active site are functionally important. *Biochem. Biophys. Res. Commun.* **283,** 621–626.

Xiao, X., Wolfe, S., and Demain, A. L. (1991). Purification and characterization of cephalosporin 7α-hydroxylase from *S. clavuligerus*. *Biochem. J.* **280,** 471–474.

Yu, H., Serpe, E., Romero, J., Coque, J. J., Maeda, K., Oelgeschläger, M., Hintermann, G., Liras, P., Martín, J. F., and Demain, A. L. (1994). Possible involvement of the lysine epsilon-aminotransferase gene (*lat*) in the expression of the genes encoding ACV synthetase (*pcbAB*) and isopenicillin N synthase (*pcbC*) in *Streptomyces clavuligerus*. *Microbiology* **140,** 3367–3377.

Zhang, J., Wolfe, S., and Demain, A. L. (1992). Biochemical studies on the activity of delta-(L-alpha-aminoadipyl)-L-cysteinyl-D-valine synthetase from *Streptomyces clavuligerus*. *Biochem. J.* **283,** 691–698.

CHAPTER SEVENTEEN

Siderophore Biosynthesis: A Substrate Specificity Assay for Nonribosomal Peptide Synthetase-Independent Siderophore Synthetases Involving Trapping of Acyl-Adenylate Intermediates with Hydroxylamine

Nadia Kadi *and* Gregory L. Challis

Contents

1. Introduction	432
2. NRPS-Dependent Pathways for Siderophore Biosynthesis	434
3. NRPS-Independent Pathway for Siderophore Biosynthesis	442
4. Hybrid NRPS/NIS Pathway for Petrobactin Biosynthesis	448
5. Hydroxamate-Formation Assay for NIS Synthetases	450
5.1. Assay principle	451
5.2. Buffers, reagents, and others materials	452
5.3. Assay procedure	453
5.4. Results	453
5.5. Technical notes	453
References	455

Abstract

Siderophores are an important group of structurally diverse natural products that play key roles in ferric iron acquisition in most microorganisms. Two major pathways exist for siderophore biosynthesis. One is dependent on nonribosomal peptide synthetase (NRPS) multienzymes. The enzymology of several NRPS-dependent pathways to structurally diverse siderophores has been intensively studied for more than 10 years and is generally well understood. The other major pathway is NRPS-independent. It relies on a novel family of synthetase enzymes that until recently has received very little attention. Over the last 2 years, these enzymes have begun to be intensively investigated and several examples have

Department of Chemistry, University of Warwick, Coventry, United Kingdom now been characterized. In this article, we give an overview of the enzymology of NRPS-dependent and NRPS-independent pathways for siderophore biosynthesis, using selected examples to highlight key features.

An important facet of many studies of the enzymology of siderophore biosynthesis has been to investigate the substrate specificity of the synthetase enzymes involved. For NRPS-dependent pathways, the ATP–pyrophophate exchange assay has been widely used to investigate the substrate specificity of adenylation domains within the synthetase multienzymes. This assay is ineffective for NRPS-independent siderophore (NIS) synthetases, probably because pyrophosphate is not released from the enzyme after the carboxylic acid substrate and ATP react to form an acyl adenylate. An alternative assay for enzymes that form acyl adenylates involves trapping of the activated carboxyl group with hydroxylamine to form a hydroxamic acid that can be converted to its ferric complex and detected spectrophotometrically. This assay has not been widely used for NRPS adenylation domains. Here, we show that it is an effective assay for examining the carboxylic acid substrate specificity of NIS synthetases. Application of the assay to the type B NIS synthetase AcsA shows that it is selective for α-ketoglutaric acid, confirming a bioinformatics-based prediction of the substrate specificity of this enzyme.

1. INTRODUCTION

Iron is an essential element for all living organisms, including bacteria. Iron is predominantly present in its ferric form in our aerobic atmosphere. However, the aqueous solubility of ferric iron is very low at neutral pH and in many environments. As a consequence, bioavailable iron is not sufficient to sustain bacterial growth. To acquire iron from their environment, bacteria have evolved a number of different strategies. The most common is siderophore-mediated iron uptake (Matzanke, 2005).

Siderophores are high-affinity ferric iron chelators that are produced by both prokaryotic and eukaryotic microbes in response to iron deficiency. They are secreted into the extracellular environment where they scavenge ferric iron from insoluble complexes or remove it from host binding proteins. The resulting ferric–siderophore complexes are recognized by specific cell surface receptors. This triggers active transport of the complexes into the microbial cell via membrane-spanning pores (Miethke and Marahiel, 2007).

The last 10 years have witnessed dramatic advances in our understanding of siderophore biosynthesis, and the recognition and transport of ferric–siderophore complexes. More than 500 structurally diverse siderophores have been isolated and characterized. This raises an intriguing question: why is there so much structural diversity among a group of metabolites that perform the same biological function (i.e., sequestration of ferric iron and its transport into microbial cells)? It has been suggested that intense competition between different microbial species, or between microbes and their hosts, for ferric iron has driven the evolution of structural diversity in siderophores and corresponding receptors that can specifically recognize their ferric complexes (Challis and Hopwood, 2003). Siderophore-mediated iron acquisition is known to be an important virulence determinant for many pathogenic microbes (Fischbach *et al.*, 2006). This fact has stimulated considerable recent interest in siderophore biosynthetic pathways as targets for antimicrobial intervention (Miethke and Marahiel, 2007; Quadri, 2007).

Despite the variety and complexity of their structures, siderophores are unified by a common property—their capacity to form highly stable multidentate (often hexadentate) complexes with ferric iron. This property results from the diverse range of ferriphilic ligands found in these natural products. Several siderophores contain one, two, or three bidentate oxygen ligands of the same type, such as hydroxamates (e.g., as in desferrioxamine E), catecholates (e.g., as in enterobactin), or α-hydroxycarboxylates (e.g., as in achromobactin) (Fig. 17.1; Miethke and Marahiel, 2007). Others contain combinations of these bidentate ligands within a single molecule. Yet other siderophores employ a combination of oxygen and nitrogen heterocycle ligands for ferric iron chelation (e.g., pyochelin) (Fig. 17.1; Miethke and Marahiel, 2007).

Many hydroxamate-, catecholate-, and nitrogen heterocycle-based siderophores are assembled from amino and/or aryl acid building blocks by nonribosomal peptide synthetases (NRPSs) (Crosa and Walsh, 2002), large multifunctional enzymes that are also responsible for the biosynthesis of peptide antibiotics (Felnagle *et al.*, 2008). NRPSs contain multiple catalytic domains usually within a single or a few polypeptide chains that collectively function as an assembly line for a single structurally complex product. The enzymology of NRPS assembly lines has been intensively studied for several decades and is covered in several chapters in this volume.

On the other hand, siderophores containing α-hydroxycarboxylates and many other hydroxamate-based siderophores are assembled by the NRPS-independent siderophore (NIS) pathway (Challis, 2005). Comparatively, little was known about the enzymology of NIS biosynthetic pathways until very recently.

Here, we give an overview of current knowledge of the enzymology of siderophore biosynthesis by the NRPS-dependent and NIS pathways, and describe a useful substrate specificity assay for NIS synthetases (the key enzymes in NRPS-independent pathways), which overcomes problems associated with investigating the substrate specificity of these enzymes using the well-established ATP–pyrophosphate exchange assay (Santi *et al.*, 1974) that has been widely used to investigate NRPS substrate specificity.

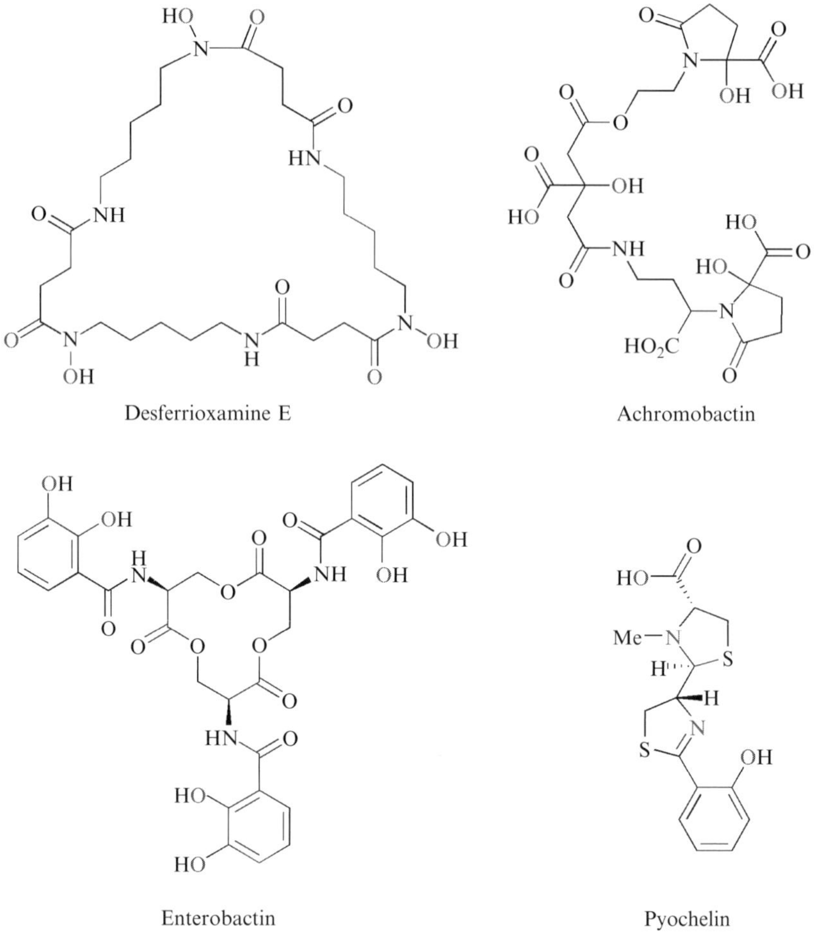

Figure 17.1 Structures of siderophores containing different types of ferric-iron-chelating functional groups, that is, hydroxamate (desferrioxamine E), α-hydroxycarboxylate (achromobactin), catecholate (enterobactin), and nitrogen heterocycle/phenolate/carboxylate (pyochelin). The atoms that bind to Fe^{3+} in each structure are highlighted in gray.

2. NRPS-DEPENDENT PATHWAYS FOR SIDEROPHORE BIOSYNTHESIS

NRPSs can assemble peptides of a wide structural diversity and broad biological activity. They have a distinct modular structure in which each module is responsible for the recognition, activation, and sometimes modification of each amino (or other carboxylic acid) incorporated into the final peptide product (Lautru and Challis, 2004). The size and sequence of a nonribosomal peptide are usually determined by the number and order of modules in the NRPS(s) that catalyze its assembly. A minimal NRPS chain extension module contains three domains: the adenylation (A) domain that specifically recognizes the substrate of the module and catalyzes adenylation of its carboxyl group; the thiolation (T) domain (also known as the peptidyl carrier protein (PCP) domain), which utilizes the terminal thiol of a post-translationally installed phosphopantetheine arm to capture the activated carboxyl group of the adenylate; and the condensation (C) domain, which catalyzes acylation of the resulting thioester with the activated acyl group attached to the T domain in the upstream module. In some NRPS modules, the C domain is replaced by a heterocyclization (Cy) domain that catalyzes heterocycle formation by reaction of β-aminoalcohol or β-aminothiol groups in the substrate attached to the T domain in the same module with the thioester attached to the T domain of the upstream module. The minimal chain-initiation module contains only A and T domains. The growing chain is covalently bound to the T domains in successive modules throughout the assembly process. An additional thioesterase (TE) domain is usually present in the final module. This domain catalyzes autoacylation of an active site Ser residue with the fully assembled chain attached to the T domain. Hydrolysis or cyclization results in release of the assembled chain from the NRPS. Some TE domains catalyze oligomerization of the fully assembled chains prior to release by cyclization. In some NRPSs, the TE domain is replaced by an alternative domain for chain release, for example, a reductase (R) domain that catalyzes NADH-dependent cleavage of the acyl thioester attached to the T domain. Additional epimerization (E) or methyltransferase (MT) domains that modify thioesters bound to the T domain prior to condensation with the substrate attached to the upstream module are sometimes found within a module. Intermediates in the assembly process bound to T domains can also be modified in *trans* by separately-encoded tailoring enzymes.

Siderophores assembled by NRPSs can be classified into three major groups according to the type of iron-binding ligand they contain: catecholates (e.g., enterobactin, myxochelin, bacillibactin); hydroxamates (e.g., coelichelin, exochelin MS); and nitrogen heterocycles (e.g., yersiniabactin, pyochelin—these utilize phenolates and carboxylates as additional ligands for ferric iron). Several NRPS-derived siderophores, however, contain combinations of these different ligand types. In the following paragraphs, we summarize the state of knowledge of enterobactin, coelichelin, and pyochelin biosynthesis as examples of archetypal catecholate, hydroxamate, and nitrogen-heterocycle siderophores (Miethke and Marahiel, 2007).

The discovery of enterobactin (also known as enterochelin) as an extracellular metabolite of *Escherichia coli* was reported in 1970. Enterobactin is arguably the best-characterized catecholate siderophore (Raymond et al., 2003). Its biosynthesis and the uptake of its ferric complex by various bacteria have been studied, and its total chemical synthesis has been reported (Raymond et al., 2003). The trilactone moiety of enterobactin preorganizes its three catechol moieties for iron-binding. Thus the Tris–catecholate ferric-enterobactin complex is extraordinarily stable (Raymond et al., 2003). Chorismic acid, a precursor of aromatic amino acids, is converted to the 2,3-dihydroxybenzoic acid (2,3-DHB) building block for enterobactin assembly by EntA, the N-terminal domain of EntB and EntC (Raymond et al., 2003). The stand-alone EntE adenylation domain catalyzes formation of an adenylate from 2,3-DHB and ATP and the subsequent acylation of the phosphopantetheine thiol of the T domain within the bifunctional EntB protein with the acyl adenylate (Fig. 17.2; Gehring et al., 1997). EntE and the T domain of EntB together constitute an NRPS chain-initiation module. The EntF protein contains a chain-extension module consisting of C, A, and T domains and a TE domain at its C-terminus. The A domain catalyzes acylation of the T domain with L-Ser via an acyl adenylate and the C domain catalyzes formation of an amide bond between the L-Ser thioester and the 2,3-DHB-EntB thioester to form 2,3-DHB-L-Ser attached to the T domain of EntF (Fig. 17.2; Gehring et al., 1998). This intermediate is transacylated onto the active site Ser residue of the TE domain and a second molecule of 2,3-DHB-L-Ser is assembled on the EntF T domain. The hydroxyl group in this intermediate is acylated with the 2,3-DHB-L-Ser moiety covalently bound to the active site of the TE domain and the resulting (2,3-DHB-L-Ser)$_2$ thioester is transferred back onto the active site Ser residue of the TE domain. A third molecule of 2,3-DHB-L-Ser is assembled on the EntF T domain and is joined to the TE-bound (2,3-DHB-L-Ser)$_2$ intermediate to create a (2,3-DHB-L-Ser)$_3$ intermediate that is released from the NRPS by lactonization (Fig. 17.3; Shaw-Reid et al., 1999).

Pyochelin is a thiazoline/thiazolidine-containing siderophore originally isolated from *Pseudomonas aeruginosa* that utilizes a combination of phenoxide, carboxylate, and heterocyclic nitrogen ligands to chelate ferric iron (Cox et al., 1981). Pyochelin is assembled from two molecules of cysteine and a molecule of salicylic acid by the PchD, PchE, PchF, and PchG enzymes. PchD is a stand-alone A domain that catalyzes acylation of the N-terminal T domain of PchE with salicylic acid via the salicyl adenylate. The A domain in PchE catalyzes acylation of the C-terminal T domain and the Cy domain catalyzes condensation of the cysteinyl thioester with the salicyl thioester (Quadri et al., 1999). The α-carbon of the resulting thioester is proposed to undergo epimerization, catalyzed by an MT-like domain embedded within the A domain of PchE (Patel et al., 2003). The Cy domain then catalyzes cyclodehydration to form a 2-hydroxyphenyl-thiazolinyl thioester intermediate. The A domain of PchF catalyzes *cis* loading of a molecule of cysteine onto the T domain and the Cy domain of PchF catalyzes amide bond formation between the cysteinyl thioester and the 2-hydroxyphenyl-thiazolinyl thioester and subsequent cyclodehydration to form a bis-thiazolinyl thioester intermediate. At this juncture PchG

Siderophore Biosynthesis

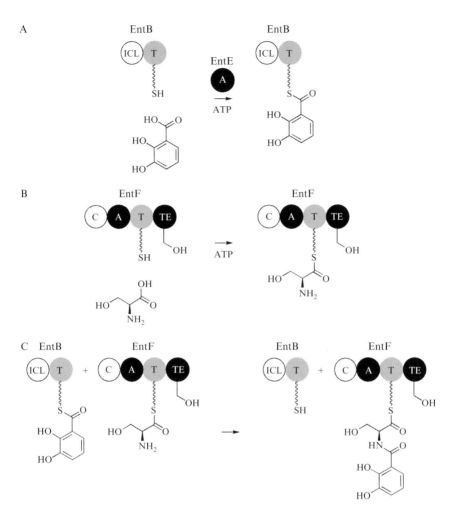

Figure 17.2 Reactions catalyzed by EntE and EntF enzymes in assembly of an L-N-(2,3-dihydroxybenzoyl)serinyl thioester intermediate in enterobactin biosynthesis. (A) EntE catalyzes adenylation of 2,3-DHB and subsequent trapping of the resulting adenylate with the T domain of EntB. (B) The EntF A domain catalyzes loading of L-Ser onto the EntF T domain via an aminoacyl adenylate intermediate. (C) The EntF C domain catalyzes transfer of the 2,3-DHB group from EntB onto the amino group of the L-Ser residue attached to the T domain of EntF.

catalyzes NADPH-dependent reduction of the second-formed thiazolines to the corresponding thiazolidine and an MT domain embedded within the PchF A domain catalyzes S-adenosylmethionine (SAM)-dependent N-methylation of the thiazolidine (Patel and Walsh, 2001). Hydrolysis of the thioester linkage in the resulting intermediate, which is catalyzed by the TE domain of PchF, yields pyochelin (Fig. 17.4).

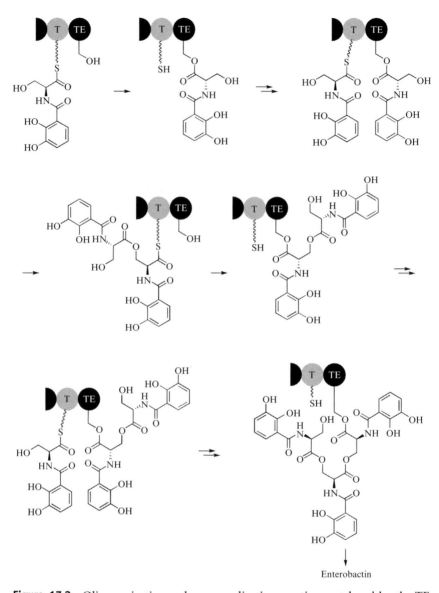

Figure 17.3 Oligomerization and macrocyclization reactions catalyzed by the TE domain of EntF in the assembly of enterobactin from the L-*N*-(2,3-dihydroxybenzoyl) serinyl thioester intermediate. Transfer of the 2,3-DHB-Ser group from the T domain to the active Ser residue of the TE domain allows a second 2,3-DHB-Ser moiety to be assembled on the T domain. The TE domain catalyzes acylation of the thioester with the ester and subsequent transfer of the dimeric thioester back onto the active site Ser residue of the TE domain. A third 2,3-DHB-Ser moiety is assembled on the T domain and the trimeric TE-bound intermediate is formed by an analogous process to that involved in formation of the dimeric intermediate. Finally, the TE catalyzes release of the mature enterobactin molecule via macrolactonization.

Siderophore Biosynthesis

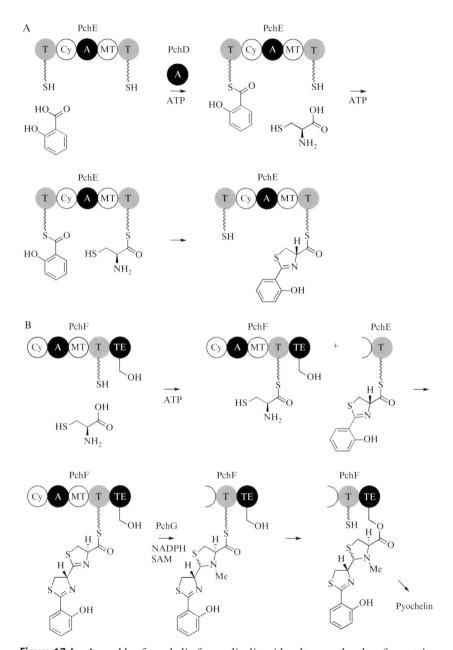

Figure 17.4 Assembly of pyochelin from salicylic acid and two molecules of L-cysteine catalyzed by the PchD, PchE, PchF, and PchG enzymes. (A) The stand-alone PchD A domain catalyzes *trans* acylation of the N-terminal T domain of PchE with salicylic acid and the internal A domain of PchE catalyzes *cis* acylation of the C-terminal T domain of PchE. Both reactions are ATP-dependent and involve adenylate intermediates. The Cy domain catalyzes *N*-acylation of the L-Cys thioester with the salicyl thioester and

Coelichelin is a Tris–hydroxamate tetrapeptide siderophore produced by *Streptomyces coelicolor* (Fig. 17.5). It was recently discovered by mining the genome sequence of *S. coelicolor* for novel nonribosomal peptide biosynthetic gene clusters (Lautru *et al.*, 2005). The *cchH* gene within the novel gene cluster encodes a trimodular NRPS, containing epimerization domains within modules 1 and 2, but lacking a thioesterase domain or an equivalent domain for release of the fully assembled peptide chain in module 3. A predictive structure-based model of amino acid recognition by NRPS A domains was applied to predict the likely substrate of each module of the NRPS (Challis and Ravel, 2000). Thus module 1 was predicted to recognize L-N5-formyl-N5-hydroxyornithine (L-fhOrn), module 2 was predicted to recognize L-Thr, and module 3 was predicted to recognize L-N5-hydroxyornithine (L-hOrn). Based on these analyses, this gene cluster was proposed to direct production of the tripeptide D-fhOrn-D-allo-Thr-L-hOrn (there are two possible isomers of this tripeptide, because L-hOrn contains two amino groups and, in principle either could condense with the thioester attached to the T domain of the upstream module) (Challis and Ravel, 2000). It was therefore a surprise to find that the product of this gene cluster was a tetrapeptide containing two D-fhOrn residues along with one D-allo-Thr and one L-hOrn residue (Lautru *et al.*, 2005). Gene disruption experiments confirmed the role of the NRPS-encoding *cchH* gene in the biosynthesis of coelichelin and also implicated the *cchJ* gene, which encodes a protein that is similar to known esterases, in this process (Lautru *et al.*, 2005). Heterologous expression of the gene cluster in *Streptomyces fungicidicus* (a coelichelin nonproducer) confirmed that CchH is the only NRPS required for coelichelin biosynthesis (Lautru *et al.*, 2005). On the basis of the above results a model for coelichelin biosynthesis was proposed involving initial assembly of the tripeptide D-fhOrn-D-allo-Thr-L-hOrn on the module 3 T domain, utilizing conventional NRPS enzymatic logic (Fig. 17.5). Then, in a break from the textbook enzymatic logic, a second molecule of L-fhOrn is proposed to be activated and epimerized by module 1, and condensed with the α-amino group of the L-hOrn residue in the tripeptide attached to the module 3 T domain (Fig. 17.5; Lautru *et al.*, 2005). This process involves iterative module use (as has been observed for other NRPSs—see e.g., enterobactin above) as well as module skipping, which is an extremely rare process that has, to the best of our knowledge, only been implicated in the

---

subsequent cyclodehydration to form the thiazoline ring. The MT-like domain catalyzes epimerization of the Cys-derived stereocenter during or after heterocycle formation. (B) The A domain of PchF catalyzes ATP-dependent acylation of the downstream T domain with L-Cys. The Cy domain of PchF catalyzes condensation of the thioesters bound to PchE and PchF to generate a bis-thiazolinyl thioester. PchG catalyzes reduction of one of the thiazolines to a thiazolidine and the MT domain of PchF catalyzes N-methylation. TE-domain mediated thioester hydrolysis yields pyochelin.

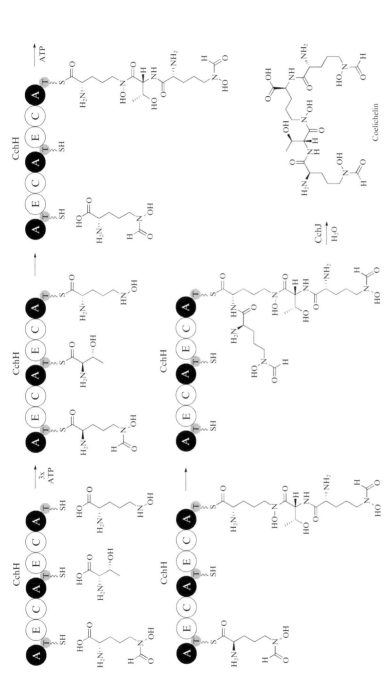

Figure 17.5 Proposed pathway for assembly of the tetrapeptide siderophore coelichelin by the trimodular NRPS constituted by the CchH and CchJ proteins, involving an unusual example of module skipping. L-fhOrn, L-Thr, and L-hOrn are loaded onto the three T domains of CchH by the adjacent A domains. The E domains catalyze epimerization of L-fhOrn and L-Thr to D-fhOrn and D-allo-Thr, respectively. The C domains catalyze assembly of a D-fhOrn–D-allo-Thr–L-hOrn thioester intermediate on the C-terminal T domain of CchH. The N-terminal A domain loads a second unit of L-fhOrn onto the adjacent T domain and, after epimerization, one of the two C domains catalyzes condensation of the resulting D-fhOrn thioester with the tripeptidyl thioester attached to the C-terminal T domain of CchH. The CchJ putative esterase then catalyzes hydrolytic release of the tetrapeptide from the T domain to yield coelichelin.

biosynthesis of two other nonribosomal peptides (Dimise et al., 2008; Wenzel et al., 2006). It is proposed that the CchJ esterase-like protein catalyzes hydrolytic release of the fully assembled tetrapeptide from the T domain within module 3 of CchH (Lautru et al., 2005).

The examples discussed above serve to highlight the catalytic diversity and complexity inherent in the assembly of siderophores by NRPSs. Despite this, a common feature of all NRPS-catalyzed siderophore assembly processes is the ATP-dependent adenylation of amino (and other) acids catalyzed by the A domain. Since this reaction is largely responsible for selecting the correct substrates for siderophore assembly from the complex mixture of amino and other acids present in the cellular pool, a significant feature of many studies of the enzymology of nonribososomal peptide biosynthesis has been to investigate the substrate specificity of A domains within different NRPS systems. In the overwhelming majority of such investigations, the same assay has been used. This is the so-called ATP–pyrophophate (ATP-PP$_i$) exchange assay, originally developed to probe the substrate specificity of tRNA synthetases (Santi et al., 1974), which measures the incorporation of radiolabeled pyrophosphate into the ATP via attack on the acyl-adenylate formed by the A domain in the absence of a T domain thiol to trap the adenylate.

3. NRPS-Independent Pathway for Siderophore Biosynthesis

The pathway to aerobactin was the first NIS biosynthetic pathway to be reported, by Neilands and coworkers in the 1980s (Challis, 2005; De Lorenzo and Neilands, 1986; De Lorenzo et al., 1986). Aerobactin has been reported as a metabolic product of several different bacterial genera, including *Vibrio*, *Yersinia*, *Salmonella*, *Aerobacter*, and *Escherichia*, but its biosynthetic pathway has been most intensively studied in *E. coli*, where it is known as a pathogenicity factor (Challis, 2005). The chelating α-hydroxycarboxylate and hydroxamate groups in aerobactin allow it to form a strong hexadentate complex with ferric iron. Aerobactin is assembled from a molecule of citric acid and two molecules each of L-lysine, acetyl-CoA and molecular oxygen by the enzymes encoded by the *iucABCD* genes (Fig. 17.6; Challis, 2005). These genes are clustered with the *iutA* gene, which encodes the ferric-aerobactin outer membrane receptor, on the pCoIV-K30 plasmid of *E. coli*.

The first enzyme involved in aerobactin biosynthesis is IucD, a FADH$_2$-dependent monooxygenase that converts L-lysine to L-N6-hydroxylysine using molecular oxygen as a cosubstrate (Challis, 2005). The conversion of L-N6-hydroxylysine to L-N6-acetyl-N6-hydroxylysine (AHL) is catalyzed by IucB, an acetyl-CoA-dependent acyltransferase (Challis, 2005). Genetic evidence suggests that IucA is a synthetase that catalyzes activation of one of

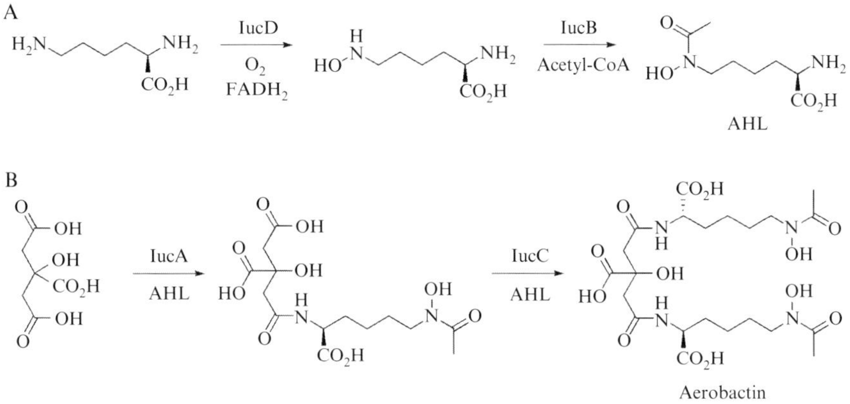

Figure 17.6 Proposed pathway for aerobactin biosynthesis. (A) Conversion of L-lysine to N6-acetyl-N6-hydroxylysine (AHL) catalyzed by IucD and IucB. (B) Putative functions of IucA and IucB deduced from the structures of aerobactin-related metabolites that accumulated in mutants missing the genes encoding these proteins.

the prochiral carboxyl groups of citric acid and subsequent condensation of the activated carboxyl group with L-N6-acetyl-N6-hydroxylysine (Challis, 2005). Finally, genetic evidence also suggests that IucC (an IucA homolog) catalyzes the condensation of a second molecule of L-N6-acetyl-N6-hydroxylysine with the product of the IucA-catalyzed reaction via an activated carboxylic acid intermediate (Challis, 2005). However, the proposed roles of IucA and IucC remain to be proven by biochemical experiments with the purified proteins.

In recent years, it has become apparent that several structurally diverse siderophores are assembled by NIS biosynthetic pathways involving similar enzymes to those involved in aerobactin biosynthesis (Challis, 2005). Indeed, the presence of one or more genes encoding a homolog of IucA/IucC has become recognized as a hallmark of NIS biosynthetic gene clusters. A phylogenetic analysis of the IucA/IucC-like putative synthetases suggested that they belong to three main types (Challis, 2005). The type A enzymes are exemplified by IucA and were proposed to catalyze activation of one of the prochiral carboxyl groups of citric acid by reaction with a nucleotide triphosphate (NTP), followed by condensation of the resulting activated carboxyl group with an amine or alcohol to furnish the corresponding amide or ester, respectively. The type B enzymes are proposed to catalyze an analogous reaction to the type A enzymes, except using α-ketoglutaric acid as a substrate (the carboxyl group attached to C-4 is proposed to undergo activation and subsequent condensation with amine). The type C enzymes are exemplified by IucC and are proposed to catalyze NTP-dependent activation of carboxyl groups in monoamide/monoester derivatives of citric acid or succinic acid, followed by condensation of the activated carboxyl group with an amine or alcohol. At the time this model was proposed, only very limited experimental data from genetic studies of the aerobactin pathway and a handful of other NIS pathways were available. In the three years since publication of this model several biochemical studies of purified recombinant NIS synthetases have been reported (Kadi et al., 2007, 2008a,b; Oves-Costales et al., 2007, 2008, 2009; Schmelz et al., 2009). Some generalizations can be drawn from these investigations. All NIS synthetases studied to date used ATP as a cosubstrate and appear to catalyze adenylation of carboxyl groups in their substrates. Two type A synthetases and four type C synthetases have been investigated to date and in each case the enzymes were shown to possess the substrate specificities predicted by the phylogenetic model. AsbA and AcsD, the two type A synthetases studied, both appear to catalyze desymmetrization of citric acid with a high degree of enantioselectivity, suggesting that they may prove useful as biocatalysts for asymmetric synthesis (Schmelz et al., 2009; Oves-Costales et al., 2009). No biochemical investigations of purified type B enzymes have yet been reported. Here, we describe the first investigation of the substrate specificity of a type B enzyme. It remains to be seen whether all such enzymes are specific for α-ketoglutarate, as predicted, or whether they also possess other substrate specificities.

The first example of a type C NIS synthetase to be biochemically characterized was DesD from *S. coelicolor* (Kadi et al., 2007). A cluster of four genes (*desABCD*) was proposed, on the basis of genetic studies, to direct the biosynthesis of desferrioxamines E and B in *S. coelicolor* (Barona-Gómez et al., 2004). The macrocyclic structure of desferrioxamine E preorganizes it for high-affinity ferric iron binding via the three bidentate hydroxamate ligands. DesA, DesB, and DesC were proposed to catalyze the conversion of L-lysine to *N*-hydroxy-*N*-succinylcadeverine (HSC) via pyridoxal-phosphate-mediated decarboxylation, $FADH_2$ and O_2-dependent monooxygenation, and succinyl-CoA-dependent acylation reactions, respectively (Barona-Gómez F. et al., 2004; Fig. 17.7). Purified recombinant DesD was shown to catalyze the ATP-dependent conversion of chemically synthesized HSC to desferrioxamines G1 and E. Kinetic studies implicated desferrioxamine G1 as a free intermediate in the assembly of desferrioxamine E and mass spectrometric data also suggested that a dimer of HSC was a transient free intermediate in the formation of desferrioxamine G1 (Kadi et al., 2007; Fig. 17.7). The transformation of HSC to desferrioxamine E by DesD parallels the transformation of L-serine and 2,3-DHB-EntB to enterobactin catalyzed by EntF. Both involve iterative use of the active sites in the enzymes. However, the mechanisms by which these transformations are achieved are strikingly different. EntF is a multienzyme that exclusively employs covalent linkages between intermediates and the enzyme throughout the assembly process. On the other hand, DesD appears to have only a single active site and freely dissociable intermediates are involved in the reactions it catalyzes.

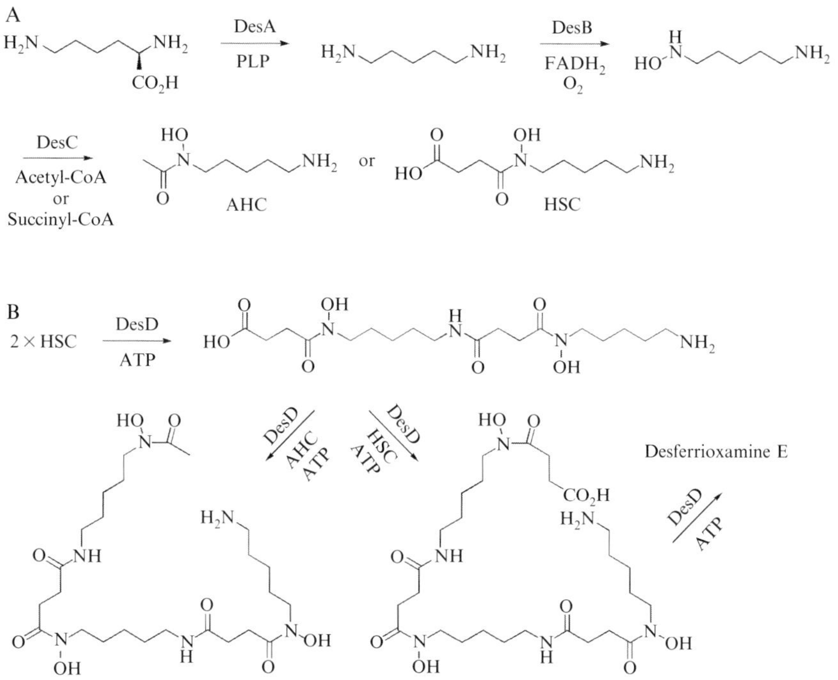

Figure 17.7 Pathway for the biosynthesis of deferrioxamines B, G1, and E. (A) Reactions catalyzed by the DesA and DesB enzymes in the conversion of L-lysine to N-hydroxycadaverine. DesC is proposed to catalyze the acylation of N-hydroxycadaverine with acetyl-CoA or succinyl-CoA to yield N-acetylcadaverine or N-succinylcadaverine, respectively. (B) DesD catalyzes adenylation of the HSC carboxyl group and subsequent condensation with a second molecule of HSC to form a dimer, followed by adenylation of the carboxyl group in the dimer and subsequent condensation with AHC to form desferrioxamine B or with a molecule of HSC to form desferrioxamine G1. DesD-catalyzed adenylation of the carboxyl group in desferrioxamine G1, followed by acylation of the amino group, affords the macrocycle desferrioxamine E. The dimer of HSC and desferrioxamine G1 are transient, freely dissociable intermediates in the assembly of desferrioxamine E by DesD.

More recently, biochemical characterization of two further type C NIS synthetases that catalyze analogous reactions to those of DesD have been reported. The first is encoded by *pubC*, which resides within the *pubABC-putAB* gene cluster that is conserved in all sequenced *Shewanella* species, and is proposed to direct production of the macrocyclic bis-hydroxamate putrebactin (Ledyard and Butler, 1997) and the uptake/utilization of its ferric complex (Kadi et al., 2008a). Purified recombinant PubC catalyzes the ATP-dependent conversion of chemically synthesized N-hydroxy-N-succinylputrescine (HSP) to putrebactin (Kadi et al., 2008a; Fig. 17.8). PubA and PubB are proposed to catalyze the assembly of HSP from putrescine (the decarboxylation product of ornithine) via analogous

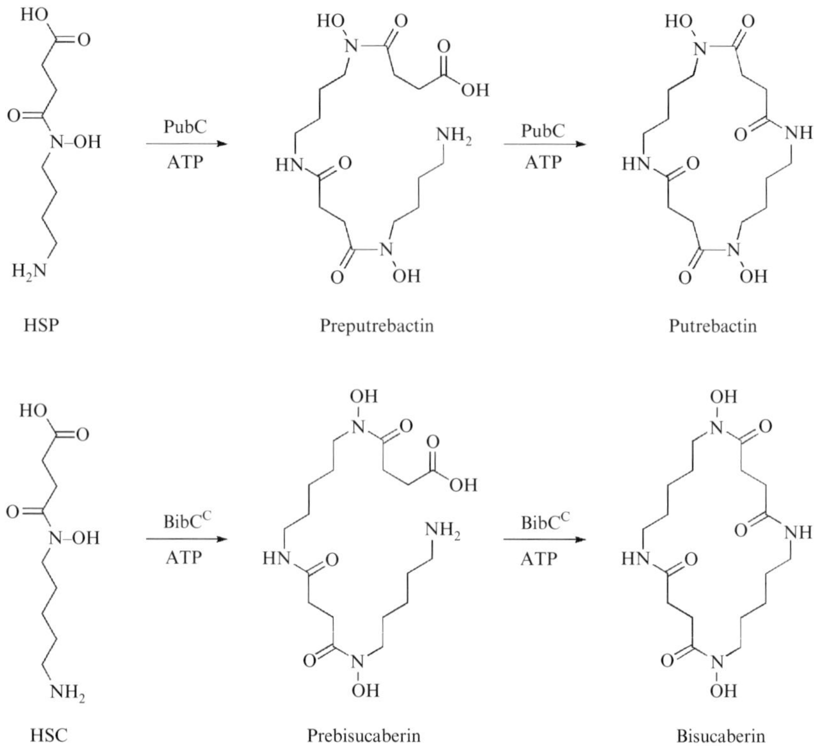

Figure 17.8 ATP-dependent dimerization and subsequent macrocylization reactions catalyzed by the PubC enzyme and the C-terminal domain of BibC in putrebactin and bisucaberin biosynthesis, respectively. Acyl adenylate intermediates are involved in both reactions and preputrebactin/prebisucaberin are transient, freely dissociable intermediates in macrocycle assembly.

chemistry to the conversion of cadaverine to HSC by DesB and DesC (Kadi et al., 2008a). Interestingly, the other type C NIS synthetase that functions analogously to DesD is a domain within a multienzyme encoded by the *bibC* gene in the *bibABCbitABCDE* gene cluster of *Vibrio salmonicida* (Kadi et al., 2008b), a fish pathogen that has been reported to produce the macrocyclic bis-hydroxamate siderophore bisucaberin (Winkelmann et al., 2002). As anticipated, the purified recombinant C-terminal domain of BibC catalyzes the ATP-dependent conversion of HSC to bisucaberin (Kadi et al., 2008b; Fig. 17.8). BibA, BibB, and the N-terminal domain of BibC show significant sequence similarity to DesA, DesB, and DesC, respectively, and are consequently hypothesized to catalyze analogous reactions in the assembly of HSC (Kadi et al., 2008b). A common feature in the reactions catalyzed by PubC and the BibC C-terminal domain is the involvement of uncyclized dimers of HSP/HSC as transient free intermediates (Fig. 17.8). In both cases, the isolated intermediates are converted to the corresponding natural products when incubated with the appropriate enzyme and ATP (Kadi et al., 2008a,b).

Very recently, structural and biochemical studies of AcsD, a type A NIS synthetase from *Pectobacterium chrsyanthemi* (*Dickeya dadantii*) involved in achromactin biosynthesis (Franza et al., 2005), have been reported (Schmelz et al., 2009). The *acsD*, *acsA*, and *acsC* genes within the achromobactin biosynthetic gene cluster encode type A, type B, and type C NIS synthetases, respectively (Challis, 2005). These three enzymes are proposed to catalyze assembly of achromobactin from two molecules of α-ketoglutarate and one molecule each of citric acid, ethanolamine and L-2,4-diaminobutyrate (DABA) (Challis, 2005). Activity assays with recombinant AcsD showed that it catalyzes stereospecific adenylation of one of the two prochiral carboxyl groups of citric acid (Schmelz et al., 2009). The enzyme showed no activity with α-ketoglutaric acid, as predicted for a type A enzyme. The activated carboxyl group resulting form the adenylation reaction can be efficiently trapped with L-serine (and analogues) to afford the corresponding ester, which undergoes (presumably uncatalyzed) intramolecular rearrangement to the amide (Schmelz et al., 2009; Fig. 17.9). The adenylate intermediate could not be efficiently trapped with DABA. The structural studies revealed that NIS synthetases possess a new fold for catalysis of acyl adenylate formation. They also established the molecular basis for ATP and citric acid binding, and provided an explanation for the high degree of enantioselectivity observed in the AcsD-catalyzed reaction (Schmelz et al., 2009). Here, we show that AcsA is specific for

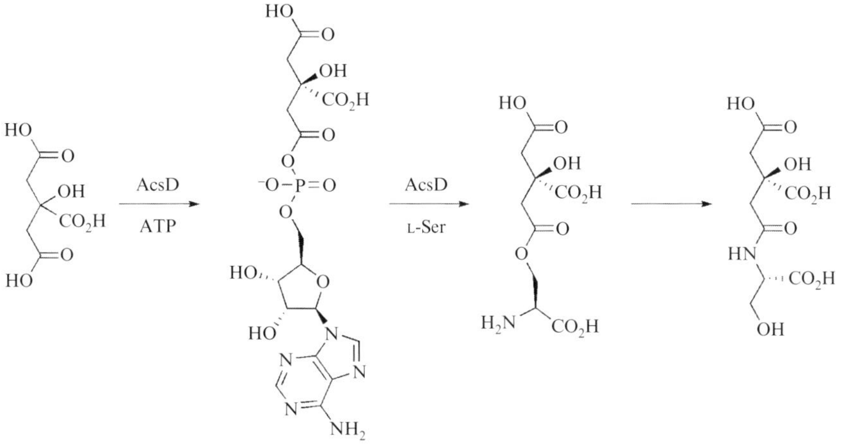

Figure 17.9 AcsD-catalyzed desymmetization of citric acid by selective adenylation of one of the two prochiral carboxyl groups. The adenylate is trapped by L-serine. The ester resulting from trapping with L-Ser undergoes rapid (presumably uncatalyzed) intramolecular rearrangement of the amide, which is stable and can thus be purified and fully characterized.

α-ketoglutarate as its carboxylic acid substrate. Further investigations will be required to determine the amine substrate specificity of AcsA. The substrate specificity of AcsC will also need to be investigated to elucidate the intermediates and likely order of reactions in achromobactin biosynthesis. It is anticipated that AcsC catalyzes the condensation of O-citryl-ethanolamine with DABA and that AscA catalyzes iterative acylation of the two amino groups in the product of the AscC-catalyzed reaction with two successive molecules of α-ketoglutarate. However, an alternative pathway in which AcsA acts prior to AcsD and AcsC cannot be ruled out at this stage.

The remaining two NIS synthetases that have been biochemically characterized to date form part of a unique hybrid NRPS/NIS pathway for siderophore biosynthesis. They are discussed in the following section.

4. Hybrid NRPS/NIS Pathway for Petrobactin Biosynthesis

Petrobactin contains an unusual combination of two catecholate ligands and an α-hydroxycarboxylate ligand for ferric-iron binding (Fig. 17.10). It was originally isolated from the marine bacterium *Marinobacter hydrocarbonoclasticus* (Bergeron *et al.*, 2003), but more recently it has also been identified as the pathogenicity-conferring siderophore of the human pathogen *Bacillus anthracis* (Koppisch *et al.*, 2005; Wilson *et al.*, 2006). Genetic studies have implicated the *B. anthracis asbABCDEF* gene cluster in petrobactin biosynthesis (Lee *et al.*, 2007). AsbF catalyzes conversion of chorismic acid to 3,4-dihydroxybenzoic acid (3,4-DHB)—one of the building blocks from which petrobactin is assembled (Fox *et al.*, 2008; Pfleger *et al.*, 2008). The *asbA* and *asbB* genes encode type A and type C NIS synthetases, respectively (Challis, 2005). On the other hand, *asbC* and *asbD* encode stand-alone A and T domains, respectively (Pfleger *et al.*, 2007). Thus the petrobactin biosynthetic assembly line represents a unique example of a hybrid NRPS/NIS system. Analysis of petrobactin-related metabolites that accumulate in mutants lacking each of the *asbABCDEF* genes, coupled with biochemical studies of the AsbC, AsbD, and AsbE enzymes, led Sherman and coworkers to suggest a pathway for petrobactin biosynthesis involving AsbC-catalyzed ATP-dependent loading of 3,4-DHB onto the AsbD T domain, followed by AsbE-catalyzed acylation of N1 of spermidine with 3,4-DHB-AsbD (Lee *et al.*, 2007; Pfleger *et al.*, 2007). This suggested that N1-(3,4-DHB) spermidine is a key intermediate in petrobactin biosynthesis. However, studies of purified recombinant AsbA showed that it was able to catalyze efficient ATP-dependent condensation of citric acid with spermidine (Fig. 17.10), but not its N1-(3,4-DHB) derivative (Oves-Costales *et al.*, 2007). Likewise, purified recombinant AsbB catalyzes efficient condensation of N8-citryl-spermidine or its N1-3,4-DHB

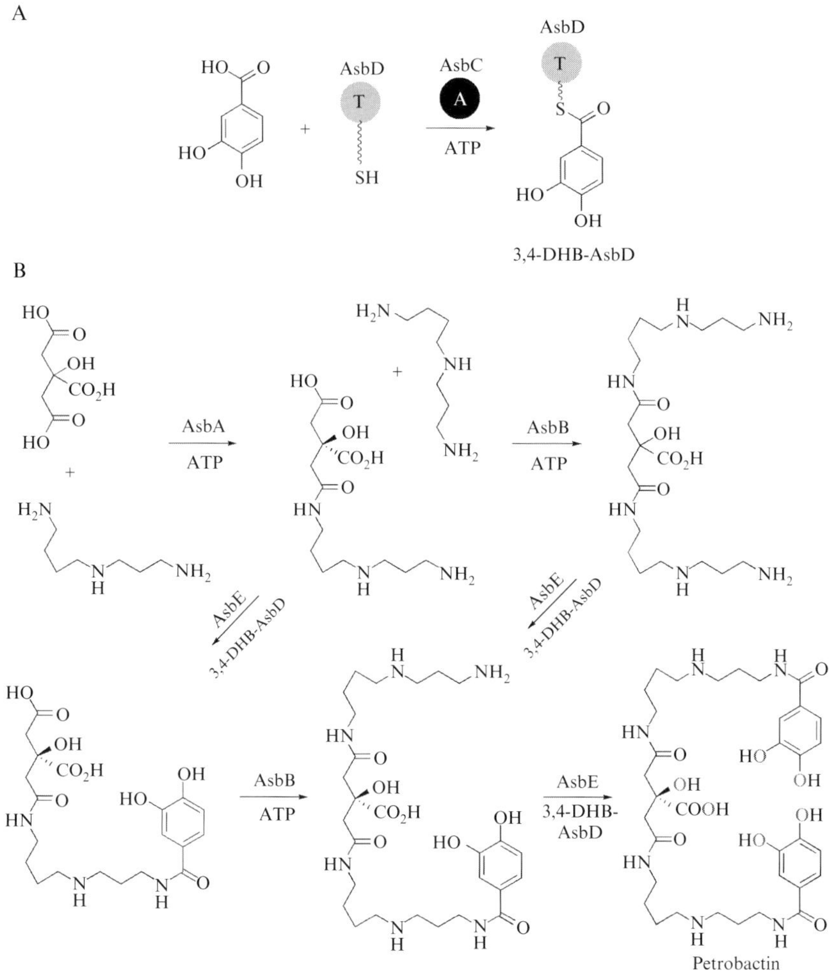

Siderophore Biosynthesis

Figure 17.10 Assembly of petrobactin by a unique hybrid NRPS-dependent/NRPS-independent pathway. (A) The stand-alone adenylation domain AsbC catalyzes adenylation of the carboxyl group in 3,4-DHB and subsequent acylation of the phosphopantetheine thiol of the stand-alone T domain AsbD. (B) AsbA catalyzes desymmetrization of citric acid by adenylation of one of its two prochiral carboxyl groups (analogous to the reaction catalyzed by AcsD) and subsequent acylation of N8 of spermidine. AsbB catalyzes analogous ATP-dependent condensation of N8-citryl-spermidine with a second molecule of spermidine to form the symmetrical bis-amide of citric acid. AsbE is proposed to catalyze acylation of N1 of each of the spermidine residues within the symmetrical bis-amide with two molecules of 3,4-DHB-AsbD to form petrobactin. It is hypothesized that AsbE could also catalyze acylation of N1 of the spermidine residue in N8-citryl-spermidine with 3,4-DHB-AsbD because AsbB can also catalyze efficient condensation of N1-(3,4-dihydroxybenzoyl)-N8-citryl-spermidine with N8 of spermidine. Note that AsbE does not show any sequence similarity to NRPS C domains.

derivative with spermidine (Fig. 17.10), but not the N1-3,4-DHB derivative of spermidine (Oves-Costales *et al.*, 2008). Thus, it can be concluded that AsbA and AsbB function as predicted for type A and type C NIS synthetases, and that they act prior to the NRPSs in petrobactin biosynthesis (Oves-Costales *et al.*, 2008).

5. Hydroxamate-Formation Assay for NIS Synthetases

An important aspect of early studies of the enzymology of NIS synthetases has been to investigate the substrate specificity in the carboxylic acid adenylation reaction. In the NRPS field, the ATP–PP$_i$ exchange assay has been widely employed for this purpose. Here, we show by comparison with a published control reaction (May *et al.*, 2001) (adenylation of salicylic acid by the DhbE A domain involved in bacillibactin biosynthesis) that the ATP–PP$_i$ exchange assay is ineffective for NIS synthetases (Fig. 17.11). An explanation for this observation has come from recently published X-ray crystallographic studies of AcsD, a type A NIS synthetase (Schmelz *et al.*, 2009). In order for the ATP–PP$_i$ exchange assay to be effective, PP$_i$ must be released from the active site after adenylate formation (to allow exchange with radiolabeled PP$_i$). It is quite clear from the structure of AcsD with ATP bound that this cannot occur upon adenylate formation in NIS synthetases because the PP$_i$ is buried in a positively charged pocket (Schmelz *et al.*, 2009). Thus, it cannot be released unless the adenylate is released or decomposed by nucleophilic attack of the amine or alcohol substrate of the synthetase.

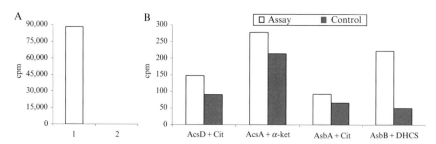

Figure 17.11 Results of ATP–PP$_i$ exchange assay with various adenylating enzymes involved in siderophore biosynthesis. (A) A very high level of activity is observed for the NRPS A domain DhbE with salicylic acid (1) relative to a boiled enzyme control (2). (B) Negligible levels of activity relative to boiled enzyme controls are observed for the NIS synthetases AcsD, AcsA, AsbA, AsbB with their cognate substrates (cit = citrate; α-ket = α-ketoglutarate; DHCS = N1-(3,4-dihydroxy-benzoyl)-N8-citryl-spermidine). This is attributed to the inability of NIS synthetases to release PP$_i$ after adenylate formation, thus preventing incorporation of radiolabeled PP$_i$ into ATP.

The failure of the ATP–PP$_i$ exchange assay with NIS synthetases motivated us to seek an alternative assay for probing carboxylic acid substrate specificity. An assay that was developed in the 1950s for detecting adenylate formation caught our attention. In 1945, a colorimetric "micromethod" for determination of acyl phosphates was published by Lipmann and Tuttle (1945). It was based on the formation of hydroxamic acids from hydroxylamine and acyl phosphates. Trivalent iron was used to form an orange-brown to purplish-brown complex with the hydroxamic acid. The absorbance maxima of the iron complexes are between 540 and 480 nm, at which ferric chloride does not absorb. In 1955, Hoagland suggested a mechanism for amino acid activation in rat liver involving formation of an aminoacyl adenylate. The addition of high concentrations of hydroxylamine trapped the activated amino acid (Hoagland, 1955). Later in the same year, Berg confirmed the presence of an enzyme-bound intermediate in the acetate-activating system from yeast (Berg, 1955). This enzyme can catalyze the incorporation of ^{32}PP$_i$ into ATP in the presence of acetate by forming an acetyl adenylate intermediate. Acetyl-CoA is produced when coenzyme A (CoA) is present in the reaction mixture. The following year, Moyed and Lipmann and De Moss *et al.* confirmed the proposal by Hoagland and Berg that acyl-adenylates are intermediates in these reactions (DeMoss *et al.*, 1956; Moyed and Lipmann, 1956). Lipmann and Tuttle used an organism from soil that could grow on propionate as a carbon source. Extracts of this organism contained an enzyme that could catalyze the conversion of propionate to propionyl-CoA via a propionyl adenylate intermediate. Over the years, the original hydroxylamine-trapping assay developed by Lipmann and Tuttle has been modified and used to detect various other kinds of activated carboxylic acid intermediates in enzyme-catalyzed reactions. However, unlike the ATP–PP$_i$ exchange assay, it has not been widely employed for the detection of acyl-adenylate intermediates. Here, we report that the hydroxylamine-trapping assay is useful for detecting the formation of acyl adenylate intermediates formed by NIS synthetases.

5.1. Assay principle

The hydroxylamine-trapping assay is used to examine the specificity of a synthetase towards a variety of carboxylic acids in the absence of the cognate amine or alcohol (note that it is often difficult to predict a priori the nature of this nucleophilic species from the structure of the siderophore under investigation—see the examples of achromobactin and petrobactin above). Hydroxylamine is added as a highly reactive surrogate for the missing amine/alcohol. If an acyl adenylate is formed, reaction with hydroxylamine leads to formation of a hydroxamic acid, which can easily be detected by spectrophotometry after addition of ferric iron (Fig. 17.12).

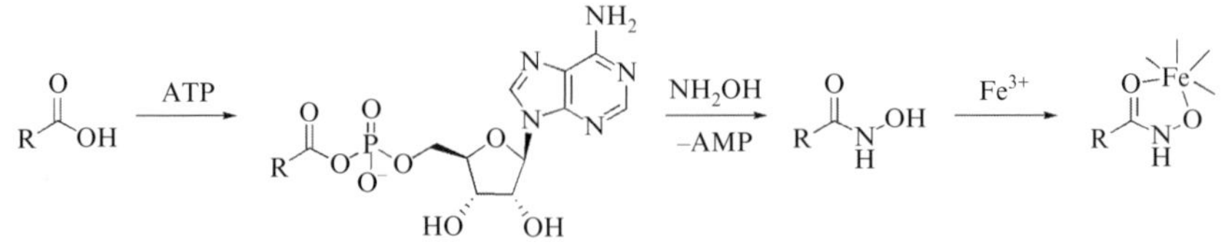

Figure 17.12 Interception of enzyme-bound acyl adenylates with hydroxylamine to form hydroxamic acids, which can be detected spectrophotometrically by conversion to a ferric complex.

5.2. Buffers, reagents, and others materials

- 50 mM Tris buffer, pH 8; stored at 4 °C.
- 300 mM MgCl$_2$ solution buffered with 25 mM Tris buffer (pH 8); stored at 4 °C.
- 30 mM aqueous solution of ATP; freshly prepared and stored on ice for no more than 1 day.
- 2 M hydroxylamine in 3.5 M NaOH (the pH is adjusted to 8 with concentrated HCl); stored at 4 °C.
- N1-(3,4-dihydroxybenzoyl)-N8-citryl-spermidine (DHCS) and N8-citryl-spermidine are chemically synthesized according to published procedures (Oves-Costales et al., 2007, 2008).
- 60 mM solutions of each carboxylic acid (citric acid; propionic acid; DL-lactic acid; mucic acid; L-glutamic acid 5-methyl ester; L-1,4-diaminobutyric acid; L-malic acid; L-tartaric acid; malonic acid; α-ketoglutaric acid; L-glutamic acid; levulinic acid; 1,3-acetone dicarboxylic acid; oxaloacetic acid; DHCS; N8-citryl-spermidine) are prepared in 25 mM Tris buffer. The pH is adjusted to 8 with 1 M HCl or 1 M NaOH solutions; stored at 4 °C.
- Stopping solution: A solution containing 10% (w/v) FeCl$_3$·6H$_2$O and 3.3% TCA is made up in 0.7 M HCl; stored at room temperature.
- Enzyme solutions: Each enzyme (AcsD, AsbA, AsbB, and AcsA) is purified according to published procedures (Oves-Costales et al., 2007, 2008; Schmelz et al., 2009). The aliquots are stored at −80 °C. 50 μM solutions of each enzyme in 25 mM Tris buffer are prepared fresh and stored on ice for no more than 1 day.
- Enzyme control solutions: Each 50 μM enzyme solution is inactivated by boiling for 10 min at 100 °C prior to use.
- 1.5 mL plastic spectrophotometric cuvettes are used with a Beckman Coulter DU 7400 UV–vis spectrometer, to measure absorbance at 540 nm.

5.3. Assay procedure

On the day, add to a 1.5 mL microcentrifuge tube, 150 μL Tris buffer, 15 μL $MgCl_2$ solution, 22.5 μL ATP solution, 22.5 μL hydroxylamine solution, and 15 μL of a carboxylic acid solution (for the assay carried out with AcsD and AcsA), or 7.5 μL of a carboxylic acid solution (for assays carried out with AsbA and AsbB). Make up to a total volume of 300 μL with distilled water. For each carboxylic acid an assay reaction and a control reaction are prepared.

1. Initiate the reaction by adding 15 μL of 50 μM enzyme solution to the assay mixture, and 15 μL of enzyme control solution to the control mixture. This yields the following final concentrations for each assay: 25 mM Tris buffer (pH 8), 15 mM $MgCl_2$, 2.25 mM ATP, 150 mM hydroxylamine, 3 mM carboxylic acid (for assay performed with AcsD and AcsA enzymes), or 1.5 mM carboxylic acid (for assay performed with AsbA and AsbB enzymes), 2.5 μM enzyme. Cap tubes and incubate at 37 °C for 1 h.
2. Add 300 μL of stopping solution to each microcentrifuge tube.
3. Centrifuge the microcentrifuge tubes for 5 min at 14,000 rpm to pellet the precipitated enzymes.
4. Transfer the supernatants to the cuvettes.
5. Immediately measure the absorbance for each cuvette at 540 nm. (Note: Do not delay these measurements because the color can fade quickly.)

5.4. Results

Figure 17.13 shows normalized values of the absorbance for the assay reaction minus the absorbance for the corresponding control reaction for each enzyme with a variety of carboxylic acids. Published data confirm the results obtained in this assay (Oves-Costales *et al.*, 2007, 2008; Schmelz *et al.*, 2009). AcsD has been shown to catalyze condensation of citric acid with L-serine; AsbA has been shown to catalyze condensation of citric acid with spermidine; and AsbB has been shown to catalyze condensation of DHCS or N8-citryl-spermidine with spermidine. The results for AcsA provide the first experimental verification of the prediction (Challis, 2005) that type B enzymes catalyze selective adenylation of α-ketoglutaric acid.

5.5. Technical notes

For less active enzymes, better results can be obtained by carrying out the assays with higher enzyme concentrations. For obvious reasons, the assay cannot be used with hydroxamic acid-containing substrates such as those utilized by DesD, PubC, and the BibC C-terminal domain.

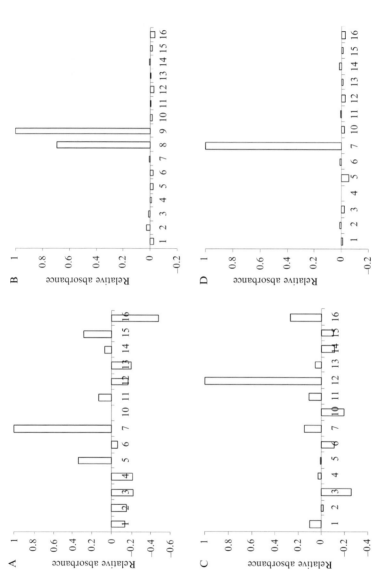

Figure 17.13 Results of the hydroxylamine-trapping assay: (A) with AcsD; (B) with AsbB; (C) with AcsA; and (D) with AsbA. The carboxylic acids investigated as substrates were as follows: 1, propionic acid; 2, DL-lactic acid; 3, mucic acid; 4, L-glutamic acid 5-methyl ester; 5, L-1,4-diaminobutyric acid; 6, L-malic acid; 7, citric acid; 8, DHCS; 9, N8-citryl-spermidine; 10, L-tartaric acid; 11, malonic acid; 12, α-ketoglutaric acid; 13, L-glutamic acid; 14, levulinic acid; 15, 1,3-acetone dicarboxylic acid; 16, oxaloacetic acid.

REFERENCES

Barona-Gómez, F., Wong, U., Giannakopulos, A. E., Derrick, P. J., and Challis, G. L. (2004). Identification of a cluster of genes that directs desferrioxamine biosynthesis in *Streptomyces coelicolor* M145. *J. Am. Chem. Soc.* **126,** 16282–16283.

Berg, P. (1955). Participation of adenyl-acetate in the acetate-activating system. *J. Am. Chem. Soc.* **77,** 3163–3164.

Bergeron, R. J., Huang, G., Smith, R. E., Bharti, N., McManis, J. S., and Butler, A. (2003). Total synthesis and structure revision of petrobactin. *Tetrahedron* **59,** 2007–2014.

Challis, G. L. (2005). A widely distributed bacterial pathway for siderophore biosynthesis independent of nonribosomal peptide synthetases. *Chem. Bio. Chem* **6,** 1–11.

Challis, G. L., and Hopwood, D. A. (2003). Synergy and contingency as driving forces for the evolution of multiple secondary metabolite production in *Streptomyces* species. *Proc. Natl. Acad. Sci. USA* **100,** 14555–14561.

Challis, G. L., and Ravel, J. (2000). Coelichelin, a new peptide siderophore encoded by the *Streptomyces coelicolor* genome: Structure prediction from the sequence of its non-ribosomal peptide synthetase. *FEMS Microbiol. Lett.* **187,** 111–114.

Cox, C. D., Rinehart, K. L., Moore, M. L., and Cook, J. C. (1981). Pyochelin: Novel structure of an iron-chelating growth promoter for *Pseudomonas aeruginosa*. *Proc. Natl. Acad. Sci. USA* **78,** 4256–4260.

Crosa, J. H., and Walsh, C. T. (2002). Genetics and assembly line enzymology of siderophore biosynthesis in bacteria. *Microbiol. Mol. Biol. Rev.* **66,** 223–249.

De Lorenzo, V., Bindereif, A., Paw, B. H., and Neilands, J. B. (1986). Aerobactin biosynthesis and transport genes of plasmid ColV-K30 in *Escherichia coli* K-12. *J. Bacteriol.* **165,** 570–578.

De Lorenzo, V., and Neilands, J. B. (1986). Characterization of *iucA* and *iucC* genes of the aerobactin system of plasmid ColV-K30 in *Escherichia coli*. *J. Bacteriol.* **167,** 350–355.

DeMoss, J. A., Genuth, S. M., and Novelli, G. D. (1956). The enzymatic activation of amino acids via their acyl-adenylate derivatives. *Proc. Natl. Acad. Sci. USA* **42,** 325–332.

Dimise, E. J., Widboom, P. F., and Bruner, S. D. (2008). Structure elucidation and biosynthesis of fuscachelins, peptide siderophores from the moderate thermophile *Thermobifida fusca*. *Proc. Natl. Acad. Sci. USA* **105,** 15311–15316.

Felnagle, E. A., Jackson, E. E., Chan, Y. A., Podevels, A. M., Berti, A. D., McMahon, M. D., and Thomas, M. G. (2008). Nonribosomal peptide synthetases involved in the production of medically relevant natural products. *Mol. Pharm.* **5,** 191–211.

Fischbach, M. A., Lin, H., Liu, D. R., and Walsh, C. T. (2006). How pathogenic bacteria evade mammalian sabotage in the battle for iron. *Nat. Chem. Biol.* **2,** 132–138.

Fox, D. T., Hotta, K., Kim, C.-Y., and Koppisch, A. T. (2008). The missing link in petrobactin biosynthesis: asbF encodes a (−)-3-dehydroshikimate dehydratase. *Biochemistry* **47,** 12251–12253.

Franza, T., Mahé, B., and Expert, D. (2005). *Erwinia chrysanthemi* requires a second iron transport route dependent of the siderophore achromobactin for extracellular growth and plant infection. *Mol. Microbiol.* **55,** 261–275.

Gehring, A. M., Bradley, K. A., and Walsh, C. T. (1997). Enterobactin biosynthesis in *Escherichia coli*: Isochorismate lyase (EntB) is a bifunctional enzyme that is phosphopantetheinylated by EntD and then acylated by EntE using ATP and 2,3-dihydroxybenzoate. *Biochemistry* **36,** 8495–8503.

Gehring, A. M., Mori, I., and Walsh, C. T. (2008). Reconstitution and characterization of the *Escherichia coli* enterobactin synthetase from EntB, EntE, and EntF. *Biochemistry* **37,** 2648–2659.

Hoagland, M. B. (1955). An enzymic mechanism for amino acid activation in animal tissues. *Biochim. Biophys. Acta* **16,** 288–289.

Kadi, N., Arbache, S., Song, L., Oves-costales, D., and Challis, G. L. (2008a). Identification of a gene cluster that directs putrebactin biosynthesis in *Shewanella* species: PubC catalyses cyclodimerization of *N*-hydroxy-*N*-succinylputrescine. *J. Am. Chem. Soc.* **130**, 10458–10459.

Kadi, N., Oves-Costales, D., Barona-Gómez, F., and Challis, G. L. (2007). A new family of ATP-dependent oligomerization-macrocyclization biocatalysts. *Nat. Chem. Biol.* **3**, 652–656.

Kadi, N., Song, L., and Challis, G. L. (2008b). Bisucaberin biosynthesis: An adenylating domain of the BibC multi-enzyme catalyzes cyclodimerization of *N*-hydroxy-*N*-succinylcadaverine. *Chem. Commun.* 5119–5121.

Koppisch, A. T., Browder, C. C., Moe, A. L., Shelley, J. T., Kinkel, B. A., Hersman, L. E., Iyer, S., and Ruggiero, C. E. (2005). Petrobactin is the primary siderophore synthesized by *Bacillus anthracis* Str. Sterne under conditions of iron starvation. *Biometals* **18**, 577–585.

Lautru, S., and Challis, G. L. (2004). Substrate recognition by nonribosomal peptide synthetase multi-enzymes. *Microbiology* **150**, 1629–1636.

Lautru, S., Deeth, R. J., Bailey, L. M., and Challis, G. L. (2005). Discovery of a new peptide natural product by *Streptomyces coelicolor* genome mining. *Nat. Chem. Biol.* **1**, 265–269.

Ledyard, K. M., and Butler, A. (1997). Structure of putrebactin, a new dihydroxamate siderophore produced by *Shewanella putrefaciens*. *J. Biol. Inorg. Chem.* **2**, 93–97.

Lee, J. Y., Janes, B. K., Passalacqua, K. D., Pfleger, B. F., Bergman, N. H., Liu, H., Hakansson, K., Somu, R. V., Aldrich, C. C., Cendrowski, S., Hanna, P. C., and Sherman, D. H. (2007). Biosynthetic analysis of the petrobactin siderophore pathway from *Bacillus anthracis*. *J. Bacteriol.* **189**, 1698–1710.

Lipmann, F., and Tuttle, L. C. (1945). A specific micromethod for the determination of acyl phosphates. *J. Biol. Chem.* **159**, 21–28.

Matzanke, B. F. (2005). Iron transport: Siderophores. *In* "Encyclopedia of Inorganic Chemistry" (R. B. King, ed.), 2nd ed., Vol. IV, pp. 2619–2646. Wiley, London.

May, J. J., Wendrich, T. M., and Marahiel, M. A. (2001). The *dhb* operon of *Bacillus subtilis* encodes the biosynthetic template for the catecholic siderophore 2,3-dihydroxybenzoate-glycine-threonine trimeric ester bacillibactin. *J. Biol. Chem.* **276**, 7209–7217.

Miethke, M., and Marahiel, M. A. (2007). Siderophore-based iron acquisition and pathogen control. *Microbiol. Mol. Biol. Rev.* **71**, 413–451.

Moyed, H. S., and Lipmann, F. (1956). Studies on the adenosine triphosphate-propionate reaction in extracts of an unidentified bacterium. *J. Bacteriol.* **73**(1), 117–121.

Oves-Costales, D., Kadi, N., Fogg, M. J., Song, L., Wilson, K. S., and Challis, G. L. (2007). Enzymatic logic of anthrax stealth siderophore biosynthesis: AsbA catalyzes ATP-dependent condensation of citric acid and spermidine. *J. Am. Chem. Soc.* **129**, 8416–8417.

Oves-Costales, D., Kadi, N., Fogg, M. J., Song, L., Wilson, K. S., and Challis, G. L. (2008). Petrobactin biosynthesis: AsbB catalyzes condensation of spermidine with N^8-citryl-spermidine and its N^1-(3,4-dihydroxybenzoyl) derivative. *Chem. Commun.* 4034–4036.

Oves-Costales, D., Song, L., and Challis, G. L. (2009). The petrobactin biosynthetic enzyme AsbA has relaxed substrate specificity and catalyzes enantioselective desymmetrisation of citric acid. *Chem. Commun.* Doi:10.1039/6823147h.

Patel, H. M., Tao, J., and Walsh, C. T. (2003). Epimerization of an L-cysteinyl to a D-cysteinyl residue during thiazoline ring formation in siderophore chain elongation by pyochelin synthetase from *Pseudomonas aeruginosa*. *Biochemistry* **42**, 10514–10527.

Patel, H. M., and Walsh, C. T. (2001). *In vitro* reconstitution of the *Pseudomonas aeruginosa* nonribosomal peptide synthesis of pyochelin: Characterization of backbone tailoring thiazoline reductase and *N*-methyltransferase activities. *Biochemistry* **40**, 9023–9031.

Pfleger, B. F., Kim, Y., Nusca, T. D., Maltseva, N., Lee, J. Y., Rath, C. M., Scaglione, J. B., Janes, B. K., Anderson, E. C., Bergman, N. H., Hanna, P. C., Joachimiak, A., *et al.*

(2008). Structural and functional analysis of AsbF: Origin of the stealth 3,4-dihydroxybenzoic acid subunit for petrobactin biosynthesis. *Proc. Natl. Acad. Sci. USA* **105,** 17133–17138.

Pfleger, B. F., Lee, J. Y., Somu, R. V., Aldrich, C. C., Hanna, P. C., and Sherman, D. H. (2007). Characterization and analysis of early enzymes for petrobactin biosynthesis in *Bacillus anthracis*. *Biochemistry* **46,** 4147–4157.

Quadri, L. E. N. (2007). Strategic paradigm shifts in the antimicrobial drug discovery process of the 21st century. *Infect. Disord. Drug Targets* **7,** 230–237.

Quadri, L. E. N., Keating, T. A., Patel, H. M., and Walsh, C. T. (1999). Assembly of the *Pseudomonas aeruginosa* nonribosomal peptide siderophore pyochelin: *In vitro* reconstitution of Aryl-4,2-bisthiazoline synthetase activity from PchD, PchE, and PchF. *Biochemistry* **38,** 14941–14954.

Raymond, K. N., Dertz, E. A., and Kim, S. S. (2003). Enterobactin: An archetype for microbial iron transport. *Proc. Natl. Acad. Sci.* **100,** 3584–3588.

Santi, D. V., Webster, R. W., Jr., and Cleland, W. W. (1974). Kinetics of aminoacyl-tRNA synthetases catalyzed ATP–PPi exchange. *Methods Enzymol.* **29,** 620–627.

Schmelz, S., Kadi, N., MacMahon, S. A., Song, L., Oves-Costales, D., Oke, M., Liu, H., Johnson, K. A., Carter, L. G., Botting, C. H., White, M. F., Challis, G. L., and Naismith, J. H. (2009). AcsD catalyzes enantioselective citrate desymmetrization in siderophore biosynthesis. *Nat. Chem. Biol.* **5,** 174–182.

Shaw-Reid, C. A., Kelleher, N. L., Losey, H. C., Gehring, A. M., Berg, C., and Walsh, C. T. (1999). Assembly line enzymology by multimodular nonribosomal peptide synthetases: The thioesterase domain of *E. coli* EntF catalyzes both elongation and cyclolactonization. *Chem. Biol.* **6,** 385–400.

Wenzel, S. C., Meiser, P., Binz, T. M., Mahmud, T., and Müller, R. (2006). Nonribosomal peptide biosynthesis: Point mutations and module skipping lead to chemical diversity. *Angew. Chem. Int. Ed.* **45,** 2296–2301.

Wilson, M. K., Abergel, R. J., Raymond, K. N., Arceneaux, J. E. L., and Byers, B. R. (2006). Siderophores of *Bacillus anthracis*, *Bacillus cereus*, and *Bacillus thuringiensis*. *Biochem. Biophys. Res. Commun.* **348,** 320–325.

Winkelmann, G., Schmid, D. G., Nicholson, G., Jung, G., and Colquhoun, D. J. (2002). Bisucaberin—A dihydroxamate siderophore isolated from *Vibrio salmonicida*, an important pathogen of farmed *Atlantic salmon* (*Salmo salar*). *Biometals* **15,** 153–160.

CHAPTER EIGHTEEN

Molecular Genetic Approaches to Analyze Glycopeptide Biosynthesis

Wolfgang Wohlleben,* Evi Stegmann,* and Roderich D. Süssmuth[†]

Contents

1. Structural Classification of Glycopeptide Antibiotics	460
2. Methods for Analyzing Glycopeptide Biosynthesis	462
2.1. Chemical methods	462
2.2. Genetic methods	462
3. Investigation of Glycopeptide Biosynthetic Steps	466
3.1. The biosynthesis of amino acid building blocks	466
3.2. Biosynthesis of carbohydrate building blocks	470
3.3. Peptide assembly by NRPSs	471
3.4. Oxidative side chain cyclization by P450-type monooxygenases	473
3.5. Time point of halogenation and side chain cyclization in glycopeptide biosynthesis	475
4. Regulation, Self-Resistance, and Excretion	477
5. Linking Primary and Secondary Metabolism	478
6. Approaches for the Generation of New Glycopeptides	479
Acknowledgments	480
References	480

Abstract

The glycopeptide antibiotics vancomycin and teicoplanin are used in the hospital as drugs of last resort to combat resistant Gram-positive pathogens, in particular methicillin-resistant *Staphylococcus aureus*. All glycopeptides consist of a heptapeptide backbone in which the aromatic residues are connected to form a rigid cup-shaped structure required to stably interact with the D-Ala-D-Ala terminus of bacterial cell wall precursors. Structural diversity is generated by variations in the composition of the backbone, preferably at amino acid positions 1 and 3, and by different glycosylation, methylation, and chlorination patterns. The identification of several glycopeptide biosynthesis gene clusters, the

* Institut für Mikrobiologie, Mikrobiologie/Biotechnologie, Universität Tübingen, Tübingen, Germany
† Institut für Chemie, Technische Universität Berlin, Berlin, Germany

development of genetic techniques to manipulate at least some of the producing actinomycetes, and subsequent molecular analysis enabled the elucidation of their biosynthetic pathways. This led to biochemical methods being combined with molecular genetic techniques and analytical chemistry. Knowledge of the biosynthesis made it possible to apply different approaches for the generation of novel glycopeptide derivatives by mutasynthesis, precursor-directed biosynthesis, and genetic engineering.

1. Structural Classification of Glycopeptide Antibiotics

Over the past decade, glycopeptide antibiotics have been a field of interest for researchers of several disciplines. The most prominent member of the glycopeptide antibiotic family is vancomycin, which was discovered in the mid-1950s (for a review, see Williams and Bardsley, 1999). Vancomycin is mainly used in hospitals for the treatment of infections caused by (multi-)resistant Gram-positive pathogens, in particular by methicillin resistant *Staphylococcus aureus* (MRSA). In Europe, teicoplanin, a lipoglycopeptide, is also in clinical use to combat these pathogens. Furthermore, derivatives of the naturally synthesized compounds, such as dalbavancin, telavancin, and oritavancin, have successfully passed clinical trials and exhibit improved properties in treating vancomycin resistant enterococci and staphylococci (Arias and Murray, 2008).

Glycopeptide antibiotics have been grouped according to their structural features (Yao and Crandall, 1994) into five different subclasses (Fig. 18.1). Common to all glycopeptide antibiotics is a high content of aromatic amino acids (AA) structurally related to tyrosine and phenylglycine. All glycopeptide antibiotics have a heptapeptide backbone, with the primary sequence ^{7}Dpg-^{6}Hty-^{5}Hpg-^{4}Hpg-^{3}Xxx-^{2}Hty/Tyr-^{1}Yyy, which is cyclized in the aromatic side chains (Dpg, 3,5-dihydroxyphenylglycine; Hty, β-hydroxytyrosine; Hpg, 4-hydroxyphenylglycin). Further modifications arise by the attachment of chlorine atoms and sugar residues. The side chain cyclization confers on the otherwise flexible molecule a conformational rigidity which is essential for their antibiotic activity. The main criteria for a structural classification are variations in AA1 and AA3 of the heptapeptide backbone. Until now, comprehensive genetic and biochemical analyses of glycopeptide biosynthesis are limited to the so-called type I-glycopeptide antibiotics (vancomycin-like) and to lipoglycopeptides (teicoplanin-like). Therefore, this article is restricted to these two classes.

In type I-glycopeptides, AA positions 1 and 3 are commonly occupied by aliphatic amino acids with variously N-methylated (R)-^{1}Leu and (S)-^{3}Asn. They form three macrocycles with the AB-biaryl ring

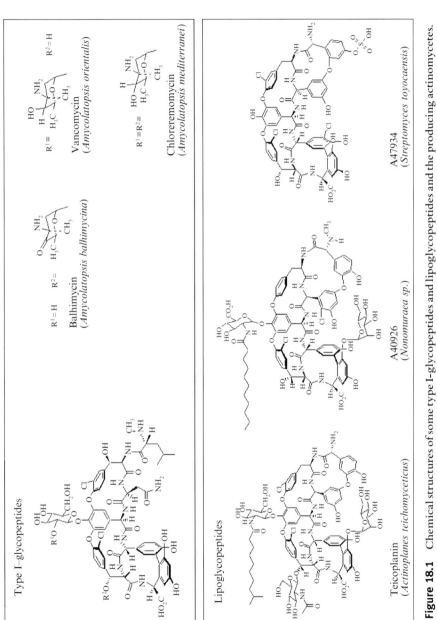

Figure 18.1 Chemical structures of some type I-glycopeptides and lipoglycopeptides and the producing actinomycetes.

(^{7}Dpg-^{5}Hpg), the C-O-D (^{6}Hty-^{4}Hpg) and the D-O-E (^{4}Hpg-^{2}Hty)-diarylethers. In teicoplanin-like lipoglycopeptides, ^{1}Leu and ^{3}Asn are replaced by ^{1}Hpg and ^{3}Dpg, which form an additional macrocycle (F-O-G ring). A further difference is a long-chain acyl group covalently attached to a carbohydrate residue.

The glycopeptide producers belong to different genera of the order *Actinomycetales*. Whereas all producers of type I-glycopeptides are members of the genus *Amycolatopsis*, members of the family *Pseudonocardiaceae*, producers of the teicoplanin-type compounds, are quite distantly related: The A40926 (dalbavancin)-synthesizing strain is a member of the genus *Nonomuraea* (formerly *Planobispora*) in the family *Streptosporangiaceae*; teicoplanin is made by an *Actinoplanes* species (family *Micromonosporaceae*); and the A47934 producer is a *Streptomyces* species (family *Streptomyceaceae*).

All glycopeptides target the N-acyl-D-Ala-D-Ala peptides of peptidoglycan precursors and thus block cell wall biosynthesis in Gram-positive bacteria. Resistance to glycopeptides has emerged in particular in enterococci, but also increasingly in staphylococci. The major resistance mechanism is due to a reprogramming of peptidoglycan precursor biosynthesis, resulting in N-acyl-D-Ala-D-Lac precursors to which vancomycin binds with 1000-fold lower affinity.

2. Methods for Analyzing Glycopeptide Biosynthesis

2.1. Chemical methods

Before the establishment of molecular genetic and protein biochemical methods, labeling studies were the preferred tools to obtain information on biosynthetic assembly. Data from Hammond *et al.* (1982, 1983), showing that Hty und Hpg were derived from tyrosine metabolism and that Dpg was obtained from acetate via a polyketide mechanism, were the prerequisite for the assignment of the respective biosynthetic genes.

For a comprehensive overview of chemical approaches in glycopeptide research, see Chapter 19, in this volume.

2.2. Genetic methods

2.2.1. Identification of glycopeptide biosynthetic gene clusters

In order to apply genetic techniques for the elucidation of glycopeptide biosynthesis different biosynthetic gene clusters have been cloned and sequenced. To date, the gene clusters encoding the biosynthesis of chloroeremomycin (*cep*) (van Wageningen *et al.*, 1998), balhimycin (*bal*) (Pelzer *et al.*, 1999; Shawky *et al.*, 2007), A47934 (*sta*) (Pootoolal *et al.*,

2002), A40926 (*dbv*) (Sosio *et al.*, 2003), and teicoplanin (*tei; tcp*) (Li *et al.*, 2004; Sosio *et al.*, 2004, respectively) are available. In all cases (except *tei*), cosmid libraries were used to identify the gene clusters using either probes deduced from consensus sequences of peptide synthetase, oxygenase or glycosyltransferase genes or heterologous probes from other glycopeptide clusters. For *tei* the respective cosmids were identified by randomly end-sequencing 500 cosmids of a genomic library.

2.2.2. Systems for the genetic manipulation of glycopeptide-producing bacteria

For the analysis of glycopeptide biosynthesis a combination of genetics and analytical chemistry has proven to be very effective. Following the generation of defined blocked mutants the chemical structures of the accumulated biosynthetic intermediates can be determined. This approach, however, requires methods for the genetic manipulation of the producers.

Not all glycopeptide producers can easily be manipulated by standard techniques. For all of them host–vector systems had to be newly developed or existing tools had to be adapted to the specific producer strains.

For *Streptomyces*, in particular for the model strains *Streptomyces coelicolor* and *Streptomyces lividans*, genetic tools are available (Kieser *et al.*, 2000) that can, after modification, be applied to other *Streptomyces* strains, as shown for *Streptomyces toyocaensis* (Matsushima and Baltz, 1996), the A47934 producer. DNA can be introduced either by PEG-mediated protoplast transformation or by conjugation of mobilizable vectors from *Escherichia coli* using specific donor strains (Flett *et al.*, 1997). In order to generate mutations in the producer, gene replacement approaches using either nonreplicative or temperature-sensitive vectors can be employed (Hillemann *et al.*, 1991; Muth *et al.*, 1999). After establishing a cosmid library, the Redirect Technology (Gust *et al.*, 2004) can also be applied to *S. toyocaensis* (Lamb *et al.*, 2006). However, the efficiency and the number of available tools are still much lower than for bacteria such as *E. coli* or bacilli, making genetic experiments time-consuming.

The situation is even worse for other glycopeptide producers. For the balhimycin producer, *Amycolatopsis balhimycina* (Nadkarni *et al.*, 1994; Wink *et al.*, 2003), integrative vectors and a transformation system have been developed (Pelzer *et al.*, 1997). Gene inactivation experiments can be performed with only nonreplicative vectors (pSP1, pSP2) containing markers (chloramphenicol and erythromycin resistance genes) for selection in *A. balhimycina*.

For the transfer of DNA, a modified version of the direct transformation protocol (first described by Madon and Hütter, 1991) delivers the highest efficiencies (Pelzer *et al.*, 1997). Essentially, washed mycelium is incubated with plasmid DNA in the presence of PEG 1000, plated on regeneration medium and then overlaid with antibiotic-containing soft agar to select for the presence of the plasmid. A further gene transfer system has been described for *Nonomuraea* (Stinchi et al., 2003): For this strain, intergeneric conjugation from *E. coli* can be used to introduce foreign DNA.

Since the gene clusters include large multicistronic operons (see Section 2.2.4), gene inactivation must use in-frame deletion mutants to avoid any polar effect on downstream genes. The standard procedure, therefore, comprises two steps. First, a nonreplicative plasmid carrying the gene of interest with an internal in-frame deletion is transferred into *A. balhimycina* and recombinants are selected which harbor (after a single crossover) the integrated plasmid. In a second step, this plasmid is excised (via a second crossover event), resulting in a deletion. Due to the low transformation and recombination frequency of the (nonstreptomycete) glycopeptide producers a direct selection for double crossover events normally fails, whereas single crossovers leading to plasmid integration provide recombinants. To select the subsequent second crossover a "stress protocol" was developed which facilitates the isolation of the mutants. The decisive step in this protocol is an ultra-sound treatment combined with incubation at elevated temperature, or alternatively protoplast formation and regeneration.

Expression experiments in the glycopeptide-producing strains are routinely performed with integrative plasmids such as pSET (Bierman et al., 1992) and its derivatives such as pSETermE (Stegmann et al., 2006a), utilizing the ΦC31 attachment site present in most actinomycete chromosomes (see Chapter 15 in Volume 459).

Approaches to express the clusters in a heterologous host that can be more efficiently manipulated require the cloning and transfer of large contiguous pieces (~70 kb) of DNA. The construction of suitable vectors (Sosio et al., 2000, 2001) as well as the generation of libraries of producing strains that include clones with large inserts (>150 kb) (Alduina et al., 2003) has been reported. However, until now, the heterologous expression of a complete glycopeptide biosynthetic gene cluster was not successful.

2.2.3. Heterologous over-expression of glycopeptide biosynthetic genes

The known DNA sequences of glycopeptide gene clusters provide the possibility to easily express individual genes in a heterologous background. The cloning of biosynthetic genes and subsequent protein fusions (with N- or C-terminal His-tags or with the maltose binding protein) enable a facilitated purification in *E. coli*. However, expression of actinomycete proteins in *E. coli* is sometimes difficult to achieve; often because of the different codon usage due to the high GC-content (65–75%) of actinomycete genes. This can be overcome either by using *E. coli* Rosetta strains (Novagen©), which are equipped with genes for those t-RNAs which are rarely utilized in *E. coli* but frequently used in actinomycetes. An alternative approach uses *S. lividans* as an expression host. Different vectors carrying either strong constitutive promoters such as permE* (Bibb et al., 1985) or inducible promoters such as the thiostrepton-inducible ptipA (Murakami *et al.*, 1989) or the tetracycline-inducible ptet (Rodríguez-García *et al.*, 2005) are available. Useful expression vectors are, for example, pIJ4123, pIJ6021 or pIJ8600 (Kieser *et al.*, 2000), and pGM190. The shuttle vector pGM190 (Fig. 18.2) is based on the medium copy-number plasmid pSG5 (Muth *et al.*, 1988, 1989) and is particularly stable, probably because of the presence of the minus-origin of replication (Reuther *et al.*, 2006b) and of a gene arrangement which resulted in the same direction of transcription for all plasmid genes. The vector possesses a broad host range (all *Streptomyces* strains tested thus far) and has successfully been applied for the expression of various functions (e.g., Mazza *et al.*, 2006; Reuther *et al.*, 2006a).

An alternative expression system for *Streptomyces* is based on the *E. coli* T7 system, which was modified by J. Altenbuchner for streptomycetes. The gene of interest can be cloned in *E. coli* into vectors of the pRSET series (Invitrogen life technologies, Carlsbad, USA) and then fused with a *Streptomyces* vector, preferably pGM9 (Wohlleben and Muth, 1993). The resulting shuttle plasmids are used to transform *S. lividans* T7 in which the T7

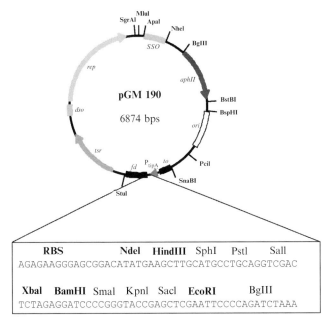

Figure 18.2 Restriction map of the expression vector pGM190. pGM190 is based on the replicon of pSG5 (Muth *et al.*, 1988), consisting of a replication gene (*rep*) and origins for single-strand (*sso*) and double-strand (*dso*) replication as well as replication functions for *E. coli* (*ori*). It carries resistance genes against kanamycin/neomycin (*aphII*) and thiostrepton (*tsr*) for selection and an expression cassette composed of the thiostrepton-inducible *tipA* promoter (P_{tipA}) followed by a multiple cloning site shown in detail (singular restriction sites are indicated in bold). The expression cassette is flanked by termination sequences (*fd*, *to*).

polymerase gene is integrated into the chromosome and can be expressed under the control of the thiostrepton-inducible promoter (Fischer, 1996). Thus, a number of proteins which could not be expressed in E. coli were successfully purified from S. lividans (e.g., Pfeifer et al., 2001).

To date, both methods, over-expression of fusion enzymes, combined with substrate conversion studies, as well as gene inactivation and chemical characterization of biosynthetic intermediates have been complementarily applied in order to study the biosynthesis of glycopeptide antibiotics. From the current viewpoint, the best investigated biosynthetic systems are those for chloroeremomycin (*Amycolatopsis orientalis*), investigated mainly by protein over-expression, and balhimycin (*A. balhimycina*), mainly investigated by gene inactivation.

2.2.4. In silico analysis of glycopeptide biosynthetic gene clusters

The functions of most biosynthetic genes have first been suggested by sequence comparisons and subsequently confirmed or disproved by genetic and biochemical analysis.

The sequences of the five different clusters (due to their high identity *tei* and *tcp* will be considered as one cluster) have been used for extensive comparisons (Donadio et al., 2005) in which the complestatin gene cluster (Chiu et al., 2001) can also be included (Fig. 18.3). Whereas the similarity between corresponding genes is high (up to 97%), the clusters show only a limited degree of synteny (Donadio et al., 2005), except for the *cep* and *bal* clusters, which are both responsible for the synthesis of a vancomycin-like glycopeptide. The regions of conserved synteny include five functional cassettes for the following biosynthetic steps: synthesis of the nonproteinogenic amino acids Hpg and Dpg; nonribosomal peptide synthesis of AA1–AA3 and AA4–AA7; and formation of the cross-links between the aromatic amino acids (Donadio et al., 2005). These functional cassettes also often comprise the noncoding upstream regions. Sequence comparisons with published genomic data revealed that some of these cassettes can also be found in other actinomycetes like *Frankia*, which are not known to be potent secondary metabolite producers (Weber et al., 2009). In these bacteria, the cassettes are obviously not part of glycopeptide gene clusters. From these findings one may speculate that the acquisition of such cassettes may have played an important role in the evolution of secondary metabolite gene clusters (Donadio et al., 2005).

3. INVESTIGATION OF GLYCOPEPTIDE BIOSYNTHETIC STEPS

3.1. The biosynthesis of amino acid building blocks

Type I-glycopeptides antibiotics consist of five different amino acids. These are (R)-^{1}Leu in its various N-methylated forms and (S)-^{3}Asn as the most important representatives in these positions. The aromatic amino acids form the aglycone with $^{4/5}$Hpg (2x), $^{2/6}$Tyr/Hty (2x), and ^{7}Dpg (1x). Among

Genetics of Glycopeptide Biosynthesis

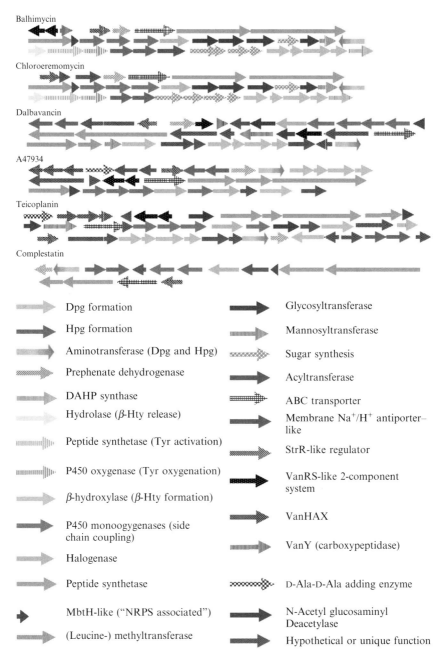

Figure 18.3 Comparison of the organization of biosynthetic gene clusters for balhimycin, chloroeremomycin, dalbavancin (A40296), A47934, teicoplanin, and complestatin, modified according to Donadio *et al.* (2005); arrows are not in scale. (See Color Insert.)

lipoglycopeptide antibiotics, structural diversity of building blocks is reduced towards aromatic amino acids, and the aglycone is solely composed of Tyr/Hty (1/1), Dpg (2x), and Hpg (3x). The genes for the biosynthesis of the nonproteinogenic amino acids which are specifically provided as part of the glycopeptide biosynthesis are located in the respective glycopeptide gene clusters.

3.1.1. Biosynthesis of β-hydroxytyrosine (AA2 and AA6)

Apart from the proteinogenic amino acids serine and threonine, other β-hydroxy amino acids are rarely found in nature with the exception of peptidic secondary metabolites. Secondary metabolites bearing β-hydroxyamino acids are novobiocin, plusbactin, nikkomycin, and CDA and all glycopeptide antibiotics (Chen et al., 2002). The Htys of vancomycin-type glycopeptides are diastereomers with $(2R,3R)$-^{2}Cht/Hty and $(2S,3R)$-^{6}Cht configurations (Cht, 3-chloro-β-hydroxytyrosine). In contrast, in teicoplanin and most other lipoglycopeptide antibiotics ^{2}Hty is replaced by ^{2}Tyr. The genes involved in Hty biosynthesis were first identified by mutational analysis in *A. balhimycina*: Gene inactivation by in-frame deletion of the genes *bhp*, *bpsD*, and *oxyD* delivered null mutants whose ability to produce balhimycin can be restored by adding β-hydroxytyrosine to the medium (Puk et al., 2002, 2004).

An elegant way to prove that all three genes are involved in the same step is by cofeeding experiments: no combination of *bhp*, *bpsD*, and *oxyD* mutants led to "complementation", whereas the cocultivation of one of these mutants together with a *pgat* mutant (see below) resulted in antibiotic biosynthesis. A further indication of a functional interplay came from reverse transcription-PCR (RT-PCR) experiments which proved that all three genes lie in one operon, that is, they are concomitantly transcribed and expressed (Puk et al., 2002, 2004). Until now, there do not exist protein expression studies of all genes for Hty biosynthesis of glycopeptide antibiotics. From sequence homologies of *bpsD*/*oxyD* to *novH*/*novI* genes of the novobiocin biosynthetic gene cluster (Chen and Walsh, 2001) a possible function can be assumed: BpsD is a single peptide synthetase module with AT-domain structure at which Tyr is activated. OxyD is an enzyme with homologies to cytochrome P450-dependent monooxygenases introducing the hydroxy-function. Bhp has similarities to a perhydrolase/"haloperoxidase," but a function as haloperoxidase could be excluded (Puk et al., 2002). Bhp acts as a thioesterase with substrate specificity for PCP-bound Hty (Mulyani et al., 2009).

Surprisingly, in lipoglycopeptide antibiotics the biosynthesis of Htys occurs in a different way (Stinchi et al., 2006). After inactivation of *orf28* (encoding a putative β-hydroxylase in A40926 biosynthesis), the mutant produced a desoxy-derivative of the lipoglycopeptide. Several pieces of evidence indicate that hydroxylation in lipoglycopeptide biosynthesis occurs after tyrosine activation by the nonribosomal peptide synthetase at the NRPS complex, namely the formation of an altered peptide in the *orf28* mutant, the inability to rescue A40926 biosynthesis upon feeding free Hty, and the lack of homologues of *bpsD, oxyD,* and *bhp* in the A40926 cluster (Stinchi *et al.*, 2006).

3.1.2. Biosynthesis of hydroxyphenylglycine (AA4 and AA5)

The nonproteinogenic amino acid 4-hydroxyphenylglycine (Hpg) is a main building block of all glycopeptide antibiotics. The biosynthesis of Hpg was investigated by heterologous overexpression of the putative genes from the chloroeremomycin-producer *Amycolatopsis mediterranei* in *E. coli*. The *in vitro* substrate conversions showed that the biosynthesis is based on four enzymes (ORF1/17/21/22^{Cep} and Pdh/HmaS/HmO/Pgat^{Bal}). The product of *orf1*, which is a prephenate dehydrogenase, Pdh, presumably provides sufficient amounts of 4-hydroxyphenylpyruvate (4-HPP) from primary metabolism. The step leading from 4-HPP to 4-hydroxymandelate (4-HMA) is performed by the hydroxymandelate synthase, HmaS (Orf21Cep), an iron-dependant decarboxylating dioxygenase (Choroba *et al.*, 2000; Li *et al.*, 2001). Oxidation of 4HMA was performed with ORF22/HmO coding for a flavin-dependent oxidase (Hubbard *et al.*, 2000). Finally, a transamination reaction (catalyzed by Pgat^{Bal} or Orf17^{Cep}) with PLP as a cofactor occurs.

3.1.3. Biosynthesis of 3,5-dihydroxyphenylglycine (AA7)

Glycopeptide antibiotics contain Dpg as the C-terminal amino acid, which forms the AB-ring. In addition, Dpg is a constituent in the 3-position of lipoglycopeptides, forming the F-*O*-G-ring. The biosynthesis is encoded by five genes *dpgA/B/C/D*^{bal} (*orf27–30*^{Cep}) and *pgat*^{bal} (Orf17^{Cep}) of which *dpgA-D* are grouped in an operon-like structure. DpgA closely resembles plant chalcon/stilbene synthases of the type III PKS family (Pfeifer *et al.*, 2001). Gene inactivation of *dpgA* resulted in loss of glycopeptide biosynthesis, restored upon feeding of 3,5-dihydroxyphenylacetic acid (DHPA) (Pfeifer *et al.*, 2001). Expression of DpgA in the heterologous host *S. lividans* resulted in production of DHPA. Incubations with radioactive malonyl-CoA (Pfeifer *et al.*, 2001) as well as heterologous expression of Orf27^{Cep} (DpgA) in *E. coli* and *in vitro* characterization (Li *et al.*, 2001) proved that only malonyl-CoA was used as the starter and extender unit for DHPA-production. Coexpression of DpgA-D in *S. lividans* established the synthesis of 3,5-dihydroxyphenylglyoxylate (GC–MS-analysis). Inactivation of the aminotransferase gene *pgat* could only be overcome by concomitant feeding of Hpg and Dpg. The sole supplementation of either one of these amino acids resulted in lack of glycopeptide production (Pfeifer *et al.*, 2001). From these results a biosynthetic scheme was proposed going from dihydroxyphenylacetic acid to mandelic acid, glyoxylate, and finally Dpg.

The interplay all four enzymes, DpgA-D (Orf27–30^{Cep}), expressed as N-terminally His-tagged proteins in *E. coli*, was also investigated *in vitro*. Chen *et al.* (2001) observed that substrate conversion of malonyl-CoA by DpgA was extremely slow and was significantly accelerated upon addition of DpgB to the incubation mixture. Addition of DpgD resulted in further increase of DHPA-CoA production. DpgC was found to oxidize DHPA to the glyoxylate in the presence of oxygen.

The properties of DpgA have been characterized by mutational studies in more detail (Tseng *et al.*, 2004), resulting in a proposal for an alternative noncovalent malonyl-CoA condensing mechanism.

3.2. Biosynthesis of carbohydrate building blocks

The biosynthetic pathways of many carbohydrate residues attached to glycopeptide aglycones are often found in other antibiotics or may even come from primary metabolism (e.g., mannose and galactose). The carbohydrates vancosamine, *epi*-vancosamine, and *oxo*-vancosamine are characteristic representatives of the type I-glycopeptide antibiotics. Their biosynthetic genes are present in vancomycin, chloroeremomycin, and balhimycin producers and the structures at the same time represent the differentiating criterion for these glycopeptides (Fig. 18.1). Five genes (*orf14* and *orf23–26; evaABCDE*) were assigned to the biosynthesis of 4-L-*epi*-vancosamine in chloroeremomycin (van Wageningen *et al.*, 1998) and dehydro-vancosamine in balhimycin (Pelzer *et al.*, 1999), respectively. These functionally related genes are not colocalized in the gene clusters (Fig. 18.3). Investigation of 4-L-*epi*-vancosamine biosynthesis gave insights into the principal steps relevant to chloroeremomycin.

By heterologous expression of the *evaA-E* genes of the *cep* cluster in *E. coli*, Chen *et al.* (2000) were able to reconstitute the pathway from TDP-glucose to TDP-L-epivancosamine, which reasonably can be extrapolated to the biosyntheses of vancosamine and *oxo*-vancosamine. For the balhimycin producer *A. balhimycina* a naturally occurring deletion in the *evaE*-homologue gene results in biosynthetic arrest at the stage of *oxo*-vancosamine.

The biosynthetic origin of these vancosamine-like sugars is glucose, which is activated as uridine or thymidine diphospho-glucose. Ho *et al.* (2006) assumed that the nucleotidyltransferase responsible for the activation is Cep15. However, in-depth analysis of Cep15 revealed that it functions as an *N*-acetylglucosaminyl deacetylase and that the activated glucose is taken from primary metabolism (Truman *et al.*, 2006, 2007). Surprisingly, the gene clusters for vancomycin-like glycopeptides include this gene, although vancomycin carries no sugar derived from glucosamine. Inspection of the gene by deletion and site-directed mutagenesis revealed that it can be inactivated in the *bal* cluster with no change in phenotype. Comparison with the lipoglycopeptide counterparts led to the identification of one residue which is mutated in Cep15, thus abolishing deacetylase activity (Truman et al., 2008). By reversion of this mutation (N164H) its function could be reactivated.

3.3. Peptide assembly by NRPSs

The glycopeptide aglycone is composed of a heptapeptide backbone from which side chain cross-links of the aromatic amino acids are the characteristic structural features uniquely represented by this compound class (Fig. 18.1). DNA sequence analysis of the chloroeremomycin biosynthetic gene cluster (van Wageningen et al., 1998) as well as that of the balhimycin gene cluster (Pelzer et al., 1999) revealed four NRPS proteins (BpsA/B/C/D) with eight modules (M) for amino acid activation (A) and thiolation (T). Additionally, condensation (C) domains, epimerisation (E) domains, and one thioesterase (Te) domain complete the domain organization. Gene inactivation studies showed that BpsD is involved in the biosynthesis of β-hydroxytyrosine (see Section 3.1.1), thus acting as an independent NRPS module (Puk et al., 2004).

Whereas the peptide synthetase modules in the biosynthesis of the vancomycin-like glycopeptides are organized into three enzymes, BpsA (M1–3), BpsB (M4–6), and BpsC (M7), those involved in lipoglycopeptide synthesis are organized into four enzymes, such as TeiA (M1–2), TeiB (M3), TeiC (M4–6), and TeiD (M7) (Donadio et al., 2005; Li et al., 2004).

In silico analysis of the organization of the peptide synthetases based on the consensus sequences of the different domains (Konz and Marahiel, 1999) led to a domain prediction (Recktenwald et al., 2002) which is in accordance with the chemical structure of the linear peptide, with a few exceptions: For the first amino acid, neither an epimerization nor a methylation domain is present, and an additional condensation-like domain (X-domain) is found in Module 7 (Rausch et al., 2007; Recktenwald et al., 2002). Since only a few NRPSs activating Hpg or Dpg have been sequenced, the possibility to precisely predict the amino acid specificity of the A-domains (Challis et al., 2000; Stachelhaus et al., 1999) is limited; however, the basic chemical structure could be assigned (Rausch et al., 2005).

Confirmation that the in silico predictions are correct were obtained from over-expression of A-domains in S. lividans (Recktenwald et al., 2002), combined with ATP/PP$_i$ exchange assays.

Module 1: In type I-glycopeptide antibiotics, (R)-MeLeu is the most common amino acid at position AA1 of the aglycone. Module 1 has an A–T-domain structure devoid of an epimerisation (E) domain or an N-methylation (M) domain. Heterologous over-expression of the first A-domain of Cep1 (van Wageningen et al., 1998) in E. coli, followed by an activation assay, showed the preference of (S)-Leu over (R)-Leu and N-MeLeu in the (R)- or

(S)-configuration as a substrate (Trauger and Walsh, 2000). Thus, a racemase or an in *trans* working E-domain (ORF29/30^{Cep}) was postulated for its biosynthesis. The characterization of linear heptapeptide intermediates from oxygenase mutants (Süssmuth *et al.*, 1999) and heterologous over-expression of an *N*-methyltransferase (*mtfA, orf16, A. orientalis*) gene in *E. coli* (O'Brien *et al.*, 2000) confirmed that *N*-methylation is a late biosynthetic step (see Section 3.6.1).

Modules 2 and 6: These modules code for the assembly of Hty into the heptapeptide chain. Module 2 has a C–A–T–E-domain organization, in contrast to module 6, which shows only a C–A–T-domain organization. ATP/PP$_i$-assays for the A-domain of module 6 showed only a weak selectivity for Hty, possibly because of the use of racemic mixtures (Recktenwald *et al.*, 2002). The configurations of the Htys in the aglycone are ($2R,3R$)-^{2}Cht and ($2S,3R$)-^{6}Cht. Since only module 2 has an epimerization domain, it can be concluded that the configuration of Hty activated by the A-domain is $2S,3R$ for both modules.

Module 3: This module has a C–A–T-domain structure and, although not specifically shown, likely activates (S)-Asn.

Modules 4 and 5: The modules both display C–A–T–E domain structure and are responsible for the activation and condensation of Hpg. ATP/PP$_i$-exchange experiments with over-expressed A-domains of modules 4 and 5 showed activation of (S)-Hpg (Recktenwald *et al.*, 2002). Thus, the E-domains perform the epimerization at the stage of the tetra- and pentapeptide into the (R)-configurations present in the aglycone.

Module 7: The last module in the peptide assembly line has a C–A–T–X–Te-domain organization. ATP/PP$_i$-exchange experiments with over-expressed A-domain showed (S)-Dpg to be the activated amino acid (Recktenwald *et al.*, 2002). The X-domain shows some sequence similarity to C-domains (Rausch *et al.*, 2007). A mutant with a deletion of the X-domain resembles an *oxyA* mutant (see Section 3.4.1), which led to the speculation that this domain is required for interaction of the oxygenase with the NRPS (Stegmann, unpublished data). The Te-domain codes for the thioesterase responsible for the cleavage of the processed heptapeptide from the NRPS complex.

In all clusters, downstream from the peptide synthetase encoding M7, an *mbtH*-like gene has been identified. Similar genes have been found in almost all clusters involved in NRPS biosynthesis (Stegmann *et al.*, 2006b). A gene inactivation in *A. balhimycina* did not reveal any phenotype, suggesting that *mbtH* does not have a specific function in balhimycin biosynthesis. However, from analysis of *S. coelicolor*, which has two *mbtH*-like genes, it can be concluded that MbtH proteins from different clusters can substitute for each other (Lautru *et al.*, 2007; Wolpert *et al.*, 2007). Since *A. balhimycina* includes a second *mbtH*-like gene it cannot be excluded that a MbtH-specific function is required in glycopeptide biosynthesis (Stegmann *et al.*, 2006b).

3.4. Oxidative side chain cyclization by P450-type monooxygenases

3.4.1. Vancomycin-type glycopeptides

In glycopeptide biosynthesis, the three oxidative side chain cyclizations are central steps of glycopeptide biosynthesis. From sequence homology comparisons, it was assumed that three P450-type monooxygenases, OxyA/B/C, were involved in aglycone assembly (Bischoff et al., 2001b).

(RT)-PCR analysis revealed that the *oxyABC* genes in *A. balhimycina* are cotranscribed (Stegmann et al., 2006a). Heterologous complementation of *A. balhimycina oxy* mutants furthermore proved the functional equivalence of the oxygenases from the balhimycin and vancomycin producers (Stegmann et al., 2006a).

The breakthrough in understanding a considerable portion of these biosynthetic steps came from the generation of oxygenase gene inactivation mutants in *A. balhimycina*, followed by structure elucidation of peptide intermediates expressed in the culture medium (Bischoff et al., 2001b; Süssmuth et al., 1999).

The characterization of a linear hexa- and heptapeptide from an *A. balhimycina* mutant in which transcription of the oxygenase genes is interrupted verified the participation of the oxygenases in side chain cross-linking. In order to assign the three cross-link formations to the three *oxy* genes, in-frame deletion mutants of each gene were constructed (Stegmann et al., 2006a) (Fig. 18.4) and the corresponding balhimycin intermediates were isolated and structurally elucidated (Bischoff et al., 2001a,b).

The culture filtrates from the gene inactivation mutant *oxyA* revealed a monocyclic heptapeptide with the C-O-D ring closed (Bischoff et al., 2001b). As a consequence, one of the intact oxygenases, OxyB or OxyC, was responsible for C-O-D-ring formation. By subsequent analysis of inactivation mutants of all three oxygenase genes, *oxyA/B/C*, yielding linear, monocyclic, and bicyclic peptide intermediates, the order of the cyclization steps could finally be resolved (Bischoff et al., 2001a) (Fig. 18.4). The sequential aglycone assembly performed by the oxygenases was devised as (1) OxyB—C-O-D-ring (2) OxyA—D-O-E-ring, and (3) OxyC—AB-ring. These results showed that there exists a strict sequence in side chain cyclization, and that a defect of one oxygenase cannot be by-passed or compensated in any way by the other oxygenases.

In order to characterize the enzymatic conversion of heptapeptide substrates by the oxygenases, extensive biochemical and structural investigations were performed (Geib et al., 2008; Pylypenko et al., 2003; Woithe et al., 2007, 2008; Zerbe et al., 2002, 2004) using the vancomycin oxygenases. (For a detailed description see Chapter 19, in this volume.)

The fact that no substrate conversion was obtained with the above-mentioned over-expressed P450-monooxygenases using free peptide

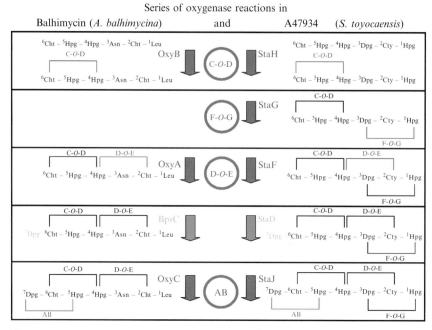

Figure 18.4 Order of the oxygenase reactions (in red) in vancomycin-like glycopeptides and in lipoglycopeptides. Before the final cross-link reaction takes place it is hypothesized that the peptide synthetase, BpsC or StaD, respectively, catalyzes the addition of the seventh amino acid (in green). (See Color Insert.)

substrates, as well as the detection of hexapeptides next to heptapeptides in comparable amounts in the culture filtrates of oxygenase mutants (Bischoff et al., 2001a,b; Süssmuth et al., 1999), raised doubts whether a cytoplasmic cyclization of free heptapeptide can occur. As a consequence, NRPS-bound heptapeptides, but also NRPS-bound hexapeptides, were considered as possible substrates of oxygenases OxyA/B/C. A cyclization prior to a hexapeptide stage could be excluded since the C-O-D-ring of ^{4}Hpg and ^{6}Cht represents the first ring to be closed.

In order to shed more light on the biosynthetic assembly, two *A. balhimycina* mutants, both inactivated in heptapeptide formation, were examined for their potential to produce glycopeptide biosynthesis intermediates. The *dpgA* deletion mutant cannot synthesize the nonproteogenic amino acid ^{7}Dpg (Pfeifer et al., 2001); the *bpsC* mutant has a deletion in the first condensation-domain of BpsC, thus disabling the condensation of a hexapeptide precursor with the amino acid ^{7}Dpg (Recktenwald et al., 2002). Therefore, in both mutants the biosynthesis stops at the stage of the hexapeptide. Culture filtrates of both mutants showed no antibiotic activity against an indicator strain of *Bacillus subtilis* and were analyzed by a subsequent screening with HPLC–ESI–MS. The peptides characterized from the *dpgA* mutant and the *bpsC* mutant both suggested an oxidative cyclization on the NRPS template during peptide formation (Bischoff *et al.*, 2005). These results set the stage for *in vitro* experiments by the Robinson group. Zerbe *et al.* (2004) analyzed OxyB-catalyzed reactions with different substrates and found that only the peptidyl-PCP derivative showed significant turnover rates with OxyB, whereas the SNAc-ester did not.

3.4.2. Side chain cyclization in lipoglycopeptides

The lipoglycopeptides contain four side chain cross-links (Fig. 18.1). It was tempting to speculate that the lipoglycopeptide-specific F-O-G cross-link between AA1 and AA3 is carried out by the oxygenase encoded by the oxygenase gene located near the homologues of *oxyABC*; which in the case of A47934 is *staG* (Hadatsch *et al.*, 2007; Pootoolal *et al.*, 2002). Following the construction of in-frame deletion mutants of the A47934 producer *S. toyocaensis* the structures of the accumulated intermediates were determined, and complementation studies with the respective vancomycin oxygenase genes as well as analyzes of A47934 intermediates present in the wild type were performed (Hadatsch *et al.*, 2007). Surprisingly, the additional ring closure in A47934, which is indeed catalyzed by StaG, is the second oxygenase reaction (Fig. 18.4) and, in contrast to the synthesis of vancomycin-type glycopeptides, oxygenase reactions in A47934 oxygenase mutants can be skipped to a certain degree.

3.5. Time point of halogenation and side chain cyclization in glycopeptide biosynthesis

Vancomycin is chlorinated at ^{2}Hty and ^{6}Hty, which represents a commonly found chlorination pattern of glycopeptide antibiotics. Antibiotic activity decreases in the absence of either one of the two chlorines by a factor of 2–10-fold, probably due to a decreased dimerization tendency of glycopeptide antibiotics (Gerhard *et al.*, 1993).

The halogenase BhaA ($bhaA^{bal}$, $orfX^{cep}$) shows homologies to flavin-dependent halogenases of other secondary metabolic gene clusters and thus the halogenating function in glycopeptide biosynthesis was assigned to BhaA. This was proved by inactivation of *bhaA* from *A. balhimycina*, which led to dechlorobalhimycin in the culture filtrates of this mutant (Puk *et al.*, 2002). However, supplementation of β-hydroxytyrosine biosynthesis mutants (*oxyD* and *bpsD*) with Hty fully restored antibiotic production and halogenation. In contrast, Cht was not accepted as a substrate of these mutants, indicating that Cht is not recognized by the A-domains of M2 and M6. A rejection of Cht by cellular uptake systems of these mutants was experimentally excluded. These results suggest a chlorination event at a time point later than Hty biosynthesis during peptide assembly on the NRPS, putatively between adenylation of Hty by the multimodular NRPSs BpsA and BpsB and coupling to the growing peptide chain. This assumption was further confirmed by analyzing a mutant in which BpsA was engineered by replacing the C-domain of M3 for the Te domain. The product of the mutant is a chlorinated dipeptide, proving that chlorination occurs at the NRPS and not after completion of peptide formation (Kittel et al., 2009).

The current picture on the time point of side chain cyclization, as well as on the time point of chlorination, implies a much more complex biosynthesis than previously assumed, with halogenation and side chain cyclization occurring on the NRPS complex (Fig. 18.5).

3.5.1. Methylation of glycopeptide antibiotics

The isolation of variously cyclized peptide intermediates from oxygenase mutants indicated that N-methylation is a late step in glycopeptide biosynthesis (Bischoff et al., 2001a; Süssmuth et al., 1999). Enzyme assays with over-expressed methyltransferase of chloroeremomycin-producer A. orientalis supported this hypothesis (O'Brien et al., 2000).

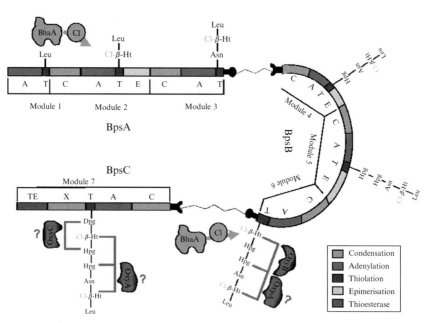

Figure 18.5 Composition of the multienzyme complex catalyzing peptide backbone formation and its simultaneously cross-linking by the oxygenases (OxyA, B, C) and its chlorination by BhaA. The supposed time points of cross-linking by OxyA and OxyC have to be definitively proven. (See Color Insert.)

3.5.2. Glycosylation of glycopeptide antibiotics

Whereas only a limited number of different amino acids occur in glycopeptides (see Sections 1 and 3.1) a broad structural diversity for glycosyl residues is found. Apart from more commonly found carbohydrates, vancosamine sugars are important representatives, particularly for type I-glycopeptide antibiotics.

The chloroeremomycin biosynthetic gene cluster shows three glycosyltransferases GtfA/B/C (*orf11/12/13*) downstream of the oxygenase genes (van Wageningen *et al.*, 1998). The same gene arrangement is found in the balhimycin biosynthetic gene cluster with *bgtfA/B/C* (Pelzer *et al.*, 1999), whereas in the vancomycin biosynthetic gene cluster only two glycosyltransferase genes (*gftD* and *gtfE*) are present.

Mutational analysis of the balhimycin *bgtf* genes revealed that glycosylation (similarly to oxidative cross-linking) occurs in a defined order (Stegmann *et al.*, 2005): First glucose is attached to AA4 by BgtfB, then *oxo*-vancosamine is transferred to AA5 by BgtfA, and finally BgtfC attaches *oxo*-vancosamine to glucose, a reaction which only happens with low efficiency.

Early investigations by the Baltz group were directed towards assignment of the glycosyltransferase functions and substrate conversions in order to generate novel hybrid glycopeptide antibiotics (Solenberg *et al.*, 1997). In recent years, glycosyltransferases from different glycopeptide pathways have been structurally characterized in order to understand the transferase reaction on a molecular level (for a review, see Sattely *et al.*, 2008), opening the possibility for the generation of glycopeptides with altered glycosylation patterns.

4. REGULATION, SELF-RESISTANCE, AND EXCRETION

Almost all antibiotic biosynthetic gene clusters include gene(s) the product(s) of which control transcriptional regulation of the structural genes. (See Chapter 4, in this volume.) Sequence analysis of all glycopeptide clusters revealed a gene for a StrR-like regulator that was first described in the streptomycin gene cluster (Retzlaff and Distler, 1995). Inactivation of the respective gene (*bbr*) in the balhimycin cluster resulted in a nonproducing mutant, which is in accordance with its presumed function as a transcriptional activator (Stegmann, unpublished data). Following overexpression as a His_6-Tag fusion protein in *E. coli*, five promoter regions were identified in gel retardation experiments to which His_6-Bbr binds (Shawky *et al.*, 2007). Together with the results of RT-PCR experiments it was possible to define the complete operon structure. In contrast, the Bbr ortholog Dbv4, which is a phosphate-controlled regulator of A40926 biosynthesis, was shown to have two binding sites in the corresponding cluster (Alduina *et al.*, 2007).

Further putative regulator genes are represented by two genes whose products resemble the two-component system VanRS, known to be essential for activation of the vancomycin resistance genes *vanHAX* in the resistance transposon Tn*1546* (Arthur et al., 1993) and in *S. coelicolor* (Hong et al., 2004). Since a *vanR* deletion mutant of *A. balhimycina* is not impaired in balhimycin production and is not affected in resistance, the role of these genes is unclear (Schäberle et al., 2009).

In contrast to almost all known antibiotic gene clusters, the clusters for vancomycin-like glycopeptides do not include an obvious resistance gene. Whereas the *sta* und *tcp* cluster harbor homologues of the *vanHAX* genes, such genes are not present in the other clusters. Screening of an *A. balhimycina* cosmid library revealed *vanHAX*-like genes at other positions in the chromosome. The functionality of these genes was demonstrated by heterologous expression in the vancomycin-sensitive *S. coelicolor* Müller.

A possible candidate for resistance is a *vanY* homologue, which was identified only in the *bal* and *dbv* clusters. Over-expression and subsequent enzyme assays documented its function as a DD-carboxypeptidase, which agrees with the function of its counterpart in enterococcal resistance transposons. However, the gene itself is unable to confer resistance; it only increases the resistance level in the presence of transcribed *vanHAX* genes, as demonstrated by expression of *vanY* in *S. coelicolor* and in a *S. coelicolor vanRS* mutant (Schäberle et al., 2009).

A common antibiotic resistance mechanism is an active export out of the bacterial cell. There are two genes in all clusters which encode putative transporters. For one of them (*orf2* in *bal*) a function as a membrane ion antiporter (Na^+/H^+ transporter) was deduced. But again, gene inactivation ruled out the involvement of this gene in resistance, since the mutant does not exhibit any obvious phenotype (Schäberle et al., 2009). The second transporter (Tba in *bal*) belonging to the ABC transporter family plays a role in balhimycin excretion, as demonstrated by mutational analysis and by measuring the intra- and extracellular balhimycin concentration in the wild type and a *tba* mutant. However, the resistance level is not affected, indicating that *tba* is not required for self-resistance (Menges et al., 2007).

5. Linking Primary and Secondary Metabolism

Four or five amino acids of the heptapeptide are directly derived from the shikimate pathway for vancomycin or teicoplanin, respectively. It is therefore tempting to speculate that during secondary metabolite synthesis the precursor supply of Hty, Tyr, and Hpg may be a limiting factor. Inspection of the gene clusters revealed one or two genes for enzymes of the shikimate branch. One gene, *pdh* encoding a prephenate dehydrogenase, is present in all clusters, while the vancomycin-type clusters contain additionally a 3-desoxyarabino-heptulosonate-7-P synthase gene (*dahp*). These genes are apparently specific for glycopeptide production since their expression is under the transcriptional control of Bbr and a second copy of these genes, most likely responsible for the enzymes of primary metabolism, is present outside of the *bal* cluster.

In order to analyze their roles in balhimycin production the genes were over-expressed and antibiotic production was determined after growth on solid medium and cultivation in a fermenter. Whereas *dahp* over-expression resulted in a significant over-production (twofold improvement), *pdh* over-expression did not lead to a measurable improvement in yield (Thykaer *et al.*, 2009). This may indicate that Dahp is indeed required to optimize the precursor supply and a higher concentration of this enzyme can channel the flux in the direction of balhimycin synthesis.

6. APPROACHES FOR THE GENERATION OF NEW GLYCOPEPTIDES

The molecular genetic and biochemical analyses of glycopeptide synthesis provide the basis for the engineering of the biosynthesis, in several ways.

1. By cultivating the balhimycin producers in media without chloride salts it could be shown that the halogenase BhaA also accepts bromine at both positions of the molecule. This offers the opportunity to direct the synthesis to balhimycin with either chlorine or bromine or chlorine and bromine by varying the composition of the culture medium (Bister *et al.*, 2003).
2. The glycosyltransferases were used to generate glycopeptides with novel glycosylation patterns *in vivo* and *in vitro* (Baltz, 2002; Solenberg *et al.*, 1997). In order to create a broad spectrum of differently glycosylated glycopeptides the supply of activated sugars is the bottleneck. By using a galactokinase from *E. coli* (Yang *et al.*, 2005) it was possible to obtain nucleotide-activated sugars which subsequently were used in an *in vitro* approach to glycosylate the vancomycin aglycone. This "glycorandomization" approach delivered novel mono- and di-glycosylated derivatives (Fu *et al.*, 2003).
3. The mutants defective in the biosynthesis of precursors are null-mutants and can be complemented by feeding the respective precursors (see Section 3.1). This was shown in the case of Dpg synthesis for the *dpgA* mutant (Pfeifer *et al.*, 2001) and in the case of Hty for the *bhp*, *oxyD*, and *bpsD* mutants (Puk *et al.*, 2004). Such mutants are ideal recipients for

precursor-directed biosynthesis. By feeding derivatives of the natural building block, glycopeptides were generated in which either Hty (Weist *et al.*, 2002) or Dpg (Weist *et al.*, 2004) was replaced. In this way, a rather broad substrate specificity of the biosynthetic enzymes has been demonstrated, leading to a large number of new compounds, many of them showing antibiotic activity.

4. Knowledge of the organization of the NRPS enzymes allows the targeted exchange ("module swapping") of modules to generate new peptide backbones (Schneider *et al.*, 1998). Alternatively, the selectivity-conferring code for NRPSs can be used to change the specificity of A-domains by site-directed mutagenesis (Eppelmann *et al.*, 2002). For glycopeptides, such experiments have not been reported yet. However, the insertion of an additional Hpg-specific module between M4 and M5 by manipulating *bpsB* has been successful (Butz *et al.*, 2008) and led to the biosynthesis of linear and monocyclic octapeptides.

5. The high conservation of glycopeptide biosynthetic enzymes offers the opportunity to isolate new glycopeptide producers by screening gene libraries with DNA probes. Using halogenase-specific probes, new glycopeptide producer were identified and shown to synthesize novel glycopeptides (Hornung *et al.*, 2007). The identified halogenase genes can also be used to generate new glycopeptide derivatives by complementation of halogenase mutants such as the *bhaA* mutant of *A. balhimycina*.

ACKNOWLEDGMENTS

Research on glycopeptides in the authors' laboratories was supported by the EU (MEGA-TOP (to W.W.) and COMBIGTOP (to W.W. and R.S.)) and the DFG (SFB 766 to W. W.). We are grateful to all members of the COMBIGTOP consortium for sharing results prior to publication and for fruitful discussions.

REFERENCES

Alduina, R., De Grazia, S., Dolce, L., Salerno, P., Sosio, M., Donadio, S., and Puglia, A. M. (2003). Artificial chromosome libraries of *Streptomyces coelicolor* A3(2) and *Planobispora rosea*. *FEMS Microbiol. Lett.* **218**, 181–186.

Alduina, R., Lo Piccolo, L., D'Alia, D., Ferraro, C., Gunnarsson, N., Donadio, S., and Puglia, A. M. (2007). Phosphate-controlled regulator for the biosynthesis of the dalbavancin precursor A40926. *J. Bacteriol.* **189**, 8120–8129.

Arias, C. A., and Murray, B. E. (2008). Emergence and management of drug-resistant enterococcal infections. *Expert Rev. Anti Infect. Ther.* **6**, 637–655.

Arthur, M., Molinas, C., Depardieu, F., and Courvalin, P. (1993). Characterisation of Tn*1546*, a Tn*3*-related transposon conferring glycopeptide resistance by synthesis of depsipeptide peptidoglycan precursors in *Enterococcus faecium* BM4147. *J. Bacteriol.* **175**, 117–127.

Baltz, R. H. (2002). Combinatorial glycosylation of glycopeptide antibiotics. *Chem. Biol.* **9,** 1268–1270.

Bibb, M. J., Janssen, G. R., and Ward, J. M. (1985). Cloning and analysis of the promoter region of the erythromycin resistance gene (*ermE*) of *Streptomyces erythreus*. *Gene* **41,** 357–368.

Bierman, M., Logan, R., O'Brien, K., Seno, E. T., Rao, R. N., and Schoner, B. E. (1992). Plasmid cloning vectors for the conjugal transfer of DNA from *Escherichia coli* to *Streptomyces spp*. *Gene* **116,** 43–49.

Bischoff, D., Bister, B., Bertazzo, M., Pfeifer, V., Stegmann, E., Nicholson, G. J., Keller, S., Pelzer, S., Wohlleben, W., and Süssmuth, R. D. (2005). The biosynthesis of vancomycin-type glycopeptide antibiotics—A model for oxidative side-chain cross-linking by oxygenases coupled to the action of peptide synthetases. *ChemBioChem* **6,** 2267–2272.

Bischoff, D., Pelzer, S., Bister, B., Nicholson, G. J., Stockert, S., Schirle, M., Wohlleben, W., Jung, G., and Süssmuth, R. D. (2001a). The biosynthesis of vancomycin-type glycopeptide antibiotics—The order of cyclisation steps. *Angew. Chem. Int. Ed.* **40,** 4688–4691.

Bischoff, D., Pelzer, S., Höltzel, A., Nicholson, G., Stockert, S., Wohlleben, W., Jung, G., and Süssmuth, R. (2001b). The biosynthesis of vancomycin-type glycopeptide antibiotics—New insights into the cyclisation steps. *Angew. Chem. Int. Ed.* **40,** 1693–1696.

Bister, B., Bischoff, D., Nicholson, G. J., Stockert, S., Wink, J., Brunati, C., Donadio, S., Pelzer, S., Wohlleben, W., and Süssmuth, R. D. (2003). Bromobalhimycin and chlorobromobalhimycins—Illuminating the potential of halogenases in glycopeptide antibiotic biosyntheses. *ChemBioChem* **4,** 658–662.

Butz, D., Schmiederer, T., Hadatsch, B., Wohlleben, W., Weber, T., and Süssmuth, R. D. (2008). Module extension of a non-ribosomal peptide synthetase of the glycopeptide antibiotic balhimycin produced by *Amycolatopsis balhimycina*. *ChemBioChem* **9,** 1195–1200.

Challis, G. L., Ravel, J., and Townsend, C. A. (2000). Predictive, structure-based model of amino acid recognition by nonribosomal peptide synthetase adenylation domains. *Chem. Biol.* **7,** 211–224.

Chen, H., Hubbard, B. K., O'Connor, S., and Walsh, C. (2002). Formation of β-hydroxy histidine in the biosynthesis of nikkomycin antibiotics. *Chem. Biol.* **9,** 103–112.

Chen, H., Thomas, M. G., Hubbard, B. K., Losey, H. C., Walsh, C. T., and Burkart, M. D. (2000). Deoxysugars in glycopeptide antibiotics: Enzymatic synthesis of TDP-L-epivancosamine in chloroeremomycin biosynthesis. *Proc. Natl. Acad. Sci. USA* **97,** 11942–11947.

Chen, H., Tseng, C. C., Hubbard, B. K., and Walsh, C. T. (2001). Glycopeptide antibiotic biosynthesis: Enzymatic assembly of the dedicated amino acid monomer (S)-3,5-dihydroxyphenylglycine. *Proc. Natl. Acad. Sci. USA* **98,** 14901–14906.

Chen, H., and Walsh, C. T. (2001). Coumarin formation in novobiocin biosynthesis: β-Hydroxylation of the aminoacyl enzyme tyrosyl-S-NovH by a cytochrome P450 NovI. *Chem. Biol.* **8,** 301–312.

Chiu, H. T., Hubbard, B. K., Shah, A. N., Eide, J., Fredenburg, R. A., Walsh, C. T., and Khosla, C. (2001). Molecular cloning and sequence analysis of the complestatin biosynthetic gene cluster. *Proc. Natl. Acad. Sci. USA* **98,** 8548–8553.

Choroba, O. W., Williams, D. H., and Spencer, J. B. (2000). Biosynthesis of the vancomycin group of antibiotics: Involvement of an unusual dioxygenase in the pathway to (S)-4-hydroxyphenylglycine. *J. Am. Chem. Soc.* **122,** 5389–5390.

Donadio, S., Sosio, M., Stegmann, E., Weber, T., and Wohlleben, W. (2005). Comparative analysis and insights into the evolution of gene clusters for glycopeptide biosynthesis. *Mol. Gen. Genet.* **274,** 40–50.

Eppelmann, K., Stachelhaus, T., and Marahiel, M. A. (2002). Exploitation of the selectivity-conferring code of nonribosomal peptide synthetases for the rational design of novel peptide antibiotics. *Biochemistry* **41**, 9718–9726.

Fischer, J. (1996). Entwicklung eines regulierbaren Expressionssystems zur effizienten Synthese rekombinanter Proteine in *Streptomyces lividans*. PhD thesis, University of Stuttgart, Germany.

Flett, F., Mersinias, V., and Smith, C. P. (1997). High efficiency intergeneric conjugal transfer of plasmid DNA from *Escherichia coli* to methyl DNA-restricting streptomycetes. *FEMS Microbiol. Lett.* **155**, 223–229.

Fu, X., Albermann, C., Jiang, J., Liao, J., Zhang, C., and Thorson, J. S. (2003). Antibiotic optimization via *in vitro* glycorandomization. *Nat. Biotechnol.* **21**, 1467–1469.

Geib, N., Woithe, K., Zerbe, K., Li, D. B., and Robinson, J. A. (2008). New insights into the first oxidative phenol coupling reaction during vancomycin biosynthesis. *Bioorg. Med. Chem. Lett.* **18**, 3081–3084.

Gerhard, U., Mackay, J. P., Maplestone, R. A., and Williams, D. H. (1993). The role of the sugar and chlorine substituents in the dimerization of vancomycin antibiotics. *J. Am. Chem. Soc.* **115**, 232–237.

Gust, B., Chandra, G., Jakimowicz, D., Yuqing, T., Bruton, C. J., and Chater, K. F. (2004). Lambda red-mediated genetic manipulation of antibiotic-producing *Streptomyces*. *Adv. Appl. Microbiol.* **54**, 107–128.

Hadatsch, B., Butz, D., Schmiederer, T., Steudle, J., Wohlleben, W., Süssmuth, R. D., and Stegmann, E. (2007). The biosynthesis of teicoplanin-type glycopeptide antibiotics: Assignment of p450 mono-oxygenases to side chain cyclisations of glycopeptide A47934. *Chem. Biol.* **14**, 1078–1089.

Hammond, S. J., Williams, D. H., and Nielsen, R. C. (1983). The biosynthesis of ristocetin. *J. Chem. Soc. Chem. Commun.* 116–117.

Hammond, S. J., Williamson, M. P., Williams, D. H., Boeck, L. D., and Marconi, G. G. (1982). On the biosynthesis of the antibiotic vancomycin. *J. Chem. Soc. Chem. Commun.* 344–346.

Hillemann, D., Pühler, A., and Wohlleben, W. (1991). Gene disruption and gene replacement in *Streptomyces* via single stranded DNA transformation of integration vectors. *Nucleic Acids Res.* **19**, 727–731.

Ho, J. Y., Huang, Y. T., Wu, C. J., Li, Y. S., Tsai, M. D., and Li, T. L. (2006). Glycopeptide biosynthesis: Dbv21/Orf2 from dbv/tcp gene clusters are N-Ac-Glm teicoplanin pseudoaglycone deacetylases and Orf15 from cep gene cluster is a Glc-1-P thymidyltransferase. *J. Am. Chem. Soc.* **128**, 13694–13695.

Hong, H. J., Hutchings, M. I., Neu, J. M., Wright, G. D., Paget, M. S., and Buttner, M. J. (2004). Characterisation of an inducible vancomycin resistance system in *Streptomyces coelicolor* reveals a novel gene (*vanK*) required for drug resistance. *Mol. Microbiol.* **52**, 1107–1121.

Hornung, A., Bertazzo, M., Dziarnowski, A., Schneider, K., Welzel, K., Wohlert, S. E., Holzenkämpfer, M., Nicholson, G. J., Bechthold, A., Süssmuth, R. D., Vente, A., and Pelzer, S. (2007). A genomic screening approach to the structure-guided identification of drug candidates from natural sources. *ChemBioChem* **8**, 757–766.

Hubbard, B. K., Thomas, M. G., and Walsh, C. T. (2000). Biosynthesis of L-p-hydroxyphenylglycine, a non-proteinogenic amino acid constituent of peptide antibiotics. *Chem. Biol.* **7**, 931–942.

Kieser, T., Bibb, M. J., Buttner, M. J., Chater, K. F., and Hopwood, D. A. (2000). Practical *Streptomyces* Genetics. John Innes Foundation, Norwich, UK.

Kittel, C., Butz, D., Süssmuth, R. D., Wohlleben, W., and Stegmann, E. (2009). *In vivo* analysis of the balhimycin biosynthesis: The FADH2-dependent halogenase BhaA chlorinates β-hydroxytyrosine covalently bound to the nonribosomal peptide synthetase BpsA (Submitted for publication).

Konz, D., and Marahiel, M. A. (1999). How do peptide synthetases generate structural diversity? *Chem. Biol.* **6,** R39–R48.

Lamb, S. S., Patel, T., Koteva, K. P., and Wright, G. D. (2006). Biosynthesis of sulfated glycopeptide antibiotics by using the sulfotransferase StaL. *Chem. Biol.* **13,** 171–181.

Lautru, S., Oves-Costales, D., Pernodet, J. L., and Challis, G. L. (2007). MbtH-like protein-mediated cross-talk between non-ribosomal peptide antibiotic and siderophore biosynthetic pathways in *Streptomyces coelicolor* M145. *Microbiology* **153,** 1405–1412.

Li, T.-L., Choroba, O. W., Charles, E. H., Sandercock, A. M., Williams, D. H., and Spencer, J. B. (2001). Characterisation of a hydroxymandelate oxidase involved in the biosynthesis of two unusual amino acids occurring in the vancomycin group of antibiotics. *J. Chem. Soc. Chem. Commun.* 1752–1753.

Li, T.-L., Huang, F., Haydock, S. F., Mironenko, T., Leadlay, P. F., and Spencer, J. B. (2004). Biosynthetic gene cluster of the glycopeptide antibiotic teicoplanin. Characterisation of two glycosyltransferases and the key acyltransferase. *Chem. Biol.* **11,** 107–119.

Madoń, J., and Hütter, R. (1991). Transformation system for *Amycolatopsis* (Nocardia) *mediterranei*: Direct transformation of mycelium with plasmid DNA. *J. Bacteriol.* **173,** 6325–6331.

Matsushima, P., and Baltz, R. H. (1996). A gene cloning system for '*Streptomyces toyocaensis*'. *Microbiology* **142,** 261–267.

Mazza, P., Noens, E. E., Schirner, K., Grantcharova, N., Mommaas, A. M., Koerten, H. K., Muth, G., Flärdh, K., van Wezel, G. P., and Wohlleben, W. (2006). MreB of *Streptomyces coelicolor* is not essential for vegetative growth but is required for the integrity of aerial hyphae and spores. *Mol. Microbiol.* **60,** 838–852.

Menges, R., Muth, G., Wohlleben, W., and Stegmann, E. (2007). The ABC transporter Tba of *Amycolatopsis balhimycina* is required for efficient export of the glycopeptide antibiotic balhimycin. *Appl. Microbiol. Biotechnol.* **77,** 125–134.

Mulyani, S., Wohlleben, W., Süssmuth, R. D., and van Peé, K.-H. (2009). Identification of Bhp as a thioesterase in the formation of β-hydroxytyrosine during balhimycin biosynthesis in *Amycolatopsis balhimycina* (Submitted for publication).

Murakami, T., Holt, T. G., and Thompson, C. J. (1989). Thiostrepton-induced gene expression in *Streptomyces lividans*. *J. Bacteriol.* **171,** 1459–1466.

Muth, G., Brolle, D. F., and Wohlleben, W. (1999). Genetics of *Streptomyces*. In "Manual of Industrial Microbiology and Biotechnology" (A. L. Demain and J. E. Davies, eds.), pp. 353–367. ASM Press, Washington, DC.

Muth, G., Nussbaumer, B., Wohlleben, W., and Pühler, A. (1989). A vector system with temperature-sensitive replication for gene disruption and mutational cloning in Streptomycetes. *Mol. Gen. Genet.* **219,** 341–348.

Muth, G., Wohlleben, W., and Pühler, A. (1988). The minimal replicon of the *Streptomyces ghanaensis* plasmid pSG5 identified by subcloning and Tn5 mutagenesis. *Mol. Gen. Genet.* **211,** 424–429.

Nadkarni, S. R., Patel, M. V., Chatterjee, S., Vijayakumar, E. K., Desikan, K. R., Blumbach, J., and Ganguli, B. N. (1994). Balhimycin, a new glycopeptide antibiotic produced by *Amycolatopsis* sp. Y-86,21022. *J. Antibiot.* **47,** 334–341.

O'Brien, D. P., Kirkpatrick, P. N., O'Brien, S. W., Staroske, T., and Richardson, T. I. (2000). Expression and assay of an *N*-methyltransferase involved in the biosynthesis of a vancomycin group antibiotic. *Chem. Commun.* **2000,** 103–104.

Pelzer, S., Reichert, W., Huppert, M., Heckmann, D., and Wohlleben, W. (1997). Cloning and analysis of a peptide synthetase gene of the balhimycin producer *Amycolatopsis mediterranei* DSM5908 and development of a gene disruption/replacement system. *J. Biotechnol.* **56,** 115–128.

Pelzer, S., Süssmuth, R. D., Heckmann, D., Recktenwald, J., Huber, P., Jung, G., and Wohlleben, W. (1999). Identification and analysis of the balhimycin biosynthetic gene cluster and its use for manipulating glycopeptide biosynthesis in the producing organism *Amycolatopsis mediterranei* DSM5908. *Antimicrob. Agents Chemother.* **43,** 1565–1573.

Pfeifer, V., Nicholson, G. J., Ries, J., Recktenwald, J., Schefer, A. B., Shawky, R. M., Schröder, J., Wohlleben, W., and Pelzer, S. (2001). A polyketide synthase in glycopeptide biosynthesis: The biosynthesis of the non-proteinogenic amino acid (S)-3,5-dihydroxyphenylglycine. *J. Biol. Chem.* **276,** 38370–38377.

Pootoolal, J., Thomas, M. G., Marshall, C. G., Neu, J. M., Hubbard, B. K., Walsh, C. T., and Wright, G. D. (2002). Assembling the glycopeptide antibiotic scaffold: The biosynthesis of A47934 from *Streptomyces toyocaensis* NRRL15009. *Proc. Natl. Acad. Sci. USA* **99,** 8962–8967.

Puk, O., Bischoff, D., Kittel, C., Pelzer, S., Weist, S., Stegmann, E., Süssmuth, R., and Wohlleben, W. (2004). Biosynthesis of chloro-β-hydroxytyrosine, a non-proteinogenic amino acid of the peptidic backbone of vancomycin-type glycopeptide antibiotics. *J. Bacteriol.* **186,** 6093–6100.

Puk, O., Huber, P., Bischoff, D., Recktenwald, J., Jung, G., Süssmuth, R. D., Van Pee, K.-H., Wohlleben, W., and Pelzer, S. (2002). Glycopeptide biosynthesis in *Amycolatopsis mediterranei*: Function of a halogenase and a haloperoxidase/perhydrolase. *Chem. Biol.* **9,** 225–235.

Pylypenko, O., Vitali, F., Zerbe, K., Robinson, J. A., and Schlichting, I. (2003). Crystal structure of OxyC, a cytochrome P450 implicated in an oxidative C–C coupling reaction during vancomycin biosynthesis. *J. Biol. Chem.* **278,** 46727–46733.

Rausch, C., Hoof, I., Weber, T., Wohlleben, W., and Huson, D. H. (2007). Phylogenetic analysis of condensation domains in NRPS sheds light on their functional evolution. *BMC Evol. Biol.* **7,** 78.

Rausch, C., Weber, T., Kohlbacher, O., Wohlleben, W., and Huson, D. H. (2005). Specificity prediction of adenylation domains in nonribosomal peptide synthetases (NRPS) using transductive support vector machines (TSVMs). *Nucleic Acids Res.* **33,** 5799–5808.

Recktenwald, J., Shawky, R., Puk, O., Pfennig, F., Keller, U., Wohlleben, W., and Pelzer, S. (2002). Nonribosomal biosynthesis of vancomycin-type antibiotics: A heptapeptide backbone and eight peptide synthetase modules. *Microbiology* **148,** 1105–1118.

Retzlaff, L., and Distler, J. (1995). The regulator of streptomycin gene expression, StrR, of *Streptomyces griseus* is a DNA binding activator protein with multiple recognition sites. *Mol. Microbiol.* **18,** 151–162.

Reuther, J., Gekeler, C., Tiffert, Y., Wohlleben, W., and Muth, G. (2006a). Unique conjugation mechanism in mycelial streptomycetes: A DNA-binding ATPase translocates unprocessed plasmid DNA at the hyphal tip. *Mol. Microbiol.* **61,** 436–446.

Reuther, J., Wohlleben, W., and Muth, G. (2006b). Modular architecture of the conjugative plasmid pSVH1 from *Streptomyces venezuelae*. *Plasmid* **55**(3), 201–209.

Rodríguez-García, A., Combes, P., Pérez-Redondo, R., Smith, M. C., and Smith, M. C. (2005). Natural and synthetic tetracycline-inducible promoters for use in the antibiotic-producing bacteria *Streptomyces*. *Nucleic Acids Res.* **33,** e87.

Sattely, E. S., Fischbach, M. A., and Walsh, C. T. (2008). Total biosynthesis: *In vitro* reconstitution of polyketide and nonribosomal peptide pathways. *Nat. Prod. Rep.* **25,** 757–793.

Schäberle, T., Vollmer, W., Frasch, H.-J., Hüttl, S., Röttgen, M., von Thaler, A.-K., Wohlleben, W., and Stegmann, E. (2009). Self-resistance and cell wall composition in the glycopeptide producer *Amycolatopsis balhimycina*. (Submitted for publication).

Schneider, A., Stachelhaus, T., and Marahiel, M. A. (1998). Targeted alteration of the substrate specificity of peptide synthetases by rational module swapping. *Mol. Gen. Genet.* **257,** 308–318.

Shawky, R. M., Puk, O., Wietzorrek, A., Pelzer, S., Takano, E., Wohlleben, W., and Stegmann, E. (2007). The border sequence of the balhimycin biosynthesis gene cluster from *Amycolatopsis balhimycina* contains *bbr*, encoding a StrR-like pathway-specific regulator. *J. Mol. Microbiol. Biotechnol.* **13,** 76–88.

Solenberg, P. J., Matsushima, P., Stack, D. R., Wilkie, S. C., Thompson, R. C., and Baltz, R. H. (1997). Production of hybrid glycopeptide antibiotics *in vitro* and in *Streptomyces toyocaensis*. *Chem. Biol.* **4,** 195–202.

Sosio, M., Bossi, E., and Donadio, S. (2001). Assembly of large genomic segments in artificial chromosomes by homologous recombination in *Escherichia coli*. *Nucleic Acids Res.* **29,** E37.

Sosio, M., Giusino, F., Cappellano, C., Bossi, E., Puglia, A. M., and Donadio, S. (2000). Artificial chromosomes for antibiotic-producing actinomycetes. *Nat. Biotechnol.* **18,** 343–345.

Sosio, M., Kloosterman, H., Bianchi, A., De Vreugd, P., Dijkhuizen, L., and Donadio, S. (2004). Organisation of the teicoplanin gene cluster in *Actinoplanes teichomyceticus*. *Microbiology* **150,** 95–102.

Sosio, M., Stinchi, S., Beltrametti, F., Lazzarini, A., and Donadio, S. (2003). The gene cluster for the biosynthesis of the glycopeptide antibiotic A40926 by *Nonomuraea* species. *Chem. Biol.* **10,** 541–549.

Stachelhaus, T., Mootz, H. D., and Marahiel, M. (1999). The specificity-conferring code of adenylation domains in nonribosomal peptide synthetases. *Chem. Biol.* **7,** 211–224.

Stegmann, E., Bischoff, D., Kittel, C., Pelzer, S., Puk, O., Recktenwald, J., Weist, S., Süssmuth, R., and Wohlleben, W. (2005). Precursor-directed biosynthesis for the generation of novel glycopetides. *Ernst Schering Res. Found. Workshop* **51,** 215–232.

Stegmann, E., Pelzer, S., Bischoff, D., Puk, O., Stockert, S., Butz, D., Zerbe, K., Robinson, J., Süssmuth, R. D., and Wohlleben, W. (2006a). Genetic analysis of the balhimycin (vancomycin-type) oxygenase genes. *J. Biotechnol.* **124,** 640–653.

Stegmann, E., Rausch, C., Stockert, S., Burkert, D., and Wohlleben, W. (2006b). The small MbtH-like protein encoded by an internal gene of the balhimycin biosynthetic gene cluster is not required for glycopeptide production. *FEMS Microbiol. Lett.* **262,** 85–92.

Stinchi, S., Azimonti, S., Donadio, S., and Sosio, M. (2003). A gene transfer system for the glycopeptide producer *Nonomuraea* sp. ATCC39727. *FEMS Microbiol. Lett.* **225,** 53–57.

Stinchi, S., Carrano, L., Lazzarini, A., Feroggio, M., Grigoletto, A., Sosio, M., and Donadio, S. (2006). A derivative of the glycopeptide A40926 produced by inactivation of the beta-hydroxylase gene in *Nonomuraea sp*. ATCC39727. *FEMS Microbiol. Lett.* **256,** 229–235.

Süssmuth, R., Pelzer, S., Nicholson, G., Walk, T., Wohlleben, W., and Jung, G. (1999). New advances in the biosynthesis of glycopeptide antibiotics of the vancomycin type from *Amycolatopsis mediterranei*. *Angew. Chem. Int. Ed.* **38,** 1976–1979.

Thykaer, J., Nielsen, J., Wohlleben, W., Lantz, A. E., and Stegmann, E. (2009). Bridging primary and secondary metabolism by increasing the precursor supply results in improved glycopeptide production. (Submitted for publication).

Trauger, J. W., and Walsh, C. T. (2000). Heterologous expression in *Escherichia coli* of the first module of the nonribosomal peptide synthetase for chloroeremomycin, a vancomycin-type glycopeptide antibiotic. *Proc. Natl. Acad. Sci. USA* **97,** 3112–3117.

Truman, A. W., Fan, Q., Röttgen, M., Stegmann, E., Leadlay, P. F., and Spencer, J. B. (2008). The role of cep15 in the biosynthesis of chloroeremomycin: Reactivation of an ancestral catalytic function. *Chem. Biol.* **15,** 476–484.

Truman, A. W., Huang, F., Llewellyn, N. M., and Spencer, J. B. (2007). Characterisation of the enzyme BtrD from *Bacillus circulans* and revision of its functional assignment in the biosynthesis of butirosin. *Angew. Chem. Int. Ed.* **46,** 1462–1464.

Truman, A. W., Robinson, L., and Spencer, J. B. (2006). Identification of a deacetylase involved in the maturation of teicoplanin. *ChemBioChem* **7,** 1670–1675.

Tseng, C. C., McLoughlin, S. M., Kelleher, N. L., and Walsh, C. T. (2004). Role of the active site cysteine of DpgA, a bacterial type III polyketide synthase. *Biochemistry* **43**, 970–980.

Van Wageningen, A. M., Kirkpatrick, P. N., Williams, D. H., Harris, B. R., Kershaw, J. K., Lennard, N. J., Jones, M., Jones, S. J., and Solenberg, P. J. (1998). Sequencing and analysis of genes involved in the biosynthesis of a vancomycin group antibiotic. *Chem. Biol.* **5**, 155–162.

Weber, T., Rausch, C., Lopez, P., Hoof, I., Gaykova, V., Huson, D., and Wohlleben, W. (2009). CLUSEAN: A computer-based framework for the automated analysis of bacterial secondary metabolite biosynthetic gene clusters. *J. Biotechnol.* **140**, 13–17.

Weist, S., Bister, B., Puk, O., Bischoff, D., Nicholson, G., Stockert, S., Wohlleben, W., Jung, G., and Süssmuth, R. D. (2002). Fluorobalhimycin—A new chapter in glycopeptide antibiotic research. *Angew. Chem. Int. Ed.* **41**, 3383–3385.

Weist, S., Kittel, C., Bischoff, D., Bister, B., Pfeifer, V., Nicholson, G. J., Wohlleben, W., and Süssmuth, R. D. (2004). Mutasynthesis of glycopeptide antibiotics: Variations of vancomycin's AB-ring amino acid 3,5-dihydroxyphenylglycine. *J. Am. Chem. Soc.* **126**, 5942–5943.

Williams, D. H., and Bardsley, B. (1999). The vancomycin group of antibiotics and the fight against resistant bacteria. *Angew. Chem. Int. Ed.* **38**, 1172–1193.

Wink, J. M., Kroppenstedt, R. M., Ganguli, B. N., Nadkarni, S. R., Schumann, P., Seibert, G., and Stackebrandt, E. (2003). Three new antibiotic producing species of the genus Amycolatopsis, *Amycolatopsis balhimycina* sp. nov., *A. tolypomycina* sp. nov., *A. vancoresmycina* sp. nov., and description of *Amycolatopsis keratiniphila* subsp. *keratiniphila* subsp. nov. and *A. keratiniphila* subsp. *nogabecina* subsp. nov. *Syst. Appl. Microbiol.* **26**, 38–46.

Wohlleben, W., and Muth, G. (1993). *Streptomyces* plasmid vectors. *In* "Plasmids a Practical Approach" (K. G. Hardy, ed.), pp. 147–175. Oxford University Press, Oxford.

Woithe, K., Geib, N., Meyer, O., Wörtz, T., Zerbe, K., and Robinson, J. A. (2008). Exploring the substrate specificity of OxyB, a phenol coupling P450 enzyme involved in vancomycin biosynthesis. *Bioorg. Med. Chem. Lett.* **18**, 3081–3084.

Woithe, K., Geib, N., Zerbe, K., Li, D. B., Heck, M., Fournier-Rousset, S., Meyer, O., Vitali, F., Matoba, N., Abou-Hadeed, K., and Robinson, J. A. (2007). Oxidative phenol coupling reactions catalyzed by OxyB: A cytochrome P450 from the vancomycin producing organism. Implications for vancomycin biosynthesis. *J. Am. Chem. Soc.* **129**, 6887–6895.

Wolpert, M., Gust, B., Kammerer, B., and Heide, L. (2007). Effects of deletions of *mbtH*-like genes on clorobiocin biosynthesis in *Streptomyces coelicolor*. *Microbiology* **153**, 1413–1423.

Yang, J., Fu, X., Liao, J., Liu, L., and Thorson, J. (2005). Structure-based engineering of galactokinase as a first step toward *in vivo* glycorandomization. *Chem. Biol.* **12**, 657–664.

Yao, R. C., and Crandall, L. W. (1994). Glycopeptides: Classification, occurrence, and discovery. *In* "Glycopeptide Antibiotics" (R. Nagarajan, ed.), pp. 1–28. Marcel Dekker, New York, USA.

Zerbe, K., Pylypenko, O., Vitali, F., Zhang, W., Rouset, S., Heck, M., Vrijbloed, J. W., Bischoff, D., Bister, B., Süssmuth, R. D., Pelzer, S., Wohlleben, W., *et al.* (2002). Crystal structure of OxyB, a cytochrome P450 implicated in an oxidative phenol coupling reaction during vancomycin biosynthesis. *J. Biol. Chem.* **277**, 47476–47485.

Zerbe, K., Woithe, K., Li, D. B., Vitali, F., Bigler, L., and Robinson, J. A. (2004). An oxidative phenol coupling reaction catalyzed by OxyB, a cytochrome P450 from the vancomycin-producing microorganism. *Angew. Chem. Int. Ed.* **43**, 6709–6713.

CHAPTER NINETEEN

In Vitro Studies of Phenol Coupling Enzymes Involved in Vancomycin Biosynthesis

Dong Bo Li, Katharina Woithe, Nina Geib, Khaled Abou-Hadeed, Katja Zerbe, *and* John A. Robinson

Contents

1. Introduction	488
2. Peptide Synthesis	491
3. Peptide Thioesters	499
3.1. Synthesis of hexapeptide *S*-phenyl thioester (1-SPh)	499
3.2. Synthesis of hexapeptide-SCoA thioester (1-SCoA)	500
3.3. Synthesis of hexapeptide-PCP conjugate (1-S-PCP)	501
4. *In Vitro* Assays with OxyB	502
5. Production and Purification of Enzymes	503
5.1. Production and purification of OxyB	503
5.2. Production and purification of spinach ferredoxin	504
5.3. Production and purification of PCP domain	505
5.4. Production and purification of *E. coli* NADPH-flavodoxin reductase	506
References	507

Abstract

Oxidative phenol cross-linking reactions play a key role in the biosynthesis of glycopeptide antibiotics such as vancomycin. The vancomycin aglycone contains three cross-links between aromatic amino acid side-chains, which stabilize the folded backbone conformation required for binding to the target D-Ala-D-Ala dipeptide. At least the first cross-link is introduced into a peptide precursor whilst it is still bound as a thioester to a peptide carrier protein (PCP) domain (also called a thiolation domain) within the nonribosomal peptide synthetase. We described here methods for the solid-phase synthesis of

Institute of Organic Chemistry, University of Zürich, Zürich, Switzerland peptides and their coupling to PCP domains, which may be useful for *in vitro* studies of cross-linking and related tailoring reactions during nonribosomal glycopeptide antibiotic biosynthesis.

1. Introduction

One of the strategies used in Nature to create conformationally constrained peptides involves oxidative cross-linking of aromatic amino acid side-chains. Examples are found in the glycopeptide antibiotics, such as vancomycin and teicoplanin (Fig. 19.1), which contain biaryl ether

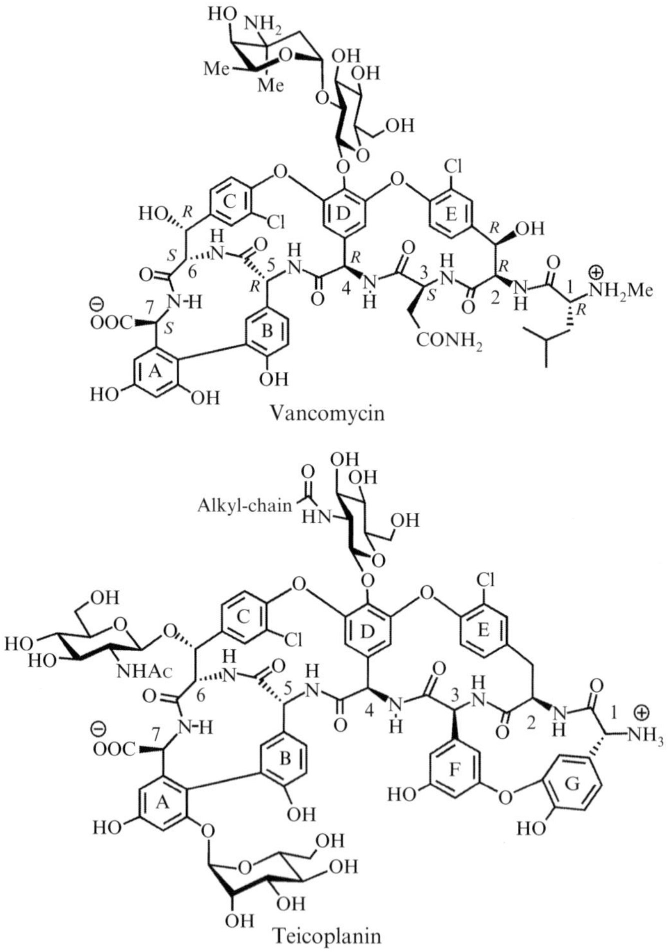

Figure 19.1 Structures of vancomycin and teicoplanin.

(C–O–C) as well as direct biaryl (C–C) cross-links (Hubbard and Walsh, 2003). These are essential for the creation within the glycopeptide scaffold of a binding site for the dipeptide D-Ala-D-Ala found in intermediates of bacterial peptidoglycan biosynthesis. Such aromatic C–O–C and C–C cross-links, however, occur more widely in Nature, for example, in the closely related bacterial metabolites complestatin (Seto *et al.*, 1989), kistamicin (Naruse *et al.*, 1993), and chloropeptin (Matsuzaki *et al.*, 1994), as well as in peptidic natural products such as K-13 (Yasuzawa *et al.*, 1987), bouvardin (Jolad *et al.*, 1977), and the related plant metabolites RA-I-XVI (Hitotsuyanagi *et al.*, 2004), OF4949 (Sano *et al.*, 1986), the arylomycins (Holtzel *et al.*, 2002), the biphenomycins (Uchida *et al.*, 1985), the TMC-95 family (Kohno *et al.*, 2000), and RP-664536 (Helynck *et al.*, 1998).

Oxidative phenol coupling reactions also play a key role in plant alkaloid biosynthesis, especially of benzylisoquinoline alkaloids such as morphine (Ziegler and Facchini, 2008). The first report of the purification and cloning of a cytochrome P450 enzyme catalyzing a regio- and stereospecific oxidative C–O–C phenol coupling was of berbamunine synthase from *Berberis stolonifera* (Kraus and Kutchan, 1995). A related P450 enzyme catalyzing a C–C coupling was cloned recently from *Coptis japonica*, which converts (S)-reticuline into (S)-corytuberine (Ikezawa *et al.*, 2008). These plant enzymes, however, appear to be membrane-bound proteins. In contrast, several soluble microbial P450 enzymes have also been cloned and studied recently that catalyze phenol (and related) coupling reactions. These include a flaviolin oxidase (Zhao *et al.*, 2005), a tetrahydroxynaphthalene oxidase (Funa *et al.*, 2005), an indole–indole coupling enzyme (StaP) involved in staurosporine biosynthesis (Howard-Jones and Walsh, 2007; Makino *et al.*, 2007) and, last but not least, the enzyme OxyB, which catalyzes the first cross-linking step in the biosynthesis of vancomycin and related glycopeptides (Woithe *et al.*, 2007; Zerbe *et al.*, 2004).

The three oxidative cross-linking reactions during vancomycin biosynthesis are catalyzed in a defined order by three closely related cytochrome P450 enzymes, called OxyA, OxyB, and OxyC (following the order of their genes in the bacterial chromosome). The first coupling reaction occurs between the phenol rings in residues-4 and -6 (the C-O-D ring), catalyzed by OxyB, the second aryl-ether bridge is formed between side chains of residues-2 and -4 (D-O-E ring) by OxyA, and the last biaryl coupling between the aromatic side chains of residues-5 and -7 is carried out by OxyC (AB rings, Fig. 19.1) (Bischoff *et al.*, 2001a,b; Sussmuth *et al.*, 1999). The crystal structures of OxyB (CYP165B3) and OxyC (CYP165C4) from the vancomycin producer *Amycolatopsis orientalis* have been reported in substrate-free forms (Pylypenko *et al.*, 2003; Zerbe *et al.*, 2002), confirming that these proteins indeed contain a fold and heme environment typical of P450 enzymes.

The heptapeptide backbones of glycopeptides are constructed by the action of large nonribosomal peptide synthetase (NRPS) multidomain proteins (NRPS-1–3, Fig. 19.2). According to the so-called thiotemplate mechanism of assembly, peptidic intermediates remain bound as C-terminal thioesters to pantetheinyl groups attached to peptide carrier protein (PCP) domains (also called thiolation, or T domains) within the NRPS assembly line (Grünewald and Marahiel, 2006). (See Chapter 13, in this volume.) Each module in the NRPS is responsible for the incorporation of one amino acid, and contains at least a PCP domain and an amino acid activating (A) domain, usually a condensation (C) domain, and sometimes also an epimerization (E) domain amongst others. Seven modules are required to assemble glycopeptide heptapeptide backbones which, according to bioinformatic analyses, are distributed between three (in a 3:3:1 distribution) or four (in a 2:1:3:1 distribution) NRPS subunits for the vancomycin-like and teicoplanin/complestatin-type glycopeptides, respectively (Donadio et al., 2005). Fused to module 7 is a domain (X) of unknown function and a thioesterase (TE) domain. Only recently has it become clear that, whilst being moved along and remaining covalently anchored to the NRPS assembly line, the peptide intermediates may also be transformed by other enzymes acting in *trans*. This has been demonstrated *in vitro* for OxyB in vancomycin-like glycopeptides (Woithe et al., 2007; Zerbe et al., 2004), and may well also be true for the other cross-linking enzymes, and the halogenases, which introduce Cl atoms into aromatic residues during glycopeptide biosynthesis.

For *in vitro* studies of glycopeptide cross-linking and halogenase enzymes, access is required to suitable substrates, namely peptides linked as thioesters to fragments of the NRPS, in particular, isolated recombinant PCP domains (Fig. 19.2). The production of such peptide–PCP conjugates

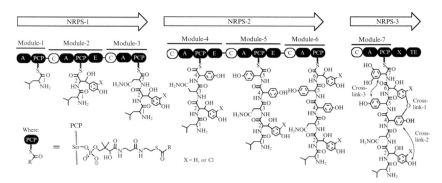

Figure 19.2 The NRPS assembly line for the biosynthesis of vancomycin-like antibiotics deduced by bioinformatic analyses. The domain structures of modules 1–7 are shown. Peptide intermediates likely to be bound to PCPs at various stages of assembly are also shown.

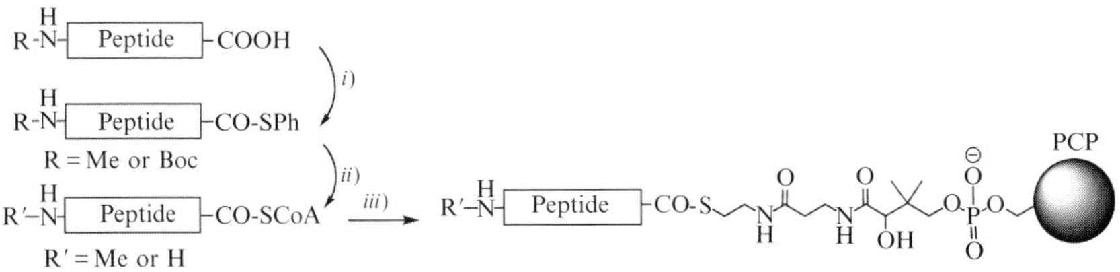

Figure 19.3 Synthetic approach to peptide–PCP thioester substrates.

Reagents: *i*) PhSH, coupling reagent; *ii*) If necessary TFA, then CoA-SH; *iii*) Sfp + apo-PCP

is often not straightforward, due to the stereochemical and chemical lability of these molecules. The approach described here involves first the solid-phase synthesis of a suitable peptide (Fig. 19.3).

The required synthetic peptides typically contain hydroxyphenylglycine (Hpg) and β-hydroxytyrosine (Bht) derivatives, which are often rapidly epimerized and/or degraded upon treatment with acids and bases. Thus, the peptides cannot be synthesized efficiently using standard Fmoc- and Boc-methods of solid-phase peptide synthesis. Next, the peptide should be activated at the C-terminus, first as an *S*-phenyl and then an *S*-CoA thioester. With the peptide-SCoA thioester, the broad specificity pantetheinyl transferase Sfp from *Bacillus subtilis* (Quadri *et al.*, 1998) can be used to transfer the entire peptide-pantetheinyl portion to the active site Ser residue of a recombinant apo-PCP domain. Finally, a suitable fragment of the NRPS comprising an intact PCP domain must be produced in the apo-form. Although powerful general methods exist for producing large amounts of protein in the cytoplasm of *Escherichia coli*, the expression protocol used and the choice of N- and C-termini for the PCP domain can have a major influence on its folding and solubility in aqueous buffers. Below, we expand on one approach for synthesizing such molecules, which has been validated recently in the production of a variety of peptide–PCP conjugates for *in vitro* studies of glycopeptide antibiotic cross-linking enzymes (Geib *et al.*, 2008; Woithe *et al.*, 2007, 2008).

2. Peptide Synthesis

As a typical example, we focus here on the model hexapeptide **1** (Fig. 19.4) and its use for *in vitro* assays with OxyB. The methods described here for the production of **1** and its PCP-bound form (**1-S-PCP**) can, however, be readily adapted for the production of a variety of other related peptide–PCP conjugates. In the interests of ease of synthesis, β-hydroxytyrosine is incorporated only at position-6 of this hexapeptide, although the biosynthetic logic (Fig. 19.2) dictates that β-hydroxytyrosine, or perhaps

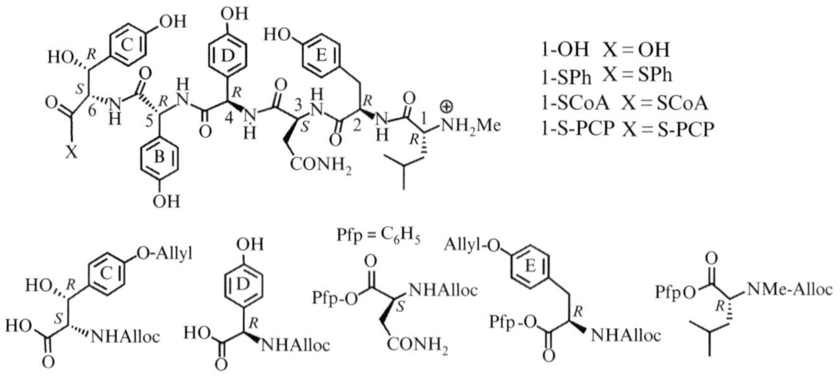

Figure 19.4 The model hexapeptide 1 and protected amino acids required for its synthesis.

m-chloro-β-hydroxytyrosine, should be present at both positions-2 and -6. In earlier studies, model substrates containing tyrosine at both positions-2 and -6 were produced and used for *in vitro* studies of OxyB (Geib *et al.*, 2008; Woithe *et al.*, 2007, 2008; Zerbe *et al.*, 2004). Also for synthetic convenience, the peptide **1** has an *N*-methylated D-Leu at the N-terminus. Experience has shown that an unprotected N-terminal *N*-methylamino group is compatible with thioester formation at the C-terminus, under the conditions used. Note, however, that *N*-methylation occurs late in the vancomycin biosynthetic pathway, after aglycone formation is complete (O'Brien *et al.*, 2000), and that the *N*-methyl group does not influence to any significant extent the cross-linking reaction catalyzed by OxyB (Woithe *et al.*, 2007, 2008).

If peptides are required with an unmethylated N-terminus, then the N-terminal primary amino group should be protected, for example, with a *t*-butoxycarbonyl (Boc) group, during thioester formation at the C-terminus (Fig. 19.3). The Boc group can be removed from the peptide after formation of an *S*-phenyl thioester by brief treatment with trifluoroacetic acid (TFA) (Woithe *et al.*, 2007).

The free linear peptide **1-OH** is first prepared by solid-phase peptide synthesis. The assembly of **1-OH** can be achieved efficiently using *N*-allyloxycarbonyl (Alloc)-protected amino acids by solid-phase synthesis on chlorotrityl chloride (CT) resin (Freund and Robinson, 1999; Li and Robinson, 2005). Where necessary, *N*-Alloc and side-chain *O*-allyl protecting groups can be removed using Pd-catalysis under neutral conditions. Here, an effort has been made to avoid as far as possible side-chain protecting groups, since some (e.g., benzyl groups) can be very difficult to remove cleanly from such labile peptides. The side-chain protection of Asn is also avoided, but then exposure to the normal activating reagents (e.g., HBTU/HOBt) can lead to side reactions in the Asn side-chain amide, so this and the amino acids subsequently added to the peptide on the resin are coupled as pentafluorophenyl (Pfp) esters. When assembly is complete, only minimal exposure of the resin to acid is needed (0.6% TFA in CH_2Cl_2) to release the fully assembled peptide (**1-OH**) into solution, which can be purified and used for further studies. Hence, the protected amino acids required for the synthesis of **1-OH** are those shown in Fig. 19.4. The synthesis of peptide **1-OH** is described in detail below.

All synthetic transformations must be performed in a well-ventilated fume hood designed for organic synthesis use, and carried out only by suitably trained personnel. Avoid direct contact with all chemicals used.

Step 1. Synthesis of amino acid 2. The protected amino acid (Bht) **2** can be prepared using Evans aldol chemistry, as shown in Fig. 19.5 (Evans and Weber, 1986, 1987). The allyl group was chosen for protection of the phenol, since it can be removed under neutral conditions with Pd-catalysis during cleavage of the *N*-Alloc group.

Compound **9** was prepared as follows. Stannous triflate (5.0 g, 12 mmol) and dry tetrahydrofuran (THF) (70 ml) were cooled to $-78\ °C$ under argon. *N*-Ethylpiperidine (2.5 ml, 18 mmol) was added slowly, and stirred for 5 min. The oxazolidinone **7** (2.76 g, 10 mmol) was added in dry THF at $-78\ °C$, and the mixture was stirred for 1.5 h. The aldehyde **8** was then added, and the mixture was stirred at $-78\ °C$ for 2 h. The reaction was quenched by addition of sodium phosphate buffer, pH 7, filtered through celite, and the aqueous phase was extracted with CH_2Cl_2. The organic extract was washed with 1N aqueous sodium bisulphate, dried over anhydrous sodium sulfate, and concentrated *in vacuo*. The product was purified by flash chromatography on silica gel (*n*-hexane:EtOAc, 2:1, R_f 0.3) to give **9** as a white foam (3.39 g, 77%). ^{1}H NMR (300 MHz, CDCl$_3$) δ 2.95

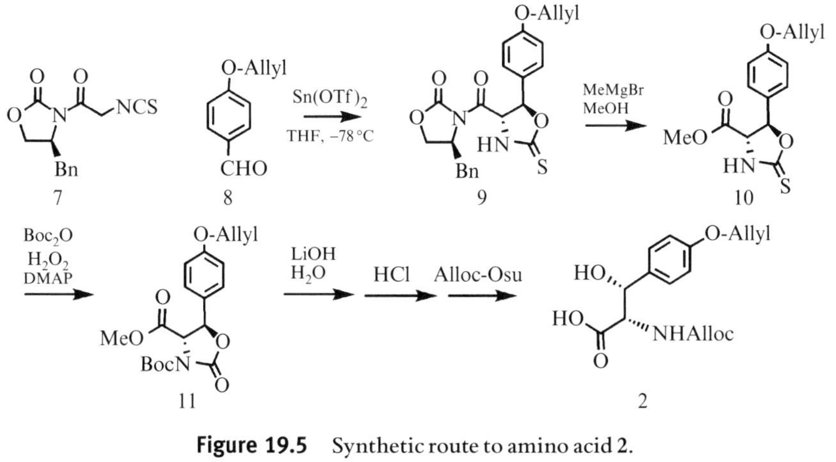

Figure 19.5 Synthetic route to amino acid 2.

(1H, dd), 3.27 (1H, dd), 4.35 (2H, m), 4.57 (2H, d), 4.78 (1H, m), 5.06 (1H, d), 5.30 (1H, d), 5.45 (1H, d), 6.05 (1H, m), 6.38 (1H, d), 6.95 (2H, d) 7.20 (2H, d), 7.35 (2H, m), 7.40 (3H, m), 7.75 (1H, s).

To the aldol adduct **9** (3.39 g, 7.74 mmol) in anhydrous MeOH (20 ml) and CH_2Cl_2 (20 ml) at 0 °C was added a suspension formed by the addition of MeMgBr (2.66 ml, 8.51 mmol) in dry diethylether (3.2 M) to anhydrous MeOH (10 ml) (Evans and Weber, 1987). After stirring for 10 min, the reaction was quenched by addition of 1N aqueous sodium bisulfate. The volatile solvent was removed *in vacuo*, and the aqueous phase was extracted three times with CH_2Cl_2. The combined organic phase was then dried over anhydrous sodium sulfate and evaporated *in vacuo*. The product was purified by flash chromatography on silica gel (n-hexane:EtOAc, 2:1, R_f 0.3) to give **10** as a light yellow oil (1.72 g, 76%). ^{1}H NMR (300 MHz, $CDCl_3$) δ 3.98 (3H, s), 4.60 (1H, d), 4.67 (2H, m), 5.40 (1H, m), 5.52 (1H, m), 6.01 (1H, d), 6.15 (1H, m), 7.06 (2H, d), 7.42 (2H, d), 7.72 (1H, br. s).

To the methyl ester **10** (1.72 g, 5.87 mmol) in dry CH_2Cl_2 (20 ml) was added Boc_2O (1.41 g, 6.45 mmol) and 4-(dimethylamino)-pyridine (36 mg, 0.29 mmol). The mixture was stirred at room temperature for 1 h. After cooling to 0 °C, a 30% solution of H_2O_2 (5 ml) and formic acid (5 ml) were added and the mixture was stirred vigorously for 30 min. Then 1N aqueous sodium bisulphate was added, the organic layer was separated, and the aqueous phase was extracted three times with CH_2Cl_2. The organic phases were combined, dried over anhydrous sodium sulfate, and concentrated *in vacuo*. The product was purified by flash chromatography on silica gel (n-hexane:EtOAc, 2:1, R_f 0.5) to give **11** as a colorless oil (1.12 g, 51%). ^{1}H NMR (300 MHz, $CDCl_3$) δ 1.62 (9H, s), 4.00 (3H, s), 4.69 (2H, m), 4.77 (1H, d), 5.42 (1H, m), 5.43 (1H, d), 5.55 (1H, m), 6.18 (1H, m), 7.10 (2H, d), 7.42 (2H, d).

To compound **11** (1.12 g, 2.98 mmol) dissolved in dioxane (20 ml) at room temperature was added a 2N aqueous LiOH solution (7.45 ml, 14.9 mmol) and the resulting mixture was stirred overnight. The reaction was quenched by addition of 1N aqueous sodium bisulfate, and extracted three times with CH_2Cl_2. The combined organic extracts were dried over anhydrous sodium sulfate and concentrated *in vacuo*. The product was purified by flash chromatography on silica (CH_2Cl_2:MeOH:AcOH, 100:1:0.5) to give the product as a colorless oil (660 mg, 66%). ^{1}H NMR (300 MHz, DMSO) δ 1.30 (9H, s), 3.30 (OH), 4.18 (1H, m), 4.55 (2H, d), 5.01 (1H, s), 5.25 (1H, m), 5.40 (1H, m), 6.03 (1H, m), 6.29 (1H, d), 6.90 (2H, m), 7.27 (2H, d), 12.15 (1H, br. s). This material (560 mg, 1.67 mmol) was dissolved in 4N HCl in dry dioxane (5 ml). After stirring for 1 h at room temperature, the solution was evaporated to dryness *in vacuo* to afford the amino acid hydrochloride salt.

In preparation for the formation of **2**, allyloxycarbonyl-N-hydroxysuccinimide ester (Alloc-OSu) was prepared (Hayakawa *et al.*, 1986). Thus, to allyloxycarbonyl chloride (170 μl, 1.6 mmol) and N-hydroxysuccinimide (184 mg, 1.6 mmol) in dry THF (2 ml) was added dropwise over 5 min triethylamine (223 μl, 1.6 mmol). After stirring for 30 min at room temperature, the solution was filtered, and the clear filtrate was evaporated to dryness *in vacuo* to afford Alloc-OSu, which was used immediately without further purification.

The amino acid–HCl salt from above (1.67 mmol) and sodium bicarbonate (335 mg, 4.1 mmol) in a mixture of water (5 ml) and acetone (5 ml) was added to the freshly prepared Alloc-OSu, and the mixture was stirred at room temperature overnight. The volatiles were removed *in vacuo*, then the pH was adjusted to 2 with dil. HCl and the remaining aqueous solution was extracted with EtOAc. The organic extracts were dried over anhydrous sodium sulfate and evaporated to dryness *in vacuo* to afford **2** (320 mg, 60%) as a white solid. ^{1}H NMR (300 MHz, DMSO) δ 3.30 (OH), 4.20 (1H, m), 4.38 (2H, m), 4.55 (2H, m), 4.70 (1H, d), 5.10–5.30 (3H, m), 5.40 (1H, d), 5.80 (1H, m), 6.05 (1H, m), 6.90 (2H, d), 6.95 (1H, s), 7.30 (2H, d), 12.7 (1H, br. s).

*Step 2. Synthesis of amino acid **3***. To a stirred solution of D-4-hydroxyphenylglycine (3.34 g, 20 mmol) and sodium bicarbonate (2.4 g, 14 mmol) in water/acetone (1:1, 80 ml) was added freshly prepared Alloc-OSu (3.98 g, 20 mmol) (see above) (Paquet, 1982). The mixture was stirred for 5 h. The pH was then adjusted to 2 with dil. HCl, and after removal of acetone *in vacuo*, the aqueous layer was extracted with EtOAc. The extracts were dried over anhydrous sodium sulfate, and evaporated to afford an oil, which was crystallized from ether/hexane to give a white solid (3.88 g, 77%). ^{1}H NMR (300 MHz, DMSO) δ 3.35 (OH), 4.50 (2H, d), 5.00 (1H, d), 5.15 (1H, d), 5.30 (1H, d), 5.90 (1H, m), 6.73 (2H, d), 7.20 (2H, d), 7.85 (1H, d), 9.45 (1H, br. s).

*Step 3. Synthesis of amino acid **4***. Alloc-Asn-OH was first prepared from Asn using the procedure described in step 2. The pentafluorophenyl (-OPfp) ester was then prepared as follows. To a stirred solution of Alloc-Asn-OH (1.3 g, 6 mmol) and pentafluorophenol (3.68 g, 20 mmol) in dry dioxane (20 ml) at 0 °C was added dicyclohexylcarbodiimide (1.51 g, 7.2 mmol). The mixture was stirred at 0 °C for 2 h and at room temperature for 3 h. The urea was then filtered off, solvent was removed *in vacuo*, the residue was triturated with *n*-hexane, and the resulting solid was recrystallized from *n*-hexane/EtOAc to give the product as a white solid (1.41 g, 61%). ^{1}H NMR (300 MHz, DMSO) δ 2.45 (2H, m), 4.28 (1H, m), 4.43 (2H, d), 5.10 (1H, d), 5.23 (1H, d), 5.85 (1H, m), 6.85 (1H, br. s), 7.30 (1H, br. s), 7.32 (1H, d).

*Step 4. Synthesis of amino acid **5***. The amino acid **5** is prepared in five steps from D-tyrosine. First, acetyl chloride (5.6 ml, 72 mmol) was added dropwise to methanol (30 ml) and stirred at 0 °C for 15 min, then D-Tyr (4.73 g, 26 mmol) was added in portions and the mixture was refluxed for 3 h. The solvent was then evaporated *in vacuo* to give a white solid, which was used without purification. ^{1}H NMR (300 MHz, D_2O) δ 3.42 (2H, m), 4.07 (3H, s), 4.60 (1H, m), 7.10 (2H, d), 7.40 (2H, d).

Second, the D-Tyr-OMe·HCl salt prepared as above (4.47 g, 19 mmol) was dissolved in acetone/water (1:1, 40 ml) and sodium bicarbonate (4.0 g, 48 mmol) was added, followed by freshly prepared Alloc-OSu (3.98 g, 20 mmol, prepared as above). The mixture was stirred overnight at room temperature. The pH was then adjusted to 2–3 with dil. HCl, the acetone was removed *in vacuo*, and the aqueous phase was extracted three times with EtOAc. The organic extracts were dried over anhydrous sodium sulfate and evaporated to dryness *in vacuo*, to give a colorless oil, which was used without purification (TLC, silica *n*-hexane:EtOAc (1:1) showed a single component, R_f 0.75).

Third, the Alloc-D-Tyr-OMe (5.31 g, 19 mmol) was dissolved in dry DMF (30 ml) and potassium carbonate (5.32 g, 38 mmol), allyl bromide (2.5 ml, 24 mmol), and *n*-Bu_4NBr (1.9 g, 5 mmol) were added. The mixture was stirred for 24 h at room temperature. Then 1N sodium bisulfate was added and the mixture was extracted three times with EtOAc. The organic extracts were washed with brine, dried over anhydrous sodium sulfate, and evaporated to dryness *in vacuo*. The resulting oil (5.8 g, 96%) was used without purification (TLC, silica *n*-hexane:EtOAc (2:1) showed a single component, R_f 0.3). ^{1}H NMR (300 MHz, acetone) δ 3.13 (1H, dd), 3.28 (1H, dd), 3.89 (3H, s), 4.63 (1H, m), 4.68 (2H, m), 4.75 (2H, m), 5.32 (1H, m), 5.45 (1H, m), 5.48 (1H, m), 5.60 (1H, m), 6.08 (1H, m), 6.25 (1H, m), 6.62 (1H, br.d), 7.08 (2H, d), 7.39 (2H, d).

Fourth, the Alloc-D-Tyr(allyl)-OMe (5.8 g, 18 mmol) was dissolved in THF:water (1:1, 30 ml) and LiOH·H_2O (1.01 g, 24 mmol) was added with stirring at 0 °C overnight. The pH was then adjusted to 2–3 with dil. HCl, the THF was removed *in vacuo*, and the aqueous phase was extracted three times with CH_2Cl_2. The organic extracts were dried over anhydrous sodium sulfate and evaporated to dryness *in vacuo*, to give a colorless oil (5.47 g, 99%). ^{1}H NMR (300 MHz, DMSO) δ 2.90 (1H, m), 3.10 (1H, m), 3.40 (br. OH), 4.21 (1H, m), 4.55 (2H, m), 4.67 (2H, m), 5.27 (1H, m), 5.35 (2H, m), 5.39 (1H, m), 5.50 (1H, m), 5.95 (1H, m), 6.15 (1H, m), 6.97 (2H, d), 7.29 (2H, d), 7.60 (1H, d).

Fifth, the Alloc-D-Tyr(allyl)-OH (5.2 g, 17 mmol) and C_6F_5-OH (3.35 g, 18 mmol) in dry dioxane (20 ml) at 0 °C was treated with dicyclohexylcarbodiimide (3.7 g, 18 mmol). After stirring 1 h at 0 °C, and another 1 h at room temperature, the mixture was filtered, and the filtrate was concentrated *in vacuo*. The resulting oil was crystallized from *n*-hexane and diethyl ether to give the product **5** as a white solid (6.2 g, 76%). ^{1}H NMR (300 MHz, $CDCl_3$) δ 3.50 (2H, m), 4.82 (2H, m), 4.88 (2H, m), 5.21 (1H, m), 5.40 (1H, br. d), 5.50 (1H, d), 5.53 (1H, m), 5.60 (1H, m), 5.68 (1H, m), 6.18 (1H, m), 6.33 (1H, m), 7.17 (2H, d), 7.41 (2H, d).

Step 5. Synthesis of amino acid **6**. To *N*-Methyl-D-leucine (184 mg, 1.27 mmol) in dioxane:water (1:1, 10 ml) was added sodium bicarbonate (1.07 g, 12.7 mmol) and Alloc-chloride (1.35 ml, 12.7 mmol) and the mixture was stirred overnight at room temperature. The volatiles were then removed *in vacuo*, and the aqueous phase (pH 8) was extracted with diethyl ether. The aqueous phase was then adjusted to pH 2–3 with dil. HCl and was again extracted with ether. The ether extract was dried over anhydrous sodium sulfate and evaporated to dryness *in vacuo* to give Alloc-*N*-methyl-D-Leu-OH (270 mg, 93%). This material was dissolved in dry dioxane (5 ml) and pentafluorophenol (217 mg, 1.18 mmol) was added followed by dicyclohexylcarbodiimide (243 mg, 1.18 mol) and the mixture was stirred at 0 °C for 1 h, then at room temperature for 2 h. The urea was removed by filtration and the filtrate was evaporated to dryness *in vacuo*, to afford the product as an oil, which was used without further purification. ^{1}H NMR (300 MHz, CDCl$_3$) δ 0.90 (6H, m), 1.51–1.82 (3H, m), 2.80 (3H, s), 2.90 (1H, m), 4.08 (2H, m), 5.20 (2H, m), 5.58 (1H, m).

Step 6. Peptide assembly

Step 6A. Loading residue-6 on the resin. The amino acid **2** (320 mg, 1 mmol) was dissolved in a mixture of dry CH$_2$Cl$_2$ and dry DMF (19:1, 15 ml) and dry *N*-methylmorpholine (510 μl, 4.7 mmol) was added. This mixture was added to freshly activated and dried CT-resin (1.2 g, 1.56 mmol/g, *Novabiochem*) and stirred at room temperature overnight. Methanol (5 ml) was then added and the mixture was agitated for 10 min. The resin was filtered and washed successively with DMF (4 × 25 ml), MeOH (4 × 25 ml) and CH$_2$Cl$_2$ (4 × 25 ml). The presence of amino acid on the resin was checked by treating a few resin beads with TFA (0.6%, v/v) in CH$_2$Cl$_2$ (100 μl), and analysis by HPLC–MS.

Step 6B. Deprotection. To this resin (600 mg) under argon in the dark was added a solution of Pd(PPh$_3$)$_4$ (200 mg, 0.7 mmol) and PhSiH$_3$ (1.3 ml, 10.6 mmol) in CH$_2$Cl$_2$ (15 ml), and the mixture was agitated for 3 h. The resin was then filtered and washed successively with DMF (4 × 25 ml), CH$_2$Cl$_2$ (4 × 25 ml), and DMF (4 × 25 ml). A few resin beads were treated with TFA (0.6%, v/v) in CH$_2$Cl$_2$ (100 μl), and analysis by HPLC–MS was performed to ensure that deprotection was complete.

Step 6C. Coupling residue-5. To the resin was added **3** (251 mg, 1 mmol), diisopropylcarbodiimide (DIC) (126 mg, 1 mmol) and 1-hydroxybenzotriazole (HOBt) (306 mg, 2 mmol) in dry DMF (10 ml) and the mixture was agitated overnight at room temperature. The resin was then filtered and washed with DMF (4 × 25 ml) and CH$_2$Cl$_2$ (4 × 25 ml). A few resin beads were treated with TFA (0.6%, v/v) in CH$_2$Cl$_2$ (100 μl), and analysis by HPLC–MS was performed to ensure that coupling was >95% complete. The Alloc group was then removed using the procedure in step 6B.

Step 6D. Coupling residue-4. This step was carried out as described for step 6C.

Step 6E. Coupling residue-3. For this step, the freshly prepared amino acid derivative **4** was used (see above). The compound **4** (382 mg, 1 mmol) and HOBt (306 mg, 2 mmol) in dry DMF (10 ml) was added to the resin, and the mixture was agitated overnight at room temperature. The resin was then filtered and washed with DMF (4 × 25 ml) and CH_2Cl_2 (4 × 25 ml). A few resin beads were treated with TFA (0.6%, v/v) in CH_2Cl_2 (100 μl), and analysis by HPLC–MS was performed to check that coupling was > 95% complete. At this stage, it is necessary to use nBu_3SnH instead of $PhSiH_3$ as hydride donor in the deprotection step. Thus, under argon in the dark a solution of $Pd(PPh_3)_4$ (140 mg, 0.49 mmol) and nBu_3SnH (2 ml, 7.43 mmol) in CH_2Cl_2 (15 ml) was added to the resin and the mixture was agitated for 3 h at room temperature. The resin was then filtered and washed successively with DMF (4 × 25 ml), CH_2Cl_2 (4 × 25 ml), and DMF (4 × 25 ml). A few resin beads were treated with TFA (0.6%, v/v) in CH_2Cl_2 (100 μl), and analysis by HPLC–MS was performed to ensure that deprotection was complete.

Step 6F. Coupling residue-2. For this step, the freshly prepared amino acid derivative **5** was used (see above). To the resin was added compound **5** (473 mg, 1 mmol) and HOBt (153 mg, 1 mmol) in dry DMF (10 ml), and the mixture was agitated overnight at room temperature. The resin was then filtered and washed with DMF (4 × 25 ml) and CH_2Cl_2 (4 × 25 ml). A few resin beads were treated with TFA (0.6%, v/v) in CH_2Cl_2 (100 μl), and analysis by HPLC–MS was performed to check that coupling was >95% complete. Removal of the *N*-Alloc and *O*-allyl groups was performed using the procedure described in step 6E, with nBu_3SnH as hydride donor. Then, a few resin beads were treated with TFA (0.6%, v/v) in CH_2Cl_2 (100 μl), and analysis by HPLC–MS was performed to ensure that deprotection was complete.

Step 6G. Coupling residue-1. For this step, the freshly prepared amino acid derivative **6** was used (see above). To the resin was added compound **6** (435 mg, 1.1 mmol) and HOBt (168 mg, 1.1 mmol) in dry DMF (10 ml), and the mixture was agitated overnight at room temperature. The resin was then filtered and washed with DMF (4 × 25 ml) and CH_2Cl_2 (4 × 25 ml). A few resin beads were treated with TFA (0.6%, v/v) in CH_2Cl_2 (100 μl), and analysis by HPLC–MS was performed to ensure that coupling was >95% complete. Removal of the *N*-Alloc was performed using the procedure described in step 6E, with nBu_3SnH as hydride donor.

Step 6H. Cleavage from the resin and purification. A solution of TFA in CH_2Cl_2 (0.6% v/v, 20 ml) was added to the resin with agitation for 5 min at room temperature. The resin was then filtered, and the cleavage procedure was repeated four more times. The resin was then washed with methanol, and finally all organic filtrates were combined and evaporated to dryness *in vacuo*. The hexapeptide **1-OH** (26 mg, >95% purity) was then obtained by preparative reverse-phase HPLC using a preparative C18 column and a

Table 19.1 ^{1}H NMR Chemical shift assignments for peptide **1-OH** (500 MHz at 300 K in DMSO-d_6)

Residue	NH	C(α)-H	C(β)-H	Others
Leu^1	8.64	3.58	1.51, 1.38	C(γ)H = 1.49, C(δ)H$_3$ = 0.79, 0.84, NMe = 1.94
Tyr^2	8.82	4.76	2.95, 2.56	C(δ)H = 7.04, C(ε)H = 6.62, OH = 9.32
Asn^3	8.45	4.67	2.45, 2.31	N(δ)H = 7.30, 6.93
Hpg^4	8.06	5.51	–	C(γ)H = 7.16, C(δ)H = 6.66, OH = 9.32
Hpg^5	8.72	5.54	–	C(γ)H = 7.01, C(δ)H = 6.62, OH = 9.18
Bht^6	8.10	4.26	4.94, OH = 5.51	C(δ)H = 6.91, C(ε)H = 6.51, OH = 9.20

gradient of 5–35% MeCN in H$_2$O + 0.1% TFA. ES-MS found m/z 900.5 ([M + H]$^+$). ^{1}H NMR (500 MHz, DMSO, 300 K) chemical shift assignments were obtained from 1D and 2D ^{1}H NMR spectra and are given in Table 19.1.

3. Peptide Thioesters

The synthesis of the hexapeptide S-phenyl thioester **1-SPh**, hexapeptide-SCoA thioester **1-SCoA**, as well as the hexapeptide–PCP conjugate **1-S-PCP** was achieved using methods described previously (Vitali et al., 2003; Woithe et al., 2007; Zerbe et al., 2004).

3.1. Synthesis of hexapeptide S-phenyl thioester (1-SPh)

The hexapeptide **1-OH** (2 mg) in freshly distilled DMF (1 ml) was stirred with PyBOP (1.2 eqiv), DIEA (1.2 eqiv), and thiophenol (2.4 eqiv) under nitrogen atmosphere. The reaction was monitored by LC/MS and was complete after 10 min. After lyophilization, the resulting hexapeptide S-phenyl thioester (**1-SPh**) was purified by semipreparative reverse-phase HPLC (C$_{18}$ Vydac 218TP1010, 250/10, pore diameter 300 Å, particle size

Table 19.2 ^{1}H NMR Chemical shift assignments for peptide **1-SPh** (500 MHz at 300 K in DMSO-d$_6$)

Residue	NH	C(α)-H	C(β)-H	Others
Leu^1	–	3.56	1.50, 1.37	C(γ)H = 1.50, C(δ)H$_3$ = 0.77, 0.83, NMe = 1.96
Tyr^2	8.78	4.77	2.96, 2.57	C(δ)H = 7.04, C(ε)H = 6.63, OH = 9.15
Asn^3	8.44	4.68	2.47, 2.32	N(δ)H = 7.28, 6.91
Hpg^4	8.03	5.58	–	C(γ)H = 7.18, C(δ)H = 6.58, OH = 9.25
Hpg^5	8.85	5.78	–	C(γ)H = 7.02, C(δ)H = 6.64, OH = 9.32
Bht^6	8.80	4.47	5.05, OH = 5.74	C(δ)H = 7.02, C(ε)H = 6.57, OH = 9.26
CO.SPh	–	–	–	7.46, 7.26

5 μm) with a gradient of 5–100% MeCN + 0.1% TFA in water + 0.1% TFA in five column volumes at a flow rate of 5 ml/min. The corresponding peptide S-phenyl thioester was obtained in ≈90% yield. ES-MS m/z = 992.4 ([M + H]$^+$). Chemical shift assignments were obtained from 1D and 2D ^{1}H NMR spectra and are given in Table 19.2.

3.2. Synthesis of hexapeptide-SCoA thioester (1-SCoA)

The peptide S-phenyl thioester **1-SPh** (2 mg) and coenzyme A (4 eqiv) were stirred in sodium phosphate buffer (4 ml, 50 mM, pH 8.5). The reaction was complete after 120 min, as monitored by LC/MS and the product was purified by semipreparative HPLC on a reverse phase C$_{18}$ column (Vydac 218TP1010, 250/10, pore diameter 300 Å, particle size 5 μm) using a gradient of 45–100% MeCN in 20 mM NH$_4$AcO in five column volumes, with a flow rate of 5 ml/min and subsequent desalting using a gradient of 25–40% MeCN + 0.1% TFA in water + 0.1% TFA in five column volumes at 5 ml/min. The corresponding hexapeptide-SCoA thioester **1-SCoA** was recovered in 40–60% yield. MALDI-MS m/z 1649.0 ± 0.5 ([M + H]$^+$). Chemical shift assignments were obtained from 1D and 2D ^{1}H NMR spectra and are given in Table 19.3.

Table 19.3 ¹H NMR Chemical shift assignments for peptide **1-SCoA** (600 MHz at 300 K in DMSO-d_6)

Residue	NH	C(α)-H	C(β)-H	Others[a]
Leu¹	–	3.58	1.50, 1.47	C(γ)H = 1.47, C(δ)H$_3$ = 0.75, 0.83, NMe = 1.98
Tyr²	8.84	4.74	2.94, 2.58	C(δ)H = 7.04, C(ε)H = 6.633
Asn³	8.49	4.64	2.48, 2.32	N(δ)H = 7.31, 6.91
Hpg⁴	8.01	5.50	–	C(γ)H = 7.16, C(δ)H = 6.66
Hpg⁵	8.74	5.71	–	C(γ)H = 7.00, C(δ)H = 6.63
Bht⁶	8.60	4.34	5.03	C(δ)H = 6.98, C(ε)H = 6.54
COSCoA	–	–	–	1 = 2.75, 2.85, 2 = 3.09, 3 = 8.19, 4 = 2.29, 5 = 3.30, 6 = 7.78, 7 = 3.43, 8 = 0.74, 9 = 0.96, 10 = 3.80, 11 = 4.10, 12 = 4.35, 13 = 4.89, 14 = 4.72, 15 = 5.94, 16 = 8.42, 17 = 8.17

[a] The numbering used for assignments in the CoA are shown below:

3.3. Synthesis of hexapeptide-PCP conjugate (1-S-PCP)

The reaction mixture containing the following components at the concentrations given [apo-PCP (120 μM), hexapeptidyl-CoA thioester **1-SCoA** (100 μM), *B. subtilis* Sfp (5 μM) and MgCl$_2$ (50 mM) in Tris/HCl buffer (50 mM, pH 7.5)], were incubated at 37 °C for 30 min. The reaction was monitored by analytical HPLC (Vydac C18 218TP54 column, 250 × 4.6 mm, pore diameter 300 Å, particle size 5 μm) using a gradient of 5–100% MeCN + 0.1% TFA in water + 0.1% TFA over four column volumes at 1 ml/min. The conversion proceeded quantitatively. The resulting peptide–PCP conjugate **1-S-PCP** was isolated by HPLC chromatography and the identity of the peptide–PCP conjugate was confirmed by MALDI-MS (**1-S-PCP**: m/z 11579 ± 2 [M + H]$^+$, calc. mass: 11579).

Upon complete reaction, the mixture was dialysed against TRIS/HCl buffer (50 mM, pH 7.5), and the resulting solution containing the peptide–PCP conjugate **1-S-PCP** was used in assays without further purification.

4. IN VITRO ASSAYS WITH OxyB

Assays with OxyB were carried out essentially as described earlier (Woithe et al., 2007; Zerbe et al., 2004). The assay contained the following components at the concentrations given: peptide–PCP-conjugate **1-S-PCP** (80 μM), OxyB (8 μM), recombinant spinach ferredoxin (20 μM), E. coli flavodoxin reductase (10 μM), NADPH (1 mM), glucose-6-phosphate (1 mM), glucose-6-phosphate-dehydrogenase (0.5 U) in TRIS/HCl buffer (50 mM, pH 7.5). Incubation at 30 °C for 60 min was followed by addition of 1/10 volume of a 25% v/v aqueous hydrazine solution. Further incubation at 30 °C for 30 min yielded the peptide hydrazides, which were separated from proteins by solid-phase extraction. The peptidic fraction was analyzed by analytical HPLC (see Fig. 19.6) on a reverse-phase C18 Zorbax Eclipse XDB

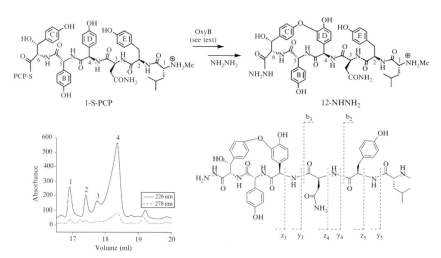

Figure 19.6 The reaction catalyzed by OxyB using **1-S-PCP** as substrate. Shown bottom left is a section of an HPLC chromatogram of the product mixture. By ES-MS peaks 1 and 2 correspond to epimeric linear peptide hydrazides, whereas peaks 3 and 4 correspond to epimeric monocyclic peptide hydrazides. Bottom right shows the fragment ions observed for peak 4 upon ES-MS/MS analysis.

Table 19.4 ^{1}H NMR Chemical shift assignments for peptide **12-NHNH$_2$** (600 MHz at 310 K in DMSO-d$_6$)

Residue	NH	C(α)-H	C(β)-H	Others
Leu^1	8.62	3.58	1.47, 1.37	C(γ)H = 1.47, C(δ)H$_3$ = 0.74, 0.80, NMe = 1.96
Tyr^2	8.77	4.77	2.97, 2.59	C(δ)H = 7.04, C(ε)H = 6.62, OH = 9.12
Asn^3	8.47	4.73	2.57, 2.38	N(δ)H = 7.30, 6.91
Hpg^4	7.94	5.36	–	C(γ1)H = 6.23, C(γ2)H = 6.64, C(δ)H = 6.78, OH = 9.32
Hpg^5	8.71	5.27	–	C(γ)H = 7.19, C(δ)H = 6.72, OH = 9.40
Bht^6	8.65	4.67	5.30	C(δ1)H = 6.94, C(δ2)H = 7.56, C(ε1)H = 7.24, C(ε2)H = 7.12
CONHNH$_2$	–	–	–	7.15, 7.06, 6.98

column eluting with a gradient of 5–40% acetonitrile + 0.1% TFA in water + 0.1% TFA in 8 column volumes with a flow rate of 1 ml/min.

The major product isolated from the assay (Fig. 19.6) was the monocyclic peptide hydrazide (**12-NHNH$_2$**), as confirmed by high-resolution MS (exact mass calculated for C$_{45}$H$_{53}$N$_9$O$_{12}$Na: 934.3711 ([M + Na]$^+$), m/z found 934.3709), ESI-MS/MS and 1D and 2D COSY, TOCSY, and NOESY spectra. ^{1}H NMR chemical shift assignments are given in Table 19.4. The occurrence of epimeric products during peptide S-PCP-thioester formation and/or the OxyB assay and work-up has been documented in earlier studies (Geib *et al.*, 2008; Woithe *et al.*, 2008).

5. Production and Purification of Enzymes

5.1. Production and purification of OxyB

Recombinant His$_6$-tagged OxyB was produced and purified as described earlier (Zerbe *et al.*, 2002). Thus, recombinant His$_6$-tagged OxyB was produced in *E. coli* BL21(DE3)pLysS, transformed with a pET14b-derived plasmid (pOCI1047) containing the *oxyB* gene. TB medium (400 ml) with ampicillin (100 µg/ml) and chloramphenicol (34 µg/ml) was inoculated with a preculture and incubated at 24 °C with shaking at 200 rpm. δ-Aminolevulinic acid (8 mg) was added at $A_{600} = 0.5$. Cultures were induced with IPTG at $A_{600} = 1$ and shaken at 22 °C. At 20 h postinduction a second portion of δ-aminolevulinic acid (8 mg) was added and the cultures were harvested by centrifugation after another 24 h.

The cells were disrupted in phosphate buffer (50 mM potassium phosphate, pH 7.4, 10% glycerol, 1 mM DTT, 1 mM PMSF, 1 mM benzamidine, 0.1% β-mercaptoethanol) by sonication. The *E. coli* cell-free extract containing His_6-OxyB was applied to a Ni-NTA column (0.5 cm × 5 cm, Ni^{2+} NTA superflow, *Qiagen*) pre-equilibrated with buffer A (potassium phosphate, pH 7.4, 50 mM, with KCl (300 mM), and imidazole (20 mM)), at a flow rate of 2 ml/min. After washing (20 column volumes of buffer A) the His_6-tagged protein was eluted in buffer A with 300 mM imidazole, and dialyzed against buffer B (Tris–HCl (50 mM), pH 7.5, and EDTA (0.5 mM)). The protein was then chromatographed on MonoQ (HR10/10, *GE Healthcare*) in buffer B at a flow rate of 2 ml/min, and eluted with a linear gradient of 0–0.4 M KCl over 100 ml with a flow rate of 1.5 ml/min. His_6-OxyB eluted with 0.17 M KCl, and showed a single band by SDS-PAGE, gave an electrospray mass spectrum consistent with the expected mass, and the correct N-terminal sequence by Edman degradation. The concentration of the purified P450 enzyme was determined spectrophotometrically from CO difference UV spectra using dithionite-reduced heme and an extinction coefficient of 91 mM^{-1} cm^{-1} (Omura and Sato, 1964).

The amino acid sequence of the His_6-tagged OxyB (His-tag underlined) is shown below.

```
1     GSSHHHHHHS SGLVPRGSHM SEDDPRPLHI RRQGLDPADE LLAAGALTRV
51    TIGSGADAET HWMATAHAVV RQVMGDHQQF STRRRWDPRD EIGGKGIFRP
101   RELVGNLMDY DPPEHTRLRR KLTPGFTLRK MQRMAPYIEQ IVNDRLDEME
151   RAGSPADLIA FVADKVPGAV LCELVGVPRD DRDMFMKLCH GHLDASLSQK
201   RRAALGDKFS RYLLAMIARE RKEPGEGMIG AVVAEYGDDA TDEELRGFCV
251   QVMLAGDDNI SGMIGLGVLA MLRHPEQIDA FRGDEQSAQR AVDELIRYLT
301   VPYSPTPRIA REDLTLAGQE IKKGDSVICS LPAANRDPAL APDVDRLDVT
351   REPIPHVAFG HGVHHCLGAA LARLELRTVF TELWRRFPAL RLADPAQDTE
401   FRLTTPAYGL TELMVAW*
```

5.2. Production and purification of spinach ferredoxin

Recombinant His_6-tagged spinach ferredoxin was produced and purified as described earlier (Woithe *et al.*, 2007). Thus, a plasmid containing the His_6-tagged ferredoxin gene (pOCI829) was introduced into *E. coli* BL21(DE3) pLysS (*Novagen*). The spinach ferredoxin was produced in LB medium containing ampicillin (100 µg/ml), chloramphenicol (34 µg/ml), and

FeSO$_4$ (10 μM) at 37 °C. The cultures were induced at $A_{600} = 1.4$ with IPTG (0.1 mM) and further cultivated at 30 °C for 4 h. Cell pellets were resuspended in phosphate buffer (20 ml, 50 mM, pH 7.4, 300 mM KCl, 1 mM DTT, 1 mM PMSF, 1 mM benzamidine, 0.1% 2-mercaptoethanol, 10% glycerol) and disrupted by sonication. After centrifugation (40,000×g for 45 min), the supernatant was loaded onto a 7 ml Ni-NTA superflow column (Qiagen) and equlibrated with 20 ml buffer A (50 mM phosphate buffer pH 7.4, 300 mM KCl, 20 mM imidazole). After washing with 40 ml buffer A, recombinant spinach ferredoxin was eluted with 10 ml buffer A containing 300 mM imidazole. Fractions containing the red spinach ferredoxin were dialyzed against 50 mM Tris–HCl buffer pH 7.5, 10% glycerol, flash-frozen and stored at −80 °C.

The molecular weight of the recombinant spinach ferredoxin was analyzed by ES-MS (pH 9 (+ [2Fe-2S] cluster) m/z obs. 12818 ± 1 [M + H$^+$], calc. 12818; pH 3 (with loss of the [2Fe-2S] cluster) m/z obs. 12645 ± 1 [M + H$^+$], calc. 12646). The UV spectrum in Tris–HCl buffer (50 mM, pH 7.5) showed absorption maxima (λ_{max}) at 273, 328, 420, and 465 nm, in agreement with the literature values for plant [2Fe-2S] ferredoxins (Armengaud et al., 2000; Crispin et al., 2001; Tagawa and Arnon, 1962). The solution concentration was determined by UV using $\varepsilon_{422\ nm} = 9700$ cm^{-1} M^{-1} (Crispin et al., 2001).

The amino acid sequence of the His$_6$-spinach ferredoxin is shown below, with the His-tag underlined:

```
1   GSSHHHHHHS SGLVPRGSHM AAYKVTLVTP TGNVEFQCPD DVYILDAAEE
51  EGIDLPYSCR AGSCSSCAGK LKTGSLNQDD QSFLDDDQID EGWVLTCAAY
101 PVSDVTIETH KEEELTA
```

5.3. Production and purification of PCP domain

A recombinant His$_6$-tagged PCP domain from module-7 of the vancomycin NRPS (vpsC) from A. orientalis (residues 967–1043, including a Cys979Ser mutation, to avoid dimer formation through oxidation) was produced and purified as described earlier (Woithe et al., 2007). Plasmid pOCI865 was introduced into E. coli Rosetta2(DE3)pLysS (Novagen). An overnight preculture (20 ml) was used to inoculate TB medium (400 ml) containing ampicillin (100 μg/ml) and chloramphenicol (34 μg/ml), with incubation at 37 °C and shaking at 200 rpm. The cultures were induced at $A_{600} = 1.2$ with IPTG (0.1 mM) and further cultivated at 22 °C for 6 h. Cell pellets were resuspended in 20 ml phosphate buffer (50 mM, pH 7.4, 1 mM DTT, 1 mM PMSF, 1 mM benzamidine, 0.1% 2-mercaptoethanol, 10% glycerol) and disrupted by sonication (three bursts of 60 s). After centrifugation (40,000×g for 45 min), the supernatant was loaded onto a 7 ml Ni-NTA superflow column (Qiagen) and equlibrated with 20 ml buffer A (50 mM phosphate buffer pH 7.4, 300 mM KCl, 20 mM imidazole). After washing with 40 ml buffer A, recombinant

PCP was eluted with 10 ml buffer A containing 300 mM imidazole. Eluted protein fractions were purified further by anion exchange chromatography (MonoQ HR10/10, *GE Healthcare*) in buffer B (20 mM diethanolamine pH 8.5, gradient: 0–0.5 M KCl in 15 cv). The proteins eluted at 0.28 M KCl. The isolated PCP was dialysed against 50 mM Tris–HCl buffer pH 7.5, 10% glycerol, flash-frozen and stored at $-80\,^{\circ}$C.

The concentration of purified PCP was measured spectrophotometrically at 562 nm using a colorimetric BCA protein assay (*Pierce*, extinction coefficient $\varepsilon_{562\,nm} = 6833$ cm^{-1} M^{-1} determined by quantitative amino acid analysis). The purified protein showed a single band by SDS-PAGE and gave an electrospray mass spectrum (ES-MS) corresponding to the loss of N-terminal methionine (m/z obs. [M + H]$^{+}$ 10360 ± 2, calc. 10358).

The sequence of the PCP, which includes residues 967–1043 of VpsC and a His$_6$-tag (underlined) and a Cys-to-Ser mutation (underlined), is shown below.

```
 1   GSSHHHHHHS SGLVPRGSHM EKAPENETEK VLSALFAEIL SVDQVGVDDA
51   FQDLGGSSAL AMRLVARIRE ELGEDLPIRQ LFSSPTPAGL ARALAAK
```

5.4. Production and purification of *E. coli* NADPH-flavodoxin reductase

The production and purification of His$_6$-tagged *E. coli* NADPH-flavodoxin reductase (eco-FlvR) was performed as described previously (Woithe *et al.*, 2007). The plasmid pOCI819 containing the required gene was introduced into *E. coli* BL21(DE3)pLysS cells (*Novagen*). His$_6$-tagged eco-FlvR was produced in LB medium containing ampicillin (100 μg/ml) and chloramphenicol (34 μg/ml) at 37 °C. The cultures were induced at $A_{600} = 0.6$ with IPTG (0.1 mM) and further cultivated at 37 °C for 4 h.

The cell pellet was resuspended in 20 ml phosphate buffer (50 mM, pH 7.4, 1 mM DTT, 1 mM PMSF, 1 mM benzamidine, 0.1% 2-mercaptoethanol, 10% glycerol) and disrupted by sonication (three bursts of 60 s). After centrifugation (40,000×g for 45 min) the supernatant was loaded onto a 7 ml Ni-NTA superflow column (*Qiagen*) and equilibrated with 20 ml buffer A (50 mM phosphate buffer pH 7.4, 300 mM KCl, 20 mM imidazole). After washing with 40 ml buffer A, the recombinant eco-FlvR was eluted from the column with 10 ml buffer A containing 300 mM imidazole. Fractions containing bright yellow eco-FlvR were dialysed against 50 mM Tris–HCl buffer pH 7.5, 10% glycerol, flash-frozen and stored at $-80\,^{\circ}$C.

The UV–vis absorbance spectrum of the protein showed the typical maxima at 400, 456, and 483 nm (Jenkins and Waterman, 1994). The flavodoxin reductase concentration was determined using the absorbance of bound FAD at 456 nm ($\varepsilon = 7100$ M^{-1} cm^{-1}) (Fujii and Huennekens, 1974). The purified protein showed a single band by SDS-PAGE and gave an ES-MS corresponding to the loss of N-terminal methionine (m/z obs. [M + H]$^{+}$ 29776 ± 6, calc. 29783).

REFERENCES

Armengaud, J., Gaillard, J., and Timmis, K. N. (2000). A second [2Fe-2S] ferredoxin from *Sphingomonas* sp. strain RW1 can function as an electron donor for the dioxin dioxygenase. *J. Bacteriol.* **182**, 2238–2244.

Bischoff, D., Pelzer, S., Bister, B., Nicholson, G. J., Stockert, S., Schirle, M., Wohlleben, W., Jung, G., and Süssmuth, R. D. (2001a). The biosynthesis of vancomycin-type glycopeptide antibiotics—The order of the cyclization steps. *Angew. Chem. Int. Ed.* **40**, 4688–4691.

Bischoff, D., Pelzer, S., Höltzel, A., Nicholson, G. J., Stockert, S., Wohlleben, W., Jung, G., and Sussmuth, R. D. (2001b). The biosynthesis of vancomycin-type glycopeptide antibiotics—New insights into the cyclization steps. *Angew. Chem. Int. Ed.* **40**, 1693–1696.

Crispin, D. J., Street, G., and Varey, J. E. (2001). Kinetics of the decomposition of [2F-2S] ferredoxin from spinach: Implications for iron bioavailability and nutritional status. *Food Chem.* **72**, 355–362.

Donadio, S., Sosio, M., Stegmann, E., Weber, T., and Wohlleben, W. (2005). Comparative analysis and insights into the evolution of gene clusters for glycopeptide antibiotic biosynthesis. *Mol. Genet. Genomics* **274**, 40–50.

Evans, D. A., and Weber, A. E. (1986). Asymmetric glycine enolate aldol reactions: Synthesis of cyclosporin's unusual amino acid, MeBmt. *J. Am. Chem. Soc.* **108**, 6757–6761.

Evans, D. A., and Weber, A. E. (1987). Synthesis of the cyclic hexapeptide echinocandin D. New approaches to the asymmetric synthesis of β-hydroxy alpha-amino acids. *J. Am. Chem. Soc.* **109**, 7151–7157.

Freund, E., and Robinson, J. A. (1999). Solid-phase synthesis of a putative heptapeptide intermediate in vancomycin biosynthesis. *Chem. Commun.* 2509–2510.

Fujii, K., and Huennekens, F. M. (1974). Activation of methionine synthetase by a reduced triphosphopyridine nucleotide-dependent flavoprotein system. *J. Biol. Chem.* **249**, 6745–6753.

Funa, N., Funabashi, M., Ohnishi, Y., and Horinouchi, S. (2005). Biosynthesis of hexahydroxyperylenequinone melanin via oxidative aryl coupling by cytochrome P-450 in *Streptomyces griseus*. *J. Bacteriol.* **187**, 8149–8155.

Geib, N., Woithe, K., Zerbe, K., Li, D. B., and Robinson, J. A. (2008). New insights into the first oxidative phenol coupling reaction during vancomycin biosynthesis. *Bioorg. Med. Chem. Lett.* **18**, 3081–3084.

Grünewald, J., and Marahiel, M. A. (2006). Chemoenzymatic and template-directed synthesis of bioactive macrocyclic peptides. *Microbiol. Mol. Biol. Rev.* **70**, 121–146.

Hayakawa, Y., Kato, H., Uchiyama, M., Kajino, H., and Noyori, R. (1986). Allyloxycarbonyl group: A versatile blocking group for nucleotide synthesis. *J. Org. Chem.* **51**, 2400–2402.

Helynck, G., Dubertret, C., Frechet, D., and Leboul, J. (1998). Isolation of RP 66453, a new secondary peptide metabolite from *Streptomyces* sp. useful as a lead for neurotensin antagonists. *J. Antibiot.* **51**, 512–514.

Hitotsuyanagi, Y., Ishikawa, H., Hasuda, T., and Takeya, K. (2004). Isolation, structural elucidation, and synthesis of RA-XVII, a novel hexapeptide from *Rubida cordifolia*, and the effect of side chain at residue 1 upon the conformation and cytotoxic activity. *Tet. Lett.* **45**, 935–938.

Holtzel, A., Schmid, D. G., Nicholson, G. J., Stevanovic, S., Schimana, J., Gebhardt, K., Fiedler, H.-P., and Jung, G. (2002). Arylomycins A and B, new biaryl-bridged lipopeptide antibiotics produced by *Streptomyces* sp. Tu 6075. II. Structure elucidation. *J. Antibiot.* **55**, 571–577.

Howard-Jones, A. R., and Walsh, C. T. (2007). Nonenzymatic oxidative steps accompanying action of the cytochrome P450 enzymes StaP and RebP in the biosynthesis of staurosporine and rebeccamycin. *J. Am. Chem. Soc.* **129,** 11016–11017.

Hubbard, B. K., and Walsh, C. T. (2003). Vancomycin assembly: Nature's way. *Angew. Chem. Int. Ed.* **42,** 730–765.

Ikezawa, N., Iwasa, K., and Sato, F. (2008). Molecular cloning and characterization of CYP80G2, a cytochrome p450 that catalyzes an intramolecular C–C phenol coupling of (S)-reticuline in magnoflorine biosynthesis, from cultured *Coptis japonica* cells. *J. Biol. Chem.* **283,** 8810–8821.

Jenkins, C. M., and Waterman, M. R. (1994). Flavodoxin and NADPH-flavodoxin reductase from *Escherichia coli* support bovine cytochrome P450c17 hydroxylase activities. *J. Biol. Chem.* **269,** 27401–27408.

Jolad, S. D., Hoffman, J. J., Torrance, S. J., Wiedhopf, R. M., Cole, J. R., Arora, S. K., Bates, R. B., Gargiulo, R. L., and Kriek, G. R. (1977). Bouvardin and deoxybouvardin, antitumor cyclic hexapeptides from *Bouvardia ternifolia* (Rubiaceae). *J. Am. Chem. Soc.* **99,** 8040–8044.

Kohno, J., Koguchi, Y., Nishio, M., Nakao, K., Kuroda, M., Shimizu, R., Ohnuki, T., and Komatsubara, S. (2000). Structures of TMC-95A-D: Novel proteasome inhibitors from *Apiospora montagnei* Sacc. TC 1093. *J. Org. Chem.* **65,** 990–995.

Kraus, F., and Kutchan, T. M. (1995). Molecular cloning and heterologous expression of a cDNA encoding berbamunine synthase, a C–O phenol-coupling cytochrome P450 from the higher plant *Berberis stolonifera*. *Proc. Natl. Acad. Sci. USA* **92,** 2071–2075.

Li, D. B., and Robinson, J. A. (2005). An improved solid-phase methodology for the synthesis of putative hexa- and heptapeptide intermediates in vancomycin biosynthesis. *Org. Biomol. Chem.* **3,** 1233–1239.

Makino, M., Sugimoto, H., Shiro, Y., Asamizu, S., Onaka, H., and Nagano, S. (2007). Crystal structures and catalytic mechanism of cytochrome P450 StaP that produces the indolocarbazole skeleton. *Proc. Natl. Acad. Sci. USA* **104,** 11591–11596.

Matsuzaki, K., Ikeda, H., Ogino, T., Matsumoto, A., Woodruff, H. B., Tanaka, H., and Omura, S. (1994). Chloropeptins I and II, novel inhibitors against gp120-CD4 binding from *Streptomyces* sp. *J. Antibiot.* **47,** 1173–1174.

Naruse, N., Oka, M., Konishi, M., and Oki, T. (1993). New antiviral antibiotics, kistamicins A and B II. Structure determination. *J. Antibiot.* **46,** 1812–1818.

O'Brien, D. P., Kirkpatrick, P. N., O'Brien, S. W., Staroske, T., Richardson, T. I., Evans, D. A., Hopkinson, A., Spencer, J. B., and Williams, D. H. (2000). Expression and assay of an N-methyltransferase involved in the biosynthesis of a vancomycin group antibiotic. *Chem. Comm.* 103–104.

Omura, T., and Sato, R. (1964). The carbon monoxide-binding pigment of liver microsomes I. Evidence for its hemoprotein nature. *J. Biol. Chem.* **239,** 2370–2378.

Paquet, A. (1982). Introduction of 9-fluorenylmethyloxycarbonyl, trichloroethoxycarbonyl, and benzyloxycarbonyl amine protecting groups into O-unprotected hydroxyamino acids using succinimidyl carbonates. *Can. J. Chem.* **60,** 976–980.

Pylypenko, O., Vitali, F., Zerbe, K., Robinson, J. A., and Schlichting, I. (2003). Crystal structure of OxyC, a cytochrome P450 implicated in an oxidative C–C coupling reaction during vancomycin biosynthesis. *J. Biol. Chem.* **278,** 46727–46733.

Quadri, L. E. N., Weinreb, P. H., Lei, M., Nakano, M. M., Zuber, P., and Walsh, C. T. (1998). Characterization of Sfp, a *Bacillus subtilis* phosphopantetheinyl transferase for peptidyl carrier protein domains in peptide synthetases. *Biochemistry* **37,** 1585–1595.

Sano, S., Ikai, K., Katayama, K., Takesato, K., Nakamura, T., Obayashi, A., Ezure, Y., and Enomoto, H. (1986). OF4949, New inhibitors of aminopeptidase B. *J. Antibiot.* **39,** 1685–1696.

Seto, H., Fujioka, T., Furihata, K., Kaneko, I., and Takahashi, S. (1989). Structure of complestatin, a very strong inhibitor of protease activity of complement in the human complement system. *Tet. Lett.* **30,** 4987–4990.

Sussmuth, R. D., Pelzer, S., Nicholson, G., Walk, T., Wohlleben, W., and Jung, G. (1999). New advances in the biosynthesis of glycopeptide antibiotics of the vancomycin type from *Amycolatopsis mediterranei*. *Angew. Chem. Int. Ed.* **38,** 1976–1979.

Tagawa, K., and Arnon, D. I. (1962). Ferredoxins as electron carriers in photosynthesis and in the biological production and consumption of hydrogen gas. *Nature* **484,** 537–543.

Uchida, I., Shigematsu, N., Ezaki, M., Hashimoto, M., Aoki, H., and Imanaka, H. (1985). Biphenomycins A and B, novel peptide antibiotics II. Structural elucidation of biphenomycins A and B. *J. Antibiot.* **38,** 1462–1468.

Vitali, F., Zerbe, K., and Robinson, J. A. (2003). Production of vancomycin aglycone conjugated to a peptide carrier domain derived from a biosynthetic non-ribosomal peptide synthetase. *Chem. Comm.* 2718–2719.

Woithe, K., Geib, N., Meyer, O., Wörtz, T., Zerbe, K., and Robinson, J. A. (2008). Exploring the substrate specificity of OxyB, a phenol coupling P450 enzyme involved in vancomycin biosynthesis. *Org. Biomol. Chem.* **6,** 2861–2867.

Woithe, K., Geib, N., Zerbe, K., Li, D. B., Heck, M., Fournier-Rousset, S., Meyer, O., Vitali, F., Matoba, N., Abou-Hadeed, K., and Robinson, J. A. (2007). Oxidative phenol coupling reactions catalyzed by OxyB: A cytochrome P450 from the vancomycin producing organism. Implications for vancomycin biosynthesis. *J. Am. Chem. Soc.* **129,** 6887–6895.

Yasuzawa, T., Shirahata, K., and Sano, H. (1987). K-13, a novel inhibitor of angiotensin I converting enzyme produced by *Micromonospora halophytica* subsp. *exilisia*. *J. Antibiot.* **40,** 455–458.

Zerbe, K., Pylypenko, O., Vitali, F., Zhang, W. W., Rouse, S., Heck, M., Vrijbloed, J. W., Bischoff, D., Bister, B., Süssmuth, R. D., Pelzer, S., Wohlleben, W., *et al.* (2002). Crystal structure of OxyB, a cytochrome P450 implicated in an oxidative phenol coupling reaction during vancomycin biosynthesis. *J. Biol. Chem.* **277,** 47476–47485.

Zerbe, K., Woithe, K., Li, D. B., Vitali, F., Bigler, L., and Robinson, J. A. (2004). An oxidative phenol coupling reaction catalyzed by OxyB, a cytochrome P450 from the vancomycin-producing microorganism. *Angew. Chem. Int. Ed.* **43,** 6709–6713.

Zhao, B., Guengerich, F. P., Bellamine, A., Lamb, D. C., Izumikawa, M., Lei, L., Podust, L. M., Sundaramoorthy, M., Kalaitzis, J. A., Reddy, L. M., Kelly, S. L., Moore, B. S., *et al.* (2005). Binding of two flaviolin substrate molecules, oxidative coupling, and crystal structure of *Streptomyces coelicolor* A3(2) cytochrome P4450 158A2. *J. Biol. Chem.* **280,** 11599–11607.

Ziegler, J., and Facchini, P. J. (2008). Alkaloid biosynthesis: Metabolism and trafficking. *Annu. Rev. Plant Biol.* **59,** 735–769.

CHAPTER TWENTY

Biosynthesis and Genetic Engineering of Lipopeptides in *Streptomyces roseosporus*

Richard H. Baltz

Contents

1. Introduction	512
2. Biosynthesis and Genetic Engineering of Daptomycin in *S. rosesoporus*	514
2.1. Cloning of the daptomycin biosynthetic genes and deductions on biosynthesis	514
2.2. Heterologous expression of the daptomycin gene cluster in *S. lividans*	517
2.3. Construction of deletion mutants of *S. roseosporus* and expression vectors	517
2.4. Functions of *dptGHIJ* genes	520
2.5. Reconstitution of the daptomycin biosynthetic pathway by ectopic expression of individual genes or groups of genes	521
3. Sources of Genes for Combinatorial Biosynthesis	523
3.1. The A54145 biosynthetic genes from *Streptomyces fradiae* and the CDA genes from *Streptomyces coelicolor*	523
4. Genetic Engineering of Novel Lipopeptides	524
4.1. Gene replacement	524
4.2. Module and multidomain replacements	525
4.3. Combinatorial biosynthesis	526
4.4. *In vitro* antibacterial activities of novel lipopeptides	526
5. Concluding Remarks	527
Acknowledgments	528
References	528

Abstract

Daptomycin is an acidic cyclic lipopeptide antibiotic approved for treatment of infections caused by Gram-positive pathogens, including *Staphylococcus aureus* strains resistant to other antibiotics. Daptomycin biosynthesis is carried out by a giant multisubunit, multienzyme nonribosomal peptide synthetase (NRPS). The daptomycin (*dpt*) biosynthetic genes have been cloned in a bacterial artificial chromosome (BAC) vector, sequenced, and expressed in *Streptomyces lividans*. Several of the *dpt* genes, including the three NRPS genes, are transcribed as a lengthy polycistronic message. The daptomycin-producing strain, *Streptomyces roseosporus*, can be genetically manipulated, and a number of deletion mutants encompassing one or more of the *dpt* genes have been constructed. Several of the *dpt* genes have been expressed from ectopic chromosomal loci (ϕC31 or IS*117 attB* sites) under the transcriptional control of the strong constitutive *ermEp** promoter, and recombinant strains produced high levels of lipopeptides, thus establishing a *trans*-complementation system for combinatorial biosynthesis. A number of hybrid NRPS subunits have been generated by λ-Red-mediated recombination, and combinatorial libraries of lipopeptides have been generated by NRPS subunit exchanges, module exchanges, multidomain exchanges, deletion mutagenesis, and multiple natural lipidations, using the ectopic *trans*-complementation system in *S. roseosporus*.

1. Introduction

The cyclic acidic lipopeptide antibiotic complex, A21978C, was discovered by scientists at Eli Lilly and Company by screening soil samples for actinomycetes that produce antibiotics active against Gram-positive pathogens (Debono *et al.*, 1987). A21978C is composed of several lipopeptide factors containing different lipid side chains. In the three major factors, referred to as $A21978C_1$, $A21978C_2$, $A21978C_3$, these are *anteiso*-undecanoate, *iso*-dodecanoate, and *anteiso*-tridecanoate (Fig. 20.1). The lipid side chains can be removed by a deacylase produced by *Actinoplanes utahensis*, leaving the tridecapeptide (consisting of 13 amino acid residues) with a free amino group at the N-terminus of Trp_1 amenable to reacylation by different fatty acids (Baltz *et al.*, 2005). Lilly carried out a chemical modification program, and found that the derivative containing an *N*-decanoyl group (subsequently named daptomycin) had the best efficacy against Gram-positive pathogens, coupled with low toxicity (Baltz *et al.*, 2005; Debono *et al.*, 1988). Daptomycin is currently produced commercially by a process that includes feeding the decanoic acid side chain during fermentation (Baltz *et al.*, 2005; Huber *et al.*, 1988).

Lilly initiated clinical studies with daptomycin in the late 1980s to early 1990s. Clinical trials were subsequently abandoned when muscle toxicity was observed in a phase I safety study aimed at increasing the dose from 3 to 4 mg/kg twice per day to treat staphylococcal endocarditis (Eisenstein *et al.*, 2009). In 1997, Cubist Pharmaceuticals licensed daptomycin, and discovered that the muscle toxicity could be avoided by dosing at higher levels once per day (Eisenstein *et al.*, 2009). This enabled Cubist to carry out

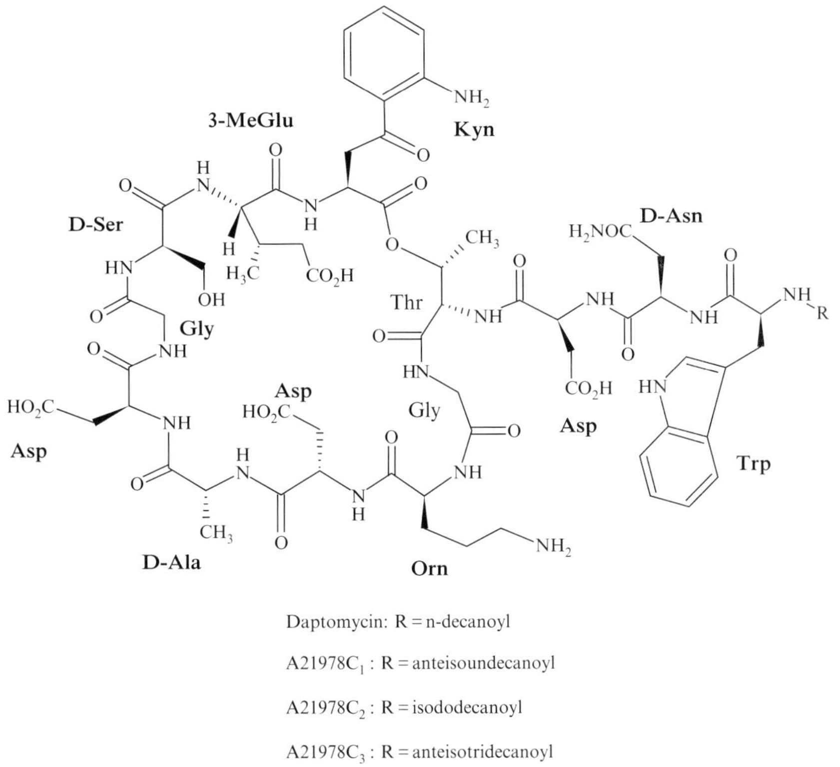
3-MeGlu
Kyn
D-Ser
D-Asn
Thr
Gly
Asp
Asp
Gly
Asp
Trp
D-Ala
Orn
Daptomycin: R = n-decanoyl
A21978C1 : R = anteisoundecanoyl
A21978C2 : R = isododecanoyl
A21978C3 : R = anteisotridecanoyl

Figure 20.1 Structures of A21978C factors and daptomycin. (Figure reproduced from Baltz *et al.*, 2005.)

clinical trials and gain FDA approval for skin and skin structure infections caused by various Gram-positive pathogens (Arbeit *et al.*, 2004), and for bacteremia and right-sided endocarditis caused by *Staphylococcus aureus* strains, including those resistant to methicillin (MRSA) (Fowler *et al.*, 2006). One shortcoming of daptomycin is that it failed to show noninferiority in clinical trials to treat community-acquired pneumonia (Pertel *et al.*, 2008). It appears that daptomycin becomes sequestered in lung surfactant, thus interfering with its availability to treat *Streptococcus pneumoniae* (Silverman *et al.*, 2005), even though it is very active against this pathogen *in vitro* (Baltz *et al.*, 2005). Improved activity in the lung presents a possible target for an improved derivative of daptomycin.

The possibilities for modifying daptomycin by medicinal chemistry via readily accessible routes are limited to exchanges of the lipid tail and modification of the δ-amino group of Orn_6 (Fig. 20.1; Baltz, 2008; Baltz *et al.*, 2005). Since daptomycin is synthesized by an NRPS mechanism (Baltz, 2008; Baltz *et al.* 2005; Fischbach and Walsh, 2006; Sieber and

Marahiel, 2005), genetic engineering of the multisubunit multienzyme NRPS is an attractive approach to generate new derivatives for evaluation.

2. BIOSYNTHESIS AND GENETIC ENGINEERING OF DAPTOMYCIN IN *S. ROSESOPORUS*

In order to genetically engineer the biosynthesis of a complex antibiotic like daptomycin to generate novel derivatives, a number of technical issues need to be addressed. First, the genes need to be cloned, sequenced, and functionally characterized. Second, gene expression host(s) need to be developed. These could be heterologous hosts that do not normally produce daptomycin, or homologous hosts containing deletion mutations encompassing different sets of daptomycin biosynthetic genes. Third, vectors for the stable expression of one or more of the daptomycin genes in ectopic loci under the control of an appropriate promoter are required. Fourth, a facile method to engineer the daptomycin NRPS genes in *Escherichia coli* is advantageous to facilitate the process. Fifth, a method to transfer the cloned and engineered genes on plasmids from *E. coli* to the chosen streptomycete expression host is required.

In this chapter, I discuss how these technical issues have been successfully addressed in an industrial strain, *Streptomyces roseosporus*. This serves as an example of how the biosynthesis of complex peptides can be exploited to generate novel compounds not attainable, or if attainable, not scalable by medicinal chemistry. It is not so much the details of engineering the daptomycin gene cluster *per se* that may be of highest interest to the scientific community, but the strategy and generality of the approaches. As such, I do not include many specific details, but general concepts that have proven to work. The concepts are not exhaustive, and can undoubtedly be improved upon. In particular, there are not yet many examples of highly productive module or domain exchanges in the *S. roseosporus* system, or in other streptomycete systems for that matter. Therefore, this system should serve as a harbinger for better things to come in the future.

2.1. Cloning of the daptomycin biosynthetic genes and deductions on biosynthesis

Initially, in work carried out at Eli Lilly and Company in the early 1990s, the daptomycin genes were localized to one end of the linear chromosome of *S. roseosporus* by transposon mutagenesis and subsequent insertional mutagenesis with cloned fragments of the NRPS genes (McHenney and Baltz, 1996; McHenney *et al.*, 1998). The genes were cloned in a cosmid vector, and partial a sequence was obtained, confirming the presence of

NRPS sequences. After the licensing of daptomycin from Lilly, Cubist cloned the complete daptomycin gene cluster in pStreptoBAC V (Miao et al., 2005). Importantly, this cloning vector was derived from the original BAC vector based upon the single copy E. coli F-factor that allows for stable maintenance of large inserts in E. coli (Shizuya et al., 1992). pStreptoBAC V has three key features for cloning genes from Streptomyces and other actinomycetes: (i) sacB for selection of insertions; (ii) oriT (in cis) for conjugal transfer from E. coli strains harboring plasmids RK2 or RP4 that express transfer functions in trans to drive conjugal transfer into actinomycetes (Baltz, 1998, 2006a,b; Bierman et al., 1992); and (iii) ϕC31 att/int functions for site-specific integration into the ϕC31 attB site in streptomycetes (Baltz, 1998, 2006a,b; Bierman et al., 1992). The plasmid carrying a 128-kb insert containing the complete daptomycin gene cluster was designated pCV1 (Miao et al., 2005; Table 20.1).

The daptomycin biosynthetic gene cluster is shown in Fig. 20.2. Daptomycin is assembled by a NRPS that contains three giant multimodular multienzyme subunits, DptA, DptBC, and DptD, encoded by the dptA, dptBC, and dptD genes. Just upstream of the NRPS genes are the dptE and dptF genes which encode an acyl-CoA ligase (DptE) and an acyl carrier protein (DptF). DptA has five modules which catalyze the binding, activation, and incorporation of the first five amino acids; DptBC has six modules; and DptD has two (Figs. 20.2 and 20.3). The first module of DptA has a special C^{III} condensation domain that is likely to interact with the DptE and DptF proteins to initiate lipopeptide assembly by coupling the long-chain fatty acids to the N-terminus of Trp_1. The daptomycin cluster also contains genes encoding a methyltransferase involved in the biosynthesis of $3mGlu_{12}$ (dptI) and a tryptophan 2,3-dioxygenase (dptJ) involved in biosynthesis of Kyn_{13}, discussed below. All of the dpt genes shown in Fig. 20.2 are transcribed in the same direction, probably as a single giant mRNA (Coëffet-Le Gal et al., 2006) which is up-regulated several fold in a strain selected for improved daptomycin production relative to the wild type strain (Rhee and Davies, 2006).

Several of the NRPS modules have typical CAT structures with condensation (C), adenylation (A), and thiolation (T) domains (Fischbach and Walsh, 2006). Three of the modules (positions 2, 8, and 11) have CATE structures containing epimerase (E) domains that bind and activate L-amino acids and then convert them to D-isomers (Fig. 20.3). The C domains following CAT modules (with one exception in the daptomycin NRPSs) are specialized to accept L-amino acids, and have been referred to as ^LC_L, because they accept an incoming peptide with a proximal L-amino acid ready to couple with a downstream L-amino acid during peptide elongation. C domains following CATE modules, and following the $CA_{Gly5}T$ module, have specialized C^{II} sequences, also known as ^DC_L, (Fischbach and Walsh, 2006; Sattely et al., 2008) that couple growing peptides containing proximal

Table 20.1 Strains and plasmids

Strains and plasmids	Relevant characteristics	Reference
S. roseosporus		
UA343	A21978C producer	NRRL 15998
UA117	UA343 *rpsL7* (Sm^R)	Miao et al. (2006b)
UA378	UA117 Δ*dptD::ermE*	Miao et al. (2006b)
UA474	UA117 Δ*dptA* Δ*dptD::ermE*	Coëffet-Le Gal et al. (2006)
UA431	UA117 Δ*dptEFABCDGHIJ*	Nguyen et al. (2006b)
KN100	UA117 Δ*dptBCD*	Nguyen et al. (2006b)
KN125	UA117 Δ*dptBCDGHIJ*	Nguyen et al. (2006b)
Plasmids		
pRHB538	*rpsL rep*^{ts} Am^R	Hosted and Baltz (1997)
pHM11a	Hm^R *oriT ermEp* att*/int*^{IS117}	Motamedi et al. (1995)
pRB04	pHM11a::*dptD*	Miao et al. (2006b)
pMF26	pHM11a::*cdaPS3*	Miao et al. (2006b)
pMF30	pHM11a::*lptD*	Miao et al. (2006b)
pKN33	pHM11a::*dptGHI*	Nguyen et al. (2006a)
pKN34	pHM11a::*dptGHIJ*	Nguyen et al. (2006a)
pKN37	pHM11a::*dptGIJ*	Nguyen et al. (2006a)
pKN38	pHM11a::*dptGHJ*	Nguyen et al. (2006a)
pStreptoBAC V	BAC *oriT att/int*$^{\phi C31}$ Am^R	Miao et al. (2005)
pCV1	pStreptoBAC V::128 kb (*dpt*)	Miao et al. (2005)
pDA300	pStreptoBAC V::*dptMNEFABCDGHIJ*	Nguyen et al. (2006a)
pDR2153	pDA300 Δ*D-Ala8-CAT::D-Ser-CAT*	Nguyen et al. (2006a)
pDR2155	pDR2153 Δ*dptDGHIJ*	Nguyen et al. (2006a)
pDR2158	pDA300 Δ*D-Ser11-CAT::D-Ala-CAT*	Nguyen et al. (2006a)
pDR2160	pDR2158 Δ*dptDGHIJ*	Nguyen et al. (2006a)
pLT01	pStreptoBAC V::*dptPMNEFA*	Coëffet-Le Gal et al. (2006)
pLT03	pStreptoBAC V::*dptEF ermEp* *dptA*	Coëffet-Le Gal et al. (2006)
pKN24	pStreptoBAC V::*dptBC*	Nguyen et al. (2006a)
pKN25	pStreptoBAC V::*dptBCD*	Nguyen et al. (2006a)
pKN26	pStreptoBAC V::*dptBCDGHIJ*	Nguyen et al. (2006a)

(*continued*)

Table 20.1 (continued)

Strains and plasmids	Relevant characteristics	Reference
pKN65	pKN24 ΔD-Ala8-CAT:: D-Asn-CAT	Nguyen et al. (2006a)
pKN66	pKN24 ΔD-Ala8-CATE:: D-Asn-CATE	Nguyen et al. (2006a)
pKN67	pKN24 ΔD-Ser11-CAT:: D-Asn-CAT	Nguyen et al. (2006a)
pKN68	pKN24 ΔD-Ser11-CATE:: D-Asn-CATE	Nguyen et al. (2006a)
pKN69	pKN24 ΔAla8-CAT:: D-Asn-CAT	Nguyen et al. (2006a)

D-amino acids to a downstream L-amino acid. The special case of the $CA_{Gly5}T$ module being followed by a C^{II} (^DC_L) condensation domain may be a historic vestige of an ancient lipopeptide pathway that had a D-amino acid at this position. Indeed, the phylogenetically related cyclic lipopeptide pathways, CDA and friulimicin, have D-amino acids at this position of the 10-membered ring (Baltz, 2008). The distinctions between the three different catalytic capabilities of the C domains identified in the daptomycin NRPSs need to be taken into account when designing genetic engineering strategies.

2.2. Heterologous expression of the daptomycin gene cluster in *S. lividans*

When pCV1 was introduced into *S. lividans* and inserted in the chromosomal ϕC31 *attB* site, the recombinant produced $A21978C_{1-3}$ in the same ratios as produced in *S. roseosporus* (Baltz, 2008; Miao *et al.*, 2005; Penn *et al.*, 2006), proving that all of the genes required for daptomycin production, including genes dedicated to regulation, resistance, and transport, were present on pCV1. Some of those additional genes have been tentatively assigned, mainly through bioinformatics (Baltz, 2008; Miao *et al.*, 2005). After minimal fermentation medium development, *S. lividans* (pCV1) produced 55 mg/l of A21987C factors. This provided a potential heterologous host system for engineering the daptomycin pathway, but was confounded by the coproduction of calcium-dependent antibiotic (CDA) (Penn *et al.*, 2006).

2.3. Construction of deletion mutants of *S. roseosporus* and expression vectors

Chromosomal deletions of single or multiple genes were generated in *S. roseosporus* by using positive selection for deletion mutations by double crossovers. Plasmid pRHB538 (Hosted and Baltz, 1997) is temperature-sensitive

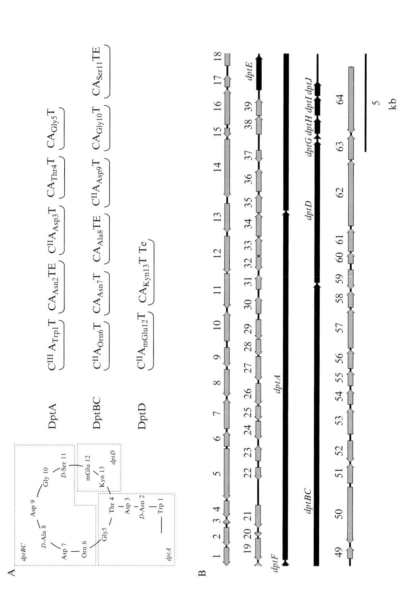

Figure 20.2 Organization of daptomycin biosynthetic genes and NRPS subunit structures. (A) The module and domain organization of the DptA, DptBC, and DptD subunits of the NRPS, encoded by *dptA*, *dptBC*, and *dptD*. (B) Organization of *dpt* genes within a 128-kb segment of *S. roseosporus* chromosomal DNA cloned in pCV1 (Table 20.1). (Figure reproduced from Miao *et al.*, 2005.)

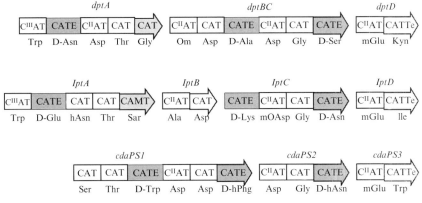

Figure 20.3 Organization and structure of the NRPS genes for daptomycin, A54145 and CDA biosynthesis. Condensation domains, C, C^{II}, and C^{III}, generally couple L-amino acids, D-amino acids, and long-chain fatty acids, respectively. Dark shading highlights the locations of CATE modules that incorporate D-amino acids. Light shading highlights where achiral amino acids, Gly or sarcosine (Sar), are followed by C^{II} condensation domains. (Figure modified from Baltz, 2008.)

for replication (*rep*^{ts}), has a cloned *rpsL* gene encoding dominant streptomycin sensitivity (Sm^S), and has an apramycin resistance gene (Am^R) for selection. This vector can be used for selection for double crossovers in Sm^R streptomycetes. *S. roseosporus* UA117 has a recessive *rpsL* mutation that expresses a Sm^R phenotype. At least 1 kb each of DNA flanking the 5′ and 3′ ends of the gene(s) to be deleted was cloned flanking the *ermE* erythromycin resistance gene from *Saccharopolyspora erythraea* as *dpt-ermE-dpt* cassettes. Several mutants were also constructed by in-frame deletions without the insertion of the *ermE* gene. Transformants were first selected for Am^R, then Am^S, Sm^R recombinants containing double crossovers were obtained after additional growth to cure the freely replicating plasmid. Deletions were confirmed by PCR. By using this method, a number of different strains, including one deleted for all of the daptomycin genes discussed above, were generated (Coëffet-Le Gal *et al.*, 2006; Miao *et al.*, 2006b; Nguyen *et al.*, 2006a,b; Table 20.1; Fig. 20.4). As expected, none of the mutants deleted for one or more NRPS genes produced A21987C factors.

Other specific deletion mutants (recombinants) were constructed by introducing different sets of genes into strains KN100 (Δ*dptBCD*) or KN125 (Δ*dptBCDGHIJ*) by conjugation and site-specific integration of plasmids (Nguyen *et al.*, 2006a; Table 20.1; Fig. 20.4). For these studies, pCV1 was trimmed down to contain only the *dptBCD* genes (pKN25) or the *dptBCDGHIJ* genes (pKN26), under the transcription control of the *ermEp*★ promoter (Bibb *et al.*, 1985) and under the translational control of an optimized ribosome binding site (Motamedi *et al.*, 1995). These plasmids

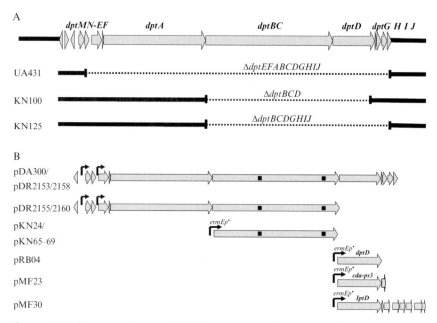

Figure 20.4 Daptomycin genes, deletion mutants and expression plasmids. (A) Deletions of *dpt* genes in mutants used for combinatorial biosynthesis. (B) Plasmids used in combinatorial biosynthesis studies. Filled squares indicate positions of module and multidomain exchanges at positions 8 and 11. Bent arrows show positions of natural or inserted promoters. (Figure reproduced from Nguyen *et al.*, 2006b.)

were generated by λ-Red-mediated recombination in *E. coli* (Datsenko and Wanner, 2000). Plasmid pKN25 was introduced into *S. roseosporus* KN125 by conjugation and site-specific insertion into the ϕC31 *attB* site to generate strain NK189 ($\Delta dptGHIJ$). Individual *dptGHIJ* genes and combinations were amplified by PCR using high fidelity *PfuTurbo* DNA polymerase, subcloned in plasmid pHM11a (Motamedi *et al.*, 1995), and introduced into *S. roseosporus* KN189 by conjugation and site-specific insertion into the IS*117* integration site. This generated a number of different deletion mutants (Table 20.1) to assess the functions of the *dptGHIJ* genes.

2.4. Functions of *dptGHIJ* genes

The functions of the *dptGHIJ* genes were assessed by constructing mutants deleted in-frame for each of the genes individually or in combination (Nguyen *et al.*, 2006a). Briefly, *dptG* and *dptH* are dispensable, but both are required for maximum lipopeptide yield. The function of *dptG* is not known, but it has counterparts in other NRPS pathways (Baltz, 2008). The *dptH* gene encodes what appears to be an editing thioesterase. The *dptI* gene encodes a methyltransferase involved in the biosynthesis of 3mGlu (Baltz, 2008; Mahlert et al., 2007). Deletion of the *dptI* gene caused the production of lipopeptides containing Glu_{12} substituted for $3mGlu_{12}$, and reduction in total lipopeptides by ~60% (Nguyen et al., 2006a). The *dptJ* gene encodes a tryptophan-2, 3-dioxygenase (TDO) involved in the conversion of Trp to Kyn. Deletion of *dptJ* caused ~50% reduction in $A21978C_{1-3}$ yield, indicating that a second TDO from primary metabolism can partially compensate for the loss of *dptJ* function (Baltz, 2008). The inclusion or deletion of the *dptI* gene has been used in combinatorial biosynthesis studies to insert 3mGlu or Glu at position 12 (see below).

2.5. Reconstitution of the daptomycin biosynthetic pathway by ectopic expression of individual genes or groups of genes

The finding that the daptomycin gene cluster is likely to be transcribed as one lengthy polycistronic message posed a challenge to set up a system for combinatorial biosynthesis. In principle, it would be expeditious to be able to express each of the NRPS genes separately, so that any engineering changes altering one subunit could be easily combined with changes in the other subunits. When combinatorial biosynthesis work was initiated at Cubist, we had access to two types of efficient site-specific integration vectors, integrating in the ϕC31 *attB* site (Bierman et al., 1992; Miao et al., 2005), and in the IS*117* site (Motamedi et al., 1995). Our initial design was to leave one NRPS gene in the native daptomycin gene cluster, and to express the other two from ectopic loci provided by ϕC31 and the IS*117 attB* sites. Alternatively, we could express two NRPS genes from one ectopic locus, and combinations of other genes from the other locus. We also chose to express the genes from ectopic sites using the *ermEp*★ promoter and an optimized ribosome binding site upstream of the first gene to be expressed. The choice of the *ermEp*★ promoter was influenced by its successful application for high-level expression of a hybrid glycopeptide (Solenberg et al., 1997) and other hybrid secondary metabolites (Baltz, 2006a,b).

Table 20.2 summarizes the relative production of A21978C factors by recombinant strains with the daptomycin genes expressed from different chromosomal sites. Expression of *dptD* from the IS*117* locus under transcriptional control of *ermEp*★ gave nearly 100% of the control $A21978C_{1-3}$ yield. Similarly, both *dptA* and *dptD* were expressed from two different ectopic loci, yielding 79% of control lipopeptides. Other constructions also gave about 100% of control yields, including expression of *dptBC* and *dptD* from different ectopic loci, expression of *dptBC* and *dptD* as a di-cistron from the ϕC31 locus and *dptGHIJ* from the IS*117* locus, and expression of

Table 20.2 Ectopic expression of genes involved in daptomycin biosynthesis[a]

| Expression locus | | | | Lipopeptide | |
dptA	dptBC	dptD	dptGHIJ	yield (%)	Reference
Dpt-n	Dpt-n	Dpt-n	Dpt-n	100	Miao et al. (2006b)
Dpt-n	Dpt-n	IS*117*	Dpt–*n*	96	Miao et al. (2006b)
φC31	Dpt-n	IS*117*	Dpt–*n*	79	Coëffet-Le Gal et al. (2006)
Dpt-n	φC31	IS*117*	Dpt–*n*	9	Coëffet-Le Gal et al. (2006)
Dpt-e	φC31	IS*117*	Dpt–*n*	50	Coëffet-Le Gal et al. (2006)
Dpt-n	φC31	IS*117*	Dpt-n	~100	Nguyen et al. (2006b)
φC31	φC31	φC31	φC31	~100	Nguyen et al. (2006b)
Dpt-n	φC31	φC31	IS*117*	107	Nguyen et al. (2006a)

[a] Dpt-n indicates expression from the natural daptomycin biosynthetic gene locus and *dpt* promoter; Dpt–*n* indicates that the natural *dptD* locus has been changed by insertion of the *ermE* gene, potentially influencing the expression of the downstream *dptGHIJ* genes. Dpt-e indicates that the *ermEp*★ promoter was inserted upstream in the *dptA* gene in the native *dpt* locus. Genes inserted on plasmids at the φC31 and IS*117 attB* insertion sites were expressed from the *ermEp*★ promoter.

all *dpt* genes as a cluster from the φC31 locus. One construction, leaving the *dptA* gene at its normal chromosomal position under transcriptional control of its native promoter, and expressing the *dptBC* and *dptD* genes from different ectopic loci under transcriptional control of *ermEp*★, gave only 9% of control. Expression was improved to 50% of control by inserting the *ermEp*★ promoter in front of *dptA*. Perhaps in this case the stoichiometry, or timing of expression, of NRPS subunits was out of balance with *dptA* being expressed from the native promoter.

These experiments demonstrated that the daptomycin biosynthetic genes need not be transcribed as a multicistron for good expression; that the strong constitutive *ermEp*★ promoter works very well to drive expression of *dpt* genes expressed ectopically, or from the native *dpt* locus; and that the IS*117*- and φC31-based plasmids are very effective for the stable insertion of different *dpt* genes or sets of genes into the chromosome outside of the main *dpt* cluster. This work established a robust ectopic *trans*-complementation system for combinatorial biosynthesis.

Figure 20.4 shows some of the strains and expression plasmids used for combinatorial biosynthesis. Strain UA431 is deleted for *dptEFABCDGHIJ* genes, and this strain can be used with pDA300 to engineer specific genes by λ-Red-mediated recombination in *E. coli*. Strain KN100 is deleted for *dptBC* and *dptD*, and can be used with pKN24 and engineered derivatives of *dptBC* and insert them into the φC31 *attB* site. This combination can be coupled with insertion of plasmids containing *dptD*, *cdaPS3*, or *lptD* in the IS*117 attB* site. Finally strain KN125 is deleted for *dptBCDGHIJ*, and can be used to make changes in *dptBC*, *dptD*, and couple them with specific deletions of different combinations of *dptG*, *dptH*, *dptI*, and *dptJ*. Examples of these applications are discussed below.

3. Sources of Genes for Combinatorial Biosynthesis

There are many NRPS genes potentially available from literature citations. However, two particular gene clusters, those of A54145 and CDA, were of particular interest because they encode acidic lipopeptide antibiotics distantly related to daptomycin, and share a number of common features (Baltz, 2008; Baltz *et al.*, 2005, 2006).

3.1. The A54145 biosynthetic genes from *Streptomyces fradiae* and the CDA genes from *Streptomyces coelicolor*

Like daptomycin, A54145 is produced as a complex of cyclic lipo-tridecapeptide antibiotics with 10-membered ring structures (Boeck *et al.*, 1990). The A54145 (*lpt*) biosynthetic gene cluster has been cloned and sequenced (Miao *et al.*, 2006a), and the stereochemistry has been predicted to be the same as in daptomycin, based upon the location of E domains in the four NRPS genes (Fig. 20.3). The CDA genes have also been cloned as part of the *S. coelicolor* genome sequencing project, and the genes analyzed (Hojati *et al.*, 2002). Of particular note are three: (i) The genes encoding the final subunits of the three pathways are di-modular, all starting with the rare 3mGlu module but ending with CATTe modules for Kyn, Trp, or Ile. The subunits appear to have diverged from a common ancestral NRPS subunit (Baltz, 2008). (ii) The CATE modules are generally located at the same relative positions (2, 8, and 11) in daptomycin and A54145 NRPSs. The CDA pathway lacks the first CATE because the peptide is truncated by two exocyclic amino acids relative to the others. CDA also has a CATE module at position 5 relative to the other two. Interestingly, the *dpt* and *lpt* pathways have achiral amino acids (Gly and Sar) at position 5, followed by a C^{II} condensation domain that normally follows CATE modules. The D-amino acid in CDA and achiral amino acids in the *dpt* and *lpt* pathways at position 5, along with the other D-amino acids at position 8 and 11, may facilitate folding of the respective molecules into their active forms (Baltz, 2008; Baltz *et al.*, 2005). (iii) The DXDG configuration at positions 7–10 relative to *dpt* and *lpt* are conserved Ca^{++} binding sites that must be maintained for antibiotic activity in combinatorial biosynthesis studies (Baltz, 2008; Kopp *et al.*, 2006).

4. Genetic Engineering of Novel Lipopeptides

4.1. Gene replacement

The apparent similarities of the terminal subunits of the NRPSs of daptomycin, A54145 and CDA made them attractive candidates for whole gene/subunit exchanges. Even though the DptD, LptD, and CdaPS3 proteins share only ~50% amino acid identities, and the conserved 3mGlu modules only 53–57% identities, these proteins appear to have maintained some level of conservation of interpeptide docking sequences (needed to interact with the DptBC protein) at their N-termini (Baltz, 2008). The simple test was to insert the *lptD* or *cdaPS3* genes into an insertion vector (pHM11a; Table 20.1) under the transcriptional control of *ermEp**, then introduce the plasmids into a strain deleted for *dptD* (UA378; Table 20.1). Recombinants containing the *lptD* and *cdaPS3* genes produced the expected hybrid lipopeptide antibiotics at about 25 and 50% of control levels (Table 20.3). The yields were improved to 40 and 69%, respectively, by inserting the genes into a strain containing the *dptA* gene expressed from the *ermEp** promoter at the ϕC31 *attB* site. These results set the stage for testing module exchanges in combination with the subunit exchanges.

Table 20.3 Production of hybrid lipopeptides by subunit exchanges for DptD[a]

| Expression locus | | | | | Lipopeptide | |
dptA	dptBC	dptD	lptD	cdaPS3	yield (%)	Reference
Dpt-n	Dpt-n	Dpt-n	—	—	100	Miao et al. (2006a,b)
Dpt-n	Dpt-n	IS*117*	—	—	96	Miao et al. (2006a,b)
Dpt-n	Dpt-n	—	IS*117*	—	28, 25	Coëffet-Le Gal et al. (2006); Miao et al. (2006a,b)
ϕC31	Dpt-n	—	IS*117*	—	40	Coëffet-Le Gal et al. (2006)
Dpt-n	Dpt-n	—	—	IS*117*	49, 50	Coëffet-Le Gal et al. (2006); Miao et al. (2006b)
ϕC31	Dpt-n	—	—	IS*117*	69	Coëffet-Le Gal et al. (2006)

[a] The same promoters were used as in Table 20.2. The *lptD* and *cdaPS3* genes encode the final two modules, $CA_{3mGlu}T$-$CA_{Ile}TTe$ and $CA_{3mGlu}T$-$CATrpTTe$, in the A54145 and CDA biosynthetic pathways, and the recombinant strains produce derivatives of daptomycin containing Ile and Trp, respectively, instead of Kyn at position 13.

4.2. Module and multidomain replacements

Since the *dptD* gene could be replaced in *trans* by the *lptD* or *cdaPS3* genes to produce A21978C$_{1-3}$ derivatives containing Ile$_{13}$ or Trp$_{13}$, respectively, it was of interest to see if the same compounds could be generated by heterologous CAT::CATTe module fusions. In these studies the DptD 3mGlu$_{12}$ module was initially fused to the DptD Kyn$_{13}$ module in the inter-module peptide linker, making three different double-amino acid substitutions (to insert a restriction site), a four-amino acid insertion and a four-amino acid deletion. All recombinants produced control amounts of A21978C$_{1-3}$ (Doekel *et al.*, 2008; Table 20.4), indicating that there is substantial flexibility in the inter-module peptide linker structure. Module fusions between DptD 3mGlu$_{12}$ and LptD Ile$_{13}$ or CdaPS3 Trp$_{13}$ were generated, and the recombinant strains produced the expected A21978C$_{1-3}$ derivatives containing Ile$_{13}$ or Trp$_{13}$ in higher yields than observed with the whole-subunit exchanges (Table 20.4). This may be due to more efficient protein–protein interactions between the homologous DptBC and DptD inter-peptide docking sites. To further explore the possibility of making amino acid substitutions at position 13, the CAT and CA domains from the CA$_{D-Asn11}$TE module from LptC were exchanged for the CAT and CA domains from the CA $_{Kyn13}$TTe module of DptD. The recombinant strain containing the CAT::***CA***::TTe double fusion produced A21978C$_{1-3}$ derivatives containing Asn$_{13}$ at about 43% of control. The strain containing the CAT::***CAT***::Te double fusion produced no lipopeptides, indicating that it is important to maintain the native configuration of TTe, as anticipated from structural studies on the EntF TTe (Zhou *et al.*, 2006).

Table 20.4 Lipopeptide production by recombinants containing module fusions in DptD[a]

DptD structure	Lipopeptide yield (% of control)
CA$_{3mGlu}$T::CA$_{Kyn}$TTe	100 ± 8
CA$_{3mGlu}$T::***CA$_{Ile}$TTe***	76 ± 8
CA$_{3mGlu}$T::***CA$_{Trp}$TTe***	119 ± 8
CA$_{3mGlu}$T::***CA$_{Asn}$T***::Te	0
CA$_{3mGlu}$T::***CA$_{Asn}$***::TTe	~43

[a] The module fusions in the inter-module T–C linker and inter-domain A–T and T–Te linkers are represented as (::). The homologous CA$_{3mGlu}$T::CA$_{Kyn}$TTe fusions were carried out in five different ways, each of which had amino acid substitutions to insert a restriction site, and two contained either an insertion or deletion of four amino acids. All five recombinants produced control yields (Doekel *et al.*, 2008). For comparison, the data from the five different homologous fusions were averaged to compare with the heterologous fusions (shown in bold italics). The *CA$_{Ile}$TTe* and *CA$_{Trp}$TTe* modules are from LptD (Miao *et al.*, 2006a) and CdaPS3 (Hojati *et al.*, 2002). The *CA$_{Asn}$T* tri-domain is from the LptC CA$_{Asn11}$TE module.

Exchanges of CAT domains have been carried out at CATE modules in DptBC to generate amino acid substitutions at positions 8 and 11 (see Figs. 20.3 and 20.4). Recombinants produced the expected novel A21978C$_{1-3}$ derivatives in yields ranging from 10 to 50% of control (Table 20.5). CATE exchanges inserting the complete Asn$_{11}$ module at positions 8 or 11 gave lower yields than the comparable CAT multidomain exchanges (Nguyen et al., 2006b).

4.3. Combinatorial biosynthesis

The multidomain exchanges at positions 8 and 11 were coupled with changes at positions 12 (deletion of *dptI*) and 13 (subunit exchanges), and natural variations in lipid tails to generate multiple combinations using the ectopic *trans*-complementation system outlined in Fig. 20.4. Many of the combinations yielded the expected novel lipopeptides. Recombinants with multiple changes generally produced lower amounts of product than those with single changes, with yields ranging from about 1.5 to 28% of control (Nguyen et al., 2006b). These yields might be improved by site specific mutagenesis, as has been demonstrated in another heterologous NRPS system (Fischbach et al., 2007).

4.4. *In vitro* antibacterial activities of novel lipopeptides

The antibacterial activities of a representative sample of novel lipopeptides are shown in Table 20.6. Antibacterial MICs against *S. aureus* ATTC 29213 ranged from 1 to 128 μg/ml compared with 0.5–1.0 μg/ml for daptomycin.

Table 20.5 Production of novel lipopeptides by recombinants containing multidomain exchanges in DptBC[a]

DptBC structure	Lipopeptide yield (% of control)
CA$_{Orn6}$T-CA$_{Asp7}$T-CA$_{D-Ala8}$TE-CA$_{Asp9}$T-CA$_{Gly}$T-CA$_{D-Ser11}$TE	100
CA$_{Orn6}$T-CA$_{Asp7}$T-CA$_{D-Ala8}$TE-CA$_{Asp9}$T-CA$_{Gly}$T::***CA**$_{D-Ala8}$**T***::E	50
CA$_{Orn6}$T-CA$_{Asp7}$T-CA$_{D-Ala8}$TE-CA$_{Asp9}$T-CA$_{Gly}$T::***CA**$_{D-Asn11}$**T***::E	17
CA$_{Orn6}$T-CA$_{Asp7}$T::***CA**$_{D-Ser11}$**T***::E-CA$_{Asp9}$T-CA$_{Gly}$T-CA$_{D-Ser11}$TE	18
CA$_{Orn6}$T-CA$_{Asp7}$T::***CA**$_{D-Asn11}$**T***::E-CA$_{Asp9}$T-CA$_{Gly}$T-CA$_{D-Ser11}$TE	10

[a] The fusion sites in the T–C and T–E linkers are shown as (::), and the heterologous domains are in ***bold italics***. (Data from Nguyen et al., 2006b and Gu et al., 2007)

Table 20.6 Antibacterial activities of daptomycin analogs[a]

Position 8	Position 11	Position 12	Position 13	MIC (µg/ml)[b]
D-Ala	D-Ser	3mGlu	Kyn	0.5–1.0
D-Ala	D-Ser	*Glu*	Kyn	8
D-Ala	D-Ser	3mGlu	*Trp*	1
D-Ala	D-Ser	3mGlu	*Ile*	4
D-Ala	D-Ser	3mGlu	*Asn*	128
D-Ser	D-Ser	3mGlu	Kyn	1
D-Asn	D-Ser	3mGlu	Kyn	8
D-Asn	D-Ser	3mGlu	*Ile*	16
D-Asn	D-Ser	*Glu*	Kyn	128
D-Ala	*D-Ala*	3mGlu	Kyn	0.5–1.0
D-Ala	*D-Asn*	3mGlu	Kyn	1
D-Ala	*D-Asn*	3mGlu	*Ile*	4
D-Ala	*D-Asn*	*Glu*	Kyn	32
D-Lys	*D-Asn*	3mGlu	Kyn	1

[a] The first row represents daptomycin. Variations from daptomycin at amino acid positions 8, 11, 12, or 13 are shown in **bold italics**. All of the daptomycin analogs have the A21978C$_1$ lipid side chain (*anteiso*-undecanate).

[b] MIC, minimal inhibitory concentration against *S. aureus* ATTC 29213. (Data from Gu *et al.*, 2007; Miao *et al.*, 2006b; and Nguyen *et al.*, 2006b.)

Five compounds were essentially as active as daptomycin. Amino acid substitutions at position 11 had little effect on antibacterial activity, whereas substitutions at position 12 and 13 influenced antibacterial activities dramatically. Substitution of Glu for 3mGlu, either singly or in combination with other changes, generally led to about 8–16-fold reduced activity. Substitution of Trp for Kyn at position 13 was well tolerated, whereas substitution of Asn caused a 128-fold increase in MIC. Interestingly, the combination of D-Lys$_8$ and D-Asn$_{11}$ substitutions, which changed the overall charge of the molecule, did not cause a significant change in MIC. These modifications represent a fraction of possible modifications of A21978C to generate daptomycin analogs. For instance, no module or multidomain exchanges in the DptA subunit have yet been reported.

5. Concluding Remarks

Some practical findings from this body of work are as follows. (i) An industrial production strain for daptomycin fermentation, *S. roseosporus*, can be manipulated genetically by a number of methods. Most importantly, plasmid DNA can be introduced by conjugation from *E. coli*, driven by RP4/RK2 functions in *trans* and *oriT* in *cis*. This has been demonstrated in other systems, and should be useful in many actinomycetes (Baltz, 1998, Baltz, 2006a,b). (ii) Chromosomal deletions can be selected using pRHB538, which carries a temperature-sensitive replicon and a marker (*rpsL*) for direct selection of double crossover events. (iii) pStreptBAC V is a useful BAC vector for cloning very large inserts of streptomycete DNA, engineering the inserts in *E. coli*, transferring genes back into streptomycetes by conjugation, and stable insertion of cloned genes into the chromosome at the ϕC31 *attB* site. (iv) pHM11a is another useful vector for cloning in *E. coli*, transfer to streptomycetes by conjugation and stable insertion into the chromosome at the IS*117 attB* site. It can be used in combination with ϕC31-vectors for combinatorial biosynthesis. (v) Genetic engineering of large, complex gene clusters in *E. coli* is greatly accelerated by the use of λ-Red-mediated recombination. (vi) Whole subunit *trans*-complementation can work in some special cases, but module exchanges and exchanges of multidomains are preferable. (vii) Successful module fusions have been made between CAT and CATTe. (viii) Successful multidomain exchanges have been made with CAT or CA from CATE modules. (ix) The special interaction of the T domain adjacent to a Te domain should be incorporated into combinatorial biosynthetic design by keeping them together as a TTe di-domain, as anticipated from structural studies with EntF (Zhou *et al.*, 2006). (x) The Te domains from DptD, LptD, and CdaPS3 can cyclize a number of different linear peptides, as has also been shown in chemoenzymatic studies (Grünewald *et al.*, 2004; Kopp *et al.*, 2006). (xi) Novel lipopeptide fermentation yields, in some cases, were surprisingly high, indicating that fermentation scale-up may be possible. In other cases, where product yields were relatively low, we might anticipate improving the yields by site-directed mutagenesis, as has been demonstrated in another heterologous NRPS system (Fischbach *et al.*, 2007).

ACKNOWLEDGMENTS

I thank my colleagues at Eli Lilly and Company and Cubist Pharmaceuticals who have contributed to this project. Particular thanks go to Margaret McHenney and Tom Hosted for early work at Lilly, and to Vivian Miao, Kien Nguyen, Dylan Alexander, Marie Coëffet-Le Gal, Daniel Ritz, Steve Wrigley, Jian-Qiao Gu, Min Chu, and Paul Brian for many contributions at Cubist. I thank Cubist Pharmaceuticals for supporting the research.

REFERENCES

Arbeit, R. D., Maki, D., Tally, F. P., Campanaro, E., and Eisenstein, B. I. (2004). The safety and efficacy of daptomycin for the treatment of complicated skin and skin-structure infections. *Clin. Infect. Dis.* **38**, 1673–1681.
Baltz, R. H. (1998). Genetic manipulation of antibiotic-producing *Streptomyces*. *Trends Microbiol.* **6**, 76–83.

Baltz, R. H. (2006a). Combinatorial biosynthesis of novel antibiotics and other secondary metabolites in actinomycetes. *SIM News* **56,** 148–160.

Baltz, R. H. (2006b). Molecular engineering approaches to peptide, polyketide and other antibiotics. *Nat. Biotechnol.* **24,** 1533–1540.

Baltz, R. H. (2008). Biosynthesis and genetic engineering of lipopeptide antibiotics related to daptomycin. *Curr. Top. Med. Chem.* **8,** 618–638.

Baltz, R. H., Miao, V., and Wrigley, S. W. (2005). Natural products to drugs: Daptomycin and related lipopeptide antibiotics. *Nat. Prod. Rep.* **22,** 717–741.

Baltz, R. H., Brian, P., Miao, V., and Wrigley, S. K. (2006). Combinatorial biosynthesis of lipopeptide antibiotics in *Streptomyces roseosporus*. *J. Ind. Microbiol. Biotechnol.* **33,** 66–74.

Bibb, M. J., Janssen, G. R., and Ward, J. M. (1985). Cloning and analysis of the promoter region of the erythromycin resistance gene (*ermE*) of *Streptomyces erythraeus*. *Gene* **38,** 215–226.

Bierman, M., Logan, R., O'Brien, K., Seno, E. T., Rao, R. N., and Schoner, B. E. (1992). Plasmid cloning vectors for conjugal transfer from *Escherichia coli* to *Streptomyces ssp*. *Gene* **116,** 43–49.

Boeck, L. D., Papiska, H. R., Wetzel, R. W., Mynderse, J. S., Fukuda, D. S., Mertz, F. P., and Berry, D. M. (1990). A54145, a new lipopeptide antibiotic complex: Discovery, taxonomy, fermentation and HPLC. *J. Antibiot.* **43,** 587–593.

Coëffet-Le Gal, M.-F., Thurson, L., Rich, P., Miao, V., and Baltz, R. H. (2006). Complementation of daptomycin *dptA* and *dptD* deletion mutations *in-trans* and production of hybrid lipopeptide antibiotics. *Microbiology* **152,** 2993–3001.

Datsenko, K. A., and Wanner, B. (2000). One-step inactivation of chromosomal genes in *Escherichia coli* K-12 using PCR products. *Proc. Natl. Acad. Sci. USA* **97,** 6640–6645.

Debono, M., Abbott, B. J., Molloy, R. M., Fukuda, D. S., Hunt, A. H., Daupert, V. M., Counter, F. T., Ott, J. L., Carrell, C. B., Howard, L. C., Boeck, L. D., and Hamill, R. L. (1988). Enzymatic and chemical modifications of lipopeptide antibiotic A21978C: The synthesis and evaluation of daptomycin (LY146032). *J. Antibiot.* **41,** 1093–1105.

Debono, M., Barnhart, M., Carrell, C. B., Hoffmann, J. A., Occolowitz, J. L., Abbott, B. J., Fukuda, D. S., Hamill, R. L., Biemann, K., and Herlihy, W. C. (1987). A21978C, a complex of new acidic peptide antibiotics: Isolation, chemistry, and mass spectral structure elucidation. *J. Antibiot.* **40,** 761–777.

Doekel, S., Coëffet-Le Gal, M.-F., Gu, J.-Q., Chu, M., Baltz, R. H., and Brian, P. (2008). *Microbiology* **154,** 2872–2880.

Eisenstein, B. I., Oleson, F. B., Jr., and Baltz, R. H. (2009). Daptomycin: From the mountain to the clinic with the essential help from Francis Tally, MD. *Clin. Inf. Dis.* in press.

Fischbach, M. A., Lai, J. R., Roche, E. D., Walsh, C. T., and Liu, D. R. (2007). Directed evolution can rapidly improve the activity of chimeric assembly-line enzymes. *Proc. Nat. Acad. Sci. USA* **104,** 11951–11956.

Fischbach, M. A., and Walsh, C. T. (2006). Assembly-line enzymology for polyketide and nonribosomal peptide antibiotics: Logic, machinery, and mechanisms. *Chem. Rev.* **106,** 3468–3496.

Fowler, V. G., Boucher, H. W., Corey, G. R., Abrutyn, E., Karchmer, A. W., Rupp, M. E., Levine, D. P., Chambers, H. F., Tally, F. P., Vigliani, G. A., Cabell, C. H., Link, A. S., *et al.* (2006). Daptomycin versus standard therapy for bacteremia and endocarditis caused by *Staphylococcus aureus*. *N. Engl. J. Med.* **355,** 653–655.

Grünewald, J., Sieber, S. A., Mahlert, C., Linne, U., and Marahiel, M. A. (2004). Synthesis and derivation of daptomycin: A chemoenzymatic route to acidic lipopeptide antibiotics. *J. Am. Chem. Soc.* **126,** 17025–17031.

Gu, J. -Q., Nguyen, K. T., Gandhi, C., Rajgarhia, V., Baltz, R. H., Brian, P., and Chu, M. (2007). Structural characterization of daptomycin analogues A21978C$_{1-3}$(D-Asn$_{11}$) produced by a recombinant *Streptomyces roseosporus* strain. *J. Nat. Prod.* **70**, 233–240.

Hojati, Z., Milne, C., Harvey, B., Gordon, L., Borg, M., Flett, F., Wilkinson, B., Sidebottom, P. J., Rudd, B. A. M., Hayes, M. A., Smith, C. P., and Micklefield, J. (2002). Structure, biosynthetic origin, and engineered biosynthesis of calcium-dependent antibiotics from *Streptomyces coelicolor*. *Chem. Biol.* **9**, 1175–1187.

Hosted, T. J., and Baltz, R. H. (1997). Use of *rpsL* for dominance selection and gene replacement in *Streptomyces roseosporus*. *J. Bacteriol.* **179**, 180–186.

Huber, F. M., Pieper, R. L., and Tietz, A. J. (1988). The formation of daptomycin by supplying decanoic acid to *Streptomyces roseosporus* cultures producing the antibiotic complex A21978C. *J. Biotechnol.* **7**, 283–292.

Kopp, F., Grünewald, J., Mahlert, C., and Marahiel, M. A. (2006). Chemoenzymatic design of acidic lipopeptide hybrids: New insights into the structure–activity relationship of daptomycin and A54145. *Biochemistry* **45**, 10474–10481.

Mahlert, C., Kopp, F., Thirlway, J., Micklefield, J., and Marahiel, M. A. (2007). Stereospecific enzymatic transformation of a-ketoglutarate to (2S,3R)-3-methyl glutamate during acidic lipopeptide biosynthesis. *J. Am. Chem. Soc.* **129**, 12011–12018.

McHenney, M. A., and Baltz, R. H. (1996). Gene transfer and transposition mutagenesis in *Streptomyces roseosporus*: Mapping of insertions that influence daptomycin or pigment production. *Microbiology* **142**, 2363–2373.

McHenney, M. A., Hosted, T. J., DeHoff, B. S., Rosteck, P. R., Jr., and Baltz, R. H. (1998). Molecular cloning and physical mapping of the daptomycin gene cluster from *Streptomyces roseosporus*. *J. Bacteriol.* **180**, 143–151.

Miao, V., Coëffet-Le Gal, M.-F., Brian, P., Brost, R., Penn, J., Whiting, A., Martin, S., Ford, R., Parr, I., Bouchard, M., Silva, C. J., Wrigley, S. W., and Baltz, R. H. (2005). Daptomycin biosynthesis in *Streptomyces roseosporus*: Cloning and analysis of the gene cluster and revision of peptide stereochemistry. *Microbiology* **151**, 1507–1523.

Miao, V., Brost, R., Chapple, J., She, K., Coëffet-Le Gal, M. -F., and Baltz, R. H. (2006a). The lipopeptide antibiotic A54145 biosynthetic gene cluster from *Streptomyces roseosporus*. *J. Ind. Microbiol. Biotechnol.* **33**, 129–140.

Miao, V., Coëffet-Le Gal, M.-F., Nguyen, K., Brian, P., Penn, J., Whiting, A., Steele, J., Kau, D., Martin, S., Ford, R., Gibson, T., Bouchard, M., et al. (2006b). Genetic engineering in *Streptomyces roseoporus* to produce hybrid lipopeptide antibiotics. *Chem. Biol.* **13**, 269–276.

Motamedi, H., Shafiee, A., and Cai, S. J. (1995). Integrative vectors for heterologous gene expression in *Streptomyces spp*. *Gene* **160**, 25–31.

Nguyen, K., Kau, D., Gu, J.-Q., Brian, P., Wrigley, S. K., Baltz, R. H., and Miao, V. (2006a). A glutamic acid 3-methyltransferase encoded by an accessory gene locus important for daptomycin biosynthesis in *Streptomyces roseosporus*. *Mol. Microbiol.* **61**, 1294–1307.

Nguyen, K., Ritz, D., Gu, J.-Q., Alexander, D., Chu, M., Miao, V., Brian, P., and Baltz, R. H. (2006b). Combinatorial biosynthesis of lipopeptide antibiotics related to daptomycin. *Proc. Natl. Acad. Sci. USA* **103**, 17462–17467.

Penn, J., Li, X., Whiting, A., Latif, M., Gibson, T., Silva, C. J., Brian, P., Davies, J., Miao, V., Wrigley, S. W., and Baltz, R. H. (2006). Heterologous production of daptomycin in *Streptomyces lividans*. *J. Ind. Microbiol. Biotechnol.* **33**, 121–128.

Pertel, P. E., Bernardo, P., Fogerty, C., Matthews, P., Northland, R., Benvenuto, M., Thorne, G. M., Luperchio, S. A., Arbeit, R. D., and Alder, J. (2008). Effects of prior effective therapy on the efficacy of daptomycin and ceFtriaxone for the treatment of community-acquired pneumonia. *Clin. Infect. Dis.* **46**, 1142–1151.

Rhee, K.-H., and Davies, J. (2006). Transcription analysis of daptomycin biosynthetic genes in *Streptomyces roseosporus*. *J. Microbiol. Biotechnol.* **16,** 1841–1848.

Sattely, E. S., Fischbach, M. A., and Walsh, C. T. (2008). Total biosynthesis: *In vitro* reconstruction of polyketide and nonribosomal peptide pathways. *Nat. Prod. Rep.* **25,** 757–793.

Shizuya, H., Birren, B., Kim, U.-J., Mancino, V., Slepak, T., Tachiiri, Y., and Simon, M. (1992). Cloning and stable maintenance of 300-kilobase fragments of human DNA in *Escherichia coli* using as F-factor-based vector. *Proc. Nat. Acad. Sci. USA* **89,** 8794–8797.

Sieber, S. A., and Marahiel, M. A. (2005). Molecular mechanisms underlying nonribosomal peptide synthesis: Approaches to new antibiotics. *Chem. Rev.* **105,** 715–738.

Silverman, J. A., Morton, L. I., Vanpraagh, A. D., Li, T., and Alder, J. (2005). Inhibition of daptomycin by pulmonary surfactant. *J. Infect. Dis.* **191,** 2149–2152.

Solenberg, P. J., Matsushima, P., Stack, D. R., Wilkie, S. C., Thompson, R. C., and Baltz, R. H. (1997). Production of hybrid glycopeptide antibiotics *in vitro* and in *Streptomyces toyocaensis*. *Chem. Biol.* **4,** 195–202.

Zhou, Z., Lai, J. R., and Walsh, C. T. (2006). Interdomain communication between the thiolation and thioesterase domains of EntF explored by combinatorial mutagenesis and selection. *Chem. Biol.* **8,** 869–879.

CHAPTER TWENTY-ONE

IN VITRO STUDIES OF LANTIBIOTIC BIOSYNTHESIS

Bo Li,[1] Lisa E. Cooper,[1] *and* Wilfred A. van der Donk

Contents

1. Introduction	534
2. Mining Microbial Genomes for Novel Lantibiotics	537
2.1. Overview	537
3. Expression and Purification of Lantibiotic Precursor Peptides (LanAs)	538
3.1. Overview	538
3.2. Protocol for heterologous overexpression and purification of LanA peptides	539
3.3. Mutagenesis and construction of synthetic analogs of LanA peptides	541
4. Expression, Purification, and Assay of LanM Enzymes	542
4.1. Overview	542
4.2. Heterologous expression and purification of LanM proteins	543
4.3. Dehydration and cyclization activity by LanM proteins	544
4.4. Order of dehydration and cyclization	546
4.5. CNBr cleavage	547
5. Expression, Purification, and Assays of LanC Cyclases	547
5.1. Overview	547
5.2. Overexpression and purification of wild-type NisC and mutants	549
6. The Protease Domain of Class II Lantibiotic Transporters	552
6.1. Protocol for cleavage of LctA substrates by LctT150	552
7. Additional Posttranslational Modifications in Lantibiotics	553
References	554

Abstract

The lantibiotics are ribosomally synthesized and posttranslationally modified peptide antibiotics containing the thioether crosslinks lanthionine (Lan) and 3-methyllanthionine (MeLan) and typically also the dehydroamino acids

Departments of Chemistry and Biochemistry and the Howard Hughes Medical Institute, University of Illinois, Urbana, Illinois, USA
[1] These authors contributed equally to this work

dehydroalanine (Dha) and (Z)-dehydrobutyrine (Dhb). These modifications are formed by dehydration of Ser/Thr residues to produce the Dha and Dhb structures, and subsequent conjugate additions of Cys residues onto the unsaturated amino acids to form thioether rings (Lan and MeLan). Several of the enzymatic reactions involved in lantibiotic biosynthesis have been reconstituted *in vitro* in recent years and these systems as well as a general overview of lantibiotic biosynthesis are discussed in this chapter.

1. Introduction

Lantibiotics are ribosomally synthesized, posttranslationally modified, lanthionine-containing antimicrobial peptides (Chatterjee *et al.*, 2005b). The lanthionine structural motif is composed of two alanine residues connected at their β-carbons by a thioether linkage (Fig. 21.1). The *meso*-stereochemistry of lanthionines has been established by NMR studies for several but not all lantibiotics. In addition to lanthionines and/or methyl substituted lanthionines, lantibiotics typically contain the unsaturated amino acids 2,3-dehydroalanine (Dha) and (Z)-2,3-dehydrobutyrine (Dhb) (Fig. 21.1) (Chatterjee *et al.*, 2005b). The topology of the lanthionine rings shows great diversity among family members. For example, the best studied lantibiotic nisin contains three consecutive rings followed by two overlapping rings and has an elongated structure, whereas cinnamycin consists of four intertwined rings and its structure is globular (Fig. 21.2).

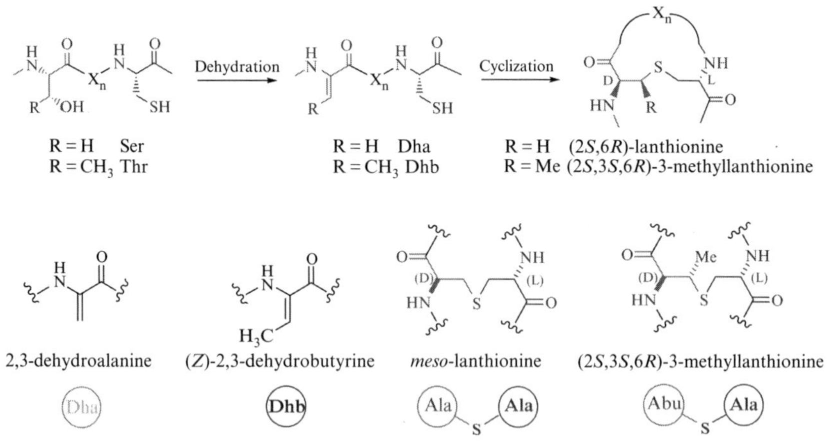

Figure 21.1 Posttranslational modifications that introduce the four characteristic structural motifs found in lantibiotics. The shorthand notation that will be used in other figures in this review is listed below each structure.

Lantibiotic Biosynthetic Enzymes 535

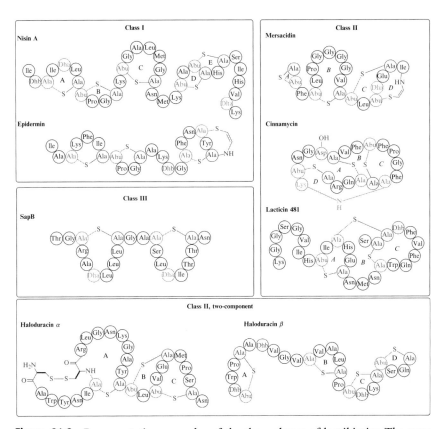

Figure 21.2 Representative examples of the three classes of lantibiotics. The same color-coding and shorthand notation is used as defined in Figs. 21.1 and 21.4. For Lan and MeLan structures, the segments derived from Ser/Thr are in red whereas those derived from Cys are in blue. The ring lettering is shown for some members and is typically alphabetical from the N- to C-terminus. (See Color Insert.)

These rings impose conformational constraints on lantibiotic peptides that are important for biological activity.

Nisin, widely used as a food preservative (Delves-Broughton *et al.*, 1996), is active against a range of food-borne pathogens, including *Clostridium* and *Listeria* species. Mode of action studies revealed that its N-terminus binds to and mislocalizes lipid II (Breukink *et al.*, 1999; Brötz *et al.*, 1998; Hasper *et al.*, 2006), an essential intermediate in cell wall biosynthesis, whereas the C-terminus is involved in pore formation (Wiedemann *et al.*, 2001). The globular lantibiotic mersacidin is too short to span the lipid bilayer and, although it binds to lipid II and inhibits transglycosylation during cell wall biosynthesis, it does not form pores (Brötz *et al.*, 1995, 1997; Hasper *et al.*, 2006). Other lantibiotic-like molecules have additional biological activities, including binding to phosphatidylethanolamine by cinnamycin (Fredenhagen *et al.*, 1990; Märki *et al.*, 1991), and promotion of aerial mycelium formation in streptomycetes by SapB (Kodani *et al.*, 2004).

The biosynthesis of lantibiotics has received much attention after it was demonstrated that their precursors, designated as LanAs, are synthesized ribosomally (Schnell *et al.*, 1988). LanAs are composed of a leader peptide region and a structural region (also called propeptide) with the structural region undergoing modifications by lantibiotic biosynthetic enzymes. In all lantibiotics, serine and threonine residues in the structural region are dehydrated to Dha and Dhb, respectively (Fig. 21.1). Subsequent intramolecular Michael-type addition of cysteine residues to the unsaturated side-chain of Dha or Dhb yields lanthionines and methyllanthionines (Fig. 21.1). In class I lantibiotics, such as nisin and epidermin (Fig. 21.2), the dehydration step is catalyzed by a designated dehydratase, LanB, and the cyclization step by a cyclase, LanC (Jung, 1991). A bifunctional enzyme, LanM, is responsible for both steps in the biosynthesis of class II lantibiotics (Siezen *et al.*, 1996), which include lacticin 481, mersacidin, cinnamycin, and the two-component lantibiotic haloduracin (Fig. 21.2). The last is a member of a growing group of lantibiotics that require the synergistic activity of two lantibiotics for antimicrobial activity (Ryan *et al.*, 1999). The biosynthetic machinery of two-component lantibiotics utilizes two different gene-encoded precursor peptides, each of which is modified by a separate LanM enzyme. For most lantibiotics, an ATP-binding cassette (ABC)-type transporter, LanT, transports the modified peptides out of the producing cells. LanT transporters for class II lantibiotics contain a protease domain and the removal of the leader peptide occurs concomitantly with transport (Fig. 21.3). Class I lantibiotics, on the other hand, utilize a dedicated protease, LanP, for the cleavage of the leader peptide. A third class of lantibiotic-like molecules was recently introduced as peptide-derived lanthionine-containing molecules that lack significant antimicrobial activities and have distinct biosynthetic enzymes (Willey and van der Donk, 2007). For instance, SapB (Kodani *et al.*, 2004) (Fig. 21.2) and SapT (Kodani *et al.*, 2005) are produced by *Streptomyces* species and serve as biosurfactins required for the formation of aerial hyphae during sporulation.

The structure–function relationship for many lantibiotics, as well as their biosynthetic enzymes, were first mapped via *in vivo* mutagenesis of precursor genes resulting in production of analogs using methods described in Chapter 22 of this volume. With respect to understanding the biosynthetic enzymes, drawbacks of the *in vivo* approach include the often decreased or aborted production of mutants, and the associated difficulty to determine which step of the biosynthesis is impaired. More recently, the activities of many biosynthetic enzymes have been reconstituted *in vitro*, which has enabled detailed investigations of their mechanisms, and exploration of structure–function relationships with a more diverse pool of

Lantibiotic Biosynthetic Enzymes 537

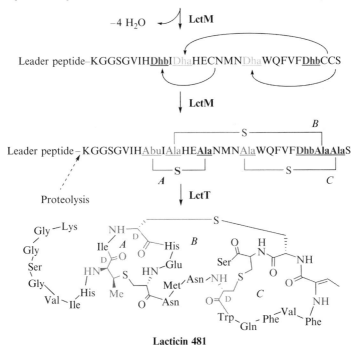

Figure 21.3 Biosynthesis of lacticin 481 as a representative example of the posttranslational maturation process of lantibiotics. Following ribosomal synthesis, LctM dehydrates serine and threonine residues in the structural region of the prepeptide LctA, and subsequently catalyzes intramolecular addition of cysteine residues onto the dehydro amino acids in a stereo- and regioselective manner. Subsequent transport of the final product across the cell membrane by LctT and proteolytic removal of the leader sequence by the N-terminal protease domain of LctT produces the mature lantibiotic.

lantibiotic analogs. The methods used for studying the LanA, LanC, LanM, and LanT proteins *in vitro* are described in this chapter.

2. Mining Microbial Genomes for Novel Lantibiotics

2.1. Overview

Over the past decade, the availability of fully sequenced genomes has provided access to vast numbers of potential natural products that were previously undetectable using traditional culture-based approaches. As the number of genomes has increased, so has the number of newly discovered lantibiotics. Currently, more than 50 lantibiotics of varying sizes, structures, and functions have been reported. With each new lantibiotic discovered, trends in sequence conservation of the biosynthetic enzymes have become more apparent and thus identification of new lanthionine-containing compounds by genomic analysis has become increasingly straightforward.

One recent example of bioinformatics-directed discovery pertains to the novel two-component lantibiotic haloduracin. While searching for analogs of the lantibiotic mersacidin, a new gene cluster was identified in the fully sequenced genome of the alkaliphilic bacterium *Bacillus halodurans* C-125 (Lawton *et al.*, 2007; McClerren *et al.*, 2006). The biosynthetic gene products were subsequently overexpressed, purified, and assayed *in vitro* for enzymatic and biological activity, which demonstrated that this cluster encoded a lanthionine-containing two-component antibiotic (McClerren *et al.*, 2006). Structures were then proposed based on amino acid sequence homology, but recent mutagenesis studies have shown that structure prediction based on other known lantibiotic structures is perilous and must be used with caution (Cooper *et al.*, 2008). In another example, ~100 actinomycete genomes were screened for *lanA* and *lanM* genes using PCR amplification with degenerate oligonucleotides, with over 20% of the strains tested showing lantibiotic-specific genes (Dodd *et al.*, 2006). Genomic-based methodologies clearly represent powerful new tools for the identification of novel lantibiotics and their biosynthetic genes.

3. Expression and Purification of Lantibiotic Precursor Peptides (LanAs)

3.1. Overview

The lantibiotic precursor peptides consist of a structural region (19–37 residues) that is modified to the mature compound, and an N-terminal leader sequence (23–59 residues) that is not modified (Schnell *et al.*, 1988). The role of the leader peptide is still under investigation, with several functions supported by experimental data. Leader peptide removal is required for antibiotic activity, and therefore the leader peptide may keep the lantibiotic inactive inside the cell, thereby protecting the producing bacterial strain (Chen *et al.*, 2001; Corvey *et al.*, 2003; McClerren *et al.*, 2006; van der Meer *et al.*, 1994; Xie *et al.*, 2004). Furthermore, the leader peptides appear to be important for molecular recognition by the dehydration and cyclization enzymes as well as the lantibiotic-specific proteases and transporter systems (Chatterjee *et al.*, 2006; Chen *et al.*, 2001; Furgerson Ihnken *et al.*, 2008; Izaguirre and Hansen, 1997; Kluskens *et al.*, 2005; Kuipers *et al.*, 2004; Li *et al.*, 2006; Rink *et al.*, 2005, 2007; van der Meer *et al.*, 1994). The lantibiotics are therefore examples of a growing group of natural products, including microcins (Duquesne *et al.*, 2007; Roy *et al.*, 1999; Severinov *et al.*, 2007) and cyanobactins (Chapter 23), for which a leader peptide appears to direct the maturation process.

The precise mechanism of leader peptide recognition by the various lantibiotic biosynthetic enzymes is at present not understood. Sequence comparison of the class I leader peptides revealed a conserved "FNLD" motif between residues -20 to -15 and a Pro residue usually found in the -2 position that may mediate the interactions with biosynthetic enzymes (van der Meer *et al.*, 1994). In class II lantibiotic leader peptides, a preponderance of Asp and Glu residues is typically encountered, as well as a GlyGly or GlyAla/Ser cleavage recognition site. This "double glycine motif" (Håvarstein *et al.*, 1994) resides at the junction between the leader peptide and the structural region of the LanA precursor peptide. Except for cinnamycin (Widdick *et al.*, 2003), the leader peptides do not contain the recognition motifs for the secretory-dependent or twin-arginine translocation pathways.

3.2. Protocol for heterologous overexpression and purification of LanA peptides

Many lantibiotic precursor peptides have been cloned, overexpressed, and purified using a variety of different protocols. For the lantibiotic subtilin, its prepeptide, SpaS, was cloned into an intein-chitin binding domain system and then purified by chitin affinity chromatography with subsequent inteinmediated cleavage (Xie *et al.*, 2002). EpiA (epidermin precursor) and MrsA (mersacidin precursor) were purified as N-terminal maltose binding protein (MBP) fusions using amylose affinity chromatography (Kupke *et al.*, 1993; Majer *et al.*, 2002). MrsA was also expressed as an N-terminal histidine fusion peptide that was more easily purified by nickel affinity chromatography (Majer *et al.*, 2002). A slightly modified version of this methodology was later adopted for the purification of LctA (lacticin 481) (Xie *et al.*, 2004), HalA1 and HalA2 (haloduracin) (McClerren *et al.*, 2006), and NisA (nisin) (Li and van der Donk, 2007) by nickel affinity chromatography followed by reverse-phase high performance liquid chromatography (RP-HPLC) and this protocol, which appears widely applicable, is outlined here.

3.2.1. Procedure

1. *Escherichia coli* BL21 (DE3) cells are freshly transformed with the desired *lanA*-containing plasmid, usually a pET15b-derived plasmid (Novagen). In our experience, freshly transformed cells result in higher overexpression yields than using a glycerol stock of transformed cells.

2. Single colonies of transformants are grown in a 37 °C shaker for 12–15 h in 50 mL of Luria–Bertani (LB) medium supplemented with 100 μg/mL ampicillin. A 20-mL aliquot is centrifuged at 5000×g for 10 min, the spent LB medium discarded, and the cell pellet carefully resuspended in fresh LB medium. This step was found to be beneficial when using ampicillin resistant strains in order to remove the secreted β-lactamase and select for retention of the *lanA*-containing plasmid. The resuspended cells (20 mL) are added to 2 L of LB-ampicillin (100 μg/mL), and the culture is grown aerobically at 37 °C until the A_{600} is ~0.75–0.85. IPTG is added to a final concentration of 1 mM and the culture is grown aerobically at 37 °C for an additional 3 h. Cells are harvested by centrifugation at 11,900×g for 20 min at 4 °C. The medium is discarded and the cell paste (~6 g) is stored at −80 °C until use.
3. The cell paste is resuspended in ~20 mL of LanA Start Buffer (20 mM NaH_2PO_4, pH 7.5 at 25 °C, 500 mM NaCl, 0.5 mM imidazole, 20% glycerol) and lysed by sonication (35% amplitude, 4.0 s pulse, 9.9 s pause, 15 min). The sample is centrifuged at 23,700×g for 30 min at 4 °C and the supernatant is discarded as the His_6-LanA peptides are typically overexpressed insolubly. The pellet is treated once more with Start Buffer to remove any soluble proteins from the pellet. The pellet from the second wash is resuspended in ~20 mL of denaturing LanA Buffer 1 (6 M guanidine hydrochloride, 20 mM NaH_2PO_4, pH 7.5 at 25 °C, 500 mM NaCl, 0.5 mM imidazole). The insoluble portion is removed by centrifugation and the supernatant is clarified through 0.45-μm syringe filters (Corning).
4. Immobilized metal affinity chromatography (IMAC) is carried out at 25 °C using a gravity flow column, a peristaltic pump-driven column, or a HiTrap chelating HP nickel affinity column (GE Healthcare). All three methods have been successful but the last method is recommended. The HiTrap resin is charged with four column volumes (CV) of 0.1 M $NiSO_4$, washed with 2 CV of filtered water to remove excess nickel, and equilibrated with 2 CV of LanA Buffer 1. After loading the filtered sample, the resin is washed with 2 CV each of LanA Buffer 1 and LanA Buffer 2 (4 M guanidine hydrochloride, 20 mM NaH_2PO_4, pH 7.5 at 25 °C, 300 mM NaCl, 30 mM imidazole). The desired LanA peptide is eluted using 1–3 CV of LanA Elution Buffer (4 M guanidine hydrochloride, 20 mM Tris, pH 7.5 at 25 °C, 100 mM NaCl, 1 M imidazole) and stored at room temperature overnight. Storage for more than 48 h or at lower temperatures is not advised as this may result in peptide precipitation. Typically, nickel affinity chromatography is performed one day prior to further purification by preparative RP-HPLC (vide infra).
5. Imidazole removal from the His_6-LanA peptides using dialysis, size exclusion chromatography (PD-10 columns; GE Healthcare), or sample extraction products (e.g., C18 Sep-Pak; Waters Scientific or C18 Bond

Elut; Varian, Inc.) are less successful than RP-HPLC due to precipitation problems. In addition, these methods typically do not provide the binding capacity desired for large scale desalting. Therefore, desalting is most successfully achieved by preparative RP-HPLC using a Waters Delta-pak™ C4 15 μm 300 Å 25 × 100 mm PrepPak® Cartridge. Solvents for preparative RP-HPLC are solvent A (0.1% TFA in water) and solvent B (0.086% TFA in 80% acetonitrile/20% water). A gradient of 2–100% of solvent B over 45 min is executed with a flow rate of 8 mL/min. Peptides are detected by absorbance at 220 nm. In our experience, most His_6-LanA peptides elute as one or two peaks (with and without disulfide linkages), typically beginning at 50–60% solvent B. Heterogeneous disulfides often broaden the peak, sometimes to the extent that a clear peak is not visible. Prereduction with dithiothreitol (DTT) or tris(carboxyethyl)phosphine (TCEP) can improve peak shape and yields.

6. Collected fractions are lyophilized and a white fluffy solid is obtained. Typical peptide yields range from 1 to 60 mg of final dried peptide per liter of overexpressed cells depending on the desired LanA. Peptides are analyzed by matrix-assisted laser desorption time-of-flight (MALDI-TOF) mass spectrometry (MS) and the desired products are stored dry at −20 °C, under nitrogen for long-term storage.

3.3. Mutagenesis and construction of synthetic analogs of LanA peptides

Lantibiotic biosynthetic proteins in general show remarkable substrate promiscuity (Chatterjee et al., 2005b, 2006; Cotter et al., 2006b; Kluskens et al., 2005; Rink et al., 2005; Zhang and van der Donk, 2007). This characteristic has been exploited for the preparation of analogs of many lantibiotics, including nisin (Kuipers et al., 1996; van Kraaij et al., 1997), subtilin (Liu and Hansen, 1992), epidermin and gallidermin (Ottenwälder et al., 1995), cinnamycin (Chapter 22 of this volume) (Widdick et al., 2003), mersacidin (Szekat et al., 2003), and lacticin 3147 (Cotter et al., 2006a). The promiscuous nature of LanM enzymes was further illustrated in vitro by successful modification by the lacticin 481 synthetase, LctM, of a series of chimeric LanA substrates (Patton et al., 2008), nonproteinogenic amino acid-containing LanA substrates (Chatterjee et al., 2006; Zhang and van der Donk, 2007), and nonlantibiotic conformationally constrained peptides (Levengood and van der Donk, 2008).

LanA peptides have lengths amenable to semisynthetic peptide ligation techniques. For example, a range of nonproteinogenic amino acids have been incorporated into the lacticin 481 substrate peptide, LctA, by expressed protein ligation (EPL) (Chatterjee et al., 2006; Zhang and van der Donk, 2007). While this approach allowed investigation of the substrate specificity of LctM, scale-up to produce the amounts required for quantitative SAR (structure–activity-relationship) studies of lacticin 481 analogs proved difficult. Therefore, an alternative method using nonpeptidic linkers between the C- and N-terminal regions of LctA was developed (Levengood et al., 2007). The copper(I)-catalyzed [3 + 2]-cycloaddition of the LctA leader sequence with a C-terminal alkyne to the LctA structural region with an N-terminal azide proved to be a superior strategy (Levengood et al., 2007). In-depth discussions of the protocols for (semi)synthetic incorporation of nonproteinogenic amino acids into LanA peptides using EPL and cycloaddition chemistry were recently published (Levengood et al., 2007; Zhang and van der Donk, 2009).

4. Expression, Purification, and Assay of LanM Enzymes

4.1. Overview

LanM enzymes are large proteins of ~115–120 kDa (900–1200 residues) that catalyze both the dehydration and cyclization of their LanA substrates (McClerren et al., 2006; Xie et al., 2004). The N-terminal domain of LanM enzymes is responsible for dehydration and the C-terminal domain for cyclization. The cyclization domain has low but detectable sequence homology with the LanC cyclases involved in the biosynthesis of class I lantibiotics. The sequence homology includes conservation of zinc ligands that were identified first for the LanC proteins as described later in this chapter. At present, the details of the interaction between the dehydration and cyclization domains are unclear. LanM proteins have no homology with LanB dehydratases, suggesting that they did not evolve from a simple fusion of the LanB and LanC enzymes found in class I lantibiotics (Siezen et al., 1996).

Several studies have investigated the enzymatic mechanism of dehydration of the prepeptide LctA by lacticin 481 synthetase (LctM). The enzyme utilizes ATP and Mg^{2+} to phosphorylate the Ser and Thr residues that are targeted for dehydration. In a subsequent step, LctM eliminates the phosphate to produce Dha and Dhb, respectively (Chatterjee et al., 2005a). Sequence alignments have not identified a traditional ATP binding motif (You and van der Donk, 2007), but site-directed mutagenesis studies suggest a mechanism for phosphorylation similar to that of Ser/Thr kinases. If the fold of LanM enzymes indeed turns out to resemble that of kinases, then lantibiotic production may have evolved from a Ser/Thr kinase that picked up an active site base to catalyze the elimination step. Site-directed mutagenesis studies suggest that Asp259, Thr405, and Arg399 are important for this step (You and van der Donk, 2007). The procedures that have been successfully used to purify and assay three active LanM proteins (LctM and the haloduracin synthetases HalM1 and HalM2) are detailed below.

4.2. Heterologous expression and purification of LanM proteins

4.2.1. Procedure

1. *E. coli* BL21 (DE3) cells are freshly transformed with a plasmid carrying the desired *lanM* gene. His$_6$-LanM proteins have been successfully overexpressed using pET15b (Novagen) and pET28b. Freshly transformed cells result in higher protein yields than using a glycerol stock.
2. Single colony transformants are grown in a 37 °C shaker for 12–15 h in 50 mL of LB medium supplemented with either 100 μg/mL ampicillin (pET15b) or 50 μg/mL kanamycin (pET28b). A 1% inoculation of a 2 L LB-antibiotic culture is completed as described above for LanA overexpression, and the culture is grown aerobically at 37 °C until the A_{600} is ~0.6–0.8. IPTG is added to a final concentration of 0.5 (LctM) or 1 mM (HalM1, HalM2) and the culture is grown aerobically at 18 °C for an additional 20 h. Cells are harvested by centrifugation at 5000×g for 20 min at 4 °C. The cell paste (~4 g) is stored at −80 °C until use.
3. The cell paste is resuspended at 4 °C in ~20 mL of LanM Start Buffer (20 mM Tris, pH 7.6, 500 mM NaCl, 10% glycerol) and lysed by sonication (35% amplitude, 4.0 s pulse, 9.9 s pause, 15 min). The sample is centrifuged at 23,700×g for 45 min at 4 °C and the supernatant is clarified through 0.45-μm syringe filters (Corning).
4. IMAC can be completed using a HiTrap chelating HP nickel affinity column (GE Healthcare, Option 1), or nickel affinity fast protein liquid chromatography (FPLC, Option 2). In both cases, 5 mL fractions are collected and analyzed for purity by Tris–Glycine SDS-PAGE (4–20% acrylamide gradient).
    a. Option 1: A stepwise gradient of increasing imidazole concentration is used with a 5 mL HiTrap chelating HP nickel affinity column (GE Healthcare). Wash Buffers are prepared by combining the LanM Start Buffer and LanM Elution Buffer (20 mM Tris, pH 7.6, 500 mM NaCl, 1 M imidazole, 10% glycerol) to give the final imidazole concentrations desired. The resin is charged and washed with 2 CV of LanM Start buffer as described for LanA purifications. After loading the LanM protein sample, the resin is washed with 2 CV each of Wash Buffers containing 0, 25, 50, 100, 200, 500, 800, and 1000 mM imidazole. The LanM enzymes elute from the column in 2–3 fractions between 100 and 200 mM imidazole.
    b. Option 2: A continuous gradient of increasing imidazole concentration is employed using FPLC. The FPLC used in our laboratory is an

Amersham Biosciences/GE Healthcare ÄKTA system. An 8-mL MC-20 POROS (Applied Biosystems) IMAC column is charged with 0.1 M NiSO$_4$ and washed with filtered water to remove the excess nickel in accordance with the manufacturer's protocols. Solvents for FPLC are solvent A (LanM Start Buffer) and solvent B (LanM Elution Buffer). A gradient of 0–100% of solvent B over 45 min is executed with a flow rate of 4 mL/min. Protein is detected by absorbance at 220 nm. For LctM, a cation exchange column is required to obtain highly pure protein (Xie et al., 2004), but HalM1 and HalM2 are of high purity after IMAC.

5. The LanM protein eluted from the Ni-affinity column is concentrated from ~15–20 to ~2–3 mL using a centrifugal filtering device (Millipore, 30 kDa Mw cutoff) and the buffer is exchanged to LanM Final Buffer (20 mM Tris, pH 7.6, 500 mM KCl, 10% glycerol) via dialysis or PD-10 size exclusion chromatography (GE Healthcare). LanM proteins are stored at −80 °C. Protein concentrations are determined using A_{280} measurements and are calculated with theoretical extinction coefficients obtained from the ProtParam function on the ExPASy Proteomics Server (His$_6$-LctM, ε_{280} = 115,365 M^{-1} cm^{-1}; His$_6$-HalM1, ε_{280} = 122,550 M^{-1} cm^{-1}; His$_6$-HalM2, ε_{280} = 119,570 M^{-1} cm^{-1}). Typical protein yields range from 3 to 15 mg of final protein per liter of overexpressed cells. Proteins are thawed on ice before use but it is recommended that thawing and refreezing be kept to a minimum (2–3 times per aliquot). LanM protein remains active for ~6–8 months before reduction of activity is observed, which typically is manifested by increasing amounts of phosphorylation peaks in activity assays.

4.3. Dehydration and cyclization activity by LanM proteins

4.3.1. Procedure

1. Purified LanA peptides are dissolved in LanM assay buffer to a final concentration of 0.3 mg/mL and supplemented with 10 mM MgCl$_2$, 2.5–5 mM ATP, and 1–5 mM TCEP or DTT (final volume 40 μL). A variety of buffers have been utilized successfully, including 50 mM MOPS, pH 7.5, 50 mM HEPES, pH 7.5, and 50 mM Tris, pH 7.5–8.5. The pH sometimes needs to be readjusted to the desired value prior to addition of LanM as residual TFA may be present in the LanA peptide solutions after RP-HPLC. LanA peptides are incubated with 0.2 mg/mL of their corresponding LanM enzymes. Control experiments use the same assay conditions without the addition of LanM. Assay samples are incubated at 25 °C for 1–3 h and then divided into two 20-μL aliquots to monitor LanM-mediated dehydration and cyclization (vide infra).

Troubleshooting: LanA peptides are often poorly soluble in the assay conditions. Using different buffers and different pH values may improve solubility (Cooper et al., 2008). In some cases, 12.5 μg/mL BSA and/or 5 μM ZnCl$_2$ is added to optimize the reaction conditions for enzyme activity. For applications in which only the final products are desired, heterogeneous assays can be performed with the substrate slowly dissolving throughout the duration of the assay since the products are more soluble than the prepeptides. For instance, full conversion of an initially heterogeneous solution of LctA has been observed (Paul et al., 2007). Sometimes a LanA mutant (e.g., His$_6$-LctA-N39R/F45H) with better solubility properties can be engineered without compromising enzyme activity (You and van der Donk, 2007).

2. To monitor dehydration, a 20-μL aliquot is quenched by addition of 5% TFA to a final concentration of 0.5% TFA. Samples are centrifuged at 18,400×g for 3 min and then desalted using C18 ZipTips (Millipore) and eluted with 4 μL of α-hydroxyl cinnamic acid. From this solution, 3 μL is applied to the MALDI target and analyzed for dehydration by MALDI-TOF MS.

3. To monitor cyclization, a chemical modification step following the enzymatic assay is required since cyclization does not result in a change in mass. For most applications using MALDI-TOF MS, modification has been accomplished using the thiol modification agents iodoacetamide (IAA, Option 1) (Cooper et al., 2008; McClerren et al., 2006) or p-hydroxymercuribenzoic acid (PHMB, Option 2) (Li et al., 2006; Paul et al., 2007). The use of IAA or PHMB is dictated by the efficiency of the modification, which differs for each lantibiotic system. When MS capabilities include high resolution MS, 1-cyano-4-dimethylaminopyridinium tetrafluoroborate (CDAP) may be used (Miller et al., 2006). Its disadvantage is a relative small mass change upon cyanylation that is in the same range as sodium adducts, but its advantage is that CDAP derivatization can be accomplished at pH 3, thereby minimizing any noncatalyzed cyclization that can compete with derivatization. Nonenzymatic cyclization of Cys residues with Dha/Dhb amino acids is well documented (Burrage et al., 2000; Okeley et al., 2000; Toogood, 1993; Zhou and van der Donk, 2002; Zhu et al., 2003).

   a. Option 1: For LanM assays carried out in a buffer at pH 8, 1 μL of 100 mM IAA is added directly to a 20-μL assay aliquot (~5 mM final concentration) and incubated at 25 °C for 90 min in the dark. For LanM assays carried out at pH below 8, the pH must be adjusted to pH 8–9 or the sample can be concentrated to dryness and then resuspended in a buffer at pH 8–9. As a control, the linear LanA peptide is incubated under LanM assay conditions in the absence of LanM, and subsequently treated with IAA using the same protocol.

b. Option 2: A 20-μL assay aliquot is dried by centrivap (Labconco). Assay products are resuspended in 6 μL of 4 M guanidine hydrochloride (Gn-HCl) with 10 mM TCEP and incubated at 25 °C for 20 min. Saturated PHMB solution (~10 mM) is added (3 μL) and the reaction mixture is incubated at 25 °C for 2 h in the dark. Control assays with LanA (0.3 mg/mL) not treated with LanM are performed in the same way. When the pH is below 8, PHMB precipitates; adjusting the sample to pH 8–9 will clear this precipitation and increase the efficiency of modification. One advantage of PHMB is the larger change in mass (IAA adduct: +57 Da; HgAr adduct: +320 DA; or bridging Hg between two Cys: +200 Da). However, the PHMB modification reaction often does not go to completion, even in the presence of denaturants (Gn-HCl) and reducing agents (TCEP).
4. Samples are centrifuged at 18,400×g for 5 min. The supernatant is desalted using C18 ZipTips (Millipore) and eluted with 4 μL of α−hydroxyl cinnamic acid. From this solution, 3 μL is applied to the MALDI target and analyzed for the presence or absence of thiol adducts by MALDI-TOF MS.

4.4. Order of dehydration and cyclization

The LanM enzymes catalyze a remarkable array of chemical transformations. For example, HalM2 dehydrates seven Ser/Thr residues and installs four rings in HalA2. Both LctM and HalM2 have been shown to be distributive enzymes, as partially dehydrated substrate intermediates were observed at low substrate concentration for LctM (Patton et al., 2008) and at both low and high substrate concentrations for HalM2 (Lee et al., Submitted for publication). Because of the distributive behavior of LctM and HalM2, reaction intermediates can be captured and analyzed at various time points to study the order of dehydrations and cyclizations catalyzed by these two enzymes.

For HalM2, the dehydrated intermediates were fragmented by collisionally induced dissociation (CID) to locate the sites of dehydration. The first four dehydrations occurred sequentially on the N-terminal four Ser/Thr residues in an N-to-C direction. The order of dehydration for the last three Ser/Thr residues could not be clearly determined, but based on the first four dehydrations, HalM2 appeared to act in an N-to-C direction, and similar conclusions have been reached for LctM (Lee et al., Submitted for publication).

To study the directionality of cyclization, intermediates were treated at different time points with IAA as detailed above for LanM cyclization assays. By assigning the positions of alkylated Cys (uncyclized Cys) by MS–MS, the order in which Cys residues undergo cyclization could be deduced, demonstrating that the A-ring, B-ring, C-ring, and D-ring in HalA2 were formed in this order. Hence, HalM2 also catalyzed cyclization in an N-to-C direction.

The substrate selectivity with respect to the hydroxyl amino acids and cysteines has been investigated using semisynthetic LctA analogs as substrates for LctM. LctA(-24 to 20) included the leader peptide (residues -24 to -1, Fig. 21.3) and a portion of the LctA structural region (residues 1–20). LctM dehydrated Thr analogs carrying ethyl, vinyl, ethynyl, propynyl, and E-alkenyl groups at the β-carbon of serines, but propyl, isopropyl, or Z-alkenyl groups were not tolerated. The (R)-stereochemistry at the β-carbon was essential (Zhang and van der Donk, 2007).

The chemo- and regioselectivity of cyclization by LctM with Cys analogs has been examined by monitoring ring formation in LctA(-24 to 20) using cyanogen bromide (CNBr) cleavage at Met16. The observed cleavage products were used to distinguish between enzyme-catalyzed MeLan formation and nonenzymatic Lan formation. Collectively, these studies showed that substitution of Cys with D-Cys, homocysteine, or β^3-homocysteine resulted in LctM-catalyzed cyclization. However, the enzyme did not catalyze ring formation of (R)-3-methylcysteine (Zhang and van der Donk, 2009; Zhang et al., 2007).

4.5. CNBr cleavage

4.5.1. Procedure

His-LctA (-24 to 20) analogs containing nonproteinogenic amino acids are treated with LctM. A 15-μL sample of the assay is purified by C18 ZipTip (Millipore) and the peptides are eluted with 50% CH$_3$CN, 40% H$_2$O, 0.3% TFA. Eluted peptides are dried and resuspended in 10 μL of CNBr solution (50 mg/mL in 70% HCOOH). The mixture is incubated in the dark overnight at room temperature, concentrated, and resuspended in 10 μL of distilled water. A 5-μL sample is mixed with 5 μL of matrix (α-cyano-4-hydroxycinnamic acid in 50% CH$_3$CN, 40% H$_2$O, 0.3% TFA), and analyzed by MALDI-TOF MS. *Note*: CNBr is toxic.

5. Expression, Purification, and Assays of LanC Cyclases

5.1. Overview

In contrast to the full reconstitution of the biosynthesis of class II lantibiotics by the bifunctional LanM proteins, only reconstitution of the cyclization process by LanC enzymes has been achieved for class I lantibiotics.

The LanB dehydratases in class I lantibiotics are large proteins (~1000 residues) without sequence homology with other proteins in the database. Within the family, the LanB proteins only share around 25% pairwise sequence identity (Siezen et al., 1996). The first LanB expressed and partially purified was EpiB, the dehydratase for the biosynthesis of epidermin in *Staphylococcus epidermis* (Peschel et al., 1996). The *epiB* gene was introduced into a staphylococcal expression vector containing the xylose-inducible *xylA* promoter of *Staphylococcus xylosus*. Expression of the *epiB* gene in *Staphylococcus carnosus* resulted in production of the EpiB protein, but partially purified EpiB did not demonstrate *in vitro* activity towards purified EpiA (Peschel et al., 1996). Similarly, SpaB was successfully expressed in *E. coli* with the aid of the GroEL/ES chaperones, but could not modify the subtilin precursor SpaS *in vitro* (Xie et al., 2002). Recent *in vitro* expression of *nisA*, *nisB*, and *nisC* using a commercial Rapid Translation System generated a mixture of products recognized by a nisin antibody after leader peptide removal by trypsin (Cheng et al., 2007). The presence of nisin in the final product mixture was also supported by antimicrobial activity and activation of the nisin-inducible *nisF* promoter in a Green Fluorescent Protein (GFP) fluorescence-based assay. However, NisB activity was not purified and the dehydration mechanism used by this protein family remains elusive. LanB enzymes share no sequence homology with LanM proteins and may employ a different dehydration mechanism than that discussed in the LanM section.

LanC cyclases catalyze the formation of lanthionine rings in class I lantibiotics. NisC, the nisin cyclase, is the only LanC whose activity has been demonstrated *in vitro* (Li et al., 2006). The substrate for NisC, eightfold dehydrated NisA, was purified from the culture medium of an engineered *Lactococcus lactis* strain carrying a plasmid containing *nisA*, *nisB*, and *nisT* (Kuipers et al., 2004). Treatment of eightfold dehydrated NisA with NisC resulted in thioether ring formation (Li et al., 2006). Subsequent removal of the leader peptide from the product(s) generated bioactive compound. Both NisC and SpaC contain one zinc ion per polypeptide (Okeley et al., 2003), which was verified in the X-ray crystal structure of NisC (Li et al., 2006). The zinc ion in NisC is coordinated by two cysteines, one histidine, and one water molecule. All three zinc ligands are conserved among LanC proteins and in the cyclization domain of LanM proteins. In addition, several other conserved residues were found in close proximity to the zinc. These include His212, Trp 283, Asp141, Tyr285, and Arg280. The last three residues are conserved in LanC enzymes but not in LanM proteins. Mutagenesis studies of NisC have shown that the zinc ligands, as well as His212 and Asp141, are essential for the activity of NisC, whereas Tyr285 and Arg280 are not (Li and van der Donk, 2007). The methods used for the *in vitro* NisC studies are described below.

5.2. Overexpression and purification of wild-type NisC and mutants

5.2.1. Procedure

1. A 20-mL culture of *E. coli* BL21 (DE3) cells carrying a pET15b plasmid containing *nisC* is grown at 37 °C in LB containing ampicillin (0.1 mg/mL). An aliquot of 10 mL of overnight culture is used to inoculate 1 L of LB/ampicillin and grown at 37 °C. When the A_{600} of the culture reaches 0.7–0.9, protein overexpression is induced with 0.2 mM IPTG. The culture is also supplemented with 100 μM ZnCl$_2$ and allowed to grow with shaking at 18 °C for 15–20 h. The cells are harvested by centrifugation and resuspended in 15 mL of resuspension buffer (20 mM Tris, 0.5 M KCl, 10% glycerol, pH 8.3) containing protease inhibitor cocktail (Roche Bioscience). Cells are stored at −80 °C until use.
2. After sonication of the thawed, resuspended cells, the cellular debris is removed by centrifugation at 23,700×g for 30 min. The filtered supernatant is applied to a 5 mL cobalt affinity column (BD Bioscience). The column is washed with 15 CV of equilibration buffer (20 mM Tris, 0.5 M NaCl, 10% glycerol, pH 7.5) and 5 CV of wash buffer (20 mM Tris, 0.5 M NaCl, 20 mM imidazole, 10% glycerol, pH 7.5). His$_6$-tagged proteins are eluted with 5 CV of elution buffer (20 mM Tris, 0.5 M NaCl, 150 mM imidazole, 10% glycerol, pH 7.5). Elution fractions are monitored by Bradford protein assay (Biorad) and SDS-PAGE. The fractions containing NisC or its mutants are combined and concentrated by Amicon-12 (YM-10, Millipore) to 2.5–5 mL depending on protein solubility. The concentrated protein is exchanged into the equilibration buffer containing 10% glycerol using a PD-10 desalting column (2.5 mL) (GE HealthCare), and stored at −80 °C.

Removal of the His$_6$ tag with thrombin. Whereas the His$_6$-tag does not interfere with NisC activity, it is removed for other applications such as the determination of metal content. His$_6$-NisC (or one of its mutants) is incubated with biotinylated thrombin (0.25 units of biotinylated thrombin per milligram of protein, Novagen) overnight at 4 °C to cleave the His$_6$-tag. Biotinylated thrombin is then removed with streptavidin immobilized onto agarose beads (32 μL of 50% streptavidin beads slurry per unit of thrombin, Novagen). Separation of nontagged NisC from the His$_6$-tag is achieved by passing the cleavage reaction through a cobalt affinity column, followed by concentration by Amicon-4 (YM-10, Millipore). Protein concentrations are determined by Bradford protein assay, BCA (bicinchoninic acid) protein assay (Pierce), and absorbance at 280 nm using a calculated extinction coefficient value (60,740 M^{-1} cm^{-1}, ExPASy). The average of the values from these three methods has been used as the protein concentration for NisC.

Metal analysis. All flasks and beakers are rinsed with 20% HNO_3 and Milli-Q water. Slidalyzers (Pierce) used for dialysis are presoaked in 20 mM Tris buffer, pH 7.5 treated with 10 g/L Chelex-100 (Biorad). To remove the loosely bound zinc, NisC (or a mutant) is dialyzed with four changes of buffer (20 mM Tris, 300 mM NaCl, pH 7.5) pretreated with 25–50 g/L Chelex-100 over a total time of 48 h at 4 °C. The protein is then concentrated by ultrafiltration with an Amicon-4 (YM-10, Millipore) presoaked in 20 mM Tris buffer, pH 7.5 treated with 10 g/L Chelex-100. Metal content of wild-type NisC is determined by inductively coupled plasma mass spectrometry (ICP-MS). The zinc content for NisC and mutants can also be determined by a spectroscopic assay based on the absorption change at 500 nm associated with the formation of a zinc complex with 4-(2-pyridylazo)resorcinol (PAR; Sigma). A sample of 100 μL of 9 μM protein is incubated with 4 M guanidine hydrochloride to denature the protein in the presence of 20 mM Tris buffer, pH 7.5 containing 0.1 M PAR. Titration of the protein mixture with 1 mM of the thiol-modifying reagent 5,5'-dithiobis(2-nitrobenzoic acid) (DTNB; Acros) results in the gradual release of cysteine-bound metal, which forms a complex with PAR. The absorbance at 500 nm is measured, and the titration is considered complete when no further significant increase in absorbance is observed. The addition of DTNB caused a minimal background increase in absorbance at 500 nm, which is measured using an identical titration of protein sample lacking PAR and subtracted from the experimental value obtained in the presence of PAR. The concentration of Zn^{2+} is determined using a zinc atomic absorption standard treated under the same conditions to obtain a standard curve. This procedure has also been used successfully to determine the Zn content of LanM proteins.

Expression and purification of dehydrated prenisin. Dehydrated prenisin is expressed in *L. lactis* transformed with a plasmid carrying *nisA*, *nisB*, and *nisT* (Kuipers *et al.*, 2004). A culture (5 mL) is grown overnight at 25 °C in M17 broth supplemented with 0.5% glucose and erythromycin (5 μg/mL). A sample of 2 mL overnight culture is used to inoculate a 200-mL culture, which is allowed to grow at room temperature. When the optical density at 600 nm reaches 0.8, the cells are harvested by centrifugation and resuspended in an equal volume of minimal medium (for the composition, see below) supplemented with 1/1000 volume of filtered overnight culture medium of an *L. lactis* nisin-producing strain, which contains nisin, to turn on the autoinduction system for production of dehydrated prenisin (Kuipers *et al.*, 2004). The cells are harvested after growing at 25 °C overnight. The culture supernatant is filtered (0.45 μm, Corning) and ice-cold aqueous trichloroacetic acid (10% final concentration) is added to precipitate the peptide. The mixture is incubated on ice for 2 h, and centrifuged at

21,600×g for 30 min at 4 °C. The pellet is washed with 10% trichloroacetic acid and with acetonitrile (ACN), and air-dried. Resuspension in 0.1% TFA typically only partially dissolves the precipitate. The mixture is centrifuged again and the supernatant containing dehydrated prenisin is lyophilized. The lyophilized peptide is dissolved in 5 mM TCEP to reduce possible disulfide bonds and then purified by RP-HPLC to separate dehydrated prenisin from the small amount of nisin added as an inducer. Separation is achieved using a gradient of 2–100% solvent B over 45 min on a C18 column (for the content of solvents A and B, see the procedure for purification of LanA peptides). The HPLC fractions are lyophilized separately and resuspended in distilled sterile water. Mass spectrometric analysis of the purified peptide is used to confirm that the product is dehydrated prenisin. The resuspended peptides from different HPLC fractions must be tested in an agar-diffusion assay against a nisin sensitive strain, *L. lactis* NZ9000, to probe for the presence of nisin used for induction. Typically, a small amount of nisin elutes at the tail of the dehydrated prenisin product peak as shown by the bioassay, even though nisin is usually not observed in the mass spectrum. Therefore, only the dehydrated prenisin samples that do not show a zone of inhibition in the agar diffusion assay are utilized in the NisC activity assay described below.

Minimal medium is prepared by adding 20 mL of solution A (per 200 mL: 2.0 g $(NH_4)_2SO_4$, 6.0 g Na_2HPO_4, 3.0 g KH_2PO_4, 1.0 g NaCl) to 80 mL of solution B (per 800 mL: 10.0 g casamino acids, 2.0 g NaOAc, 0.08 g Asn, 0.2 g $MgCl_2$, 0.01 g $CaCl_2$, 0.6 mg $FeCl_3(7H_2O)$). Afterwards, 2.5 mL of 20% glucose and 100 μL of vitamin mix (per 100 mL: 0.01 g biotin, 0.1 g folic acid, 0.1 g riboflavin, 0.1 g nicotinic acid, 0.1 g pantothenic acid, 0.2 g pyridoxal) are added to the 100-mL mixture of solution A and B. The pH of the medium is maintained at 7.0 by 0.12 M MOPS. Solution A and the casamino acids in solution B are autoclaved. The other culture components are sterilized by filtration through 0.22-μM filters (Millipore).

Activity assay. Dehydrated prenisin (~21 μM) is treated with His_6-NisC or mutants (NisC at 0.8 or 8 μM and NisC mutants at 8 μM) in 25 mM Tris, 2.5 mM $MgCl_2$, 5 μM $ZnCl_2$ at pH 7.0 in 50 μL total volume. The reaction mixture is incubated at 25 °C for 1.5 h, and the leader peptide is removed from the prepeptide with trypsin (HPLC grade, Roche) (10 μg/mL). The resulting mixture is lyophilized and resuspended in 5 μL of distilled, sterile water. The resuspended sample is applied to GM17 agar medium seeded with the nisin sensitive strain *L. lactis* NZ9000. The plate is incubated at room temperature overnight to allow the zones of inhibition to develop. In addition to the bioassay to test for NisC activity, cyclization assays monitoring the presence of free thiols are performed using methodology similar to that described above for LanM proteins (Li *et al.*, 2006).

6. The Protease Domain of Class II Lantibiotic Transporters

During the final stages of class II lantibiotic maturation, a bifunctional LanT ABC transporter with an N-terminal peptidase domain is proposed to remove the leader sequence concomitant with export of the mature species. Amide bond hydrolysis occurs C-terminal to a "double glycine" sequence (GG, GA, or GS), a recognition motif that class II LanT enzymes share with other members of the *A*BC transporter *m*aturation and *s*ecretion (AMS) protein family (Håvarstein *et al.*, 1995). The only LanT proteolytic activity characterized *in vitro* to date is that of LctT involved in lacticin 481 biosynthesis. Because LctT is an integral membrane protein (691 amino acids), a heterologous expression system was developed for only the peptidase domain. Based on a sequence alignment of LctT with known bacteriocin AMS proteins, a possible interdomain linker region between residues 125 and 175 was identified. Consequently, an *lctT* fragment corresponding to the first 150 residues was cloned as an N-terminal His_{10}-fusion protein termed His_{10}-LctT150. Overexpression and purification of His_{10}-LctT150 yielded a protease capable of correctly removing the leader peptide from linear and LctM-modified LctA (Furgerson Ihnken *et al.*, 2008). The enzyme was further investigated for its substrate scope and by site-directed mutagenesis of the proposed catalytic residues (Cys12, His90, and Asp106), yielding activity profiles consistent with a Cys protease.

6.1. Protocol for cleavage of LctA substrates by LctT150

Residues 1–150 of *lctT* were cloned into pET16b (Novagen). Heterologous expression of His_{10}-LctT150 in *E. coli* was carried out as described for LanM by induction with 0.5 mM IPTG at $A_{600} \sim 0.7$ followed by incubation at 25 °C for an additional 5 h (500 mL LB-ampicillin culture). Purification by cobalt affinity chromatography yielded 2 mg of fusion protein per liter of cell culture. Attempts to purify His_{10}-LctT150 by nickel affinity chromatography were unsuccessful. Amide bond cleavage reaction conditions for His_{10}-LctT150 are described below.

6.1.1. Procedure

1. Purified LctA is dissolved in LctT assay buffer (50 mM Na_2HPO_4, pH 7.5 at 25 °C, 50 mM Na_2SO_4, 1 mM DTT) to a final concentration of 0.04 mg/mL. Assay components may need to be readjusted to the desired pH value prior to addition of enzyme as there may be residual TFA in the LctA peptide solutions after RP-HPLC. LctA peptides are incubated with His_{10}-LctT150 with a final enzyme concentration of

0.2 mg/mL. Cleavage assays are incubated at 25 °C for 2–4 h and then quenched by the addition of 1% TFA to a final concentration of 0.1%. Samples are desalted using C18 ZipTips (Millipore) and eluted with 4 μL of α−hydroxyl cinnamic acid. From this solution, 3 μL is applied to the MALDI target and analyzed for cleavage by MALDI-TOF MS.

Note: Peptidase activity does not require any metals or cofactors and is not inhibited by EDTA. The presence of a reducing agent (TCEP or DTT) is required for His_{10}-LctT150 activity during the assay but not during purification or storage at −80 °C.

7. ADDITIONAL POSTTRANSLATIONAL MODIFICATIONS IN LANTIBIOTICS

In addition to the enzymatic reactions discussed in this chapter that have been reconstituted *in vitro* in our laboratory, one other posttranslational modification found in certain lantibiotics has been studied *in vitro*. LanD decarboxylases catalyze the formation of *S*-aminovinyl-D-cysteine (AviCys) and aminovinyl methylcysteine (AviMeCys) found at the C-terminus of several lantibiotics, including epidermin and mersacidin (Figs. 21.2 and 21.4). The activity of EpiD, the epidermin decarboxylase that introduces AviCys, was reconstituted *in vitro* in an FMN-dependent reaction (Kupke

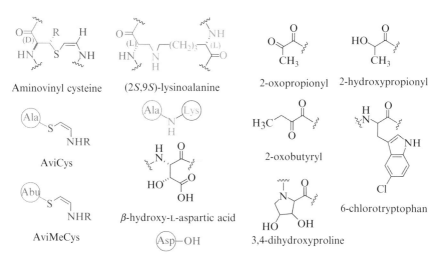

Figure 21.4 Additional posttranslational modifications that have been reported to date for the lantibiotic family (Willey and van der Donk, 2007), but for which the biosynthetic enzymes for the most part have not yet been reconstituted *in vitro*. The shorthand notation used in Fig. 21.2 is listed below each structure.

et al., 1994). On the other hand, MrsD was shown to bind FAD (Majer *et al.*, 2002). Dehydration or cyclization of EpiA was not required for EpiD modification, and the enzyme also processed the structural region of EpiA without the leader peptide (Schmid *et al.*, 2002). Subsequent NMR analysis of the EpiD decarboxylation product revealed an unusual (*Z*)-enethiol. The formation of AviCys was therefore proposed to result from the addition of the enethiol to a Dha (Kempter *et al.*, 1996), possibly catalyzed by EpiC, but this has not yet been shown experimentally. X-ray crystal structures have been solved for EpiD and an EpiD His67Asn mutant complexed with a pentapeptide (Blaesse *et al.*, 2000). His67 is conserved in all known LanD proteins and was proposed as the active-site base. For details on the methods used for the study of EpiD and MrsD the reader is referred to the references cited in this section.

Several additional posttranslational modifications have been documented for lantibiotics as shown in Fig. 21.4. At present, *in vitro* studies on the enzymes carrying out these modifications have not been reported and will most likely be the focus of future efforts in the lantibiotics field.

REFERENCES

Blaesse, M., Kupke, T., Huber, R., and Steinbacher, S. (2000). Crystal structure of the peptidyl-cysteine decarboxylase EpiD complexed with a pentapeptide substrate. *EMBO J.* **19,** 6299–6310.

Breukink, E., Wiedemann, I., van Kraaij, C., Kuipers, O. P., Sahl, H., and de Kruijff, B. (1999). Use of the cell wall precursor lipid II by a pore-forming peptide antibiotic. *Science* **286,** 2361–2364.

Brötz, H., Bierbaum, G., Markus, A., Molitor, E., and Sahl, H.-G. (1995). Mode of action of the lantibiotic mersacidin: Inhibition of peptidoglycan biosynthesis via a novel mechanism? *Antimicrob. Agents Chemother.* **39,** 714–719.

Brötz, H., Bierbaum, G., Reynolds, P. E., and Sahl, H. G. (1997). The lantibiotic mersacidin inhibits peptidoglycan biosynthesis at the level of transglycosylation. *Eur. J. Biochem.* **246,** 193–199.

Brötz, H., Josten, M., Wiedemann, I., Schneider, U., Götz, F., Bierbaum, G., and Sahl, H.-G. (1998). Role of lipid-bound peptidoglycan precursors in the formation of pores by nisin, epidermin and other lantibiotics. *Mol. Microbiol.* **30,** 317–327.

Burrage, S., Raynham, T., Williams, G., Essex, J. W., Allen, C., Cardno, M., Swali, V., and Bradley, M. (2000). Biomimetic synthesis of lantibiotics. *Chem. Eur. J.* **6,** 1455–1466.

Chatterjee, C., Miller, L. M., Leung, Y. L., Xie, L., Yi, M., Kelleher, N. L., and van der Donk, W. A. (2005a). Lacticin 481 synthetase phosphorylates its substrate during lantibiotic production. *J. Am. Chem. Soc.* **127,** 15332–15333.

Chatterjee, C., Patton, G. C., Cooper, L., Paul, M., and van der Donk, W. A. (2006). Engineering dehydro amino acids and thioethers into peptides using lacticin 481 synthetase. *Chem. Biol.* **13,** 1109–1117.

Chatterjee, C., Paul, M., Xie, L., and van der Donk, W. A. (2005b). Biosynthesis and mode of action of lantibiotics. *Chem. Rev.* **105,** 633–684.

Chen, P., Qi, F. X., Novak, J., Krull, R. E., and Caufield, P. W. (2001). Effect of amino acid substitutions in conserved residues in the leader peptide on biosynthesis of the lantibiotic mutacin II. *FEMS Microbiol. Lett.* **195,** 139–144.

Cheng, F., Takala, T. M., and Saris, P. E. (2007). Nisin biosynthesis *in vitro. J. Mol. Microbiol. Biotechnol.* **13,** 248–254.

Cooper, L. E., McClerren, A. L., Chary, A., and van der Donk, W. A. (2008). Structure–activity relationship studies of the two-component lantibiotic haloduracin. *Chem. Biol.* **15,** 1035–1045.

Corvey, C., Stein, T., Dusterhus, S., Karas, M., and Entian, K. D. (2003). Activation of subtilin precursors by *Bacillus subtilis* extracellular serine proteases subtilisin (AprE), WprA, and Vpr. *Biochem. Biophys. Res. Commun.* **304,** 48–54.

Cotter, P. D., Deegan, L. H., Lawton, E. M., Draper, L. A., O'Connor, P. M., Hill, C., and Ross, R. P. (2006a). Complete alanine scanning of the two-component lantibiotic lacticin 3147: Generating a blueprint for rational drug design. *Mol. Microbiol.* **62,** 735–747.

Cotter, P. D., Draper, L. A., Lawton, E. M., McAuliffe, O., Hill, C., and Ross, R. P. (2006b). Overproduction of wild-type and bioengineered derivatives of the lantibiotic lacticin 3147. *Appl. Environ. Microbiol.* **72,** 4492–4496.

Delves-Broughton, J., Blackburn, P., Evans, R. J., and Hugenholtz, J. (1996). Applications of the bacteriocin, nisin. *Antonie van Leeuwenhoek* **69,** 193–202.

Dodd, H., Gasson, M., Mayer, M., and Narbad, A. (2006). Identifying Lantibiotic Gene Clusters and Novel Lantibiotic Genes. Patent number: WO 2006/111743 A2 PCT/GB2006/001429. 26.

Duquesne, S., Destoumieux-Garzon, D., Zirah, S., Goulard, C., Peduzzi, J., and Rebuffat, S. (2007). Two enzymes catalyze the maturation of a lasso peptide in *Escherichia coli. Chem. Biol.* **14,** 793–803.

Fredenhagen, A., Fendrich, G., Marki, F., Marki, W., Gruner, J., Raschdorf, F., and Peter, H. H. (1990). Duramycins B and C, two new lanthionine containing antibiotics as inhibitors of phospholipase A2. Structural revision of duramycin and cinnamycin. *J. Antibiot.* **43,** 1403–1412.

Furgerson Ihnken, L. A., Chatterjee, C., and van der Donk, W. A. (2008). *In vitro* reconstitution and substrate specificity of a lantibiotic protease. *Biochemistry* **47,** 7352–7363.

Hasper, H. E., Kramer, N. E., Smith, J. L., Hillman, J. D., Zachariah, C., Kuipers, O. P., de Kruijff, B., and Breukink, E. (2006). A new mechanism of antibiotic action. *Science* **313,** 1636–1637.

Håvarstein, L. S., Diep, D. B., and Nes, I. F. (1995). A family of bacteriocin ABC transporters carry out proteolytic processing of their substrates concomitant with export. *Mol. Microbiol.* **16,** 229–240.

Håvarstein, L. S., Holo, H., and Nes, I. F. (1994). The leader peptide of colicin V shares consensus sequences with leader peptides that are common among peptide bacteriocins produced by gram-positive bacteria. *Microbiology* **140**(Pt. 9), 2383–2389.

Izaguirre, G., and Hansen, J. N. (1997). Use of alkaline phosphatase as a reporter polypeptide to study the role of the subtilin leader segment and the SpaT transporter in the posttranslational modifications and secretion of subtilin in *Bacillus subtilis* 168. *Appl. Environ. Microbiol.* **63,** 3965–3971.

Jung, G. (1991). Lantibiotics-ribosomally synthesized biologically active polypeptides containing sulfide bridges and α,β-dehydroamino acids. *Angew. Chem. Intl. Ed. Engl.* **30,** 1051–1068.

Kempter, C., Kupke, T., Kaiser, D., Metzger, J. W., and Jung, G. (1996). Thioenols from peptidyl cysteines: Oxidative decarboxylation of a 13C-labeled substrate. *Angew. Chem. Intl. Ed. Engl.* **35,** 2104–2107.

Kluskens, L. D., Kuipers, A., Rink, R., de Boef, E., Fekken, S., Driessen, A. J., Kuipers, O. P., and Moll, G. N. (2005). Posttranslational modification of therapeutic peptides by NisB, the dehydratase of the lantibiotic nisin. *Biochemistry* **44,** 12827–12834.

Kodani, S., Hudson, M. E., Durrant, M. C., Buttner, M. J., Nodwell, J. R., and Willey, J. M. (2004). The SapB morphogen is a lantibiotic-like peptide derived from the product of the developmental gene ramS in *Streptomyces coelicolor*. *Proc. Natl. Acad. Sci. USA* **101,** 11448–11453.

Kodani, S., Lodato, M. A., Durrant, M. C., Picart, F., and Willey, J. M. (2005). SapT, a lanthionine-containing peptide involved in aerial hyphae formation in the streptomycetes. *Mol. Microbiol.* **58,** 1368–1380.

Kuipers, O. P., Bierbaum, G., Ottenwälder, B., Dodd, H. M., Horn, N., Metzger, J., Kupke, T., Gnau, V., Bongers, R., van den Bogaard, P., Kosters, H., Rollema, H. S., *et al.* (1996). Protein engineering of lantibiotics. *Antonie van Leeuwenhoek* **69,** 161–169.

Kuipers, A., De Boef, E., Rink, R., Fekken, S., Kluskens, L. D., Driessen, A. J., Leenhouts, K., Kuipers, O. P., and Moll, G. N. (2004). NisT, the transporter of the lantibiotic nisin, can transport fully modified, dehydrated and unmodified prenisin and fusions of the leader peptide with non-lantibiotic peptides. *J. Biol. Chem.* **279,** 22176–22182.

Kupke, T., Kempter, C., Gnau, V., Jung, G., and Götz, F. (1994). Mass spectrometric analysis of a novel enzymatic reaction. *J. Biol. Chem.* **269,** 5653–5659.

Kupke, T., Stevanovich, S., Ottenwälder, B., Metzger, J. W., Jung, G., and Götz, F. (1993). Purification and characterization of EpiA, the peptide substrate for posttranslational modifications involved in epidermin biosynthesis. *FEMS Microbiol. Lett.* **112,** 43–48.

Lawton, E. M., Cotter, P. D., Hill, C., and Ross, R. P. (2007). Identification of a novel two-peptide lantibiotic, Haloduracin, produced by the alkaliphile *Bacillus halodurans* C-125. *FEMS Microbiol Lett.* **267,** 64–71.

Lee, M. V., Furgerson Ihnken, L. A., You, Y. O., McClerren, A. L., van der Donk, W. A., and Kelleher, N. L. (Submitted for publication).

Levengood, M. R., Patton, G. C., and van der Donk, W. A. (2007). The leader peptide is not required for posttranslational modification by lacticin 481 synthetase. *J. Am. Chem. Soc.* **129,** 10314–10315.

Levengood, M. R., and van der Donk, W. A. (2008). Use of lantibiotic synthetases for the preparation of bioactive constrained peptides. *Bioorg. Med. Chem. Lett.* **18,** 3025–3028.

Li, B., and van der Donk, W. A. (2007). Identification of essential catalytic residues of the cyclase NisC involved in the biosynthesis of nisin. *J. Biol. Chem.* **282,** 21169–21175.

Li, B., Yu, J.-P. J., Brunzelle, J. S., Moll, G. N., van der Donk, W. A., and Nair, S. K. (2006). Structure and mechanism of the lantibiotic cyclase involved in nisin biosynthesis. *Science* **311,** 1464–1467.

Liu, W., and Hansen, J. N. (1992). Enhancement of the chemical and antimicrobial properties of subtilin by site-directed mutagenesis. *J. Biol. Chem.* **267,** 25078–25085.

Majer, F., Schmid, D. G., Altena, K., Bierbaum, G., and Kupke, T. (2002). The flavoprotein MrsD catalyzes the oxidative decarboxylation reaction involved in formation of the peptidoglycan biosynthesis inhibitor mersacidin. *J. Bacteriol.* **184,** 1234–1243.

Märki, F., Hanni, E., Fredenhagen, A., and van Oostrum, J. (1991). Mode of action of the lanthionine-containing peptide antibiotics duramycin, duramycin B and C, and cinnamycin as indirect inhibitors of phospholipase A2. *Biochem. Pharmacol.* **42,** 2027–2035.

McClerren, A. L., Cooper, L. E., Quan, C., Thomas, P. M., Kelleher, N. L., and van der Donk, W. A. (2006). Discovery and *in vitro* biosynthesis of haloduracin, a new two-component lantibiotic. *Proc. Natl. Acad. Sci. USA* **103,** 17243–17248.

Miller, L. M., Chatterjee, C., van der Donk, W. A., and Kelleher, N. L. (2006). The dehydration activity of lacticin 481 synthetase is highly processive. *J. Am. Chem. Soc.* **128,** 1420–1421.

Okeley, N. M., Paul, M., Stasser, J. P., Blackburn, N., and van der Donk, W. A. (2003). SpaC and NisC, the cyclases involved in subtilin and nisin biosynthesis, are zinc proteins. *Biochemistry* **42,** 13613–13624.
Okeley, N. M., Zhu, Y., and van der Donk, W. A. (2000). Facile chemoselective synthesis of dehydroalanine-containing peptides. *Org. Lett.* **2,** 3603–3606.
Ottenwälder, B., Kupke, T., Brecht, S., Gnau, V., Metzger, J., Jung, G., and Götz, F. (1995). Isolation and characterization of genetically engineered gallidermin and epidermin analogs. *Appl. Environ. Microbiol.* **61,** 3894–3903.
Patton, G. C., Paul, M., Cooper, L. E., Chatterjee, C., and van der Donk, W. A. (2008). The importance of the leader sequence for directing lanthionine formation in lacticin 481. *Biochemistry* **47,** 7342–7351.
Paul, M., Patton, G. C., and van der Donk, W. A. (2007). Mutants of the zinc ligands of lacticin 481 synthetase retain dehydration activity but have impaired cyclization activity. *Biochemistry* **46,** 6268–6276.
Peschel, A., Ottenwälder, B., and Götz, F. (1996). Inducible production and cellular location of the epidermin biosynthetic enzyme EpiB using an improved staphylococcal expression system. *FEMS Microbiol. Lett.* **137,** 279–284.
Rink, R., Kuipers, A., de Boef, E., Leenhouts, K. J., Driessen, A. J., Moll, G. N., and Kuipers, O. P. (2005). Lantibiotic structures as guidelines for the design of peptides that can be modified by lantibiotic enzymes. *Biochemistry* **44,** 8873–8882.
Rink, R., Wierenga, J., Kuipers, A., Kluskens, L. D., Driessen, A. J. M., Kuipers, O. P., and Moll, G. N. (2007). Production of dehydroamino acid-containing peptides by *Lactococcus lactis*. *Appl. Environ. Microbiol.* **73,** 1792–1796.
Roy, R. S., Gehring, A. M., Milne, J. C., Belshaw, P. J., and Walsh, C. T. (1999). Thiazole and oxazole peptides: Biosynthesis and molecular machinery. *Nat. Prod. Rep.* **16,** 249–263.
Ryan, M. P., Jack, R. W., Josten, M., Sahl, H. G., Jung, G., Ross, R. P., and Hill, C. (1999). Extensive posttranslational modification, including serine to D-alanine conversion, in the two-component lantibiotic, lacticin 3147. *J. Biol. Chem.* **274,** 37544–37550.
Schmid, D. G., Majer, F., Kupke, T., and Jung, G. (2002). Electrospray ionization Fourier transform ion cyclotron resonance mass spectrometry to reveal the substrate specificity of the peptidyl-cysteine decarboxylase EpiD. *Rapid Commun. Mass Spectrom.* **16,** 1779–1784.
Schnell, N., Entian, K.-D., Schneider, U., Götz, F., Zahner, H., Kellner, R., and Jung, G. (1988). Prepeptide sequence of epidermin, a ribosomally synthesized antibiotic with four sulphide-rings. *Nature* **333,** 276–278.
Severinov, K., Semenova, E., Kazakov, A., Kazakov, T., and Gelfand, M. S. (2007). Low-molecular-weight posttranslationally modified microcins. *Mol. Microbiol.* **189,** 8772–8785.
Siezen, R. J., Kuipers, O. P., and de Vos, W. M. (1996). Comparison of lantibiotic gene clusters and encoded proteins. *Antonie van Leeuwenhoek* **69,** 171–184.
Szekat, C., Jack, R. W., Skutlarek, D., Farber, H., and Bierbaum, G. (2003). Construction of an expression system for site-directed mutagenesis of the lantibiotic mersacidin. *Appl. Environ. Microbiol.* **69,** 3777–3783.
Toogood, P. L. (1993). Model studies of lantibiotic biogenesis. *Tetrahedron Lett.* **34,** 7833–7836.
van der Meer, J. R., Rollema, H. S., Siezen, R. J., Beerthuyzen, M. M., Kuipers, O. P., and de Vos, W. M. (1994). Influence of amino acid substitutions in the nisin leader peptide on biosynthesis and secretion of nisin by *Lactococcus lactis*. *J. Biol. Chem.* **269,** 3555–3562.
van Kraaij, C., Breukink, E., Rollema, H. S., Siezen, R. J., Demel, R. A., De Kruijff, B., and Kuipers, O. P. (1997). Influence of charge differences in the C-terminal part of nisin on antimicrobial activity and signaling capacity. *Eur. J. Biochem.* **247,** 114–120.

Widdick, D. A., Dodd, H. M., Barraille, P., White, J., Stein, T. H., Chater, K. F., Gasson, M. J., and Bibb, M. J. (2003). Cloning and engineering of the cinnamycin biosynthetic gene cluster from *Streptomyces cinnamoneus cinnamoneus* DSM 40005. *Proc. Natl. Acad. Sci. USA* **100,** 4316–4321.

Wiedemann, I., Breukink, E., van Kraaij, C., Kuipers, O. P., Bierbaum, G., de Kruijff, B., and Sahl, H. G. (2001). Specific binding of nisin to the peptidoglycan precursor lipid II combines pore formation and inhibition of cell wall biosynthesis for potent antibiotic activity. *J. Biol. Chem.* **276,** 1772–1779.

Willey, J. M., and van der Donk, W. A. (2007). Lantibiotics: Peptides of diverse structure and function. *Annu. Rev. Microbiol.* **61,** 477–501.

Xie, L., Chatterjee, C., Balsara, R., Okeley, N. M., and van der Donk, W. A. (2002). Heterologous expression and purification of SpaB involved in subtilin biosynthesis. *Biochem. Biophys. Res. Commun.* **295,** 952–957.

Xie, L., Miller, L. M., Chatterjee, C., Averin, O., Kelleher, N. L., and van der Donk, W. A. (2004). Lacticin 481: *In vitro* reconstitution of lantibiotic synthetase activity. *Science* **303,** 679–681.

You, Y. O., and van der Donk, W. A. (2007). Mechanistic investigations of the dehydration reaction of lacticin 481 synthetase using site-directed mutagenesis. *Biochemistry* **46,** 5991–6000.

Zhang, X., Ni, W., and van der Donk, W. A. (2007). On the regioselectivity of thioether formation by lacticin 481 synthetase. *Org. Lett.* **9,** 3343–3346.

Zhang, X., and van der Donk, W. A. (2007). On the substrate specificity of the dehydratase activity of lacticin 481 synthetase. *J. Am. Chem. Soc.* **129,** 2212–2213.

Zhang, X., and van der Donk, W. A. (2009). Using expressed protein ligation to probe the substrate specificity of lantibiotic synthetases. *Methods Enzymol.* (in press).

Zhou, H., and van der Donk, W. A. (2002). Biomimetic stereoselective formation of methyllanthionine. *Org. Lett.* **4,** 1335–1338.

Zhu, Y., Gieselman, M., Zhou, H., Averin, O., and van der Donk, W. A. (2003). Biomimetic studies on the mechanism of stereoselective lanthionine formation. *Org. Biomol. Chem.* **1,** 3304–3315.

CHAPTER TWENTY-TWO

WHOLE-CELL GENERATION OF LANTIBIOTIC VARIANTS

Jesús Cortés, Antony N. Appleyard, *and* Michael J. Dawson

Contents

1. Introduction	559
2. Variant Generation	561
2.1. Lantibiotics generated by homologous expression systems	561
2.2. Lantibiotics generated by heterologous expression systems	565
2.3. Comparison between *cis* and *trans* complementation for the production of mersacidin analogues	566
2.4. Strategies for characterizing the structure and identity of lantibiotic variants	568
3. Conclusions	571
References	571

Abstract

The generation of modified lantibiotics in whole cells has proved to be of value for the investigation of the specificity of the lantibiotic-processing enzymes and their tolerance to mutations in the primary sequence of lantibiotics. The development of methods to produce new lantibiotic variants has also enabled the investigation of the structure–activity relationships of these compounds and hence an evaluation of this hitherto underexploited class of natural products as a source of potential therapeutic drug candidates. We report the methods and strategies that have been used to engineer new lantibiotic variants and practical methods to analyze libraries of new compounds with a view toward optimizing drug properties.

1. Introduction

Lantibiotics are highly posttranslationally modified peptides 19–36 amino acids long produced by Gram-positive bacteria (Chatterjee *et al.*, 2005b). Unlike many bacterial peptides they are ribosomally synthesized, but undergo posttranslational modifications, the most characteristic of which is the formation of lanthionine or methyl-lanthionine amino acids by dehydration of a serine or threonine residue and electrophilic addition of the sulfur of cysteine. Other modifications include oxidative decarboxylation of a C-terminal cysteine to yield an aminovinylcysteine moiety, which can also participate in ring formation (mersacidin; Majer et al., 2002 and epidermin; Kupke et al., 1995), conversion of serine to D-alanine (lacticin 3147; Ryan et al., 1999), hydroxylation of aspartate (cinnamycin; Widdick et al., 2003) or proline (microbisporicin; Castiglione et al., 2008), chlorination of tryptophan (microbisporicin; Castiglione et al., 2008), and sulfoxidation (actagardine; Zimmerman et al., 1995).

Because of the promising biological activities of the lantibiotics, especially their antibacterial activity, there has been much interest in modifying their structure to evaluate structure–activity relationships and to design antibiotics with improved therapeutic potential. Such studies have also been focused on understanding the exquisite control of the posttranslational processing apparatus.

Most lantibiotics are chromosomally encoded, though some biosynthetic clusters reside on plasmids (e.g., epidermin, lacticin 481) or on transposable elements (e.g., nisin). Lantibiotic biosynthetic pathways contain a number of common elements (Chatterjee et al., 2005b). The peptide gene, generically termed *lanA*, encodes the prepropeptide, which includes a cleavable leader sequence. The key dehydration and cyclization reactions are carried out either by two gene products, denoted LanB and LanC, or by a single gene product, LanM. A LanT gene product is involved in transport, either of the final mature product or the posttranslationally modified lantibiotic with its leader sequence still attached. Leader cleavage may be catalyzed by a protease, LanP, or a protease domain within LanT. In some cases there is no apparent protease within the cluster and "host" proteases are presumably recruited for this purpose (e.g., subtilin; Corvey et al., 2003). Additional genes (*lanI, lanE, lanF,* and *lanG*) may encode immunity or resistance determinants. Some lantibiotic biosynthetic clusters contain genes specific for other posttranslational modifications, such as *lanD*, responsible for the oxidative decarboxylation of the C-terminal cysteine of epidermin or mersacidin (Kupke et al., 1995; Majer et al., 2002).

Lantibiotic biosynthetic clusters are quite diverse in their regulation but several, including nisin (Kuipers et al., 1995), subtilin (Stein et al., 2002), and mersacidin (Schmitz et al., 2006), have been shown to be auto-inducible. Expression is controlled by a two-component regulation system comprising a sensor histidine kinase (LanK) and a response regulator (LanR).

Because of the ribosomally encoded nature of the lantibiotics, structural modification can be readily attempted by manipulation of the *lanA* gene in the context of the remainder of the biosynthetic pathway. A number of systems have been developed for facilitating such manipulations. Basically modification can be effected either:

1. *in cis*—by site-directed mutation or gene replacement; or
2. *in trans*—by introduction of a separate *lanA* gene, either on a plasmid or integrated into the chromosome. In some instances the native *lanA* gene is knocked out to facilitate analysis.

Modifications may be carried out either in the natural host organism or after heterologous expression in a genetically well-characterized host.

This chapter will provide a practical guide to methods which have been used successfully for manipulation of lantibiotic structure in whole-cell systems. Lantibiotic synthesis in cell-free systems will be covered elsewhere in this volume (Van der Donk). Some attempt has been made to fathom general "rules" which govern the ability of the lantibiotic biosynthetic machinery to process modified lantibiotics (Rink *et al.*, 2005). However, it should be noted that the ability of the whole-cell system to make modified lantibiotics depends on a number of factors:

- The ability of the dehydration and cyclization enzymes to tolerate modified *lanA* genes
- Tolerance of the export systems and leader-processing enzymes
- Ability of the host immunity systems to tolerate modified lantibiotics and of the modified lantibiotics to act as inducers for the biosynthetic apparatus
- In some cases specificity of other processing machinery, such as the *lanD* gene product

In at least some systems (e.g., nisin and lacticin 3147) the modification and export functions appear hugely flexible (Kluskens *et al.*, 2005; Kuipers *et al.*, 2004, 2008; Rink *et al.*, 2007), but it remains to be seen whether this extends to all lantibiotic systems. In contrast, other components of the biosynthetic machinery, for example, LanD, may be quite narrow in specificity (Kupke *et al.*, 1995).

Finally, attempts to generate modified lantibiotics have been almost exclusively addressed by modification of the *lanA* gene, but there should be potential to generate new structures also by modification of other elements of the biosynthetic pathway, both to extend the range of primary amino acid sequences which can be tolerated and to generate diversity in other structural elements, such as by heterologous combinations of processing genes. Such studies are in their infancy, but will likely increase in the coming years.

2. Variant Generation

2.1. Lantibiotics generated by homologous expression systems

2.1.1. *Cis* complementation system

A *cis* complementation system involves the substitution of the *lanA* gene for a mutant gene placed in the same *lanA* locus. This system has the advantage that there are no alterations of the gene cluster and the possible regulatory or stability sequences flanking *lanA* genes are conserved. The disadvantage of this system is that the manipulations required to generate a large number of mutants make it extremely laborious and time consuming. *Cis* complementation systems have been used to generate mutants of several lantibiotics; some examples are described below.

2.1.1.1. Subtilin The host strain *Bacillus subtilis* spaS$^-$ was created by a single-step double-crossover gene replacement, substituting *spaS* for an erythromycin resistance cassette via transformation with linear plasmid DNA (Liu and Hansen, 1992). The construction of the substitution plasmid was performed by PCR site-specific mutagenesis using the *Escherichia coli* cloning vector pTZ19U as backbone. Subtilin variants were generated using the same system but a chloramphenicol resistance cassette was used instead of erythromycin resistance, cloned upstream of the modified *spaS*, allowing for selection of the incoming DNA. The vectors used for deletion of DNA in the host and for delivery of variant *spaS* genes lacked a *Bacillus* origin of replication, so recombination was necessary for expression of the resistance marker.

2.1.1.2. Pep5 The lantibiotic Pep5 is produced by *Staphylococcus epidermidis* 5. The genes encoding the biosynthesis of this lantibiotic lie in a 20-kb plasmid. The host strain for the variant generation system was created by curing the producer strain of this plasmid (Bierbaum *et al.*, 1994). The vector used for delivering *pepA* mutants consisted of a 6.2-kb fragment encoding the whole Pep5 pathway cloned into the bifunctional *E. coli* and *Staphylococcus* vector pCU1 (Augustin *et al.*, 1992). Mutants of *pepA* were generated in a 1.4-kb *Kpn*I fragment containing *pepA* and *pepI* by oligonucleotide-assisted single-stranded DNA site-directed mutagenesis. Following mutagenesis, the fragments were substituted into the delivery plasmid and used to transform *Staphylococcus carnosus* protoplasts. Plasmids obtained from *S. carnosus* were used to transform the host strain by electroporation.

2.1.1.3. Mutacin II The host strain *Streptococcus mutans* mutA$^-$ was created by a single-step double crossover, introducing a tetracycline resistance cassette upstream of the *mutA* locus and simultaneously creating a deletion of *mutA* via transformation with linear DNA (Chen *et al.*, 1998). The construction of the deletion plasmid was performed by PCR site-specific mutagenesis using the *E. coli* cloning vector pCRII as backbone. Mutacin II variants were generated using the same system, but a kanamycin resistance cassette was used instead of tetracycline resistance cloned upstream of the modified *mutA*, allowing for selection of the incoming DNA. The vectors used for deletion of DNA in the host and for delivery of modified *mutA* genes lack a *Streptococcus* origin of replication so recombination is necessary for expression of the resistance marker.

2.1.1.4. Mersacidin This method was described by Szekat *et al.* (2003). A 1.1-kb restriction fragment of the mersacidin gene cluster containing *mrsA* in the middle of the fragment was subcloned into a phagemid suitable for oligonucleotide-assisted site-directed mutagenesis. After the mutagenesis, the fragments containing mutated *mrsA* genes were moved to the temperature-sensitive *Staphylococcus* cloning vector pTVO. The ligation mixtures were used to transform protoplasts of *S. carnosus*. Plasmids prepared from *S. carnosus* were used to transform protoplasts of the producer strain, *Bacillus* sp. HIL Y-85,54728. Integrants were selected at 45 °C and resolved after successive rounds of growth in the absence of antibiotic selection.

2.1.1.5. Lacticin 3147 The substitution of each of the amino acids in the two-component lantibiotic lacticin 3147 to alanine has been performed (Cotter *et al.*, 2006a). The host strain was generated by creating an *ltnA1* and *ltnA2* deletion on the native conjugative plasmid pMRC01 that codes for the synthesis of this lantibiotic (Cotter *et al.*, 2003). This was achieved by PCR amplification of DNA fragments containing the flanking regions of *ltnA1* and *ltnA2* joined by splicing overlap extension PCR. This PCR product was cloned into the RepA$^-$ plasmid pORI280, and the resulting plasmid introduced by electroporation into a lacticin 3147-producing strain carrying pMRC01 and pVE6007 (temperature-sensitive, RepA$^+$). The integrant was obtained at nonpermissive temperature and resolved in the absence of selective antibiotic. Mutants of *ltnA1* and *ltnA2* were produced by cloning a PCR product containing *ltnA1* and *ltnA2* and flanking regions into pORI280. This plasmid was then used as a template for oligonucleotide-assisted site-directed mutagenesis. Mutant plasmids were introduced by electroporation into the host strain and the double crossovers selected following the same procedure as for the deletion to create the host strain.

2.1.2. *Trans* complementation system

The *trans* complementation systems require inactivation of the structural *lanA* gene to avoid the formation of mixtures of compounds or to prevent the wild-type lantibiotic being preferentially synthesized when competing with a mutant gene. Inactivation can be achieved by inserting an early stop codon, by entire deletion of the *lanA* gene, or by generating a frameshift. The knockout mutant is then complemented with mutant *lanA* genes using a suitable autonomous or integrative expression vector. Examples of the use of *trans* complementation systems are described below.

2.1.2.1. Nisin Kuipers *et al.* (1992) developed a system consisting of a bifunctional *E. coli*–*Lactococcus* plasmid, pNZ9013, containing the *nisZ* gene downstream of the *Lactococcus lactis lac* promoter. Mutants were produced by PCR site-specific mutagenesis and the resulting plasmids were introduced by electroporation into the nisin producer *L. lactis* T165.5. The variants in this system are produced as mixtures with nisin A produced by the host strain.

Dodd et al. (1992) developed a system for the generation of nisin variants. The host strain, L. lactis F17332, was a $nisA^-$ strain constructed by gene replacement, substituting nisA for an erythromycin resistance cassette using the plasmid pFI1283; this is a pBR322-based plasmid containing 2 kb of the nisin cluster with the erythromycin cassette inserted in the middle of nisA, generating a polar mutant that does not express the immunity genes. After further manipulation of this strain a transposon (IS905) was identified within the erythromycin cassette and a deletion further downstream seemed to remove the polar effect observed in the initial construct, making the strain easier to transform. For generation of variants, plasmid pFI354 was constructed. This plasmid contains the 5' part of nisA, including the promoter, and lacks the sequence that codes for the last 20 amino acids of nisin. The backbone of this vector is pTG262 (Gasson and Anderson, 1985). Mutations were introduced by cloning PCR products generated by overlap extension into this plasmid. Mutant plasmids were introduced into the producing strain by electroporation.

Field et al. (2008) developed a system involving a host strain L. lactis NZ9800 ($nisA^-$) described by Kuipers et al. (1993). This strain was generated by double-crossover-assisted gene replacement and contains a 4-bp deletion in nisA; this frameshift generates a stop codon at the position corresponding to amino acid 16 of pronisin. This strain was complemented by the low copy number bifunctional E. coli–Lactococcus plasmid pPTPL (O'Driscoll et al., 2004) containing the nisA gene under the control of its own promoter (P_{nis}). Mutants of nisA were obtained by error-prone PCR and moved into pPTPL. Mutant plasmids were introduced into the producing strain by electroporation.

2.1.2.2. Gallidermin/epidermin The system described by Ottenwälder et al. (1995) consists of a host strain, S. epidermidis $epiA^-$, obtained by random mutagenesis using ethyl methanesulfonate, complemented with derivatives of pT181mcs or pCU1 (Augustin et al., 1992) containing a DNA fragment encoding gdmA or derivatives of pT181mcs containing a DNA fragment encoding epiA. Mutants of gdmA and epiA were obtained by generation of single-stranded templates and mutagenic oligonucleotides or by PCR site-specific mutagenesis. The plasmids are delivered to the host strain by electroporation.

2.1.2.3. Mersacidin The system for production of mersacidin variants consists of a host strain Bacillus sp. $mrsA^-$ constructed by double-crossover-assisted gene replacement using the temperature-sensitive vector pBT2 (Brückner, 1997) as described by Szekat et al. (2003). The mrsA gene was substituted by a mutant mrsA that contains a stop codon in the position that encodes the fourth amino acid of the leader peptide. This mutant was created by oligonucleotide-assisted site-directed mutagenesis (Dawson et al., 2005). *Bacillus* sp. *mrsA*$^-$ is complemented by a derivative of pCU1 (Augustin et al., 1992) containing an engineered *mrsA* gene with the sequence encoding mersacidin flanked by unique restriction sites, so it can be substituted by annealed oligonucleotides encoding mersacidin variants. This engineered *mrsA* was created by PCR site-specific mutagenesis. The nonmethylated mutant plasmids obtained from *E. coli* ET12567 (MacNeil et al., 1992) were delivered to the host strain by electroporation.

2.1.2.4. Actagardine The system for production of actagardine variants consists of a host strain, *Actinoplanes garbadinensis garA*$^-$, constructed by double-crossover-assisted gene replacement (Boakes et al., 2007). The *garA* deletion was created using the Redirect method described by Gust et al. (2003). The mutant *garA* encodes only the first five amino acids and the stop codon. This mutant is complemented by a derivative of pSET152 (Bierman et al., 1992) containing an engineered *garA* gene with the sequence encoding actagardine flanked by unique restriction sites so it can be substituted by annealed oligonucleotides encoding actagardine variants. The mutant plasmids were delivered to the host strain by intergeneric conjugation from *E. coli* ET12567/pUZ8002. Plasmids were integrated into the chromosome at the ΦC31 phage attachment site.

2.1.2.5. Lacticin 3147 Field et al. (2007) developed a system for generation of random libraries of lacticin 3147. The host strain for the production of variants is *L. lactis* MG1363 pOM44 (Cotter et al., 2006b). pOM44 is a derivative of the shuttle vector pCI372 containing all the genes responsible for the synthesis of lacticin 3147 except *ltnA1* and *ltnA2*. This strain was complemented by the low copy number bifunctional *E. coli–Lactococcus* plasmid pPTPL (O'Driscoll et al., 2004) containing the mutated *ltnA1* and *ltnA2* genes under the control of their own promoter. Random mutants of *ltnA1* and *ltnA2* were obtained by error-prone PCR. The template for mutagenesis was a derivative of pC1372 containing *ltnA1* and *ltnA2* under the control of their own promoter. After mutagenesis, the inserts were moved to pPTPL. Mutant plasmids were introduced into the producing strain by electroporation.

2.2. Lantibiotics generated by heterologous expression systems

2.2.1. In *cis* expression of cinnamycin

The cloning and sequencing of the cinnamycin gene cluster and a method for generating cinnamycin variants was described by Widdick et al. (2003). The cinnamycin gene cluster from *Streptomyces cinnamoneus* was inserted into in the vector pOJ436 (Bierman et al., 1992) by sequentially cloning individual restriction fragments containing parts of the cluster. The resulting plasmid was moved to *Streptomyces lividans* 1326 by protoplast transformation, showing that cinnamycin can be produced in a heterologous host. The vector containing the gene cluster integrates into the *S. lividans* chromosome at the ΦC31 phage attachment site. A mutagenesis vector containing the last fragment cloned to reconstitute the cluster containing *cinA* was engineered by PCR site-specific mutagenesis so that the cinnamycin-encoding sequence within *cinA* was substituted by a tetracycline resistance gene flanked by unique restriction sites that facilitate the cloning of annealed oligonucleotides encoding cinnamycin variants. The inserts of the plasmids encoding cinnamycin variants were used to substitute *cinA* in the expression plasmid containing the whole cluster and the mutant plasmids were used to transform *S. lividans* 1326 protoplasts.

2.2.2. In *trans* expression of nukacin ISK-1

Aso et al. (2004) developed an expression system for nukacin by sequentially cloning fragments of the nukacin gene cluster under the control of the nisin promoter (P_{nis}) into the high copy number vector pIL253 (Simon and Chopin, 1988). The promoter was engineered so that it was placed directly upstream of *nukA* by splicing overlap extension PCR. Once the whole cluster was cloned, and expression in the host strain confirmed, a deletion of *nukA* was performed by removing a restriction fragment containing *nukA* and part of *nukM* and a PCR product containing the start of nukM up to the restriction site used to remove *nukA* was introduced. This strain was complemented by cloning *nukA* into pNZ8084. This vector is a P_{nis}-based expression vector (Pascalle et al., 1996) and mutants of *nukA* were generated by PCR (Asaduzzaman et al., 2006). The host strain was *L. lactis* NZ9800 ($nisA^-$) described above (Kuipers et al., 1993).

2.3. Comparison between *cis* and *trans* complementation for the production of mersacidin analogues

Production of targeted mutants within the mersacidin gene cluster by *cis* complementation involved a double crossover to deliver *mrsA* mutants into the host strain. This process requires temperature-assisted integration, selection of the stable integrant, and resolution of the integrant (Fig. 22.1). The overall process involves mutagenesis of *mrsA* by an appropriate method. This is followed by transformation of the host strain, isolation of transformants in selective medium, growth at restrictive temperature to promote integration in the presence of selective antibiotic, and isolation of the integrant strain. Once the integrant strain has been isolated, it is grown in nonselective liquid culture for several generations at high temperature to allow for the resolution of the integrant and loss of the temperature-sensitive vector. The final culture

A *Cis* complementation

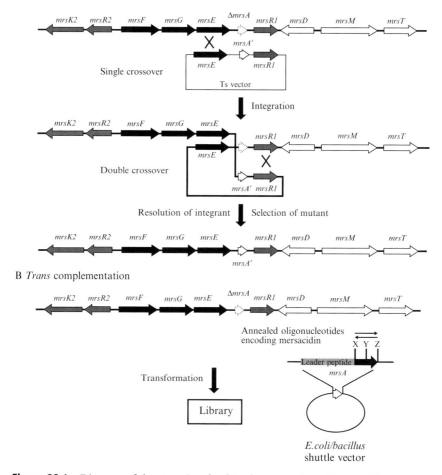

Figure 22.1 Diagram of the steps involved in the generation of mersacidin mutants by *cis* and *trans* complementation procedures. (A) *Cis* complementation involves a transformation, a double-crossover event and selection of the desired mutant from a possible 1:1 mixture with host strain. (B) *Trans* complementation involves one transformation step only.

is plated out at the appropriate colony density for replica plating and transferred to selective medium. Putative double-crossover strains need to be selected and characterized as they will be obtained as a mixture with strains that revert to the initial genotype of the host strain. Due to the steps involved in this process, this method is not suitable for the generation of large numbers of mutants or collections of mutants generated in a single mutagenesis process. For characterization of one specific mersacidin mutant, obtaining a double crossover with the minimum modification of the DNA sequences flanking the *mrsA* gene might be appropriate, but if the scope of the study is to obtain and analyze a large number of mutants, *trans* complementation allows the more rapid generation of mutant strains.

Trans complementation can be achieved by transformation of the host strain with an expression vector containing the desired *mrsA* mutant (Fig. 22.1). By using an expression vector with unique restriction sites flanking the mersacidin-encoding region of *mrsA*, it is possible to generate a large library of *mrsA* mutants in one ligation/transformation experiment. The process involves ligation of the expression vector to annealed oligonucleotides targeting the sequence within mersacidin that is the subject of study, followed by transformation of the host strain and growth of the mutants in production medium. If multiple or random changes are to be introduced, degenerate oligonucleotides can be used; the transformation mixture in this case will constitute a library of strains encoding mersacidin mutants.

2.4. Strategies for characterizing the structure and identity of lantibiotic variants

2.4.1. Variants from targeted or saturation mutagenesis libraries

In cases where the mutant lantibiotic has been derived via targeted mutagenesis of a *lanA* gene, the primary amino acid sequence of the lantibiotic is known. During the screening of a large number of broth samples LC-MS (or MALDI) is usually a suitable first-line analysis technique to determine whether a lantibiotic with the expected mass has been processed and excreted. When investigating structure–activity relationships or seeking to improve the biological activity of lantibiotics it is practical to screen broth samples for antibacterial activity. In some cases, it is necessary to analyze partially purified lantibiotic variants that have been purified using precipitation or solid-phase extraction (SPE) techniques. Variants with desirable biological properties can be isolated on a small scale (0.5–10 mg quantities) using a combination of SPE or preparative HPLC techniques to characterize the variants further or to evaluate their biological properties in detail.

Mass spectroscopy techniques are particularly useful for evaluating the extent of posttranslational modification. For instance, lantibiotics that contain free dehydroalanine residues (e.g., mersacidin) or oxidized thioethers (e.g., actagardine) can give rise to partially modified variants. In such cases, chromatographic retention times and peak areas can also be considered to evaluate variant production.

It is conceivable that a processed lantibiotic variant could contain partially cyclized lanthionine ring structures. Mass spectrometry is not particularly useful to analyze cyclization as the dehydrated and cyclized products have the same mass. However, chemical probes can be used in conjunction with mass spectroscopy to confirm that the correct number of lanthionine bridges is present (Fig. 22.2). For instance, the mutant and the wild-type

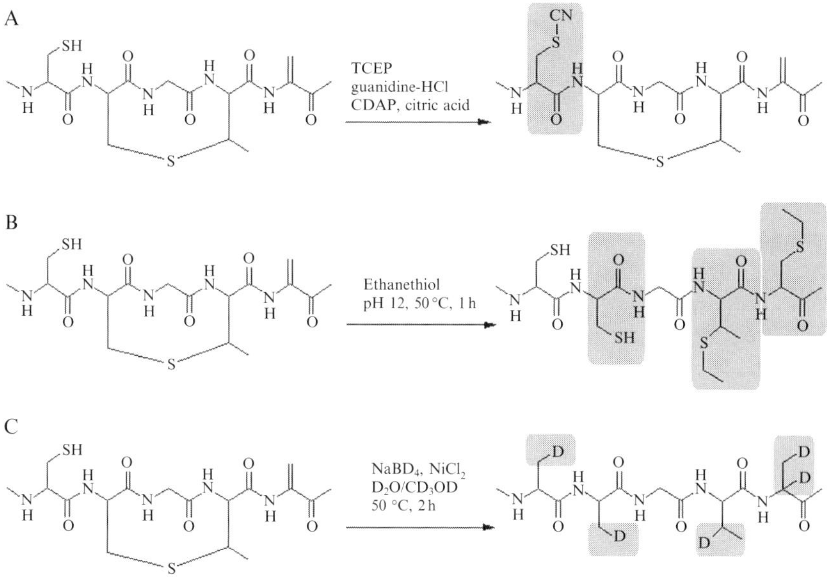

Figure 22.2 Chemical probes to aid in the elucidation of lantibiotic structures. (A) Selective cyanylation of thiols leaving lanthionine bridges intact (Chatterjee et al., 2005a; Wu and Watson, 1997). Peptide (10 nmol) in 10 μl of 0.1 M citrate buffer (pH 3.0) containing 4 M guanidine-HCl, treated with tris(2-carboxyethyl)phosphine (1 equivalent per cysteine), for 15 min at room temperature prior to addition of 1-cyano-4-dimethylaminopyridinium tetrafluoroborate (200 nmol) and the reaction left at room temperature for a further 15 min. (B) Linearization of lantibiotics with ethanethiol (Novák et al., 1996) or mercaptoethanol (Meyer et al., 1994; Zimmerman et al., 1995). Lantibiotic (10–30 nmol) treated with 30 μl of a mixture comprising ethanethiol (1.4 M), NaOH (0.5 M), in 46% aqueous ethanol and then heated at 50 °C for 2.5 h prior to addition of acetic acid (2 μl). Some lantibiotics require a longer reaction time and higher temperature (110 °C). (C) Desulfurization/reduction of lantibiotics with nickel boride under deuterating conditions (Martin et al., 2004). Dissolve the lantibiotic (0.5 mg) and nickel chloride (1.0 mg) in 2 ml of CD$_3$OD/D$_2$O (1:1). Transfer the mixture to a screw-top vial containing NaBD$_4$ (1.0 mg), creating a black Ni$_2$B precipitate. Then mix the solution for 2 h at 50 °C and remove NiB$_2$ precipitate by centrifugation.

lantibiotic can be treated with a 20-fold molar excess of 1-cyano-4-dimethylaminopyridinium tetrafluoroborate under acidic conditions at room temperature for 15 min (Chatterjee et al., 2005a; Wu and Watson, 1997). This reagent will cyanylate a free thiol but will not react with a thioether (lanthionine) bridge. If the correct number of lanthionine bridges have formed then the mass of the mutant and native lantibiotic would be similar after treatment. An increase in mass of 25 m.u. would indicate the presence of a single thiol that is not involved in bridge formation.

To identify which residues are involved in lanthionine bridge formation, the lantibiotic can be reduced using nickel boride under deuterating conditions (Martin et al., 2004). An aminobutyric acid moiety from a methyl-lanthionine bridge will be reduced with incorporation of a single deuterium atom and cysteine will be desulfurized forming alanine containing a single deuterium atom. However, a dehydrobutyrine or dehydroalanine that had not been involved in ring formation would be reduced with incorporation of two deuterium atoms (Fig. 22.2). The reduced peptide can then be sequenced using mass spectrometry to identify deuterated residues.

Modification methods such as these often require the purification of small quantities of lantibiotic, but they offer the ability to analyze variants at submilligram quantities. Although these methods will identify which residues are involved in ring formation, ultimately NMR is needed to confirm between which particular residues the rings are formed.

2.4.2. Variants from random mutagenesis libraries

Where random lantibiotic mutants have been generated the mass of the variant might not permit the identification of a variant by analysis of the mass alone. For instance, variants with identical masses could be obtained and sequence information would be required to determine the primary structure of the lantibiotic. In the random mutagenesis of a 20-mer, five residues at a time, there are 20^5 possible variants that could be produced, theoretically, and therefore a suitable screen for biological potency should be included in the experiment. For the optimization of antimicrobial properties, colonies generated via random mutagenesis can be overlaid with a suitable indicator organism such as *Micrococcus luteus*. Colonies that inhibit the growth of the marker strain can then be selected and grown in small-scale fermentations. Genomic or plasmid DNA can be recovered and the mutated *lanA* gene can be isolated using appropriate primers to amplify the gene and identify the primary sequence of the mutant lantibiotic by DNA sequencing. Variants with interesting properties can be isolated in small quantities (<10 mg) to enable further biological and structural evaluation. The primary sequence of a lantibiotic can also be determined using EDMAN degradation or Tandem MS techniques. Due to the extensive posttranslational modification of lantibiotics, it is often necessary to linearize the peptides, either by treatment with ethanethiol or by desulfurization/reduction with nickel boride as described above (Fig. 22.2).

2.4.3. Analytical strategy

To elucidate the structure of an unknown lantibiotic a number of the following techniques can be used:

1. Determine the primary sequence.
   - DNA sequence of the *lanA* gene.
   - Linearize the lantibiotic using ethanethiol or nickel boride and sequence by EDMAN degradation or MS/MS.

2. Determine the number of lanthionine bridges.
   - Treat the lantibiotic with ethanethiol to reveal all thiols, lanthionines, and free dehydroalanine or dehydrobutyrine residues. Treat the lantibiotic with 1-cyano-4-dimethylaminopyridinium tetrafluoroborate to selectively modify thiols not involved in lanthionine bridging. Compare the results of the two experiments.
   - Reduce the lantibiotic using nickel boride under deuterating conditions. This enables discrimination between a fully processed lanthionine bridge and cysteine and dehydroalanine/dehydrobutyrine residues that are not involved in bridge formation.
3. Perform NMR experiments. Several NMR experiments would be required to evaluate the structure of a lantibiotic and some of those techniques that could prove useful are listed below.
   - Establish the number and identity of amino acids using 2D ^{1}H–^{15}N HSQC experiments.
   - Determine the spin systems using 2D DQF-COSY, TOCSY, and 2D ^{1}H–^{13}C HSQC-TOCSY experiments.
   - Link scalar connectivity from side-chain protons to amide protons using 2D-DQF-COSY, TOCSY, and ^{1}H–^{15}N HSQC experiments.
   - Elucidate bridging patterns using NOESY and ^{1}H–^{13}C HMBC experiments.

3. Conclusions

The variant generation systems described above highlight the strategies that have been employed successfully to generate modified lantibiotics. The systems have enabled mutations in the primary structure of the lantibiotics. Future studies may lead to the construction of lantibiotic variants with modified ring patterns and chimeric lantibiotics. The individual methods for the generation of a library of variants will differ depending on the host system and the nature of each biosynthetic gene cluster, but the overall strategies described above should offer an insight into the approaches that can be adopted. As more lantibiotics are selected as potential drug candidates, further use of these approaches should enable the generation of even more ambitious variant libraries, particularly using random mutagenesis strategies.

REFERENCES

Asaduzzaman, S. M., Nagao, J.-I., Aso, Y., Nakayama, J., and Sonomoto, K. (2006). Lysine-oriented charges trigger the membrane binding and activity of Nukacin ISK-1. *Appl. Environ. Microbiol.* **72**(9), 6012–6017.

Aso, Y., Nagao, J.-I., Koga, H., Okuda, K.-I., Kanemasa, Y., Sashihara, T., Nakayama, J., and Sonomoto, K. (2004). Heterologous expression and functional analysis of the gene cluster for the biosynthesis of and immunity to the lantibiotic nukacin ISK-1. *J. Biosci. Bioeng.* **98**(6), 429–436.

Augustin, J., Rosenstein, R., Wieland, B., Schneider, U., Schnell, N., Engelke, G., Entian, K. D., and Gotz, F. (1992). Genetic analysis of epidermin biosynthetic genes and epidermin-negative mutants of *Staphylococcus epidermidis*. *Eur. J. Biochem.* **204**, 1149–1154.

Bierbaum, G., Reis, M., Szekat, C., and Sahl, H. G. (1994). Construction of an expression system for engineering of the lantibiotic Pep5. *Appl. Environ. Microbiol.* **60**(12), 4332–4338.

Bierman, M., Logan, R., O'Brien, K., Seno, E. T., Rao, R. N., and Schoner, B. E. (1992). Plasmid cloning vectors for the conjugal transfer of DNA from *Escherichia coli* to *Streptomyces* spp. *Gene* **116**, 43–49.

Boakes, S., Cortes Bargallo, J., and Dawson, M. J. (2007). Lantibiotic biosynthetic gene clusters from *A. garbadinensis* and *A. liguriae* WO 2007/083112.

Brückner, R. (1997). Gene replacement in *Staphylococcus carnosus* and *Staphylococcus xylosus*. *FEMS Microbiol. Lett.* **151**, 1–8.

Castiglione, F., Lazzarrini, A., Carrano, L., Corti, E., Ciciliato, I., Gastaldo, L., Candiani, P., Losi, D., Marinelli, F., Selva, E., and Parenti, F. (2008). Determining the structure and mode of action of Microbisporicin, a potent lantibiotic active against multiresistant pathogens. *Chem. Biol.* **15**, 22–31.

Chatterjee, C., Miller, L. M., Leung, Y. L., Xie, L., Yi, M., Kelleher, N. L., and van der Donk, W. A. (2005a). Lacticin 481 synthetase phosphorylates its substrate during lantibiotic production. *J. Am. Chem. Soc.* **127**, 15332–15333.

Chatterjee, C., Paul, M., Xie, L., and Van der Donk, W. A. (2005b). Biosynthesis and mode of action of lantibiotics. *Chem. Rev.* **105**, 633–683.

Chen, P., Novak, J., Kirk, M., Barnes, S., Qi, F., and Caufield, P. W. (1998). Structure–activity study of the lantibiotic mutacin II from *Streptococcus mutans* T8 by a gene replacement strategy. *Appl. Environ. Microbiol.* **64**(7), 2335–2340.

Corvey, C., Stein, T., Dusterhus, S., Karas, M., and Entian, K. D. (2003). Activation of subtilin precursors by *Bacillus subtilis* extracellular serine proteases subtilisin (AprE), WprA, and Vpr. *Biochem. Biophys. Res. Commun.* **304**, 48–54.

Cotter, P. D., Hill, C., and Ross, R. P. (2003). A food-grade approach for functional analysis and modification of native plasmids in *Lactococcus lactis*. *Appl. Environ. Microbiol.* **69**(1), 702–706.

Cotter, P. D., Deegan, L. H., Lawton, E. M., Draper, L. A., O'Connor, P. M., Hill, C., and Ross, R. P. (2006a). Complete alanine scanning of the two-component lantibiotic lacticin 3147: Generating a blueprint for rational drug design. *Mol. Microbiol.* **62**, 735–747.

Cotter, P. D., Draper, L. A., Lawton, E. M., McAuliffe, O., Hill, C., and Ross, R. P. (2006b). Overproduction of wild-type and bioengineered derivatives of the lantibiotic lacticin 3147. *Appl. Environ. Microbiol.* **72**, 4492–4496.

Dawson, M. J., Cortes Bargallo, J., Rudd, B. A. M., Boakes, S., Bierbaum, G., Hoffmann, A., and Schmitz, S. (2005). Production of mersacidin and its variants in *sigH* and/or *mrsA* negative *Bacillus* host cells WO 2005/093069.

Dodd, H. M., Horn, N., Hao, Z., and Gasson, M. J. (1992). A lactococcal expression system for engineered nisins. *Appl. Environ. Microbiol.* **58**(11), 3683–3693.

Field, D., Collins, B., Cotter, H., and Ross, R. P. (2007). A system for the random mutagenesis of the two-peptide lantibiotic lacticin 3147: Analysis of mutants producing reduced antibacterial activities. *J. Mol. Microbiol. Biotechnol.* **13**, 226–234.

Field, D., O'Connor, P. M., Cotter, P. D., Hill, C., and Ross, R. P. (2008). The generation of nisin variants with enhanced activity against specific gram-positive pathogens. *Mol. Microbiol.* **69**, 218–230.

Gasson, M. J., and Anderson, P. H. (1985). High copy number plasmid vectors for use in lactic streptococci. *FEMS Microbiol. Lett.* **30,** 193–196.

Gust, B., Challis, G. L., Fowler, K., Kieser, T., and Chater, K. F. (2003). PCR-targeted *Streptomyces* gene replacement identifies a protein domain needed for biosynthesis of the sesquiterpene soil odor geosmin. *Proc. Natl. Acad. Sci. USA* **100**(4), 1541–1546.

Kluskens, L. D., Kuipers, A., Rink, R., de Boef, E., Fekken, S., Driessen, A. J., Kuipers, O. P., and Moll, G. N. (2005). Post-translational modification of therapeutic peptides by NisB, the dehydratase of the lantibiotic nisin. *Biochemistry* **44**(38), 12827–12834.

Kuipers, O. P., Rollema, H. S., Yap, W. M. G. J., Boot, H. J., Siezen, R. J., and de Vos, W. M. (1992). Engineering dehydrated amino acid residues in the antimicrobial peptide nisin. *J. Biol. Chem.* **267**(34), 24340–24346.

Kuipers, O. P., Beerthuyzen, M. M., Siezen, R. J., and de Vos, W. M. (1993). Characterization of the nisin gene cluster *nisABTCIPR* of *Lactococcus lactis*. Requirement of expression of the *nisA* and *nisI* genes for development of immunity. *Eur. J. Biochem.* **216,** 281–291.

Kuipers, O. P., Beerthuyzen, M. M., de Ruyter, P. G., Luesink, E. J., and de Vos, W. M. (1995). Autoregulation of nisin biosynthesis in *Lactococcus lactis* by signal transduction. *J. Biol. Chem.* **270**(45), 27299–27304.

Kuipers, A., de Boef, E., Rink, R., Fekken, S., Kluskens, L. D., Driessen, A. J., Leenhouts, K., Kuipers, O. P., and Moll, G. N. (2004). NisT, the transporter of the lantibiotic nisin, can transport fully modified, dehydrated, and unmodified prenisin and fusions of the leader peptide with non-lantibiotic peptides. *J. Biol. Chem.* **279**(21), 22176–22182.

Kuipers, A., Meijer-Wierenga, J., Rink, R., Kluskens, L. D., and Moll, G. N. (2008). Mechanistic dissection of the enzyme complexes involved in biosynthesis of lacticin 3147 and nisin. *Appl. Environ. Microbiol.* **74**(21), 6591–6597.

Kupke, T., Kempter, C., Jung, G., and Götz, F. (1995). Oxidative decarboxylation of peptides catalyzed by flavoprotein EpiD. Determination of substrate specificity using peptide libraries and neutral loss mass spectrometry. *J. Biol. Chem.* **270**(19), 11282–11289.

Liu, W., and Hansen, N. (1992). Enhancement of the chemical and antimicrobial properties of subtilin by site-directed mutagenesis. *J. Biol. Chem.* **267**(35), 25078–25085.

MacNeil, D. J., Gewain, K. M., Ruby, C. L., Dezeny, G., Gibbons, P. H., and MacNeil, T. (1992). Analysis of *Streptomyces avermitilis* genes required for avermectin biosynthesis utilizing a novel integration vector. *Gene* **111**(1), 61–68.

Majer, F., Schmid, D. G., Altena, K., Bierbaum, G., and Kupke, T. (2002). The flavoprotein MrsD catalyses the oxidative decarboxylation reaction involved in formation of the peptidoglycan synthesis inhibitor mersacidin. *J. Bacteriol.* **184**(5), 1234–1243.

Martin, N. I., Spules, T., Carpenter, M. R., Cotter, P. D., Hill, C., Ross, P., and Vederas, J. J. (2004). Structural characterization of Lacticin 3147, a two-peptide lantibiotic with synergistic activity. *Biochemistry* **43,** 3049–3056.

Meyer, H. E., Heber, M., Eisermann, B., Korte, H., Metzger, J. W., and Jung, G. (1994). Sequence analysis of lantibiotics: Chemical derivatization procedures allow a fast access to complete edman degradation. *Anal. Biochem.* **223,** 185–190.

Novák, J., Kirk, M., Caufield, P. W., Barnes, S., Morrison, K., and Baker, J. (1996). Detection of modified amino acids in lantibiotic peptide Mutacin II by chemical derivatization and electrospray ionization—Mass spectroscopic analysis. *Anal. Biochem.* **236,** 358–360.

O'Driscoll, J., Glynn, F., Cahalane, O., O'Connell-Motherway, M., Fitzgerald, G. F., and van Sinderen, D. (2004). Lactococcal Plasmid pNP40 encodes a novel, temperature-sensitive restriction-modification system. *Appl. Environ. Microbiol.* **70**(9), 5546–5556.

Ottenwälder, B., Kupke, T., Brecht, S., Gnau, V., Metzger, J., Jung, G., and Götz, F. (1995). Isolation and characterization of genetically engineered gallidermin and epidermin analogs. *Appl. Environ. Microbiol.* **61,** 3894–3903.

Pascalle, G., de Ruyter, G. A., Kuipers, O. P., and de Vos, W. M. (1996). Controlled gene expression systems for *Lactococcus lactis* with the food-grade inducer nisin. *Appl. Environ. Microbiol.* **62**(10), 3662–3667.

Rink, R., Kuipers, A., de Boef, E., Leenhouts, K. J., Driessen, A. J., Moll, G. N., and Kuipers, O. P. (2005). Lantibiotic structures as guidelines for the deign of peptides that can be modified by lantibiotic enzymes. *Biochemistry* **44**(24), 8873–8882.

Rink, R., Kluskens, L. D., Kuipers, A., Driessen, A. J., Kuipers, O. P., and Moll, G. N. (2007). NisC, the cyclase of the lantibiotic nisin, can catalyze cyclization of designed non-lantibiotic peptides. *Biochemistry* **46**(45), 13179–13189.

Ryan, M. P., Jack, R. W., Josten, M., Sahl, H. G., Jung, G., Ross, R. P., and Hill, C. (1999). Extensive post-translational modification, including serine to D-alanine conversion, in the two-component lantibiotic, lacticin 3147. *J. Biol. Chem.* **274**(53), 37544–37550.

Schmitz, S., Hoffmann, A., Szekat, C., Rudd, B., and Bierbaum, G. (2006). The lantibiotic mersacidin is an autoinducing peptide. *Appl. Environ. Microbiol.* **72**(11), 7720–7727.

Simon, D., and Chopin, A. (1988). Construction of a plasmid family and its use for molecular cloning in *Streptococcus lactis*. *Biochimie* **70**, 559–566.

Stein, T., Borchert, S., Kiesau, P., Heinzmann, S., Klöss, S., Klein, C., Helfrich, M., and Entian, K. D. (2002). Dual control of subtilin biosynthesis and immunity in *Bacillus subtilis*. *Mol. Microbiol.* **44**(2), 403–416.

Szekat, C., Jack, R. W., Skutlarek, D., Farber, H., and Bierbaum, G. (2003). Construction of an expression system for site-directed mutagenesis of the lantibiotic mersacidin. *Appl. Environ. Microbiol.* **69**(7), 3777–3783.

Widdick, D. A., Dodd, H. M., Barraille, P., White, J., Stein, T. H., Chater, K. F., Gasson, M. J., and Bibb, M. J. (2003). Cloning and engineering of the cinnamycin biosynthetic gene cluster from *Streptomyces cinnamoneus cinnamoneus* DSM 40005. *Proc. Natl. Acad. Sci. USA* **100**(7), 4316–4321.

Wu, J., and Watson, J. T. (1997). A novel methodology for assignment of disulfide bond pairings in proteins. *Protein Sci.* **6**, 391–398.

Zimmerman, N., Metzger, J. W., and Jung, G. (1995). The tetracyclic lantibiotic actagardine 1H-NMR and 13C-NMR assignments and revised primary structure. *Eur. J. Biochem.* **228**(3), 786–797.

CHAPTER TWENTY-THREE

Cyanobactin Ribosomally Synthesized Peptides—A Case of Deep Metagenome Mining

Eric W. Schmidt *and* Mohamed S. Donia

Contents

1. Introduction	576
2. Some Remaining Questions	583
3. Obtaining *Prochloron* Cells and DNA	583
3.1. Identification and processing of *Prochloron*-containing ascidians	583
3.2. DNA purification from enriched *Prochloron* cells	584
3.3. DNA purification from whole animals	585
4. Chemical Analysis	586
5. Cyanobactin Gene Cloning and Identification	587
5.1. Genome mining and structure prediction	588
5.2. Cyanobactin cloning from *Prochloron* spp.	589
5.3. Cyanobactin cloning from non-*Prochloron* cyanobacteria	589
6. Heterologous Expression in *E. coli*	590
7. Deep Metagenome Mining	591
7.1. Whole-cell PCR with *Prochloron*	591
7.2. Example: Deep metagenome mining in *Prochloron*	592
8. Enzymatic Analysis of Cyanobactin Biosynthesis	593
9. Applying Deep Metagenome Mining: Pathway Engineering	593
9.1. Yeast recombination for pathway manipulation	594
Acknowledgments	595
References	595

Abstract

Deep metagenome mining is a new method for engineering natural product pathways, focusing on examining symbiotic organisms. The method has been applied to a family of compounds known as cyanobactins, which are ribosomally synthesized peptides produced by cyanobacteria. Often, these

Department of Medicinal Chemistry, University of Utah, Salt Lake City, Utah, USA cyanobacteria live symbiotically with marine animals, leading to production of natural products in whole animal samples. Here, we focus on methods to identify, clone, and study cyanobactin natural product genes from axenic organisms and metagenomic environments. The application to deep metagenome mining is described, along with other potential targets of this methodology.

1. Introduction

Genetic engineering of natural product pathways has led to the synthesis of many new small molecules and to elegant solutions to biochemical problems (Wilkinson and Mickelfield, 2007). However, significant hurdles still exist preventing rapid engineering in many cases. One solution to this problem is to look deeply at very closely related pathways to identify natural evolutionary solutions to structural diversity (Schmidt, 2005, 2008; Schmidt et al., 2000). For example, one could examine a family of closely related natural products to determine what genetic changes underlie structural differences, assuming only a few discrete mutations lead to a new structure. A good source of variability for these studies is found in certain metagenomic environments, particularly bacteria living in symbiotic relationships with animals. By observing this natural variability, it is possible to determine engineering rules for individual pathways and genes. The resulting methodology, dubbed "deep metagenome mining," was first applied to small ribosomally synthesized peptide natural products, the cyanobactins, but in principle it can be applied to any natural product type.

Deep metagenome mining does not refer to the depth of the ocean, but rather to the depth of scanning individual gene types. That is, instead of a shallow approach to genetics in which a large amount of data about different genes is sought, the deep mining approach seeks to examine numerous nearly identical genes, especially from environmental (metagenomic) samples. Discrete mutations leading to new compounds can be observed by this approach, as highlighted by the cyanobactins example described below. This stands in contrast to pathways with a greater number of mutations, in which it is not possible to determine which mutations are important to the resulting chemical structures without further experiment.

Cyanobactins are produced by many cyanobacterial strains (Fig. 23.1) (Donia et al., 2008; Ireland and Scheuer, 1980). Over 100 cyanobactins have been isolated, making them one of the biggest compound classes in cyanobacteria. While the compounds bear many similarities to the larger microcin group (Duquesne et al., 2007), key features that separate the cyanobactins from other compound classes include N–C terminal cyclization and the presence of either heterocycles or isoprenoid derivatives or both. Cyanobactins are found in many different cyanobacteria but were first

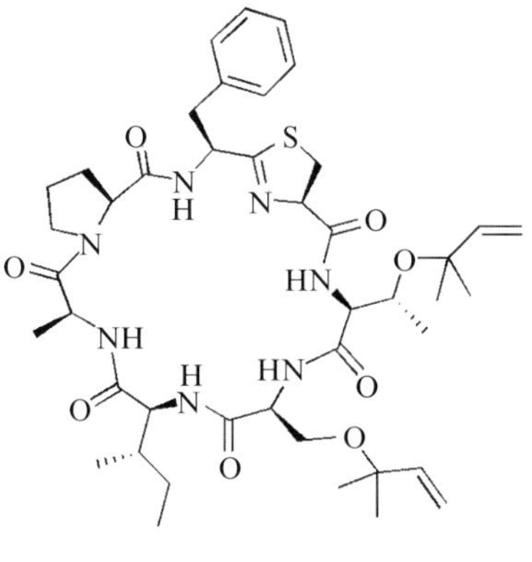

Trunkamide

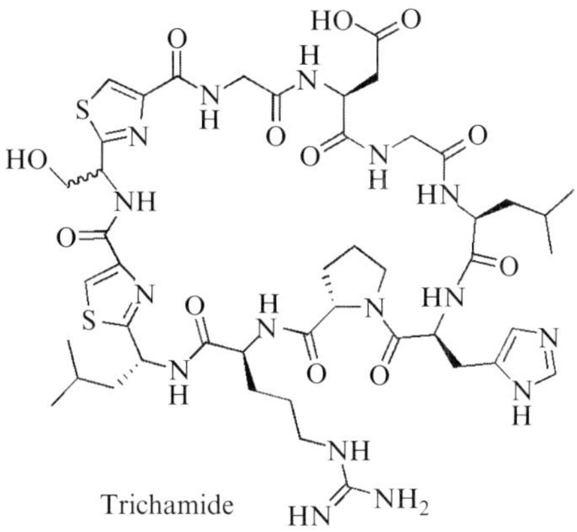

Trichamide

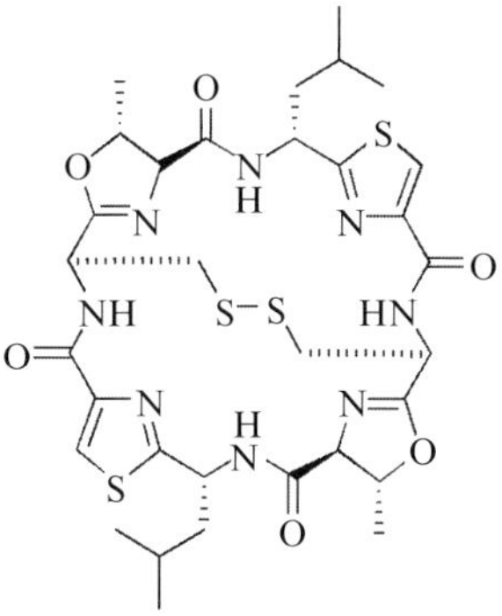

Ulithiacyclamide

Figure 23.1 Representative cyanobactin structures. Both trunkamide and ulithiacyclamide are from symbiotic *Prochloron* strains living with ascidians. Trichamide is from free-living cyanobacteria. Trunkamide exemplifies a prenylated cyanobactin, while the others are modified only by N–C cyclization and heterocyclization.

discovered in marine animals known as ascidians (Fig. 23.2). In fact, the initial isolation of ulithiacyclamide from an ascidian by Ireland and Scheuer represented the first chemical characterization of any representative of the now well-known microcin class (Ireland and Scheuer, 1980). The ascidian animals containing cyanobactins harbored specific cyanobacterial symbionts, *Prochloron* spp., leading Ireland and Scheuer to propose that these cyanobacteria could be responsible for producing the compounds found in whole animals. This prediction was validated 25 years later by genetic methods (Long *et al.*, 2005; Schmidt *et al.*, 2005). Cyanobactins were isolated mostly on the basis of their anticancer activities, which are sometimes in the low nanomolar range against human cell lines (Carroll *et al.*, 1996; Ireland and Scheuer, 1980). Not all cyanobactins exhibit this activity: the structures and the resulting activities are quite diverse.

Figure 23.2 *Prochloron*-containing didemnid ascidians from a tropical reef. The green color in these animals is due to symbiotic cyanobacteria, *Prochloron* spp. Three species of didemnid ascidians are shown, including a bluish encrusting form, a plate-like green form, and the large animal in the center of the frame, *Lissoclinum patella*. This picture was taken in the Solomon Islands by Chris Ireland (University of Utah). (See Color Insert.)

The chemical relationships defining the cyanobactin class can be clearly discerned by comparing structures and amino acid sequence alignments of known compounds, but they are most apparent in the gene clusters for cyanobactins (Fig. 23.3). These clusters always encode two paralogous serine proteases, related to PatA and PatG from the patellamide gene cluster, *pat* (Donia *et al.*, 2006, 2008; Schmidt *et al.*, 2005; Sudek *et al.*, 2006; Ziemert *et al.*, 2008). These proteases catalyze the steps required for N–C cyclization. In addition, all known gene clusters also contain homologs of *patD* and usually also *patF*, which are almost certainly responsible for heterocyclization and/or prenylation. Two further genes that are not required for synthesis when transferred to *Escherichia coli* but are conserved in cyanobacteria are *patB* and *patC*. Finally, an oxidase domain is sometimes present, either in PatG homologs or as a free-standing protein that probably oxidizes heterocycles, for example, from thiazoline to thiazole.

The cyanobactin amino acid sequences are directly encoded on a precursor peptide, PatE, and its relatives (Fig. 23.4). Usually, more than one cyanobactin structure is encoded on a single precursor peptide. The leader sequence of about 40 amino acids, which is removed by the PatA family of proteases, is relatively well conserved among cyanobactins. In addition, flanking each cyanobactin sequence are two recognition sequences, one of which is recognized by the PatA protease and the other by PatG protease. PatG catalyzes the cyclization event (Lee *et al.*, 2009).

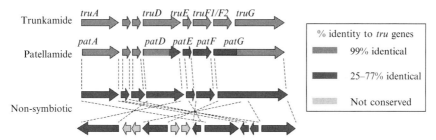

Figure 23.3 Cyanobactin gene clusters. At top is the *tru* cluster from *Prochloron*, which leads to prenylated cyanobactins such as trunkamide. Second from the top, the *pat* gene cluster from *Prochloron* is shown. This leads to heterocyclic small molecules such as ulithiacyclamide. These clusters are ~99% identical to each other (grey regions) except in the region that is important for prenylation (black region). Note that pathways from nonsymbiotic cyanobacteria have a lower level of identity to other known pathways, highlighting information that can be obtained by deep metagenome mining. Dashed lines indicate changes in gene order.

```
                                                                   87654321
>PatE1 MNKKNILPQQGQPVIRTAGQLSSQLAELSEEALGDAGLEASVTACITFCAYDGVEPSITVCISVCAYDGE
>PatE2 MNKKNILPQQGQPVIRTAGQLSSQLAELSEEALGDAGLEASVTACITFCAYDGVEPSCTLCCTLCAYDGE
>PatE3 MNKKNILPQQGQPVIRTAGQLSSQLAELSEEALGDAGLEAS:ACFPTICAYDGVEPS:FCFPTVCAYDGE
```

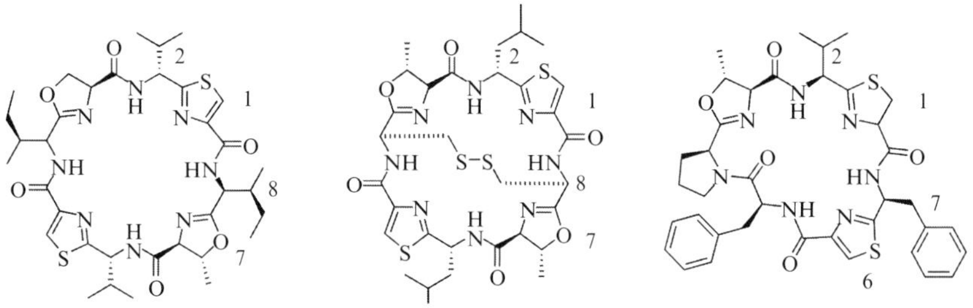

Figure 23.4 Deep metagenome mining with *pat* pathway. The three PatE peptide variants shown encode the three cyanobactin molecules in their second cassettes (numbered 1–8 above the sequences). The underlined sequences encode cyanobactin products, while sequences in bold are sequences recognized by the PatA and PatG proteases that define the boundaries of the final products. Note that the peptides are exactly identical except where they encode these different products. They are also identical at the DNA level. In addition, all three *patE* variants are flanked by identical genes encoding identical enzymes.

Our initial question in examining the cyanobactins was, "what genetic modifications are required to go from one structure to another?" This question was about both the changes to the encoded peptide sequence and posttranslational changes in functionality, such as prenylation, heterocyclization, and oxidation.

Symbiotic *Prochloron* spp. provided the ideal model to study pathway evolution. Prior to our work, there were about 60 known cyanobactins from ascidians, representing about 6% of all known ascidian natural products (MarinLit). These cyanobactins formed overlapping families of natural products that had lengths of 6–8 amino acids, variable side chains, and variable modification. First, we examined pathways that did not lead to prenylated molecules. We found that pathway gene clusters from ascidians collected across the Western Pacific were >99% identical at the DNA sequence level (Donia et al., 2006). *patE* genes were also ~99% identical to each other, except in the exact region encoding the peptide products (Table 23.1). These hypervariable regions encoded very different peptides, yet the modifying genes and intergenic regions were essentially identical. In total, we identified mutants leading to ~30 different products, in which each position was substituted at least once, with identical modifying enzymes. This is basically a natural alanine-scanning mutagenesis, since every position is substituted naturally, but the modifying enzymes were not mutated. This made it possible to identify functional mutations by simple PCR. Inspired by this, we made a completely different peptide, eptidemnamide, by genetic engineering, demonstrating conclusively that the pathway is extremely substrate tolerant.

We next examined changes leading to prenylation. To our surprise, there were no new enzyme classes in the prenylating pathways such as that for trunkamide (Fig. 23.3) (Donia et al., 2008). Prenylating and non-prenylating pathways were virtually identical (>98% DNA sequence

Table 23.1 Mutations in the PatE sequence identified by deep metagenome mining

Position	1	2	3	4	5	6	7	8
Ala	0	1	4	1	0	9	2	0
Arg	1	0	0	0	0	0	0	0
Cys	33	0	0	7	24	6	0	8
Ile	0	1	0	13	1	0	1	3
Leu	0	9	0	0	1	7	1	5
Phe	0	11	0	1	4	0	5	0
Pro	0	1	9	0	0	2	0	0
Ser	0	1	2	0	0	2	1	1
Thr	0	0	27	1	0	3	23	0
Tyr	0	0	0	1	0	0	0	0
Val	0	11	0	1	4	5	0	7
None	0	0	0	0	0	0	1	8

Position number is from the C–N terminus to be consistent with Figs. 23.3 and 23.4. The numbers in the rows refer to the number of times an individual amino acid is found in that position, out of 34 mutants identified. In all of these cases, the pathway enzymes were almost 100% identical, and mutations were only identified within the 24-nt product-coding region of the *patE*-homologous genes. Note that every site is mutated at least once, revealing broad-substrate selectivity for the pathway enzymes.

identity) except in a region encoding the C-terminus of PatD, PatF, and the oxidase region of PatG. A PatG-like oxidase was missing from the prenylated pathways; this makes sense since the prenylated molecules were not oxidized. In place of a single *patF* gene, there were two encoded PatF proteins, which were about 40% identical to each other and to the nonprenylating PatF. Thus, simply by sequence gazing, the difference between prenylating and nonprenylating pathways was easily discerned. This would not have been possible without deep metagenome mining because there are no characterized homologs of PatF. Thus, if all proteins had multiple modifications, the basis of prenylation (now explained by the change in PatD C-terminal and/or PatF) would not be clear. Deep metagenome mining allows rapid identification of functional features for genetic engineering.

Several other cyanobactin pathways from free-living cyanobacteria have been examined (Donia *et al.*, 2008; Sudek *et al.*, 2006; Ziemert *et al.*, 2008). These pathways are structurally more diverse than those found in *Prochloron* spp., including peptides 6–12 amino acids in length, with relatives containing nearly every proteinogenic amino acid. The structures, biological activities, and potential roles of diverse cyanobactins are summarized elsewhere (Donia and Schmidt, 2009).

Deep metagenome mining revealed extremely closely related pathways with clear engineering rules. The cyanobactins from free-living organisms have more divergent biosynthetic enzymes, but nonetheless useful rules can be deduced by careful comparison of sequences. In particular, the presence of certain types of amino acids at certain positions can be readily discerned (Fig. 23.5). In the natural cyanobactins known to date, the C-terminal amino acid (position 1) is always a heterocycle derived from Cys, Ser, or Thr, or the usual proteinogenic heterocycle, Pro. There is only one exception found by deep metagenome mining, although the resulting compound has not yet been isolated. Thus, position 1 might require heterocycles. Similarly, although cyanobactins have an overall high Cys content, Cys is never found in position 2. In fact, this residue is almost always hydrophobic, with a few exceptions. Thus, some selectivity exists at position 2 for aliphatic and hydrophobic groups, although the presence of Ser/Thr in some compounds suggests that the substrate selectivity may be broader at position 2 than position 1. The remaining residues are substituted by nearly every other amino acid and are thus not selective, although they contain a high proportion of aliphatic amino acids and Ser/Thr/Cys. Although these natural experiments provided a great deal of information about substrate selectivity, we were still not sure how much Trp or the more hydrophilic amino acids would be tolerated by the patellamide modifying enzymes. We selected a simple engineering approach to address these issues. The new compound, eptidemnamide, was synthesized in *E. coli* following PCR mutagenesis (Donia *et al.*, 2006). In eptidemnamide, Pro was kept in

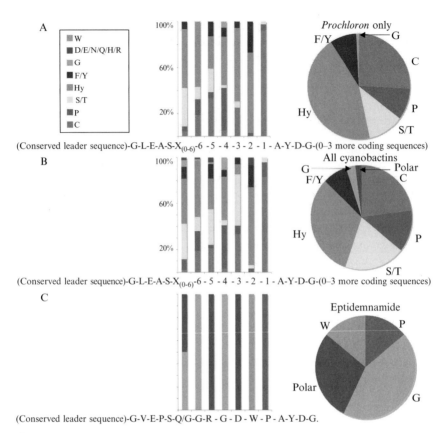

Figure 23.5 Comparison of amino acid use in cyanobactins by (A) *Prochloron*; (B) all cyanobacteria; and (C) the engineered peptide, eptidemnamide. Note that, in natural cyanobactins, either Pro or other heterocycles derived from Cys/Ser are always present in position 1. This was replicated with Pro in position 1 of eptidemnamide. In position 2, the natural molecules never contain Cys and virtually never contain Ser/Thr. In eptidemnamide, position 2 is taken by Trp, which is almost never found in natural cyanobactins. The other positions in natural cyanobactins are more varied, although they contain a high percentage of Cys, Pro, Ser/Thr, and Hy (hydrophobic amino acids Ala, Ile, Leu, and Val). Very few Gly or other more hydrophilic amino acids are found, and these amino acids are virtually never found in *Prochloron*-derived cyanobactins. In eptidemnamide, these amino acids constitute the last six amino acids. This figure illustrates the degree to which deep metagenome mining coupled with chemical analysis illuminates engineering principles. Questions that could not be answered by observing natural diversity were addressed by the construction of the artificial molecule, eptidemnamide. Finally, this figure does not include compounds predicted from sequence, but which have not yet been purified. (See Color Insert.)

position 1, Trp (which is almost never found in cyanobactins) was placed in position 2, and the remainder of the amino acids, such as Arg, Gly, Asp, and Gln, were extremely rare in cyanobactins. The engineered compound thus explored much of the sequence selectivity space not encountered by the natural molecules. Together, these studies illustrate the power of traditional engineering methods coupled with deep metagenome mining to rapidly assess design rules for natural products.

2. Some Remaining Questions

The work described above represents just the start of the application of deep metagenome mining to natural products. Numerous questions remain. Among the more pressing questions are: will similar observations be made with other pathway types, such as those to polyketides and nonribosomal peptides, or will different rules be discerned? What drives the changes described, which are still otherwise unprecedented? Finally, it should be noted that so far cyanobactin genes and compounds have been found only in cyanobacteria, where they are ubiquitous and second only to the nonribosomal peptide/polyketide group. Given the wealth of genome sequencing, the ancient origin of cyanobacteria, and the prevalence of lateral gene transfer, this is bizarre to say the least.

One possible example of deep metagenome mining in the polyketide world is found in the area of polyketides derived from mycobacteria (Pidot *et al.*, 2008). These pathogenic (i.e., symbiotic) bacteria were shown to encode different polyketide products from nearly identical genes that exhibited clear deletions or insertions. With this first example, there may be many more such systems to be found in closely related bacteria. The symbiotic pool provides an excellent source of such symbiotic bacteria (Schmidt, 2008).

The methods described below are sufficient to enable researchers to work on cyanobactins, either in ascidians or in free-living bacteria. In addition, methods for deep metagenome mining are described.

3. Obtaining *Prochloron* Cells and DNA

3.1. Identification and processing of *Prochloron*-containing ascidians

Many of these methods are derived from protocols described in a key book about *Prochloron* and references therein (Lewin and Cheng, 1989). *Prochloron* bacteria have not yet been stably cultivated. They have mostly been found as obligate symbionts of ascidians from the family *Didemnidae*. In a handful of cases, they have also been reported from other invertebrates and from stromatolites in Australia (Burns *et al.*, 2005). Over 30 different ascidian species harbor *Prochloron* (Kott, 2001), although this number is likely to be an underestimate (Hirose et al., 2008), and these ascidians are found in tropical waters throughout the Indo-Pacific and Red Sea regions. Closely related symbionts are found in Caribbean ascidians (Lafargue and Duclaux, 1979). It should be noted that *Prochloron* does not always contain the genes for cyanobactin synthesis; however, they are a relatively reliable source (Donia et al., 2006).

Prochloron-containing ascidians are extremely diverse in morphology and can be readily observed from intertidal depths to the edges of the photic zone (Fig. 23.2) (Lewin and Cheng, 1989). Patches of *Prochloron* are spinach green due to the relatively unusual presence of chlorophyll *b* in the bacteria. Sometimes, this green color has a slightly bluish tinge. Other ascidians sometimes appear to be the same color due to the presence of eukaryotic algae; however, these are readily distinguished upon close inspection because of the punctate appearance of the eukaryotes, whereas *Prochloron* is usually evenly dispersed within a patch. Conveniently, in most ascidian species, the *Prochloron* cells are readily displaced from the ascidian by squeezing the animal. Cells obtained in this way clump within a few minutes, providing good additional evidence of the presence of *Prochloron*. In addition, *Prochloron* can be readily observed even with relatively low-power field microscopes, since they are often >10 μm in size. In some cases, *Prochloron* cannot be dissociated from ascidian tissues. Most notably, they are seemingly irretrievably embedded in a mucous matrix in the extremely common ascidians, *Didemnum molle* and relatives. Fortunately, these ascidians are easy to recognize based upon gross morphology (Colin and Arneson, 1995).

3.2. DNA purification from enriched *Prochloron* cells

If *Prochloron* can be dissociated from a sample, the following protocol is recommended (Schmidt et al., 2004):

1. Collect and maintain the ascidians in leak-proof bags, maintaining temperature and keeping them wet. Transfer ascidians to larger containers containing fresh seawater as soon as feasible after collection. If possible, process immediately. However, reasonable *Prochloron* samples have been obtained from specimens kept up to $\sim$24 h if the seawater is changed and kept at a reasonable temperature.
2. Rinse the outside of the ascidians with 0.2-μm-filtered artificial seawater (ASW). ASW can be obtained from pet stores or aquarium supply stores and brought to the collection site as a dry, preweighed powder. It is conveniently dissolved in bottled water or tap water and filtered onsite.
3. Press the ascidian against the inner surface of a sterile Petri dish. Resuspend the obtained cells using a pipet equipped with an aerosol tip, pipetting up and down to disperse clumped cells. Transfer to 1.5 ml microcentrifuge tubes.

4. Pellet the cells using a lightweight microcentrifuge.
5. Further squeeze the animal sample and add more cells to the microcentrifuge tubes. Centrifuge as in step 4 and repeat as necessary to maximize cell yield.
6. Rinse cells in 1 ml ASW, resuspend, and repeat step 4. Up to 500 μl of *Prochloron* cells can be obtained from an animal that fits in a standard Petri dish. These preparations are not pure, in the sense that often spicules and other bacteria are present. However, the resulting DNA preparations originate almost entirely from *Prochloron*.
7. In contrast to other reports, we have not found *Prochloron* to be particularly delicate. To be cautious, it is worth processing samples immediately after step 6.
8. To extract DNA in the field, numerous methods are possible. We usually use the Qiagen Genomic Tip kit, following the manufacturer's instructions. Lysozyme is not required to lyse *Prochloron*, making it unnecessary to transport frozen or refrigerated samples to the site of collection. The resulting DNA is stable at room temperature for travel and is used successfully in the synthesis of fosmid libraries after up to several years of storage at 4 °C. Whole ascidian samples can also be stored in a commercial nucleic acid preservation reagent, such as RNALater, at −20 °C for DNA extraction in the laboratory.

Lightweight, field-appropriate equipment can be purchased for all of these experiments and packed into a single container for travel. Of note, these types of procedures can also be used for many other symbiotic organisms, although the details of microbial isolation differ. Also of note, no refrigeration or freezing is required for these techniques, eliminating many of the problems associated with transporting goods and reagents for field research. Seventy percent ethanol and isopropanol purchased at pharmacies are perfectly acceptable alternatives to the pure chemicals suggested in protocols.

3.3. DNA purification from whole animals

If *Prochloron* cannot be dissociated from samples, as in the case of *D. molle*, mixed samples can still be stored as above. Often, DNA can still be obtained using standard methods. However, this is problematic with highly mucous-laden animals such as *D. molle*, for which the following protocol is suggested:

1. Ascidians stored in RNALater or similar preservative are processed using this method, inspired by the large scale DNA isolation methods optimized for actinomycetes (Kieser *et al.*, 2000).
2. Resuspend dissected pieces of the animal (or preferably freeze-dried) in 5 ml of TE buffer, add 10 mg lysozyme (final conc 2 mg ml^{-1}) and swirl

to dissolve. Incubate at 30 °C for 30–60 min, monitoring every 15 min. On first sign of increased viscosity or after 1 h, proceed to step 3.
3. Add 1.2 ml 0.5 M EDTA (final conc 0.1 M), mix gently and add 0.13 ml of Proteinase K (final conc 0.2 mg ml^{-1}), mix gently and incubate at 30 °C for 5 min.
4. Add 6.5 ml 10% SDS (final conc 1%), tilt immediately and incubate at 37 °C for up to 2 h, checking for lysis. (This could be extended to overnight, depending on the speed of lysis.)
5. Add 1.2 ml of 5 M NaCl, mix, and add 1.0 ml of CTAB/NaCl solution. (10% CTAB in 0.7 M NaCl; dissolve 4.1 g NaCl in 80 ml water, slowly add 10 g CTAB [hexadecyltrimethyl ammonium bromide] while heating and stirring. Adjust final volume to 100 ml.)
6. Incubate at 65 °C for 20 min (could be extended to 1 h).
7. Perform a standard phenol/chloroform extraction, followed by an isopropanol precipitation.

The scale of the extraction can be adjusted depending on the amount of sample to be processed.

Note. It is critical to follow all relevant travel and customs regulations when transporting scientific equipment and reagents. In addition, proper local and national permits must be obtained prior to collection.

4. Chemical Analysis

Over 100 different cyanobactins have been structurally characterized, with an additional >30 identified from metagenomics experiments. Therefore, exact protocols depend upon the structural classes under consideration. The following applies to isolation of known cyanobactins from ascidians. These compounds tend to be of relatively low polarity in comparison to many from other sources. Samples stored in ethanol or isopropanol (70–100%) can be used to obtain these compounds, but it is generally better to extract fresh or frozen specimens if possible. In the procedure below, it is desirable to start with at least 50 g wet weight of ascidian if pure material is required, but the LC–MS method will work with a few mg of tissue (Donia et al., 2006):

1. Using frozen or lyophilized ascidians, cut tissue into small (∼1–8 cm^3) segments and place in at least a 10-fold excess of methanol. Leave at room temperature for several hours, decant, and extract at least twice more with fresh methanol. The last extraction step is usually left overnight. After proper extraction, the green color is almost entirely bleached from the sample.

2. Dry by rotary evaporation to remove methanol. Partition remaining water against ethyl acetate, or add water to dried samples and extract. Dry the ethyl acetate fraction and partition it between hexane and methanol. The methanol fraction will be greatly enriched for cyanobactins.
3. The dried methanol fraction can be further purified by silica or C_{18} chromatography. The exact method depends upon the compounds in question. To rapidly determine which cyanobactins are present in a sample, the following represents a universal method.
4. Prepare a small fraction of the methanol extract for LC–MS by pushing it through a plug of end-capped C_{18} resin in methanol. Dry this sample, which is pure enough for mass spectrometry.
5. In a typical LC–MS run, samples are dissolved in 1:1 water: acetonitrile and injected on a Waters Alliance 2795 HPLC, using an analytical C_{18} column (Phenomenex) and an acetonitrile/water gradient at a flow rate of 0.2 ml min^{-1}. Gradient conditions are initially 50% water, 50% acetonitrile to 1% water, 99% acetonitrile over 15 or 25 min and held at that level for another 5 min. One percent formic acid is used in the water and/or acetonitrile to enhance ionization. For high-resolution HPLC mass spectrometry (HR-HPLCMS), a Micromass Q-TOF mass spectrometer is used with a LockSpray running concurrently to ensure higher mass accuracy. This method is relatively universal for detection of cyanobactins (which are usually the major compounds in the methanol extract) from ascidians and for moderately nonpolar cyanobactins from cyanobacteria. In all cases, we use authentic standards of peptides that have been purified to homogeneity from whole ascidians and analyzed by NMR and MS to confirm the identity of cyanobactins.

It is difficult to detect more polar cyanobactins by this method because of their near coelution with salts. Other methods have been described for these compounds. In our hands, it is not helpful to use the host taxonomy in deconvoluting the masses obtained as cyanobactins are usually produced in diverse hosts. For example, trunkamide was only reported from *Lissoclinum patella* prior to our work; however, we purified it from *D. molle* (Donia *et al.*, 2008). Finally, the presence of a two-mass-unit difference in the mass spectrometric trace is usually a sign of different oxidation states (Donia *et al.*, 2006).

5. Cyanobactin Gene Cloning and Identification

Cyanobactin pathway clusters can be obtained either from whole DNA samples or from large-insert libraries (Donia *et al.*, 2006, 2008; Long *et al.*, 2005; Schmidt *et al.*, 2005). In addition, cyanobactin clusters are readily recognized in genome sequences, leading to useful structure prediction from genomes (Donia *et al.*, 2008; Sudek *et al.*, 2006; Ziemert *et al.*, 2008). Cloned pathways are useful in providing new natural products from miniscule environmental samples by heterologous expression (Donia *et al.*, 2008).

5.1. Genome mining and structure prediction

To identify cyanobactin clusters in genome sequences, it is sufficient to perform a BLAST search for homologs of PatA, PatD, PatF, or PatG. Once the contig containing this homolog is identified, areas surrounding that gene should be carefully annotated by translation in both directions and all frames. All cyanobactin genes identified to date are clustered in one locus of the genome. However, some clusters appeared to be truncated or interrupted due to insertion events, which probably inactivate the pathway.

Genome sequences can be used for accurate structure prediction, as previously demonstrated. One of the key features allowing accurate prediction is the substrate specificity of PatA and PatG homologs, which defines the N–C sequences of the actual cyanobactin products. Alignment of existing PatE sequences identifies G (hydrophobic) X A/P S, where X is usually D or E, as the recognition site for PatA homologs, while A/S F/Y D is the substrate for PatG homologs. The canonical PatG protease recognizes AYD and SYD with apparently equal affinity, while PatA protease is known to recognize GLEAS, GVEPS, and GVDAS. All known PatA and PatG relatives recognize similar sequences, even if the proteins are only ~60% identical. Thus, a peptide with a leader sequence terminating in ...GVEPSITVCISVCAYDGE can be reliably predicted to synthesize the underlined peptide. The presence of PatD and/or PatF and application of a phylogenetic analysis can be used to determine whether a heterocycle is formed. The presence or absence of an oxidase can be used to determine whether that heterocycle is oxidized. Structural hypotheses can be readily differentiated with extremely small samples by applying MS/MS techniques using FT-ICR (Sudek *et al.*, 2006). Because the amino acid sequences are directly genome-encoded, and in many cases whole-genome sequences are available, high-resolution MS/MS peptide sequencing experiments are absolutely definitive as to structure. *patE* homologs are often not predicted to encode a protein by autoannotation tools, so it pays in some cases to repeat the analysis of sequenced regions. For example, this was necessary in the cyanobactin cluster annotation of *Lyngbya aestuarii* and *Trichodesmium erythraeum*.

A caution is that individual cyanobactin gene types are often found outside of cyanobactin clusters and are probably not associated with cyanobactin synthesis. For example, a PatD homolog is found in the microcin-like pathway leading to goadsporin biosynthesis in actinomycetes (Onaka *et al.*, 2005). PatA- and PatG-like proteases are found outside of cyanobactin pathways. The oxidase domain is also found in other pathway types, including nonribosomal peptide synthetases. Thus far, we have not found PatF homologs outside of a cyanobactin gene cluster, although there is at least one cyanobactin cluster lacking PatF. Since N–C cyclization requires PatA and PatG, a cyanobactin cluster should at minimum contain homologs of both of these proteases and a PatE-like precursor peptide. This is the minimum gene set required to make cyanobactin N–C cycles, although we have never found a pathway containing just this set.

Another issue, as described above, is how to determine whether or not a pathway is prenylating. We have derived a phylogenetic model (Donia and Schmidt, in preparation) that is useful in determining whether a pathway is prenylating or heterocyclizing. In this model, modifying enzymes as well as the precursor peptide sequences are correlated to the type of posttranslational modification observed.

5.2. Cyanobactin cloning from *Prochloron* spp.

To clone pathways from DNA samples, several strategies can be attempted. Because cyanobactin pathways from *Prochloron* are so similar to each other, it is possible to clone entire, intact pathways out of ascidian samples using primers that exactly match the 5′-end of PatA and the 3′-end of PatG (Donia et al., 2008). These pathways can be directly transferred to *E. coli* for expression. A limitation to this strategy is that there is usually more than one cyanobactin pathway in a single animal. Therefore, sequencing over the PatE homolog is required to confirm that the desired pathway has been obtained. If DNA is degraded or is problematic, fragments of the desired gene sequence can be directly cloned. For example, a prenylated cyanobactin pathway can be cloned as follows:

1. Exactly matching primers internal to PatD and PatF homologs are designed and applied to the sample.
2. The resulting product is sequenced. If it encodes the desired peptide, homologous recombination can be used to make hybrids with an intact prenylating pathway for expression in *E. coli*. Methods for performing this cross are described in Section 9 below. This method has been used successfully to express the anticancer clinical candidate, trunkamide, in *E. coli* culture.

5.3. Cyanobactin cloning from non-*Prochloron* cyanobacteria

When cloning from organisms other than *Prochloron*, DNA sequences are not similar enough to use an exact-match strategy. In these cases, a degenerate priming strategy is required. To obtain whole pathways, they must then be cloned out of DNA libraries, such as cosmid or bac libraries. We have previously reported a degenerate strategy to obtain the tenuecyclamides pathway from a free-living cyanobacterium (Donia et al., 2008).

Genomic DNA from *Nostoc* sp. was cloned into the pCC1FOS fosmid vector (Epicentre) following the manufacturer's protocol. A library of about 800 colonies was constructed. Colonies were then picked into 96-well plates and stored in 20% glycerol. Using a Clustal X alignment of different members, conserved regions in cyanobactin proteases were identified. Degenerate primers were then designed using CODEHOP to specifically amplify cyanobactin proteases and not any other serine proteases. The primers used were Protease-F (TGGGCACCCTCGGCTAYGAYT-TYGG) and Protease-R (GCGTTCACGTTCCAGCCRTANADNCC). This pair of primers was used initially to amplify a ~400 bp cyanobactin protease region. Exact-match primers were then designed based on the obtained sequence and used to screen the fosmid library. Each 96-well plate was replicated in one tube by the use of a 96-well replicator immersed in the plate then in sterile water. This mixed preparation of fosmids proved to be an actual representation of every colony in the plate and was screened directly by PCR. After identifying one plate containing the desired sequence, further screening was performed to identify the appropriate single clone. This time, one mixed sample was prepared from each row and each column by mixing 10–20 μl from each well. A plasmid from the cyanobactin-containing clone was prepared and used for PCR to further prove the screening success. This clone containing the intact *10* cluster was then sequenced to 6× coverage.

6. Heterologous Expression in *E. coli*

A key advantage of many cyanobactin pathways is that they can be expressed in *E. coli*. In our hands, removal of PatA, PatD, PatE, PatF, or PatG homologs from a pathway results in no detectable product *in vivo*, while PatB and PatC homologs can be safely left out of expression constructs if desired (Donia et al., 2006). The best yields are obtained when genes are left intact in apparent operons, although placing individual promoters in front of every gene also leads to cyanobactin production in *E. coli*. There is a report that cyanobactins are extremely robustly expressed from large plasmids such as bacmids (Long et al., 2005), but we have been unable to replicate this result despite extensive investigation. It may be that certain pathways from some samples are more readily expressed in the completely native context, while most are much better expressed after some minor engineering.

The most robust expression we have obtained so far was by cloning the entire patellin biosynthetic pathway (Donia et al., 2008). PCR with exact-match primers provided the entire ~11 kb pathway, which was cloned into the pCR2.1 vector. Fortuitously, PatA was cloned in-frame with the *lacZ* gene N-terminal fragment found in the vector. Patellins were expressed from ~0.1–1 mg l^{-1}, and expression was shown to be responsive to lactose and glucose concentrations. Expression is performed as follows:

1. Due to rearrangement problems, all experiments are performed at 30 °C.
2. A freshly picked colony of Top 10 cells containing the above described vector was seeded overnight into 50 ml LB with appropriate antibiotic selection. The resulting seed culture is transferred to a 10 l fermenter in the same medium and allowed to grow for 72 h. Aeration is performed with 7.5 cc min^{-1} airflow and 250 rpm mixing.
3. After 3 days, the culture is extracted by addition of ~10 g HP20 resin. The resin is washed with water (1.5 l) and eluted with methanol (1 l). Following rotary evaporation of the methanol fraction, the residue is dissolved in water (250 ml) and extracted three times with equal amounts of ethyl acetate.
4. The resulting ethyl acetate fraction is dried and fractionated through a C$_{18}$ column. Fractions are then dried, dissolved in 1:1 acetonitrile:water and analyzed by HR-HPLCMS.

A problem that is sometimes encountered is that individual starting *E. coli* colonies produce different levels of compounds, leading to variability in production.

7. Deep Metagenome Mining

This process refers to the rapid comparison of many closely related pathways to determine functional differences that are useful in engineering. The process will be described with ascidians containing *Prochloron* as an example. In principle, this procedure could be applied to any pathway type involving symbiotic interactions.

First, obtain sample ascidians as described above. DNA can be extracted, or PCR can be performed directly on enriched *Prochloron* cells. The protocol we developed for *Prochloron* PCR works mainly for small PCR products, especially *patE* variants. Bigger PCR products require better quality DNA isolated by any of the methods described above.

7.1. Whole-cell PCR with *Prochloron*

1. Make a 10× betaine solution: add 16 µl of betaine (6.5 *M*) to 0.5 µl DMSO.

2. Add up to 10 μl *Prochloron* cells to 200 μl DMSO and grind the mixture with a tissue grinder.
3. Make serial dilution to 1/10× and 1/100× in sterile water.
4. Perform PCR using each dilution (2.5 μl) and the following components, diluted to 20 μl per reaction: 10× betaine solution (2 μl); recommended amounts of buffer and Mg, dNTP mix (0.4 μl of 10 mM solution); HiFi Taq polymerase (Invitrogen, 0.1 μl). In addition, each primer is added to give a final concentration of 1 mM.

7.2. Example: Deep metagenome mining in *Prochloron*

1. To determine regions of interest, use PCR to obtain fragments of the whole pathway using multiple overlapping exact-match primers. At this step, products must be cloned prior to sequencing. Sequence analysis of overlapping segments reveals which portions are likely to be variable and which are conserved. In the case of cyanobactins, we noticed that only the hypervariable region of PatE contained consistent mutations.
2. PCR over candidate variable regions "deeply," that is, perform PCR using multiple, closely related samples, preferably of mixed (metagenomic) origin. In the cyanobactin case, we performed PCR from a region inside PatF to a region inside PatD. In this way, we could monitor for mutations within proteins and intergenic regions easily. In addition, PatE itself was analyzed by direct PCR from samples. Although it is preferable to examine as many samples as possible, in fact numerous variable products originated from single animal samples, revealing a diverse mixed colony within a single animal. In principle, therefore, a single animal could be sufficient for some deep metagenome projects.
3. Clone products and sequence multiple isolates in 96-well plates. Use at least two PCRs per sample to minimize the probability of identifying PCR mutations as opposed to natural mutations. In the case of cyanobactins, we sequenced 12 96-well plates, identifying 30 cyanobactin pathway relatives. Because flanking sequences were available, it was possible to determine rapidly that other enzymatic changes did not contribute to product diversity.
4. For a few representative samples, sequence pathway clusters completely to ensure that the observed mutations are focused in the identified variable regions and not elsewhere.

Using this methodology, it was possible to examine numerous mutations in a short time frame. Finally, as mentioned above, comparison of prenylating and nonprenylating pathways also followed a variation of the deep metagenome mining method.

8. Enzymatic Analysis of Cyanobactin Biosynthesis

Thus far, we have reported the activities of PatA and PatG proteins (Lee *et al.*, 2009), but not yet PatD or PatF. A critical advance allowing us to perform this analysis was as follows. We experimented with many different PatE peptides, but they were very difficult to handle, often precipitating from solution and not reacting well in enzyme assays. In addition, PatE relatives were often not detected on SDS-PAGE with various stains, despite a huge abundance of the peptides as shown by tryptic digest-mass spectrometry. The advance came by using N-terminally His-tagged PatEdm, an engineered peptide containing the wholly artificial "eptidemnamide" sequence in cassette 2. This peptide was better behaved because it has only two Cys residues, rather than the four to six commonly found in these small peptides, and moreover it was readily detectable by SDS-PAGE. In addition, a key step in the successful purification was to purify under denaturing conditions in urea, followed by exhaustive dialysis against water over several days. This protocol provided the best yield of precursor.

Preparation of PatA and PatG was relatively more standard and will not be described here. As an outcome of these experiments, we were able to show that PatA cuts the N-terminal recognition sequence, while PatG cuts the C-terminus and cyclizes the peptides. This represented the first natural ribosomal peptide cyclase to be characterized *in vitro* as a pure protein.

9. Applying Deep Metagenome Mining: Pathway Engineering

As described above, it is possible to modify just the PatE hypervariable region to generate new products. Larger regions must be modified in order to switch from nonprenylating to prenylating pathways. Much remains unknown about the size limitations of the pathways. Several methods have been used to generate new molecules *in vivo*. Standard PCR mutagenesis was used to synthesize PatEdm from the natural *PatE2* gene. Whole pathways have been expressed either as an operon under control of the *lac* promoter or in individual plasmids under control of T7 promoters. To manipulate whole operons, we have used recombination in yeast and by lambdaRed (Gust *et al.*, 2003). The latter protocol uses the standard method as described. The only exception is that we use a multicopy plasmid, so the correct mutant must be obtained by various purification methods, such as restreaking followed by colony PCR or Southern hybridization. This is necessary because the lambdaRed method generates individual colonies containing mixtures of plasmids, and at least in our hands this problem is greatly exacerbated with multicopy plasmids. However, it is often technically easier to repurify these mixtures than to use single-copy vectors.

We also use the yeast recombination method for the rapid manipulation of pathways. The example described below enabled us to produce the rare compound, trunkamide, which is a potential anticancer agent occasionally found in ascidian samples. The trunkamide pathway is very closely related to that of patellin biosynthesis. Previously, we had cloned the patellin biosynthetic cluster. Thus, we rescued the trunkamide genes for heterologous expression in *E. coli* using yeast recombination. The following is a modification of a previously reported protocol (Kunes *et al.*, 1987; Sikorski and Hieter, 1989).

9.1. Yeast recombination for pathway manipulation

1. The yeast–*E. coli* shuttle vector harboring the original *tru1* cluster is linearized near the site of modification by restriction digestion.
2. The region harboring a desired modification is amplified by PCR or synthesized. In this example, we amplified the *truE2* gene with sufficient flanking regions to allow homologous recombination.
3. The two linear pieces from steps 1 and 2 are mixed and used to transform yeast following the standard high-lithium yeast transformation protocol:
4. Inoculate 0.5 ml of stationary phase cells into 10 ml of fresh YPD medium, which should give an initial O.D.600 of 1.0. Shake the culture in a 30 °C incubator for about 4 h, until the O.D.600 is about 4.0.
5. Spin cells at 3000 rpm for 5 min. Remove supernatant and resuspend cells in 5 ml of lithium solution (10 mM Tris pH 7.5, 1 mM EDTA pH 8.0, 0.1 M lithium acetate).
6. Spin again, pour off supernatant. Resuspend cells in residual lithium solution remaining in the tube. Transfer the mixture to a microcentrifuge tube.
7. Boil salmon sperm DNA (10 mg ml^{-1}) for 10 min. Place on ice before adding to yeast cells.
8. Add DNA to a microcentrifuge tube. Add 50 μl of yeast cells and 7 μl of salmon sperm DNA to the DNA solution. Mix well. Add 0.45 ml of 40% PEG solution (10 mM Tris pH 7.5, 1 mM EDTA pH 8.0, 0.1 M lithium acetate, 40% PEG 4000 or 3350) to each tube. Mix well.
9. Incubate at 30 °C for 20 min.
10. Heat-shock the cells at 42 °C for 15 min. Leave cells at room temperature thereafter.
11. Spin at 4000 rpm, resuspend in buffer TE (400 μl), and plate 100–150 μl of the mixture onto selective medium plates. Incubate at 30 °C for 2–3 days.

12. Combine at least 15–20 yeast transformant colonies and transfer to 10 mL selective media. Grow overnight until a high density is achieved, then rescue plasmids using standard yeast miniprep procedures. Lyticase incubations prior to minipreps are sometimes helpful to facilitate recovery of larger plasmids. We usually use uracil auxotrophy as a convenient yeast selection marker. Transform the isolated plasmid mixture into *E. coli*, and screen for recombinant colonies for the desired plasmids.

ACKNOWLEDGMENTS

This work was funded by NIH GM071425. We are grateful for the participation of our colleagues led by Margo Haygood (OHSU) and Jacques Ravel (U. Maryland) in these endeavors.

REFERENCES

Burns, B. P., Seifert, A., Goh, F., Pomati, F., Jungblut, A. D., Serhat, A., and Neilan, B. A. (2005). Genetic potential for secondary metabolite production in stromatolite communities. *FEMS Microbiol. Lett.* **243,** 293–301.

Carroll, A. R., Coll, J. C., Bourne, D. J., MacLeod, J. K., Zabriskie, T. M., Ireland, C. M., and Bowden, B. F. (1996). Patellins 1–6 and trunkamide A: Novel cyclic hexa-, hepta- and octa-peptides from colonial ascidians, *Lissoclinum* sp. *Aust. J. Chem.* **49,** 659–667.

Colin, P. L., and Arneson, C. (1995). "Tropical Pacific Invertebrates." Coral Reef Press, Beverly Hills, CA.

Donia, M. S., and Schmidt, E. W. (2009). Cyanobactins, ubiquitous cyanobacterial ribosomal peptide metabolites. *In* "Comprehensive Natural Products II" (in press).

Donia, M. S., Hathaway, B. J., Sudek, S., Haygood, M. G., Rosovitz, M. J., Ravel, J., and Schmidt, E. W. (2006). Natural combinatorial peptide libraries in cyanobacterial symbionts of marine ascidians. *Nat. Chem. Biol.* **2,** 729–735.

Donia, M. S., Ravel, J., and Schmidt, E. W. (2008). A global assembly line for cyanobactins. *Nat. Chem. Biol.* **4,** 341–343.

Duquesne, S., Destoumieux-Garzon, D., Peduzzi, J., and Rebuffat, S. (2007). Microcins, gene-encoded antibacterial peptides from enterobacteria. *Nat. Prod. Rep.* **24,** 708–734.

Gust, B., Challis, G. L., Fowler, K., Kieser, T., and Chater, K. F. (2003). PCR-targeted *Streptomyces* gene replacement identifies a protein domain needed for biosynthesis of the sesquiterpene soil odor geosmin. *Proc. Natl. Acad. Sci. USA* **100,** 1541–1546.

Hirose, M., Yokobori, S., and Hirose, E. (2009). Potential speciation of morphotypes in the photosymbiotic ascidian *Didemnum molle* in the Ryukyu Archipelago, Japan. *Coral Reefs* **28,** 119–126.

Ireland, C., and Scheuer, P. J. (1980). Ulicyclamide and ulithiacyclamide, two new small peptides from a marine tunicate. *J. Am. Chem. Soc.* **102,** 5688–5691.

Kieser, T., Bibb, M. J., Buttner, M. J., Chater, K. F., and Hopwood, D. A. (2000). "Practical *Streptomyces* Genetics." John Innes Foundation, Norwich.

Kott, P. (2001). The Australian Ascidiacea part 4, Aplousobranchia (3), Didemnidae. *Mem. Qld. Mus.* **4,** 1–408.

Kunes, S., Ma, H., Overbye, K., Fox, M. S., and Botstein, D. (1987). Fine structure recombination analysis of cloned genes using yeast transformation. *Genetics* **115,** 73–81.

Lafargue, F., and Duclaux, D. (1979). Premier exemple, en Atlantique tropical, d'une association symbiotique entre une ascidie didemnidae et une cyanophycée chroococcale: *Trididemnum cyanophorum* nov. sp. et *Synechocystis trididemni* nov. sp. *Ann. Inst. Oceanogr.* **55,** 163–184.

Lee, J., McIntosh, J., Hathaway, B. J., and Schmidt, E. W. (2009). Using marine natural products to discover a protease that catalyzes peptide macrocyclization of diverse substrates. *J. Am. Chem. Soc.* **131,** 2122–2124.

Lewin, R. A., and Cheng, L. (eds.), (1989). *In* "Prochloron: A Microbial Enigma." Chapman & Hall, New York.

Long, P. F., Dunlap, W. C., Battershill, C. N., and Jaspars, M. (2005). Shotgun cloning and heterologous expression of the patellamide gene cluster as a strategy to achieving sustained metabolite production. *ChemBioChem* **6,** 1760–1765.

Onaka, H., Nakaho, M., Hayashi, K., Igarashi, Y., and Furumai, T. (2005). Cloning and characterization of the goadsporin biosynthetic gene cluster from *Streptomyces* sp. TP-A0584. *Microbiology* **151,** 3923–3933.

Pidot, S. J., Hong, H., Seeman, T., Porter, J. L., Yip, M. J., Men, A., Johnson, M., Wilson, P., Davies, J. K., Leadley, P. F., and Stinear, T. P. (2008). Deciphering the genetic basis for polyketide variation among mycobacteria producing mycolactones. *BMC Genomics* **9,** 462.

Schmidt, E. W. (2005). From chemical structure to environmental biosynthetic pathways: Navigating marine invertebrate-bacteria associations. *Trends Biotechnol.* **23,** 437–440.

Schmidt, E. W. (2008). Trading molecules and tracking targets in symbiotic interactions. *Nat. Chem. Biol.* **4,** 466.

Schmidt, E. W., Obraztsova, A. Y., Davidson, S. K., Faulkner, D. J., and Haygood, M. G. (2000). Identification of the antifungal peptide-containing symbiont of the marine sponge *Theonella swinhoei* as a novel delta-proteobacterium, "*Candidatus* Entotheonella palauensis". *Mar. Biol.* **136,** 969–977.

Schmidt, E. W., Sudek, S., and Haygood, M. G. (2004). Genetic evidence supports secondary metabolic diversity in *Prochloron* spp., the cyanobacterial symbiont of a tropical ascidian. *J. Nat. Prod.* **67,** 1341–1345.

Schmidt, E. W., Nelson, J. T., Rasko, D. A., Sudek, S., Eisen, J. A., Haygood, M. G., and Ravel, J. (2005). Patellamide A and C biosynthesis by a microcin-like pathway in *Prochloron didemni*, the cyanobacterial symbiont of *Lissoclinum patella*. *Proc. Natl. Acad. Sci. USA* **102,** 7315–7320.

Sikorski, R. S., and Hieter, P. (1989). A system of shuttle vectors and yeast host strains designed for efficient manipulation of DNA in *Saccharomyces cerevisiae*. *Genetics* **122,** 19–27.

Sudek, S., Haygood, M. G., Youssef, D. T. A., and Schmidt, E. W. (2006). Structure of trichamide, a cyclic peptide from the bloom-forming cyanobacterium *Trichodesmium erythraeum*, predicted from the genome sequence. *Appl. Environ. Microbiol.* **72,** 4382–4387.

Wilkinson, B., and Mickelfield, J. (2007). Mining and engineering natural-product biosynthetic pathways. *Nat. Chem. Biol.* **3,** 379–386.

Ziemert, N., Ishida, K., Quillardet, P., Bouchier, C., Hertweck, C., Tandeau de Marsac, N., and Dittmann, E. (2008). Microcyclamide biosynthesis in two strains of *Microcystis aeruginosa*: From structure to genes and vice versa. *Appl. Environ. Microbiol.* **74,** 1791–1797.

Author Index

A

Abbondi, M., 14, 16, 31, 32, 33
Abbott, B. J., 512, 529
Abdelfattah, M. S., 283, 291, 302
Abelson, D., 49
Abergel, R. J., 448
Abou-Hadeed, K., 473, 487, 489, 490, 491, 492, 499, 502, 505, 506
Abraham, E. P., 419
Abrutyn, E., 513
Aceti, D. J., 125, 127, 128
Adamidis, T., 125, 127, 128
Adams, V., 310, 325
Admiraal, S. J., 259, 264, 381
Adrio, J. L., 30, 31, 34, 39, 41, 49, 53, 419
Agarwal, A., 14
Agarwal, S., 62
Agnihotri, G., 280
Aguirrezabalaga, I., 282, 283, 289, 290, 291, 292, 293, 297, 298, 300, 326
Aharoni, A., 327
Aharonowitz, Y., 412, 413, 414, 415, 428
Ahlert, J., 46, 120, 160
Ahmed, A., 311
Ahn, J. W., 64, 68
Aires-de-Sousa, M., 118
Akey, D. L., 193, 235
Al-Abdallah, Q., 405
Alarco, A. M., 183, 193, 203
Albang, R., 119
Albermann, C., 313, 315, 316, 317, 318, 326, 372, 479
Albermann, K., 119
Alberts, A. W., 37, 53
Alder, J., 513
Alderwick, L. J., 126
Aldrich, C. C., 448
Alduina, R., 161, 464, 477
Alemany, M. T., 413
Alexander, D. C., 408, 420, 516, 519, 520, 522, 526, 527
Ali, R. A., 128
Alici, A., 81
Allen, C., 545
Allende, N., 282, 283, 290, 291, 293, 297, 298, 300, 326
Al Nakeeb, M., 354, 359, 360
Altena, K., 539, 554, 560
Altmeyer, M. O., 81, 87
Altschul, S. F., 130, 185, 189
Amir-Heidari, B., 354, 362, 363, 364
Anderle, C., 122
Anderson, E. C., 448
Anderson, I., 421, 422
Anderson, J. B., 185
Anderson, P. H., 564
Anderson, T. B., 125, 126, 127, 131
Andersson, I., 415
Ando, A., 123
Andrade, J. M., 106
Andres, N., 95, 126
Andresen, D., 109
Andresson, O. S., 259
Ansari, M. Z., 187, 396
Aoki, H., 50, 404, 489
Appleyard, A. N., 559
Apweiler, R., 185
Arbache, S., 4, 5, 446, 447
Arbeit, R. D., 513
Arceneaux, J. E. L., 448
Arias, C. A., 460
Arias, P., 125, 126
Armau, E., 38
Armengaud, J., 505
Arneson, C., 584
Arnon, D. I., 505
Aron, Z. D., 363
Arora, S. K., 489
Arthanari, H., 343
Arthur, M., 478
Asaduzzaman, S. M., 566
Asamizu, S., 489
Aschke-Sonnenborn, U., 313
Aso, Y., 566
Atrih, A., 163, 167, 172
Attwood, T. K., 185
Auclair, K., 53
Augustin, J., 562, 564, 565
Austin, M. B., 224
Averin, O., 538, 539, 542, 544, 545
Av-Gay, Y., 412
Azimonti, S., 464

B

Bachmann, B. O., 46, 120, 160, 175, 181, 183, 193, 203, 414

Badger, J. H., 119
Bae, J., 364
Baggaley, K. H., 404
Baig, I., 279, 299, 300, 326
Bailey, L. M., 396, 440, 442
Bairoch, A., 185
Baker, B. J., 419
Baker, J., 569
Baldwin, J. E., 414, 415, 418, 427
Balibar, C. J., 224, 355
Balsara, R., 539, 548
Baltz, R. H., 4, 8, 25, 118, 120, 163, 338, 357, 358, 359, 463, 477, 479, 511, 512, 513, 514, 515, 516, 517, 518, 519, 520, 521, 522, 523, 524, 525, 526, 527, 528
Bamas-Jacques, N., 151, 152
Bamonte, R., 9
Banerjee, P. K., 12
Bañuelos, O., 415, 417
Bapat, M., 383
Baragona, S., 118
Barbazuk, W. B., 86
Barbosa, M. D., 14
Bardsley, B., 460
Barends, S., 96, 120, 122, 133
Barkallah, S., 357
Barnes, S., 562, 569
Barnhart, M., 529
Barona-Gómez, F., 444
Barr, P. J., 259, 380, 381
Barraille, P., 539, 541, 560, 565
Barredo, J. L., 419
Barrett, A. G. M., 387
Barrett, G., 108
Barry, D. J., 118
Bartels, D., 81
Bartlett, J. G., 118
Barton, G. J., 189
Barton, W. A., 326
Basu, S., 182
Bate, N., 106, 120, 289
Bateman, A., 119, 120, 130, 185
Bates, R. B., 489
Battershill, C. N., 577, 587, 590
Baumberg, S., 128
Baumgartner, R., 341
Baylis, H. A., 125
Bayona, J., 118
Bearden, D., 298, 310
Bearden, D. W., 285
Becerra, M. C., 118
Bechthold, A., 175, 183, 195, 282, 283, 284, 290, 292, 298, 300, 309, 310, 311, 313, 317, 319, 321, 322, 324, 325, 326, 327, 480
Beck, Z. Q., 49
Becker, G. W., 420
Becker, J., 13
Beckmann, R. J., 419

Bedorf, N., 34, 37, 48
Beer, L. L., 359
Beerthuyzen, M. M., 538, 539, 560, 564, 566
Beeston, N., 194
Begley, T., 385
Bekel, T., 81
Belcourt, M. F., 43
Bellamine, A., 489
Belshaw, P. J., 223, 224, 226, 259, 266, 343, 382, 539
Belt, A., 9
Beltrametti, F., 363, 366, 463
Bengtsson, M., 106
Bennett, A. E., 343
Bennett, J. W., 119
Bentley, S. D., 119, 120, 130, 161
Benvenuto, M., 513
Beppu, T., 125, 127, 128, 144, 145, 150
Berdy, J., 34, 39, 40, 49
Berg, C., 436
Berg, H., 355
Berg, P., 451
Bergeron, R. J., 448
Bergman, N. H., 448
Bernardo, P., 513
Bernhard, F., 341, 342, 362
Berry, A., 194, 201
Berry, D. M., 523
Bersos, Z., 118
Bertazzo, M., 475, 479, 480
Berti, A. D., 338, 427
Bertsch, A., 84
Besra, G. S., 126
Betlach, M. C., 259, 380, 381
Beveridge, A., 128
Bevitt, D. J., 194, 203
Bewick, M. W., 124
Beyer, S., 60, 64, 66, 79, 81
Bharti, N., 448
Bhole, D. B., 120
Bianchi, A., 463
Bibb, M. J., 93, 95, 97, 98, 100, 101, 104, 105, 106, 108, 109, 111, 113, 119, 121, 125, 126, 127, 130, 144, 145, 146, 148, 149, 151, 152, 154, 160, 161, 167, 168, 173, 174, 287, 290, 301, 385, 423, 463, 465, 539, 541, 560, 565, 585
Biemann, K., 529
Bier, F. F., 109
Bierbaum, G., 535, 539, 541, 554, 560, 562, 563, 564, 565
Bierman, M., 98, 130, 161, 289, 464, 515, 521, 565, 566
Biggins, J. B., 283, 326, 354, 372
Bigler, L., 473, 475, 489, 503
Bignell, D. R., 120, 423
Bindereif, A., 442
Bindseil, K., 282, 283, 299, 300, 325

Bingman, C. A., 302, 313, 315, 316, 317
Binns, D., 185
Binz, T. M., 175, 442
Biondi, N., 34, 40
Birren, B., 515
Bischoff, D., 25, 358, 364, 464, 468, 471, 473, 474, 475, 476, 477, 479, 480, 489, 490, 492, 499, 502
Bishop, A., 170
Bister, B., 25, 473, 474, 475, 476, 479, 480, 489, 490, 492, 499, 502
Bitto, E., 313, 315, 316, 317
Blackburn, J. M., 415, 548
Blackburn, P., 535
Blaesse, M., 554
Blake, T. J., 394
Blanchard, M., 86, 313, 324
Blanco, G., 283, 284, 285, 290, 292, 300, 302, 317, 326, 414, 423
Blasiak, L. C., 358
Blöcker, H., 81
Blondelet-Rouault, M. H., 174
Blum, M.-M., 341
Blumauerová, M., 385
Blumbach, J., 463
Boakes, S., 565
Boddy, C. N., 380, 381, 385, 387, 388, 391, 395
Bode, E., 81
Bode, H. B., 60, 63
Boeck, L. D., 354, 365, 402, 462, 512, 523
Boger, D. L., 383
Bolam, D. N., 313
Boll, R., 324
Bongers, R., 541
Boot, H. J., 563
Borchert, S., 560
Borel, J. F., 37, 50
Borg, M., 354, 362, 363, 364, 523, 525
Borghi, A., 32
Borisova, L. N., 144
Borisova, S., 382
Borisova, S. A., 282, 283, 285
Bork, P., 185
Borneman, A., 14
Börner, T., 357, 365
Borodina, I., 122
Borovok, I., 414, 415
Borriss, R., 210
Bossi, E., 161, 464
Botstein, D., 594
Bouchard, M., 354, 359, 515, 516, 517, 518, 519, 521, 522, 524, 527
Boucher, H. W., 513
Bouchier, C., 578, 581, 588
Bourne, D. J., 577
Boury, M., 381
Boutros, P. C., 110
Bovenberg, R. A., 119, 419

Bowden, B. F., 577
Bowers, R. J., 413
Bowman, J., 13
Boyce, J., 118
Braden, T., 14
Bradley, J. S., 118
Bradley, K. A., 436
Bradley, M., 545
Brakhage, A. A., 405
Bralley, P., 128
Brana, A. F., 47, 302, 317, 319, 326
Braña, A. F., 279, 282, 283, 284, 285, 287, 288, 290, 291, 292, 293, 297, 298, 299, 300, 302, 310, 321, 326, 414, 423
Brandes, G., 63
Brandi, L., 4, 14, 16, 25, 31, 32, 33
Brassington, A. M. E., 359, 363
Brecht, S., 541, 564
Breton, A., 13, 14
Breukink, E., 535, 541
Brian, P., 125, 127, 131, 354, 359, 515, 516, 517, 518, 519, 520, 521, 522, 523, 524, 525, 526, 527
Brick, P., 186, 339
Briggs, B., 354, 365
Briggs, S. L., 257
Brinkmann, V., 62
Brockman, E. R., 64
Brolle, D. F., 463
Brost, R., 354, 357, 358, 359, 515, 516, 517, 518, 521, 522, 523, 524, 525
Brötz, H., 535
Broughton, M. C., 163
Browder, C. C., 448
Brown, A. G., 53, 404
Brown, S., 119, 120, 130
Brückner, R., 564
Brunati, C., 37, 479
Bruner, L. H., 196
Bruner, S. D., 236, 345, 354, 355, 362, 442
Brunner, N. A., 14
Brunzelle, J. S., 538, 545, 548, 551
Bruton, C. J., 44, 98, 103, 163, 173, 463
Bryant, S. H., 185
Brzuszkiewicz, E., 381
Bucca, G., 108, 109, 111, 126, 130, 145, 148, 152, 359, 357
Buchholz, F., 160
Buchini, S., 327
Buchrieser, C., 381
Buckett, P. D., 259, 260, 267, 268, 269
Buffo, M. J., 278
Bull, A. T., 8, 9, 25
Bu'Lock, J. D., 30
Bulone, V., 312
Bumpus, S. B., 261, 262, 266, 269, 363
Bunet, R., 149, 156
Bunkoczi, G., 257

Burkart, M. D., 206, 208, 219, 221, 225, 226, 256, 258, 260, 265, 282, 315, 354, 355, 357, 367, 370, 470
Burkert, D., 472
Burnham, M. K., 14
Burns, B. P., 583
Burrage, S., 545
Busarow, S. B., 126
Busch, U., 9
Bushell, M. E., 123
Busti, E., 8, 9
Butler, A., 289, 445, 448
Butler, M. S., 4
Butovich, I. A., 302, 310, 326
Butterworth, D., 404
Buttner, M. J., 109, 111, 113, 126, 128, 130, 146, 149, 160, 161, 163, 167, 168, 172, 174, 287, 385, 463, 465, 478, 536, 585
Butz, D., 464, 473, 475, 476, 480
Byers, B. R., 448
Bystroff, C., 183

C

Caballero, J. L., 125
Cabell, C. H., 513
Caboche, S., 338
Caffrey, P., 194, 201, 203, 204
Cahalane, O., 564, 565
Cai, S. J., 289, 290, 516, 519, 520, 521
Calderone, C. T., 357
Calzada, J. G., 407, 413
Camalier, R. F., 62
Campanaro, E., 513
Campbell, J. A., 312
Candiani, P., 31, 32, 33, 560
Cane, D. E., 257, 259, 264, 381, 382
Cantwell, C., 419
Cantwell, C. A., 419
Cao, W. R., 48
Cappellano, C., 161, 464
Cappiello, M. G., 14
Cardno, M., 545
Cardoza, R. E., 408, 410, 411
Carlomagno, T., 63
Carney, J., 48, 380, 381, 392
Carney, J. R., 48, 354, 359, 395
Carotti, M., 25
Carpenter, L. J., 196
Carpenter, M. R., 569, 570
Carrano, L., 31, 32, 33, 468, 469, 560
Carrell, C. B., 512
Carroll, A. R., 577
Carter, L., 444, 450, 452, 453
Casaulta, M., 182
Cáslavská, J., 385
Casqueiro, J., 415, 417
Castiglione, F., 31, 32, 33, 560

Castro, J. M., 412, 413, 415, 417, 418
Caufield, P. W., 538, 562, 569
Cavaletti, L., 8, 9, 31, 32, 33, 35, 36
Cendrowski, S., 448
Cerdeno-Tarraga, A. M., 119, 120, 130, 161
Chakraburtty, R., 95, 121, 126, 146, 151, 154
Chakravarty, P. K., 383
Chalker, A. F., 14
Challis, G. L., 31, 96, 99, 119, 120, 130, 161, 163, 166, 182, 183, 186, 187, 188, 192, 196, 207, 209, 234, 396, 443, 444, 447, 448, 450, 452, 453
Challis, I. R., 194
Chamberlin, L., 163, 167, 172
Chambers, H. F., 513
Champness, W., 125, 127, 128, 131
Champness, W. C., 125, 126, 127, 128, 131
Chan, J. A., 13, 24
Chan, Y. A., 338, 355, 357, 427
Chandra, G., 44, 96, 103, 163, 171, 172, 173, 463
Chang, S., 95, 108, 111, 121
Chang, S. A., 128
Chapple, J., 357, 358, 522, 523, 524, 525
Charaniya, S., 109
Charge, A., 317, 322
Charles, E. H., 365, 469
Charnock, J. M., 418
Chary, A., 538, 545
Chater, K., 145, 148, 152
Chater, K. F., 44, 96, 97, 98, 99, 100, 101, 102, 103, 104, 105, 106, 108, 109, 111, 119, 120, 125, 126, 128, 129, 130, 146, 148, 149, 160, 161, 163, 166, 167, 168, 173, 174, 287, 385, 463, 465, 539, 541, 560, 565, 585, 593
Chatterjee, C., 534, 538, 539, 541, 542, 544, 545, 546, 548, 552, 559, 560, 569
Chatterjee, S., 463
Chen, H., 258, 259, 264, 280, 282, 315, 327, 354, 355, 357, 358, 361, 362, 365, 385, 468, 470
Chen, J., 37, 53, 95, 108, 111
Chen, L., 312, 327
Chen, M., 43, 44, 206, 208
Chen, P., 538, 562
Chen, W. J., 121
Chen, X. H., 210, 280
Cheng, F., 548
Cheng, L., 583, 584
Cheng, Y.-Q., 195, 355
Cherepanov, P. P., 169
Cheung, H. T., 391
Chin, H. S., 419
Chirgadze, N. Y., 257
Chitsaz, F., 185
Chiu, C. P., 327
Chiu, H. T., 382, 466
Choi, C. Y., 284
Chopin, A., 566

Choroba, O. W., 367, 469
Chouayekh, H., 121, 122
Chow, V. T., 13
Christenson, S. D., 371
Christianson, C. V., 236, 355
Chu, M., 516, 519, 520, 522, 525, 526, 527
Chua, S. C., 418
Chung, L., 48
Ciardelli, T. L., 383
Ciciliato, I., 16, 31, 32, 33, 560
Cironi, P., 260, 261, 263, 267, 268, 269
Clardy, J., 25, 119, 182
Clarke, K. M., 231, 260
Cleland, W. W., 433, 442
Clifton, I. J., 414, 415, 418
Clugston, S. L., 224, 266
Coëffet-Le Gal, M.-F., 354, 357, 358, 359, 515, 516, 517, 518, 519, 521, 522, 523, 524, 525, 527
Coggon, P., 37, 47
Cohen, G., 412, 413, 414, 415
Cohen, S. N., 109, 126
Cole, J., 74
Cole, J. R., 489
Cole, M., 404
Colin, P. L., 584
Coll, J. C., 577
Collado, J., 38
Collins, B., 565
Colombo, A. L., 45, 288
Colquhoun, D. J., 446
Colson, S., 122, 123, 133
Colvin, K. R., 423
Combes, P., 464, 465
Conti, E., 186, 234, 339
Cook, J. C., 436
Cooke, H. A., 355
Cooper, L. E., 533, 538, 539, 541, 542, 545, 546
Cooper, R. D., 414
Copp, J. N., 256, 259
Coque, J. J. R., 407, 408, 410, 411, 413, 417, 418, 419, 420, 421
Corey, G. R., 513
Cornish, A., 387
Corre, C., 96, 99, 364
Cortes, J., 194, 203
Cortés, J., 284, 302, 412, 413, 417, 418, 559
Cortes Bargallo, J., 565
Corti, E., 31, 32, 33, 37, 560
Corvey, C., 538, 560
Cotter, H., 564, 565
Cotter, P. D., 538, 541, 563, 565, 569, 570
Coughlin, J. M., 44
Counter, F. T., 512
Court, D. L., 161
Courvalin, P., 478
Coutinho, P. M., 312
Cox, C. D., 436

Cozzone, A. J., 126
Cragg, G. M., 40, 41, 45, 48, 53, 311
Craig, M., 123
Crandall, L. W., 460
Craney, A., 108
Cravatt, B. F., 249
Crawford, J. M., 256
Crispin, D. J., 505
Cronan, J. E., 257
Cronan, J. E., Jr., 257, 258
Crosa, J. H., 433
Crothers, D. M., 107
Crouse, G. D., 289
Crumbliss, A. L., 381
Cudic, P., 357
Cudlín, J., 385
Cui, X., 123, 132
Cullen, W. P., 387
Cullum, J., 161, 170
Cummings, R., 25, 119
Cundliffe, E., 106, 120, 289, 291
Currie, C. R., 25
Currie, S. A., 402, 404
Cutting, C., 269
Cyr, D. D., 48
Czisny, A., 160

D

Dafforn, T. R., 126
Dai, Y. F., 419
D'Alia, D., 477
Daly, C., 260, 265, 270
Damodaran, C., 300
Daran, J. M., 119
Darbon, E., 174
Das, U., 185
Datsenko, K. A., 160, 169, 520
Daugherty, L., 185
Daunert, S., 108
Daupert, V. M., 512
Dauter, Z., 187
Davidson, S. K., 576
Davies, G. J., 312, 313
Davies, J., 31, 517
Davies, J. K., 583
Davis, B. G., 313
Davis, J. T., 226
Davis, K. E. R., 8
Davison, J., 13, 14
Dawid, W., 61, 64, 66, 68
Dawson, M. J., 559, 565
Dayem, L. C., 395
DeBacker, M. D., 14
de Boef, E., 538, 541, 548, 550, 560, 561
Debono, M., 512
Decker, H., 364
Deegan, L. H., 541, 563

Deeth, R. J., 396, 440, 442
De Grazia, S., 161, 464
DeHoff, B. S., 514
de Hoogt, R., 14
Deisenhofer, J., 53
de Kruijff, B., 535, 541
de la Fuenta, J. L., 419
De Lay, N. R., 257
Deleury, E., 312
De Lorenzo, V., 442
Del Vecchio, F., 194
Delves-Broughton, J., 535
Demain, A. L., 30, 31, 34, 37, 39, 41, 49, 51, 53, 119, 405, 409, 410, 413, 419, 420, 422, 426
Demel, R. A., 541
DeMoss, J. A., 451
Denapaite, D., 161, 170
Denaro, M., 32, 33
Deng, C. H., 419
Deng, W., 196, 354, 364
Deng, Z., 129
Depardieu, F., 478
De Pascale, G., 14, 16
Derbyshire, M. K., 185
Derewenda, Z. S., 187
Dermyer, M., 14
Derrick, P. J., 444
Dertz, E. A., 435, 436
de Ruyter, G. A., 566
de Ruyter, P. G., 564, 566
Desikan, K. R., 463
Destoumieux-Garzon, D., 539, 576
DeVito, J. A., 14
de Vos, W. M., 536, 538, 539, 542, 548, 560, 561, 563, 564, 566
De Vreugd, P., 463
Dewaele, S., 14
Deweese-Scott, C., 185
Dewey, R., 38
Dezeny, G., 163, 174, 565
D'haeseleer, P., 110
Dhillon, N., 302
Diep, D. B., 552
Díez, B., 419
Dijkhuizen, L., 122, 463
Dimise, E. J., 442
Ding, W., 196, 354, 364
Ding, Y., 49
Distler, J., 477
Dittmann, E., 196, 359, 578, 581, 588
Dobrindt, U., 381
Dodd, H., 538
Dodd, H. M., 539, 541, 560, 564, 565
Doddrell, D. M., 383
Dodson, E. J., 313
Doekel, S., 341, 525
Doi-Katayama, Y., 45
Dolce, L., 161, 464

Domann, S., 298, 310
Dombrowski, A. W., 38
Dominguez, H., 174
Domling, A., 62
Donadio, S., 3, 4, 8, 9, 14, 16, 161, 194, 203, 282, 289, 363, 366, 463, 464, 466, 467, 468, 469, 471, 477, 479, 490
Dong, C., 358
Dong, S. D., 283, 354, 372
Donia, M. S., 380, 575, 576, 578, 580, 581, 584, 586, 587, 588, 589, 590
Doran, J. L., 410, 412
Dorrestein, P. C., 225, 257, 261, 262, 266, 269, 385
Dorso, K., 119
Dötsch, V., 341, 342
Dotzlaf, J. E., 417, 419, 420
Doull, J. L., 124
Doumith, M., 283, 284
Dowson, J. A., 302
Draeger, G., 280
Drager, G., 284, 298, 310, 325
Draper, L. A., 541, 563
Drapkin, P. T., 392
Drennan, C. L., 358
Dresel, D., 122, 123, 129
Driessen, A. J., 119, 538, 541, 548, 550, 561
Drueckhammer, D. G., 226
Du, L., 43, 44, 206, 208, 259
Dubertret, C., 489
Dubus, A., 418
Duclaux, D., 584
Duerfahrt, T., 383
Duitman, E. H., 362
Dunlap, W. C., 577, 587, 590
Duquenne, L., 185
Duquesne, S., 539, 576
Durkin, A. S., 86
Durr, C., 319, 321, 324, 325, 326
Durrant, M. C., 536
Dusterhus, S., 538, 560
Dworkin, M., 61, 66, 72, 73, 74, 77, 78, 79
Dyson, P., 170, 171, 172
Dziarnowski, A., 480

E

Eccleston, M., 128
Eckermann, S., 14
Eddy, S. R., 189
Edlund, P. O., 84
Edwards, D. J., 43, 44, 206, 208, 259, 354
Eguchi, T., 313, 315, 316, 319
Ehling-Schulz, M., 355, 359
Ehmann, D. E., 258, 259, 264, 383
Ehrentreich-Forster, E., 109
Ehrlich, K. C., 256
Eichner, A., 161, 170

Author Index

Eide, J., 382, 466
Eisen, J., 86
Eisen, J. A., 396, 577, 578, 587
Eisenstein, B. I., 512, 513
Eisermann, B., 569
Elander, R. P., 405
Elias-Arnanz, M., 73
Ellestad, G. A., 354, 365
Elling, L., 313, 315
Elliot, M. A., 113
Elnakady, Y. A., 62, 80
Endo, A., 37, 53
Engel, N., 48
Engelke, G., 562, 564, 565
Engels, J. W., 130
Enguita, F. J., 408, 410, 411, 420, 421
Enomoto, H., 489
Entian, K.-D., 536, 538, 560, 562, 564, 565
Epp, J. K., 302
Eppelmann, K., 341, 383, 480
Erhard, M., 359
Ertl, P., 182
Esnault, C., 122
Essen, L.-O., 206, 257, 259, 337, 343, 344, 345, 347, 354, 357
Essex, J. W., 545
Esumi, Y., 387
Etienne, G., 38
Eustaquio, A., 173, 174
Evans, D. A., 492, 493, 494
Evans, R. J., 535
Expert, D., 447
Ezaki, M., 489
Ezure, Y., 489

F

Fabbretti, A., 14, 16, 25, 31, 32, 33
Fabre, B., 38
Facchini, P. J., 489
Fajardo, A., 31
Falcone, B., 312, 327
Falk, S. P., 16
Fan, Q., 471
Fang, J., 354, 370, 371
Farber, H., 541, 563, 564
Farewell, A., 95
Farmer, P., 118
Farnet, C. M., 46, 120, 175, 183, 193, 203, 204, 210, 364
Fast, B., 14
Faulkner, D. J., 576
Faust, B., 298, 310
Fecik, R. A., 193
Fedorova, N. D., 119
Fedoryshyn, M., 321, 325, 326
Feeney, J., 391
Feifel, S. C., 355

Feitelson, J. S., 119
Fekken, S., 538, 541, 548, 550, 561
Feliciano, J. S., 108
Felnagle, E. A., 338, 433
Fendrich, G., 536
Fenical, W., 8, 354, 359
Fernández, E., 285
Fernández, F. J., 415, 417
Fernández, M. J., 285, 288
Fernandez-Moreno, M. A., 125, 126, 128
Feroggio, M., 31, 32, 33, 468, 469
Ferrari, P., 37
Ferraro, C., 477
Ferrer, J. L., 233, 234
Festin, G. M., 355
Feurer, C., 37, 50
Fichtlscherer, F., 257, 258
Ficner, R., 257, 259, 341
Fidelis, K., 182
Fiedler, H.-P., 489
Field, C., 86
Field, D., 564, 565
Fielding, E. N., 354, 368
Fielding, S., 170, 171, 172
Filippini, S., 45, 288
Finking, R., 247, 256, 257, 258, 259, 338, 341, 449
Finnan, S., 203
Firn, R. D., 396
Fischbach, M. A., 119, 184, 206, 220, 256, 338, 433, 477, 513, 515, 526, 528
Fischer, C., 283, 298, 302, 310, 326, 336
Fischer, H. P., 14
Fischer, J., 466
Fischl, A. S., 257
Fisher, D. L., 419
Fitzgerald, G. F., 564, 565
Fitzgerald, M. K., 311
Flardh, K., 123, 465
Flaxman, C. S., 126
Flecks, S., 362
Flett, F., 108, 111, 163, 354, 357, 359, 364, 463, 523, 525
Flood, E., 203
Floriano, B., 126, 127, 173
Floss, H. G., 280
Flugel, R. S., 256, 257, 258, 265, 382
Fogerty, C., 513
Fogg, M. J., 444, 448, 450, 452, 453
Folcher, M., 151, 152
Foley, T. L., 225, 258, 260, 265
Fong, J. H., 185
Fonstein, L., 45, 160, 291, 391
Fontaine, A., 338
Fontes, M., 73
Ford, R., 354, 359, 515, 516, 517, 518, 519, 521, 522, 524, 527
Forootan, A., 106

Fortin, P. D., 355, 363
Fossum, S., 14
Foster, G., 313, 327
Foster, L. A., 14
Foster, S. J., 163, 167, 172
Fournier-Rousset, S., 473, 489, 490, 491, 492, 499, 502, 505, 506
Fouts, D. E., 396
Fowler, K., 99, 163, 166, 565, 593
Fowler, V. G., 513
Fox, D. T., 448
Fox, K. R., 383
Fox, M. S., 594
Franke, P., 210
Frankel, B. A., 357
Franza, T., 447
Frasch, H.-J., 478
Frechet, D., 489
Fredenburg, R. A., 382, 466
Fredenhagen, A., 536
Freemont, P. S., 313
Freiberg, C., 14
Freitag, A., 289
Frere, J. M., 418
Frerich, A., 324
Freund, E., 492
Frey, P. A., 355
Fried, M., 107
Frisvad, J. C., 9
Fritz, C. C., 257
Frolow, F., 415
Frueh, D. P., 234, 342, 343
Fu, J., 174, 175
Fu, Q., 313, 315, 316, 317, 318, 326
Fu, X., 313, 315, 316, 317, 318, 372, 479
Fuchser, J., 325
Fudou, R., 64, 68
Fuente, J. L., 407, 408, 420
Fujii, K., 507
Fujii, T., 106, 122, 123
Fujimori, D. G., 174, 358
Fujioka, T., 489
Fukagawa, Y., 404
Fukuda, D. S., 512, 523, 529
Funa, N., 489
Funabashi, M., 489
Furgerson Ihnken, L. A., 538, 546, 552
Furihata, K., 489
Furin, J. J., 118
Furumai, T., 588
Furuya, K., 391
Futterer, K., 126

G

Gagnat, J., 174
Gaillard, H., 151, 152
Gaillard, J., 505
Gaisser, S., 279, 283, 289
Galas, D. J., 107
Galazzo, J., 392
Galgoci, A., 25, 119
Galm, U., 41, 43, 44, 46, 173, 174
Gandhi, C., 526, 527
Ganguli, B. N., 463
Ganju, L., 12
Gao, Q., 419
Garavito, R. M., 313
Garcia, R., 81, 83, 84, 86
Garcia, R. O., 59, 65, 69
García-Bernardo, J., 194, 285
García-Domínguez, M., 410, 413
Garcia-Estrada, C., 119
Gargiulo, R. L., 489
Garner, C. D., 418
Garner, M. M., 107
Gaspari, F., 34
Gasson, M. J., 538, 539, 541, 560, 564, 565
Gastaldo, L., 31, 32, 33, 560
Gatto, G. J., 282
Gaykova, V., 466
Gebhardt, K., 489
Geer, L. Y., 185
Geer, R. C., 185
Geffers, R., 63
Gehling, M., 14
Gehring, A. M., 221, 256, 257, 264, 382, 436, 539
Geib, N., 473, 487, 489, 490, 491, 492, 499, 502, 503, 505, 506
Gekeler, C., 465
Gelfand, M. S., 539
Genereux, C., 421, 422
Genuth, S. M., 451
George, N., 259, 260, 265
Geraghty, M. T., 256
Gerbaud, C., 174
Gerhard, U., 475
Gerratana, B., 414
Gershater, C. J., 146
Gerstein, M., 14
Gerth, K., 34, 37, 48, 60, 64, 66, 68, 79, 81
Gerwick, W. H., 354
Gewain, K. M., 45, 163, 174, 288, 565
Ghorbel, S., 122
Giannakopulos, A. E., 444
Gibbons, P. H., 163, 174, 565
Gibson, M., 282, 283, 298, 299, 300, 326
Gibson, T., 516, 517, 519, 522, 524, 527
Gierling, G. H., 364
Gieselman, M., 545
Gilbert, H. J., 313
Giraldes, J. W., 193, 235
Gish, W., 130, 185, 189
Giusino, F., 161, 464
Glaser, N., 62

Glaser, S. J., 324
Glazebrook, M. A., 124
Glod, F., 109
Gluckman, T. J., 392
Glund, K., 383
Glynn, F., 564, 565
Gnau, V., 364, 541, 553, 564
Goegelman, R., 404
Goff, R. D., 313, 315, 316, 317, 327, 328
Goh, K., 583
Gokhale, R. S., 187, 194, 256, 258, 264, 396
Golakoti, T., 49
Golan, D. E., 259, 260, 262, 263, 265, 267, 268, 269
Goldman, B. S., 86
Gomathinayagam, R., 300
Gomez-Escribano, J. P., 422, 423
Gong, G. L., 48
Gontang, E., 8
Gonzales, N. R., 185
González, A., 285, 288
Goo, K. S., 418
Goodfellow, M., 8, 9, 25
Goodson, T., 302
Gordon, L., 354, 357, 359, 364, 523, 525
Gorman, M., 402
Goto, T., 50
Gottelt, M., 143
Gottschalk, G., 381
Götz, F., 535, 536, 538, 539, 541, 548, 553, 560, 561, 562, 564, 565
Goulard, C., 539
Gould, S. J., 256, 354, 359
Gourzoulidou, E., 182
Gramajo, H. C., 95, 106, 126, 259, 380, 381, 392
Grammel, N., 45
Grantcharova, N., 465
Greaney, M., 419
Green, B., 128
Greener, A., 161
Greenstein, M., 355, 357, 359
Greenwell, L., 282, 283, 299, 300, 326
Grenader, A., 16, 17
Griffith, B. R., 313, 315, 316, 317, 318
Grigoletto, A., 468, 469
Grigoriadou, C., 16
Gronewold, T. M., 62
Gropl, C., 84
Gross, B., 312, 327
Gross, F., 174, 175
Gruenewald, S., 381
Grundmann, H., 118
Gruner, J., 536
Grünewald, J., 338, 346, 409, 490, 523, 528
Gu, J.-Q., 359, 516, 517, 519, 520, 521, 522, 525, 526, 527
Gualerzi, C. O., 14, 16, 25
Guan, Z., 38
Gubler, H. U., 37, 50

Guengerich, F. P., 489
Guidos, R., 118
Guilfoyle, R. A., 99
Gullón, S., 283, 291, 302
Gunatilaka, A. A., 326
Gunnarsson, N., 477
Güntert, P., 342
Guo, Y., 129
Guo, Z., 99, 280, 282
Gust, B., 44, 103, 122, 159, 163, 166, 173, 174, 175, 463, 472, 565, 593
Guthrie, E. P., 126
Gutiérrez, S., 415, 417
Gwadz, M., 185
Gwynn, M. N., 4, 118

H

Ha, S., 119, 312, 327
Haag, S., 364
Hacker, J., 381
Hackmann, K., 123, 129
Hadatsch, B., 475, 480
Hager, M. H., 41, 43, 44, 46
Hahn, M., 348
Hajdu, J., 414, 415
Hakala, J., 285, 288
Hakansson, K., 448
Hakoshima, T., 385
Hallis, T. M., 282
Halpern, A. L., 396
Haltli, B., 355, 357, 359
Hamill, R. L., 402, 512, 529
Hammond, S. J., 354, 365, 462
Hamoen, L. W., 362
Hamprecht, D., 387
Han, O., 280
Hanamoto, A., 119, 120, 130
Hancock, S. M., 327
Hanna, P. C., 448
Hannappel, E., 258
Hanni, E., 536
Hanscomb, G., 404
Hansen, D. B., 363
Hansen, J. N., 538, 539, 554
Hansen, M. E., 9
Hansen, N., 562
Hantke, K., 381
Hao, Z., 564
Hara, H., 119, 120, 130
Hara, O., 145
Hara, T., 415, 418
Hara, Y., 146
Harasym, M., 128
Harding, M. W., 37
Hargreaves, R. T., 354, 359
Härle, J., 309
Harlos, K., 418

Harper, D., 119, 120, 130, 161
Harris, B. R., 363, 462, 470, 471, 477
Harris, D. E., 119, 120, 130, 161
Harris, D. M., 119
Harris, E., 37, 53
Harvey, B., 354, 359, 362, 523, 525
Harwood, C. R., 269
Hasenkamp, R., 53
Hashimoto, M., 50, 489
Hashimoto, Y., 111
Hasper, H. E., 535
Hasuda, T., 489
Hatanaka, H., 50
Hathaway, B. J., 578, 580, 581, 584, 586, 587, 590
Hattori, M., 119, 120, 130
Haupt, C., 362
Hauser, G., 324
Haushalter, R. W., 241
Hautala, A., 285, 288
Håvarstein, L. S., 539, 552
Hayakawa, M., 9
Hayakawa, Y., 494
Hayashi, K., 588
Hayden, M. K., 118
Haydock, S., 284, 463, 471
Haydock, S. F., 194, 203, 302, 363, 366
Hayes, M. A., 354, 359, 364, 523, 525
Hayes, R. A., 118
Haygood, M. G., 576, 577, 578, 580, 581, 584, 586, 587, 588, 590
Haynes, S. W., 192, 196, 207
He, J., 196
He, M., 355, 359
He, Q., 354, 364
He, W., 235
He, X., 282
He, X. M., 278, 280
Heber, M., 569
Hecht, H. J., 194
Heck, M., 473, 489, 490, 491, 492, 499, 502, 505, 506
Heckmann, D., 365, 366, 463, 470, 471, 477
Hedrick, M. P., 383
Heidari, B. A., 354, 357
Heide, L., 122, 173, 174, 175, 289, 472
Heidelberg, J. F., 396
Heijne, W. H., 119
Heijnen, J. J., 122, 124, 125, 132, 133
Heim, J., 426
Heinzmann, S., 560
Heitmann, B., 324
Helfrich, M., 560
Helynck, G., 489
Hemscheidt, T. K., 49
Hendlin, D., 402, 404, 407
Henikoff, J. G., 162, 189, 322
Henikoff, S., 162, 322

Henkel, J., 109
Henne, A., 210
Henriksen, C. M., 409
Henrissat, B., 312
Hensgens, C. M., 414, 415
Herai, S., 111
Herdtweck, E., 62
Herlihy, W. C., 529
Hernandez, C., 311
Hernandez, S., 402
Hernández, S., 404
Herron, P., 170
Herron, P. R., 171, 172
Hersbach, G. J. M., 119
Hershberger, C. L., 122
Hersman, L. E., 448
Hertweck, C., 151, 183, 195, 196, 578, 581, 588
Hesketh, A., 93, 95, 108, 111, 121, 423
Heusser, C., 62
Hicks, L. M., 365
Hiestand, P. C., 37
Hieter, P., 594
Higashibata, H., 111
Higgens, C. E., 402
Hildebrandt, A., 84
Hill, A. M., 184
Hill, C., 536, 538, 541, 560, 563, 565, 569, 570
Hillemann, D., 463
Hillman, J. D., 535
Hintermann, G., 420
Hiraishi, A., 64, 68
Hirano, S., 128
Hirata, S., 404
Hirose, E., 584
Hirose, M., 584
Hitotsuyanagi, Y., 489
Hitzeroth, G., 210
Hixon, J., 257
Ho, J. Y., 470
Hoagland, M. B., 451
Hodges, R. L., 419
Hodgson, D. A., 34, 40, 126
Hodnick, W. F., 43
Hoehn, M. M., 402
Hoffman, C., 37, 53
Hoffman, J. J., 489
Hoffmann, A., 560, 565
Hoffmann, F. M., 311
Hoffmann, J., 284
Hoffmann, J. A., 529
Hoffmeister, D., 284, 298, 310, 313, 319, 325, 326, 327
Hofle, G., 34, 37, 48, 62, 63, 81
Hofmann, C., 324
Hofmann, H., 62
Hohenauer, T., 108
Hojati, Z., 354, 359, 364, 523, 525
Holak, T. A., 341

Holander, I. J., 426
Holbeck, S., 62
Holker, J. S. E., 354, 365
Holler, T. P., 14
Hollingshead, M., 62
Holmes, D. J., 4, 118
Holo, H., 539
Holt, T. G., 465
Höltzel, A., 473, 474, 489
Holzenkämper, M., 284, 480
Homburg, S., 381
Hong, H., 363, 365, 367, 583
Hong, H. J., 478
Hong, J. S., 284
Hong, S. K., 127, 391
Hood, J. D., 404
Hoof, I., 207, 466, 471, 472
Hook, L. A., 64
Hopke, J., 129, 130
Hopkinson, A., 492
Hopwood, D. A., 31, 96, 97, 98, 100, 101, 102, 104, 105, 106, 108, 118, 119, 120, 122, 123, 125, 127, 128, 133, 146, 149, 160, 161, 167, 168, 170, 174, 287, 385, 433, 463, 465, 585
Hori, K., 404
Horinouchi, S., 96, 113, 119, 120, 125, 126, 127, 128, 129, 130, 144, 145, 150, 151, 152, 489
Horn, N., 541, 564
Hornbogen, T., 355
Hornung, A., 480
Hosoda, J., 404
Hosted, T. J., 514, 516
Hotchkiss, G., 108, 109, 111, 126, 130, 145, 148, 152, 359, 363
Hotta, K., 379, 380, 381, 385, 387, 388, 391, 395, 448
Howard, L. C., 512
Howard-Jones, A. R., 489
Hrvatin, S., 174, 260, 261, 262, 263, 267, 268, 269, 358
Hsiao, N. H., 143, 149, 151, 156
Hsu, J. S., 418, 419
Hu, W., 13, 48
Hu, W. S., 109
Hu, Y., 175, 193, 312, 313, 327
Huang, F., 357, 359, 361, 463, 470, 471
Huang, G., 448
Huang, J., 109, 111, 126
Huang, K., 46, 120, 160
Huang, L. H., 387
Huang, Y. T., 470
Hubbard, B. K., 282, 315, 354, 355, 357, 358, 359, 361, 362, 365, 382, 385, 463, 466, 468, 469, 470, 475, 489
Hubbard, T., 182
Huber, C. G., 87
Huber, F. M., 356, 512

Huber, M. L., 302
Huber, P., 358, 365, 366, 463, 464, 468, 470, 471, 475, 477
Huber, R., 554
Huber, U., 49
Hucul, J. A., 355, 357, 359
Hudson, M. E., 536
Huennekens, F. M., 507
Hueso-Rodríguez, J. A., 13, 24
Huff, J., 37, 53
Hugenholtz, J., 535
Hughes, G., 171, 172
Hung, G., 560
Hunt, A. H., 512
Hunt, V., 37, 53
Hunter, S., 185
Hunter-Cavera, J. C., 9
Huppert, M., 366, 463
Hur, J. H., 38
Huson, D. H., 188, 207, 209, 466, 471, 472
Huson, U., 341
Hussong, R., 84
Hutchings, M. I., 478
Hutchinson, C. R., 53, 122, 126, 282, 288, 289, 291, 326
Hutchinson, R. C., 391
Hütter, R., 463
Hüttl, S., 478
Hwang, Y., 247
Hwangbo, Y., 257
Hyun, C. G., 38

I

Ichikawa, S., 383
Ichinose, K., 284, 298, 310, 313, 319, 325, 326, 327
Idborg-Bjorkman, H., 84
Igarashi, Y., 588
Iida, A., 145
Iimura, Y., 9
Iizuka, T., 64, 68
Ikai, K., 489
Ikeda, H., 111, 119, 120, 130, 489
Ikeda-Araki, A., 326
Ikenoya, M., 119, 120, 130
Ikezawa, N., 489
Imanaka, H., 50, 489
Imberty, A., 313
Inamine, E., 407
Ingolia, T. D., 420
Inventi-Solari, A., 45, 288
Ipposhi, H., 151, 152
Ireland, C. M., 576, 577, 578
Irschik, H., 34, 37, 48
Ishida, K., 578, 581, 588
Ishikawa, H., 489
Ishikawa, J., 119, 120, 130, 385

Ishikura, T., 404
Ishizuka, H., 125, 127, 128
Islam, K., 37
Istvan, E. S., 53
Ivkina, N. S., 144
Iwasa, K., 489
Iyer, S., 448
Izaguirre, G., 538
Izumi, T. A., 145
Izumikawa, M., 489

J

Jablonski, L., 385
Jack, R. W., 536, 541, 560, 563, 564
Jackson, E. E., 338, 427
Jackson, M., 402, 404
Jacobsen, J. R., 256, 259
Jacobson, G. R., 123
Jacobsson, S. P., 84
Jacques, P., 338
Jahn, D., 257, 258, 259, 338
Jakimowicz, D., 44, 103, 163, 173, 463
James, K. D., 119, 120, 130, 161
Janes, B. K., 448
Janssen, G. R., 106, 121, 130, 290, 301, 465
Janssen, P. H., 8
Jantos, J., 123, 129
Janzen, W. P., 13
Jao, C., 261, 266, 269
Jaspars, M., 577, 587, 590
Jayapal, K. P., 109
Jayasuriya, H., 38
Jefferson, M. T., 387
Jenke-Kodama, H., 196
Jenkins, C. M., 507
Jenni, S., 234
Jensen, P. R., 8, 354, 359
Jensen, S. E., 355, 408, 409, 410, 412, 413, 415, 417, 420, 423
Jessani, N., 246
Jewiarz, J., 383, 385
Jeyaraj, D. A., 182
Ji, Y., 14
Jiang, B., 14
Jiang, J., 283, 313, 326, 354, 372, 479
Jim, J., 354, 359
Jin, Q., 383
Joachimiak, A., 448
Joardar, V., 119
Jog, M. M., 120
Johnson, D. A., 43
Johnson, K., 196
Johnson, K. A., 444, 447, 450, 452, 453
Johnson, M., 583
Johnsson, K., 259, 260, 265
Johnsson, N., 259, 260, 265

Jojima, Y., 64, 68
Jolad, S. D., 489
Jonák, J., 106
Jones, C. G., 396
Jones, G. H., 128
Jones, J. J., 146
Jones, M., 365, 462, 470, 471, 477
Jones, S. J., 462, 470, 471, 477
Jones, S. J. M., 365
Joris, B., 122, 123, 133
Joseph, J. K., 118
Joseph, S. J., 8
Joshi, A., 257
Joshi, A. K., 256, 257, 258
Joshua, H., 37, 53
Josten, M., 535, 536, 560
Ju, J., 41, 43, 44, 46, 257, 357
Julien, B., 48
Jung, G., 358, 364, 365, 366, 446, 463, 464, 468, 470, 471, 472, 473, 474, 475, 476, 477, 480, 489, 536, 538, 539, 541, 553, 554, 560, 561, 564, 569
Jungblut, A. D., 583

K

Kadam, S., 66
Kadi, N., 444, 445, 446, 447, 448, 450, 452, 453
Kahan, F. M., 404
Kahan, J. S., 404
Kahne, D., 282, 283, 302, 317, 321, 323, 354, 357, 363, 372
Kaiser, D., 86, 364, 554
Kaiser, O., 81
Kaiser, R., 280
Kajino, H., 494
Kakavas, S. J., 194, 201
Kakinuma, K., 313, 315, 316, 319
Kaku, H., 123
Kalaitzis, J. A., 489
Kalman, J. R., 394
Kameyama, S., 113, 145, 151
Kamionka, A., 123
Kammerer, B., 175, 289, 472
Kaneko, I., 489
Kanemasa, Y., 566
Kang, Q., 354, 370
Kantola, J., 285, 288
Kao, C. L., 282
Kao, C. M., 109, 111, 126
Kapur, S., 236, 238
Karas, M., 538, 560
Karchmer, A. W., 513
Karoonuthaisiri, N., 109, 111, 126
Karray, F., 174
Karwowski, J. P., 66

Katayama, K., 489
Kato, H., 494
Kato, J. Y., 96
Katz, L., 48, 194, 201, 203, 282
Katz, M. L., 118
Kau, D., 359, 516, 517, 519, 520, 521, 522, 524, 527
Kauffman, S., 13
Kaur, G., 62
Kavanagh, K. L., 257
Kawamoto, S., 124
Kay, B. K., 269
Kazakov, A., 539
Kazakov, T., 539
Kealey, J. T., 259, 380, 381, 395
Keating, A. E., 343
Keating, D. H., 232
Keating, T. A., 258, 264, 343, 358, 436
Keijser, B. J., 122, 124, 125, 132, 133
Kelleher, N. L., 225, 257, 259, 260, 261, 262, 266, 267, 269, 363, 365, 436, 470, 538, 539, 542, 544, 545, 546, 569
Kellenberger, L., 279, 283
Keller, S., 475, 479
Keller, U., 45, 383, 385, 471, 472, 474
Kellner, R., 536, 538
Kelly, S. L., 489
Kempter, C., 364, 553, 554, 560, 561
Kendrew, S. G., 53, 194
Kennedy, J., 45, 53, 288, 380, 381, 395
Kerkman, R., 119, 419
Kern, B. A., 407
Kersey, P., 185
Kershaw, J. K., 365, 462, 470, 471, 477
Kerste, R., 122, 124, 125, 132, 133
Kessler, N., 339, 341
Ketela, T., 14
Khalil, M. W., 62
Khanin, R., 126
Kharel, M. K., 336
Khokhlov, A. S., 144
Khosla, C., 48, 234, 256, 258, 259, 264, 265, 380, 381, 382, 391, 466
Kidd, D., 247
Kiehl, D. E., 289
Kiel, J. A., 119
Kiesau, P., 560
Kieser, H., 119, 120, 129, 130, 160, 161, 167, 168, 174
Kieser, H. M., 125, 127, 128, 161, 170
Kieser, T., 97, 98, 100, 101, 104, 105, 106, 108, 130, 146, 149, 163, 166, 287, 385, 463, 465, 565, 585, 593
Kikuchi, H., 119, 120, 130
Kim, C.-Y., 448
Kim, H. S., 145
Kim, J. Y., 118

Kim, S. O., 38
Kim, S. S., 435, 436
Kim, S. U., 391
Kim, S. Y., 38
Kim, U.-J., 515
Kim, Y., 448
Kinashi, H., 161, 170, 385
King, T. J., 53
Kinkel, B. A., 448
Kino, T., 50
Kinoshita, H., 109, 126, 130, 145, 148, 151, 152
Kirk, M., 562, 569
Kirkpatrick, P. N., 365, 462, 470, 471, 472, 476, 477, 492
Kirschning, A., 298, 310
Kirst, H. A., 289
Kitani, S., 145, 149, 156
Kittel, C., 358, 369, 468, 471, 476, 477, 479, 480
Kittendorf, J. D., 193
Kleckner, N., 280
Klein, C., 560
Kleinkauf, H., 428
Klena, J. D., 278
Klepzig, K. D., 25
Kloosterman, H., 463
Klöss, S., 560
Kluskens, L. D., 538, 541, 548, 550, 561
Knappe, T. A., 343, 344
Kobayashi, M., 111
Kobel, H., 363
Koch, M. A., 182
Kociban, D. L., 278
Kodali, S., 25, 119
Kodani, S., 560
Koehn, F. E., 23
Koerten, H. K., 122, 123, 124, 133, 465
Koga, H., 566
Koglin, A., 256, 260, 341, 342, 343, 358
Koguchi, Y., 489
Kohlbacher, O., 188, 209, 341, 471
Kohli, R. M., 343, 345, 349
Kohno, J., 489
Kohsaka, M., 50, 404
Kojiri, K., 326
Koketsu, K., 380, 381, 385, 387, 388, 391, 395
Koki, A., 404
Kolter, R., 259, 260, 261, 262, 266, 267, 268, 269
Kolvek, S. J., 129, 130
Komatsubara, S., 489
Kong, R., 125, 127, 128
Konig, B., 230
Konishi, M., 489
Konomi, T., 404
Kontos, F., 118

Konz, D., 257, 258, 259, 338, 471
Kopp, F., 345, 346, 354, 357, 360, 362, 521, 523, 528
Koppisch, A. T., 448
Korman, T. P., 234
Kormanec, J., 122
Kornitskaia, E., 144
Korte, H., 569
Koshino, H., 387
Kosters, H., 541
Koteva, K. P., 463
Kotowska, M., 126, 148
Kott, P., 583
Koumoutsi, A., 210
Kouno, K., 404
Kovacevic, S., 417, 418, 419, 420
Kraal, B., 122, 124, 125, 132, 133
Krabben, P., 122, 124, 125, 132, 133
Kramer, N. E., 535
Kraus, F., 489
Krebs, I., 81, 83
Kreisberg-Zakarin, R., 414, 415
Kremer, L., 126
Kretz, P. L., 161
Kriek, G. R., 489
Krismer, B., 25
Kristiansen, K. N., 409
Kroening, K. D., 99
Krogh, A., 183
Kroppenstedt, R. M., 463
Krug, D., 59, 81, 82, 83, 84, 86, 174, 355
Kruger, R. G., 282, 283, 302, 357, 357
Krull, R. E., 538
Krumm, H., 260
Kryshtafovych, A., 182
Ku, J., 259
Kubicka, S., 63
Kubista, M., 106
Kucherov, H., 338
Kuczek, K., 126, 148
Kudelski, A., 50
Kuhlmann, S., 174
Kuhstoss, S., 289
Kuipers, A., 538, 541, 561
Kuipers, O. P., 535, 536, 538, 539, 541, 542, 548, 550, 560, 561, 563, 564, 566
Kumada, Y., 145
Kumagai, T., 44
Kumasaka, T., 313
Kunes, S., 594
Kunnari, T., 285, 288
Kunze, B., 62, 81
Künzel, E., 284, 285, 288, 298, 310, 325, 326
Kupke, T., 539, 541, 553, 554, 560, 561, 564
Kurnasov, O., 385
Kuroda, M., 37, 53, 489
Kuron, G., 37, 53
Kurosawa, K., 387

Kuscer, E., 194
Kutchan, T. M., 489
Kvalheim, O. M., 84
Kyung, Y. S., 109

L

LaCelle, M., 256, 258, 265, 382
La Clair, J. J., 225, 241, 258, 259, 265
Lacroix, P., 151, 152
Lafargue, F., 584
Lai, J. R., 256, 525, 526, 528
Laing, E., 108, 111
Lairson, L. L., 327
Laiz, L., 407, 412, 413, 415, 417, 418
Lam, K. S., 8, 41, 49
Lamb, D. C., 489
Lamb, S. S., 463
Lambalot, R. H., 221, 243, 256, 257, 258, 265, 382
Lancini, G., 25
Landman, O., 412, 414
Landreau, C. A., 370
Lang, C., 283
Lang, E., 66
Lange, C., 151
Lange, E., 84
Langenhan, J. M., 313
Lantz, A. E., 258, 479
Larif, M., 517
Larkin, J. M., 64
Larsen, T. O., 9
La Teana, A., 14, 16
Lau, J., 392
Laubinger, W., 383, 385
Lautru, S., 396, 434, 440, 472
Lavine, B. K., 84
Lawen, A., 50
Lawlor, E. J., 125
Lawton, E. M., 538, 541, 563
Lazzarini, A., 31, 32, 33, 35, 36, 357, 359, 463, 468, 469, 560
Leadlay, P. F., 194, 203, 279, 283, 284, 289, 302, 363, 365, 367, 463, 471, 583
Leboul, J., 489
Leclère, V., 338
Ledyard, K. M., 439
Lee, H. J., 415, 418, 419, 427
Lee, I. K., 313, 316, 317, 318
Lee, J., 358
Lee, J. Y., 448
Lee, M. V., 546
Lee, P. C., 126, 127
Leeck, C. L., 99
Leeflang, C., 119
Leenders, F., 357
Leenhouts, K., 538, 541, 548, 550, 561
Leese, R. A., 354, 365

Legendre, F., 38
Legrand, R., 283
Lei, L., 489
Lei, M., 257, 258, 491
Lei, Y., 280
Leibundgut, M., 234
Leimkuhler, C., 282, 283, 317, 321, 323, 357
Leitao, A. L., 421
Lemieux, S., 13
Lengeler, J. W., 123
Lennard, N. J., 365, 462, 470, 471, 477
Lepore, B. W., 355
Leskiw, B. K., 113, 410, 412, 413
Lesniak, J., 326
Lester, D. R., 421, 422
Leung, Y. L., 542, 569
Levengood, M. R., 541, 542
Levine, D. P., 513
Levy, S., 396
Lewandowska-Skarbek, M., 289
Lewin, R. A., 583, 584
Lewkiw, B. K., 423
Li, B., 533, 538, 539, 545, 548, 551
Li, D. B., 473, 475, 487, 489, 490, 491, 492, 499, 502, 503, 505, 506
Li, J., 355
Li, L., 196, 313, 315, 316, 318, 326
Li, R., 414
Li, S. M., 122, 173, 174, 289
Li, T.-L., 363, 365, 367, 463, 469, 470, 471, 513
Li, W., 129, 257, 357
Li, W. C., 129
Li, X., 259, 260, 265, 270, 517
Li, Y., 108
Li, Y. S., 470
Li, Y. Z., 48
Lian, W., 109
Liao, J., 313, 372, 479
Liaw, S. H., 419
Licari, P., 48, 392
Liesegang, H., 210
Lih, C. J., 109, 111, 126
Lill, R. E., 194, 289
Lin, A. J., 259, 260, 263, 265, 267, 268, 269
Lin, H., 433
Lin, L., 257
Lind, K., 106
Lindberg, A. A., 280
Lindqvist, L., 280
Ling, X. B., 23
Link, A. S., 513
Linke, D., 151
Linne, U., 339, 341, 343, 349, 354, 360, 362, 364, 409, 528
Linnenbrink, A., 317, 322
Linteau, A., 13, 14
Lipata, F., 283, 284, 326
Lipavská, H., 385

Lipman, D. J., 130, 185, 189
Lipmann, F., 451
Lipscomb, S. J., 427
Liras, P., 401, 404, 405, 407, 408, 410, 411, 412, 413, 415, 417, 418, 419, 420, 421, 422, 423
Liu, D. R., 433, 526, 528
Liu, F., 260, 265, 269, 270
Liu, H., 48, 382, 444, 447, 448, 450, 452, 453
Liu, H. W., 278, 280, 282, 283, 284, 285
Liu, L., 256, 259, 380, 381
Liu, T., 336
Liu, V. G., 14
Liu, W., 44, 46, 120, 160, 196, 257, 354, 364, 541, 562
Liu, W. C., 418
Liu, W. F., 48
Liu, X. C., 45
Liu, X. L., 48
Liu, Y., 236, 244, 246, 354, 362, 380, 381
Llewellyn, N. M., 470
Lloyd, M. D., 415, 418, 419, 427
Lodato, M. A., 536
Logan, R., 98, 130, 161, 289, 354, 359, 464, 515, 521, 565, 566
Logghe, M., 14
Löhr, F., 341, 342
Loke, P., 413, 414
Lolans, K., 118
Lombó, F., 277, 280, 282, 283, 285, 287, 288, 292, 298, 299, 300, 326
Lomovskaya, N., 45, 160, 291, 391
Long, P. F., 577, 587, 590
Loonen, I., 14
Loosli, H. R., 363
Lopez, M., 37, 53
Lopez, P., 466
López-Nieto, M. J., 412, 413
Lo Piccolo, L., 477
Lorenzana, L. M., 423
Losey, H. C., 282, 283, 313, 315, 354, 358, 365, 366, 430, 470
Losi, D., 14, 16, 31, 32, 33, 34, 40, 560
Losick, R., 302
Lovell, F. M., 354, 366
Lovett, S. T., 392
Low, L., 194
Lowden, P. A., 370
Lu, M., 280
Lu, W., 282, 283, 302, 313, 317, 321, 323, 357
Luengo, J. M., 413
Luesink, E. J., 564, 566
Lui, L., 479
Luiten, R. G. M., 124
Lünsdorf, H., 62, 80
Lunsford, R. D., 14
Luo, L., 358
Luperchio, S. A., 513
Lusong, L., 343, 349

Luzhetska, M., 310, 325
Luzhetskyy, A., 183, 195, 282, 284, 310, 311, 317, 319, 321, 322, 324, 325, 326
Lynch, S., 203
Lyubechansky, L., 413

M

Ma, H., 594
Mackay, J. P., 475
Macko, L., 14
MacLeod, J. K., 577
MacMahon, S. A., 444, 447, 450, 452, 453
MacMillan, S. V., 125, 127, 131
MacNeil, D. J., 45, 163, 174, 288, 565
MacNeil, T., 163, 174, 565
Madden, T. L., 183, 185, 385
Madduri, K., 45, 288, 407
Madon, J., 463
Maeda, A., 145
Maeda, H., 387
Maeda, K., 420, 422
Mafnas, C., 8
Magarvey, N. A., 49, 355, 357, 359
Magnusson, L. U., 95
Mahé, B., 447
Mahlert, C., 354, 357, 360, 521, 523, 528
Mahmud, T., 442
Mahr, K., 122, 123, 129
Maier, T., 234
Mainini, M., 34
Majer, F., 554, 560
Majka, J., 129
Maki, D., 513
Makino, M., 489
Maldonado, L. A., 25
Malpartida, F., 102, 119, 125, 126, 128
Maltseva, N., 448
Mancino, V., 515
Maniati, M., 118
Maniatis, A. N., 118
Mankin, A. S., 278
Mann, J., 37, 49, 50, 51
Mäntsälä, P., 285, 288
Manzoni, M., 37, 49, 50, 51
Mao, Y., 43
Maplestone, R. A., 475
Marahiel, M. A., 174, 186, 187, 188, 206, 220, 224, 256, 257, 258, 259, 264, 266, 337, 338, 339, 341, 342, 343, 344, 345, 346, 347, 349, 355, 356, 357, 358, 360, 361, 362, 382, 383, 409, 432, 433, 435, 450, 471, 480, 490, 513, 521, 523, 528
Marazzi, A., 31, 32, 33
Marchler-Bauer, A., 185
Marconi, G. G., 354, 365, 462
Marinelli, F., 29, 31, 32, 33, 34, 35, 36, 40, 560
Marki, F., 536

Marki, W., 536
Markus, A., 535
Marshall, C. G., 206, 208, 343, 358, 366, 463, 475
Marta, C., 14
Martin, C. J., 289
Martin, D. P., 226
Martin, J. F., 121, 122, 210, 405, 407, 408, 410, 411, 412, 413, 415, 417, 418, 419, 420, 421, 422, 423
Martin, J. H., 354, 365
Martin, N. I., 569, 570
Martin, S., 354, 359, 515, 516, 517, 518, 519, 521, 522, 524, 527
Martinez, A., 129, 130
Martinez, E., 128
Martinez, J. L., 31
Martinez-Fleites, C., 313
Martin-Triana, A. J., 128
Maseda, H., 111
Mata, J. M., 402
Mathew, R., 257
Matoba, N., 473, 489, 490, 491, 492, 499, 502, 505, 506
Matsumoto, A., 127, 489
Matsushima, P., 163, 463, 477, 479, 521
Matsuzaki, K., 489
Matthews, P., 513
Mattingly, C., 310, 325
Matzanke, B. F., 432
May, J. J., 339, 450
Mayer, A., 284
Mayer, M., 538
Mazodier, P., 172
Mazza, P., 9, 465
McAllister, K. A., 257
McAlpine, J. B., 66, 183, 193, 203
McAuliffe, O., 541, 565
McCafferty, D. G., 357
McCafferty, J., 270
McCallum, C. M., 162
McClerren, A. L., 538, 539, 542, 545, 546
McCormick, A., 336
McCoy, J. G., 302
McDaniel, R., 291, 326
McDowall, K. J., 128
McGahren, W. J., 354, 365
McGilvray, D., 419
McGinnis, S., 183, 385
McHenney, M. A., 514
McKenzie, N. L., 103, 117, 125, 127, 128, 131
McLoughlin, S. M., 225, 259, 260, 267, 268, 470
McMahon, A. P., 261
McMahon, M. D., 338, 427
McManis, J. S., 448
McPhail, A. T., 37, 47
Mehra, S., 109
Meier, J. L., 219, 228, 229, 243, 244, 245, 246, 260

Meier, S., 258
Meijer-Wierenga, J., 561
Meiser, P., 60, 442
Melançon, C. E., 282
Memmert, K., 62, 369, 370
Men, A., 583
Menche, D., 63
Mendez, C., 47, 310, 311, 317, 319, 321, 325, 326
Méndez, C., 279, 282, 283, 284, 285, 287, 288, 289, 290, 291, 292, 293, 297, 298, 299, 300, 302, 310, 321, 326, 414, 423
Menéndez, N., 290, 302, 321
Ménez, C., 277, 283, 302
Menges, R., 478
Menzella, H. G., 380, 381, 392
Meray, R. K., 206, 208
Mercer, A. C., 221, 248, 256, 260
Mersinias, V., 109, 126, 130, 145, 148, 152, 163, 359, 363, 463
Mertz, F. P., 523
Messer, W., 14
Messing, J., 286
Metzger, J. W., 364, 539, 541, 554, 560, 564, 569
Metzger, S., 14
Mevarech, M., 413
Meyer, H. E., 569
Meyer, O., 473, 489, 490, 491, 492, 499, 502, 503, 505, 506
Miao, V., 338, 354, 357, 358, 359, 512, 513, 515, 516, 517, 518, 519, 520, 521, 522, 523, 524, 525, 526, 527
Michel, J. M., 284, 291
Micklefield, J., 120, 260, 353, 354, 357, 359, 360, 362, 363, 364, 521, 523, 525, 576
Miethke, M., 432, 433, 435
Migita, A., 380, 381, 385, 387, 388, 391, 395
Miller, A. K., 404
Miller, D. A., 358
Miller, J. R., 417, 418, 419, 420
Miller, L. M., 542, 545, 569
Miller, M. T., 414
Miller, T. W., 404
Miller, V. P., 280
Miller, W., 130, 185, 189
Mills, J. A., 14
Mills, J. H., 260, 265, 270
Milnamow, M., 48
Milne, C., 354, 357, 359, 364, 523, 525
Milne, J. C., 539
Milon, P., 25
Minami, A., 313, 315, 316, 319
Minami, E., 123
Mincer, T. J., 8
Mirmira, R. G., 259
Mironenko, T., 363, 366, 463, 471
Mishra, K. P., 12
Mittag, M., 257, 258

Mittag, T., 341
Mittler, M., 313
Miyamoto, K., 123
Miyashita, K., 123
Mochales, S., 402, 404
Moe, A. L., 448
Moffitt, M. C., 193, 194, 200, 359
Mofid, M. R., 257, 259, 341, 342
Mohanty, D., 187, 194, 396
Mokhtarzadeh, M., 312, 327
Molinas, C., 478
Molitor, E., 535
Moll, G. N., 538, 541, 545, 548, 550, 551, 561
Molle, V., 109, 111, 126
Molloy, R. M., 512
Molnar, I., 48
Mommaas, A. M., 123, 465
Monciardini, P., 3, 8, 9
Montanini, N., 32, 33
Montavon, T. J., 355
Mooberry, S. L., 49
Moore, B. A., 354, 359
Moore, B. S., 365, 489
Moore, M. L., 436
Moore, R. E., 49
Mootz, H. D., 186, 187, 188, 257, 339, 341, 345, 347, 354, 355, 381, 382, 383, 471
Moreno, M. A., 419
Morera, S., 313
Mori, I., 430
Morikawa, M., 210
Moriya, T., 355
Morrison, K., 569
Mortensen, K. K., 389
Morton, G. O., 354, 365
Morton, L. I., 513
Moser, H., 369, 370
Motamedi, H., 289, 290, 516, 519, 520, 521
Moult, J., 182
Moura, R. S., 121
Moyed, H. S., 451
Mueller, L. V., 118
Mukherjee, J. S., 118
Mukherji, M., 427
Mulder, N. J., 185
Mulder, S., 96, 120, 122, 133
Mulichak, A. M., 313
Muller, M., 122, 123, 133
Muller, R., 174, 175, 234, 383
Müller, R., 59, 60, 62, 63, 64, 65, 66, 68, 69, 79, 80, 81, 82, 83, 84, 86, 87, 355, 442
Mulyani, S., 468
Mulzer, J., 62
Murakami, T., 355, 465
Murli, S., 395
Murray, B. E., 460
Murrell, J. M, 206, 208
Muth, G., 463, 465, 466, 469, 471, 478

Mutka, S. C., 380, 381
Muyrers, J. P., 160, 161
Myers, E. W., 130, 185, 189
Mynderse, J. S., 523

N

Nadkarni, S. R., 463
Nagano, S., 489
Nagao, J.-I., 566
Nagarajan, R., 402
Nagashima, M., 326
Nagata, Y., 123
Nair, S. K., 538, 545, 548, 551
Naismith, J. H., 362, 444, 447, 450, 452, 453
Nakagawa, T., 150
Nakaho, M., 588
Nakamura, K., 150
Nakamura, T., 489
Nakano, M. M., 257, 258, 491
Nakao, K., 489
Nakayama, J., 566
Nakayama, S., 149, 156
Nango, E., 313
Narbad, A., 538
Naruse, N., 489
Navani, N., 108
Neary, J. M., 354, 357, 360, 362, 363, 364
Nedal, A., 282
Neidle, S., 391
Neilan, B. A., 193, 194, 200, 256, 259, 365, 375, 583
Neilands, J. B., 442
Nelissen, B., 14
Nelson, J. T., 358, 577, 578, 587
Nelson, K. E., 396
Nelson, W., 396
Nes, I. F., 539, 552
Neu, J. M., 359, 463, 475, 478
Neumann, C. S., 174, 358
Neumann, P., 339
Newman, D. J., 37, 40, 41, 45, 48, 53, 311
Ng, C. P., 413
Nguyen, K. T., 151, 152, 359, 516, 517, 519, 520, 521, 522, 524, 526, 527
Nguyen, L. T., 151, 152
Ni, W., 547
Nicholson, G. J., 364, 366, 367, 368, 369, 446, 466, 469, 472, 473, 474, 475, 476, 479, 480, 489
Nickeleit, I., 63
Nie, L., 364
Nieboer, M., 419
Niederweiss, M., 122, 123, 129
Nielsen, J., 122, 258, 409, 479
Nielsen, K. F., 9
Nielsen, M. L., 259
Nielsen, R. C., 462
Nielsen, R. V., 354, 359

Niemi, J., 128
Nierman, W. C., 86
Nihira, T., 109, 126, 130, 144, 145, 146, 148, 149, 150, 151, 152, 154, 156
Nikolov, D. B., 326
Nishio, M., 489
Nishiyama, M., 50, 145
Nitschko, H., 9
Nodwell, J., 108, 128
Nodwell, J. R., 103, 117, 125, 127, 128, 131, 536
Noel, J. P., 224
Noens, E. E., 122, 123, 133, 465
Nolan, R. D., 387
Nonaka, K., 46, 120, 364
Nordsiek, G., 81
Northland, R., 513
Nothaft, H., 122, 123, 129, 133
Nougayrède, J. P., 381
Novak, J., 538, 562, 569
Novelli, G. D., 451
Novikov, V., 321, 325, 326
Nowak-Thompson, B., 48
Noyori, R., 494
Ntai, I., 193
Núñez, L. E., 414, 423
Nur-e-Alam, A. F., 321
Nur-e-Alam, C., 321
Nur-e-Alam, M., 290, 302
Nusca, T. D., 448
Nussbaumer, B., 465
Nustrom, T., 95

O

Obayashi, A., 489
Oberthür, M., 282, 283, 302, 317, 321, 323, 354, 357, 364, 370
Oberthür, W. L., 363
Obraztsova, A. Y., 576
O'Brien, D. P., 472, 476, 492
O'Brien, E., 354, 365
O'Brien, K., 98, 130, 161, 289, 464, 515, 521, 565, 566
O'Brien, S. W., 365, 472, 476, 492
Occi, J. L., 45, 288
Occolowitz, J. L., 529
Ochi, K., 121
O'Connell-Motherway, M., 564, 565
O'Connor, P. M., 541, 563, 564
O'Connor, S. E., 355, 357, 367, 377, 385, 468
Odermatt, A., 182
O'Driscoll, J., 564, 565
Oelgeschläger, M., 420
Oelkers, C., 283, 292, 300
Oestreicher, N., 174
Offenzeller, M., 369, 370
Ogino, T., 489
Oguri, H., 380, 381, 385, 387, 388, 391, 395
Oh, D.-C., 25, 354, 359

Author Index

Oh, T. J., 160
Ohnishi, Y., 96, 113, 119, 120, 126, 127, 128, 130, 145, 151, 489
Ohnuki, T., 489
Ohuchi, T., 326
Oikawa, H., 380, 381, 385, 387, 388, 391, 395
Ojanperä, I., 81, 83
Ojanperä, S., 81, 83
Oka, M., 489
Okabe, M., 404
Okakura, K., 355
Okamoto, R., 404
Okamoto, S., 150
Okamura, K., 404
Okanishi, M., 326
Oke, M., 444, 445, 446, 447, 450, 452, 453
Okeley, N. M., 539, 545, 548
Okey, A., 110
Oki, T., 489
Okuda, K.-I., 566
Okuhara, M., 50
Okumura, Y., 404
O'Kunewick, J. P., 278
Olano, C., 277, 282, 283, 289, 290, 291, 298, 302
Oleson, F. B., Jr., 512
Oliynyk, M., 203
Olsen, R. K., 383
Omura, S., 111, 119, 120, 130, 489
Omura, T., 504
Onaka, H., 113, 144, 145, 150, 151, 152, 489, 588
Onishi, M., 343, 349
Ono, M., 48
Oppermann, U., 257
O'Rourke, S., 96, 99
Osburne, M. S., 129, 130
Öster, L. M., 421, 422
Osterman, A., 385
Oswald, E., 381
Otoguro, M., 9
Ott, J. L., 512
Otten, S. L., 45
Ottenwälder, B., 539, 541, 548, 564
Overbye, K., 594
Oves-Costales, D., 444, 447, 448, 450, 452, 453
Ozanick, S. G., 355, 357

P

Paccanaro, A., 14
Pace, N. R., 396
Pacey, M. S., 387
Padmanabhan, S., 73
Paget, M. S., 163, 167, 172, 478
Painter, R., 25, 119
Paitan, Y., 34
Palissa, H., 428

Palmer, M. A., 226
Pan, K. H., 109, 126
Papageorgiou, E. A., 359
Papiska, H. R., 523
Paquet, A., 495
Paquette, J., 14
Parche, S., 123
Parenti, F., 560
Park, C., 53
Park, S. H., 280, 284
Parr, I., 354, 359, 515, 516, 517, 518, 521
Parris, K. D., 257
Parthasarathy, G., 25, 119
Pascalle, G., 566
Passalacqua, K. D., 448
Pasta, S., 257
Patallo, E. P., 283, 290, 292, 300, 302, 317, 326
Patel, D. J., 279
Patel, H. M., 358, 433, 436, 437
Patel, M. V., 463
Patel, N. R., 383
Patel, T., 463
Patterson, G. M., 49
Patton, G. C., 538, 541, 545, 546
Paul, M., 534, 538, 541, 545, 546, 548, 559, 560
Paulsen, I., 396
Paw, B. H., 442
Pawlik, K., 126, 148
Payne, D. J., 4, 118
Pearl, L. H., 391
Pearson, W. R., 189
Peduzzi, J., 539, 576
Peiru, H. M., 380, 381, 392
Pelaez, F., 38
Pelander, A., 81, 83
Pelzer, S., 284, 358, 365, 366, 367, 370, 463, 464, 466, 468, 469, 470, 471, 472, 473, 474, 475, 476, 477, 479, 480, 489, 490, 492, 499, 502
Pelzing, M., 81, 83
Peng, C., 196
Penketh, P. G., 43
Penn, J., 354, 359, 515, 516, 517, 518, 519, 521, 522, 524, 527
Pérez, M., 279, 283, 298, 299, 300, 326
Pérez-Llarena, F. J., 420
Pérez-Redondo, R., 423, 464, 465
Perham, R. N., 194, 201, 257
Perlova, O., 60, 64, 66, 79, 81, 174, 175
Pernodet, J. L., 174, 472
Perrakis, A., 415
Pertel, P. E., 513
Peschel, A., 548
Peter, H. H., 536
Peters, N. R., 311
Petersen, F., 62
Petinaki, E., 118
Petkovic, H., 194
Petrich, A. K., 410, 412, 413

Petter, R., 172
Pfeifer, B. A., 259, 381, 391, 480
Pfeifer, N., 84
Pfeifer, V., 366, 367, 368, 369, 466, 469, 474, 475, 479
Pfennig, F., 45, 383, 385, 471, 472, 474
Pfleger, B. F., 448
Phelan, V. V., 175, 193
Phillips, G. N., 302
Phillips, G. N., Jr., 313, 315, 316, 317
Phillips, J. W., 14
Picart, F., 536
Pick, H., 260
Pidot, S. J., 583
Pieper, R. L., 356, 512
Pietrokovski, S., 162
Piraee, M., 183, 193, 203
Piret, J., 128
Piret, J. M., 419
Platas, G., 38
Pliner, S. A., 144
Ploux, O., 280
Podevels, A. M., 338, 433
Podust, L. M., 489
Pohlmann, V., 354, 360, 356
Polacco, M. L., 258
Polanuyer, B., 385
Polishook, J. D., 38
Polyakov, M., 118
Pomati, F., 583
Pompliano, D. L., 4, 118
Pon, C. L., 25
Pootoolal, J., 359, 463, 475
Pope, M. K., 128
Porter, J. L., 583
Postma, P. W., 123
Poteete, A. R., 161
Powell, A., 354, 357, 359, 360, 362, 363, 364
Powers, J. C., 235
Powers, J. H., 118
Pradella, S., 60, 64, 66, 79
Prado, L., 285, 287, 288
Praphanphoj, V., 256
Praseuth, A. P., 379, 380, 381, 385, 387, 388, 391, 395
Praseuth, M. B., 379, 395
Price, B., 128
Proctor, M. R., 313
Pruess, M., 185
Puglia, A. M., 161, 464, 477
Pühler, A., 463, 465
Puk, O., 358, 463, 464, 468, 471, 472, 473, 474, 475, 477, 479, 480
Pupin, M., 338
Pylypenko, O., 473, 489, 490, 492, 499, 502

Q

Qi, F. X., 538, 562
Qiao, C., 236

Quadri, L. E., 222, 256, 257, 258, 433, 436, 491
Quail, M. A., 119, 120, 130, 161
Quan, C., 538, 539, 542, 545
Que, N. L. S., 280
Queener, S. W., 419, 420
Quesniaux, V. F., 37
Quigley, G. J., 385
Quillardet, P., 578, 581, 588
Quinn, J. P., 118
Quirós, L. M., 289

R

Rachid, S., 355
Rajgarhia, V., 526, 527
Ramaswamy, S., 415
Ramos, A., 292
Ramos, F., 413
Ramos, F. R., 412
Rance, M. J., 387
Rangan, V. S., 256, 257
Rao, N. R., 289
Rao, R. N., 98, 130, 161, 289, 464, 515, 521, 565, 566
Rapoport, I. A., 144
Raschdorf, F., 536
Rasko, D. A., 577, 578, 587
Rath, C. M., 448
Räty, K., 285, 288
Rausch, C., 188, 207, 209, 341, 466, 471, 472
Ravel, J., 181, 186, 187, 188, 209, 380, 440, 471, 576, 577, 578, 580, 581, 584, 586, 587, 588, 589, 590
Raymond, K. N., 435, 436
Raynal, M. C., 283, 284, 291
Raynham, T., 545
Reading, C., 404
Ready, S. J., 289
Reather, J., 279, 283
Rebets, Y., 183, 195
Rebuffat, S., 539, 576
Recktenwald, J., 358, 359, 366, 367, 368, 463, 464, 466, 468, 469, 470, 471, 472, 474, 475, 477, 479
Reddy, L. M., 489
Redenbach, M., 161, 170
Reed, R. R., 112
Reeves, P. R., 280
Regan, C. M., 258, 260, 265
Reichenbach, H., 34, 37, 48, 60, 61, 62, 63, 64, 65, 66, 69, 72, 73, 74, 77, 78, 79, 80, 81
Reichert, W., 366, 463
Reicke, A., 25
Reid, C. H., 161
Reid, D. G., 383
Reid, R., 256, 258, 265, 382
Reif, E., 109
Reinecke, K., 182
Reinemann, C., 112

Reinert, K., 84
Reinhardt, R., 357
Reis, M., 562
Reizer, J., 123, 132
Rembold, M., 362
Remington, K., 396
Remsing, L. L., 284, 285, 288, 326
Renner, C., 341
Retzlaff, L., 477
Reuter, K., 257, 259, 341
Reuther, J., 465
Revermann, O., 81, 82, 83, 84, 86
Revzin, A., 107
Reynolds, K. A., 37, 51
Reynolds, P. E., 535
Rezai, K., 118
Rich, A., 385
Rich, M. L., 118
Rich, P., 515, 516, 519, 522, 524
Richardson, M. A., 289
Richardson, T. I., 472, 476, 492
Riechenbach, H., 62
Riedlinger, J., 25
Ries, J., 354, 366, 367, 368, 466, 469, 474, 479
Rigali, S., 96, 120, 122, 123, 124, 125, 132, 133
Riggle, P., 125, 127, 128
Riggle, P. J., 125, 127
Rinehart, K. L., 436
Ring, M. W., 65, 69
Ringe, D., 355
Rini, J. M., 312
Rink, R., 538, 541, 548, 550, 561
Rinkel, M., 151, 152
Rippka, R., 35
Ritter, S. C., 230
Ritz, D., 516, 519, 520, 522, 526, 527
Rius, N., 422
Rivera, H., Jr., 260
Rivola, G., 45, 288
Rix, U., 284, 285, 325, 326
Roach, P. L., 414, 415
Roberts, A. A., 259
Roberts, G. A., 302
Roberts, G. C. K., 391
Roberts, S., 313
Robinson, J. A., 259, 464, 473, 475, 487, 489, 490, 491, 492, 499, 502, 503, 505, 506
Robinson, L., 470
Roche, E. D., 526, 528
Rock, Y. M., 256
Rodríguez, A. M., 283, 291, 292, 293, 297, 298, 300, 302
Rodríguez, D., 298, 302
Rodriguez, E., 380, 381, 392
Rodríguez, L., 282, 283, 290, 291, 326
Rodríguez-García, A., 404, 423, 464, 465
Roemer, T., 14
Rogov, V. V., 341, 342

Rohde, M., 8, 9
Rohr, J., 47, 279, 282, 283, 284, 285, 287, 288, 290, 291, 292, 298, 299, 300, 302, 310, 317, 319, 321, 325, 326, 336
Rolinson, G. N., 404
Roll, J. T., 291
Rollema, H. S., 538, 539, 541, 563
Rollini, M., 37, 49, 50, 51
Rollins, M. J., 409, 417
Romero, J., 420
Ron, E. Z., 34
Ronning, C. M., 86
Roongsawang, N., 210
Rose, T. M., 162, 322
Rosenberg, M., 14
Rosenstein, R., 562, 564, 565
Rosenzweig, A. C., 414
Rosovitz, M. J., 578, 580, 581, 584, 586, 587, 590
Ross, P., 569, 570
Ross, R. P., 536, 538, 541, 560, 563, 564, 565
Rost, B., 182
Rosteck, P., 419
Rosteck, P. R., Jr., 514
Roth, R. A., 196
Rothrock, J., 37, 53
Röttgen, M., 471, 478
Rouse, S., 489, 490, 492, 499, 502
Rouset, S., 473
Roy, K. L., 413
Roy, R. S., 539
Rubinstein, E., 118
Rübsamen-Waigmann, H., 14
Ruby, C. L., 163, 174, 565
Rudd, B. A., 119, 194, 354, 359, 364, 523, 525, 560, 565
Ruddock, J. C., 387
Rude, M. A., 380, 381
Ruengjitchatchawalya, M., 150
Ruger, W., 313
Ruggiero, C. E., 448
Rumbero, A., 407, 408
Rupp, M. E., 513
Rupprath, C., 313, 315
Rusch, D., 396
Russell, D. W., 104, 161, 162, 163, 166, 171, 391
Ruzicka, F. J., 355
Ryan, M. P., 536, 560
Ryding, J., 95, 108, 111, 121
Ryding, N. J., 125, 126, 127, 131

S

Sacksteder, K. A., 256
Saenger, W., 362
Sahl, H.-G., 535, 536, 560, 562
Saier, M. H., Jr., 123, 132
Sairam, M., 12
Saito, A., 123, 129

Sakai, A., 385
Sakai, H., 404
Sakaki, Y., 119, 120, 130
Sakharkar, K. R., 13
Sakharkar, M. K., 13
Sakuda, S., 145
Salah-Bey, K., 284
Salas, A. F., 321
Salas, A. P., 47, 283, 292, 300, 302, 319, 326
Salas, J. A., 47, 277, 279, 282, 283, 284, 285, 287, 288, 289, 290, 291, 292, 293, 297, 298, 299, 300, 302, 310, 311, 317, 319, 321, 325, 326, 414, 423
Salerno, P., 161, 464
Salto, F., 413
Sambrook, J., 104, 161, 162, 163, 166, 171, 391
Samel, S. A., 206, 343, 344, 345, 347
Samson, S. M., 420
Sánchez, C., 43, 44, 47, 206, 208, 259, 283, 292, 300, 302, 310, 319, 326
Sandercock, A. M., 365, 469
Sandmann, A., 63, 80, 194
Sanford, R., 74
Sano, H., 489
Sano, S., 489
Santamarta, I., 422, 423
Santer, G., 369, 370
Santi, D. V., 256, 259, 380, 381, 433, 442
Santos, R. A., 125, 127
Saris, P. E., 548
Sarubbi, E., 32, 33
Sashihara, T., 566
Sasse, F., 62, 63, 80
Sastry, M., 279
Sato, F., 489
Sato, K., 145
Sato, R., 504
Sattely, E. S., 477, 515
Sawhney, R. C., 12
Sawitzke, J. A., 161
Scaglione, J. B., 448
Scaife, W., 365
Schäberle, T., 478
Schafer, A., 257, 258, 259, 338
Schäfer, B., 341
Schaffer, A. A., 185
Schauer-Vukasinovic, V., 62
Schauwecker, F., 45, 383, 385
Scheck, M., 182
Schefer, A. B., 354, 366, 367, 368, 466, 469, 474, 479
Schegg, T. R., 258, 259, 265
Scheuer, P. J., 576, 577
Schewecke, T., 428
Schimana, J., 489
Schinke, M., 259, 260, 265, 269
Schipper, D., 419
Schirle, M., 473, 474, 476, 489
Schirner, K., 465
Schley, C., 87
Schlicht, M., 122, 123, 133
Schlichting, I., 473, 489
Schlösser, A., 123, 129
Schlumbohm, W., 383, 385
Schmelz, S., 444, 447, 450, 452, 453
Schmid, D. G., 446, 489, 554, 560
Schmidt, E. W., 182, 358, 380, 575, 576, 577, 578, 580, 581, 583, 584, 586, 587, 588, 589, 590
Schmidt, M., 363
Schmiederer, T., 355, 475, 480
Schmitt, B., 284
Schmitt-John, T., 130
Schmitz, A., 107
Schmitz, S., 560, 565
Schmoock, G., 383, 385
Schnaitman, C. A., 278
Schneider, A., 341, 480
Schneider, B., 81, 83, 84, 86
Schneider, K, 480
Schneider, U., 535, 536, 538, 562, 564, 565
Schneider-Scherzer, E., 369, 370
Schneiker, S., 81
Schnell, H. J., 175
Schnell, N., 536, 538, 562, 564, 565
Schobert, M., 257, 258, 259, 338
Schofield, C. J., 404, 414, 415, 418, 419, 427
Schönafinger, G., 343, 344
Schoner, B., 302
Schoner, B. E., 98, 130, 161, 289, 302, 464, 515, 521, 565, 566
Schorgendorfer, K., 258
Schreier, M. H., 37
Schrempf, H., 123, 129
Schröder, J., 354, 366, 367, 368, 466, 469, 474, 479
Schuffenhauer, A., 182
Schultz, A. W., 354, 359
Schultz, E. R., 162
Schultz, P. G., 260, 265, 270
Schulz, G. E., 313
Schulz-Trieglaff, O., 84
Schumacher, T., 313, 315
Schumann, P., 8, 9, 66, 463
Schupp, T., 48, 62
Schuppe-Koistinen, I., 84
Schwarzer, D., 338, 341, 345, 383
Schweizer, E., 257, 258
Scott, J. J., 25
Scrutton, N. S., 194, 201
Seehra, J., 257
Seeman, T., 583
Sehgal, S. N., 50
Seibert, G., 463
Seifert, A., 583
Seitz, H., 363

Sello, J. K., 128
Selva, E., 31, 32, 33, 560
Semenova, E., 539
Senn, H., 363
Seno, E. T., 98, 130, 161, 289, 464, 515, 521, 565, 566
Seo, M. J., 38
Serhat, A., 583
Seringhaus, M., 14
Serpe, E., 420
Serre, L., 187
Seto, H., 489
Seufert, W. H., 49
Severinov, K., 539
Seyler, R., 128
Shafiee, A., 289, 290, 516, 519, 520, 521
Shah, A. N., 382, 466
Shah, S., 48
Shapiro, S., 37
Shawky, R. M., 354, 366, 367, 368, 463, 466, 469, 471, 472, 474, 477, 479
Shaw-Reid, C. A., 436
She, K., 357, 358, 522, 523, 524, 525
Sheeler, N. L., 125, 127, 131
Sheldon, P. J., 126
Shelley, J. T., 448
Shen, B., 41, 43, 44, 46, 120, 160, 195, 196, 206, 208, 257, 259, 355, 357, 370
Shen, Y. Q., 413, 426
Shepard, E., 160
Sheridan, R. M., 289
Sherman, D. H., 43, 49, 109, 125, 127, 193, 283, 285, 448
Shetty, R. S., 108
Shevchenko, L. N., 144
Shi, J., 109, 111, 126
Shiau, C. Y., 419
Shiba, T., 119, 120, 130
Shibamoto, N., 404
Shibata, N., 414, 415
Shibuya, N., 123
Shiffman, D., 412, 413
Shigematsu, N., 489
Shimaji, M., 385
Shimauchi, Y., 404
Shimizu, R., 489
Shimizu, Y., 145
Shimkets, L., 61, 65, 66, 72, 73, 74, 78, 79
Shin, S. S., 118
Shinose, M., 119, 120, 130
Shinya, T., 123
Shipman, M., 370
Shirahata, K., 489
Shiro, Y., 489
Shizuya, H., 515
Shoop, W., 25, 119
Short, J. M., 161
Shrestha, S., 108

Sidebottom, P. J., 313, 327, 523, 525
Sieber, S. A., 220, 224, 256, 259, 266, 338, 354, 355, 513, 528
Siebring, J., 122
Siems, K., 282, 283, 299, 300
Siezen, R. J., 536, 538, 539, 541, 542, 548, 563, 564, 566
Sikorski, R. S., 594
Silakowski, B., 81, 194
Sillaots, S., 13, 14
Silva, C. J., 354, 359, 515, 516, 517, 518, 521
Silver, P. A., 260, 261, 263, 267, 268, 269
Silverman, J. A., 513
Sim, J., 415
Sim, T., 413, 414
Sim, T. S., 415, 418, 419
Simon, D., 566
Simon, M., 515
Simon, P. J., 392
Simons, F. A., 14
Sims, J. W., 358
Sindelka, R., 106
Singh, G. M., 357
Singh, S., 302, 313, 315, 316, 317
Singh, S. B., 14, 38
Sizemore, C. F., 14
Sjöback, R., 106
Sjögreen, B., 106
Skarstad, K., 14
Skelly, J. V., 391
Skutlarek, D., 541, 563, 564
Slater, S. C., 86
Slepak, T., 515
Slisz, M. L., 420
Smale, T. C., 53
Smedsgaard, J., 9
Smirnov, A., 122
Smith, C. D., 49
Smith, C. P., 108, 109, 111, 122, 126, 130, 145, 148, 152, 163, 354, 357, 359, 360, 362, 363, 364, 463, 523, 525
Smith, J. L., 193, 535
Smith, L. M., 99
Smith, M. C., 161, 173, 464, 465
Smith, R. E., 448
Smith, S., 184, 220, 256, 257, 258
Smith, T. F., 189
Smith-Spencer, W., 108
Snader, K. M., 311
Snyder, M., 14
Socci, A. R., 118
Soeding, J., 151
Sohlberg, B., 109, 111, 126
Sohng, J. K., 284
Soisson, S. M., 25, 119
Sola-Landa, A., 121
Solenberg, P. J., 365, 462, 470, 471, 477, 479, 521
Soler, G., 419

Solsbacher, J., 257, 258, 259, 338
Somers, W. S., 257
Somu, R. V., 448
Song, J., 196
Song, L., 96, 444, 446, 447, 448, 450, 452, 453
Song, W. W., 196
Sonomoto, K., 566
Sørensen, H. P., 389
Sörensen, I., 63
Soriano, A., 326
Sosio, M., 3, 4, 8, 9, 16, 161, 363, 366, 463, 464, 466, 467, 468, 469, 471, 490
Sottani, C., 37
Speers, A. E., 229, 249
Spencer, J. B., 363, 365, 367, 463, 469, 470, 471, 492
Sperandio, B., 122
Spiegelman, G. B., 31
Spiteller, D., 224, 226
Sponga, F., 32, 37
Spröer, C., 64, 66
Spröter, P., 405
Spules, T., 569, 570
Stachelhaus, T., 186, 187, 188, 259, 266, 339, 341, 343, 347, 348, 354, 355, 381, 382, 383, 471, 480
Stack, D. R., 477, 479, 521
Stackebrandt, E., 64, 463
Staffa, A., 46, 120, 160, 204, 210
Stahelin, H., 37, 50
Stahl, M., 257
Stajner, K., 385
Stanier, R. Y., 35
Stapley, E. O., 402, 404
Stapon, A., 414
Stark, W. M., 402
Starks, C. M., 48
Staroske, T., 472, 476, 492
Stasser, J. P., 548
Stassi, D., 194, 201
States, D. J., 130
Staunton, J., 194, 220, 256, 279, 283, 289
Staver, M. J., 282
Stead, J. A., 128
Steele, J., 516, 519, 522, 524, 527
Stefanelli, S., 32, 33, 37
Stegmann, E., 358, 459, 463, 464, 466, 467, 468, 471, 472, 473, 475, 476, 477, 478, 479, 490
Stehmeier, P., 381
Stein, D. B., 225
Stein, T., 363, 538, 560
Stein, T. H., 539, 541, 560, 565
Steinbacher, S., 554
Steinerová, N., 385
Steinmetz, H., 62, 63
Stephanopoulos, G., 122
Stephens, K. E., 128

Steudle, J., 475
Stevanovic, S., 489
Stevanovich, S., 539
Stewart, A. F., 160, 161, 174, 175
Stierle, A., 48
Stierle, D., 48
Stinchi, S., 4, 363, 366, 463, 464, 468, 469
Stinear, T. P., 583
Stockert, S., 325, 464, 472, 473, 474, 476, 479, 480, 489
Stocks, S. M., 123
Stoltenburg, R., 112
Stoops, J. K., 236
Stoughton, D. M., 14
Straight, P. D., 260, 263, 265, 267, 268, 269
Stratigopoulos, G., 106
Stratmann, A., 48
Strauch, E., 95, 126
Street, G., 505
Strehlitz, B., 112
Strehlow, R., 109
Strieker, M., 174, 354, 357, 358
Strieter, E. R., 342
Strobel, G., 48
Strömbom, L., 106
Strynadka, N. C., 327
Stubbs, M. T., 339, 345
Stuible, H. P., 258
Stuitje, A. R., 187
Stuiver, M. H., 123, 132
Sturm, M., 84
Stuttard, C., 124, 407
Su, Z., 369, 370
Suda, H., 326
Sudek, S., 577, 578, 580, 581, 584, 586, 587, 588, 590
Sugimoto, H., 489
Sugiyama, M., 44
Suh, J. W., 38
Sumida, N., 355
Summers, R. G., 282
Sun, J., 95
Sun, X., 48
Sunbul, M., 255, 261, 266, 269
Sundaramoorthy, M., 489
Sunga, G. N., 66
Suo, Z., 259, 264
Süssmuth, R. D., 355, 358, 366, 370, 459, 463, 464, 468, 470, 471, 472, 473, 474, 475, 476, 477, 479, 480, 489, 490, 492, 499, 502
Sutera, V. A., Jr., 392
Sutherland, J. D., 415
Suzuki, H., 119, 120, 130, 145
Svenda, M., 421, 422
Swali, V., 545
Swart, R., 87
Szekat, C., 541, 560, 562, 563, 564
Szu, P. H., 282

Author Index

T

Taal, E., 124
Tachiiri, Y., 515
Tagawa, K., 505
Taguchi, T., 321, 325, 326
Tahlan, K., 423
Taieb, F., 381
Takahashi, H., 282
Takahashi, K., 387
Takahashi, S., 489
Takala, T. M., 548
Takano, E., 95, 96, 106, 109, 126, 130, 143, 144, 145, 146, 148, 149, 151, 152, 154, 156, 463, 477
Takano, Y., 126, 127
Takatsu, T., 45
Takesato, K., 489
Takeya, K., 489
Tally, F. P., 118, 513
Tam, A., 257
Tan, Y., 357
Tanaka, A., 126, 127
Tanaka, H., 489
Tanaka, K., 128
Tandeau de Marsac, N., 578, 581, 588
Tandia, F., 14
Tang, G. L., 195, 196, 354, 364
Tang, L., 48, 291, 326
Tang, M., 354, 370
Tang, Y., 234
Tang, Y. S., 25, 119
Tang, Z., 175
Tanovic, A., 206, 234, 347
Tanzawa, K., 37, 53
Tao, J., 302, 357, 430
Tao, M., 44, 129
Taton, A., 34, 40
Taylor, H. L., 37, 47
Taylor, W. R., 189
Terwissscha van Schltinga, A. C., 418, 421, 422
Testa, G., 160
Theilgaard, H. B., 409
Thieme, K., 327
Thirlway, J., 260, 354, 357, 360, 362, 363, 364, 521
Thoma, M. G., 359
Thomae, A. W., 96, 120, 122, 133
Thomas, C. R., 123
Thomas, G. H., 256
Thomas, M. G., 282, 315, 338, 354, 355, 357, 367, 368, 372, 433, 463, 469, 470, 475
Thomas, P. M., 538, 539, 542, 545
Thomason, L. C., 161
Thompson, A., 415
Thompson, C., 172
Thompson, C. J., 130, 151, 152, 465
Thompson, R. C., 477, 479, 521
Thompson, R. H., 53
Thomson, N. R., 119, 120, 130, 161
Thorne, G. M., 513
Thorpe, H. M., 161, 173
Thorson, J., 479
Thorson, J. S., 41, 43, 44, 46, 120, 283, 302, 311, 313, 315, 316, 317, 318, 319, 324, 326, 327, 328, 354, 372 479
Thum, C., 278
Thurson, L., 515, 516, 519, 522, 524
Thykaer, J., 122, 479
Tickoo, R., 120
Tiedje, J., 74
Tiemersma, E., 118
Tietz, A. J., 362, 512
Tiffert, Y., 465
Tillet, D., 365
Tillotson, G. S., 118
Timmis, K. N., 505
Tino, M. D., 118
Tipirneni, P., 118
Tiraby, G., 38
Titgemeyer, F., 96, 120, 122, 123, 124, 125, 129, 132, 133
Tobin, M. B., 417, 418
Tokura, M., 64, 68
Tomono, A., 96
Tone, J., 387
Toney, M. D., 259
Toogood, P. L., 545
Toppo, G., 35, 36
Torkkell, S., 285, 288
Torrance, S. J., 489
Torti, F., 45
Totschnig, K., 364
Toupet, C., 48
Tovarova, I. I., 144
Towle, J. E., 128
Townsend, C. A., 186, 187, 188, 209, 256, 414, 471
Traag, B. A., 122, 124, 125, 132, 133
Traber, R., 369, 370
Traitcheva, N., 196
Tramontano, A., 182
Tran, C., 392
Trauger, J. W., 345, 382, 383, 471, 472
Tredici, M. R., 34, 40
Trefzer, A., 283, 284, 290, 292, 298, 300, 310, 317, 325, 326
Tremblay, S., 183, 193, 203
Truman, A. W., 470, 471
Tsai, M. D., 470
Tsai, R. Y. L., 112
Tsai, S. C., 184, 220
Tsai, Y. C., 418, 419

Tse, W. C., 383
Tseng, C. C., 235, 367, 368, 470
Tsuda, E., 387
Tsuji, S. Y., 259, 264
Tsuji, T., 151, 152
Tsujibo, H., 123
Tüncher, A., 405
Tuphile, K., 174
Turkenburg, J. P., 313
Turner, J. R., 163, 302
Tuttle, L. C., 445

U

Uchida, R., 313, 315, 316, 319, 489
Uchiyama, M., 494
Udwary, D. W., 256, 354, 359
Ughetto, G., 385, 391
Uguru, G. C., 128
Ulijasz, A. T., 16, 17
Ullan, R. V., 415, 417
Ullmann, D., 12
Ullrich, C., 357
Umeyama, T., 126, 127
Unligil, U. M., 312
Unversucht, S., 362
Urban, A., 14
Usui, S., 415

V

Vagstad, A. L., 256
Vaillancourt, F. H., 355, 357, 358, 362
Valdebenito, M., 381
Valegard, K., 415, 418
Van, Z., 385
van Boom, J. H., 385
van den Berg, M. A., 119
van den Bogaard, P., 541
Van der Beek, C. P., 119
van der Berg, M. A., 419
van der Donk, W. A., 533, 534, 536, 538, 539, 541, 542, 543, 544, 545, 546, 547, 548, 551, 552, 559, 560, 569
van der Marel, G. A., 385
van der Meer, J. R., 538, 539
van der Meulen, J., 124
Vandervere, P., 125, 127, 128
Van Dijck, P. W. M., 119
van Döhren, H., 428
Vandoninck, S., 14
Vanek, Z., 385
van Frank, R. M., 420
Van Horn, S. F., 14
van Keulen, G., 122
van Kraaij, C., 535, 541
Van Lanen, S. G., 41, 43, 44, 46, 120, 160, 257, 364

van Liempt, H., 428
van Lun, M., 421, 422
van Oostrum, J., 536
van Pee, K.-H., 358, 368, 464, 468, 475
Vanpraagh, A. D., 513
Van Regenmortel, M. H., 37
van Scheltinga, A. C., 415
van Sinderen, D., 564, 565
van Wageningen, A. M. A., 365, 462, 470, 471, 477
van Wezel, G. P., 96, 117, 120, 122, 123, 124, 125, 132, 133, 465
Vara, J. A., 289
Varey, J. E., 505
Varoglu, M., 43
Vater, J., 210
Vaughan, M. D., 327
Vazquez-Laslop, N., 278
Veal, L. E., 420
Vederas, J. C., 53
Vederas, J. J., 569, 570
Veillette, K., 14
Velasco, J., 419
Velicer, G. J., 81, 82, 83, 84, 86
Veltkamp, C., 124
Venema, G., 357
Vente, A., 284, 317, 322, 480
Venter, J. C., 396
Verbree, E. C., 187
Verhasselt, P., 14
Vezina, C., 50
Viaene, J., 14
Vieira, J., 286
Vigliani, G. A., 513
Vijayakumar, E. K., 463
Vijgenboom, E., 122, 124, 125, 132, 133
Viklund, J. A., 418
Vilches, C., 311
Vinciotti, V., 126
Vining, L. C., 31, 38, 124, 407, 412, 413
Virolle, M. J., 121, 122
Viru, F. A., 118
Vitali, F., 233, 259, 473, 475, 489, 490, 491, 492, 499, 502, 503, 505, 506
Vivero-Pol, L., 260
Vogel, H., 260
Voges, R., 363
Vollebregt, A. W. H., 419
Vollmer, H., 326
Vollmer, W., 478
von Döhren, H., 357, 365
von Matt, P., 62
von Mulert, U., 319, 326
von Nickisch-Rosenegk, M., 109
von Thaler, A.-K., 478
Vos, M., 81, 82, 83, 84, 86
Vosburg, D. A., 343

Author Index 623

Vrijbloed, J. W., 473, 489, 490, 492, 499, 502
Vuori, E., 81, 83

W

Wackernagel, W., 169
Wagenaar, M., 357
Wagner, B., 345
Wagner, C., 258
Wagner, G., 342, 343
Wakarchuk, W. W., 327
Wakil, S. J., 236
Waldmann, H., 182
Waldron, C., 289
Walk, T., 364, 472, 473, 474, 476, 489
Walkenhorst, J., 123, 132
Walker, D., 312, 327
Walker, S., 312, 313, 327
Wall, M. E., 37, 47
Walsh, C. T., 119, 174, 184, 206, 208, 220, 256, 257, 258, 259, 260, 262, 263, 264, 265, 266, 267, 268, 269, 282, 283, 302, 313, 315, 317, 321, 323, 338, 339, 342, 343, 345, 349, 354, 355, 357, 358, 359, 363, 365, 367, 368, 371, 372, 380, 381, 382, 383, 385, 391, 433, 436, 437, 463, 466, 468, 469, 470, 471, 472, 475, 477, 489, 491, 513, 515, 525, 526, 528, 539
Walton, J. D., 355
Wang, A. H., 385
Wang, C. C., 380, 381, 385, 387, 388, 391, 395
Wang, F., 129
Wang, J., 14, 25, 119, 175
Wang, L., 44
Wang, W. C., 418
Wang, Y., 261, 381
Wang, Y. G., 289
Wani, M. C., 37, 47
Wanner, B., 520
Wanner, B. L., 160, 169
Ward, A. C., 8, 9, 25
Ward, J. M., 106, 121, 130, 290, 465
Wardell, J. N., 123
Waring, M. J., 387, 391
Warren, P., 14
Washio, K., 210
Watanabe, K., 233, 379, 380, 381, 385, 387, 388, 391, 395
Watanabe, M., 355
Watanabe-Sakamoto, A., 326
Waterbury, J. B., 35
Waterman, M. R., 507
Waterman, M. S., 189
Watson, J. A., Jr., 311
Watson, J. T., 569
Wattanachaisaereekul, S., 258
Watve, M. G., 120

Wawrzak, Z., 313
Weaver, D., 109, 111, 126
Weber, A. E., 493, 494
Weber, C., 369
Weber, J. M., 302
Weber, M., 319, 325, 326
Weber, T., 188, 207, 209, 341, 345, 466, 467, 471, 472, 480, 490
Webster, R. W., Jr., 433, 442
Wei, C. L., 418, 419
Weigel, C., 14
Weinreb, P. H., 256, 257, 258, 491
Weinstein, M. P., 25
Weinstein, R. A., 118
Weinstock, S. F., 118
Weisblum, B., 16, 17
Weiss, H., 317, 322
Weissman, K. J., 194, 220, 234, 256, 355
Weist, S., 358, 369, 468, 471, 477, 479, 480
Weitnauer, G., 324, 325
Welle, E., 317, 322
Wellein, C., 257, 258
Welzel, K., 480
Wendrich, T. M., 450
Wendt-Pienkowski, E., 44, 160, 282
Wenger, R. M., 37
Wenzel, S. C., 174, 175, 442
Wessling-Resnick, M., 260, 267, 268, 269
Westerlaken, I., 119
Westlake, D. S., 409, 410, 412, 413, 415, 417
Weston, A. J., 289
Westpheling, J., 128
Westrich, L., 325
Wetzel, R. W., 523
Wetzel, S., 182
Weymouth-Wilson, A. C., 311
Whipple, E. B., 387
White, A. J. P., 387
White, J., 95, 121, 125, 126, 130, 539, 541, 560, 565
White, M. F., 444, 447, 450, 452, 453
White, S. W., 256
Whiteman, P., 419
Whiting, A., 354, 359, 515, 516, 517, 518, 519, 521, 522, 524, 527
Whitney, J. G., 402
Whitwam, R. E., 160
Widboom, P. F., 235, 354, 368, 442
Widdick, D. A., 539, 541, 560, 565
Wiedemann, I., 535
Wiedhopf, R. M., 489
Wieland, B., 14, 562, 564, 565
Wierenga, J., 538
Wiesblum, B., 16
Wietzorrek, A., 99, 126, 463, 477
Wilkie, S. C., 477, 479, 521
Wilkinson, B., 120, 194, 289, 313, 327, 353, 354, 357, 359, 360, 362, 363, 364, 523, 525, 576

Willems, A., 128
Willey, J. M., 536, 543
Williams, D. H., 354, 363, 365, 367, 383, 391, 394, 460, 462, 469, 470, 471, 475, 477, 492
Williams, D. J., 387
Williams, G., 545
Williams, G. J., 327, 328
Williams, S. T., 124
Williamson, M. P., 354, 363, 462
Williamson, R. T., 354, 359
Willimek, A., 122, 123, 129
Wilmotte, A., 34, 40
Wilson, K. S., 444, 448, 450, 452, 453
Wilson, M. K., 448
Wilson, P., 583
Wink, J., 479
Wink, J. M., 463
Winkelmann, G., 381, 446
Winter, J., 270
Wirtz, G., 279, 283
Wise, S. C., 14
Wit, E., 126
Withers, S. G., 327
Wittenberg, L. O., 182
Wittmann, M., 354, 362
Woese, C. R., 65, 74
Wohlert, S. E., 284, 319, 325, 326, 480
Wohlleben, W., 109, 126, 130, 145, 148, 151, 152, 188, 207, 209, 341, 358, 363, 366, 367, 370, 459, 463, 464, 465, 466, 467, 468, 469, 470, 471, 472, 473, 474, 475, 476, 477, 478, 479, 480, 489, 490, 492, 499, 502
Woithe, K., 473, 475, 487, 489, 490, 491, 492, 499, 502, 503, 505, 506
Wölcke, J., 12
Wolfe, S., 409, 410, 412, 413, 415, 417, 426
Wolpert, M., 175, 472
Wong, A., 409
Wong, E., 415
Wong, L. S., 260
Wong, U., 444
Woodnutt, G., 14
Woodruff, H. B., 402, 404, 489
Worthington, A. S., 226, 231, 236, 238, 240
Wörtz, T., 473
Wright, A. J., 119
Wright, G. D., 118, 359, 463, 475, 478
Wright, H. M., 119
Wright, L. F., 97, 119
Wrigley, S. K., 338, 354, 359, 516, 517, 519, 520, 521, 522, 523
Wrigley, S. W., 512, 513, 515, 516, 517, 518, 521, 523
Wu, C. J., 470
Wu, D., 396
Wu, J., 569
Wu, X., 257, 413

X

Xianshu, Y., 204, 210
Xiao, X., 129, 420
Xie, L., 534, 538, 539, 541, 542, 544, 548, 559, 560, 569
Xin, C., 13
Xu, Y., 108, 260, 261, 263, 267, 268, 269

Y

Yadav, G., 187, 194, 396
Yamada, Y., 144, 145, 146, 150, 151, 152, 154
Yamanaka, S., 64, 68
Yamase, H., 284
Yamashita, A., 119, 120, 130
Yamazaki, H., 96, 113
Yamazaki, T., 9
Yanagimoto, M., 145
Yanai, K., 355
Yang, C. F., 370
Yang, J., 479
Yang, M., 313
Yang, R., 44
Yang, Y. B., 418, 419
Yanko, M., 415
Yao, R. C., 460
Yao, Z., 14
Yap, W. M. G. J., 563
Yasuzawa, T., 489
Ye, X. Y., 283
Ye, Y.-Y., 354, 372
Yeh, E., 358
Yeh, W. K., 417, 419, 420
Yen, M., 260, 266, 268
Yeung, S. M., 382
Yi, M., 542, 569
Yim, G., 31
Yin, J., 255, 259, 260, 261, 262, 263, 265, 266, 267, 268, 269, 357, 358
Yin, X., 355, 357
Yin, Y., 175
Ying, X., 129
Yip, C. L., 129, 130
Yip, M. J., 583
Ylihonko, K., 285, 288
Yokobori, S., 584
Yokoyama, T., 123
Yonus, H., 339
Yoon, Y. J., 284
You, Y. O., 542, 545, 546
Young, K., 25, 119
Youssef, D. T. A., 578, 581, 588
Yu, C. A., 415
Yu, J.-P., 538, 545, 548, 551
Yu, Y., 354, 370
Yu, Z., 129

Yuceer, M. C., 25
Yuqing, T., 44, 103, 163, 173, 463

Z

Zabriskie, T. M., 355, 357, 577
Zachariah, C., 535
Zahner, H., 25, 364, 536, 538
Zanuso, G., 45, 288
Zazopoulos, E., 46, 120, 160, 175, 183, 193, 203, 204, 210, 364
Zender, S., 63
Zerbe, K., 259, 464, 473, 475, 487, 489, 490, 491, 492, 499, 502, 503, 505, 506
Zerck, A., 84
Zhan, J., 326
Zhanel, G. G., 118
Zhang, B., 14
Zhang, C., 302, 313, 315, 316, 317, 318, 326, 327, 328, 372, 479
Zhang, H., 381
Zhang, J., 122, 185, 257, 364, 409, 410
Zhang, K., 255
Zhang, L., 129, 256, 257
Zhang, L. H., 128
Zhang, W., 473
Zhang, W. W., 489, 490, 492, 499, 502
Zhang, X., 49, 541, 542, 547
Zhang, Y., 160, 161, 174, 175
Zhang, Y. M., 256
Zhang, Z., 185
Zhang, Z. H., 418
Zhao, B., 489
Zhao, C. H., 129
Zhao, G., 257
Zhao, L., 282, 283, 284, 285, 382
Zhao, Q., 354, 370
Zhao, Q. F., 196
Zhao, Z., 282
Zheng, J., 256
Zheng, Q., 354, 370
Zhou, H., 545
Zhou, X., 129
Zhou, Z., 260, 261, 263, 267, 268, 269, 525, 528
Zhu, L., 47, 283, 292, 298, 300, 302, 310, 319, 325, 326
Zhu, Y., 545
Ziegelbauer, K., 14
Ziegler, J., 489
Ziemert, N., 578, 581, 588
Zimmerman, N., 560, 569
Zimmerman, S. B., 402
Zimmermann, S., 339
Zink, D. L., 38
Zirah, S., 539
Zirkle, R., 48
Zmijewski, M. J., Jr., 354, 363
Zocher, R., 50, 355
Zotchev, S. B., 282
Zou, Y., 261, 266, 269
Zuber, P., 256, 257, 258, 264, 382, 491
Zurek, G., 81, 82, 83, 84, 86

Subject Index

A

A21987C, discovery and structure, 512–513
A40926, see Dalbavancin
A47934, see Teicoplanin
A54145, genes for combinatorial biosynthesis, 523
L-δ-α-AAA-L-cysteinyl-D-valine synthetase
    cephamycin synthesis, 409–412
    purification, 411–412
Acarbose, discovery, 36
Achromobactin, structure, 434
AcpS, see Phosphopantetheinyl transferase
Actagardine, whole-cell generation of variants, 565
Actinomycetes
    antibiotic biosynthesis regulatory genes
        heterologous overexpression and mutant alleles, 130–133
        pathway-specific regulation, 125–126
        pleiotropic regulation, 127–129
        prospects for study, 133–134
    biosynthetic gene clusters of *Streptomyces*
        overview, 160–161
        cloning and identification, 161–163, 177
        polymerase chain reaction-targeted gene replacement, 163–170
        transposon mutagenesis, 170–172
        heterologous expression, 173–174, 177
        reassembly by switching overlapping cosmid clones, 174–176
    cephamycins, see Cephamycins
    culture
        growth-dependent control mechanisms, 120–121
        morphology as determinant of productivity, 123–125
        phosphate-mediated control, 121–122
        secondary metabolite production, 31–32
        stringent control, 121
    DasR regulon and metabolism control, 122–123
    species for antibiotic production, 119–120
    strain preparation for high throughput screening
        isolation from soil, 9–10
        purification and storage, 10–11
Actinomycin
    discovery, 44–45
    structure, 42
ACVS, see L-δ-α-AAA-L-cysteinyl-D-valine synthetase
Acyl-carrier protein, see Nonribosomal peptide synthetase; Polyketide synthase
Aerobactin, biosynthesis, 442–444
Aminovinyl cysteine, lanbiotics, 553–554
Amphotericin, polyketide synthase structural element extrapolation, 203–204
Andrimid, structure, 356
Antibiotics, see also specific drugs
    CDAs, see Calcium-dependent antibiotics
    cephamycins, see Cephamycins
    cyanobactins, see Cyanobactins
    discovery barriers, 118
    glycopeptides, see Glycopeptide antibiotics
    high throughput screening, see High throughput screening
    hormonal control of bacterial production, see γ-Butyrolactones
    lanbiotics, see Lanbiotics
    lipopeptides, see Lipopeptide antibiotics
    regulatory genes, see Actinomycetes; Streptomycetes
    streptomycetes production, see Streptomycetes
Antifungal high throughput screening, see High throughput screening
Argyrin A
    mechanism of action, 62–63
    structure, 63
Atorvastatin
    discovery, 53
    mechanism of action, 52–53
    structure, 52
Azinomycin B, precursor biosynthesis, 370

B

Balhimycin
    biosynthetic gene clusters, 467
    structure, 461
Bleomycin
    discovery, 43–44
    structures, 42
(4R)-[(E)–2-Butenyl)]–4-methyl-L-threonine, biosynthesis, 369–370
γ-Butyrolactones
    antibiotic production
        regulation, 144–145
        response bioassays
            kanamycin, 148–150
            pigmented antibiotics, 146–147

γ-Butyrolactones (*cont.*)
  purification, 145–146
  receptors
    identification, 150–151
    target identification, 151–152
    electrophoretic mobility shift assay of DNA binding and probe labeling
      Cy3, 155
      digoxygenin, 155
      radiolabeling, 153–154
    types and structures, 144–145

C

Calcium-dependent antibiotics
  CDA2a structure, 361
  2,3-epoxyhexanoyl side chain biosynthesis, 363–364
  genes for combinatorial biosynthesis, 523
  tryptophan dehydrogenation in biosynthesis, 357
Calicheamicin
  discovery, 46
  structure, 42
CDAs, *see* Calcium-dependent antibiotics
Cephalosporin C
  biosynthesis, 406
  structure, 403
Cephamycins
  actinomycetes biosynthesis, 402–404
  biosynthetic pathway
    common steps in actinomyetes and fungi
      L-δ-α-AAA-L-cysteinyl-D-valine synthetase, 409–412
      deacetoxycephalosporin C hydroxylase, 419–420
      deacetoxycephalosporin C synthase, 417–419
      isopenicillin N epimerase, 415–417
      isopenicillin N synthase, 412–415
      early steps, 406–408
      overview, 405–406
    cepham nucleus tailoring in actinomycetes, 420–422
    cephamycin C
      biosynthesis, 406
      production, 404–405
      regulation of production, 422–423
      structure, 403
    structures, 403
    types, 403–405
ChIP-on-chip, *see* Chromatin immunoprecipitation/DNA microarray
Chloreremomycin
  biosynthetic gene clusters, 467
  structure, 461
Chromatin immunoprecipitation/DNA microarray, streptomycetes antibiotic

biosynthesis regulatory gene characterization, 111–112
Clavulinic acid
  production, 405
  structure, 403
Coelichelin, biosynthesis, 440, 444
Compactin, discovery, 36
Condensation domains, nonribosomal peptide synthetase, 343–344
Cyanobacteria
  culture and extraction of secondary metabolites, 35
  cyanobactins, *see* Cyanobactins
Cyanobactins
  biosynthetic enzyme analysis, 593
  chemical analysis, 586–587
  deep metagenome mining
    amino acid use comparison, 581–582
    overview, 576
    pat pathway, 578–579
    PatE mutation identification, 580
    pathway engineering
      principles, 593–594
      yeast recombination for pathway engineering, 594–595
    prenylation pathways, 580–581
    Prochloron example, 592
    prospects, 583
    whole-cell polymerase chain reaction, 591–592
  diversity, 576–577
  gene cloning and identification
    genome mining and structure prediction, 588–589
    non-*Prochloron* species, 589–590
    overview, 587–588
    Prochloron species, 589
  gene clusters, 578–579
  heterologous expression in *Escherichia coli*, 590–591
  Prochloron-containing ascidians
    DNA purification
      enriched cells, 584–585
      whole animals, 585–586
    identification and processing, 583–584
  structures, 577
Cyclosporin A
  discovery, 50–51
  precursor biosynthesis, 369–370
  structure, 50, 356
Cyptophycin
  discovery, 49
  structure, 47

D

Dalbavancin
  biosynthetic gene clusters, 467
  structure, 461

Subject Index

DAOCS, see Deacetoxycephalosporin C synthase
Daptomycin, see also Lipopeptide antibiotics
   biosynthesis and genetic engineering in
      Streptomyces rosesoporus
      antibacterial activities of analogs, 526–527
      biosynthetic genes
         cloning and analysis, 514–518
         dptGHIJ genes and functions, 520–521
         heterologous expression in *Streptomyces lividans*, 517
         nonribosomal peptide synthetase, 517–519
         deletion mutant construction, 517, 519–520
         overview, 514
         pathway reconstitution by ectopic expression of genes, 521–523
   biosynthesis overview, 356
   discovery, 512
   structure, 361, 513
Daunorubicin
   discovery, 45
   structure, 42
Deacetoxycephalosporin C hydroxylase, properties and cephamycin biosynthesis, 419–420
Deacetoxycephalosporin C synthase, properties and cephamycin biosynthesis, 417–419
Deep metagenome mining, see Cyanobactins
Deoxysugars, see also Glycosyltransferases
   aminodeoxysugars, 279
   biosynthesis
      gene cassette plasmids, 292–299
      overview, 279–282
   dideoxysugars, 279
   glycosylated compound detection, 303
   glycosylation pattern modification
      combinatorial biosynthesis, 290–292
      generating glycosylated compounds, 299–301
      heterologous gene expression, 288–290
      mithramycin modification through gene inactivation
         mtmU inactivation, 285–288
         overview, 284–285
      tailoring modifications of attached deoxysugars, 301–302
   lipopolysaccharide composition, 278
   nonribosomal peptide precursor glycosyl building blocks, 371–372
   secondary metabolite composition, 278–279
   transfer by glycosyltransferases, 283–284
Desferrioxamine E, structure, 434
3,5-Dihydroxyphenylglycine, biosynthesis analysis
   in vitro, 367–368
   in vivo, 366–367
   balhimycin moiety variation by mutasynthesis, 368–369

   overview, 365–366, 469–470
DNA microarray, see Chromatin immunoprecipitation/DNA microarray
Dorrigocin, polyketide synthase structural element extrapolation, 204
Doxorubicin
   discovery, 45
   structure, 42
Dpg proteins, 3,5-dihydroxyphenylglycine biosynthesis, 365–366

E

Echninomycin
   nonribosomal peptide synthetase heterologous expression in *Escherichia coli* for biosynthesis
      culture and biosynthesis, 390–391
      multigene assembly on expression vectors, 387–390
      open reading frame functions, 385–386
      pathway, 383–282
      self-resistance mechanism and assay, 391–392
   structure, 382
Electrophoretic mobility shift assay
   γ-butyrolactone receptor binding assay and probe labeling
      Cy3, 155
      digoxygenin, 155
      radiolabeling, 153–154
   streptomycetes antibiotic biosynthesis regulatory gene characterization, 107
EMSA, see Electrophoretic mobility shift assay
L-Enduracididine, β-hydroxylation, 357
Enniatin B, structure, 356
Enterobactin
   biosynthesis, 436–438
   discovery, 435
   structure, 434
Epidermin, whole-cell generation of variants, 564
Epothilone B
   discovery, 48
   mechanism of action, 62
   structure, 47
   structure, 63
2,3-Epoxyhexanoyl side chain, biosynthesis, 363–364
Erythromycin, polyketide synthase structural element extrapolation, 203

F

Fengycin, nonribosomal peptide synthetase structural element extrapolation, 210–211
Ferredoxin, purification of histidine-tagged protein, 505

G

Gallideremin, whole-cell generation of variants, 564
Genome mining
  deep metagenome mining, see Cyanobactins lanbiotic novel drugs, 537–538
Glycopeptide antibiotics, see also Vancomycin biosynthesis
  amino acid building block biosynthesis
    3,5-dihydroxyphenylglycine, 469–470
    4-hydroxyphenylglycine, 469
    β-hydroxytyrosine, 468–469
    overview, 466, 468
  carbohydrate building block biosynthesis, 470–471
  chemical analysis, 462
  gene cluster organization, 467
  genetic analysis
    gene cluster identification, 462–463
    genetic manipulation systems, 463–464
    overexpression of biosynthetic genes, 464–466
    in silico analysis, 466
  glycosylation, 477
  halogenation, 475–476
  methylation, 476
  regulation, 477–478
  side chain cyclization by P450-type monooxygenases
    lipoglycopeptides, 475
    vancomycin-like glycopeptides, 473–475
  excretion, 478
  linking primary and secondary metabolism, 478–479
  nonribosomal peptide synthetases and peptide assembly, 471–473
  novel glycopeptide generation approaches, 479–480
  self-resistance, 478
  structural classification, 460–462
Glycosyltransferases
  auxiliary proteins, 322–323
  bi-functional enzymes, 321
  biological relevance of sugars, 311
  classification, 312
  detection of novel glycosyltransferases, 322, 324
  glycosylation mechanism, 312
  GT-B superfamily
    characteristics, 313
    sequence alignment, 314
    X-ray crystallography, 313
  modification for novel product generation, see also Deoxysugars
    biotransformation, 326
    coexpression with deoxysugar biosynthetic genes, 326
    expression in deletion mutants, 326
    gene cluster expression, 325–326
    gene inactivation, 324–325
    heterologous expression, 325
    mutagenesis for substrate specificity alteration
      domain swapping, 328
      error-prone polymerase chain reaction, 327
      overview, 326–327
      polymerase chain reaction site-directed mutagenesis, 327
      saturation mutagenesis, 328
      sequencing, 327
    overexpression, 21
  novel product generation, 311
  pharmaceutical relevance of sugars, 311
  reactions
    classical sugar transfer, 313, 315
    sugar transfer and reverse reaction, 315–316
  substrate specificity flexibility, 315, 317–320

H

High-performance liquid chromatography
  glycosylated compound detection, 303
  pantheine analog compatibility with coenzyme A synthetic enzymes, 232
  secondary metabolite recovery from culture, 33
High throughput screening
  assay development, 14–15
  databases, 23–24
  elements of program, 6
  hit follow-up
    activity confirmation, 20
    novelty evaluation, 20–22
    overview, 19
    profiling, 22
  nomenclature, 6
  novelty in program, 5
  objectives, 4–5
  operations and costs, 24–25
  samples
    overview, 7–8
    preparation, 11–12
  screening for bacterial cell wall inhibitors, 16–19
  secondary assays, 15–16
  strains
    actinomycetes
      isolation from soil, 9–10
      purification and storage, 10–11
    isolation purification, and storage, 8–9
    overview, 7
  strategy, 5, 7
  target identification and validation, 13–14
HPLC, see High-performance liquid chromatography

Subject Index

HTF, see High throughput screening
4-Hydroxyphenylglycine, biosynthesis, 469
β-Hydroxytyrosine, biosynthesis, 468–469

I

IPNS, see Isopenicillin N synthase
Isopenicillin N epimerase
  activity assay, 417
  cephamycin biosynthesis, 415
  genes, 417
  purification from *Streptomyces clavulgerus*, 415
Isopenicillin N synthase
  cephamycin biosynthesis, 412–413
  cyclization of β-lactams, 414
  genes, 413–414
  mutagenesis, 414
  substrate specificity, 413
  X-ray crystallography, 414–415

K

Kanamycin, bioassay of γ-butyrolactone induction, 148–150

L

Lacticin 3147, whole-cell generation of variants, 563, 565
Lanbiotics
  biosynthesis overview, 536–537, 559–560
  genome mining for novel drugs, 537–538
  LanA precursors
    expression and purification
      heterologous overexpression in *Escherichia coli*, 539–540
      nickel affinity chromatography, 540–541
    leader peptide recognition by biosynthetic enzymes, 539
    mutagenesis and construction of synthetic analogs, 541–542
    overview, 536
  LanC cyclases
    functional overview, 547–548
    NisC
      assay, 551
      metal analysis, 550
      overexpression and purification, 549
      prenensin substrate preparation, 550–551
  LanM enzymes
    cyanogen bromide cleavage of products, 547
    dehydration and cyclization activity assay, 544–546
      order of reactions, 546–547
    functional overview, 542
    heterologous expression and purification, 543–544
    mechanisms, 542–543
  LanT transporter protease domain
    functional overview, 552
    LctA substrate cleavage by Lct150, 552–553
  mechanisms of action, 535
  posttranslational modifications, 534, 553–554
  regulation of expression, 560
  structural overview, 534–535
  structure–function relationship, 536
  variant generation
    cis complementation system
      lacticin 3147, 563
      mersacidin, 563, 566–568
      mutacin II, 562
      Pep5, 562
      subtilin, 562
    comparison of *cis* and *trans* complementation systems, 566–568
    heterologous expression systems
      cinnamycin, 565–566
      nukacin, 566
    overview, 560–561
    structural characterization
      analytical strategy, 570–571
      random mutagenesis library variants, 570
      saturation mutagenesis library variants, 568–570
    trans complementation system
      actagardine, 565
      epidermin, 564
      gallideremin, 564
      lacticin 3147, 565
      mersacidin, 564–568
      nisin, 563–564
Landromycin, glycosylation, 310
Lipopeptide antibiotics, see also Daptomycin A21987C, 512–513
  antibacterial activities of daptomycin analogs, 526–527
  combinatorial biosynthesis gene sources, 523
  3,5-dihydroxyphenylglycine biosynthesis
    analysis
      in vitro, 367–368
      in vivo, 366–367
    balhimycin moiety variation by mutasynthesis, 368–369
    overview, 365–366
  nonribosomal peptide precursor lipopeptide synthesis
    2,3-epoxyhexanoyl side chain biosynthesis, 363–364
    overview, 360, 362–363
  novel lipopeptide genetic engineering
    combinatorial biosynthesis, 526
    gene replacement, 524
    module and multidomain replacements, 525–526

Lovastatin
  discovery, 36, 53
  mechanism of action, 52–53
  structure, 52
Lyngbyatoxin A, structure, 356
Lysine–6-aminotransferase, cephamycin synthesis, 407

M

Mannopeptimycin δ, structure, 356
Mass spectrometry, myxobacterial metabolite screening
  checklist, 82–84
  mining of secondary metabolomes, 84–87
  overview, 81–82
Mersacidin, whole-cell generation of variants, 563–568
3-Methoxy-5-methylnaphthoic acid, biosynthesis, 370–371
(2S,3R)-Methylglutamic acid, biosynthesis analysis
  in vitro, 360
  in vivo, 359
  scheme, 358
Mevastatin
  discovery, 53
  mechanism of action, 52–53
  structure, 52
Microcystin, structure, 356
Mithramycin, glycosylation pattern modification through gene inactivation
  mtmU inactivation, 285–288
  overview, 284–285
Mitomycin C
  discovery, 41, 43
  structure, 42
MS, *see* Mass spectrometry
Mutacin II, whole-cell generation of variants, 562
Mycophenolic mofetil
  discovery, 49–50
  structure, 50
Mycosubtilin, structure, 361
Myxobacteria
  bioassays of extracts, 79–80
  culture
    incubation conditions and sample workup, 79
    maintenance and preservation, 77–79
    overview, 34–35
  extraction of secondary metabolites, 34–35
  fruiting bodies, 60–61, 67, 75–76
  isolation
    air, 74
    baiting technique, 72
    extended incubation, 74–75
    light, 73
    medium, 72–73

  pH, 74
  temperature, 73
mass spectrometry screening
  checklist, 82–84
  mining of secondary metabolomes, 84–87
  overview, 81–82
novel strain discovery
  genetic characterization, 68–71
  geographic and environmental factors, 68
  historical perspective, 66, 68
  material substrates, 65–66
  prospects, 64–65
  unculturability, 69, 71
prospects for drug discovery, 87–88
purification, 76–77
successful drug development, 62–63
swarming, 75

N

Neocarzinostatin
  discovery, 46
  structure, 42
NIS synthetase, *see* Siderophores
NisC, *see* Lanbiotics
Nisin
  mechanism of action, 535
  whole-cell generation of variants, 563–564
NMR, *see* Nuclear magnetic resonance
Nocardicin A, structure, 403
Nonribosomal peptide precursors
  amino acid types, 355, 357–359
  (4R)-[(E)-2-butenyl)]–4-methyl-L-threonine biosynthesis, 369–370
  3,5-dihydroxyphenylglycine biosynthesis analysis
    in vitro, 367–368
    in vivo, 366–367
  balhimycin moiety variation by mutasynthesis, 368–369
  overview, 365–366
  glycosyl building blocks, 371–372
  lipopeptide synthesis
    2,3-epoxyhexanoyl–3-methoxy-5-methylnaphthoic acid biosynthesis, 370–371
    overview, 360, 362–363
    side chain biosynthesis, 363–364
  (2S,3R)-methylglutamic acid biosynthesis analysis
    in vitro, 360
    in vivo, 359
    scheme, 358
  overview, 354
Nonribosomal peptide synthetase
  domains
    mechanistic and structural aspects, 339–346
    software for analysis

Subject Index   633

BLAST, 185–186
manual parsing, 186
NORINE database of nonribosomal
    peptides, 190–191
NRPS-PKS overview, 187–188
NRPSpredictor, 188
PKS/NRPS analysis website, 189–190
types and functions, 184–185, 338–339,
    382–383
domain string conversion to structural elements
caveats, 207–208
examples
    fengycin, 210–211
    penicillin, 210, 212
    ramoplanin, 210–213
modifications and stereochemistry,
    206–207
polyketide chain length and amino acid side
    chains, 206
prospects, 212–214
rendering
    consensus motif analysis, 209
    domain string determination, 208
    heuristic structure prediction, 209–210
    interprotein domain ordering, 208–209
glycopeptide antibiotic peptide assembly,
    471–473
heterologous expression in *Escherichia coli*
    echinomycin biosynthesis
        culture and biosynthesis, 390–391
        multigene assembly on expression
            vectors, 387–390
        open reading frame functions, 385–386
        pathway, 383–282
        self-resistance mechanism and assay,
            391–392
    overview, 380–381
    pathway engineering, 393–395
    prospects, 395–396
    transformant stability and assay, 392–393
mechanism studies using synthetic probes
    carrier protein posttranslational
        modification and
        phosphopantetheinyltransferase
        promiscuity, 221–222
    chemoenzymatic synthesis of probes
        carrier protein loading, 232–233
        high-performance liquid
            chromatography analysis of
            pantheine analog compatibility
            with coenzyme A synthetic
            enzymes, 232
        materials, 231
        overview, 225–226
    coenzyme A analog studies, 223–225
    overview, 220–221
    synthesis of precursors
        BODIPY-coenzyme A, 229

carrier protein analogs, 228
coenzyme A precursors, 227
overview, 225–226
pantetheine azide, 230–231
pantetheine chloroacrylamide, 231
PMB-pantetheine azide, 230
PMB-pantothenic acid, 229
minimal chemistry, 205–206, 338–339
peptide precursors, *see* Nonribosomal peptide
    precursors
phage selection for gene identification,
    261–262, 269
proteomic identification using synthetic probes
    carrier protein labeling *in vivo*
        azide-labeled carrier protein
            visualization, 249–250
        endogenous carrier proteins, 248–249
        overexpressed carrier protein, 248
        principles, 247–248
    chemoenzymatic labeling of apo-carrier
        proteins in cell lysates, 241–244
    complementary labeling of domains with
        activity-based probes, 244–246
    siderophore biosynthesis, *see* Siderophores
    structure analysis with synthetic probes
        challenges, 233–234
        overview, 220–221
        protein–protein interactions
            chemoenzymatic crosslinking of
                acyl-carrier protein and
                ketosynthase domains, 240–241
            chemoenzymatic crosslinking of
                acyl-carrier protein and partner
                enzymes, 236–240
        protein–substrate interactions, 234–236
    termination module structure, 347–349
NORINE, database of nonribosomal peptides,
    190–191
Northern blot, streptomycetes antibiotic
    biosynthesis regulatory gene
    characterization, 104
NRPS, *see* Nonribosoml peptide synthetase
Nuclear magnetic resonance, nonribosomal
    peptide synthetase termination module
    structure, 347–349

P

PCP, *see* Peptide carrier protein
PCR, *see* Polymerase chain reaction
Penicillin
    nonribosomal peptide synthetase structural
        element extrapolation, 210, 212
    penicillin G biosynthesis, 406
Pentostatin
    discovery, 45–46
    structure, 42
Pep5, whole-cell generation of variants, 562

Peptide carrier protein
  peptide–peptide carrier protein thioester synthesis
    hexapeptide S-phenyl thioester synthesis, 499–500
    hexapeptide SCoA thioester synthesis, 500–501
    overview, 490–491
    peptide carrier protein conjugation, 501–502
    peptide synthesis
      amino acid synthesis, 493–497
      model hexapeptide features, 491–493
      solid-phase synthesis, 497–499
    purification of domain from vancomycin nonribosomal peptide synthetase, 505–506
Peptidyl carrier protein domain, nonribosomal peptide synthetase, 341–343
Petrobactin, biosynthesis, 448–450
Phosphopantetheinyl transferase
  carrier protein posttranslational modification and promiscuity, 221–222
  coenzyme A conjugates for study
    Qdot conjugate preparation, 266–267
    Sfp-catalyzed protein labeling with small molecule conjugates, 267–268
    small molecule conjugate preparation, 265–266
  expression systems
    AcpS from *Escherichia coli*, 263
    Sfp
      coexpression with modules from polyketide synthase or nonribosomal peptide synthetase, 264–265
      Escherichia coli enzyme, 263–264
  function, 256
  peptide tags as surrogate substrates, 260
  Sfp
    activation of polyketide synthase or nonribosomal peptide synthetase modules *in vitro*, 265
    cell surface protein labeling, 268–269
    phage selection for polyketide synthase or nonribosomal peptide synthetase fragment identification, 269
  substrate specificity, 258–259
  superfamily subgroups, 257
  types and distribution, 257–258
Piperideine-6-carboxylate dehydrogenase
  cephamycin synthesis, 407–408
  purification from *Streptomyces clavulgerus*, 408
PKS, *see* Polyketide synthase
Polyketide synthase
  domains
    software for analysis
      BLAST, 185–186

      manual parsing, 186
      NRPS-PKS overview, 187–188
      PKS/NRPS analysis website, 189–190
    types and functions, 184–185
  functional overview, 184
  mechanism synthetic probe studies
    carrier protein posttranslational modification and phosphopantetheinyltransferase promiscuity, 221–222
    chemoenzymatic synthesis of probes
      carrier protein loading, 232–233
      high-performance liquid chromatography analysis of pantheine analog compatibility with coenzyme A synthetic enzymes, 232
      materials, 231
      overview, 225–226
    coenzyme A analog studies, 223–225
    overview, 220–221
    synthesis of precursors
      BODIPY-coenzyme A, 229
      carrier protein analogs, 228
      coenzyme A precursors, 227
      overview, 225–226
      pantetheine azide, 230–231
      pantetheine chloroacrylamide, 231
      PMB-pantetheine azide, 230
      PMB-pantothenic acid, 229
    phage selection for gene identification, 261–262, 269
    proteomic identification synthetic probe studies
      carrier protein labeling *in vivo*
        azide-labeled carrier protein visualization, 249–250
        endogenous carrier proteins, 248–249
        overexpressed carrier protein, 248
        principles, 247–248
      chemoenzymatic labeling of apo-carrier proteins in cell lysates, 241–244
      complementary labeling of domains with activity-based probes, 244–246
    structural element extrapolation
      caveats
        nonlinear enzymatic logic, 196
        polyketide synthase type limitations, 194–195
        posttranslational modifications, 196
        unprecedented domains, 196
      consensus motif analysis, 200–201
      conserved active site motifs, 193–194
      examples
        amphotericin, 203–204
        dorrigocin, 204
        erythromycin, 203
      heuristic approach, 193
      ketoreductase domain analysis, 201

Subject Index

mechanistic approach, 191–193
polyketide drawing based on domain strings
   automated parsing, 197–198
   domain identification and ordering of proteins/modules, 196–197
   interprotein domain ordering, 199–200
   manual parsing, 197
polyketide structure derivation from domain strings
   heuristic approach, 202–203
   mechanism-based approach, 202
prospects, 212–214
structure synthetic probe studies
   challenges, 233–234
   overview, 220–221
   protein–protein interactions
      chemoenzymatic crosslinking of acyl-carrier protein and ketosynthase domains, 240–241
      chemoenzymatic crosslinking of acyl-carrier protein and partner enzymes, 236–240
   protein–substrate interactions, 234–236
Polymerase chain reaction
   glycosyltransferase mutagenesis
      error-prone polymerase chain reaction, 327
      site-directed mutagenesis, 327
      screening, 322, 324
   streptomycetes antibiotic biosynthesis regulatory genes
      confirmation, 103
      reverse transcriptase-polymerase chain reaction, 106–107
   targeted gene replacement in biosynthetic gene clusters of *Streptomyces*, 163–170
PPTase, *see* Phosphopantetheinyl transferase
Pravastatin
   discovery, 53
   mechanism of action, 52–53
   structure, 52
Prenensin, preparation for NisC assay, 550–551
Primer extension assay, streptomycetes antibiotic biosynthesis regulatory gene characterization, 105
Pyochelin
   biosynthesis, 436–438
   structure, 434

R

Ramoplanin, nonribosomal peptide synthetase structural element extrapolation, 210–213
Rebeccamycin
   discovery, 46–47
   glycosylation, 310
   structure, 42, 310

S

S1 nuclease protection assay, streptomycetes antibiotic biosynthesis regulatory gene characterization, 104–105
Secondary metabolites, *see also specific drugs*
   actinomycetes culture for production, 31–32
   characteristics, 30–31
   cyanobacteria culture and extraction, 35
   deoxysugars, *see* Deoxysugars
   filamentous fungi culture for production, 32–33
   myxobacteria culture and extraction, 34–35
   recovery from culture, 33
   screening assay systems, 35–40
SELEX, *see* Systemic evolution of ligands by exponential enrichment
Sfp, *see* Phosphopantetheinyl transferase
Siderophores
   assembly, 433
   biosynthesis
      NIS synthetases
         assay of hydroxamate formation, 450–454
         types, 443–448
      nonribosomal peptide synthetase-dependent pathways, 434–442
      nonribosomal peptide synthetase-independent pathways, 442–448
      petrobactin, 448–450
   functional overview, 432
   iron complexes, 433
Simvastatin
   discovery, 53
   mechanism of action, 52–53
   structure, 52
Sirolimus
   discovery, 51
   structure, 50
Site-directed mutagenesis, *see* Glycosyltransferases
SPR, *see* Surface plasmon resonance
Statins, *see specific drugs*
Streptomycetes, antibiotic biosynthesis regulatory genes
   confirmation of nature
      homologous recombination and insertion/deletion mutagenesis, 103
      polymerase chain reaction, 103
   genome scanning, 102
   identification by mutagenesis
      cosynthesis inability, 99
      genetic complementation, 100–101
      insertional mutants, 100
      mutagenesis, 97–98
      transposon mutants, 99–100
   identification by overexpression, 101–102
   overview, 94–96
   pathway-specific regulatory gene characterization

Streptomycetes, antibiotic biosynthesis regulatory genes (cont.)
  electrophoretic mobility shift assay, 107
  Northern blot, 104
  primer extension assay, 105
  reporter gene studies, 108
  reverse transcriptase-polymerase chain reaction, 106–107
  run-off transcription analysis, 106
  S1 nuclease protection assays, 104–105
  surface plasmon resonance of DNA–protein interactions, 107–108
 pleiotropic regulatory gene characterization
  chromatin immunoprecipitation/DNA microarray analysis, 111–112
  controlled induction of expression, 111
  deletion studies, 110
  overexpression studies, 111
  overview, 108–110
  systemic evolution of ligands by exponential enrichment, 112–113
Subtilin, whole-cell generation of variants, 562
Surface plasmon resonance, streptomycetes antibiotic biosynthesis regulatory gene–protein interactions, 107–108
Syringomycin, biosynthesis, 358
Systemic evolution of ligands by exponential enrichment
 γ-butyrolactone receptor target identification, 151
 streptomycetes antibiotic biosynthesis regulatory gene characterization, 112–113

T

Tacrolimus
  discovery, 51
  structure, 50
Tandem, structure, 382
Taxol
  discovery, 47–48
  structure, 47
Teicoplanin
  biosynthetic gene clusters, 467
  structure, 461
Teicoplanin, structure, 461, 488
Thioesterase domain, nonribosomal peptide synthetase, 344–346, 435
Transposon mutagenesis, biosynthetic gene clusters of *Streptomyces*, 170–172
Trichamide, structure, 577
Triostin, structure, 382

Trunkamide, structure, 577
Tubulysin A
  mechanism of action, 62
  structure, 63

U

Ulithiacyclamide, structure, 577
Urdamycin A
  glycosylation, 310
  structure, 310
β-Ureidodehydroalanine, biosynthesis, 357

V

Vancomycin
  chlorination, 475
  nonribosomal peptide synthetases, 490
  oxidative cross-linking reactions in biosynthesis
   NADPH-favodoxin reductase purification for study, 506
   overview, 489
   OxyB
    assay, 502–503
    purification, 503–504
   peptide carrier protein domain purification from vancomycin nonribosomal peptide synthetase, 505–506
   peptide–peptide carrier protein thioester synthesis
    hexapeptide S-phenyl thioester synthesis, 499–500
    hexapeptide SCoA thioester synthesis, 500–501
    overview, 490–491
    peptide carrier protein conjugation, 501–502
   peptide synthesis
    amino acid synthesis, 493–497
    model hexapeptide features, 491–493
    solid-phase synthesis, 497–499
  side chain cyclization by P450-type monooxygenases, 473–475
  structure, 356, 461, 488
Viomycin, structure, 356

X

X-ray crystallography
  GT-B glycosyltransferases, 313
  isopenicillin N synthase, 414–415
  nonribosomal peptide synthetase termination module structure, 347–349

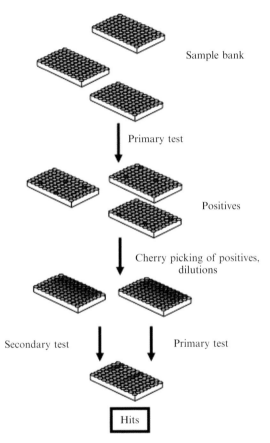

Stefano Donadio et al., Figure 1.4 The screening process. Samples meeting positivity criteria in the primary tests (red) are transferred to twin plates, diluted as appropriate and tested with primary and secondary assays for hit identification.

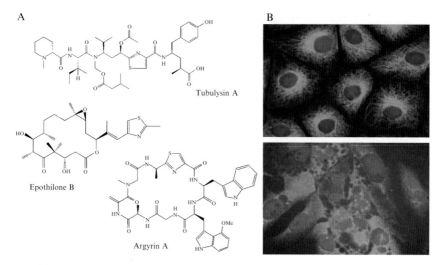

Ronald O. Garcia et al., Figure 3.2 (A) Chemical structures of some biologically active myxobacterial natural products. (B) Influence of tubulysin A on the microtubule cytoskeleton of Ptk2 potoroo cells. The cells were fixed and immunostained for tubulin. The upper picture shows microtubules of control cells (green, microtubule network; blue, cell nuclei), while the lower picture was taken after incubation with tubulysin A for 24 h Following treatment, only diffuse tubulin fluorescence remains, and the early stages of degradation of the cell nucleus are apparent. Photograph taken with permission from Sandmann et al. (2004).

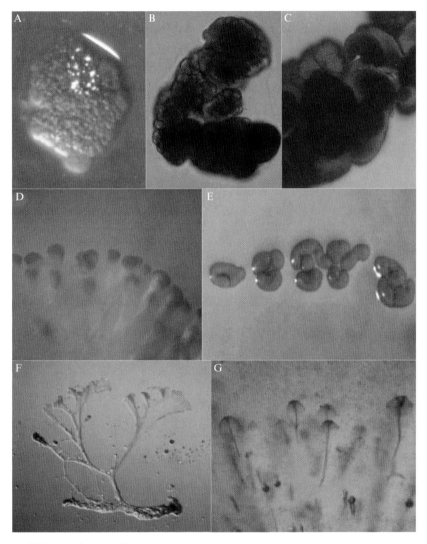

Ronald O. Garcia et al., Figure 3.3 Fruiting body morphology and swarming patterns of novel and rare myxobacterial isolates. (A) Cluster of golden sporangioles. (B) Sporangioles tightly packed in sori. (C) Mass of minute and dense sporangioles. (D) Migrating bands of swarm produced under the agar by a rare strain. (E) The bean-shaped sporangioles of *Phaselicystis flava* SBKo001^T. (F) Fan-shaped and slimy migrating colony on the surface of the agar. (G) Independently migrating colony showing comet-like structures, which are produced deep within the agar. The images in panels A and E were obtained under a dissecting microscope, while those in B and C were generated using a fluorescent laser scanning microscope. D, F, and G constitute dissecting micrographs of swarm colonies.

Ronald O. Garcia et al., Figure 3.6 Unusual deep swarming and fruiting body formation inside the agar by two novel myxobacterial strains (highlighted by white boxes), indicating their uncommon microaerophilic behavior.

Gilles P. van Wezel et al., Figure 5.1 Overproduction of the cell division activator SsgA leads to an ∼20-fold increase in the production of the red-pigmented antibiotic undecylprodigiosin during fermentation of *S. coelicolor*.

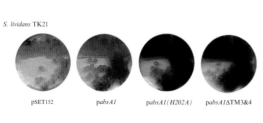

Gilles P. van Wezel et al., Figure 5.3 Activation of actinorhodin production by alleles of *absA1*. Mutants of the *S. coelicolor absA1* gene that encode proteins lacking AbsA2 kinase activity but having AbsA2∼P phosphatase activity enhance production of the blue-pigmented antibiotic actinorhodin when expressed in *S. lividans*. The strains growing on each plate are *S. lividans* TK21 bearing a control vector (pSET152) or vectors expressing wild type *absA1*, or antibiotic activating alleles *absA1(H202A)* and *absA1(delTM3 and 4)*.

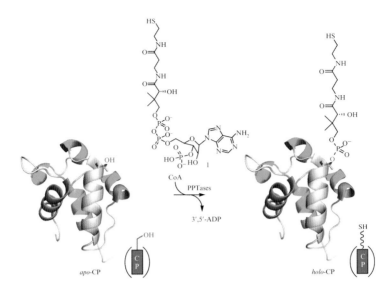

Jordan L. Meier and Michael D. Burkart, Figure 9.1 Posttranslational modification of carrier protein domains in PKS and NRPS biosynthesis. PPTase enzymes transfer 4′-phosphopantetheine from CoA to the conserved serine of carrier protein domains. The terminal thiol of this prosthetic group functions to covalently tether growing biosynthetic intermediates to the carrier protein during both polyketide and nonribosomal peptide biosynthesis. Pictured: *Escherichia coli* ACP, PDB ID 1T86.

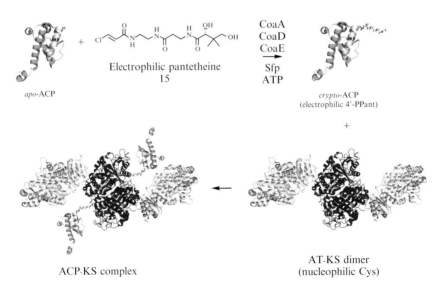

Jordan L. Meier and Michael D. Burkart, Figure 9.6 Chemoenzymatic crosslinking of ACP and KS domains. *Apo*-ACP domains (depicted here as the ACP from module 2 of the 6-deoxyerythronolide synthase) can be posttranslationally modified with electrophilic 4′-PPant arms using the one-pot chemoenzymatic synthesis to form *crypto*-ACPs. Upon addition of a nucleophilic KS domain (depicted here as the KS-AT didomain from module 3 of the 6-deoxyerythronolide synthase), which contains complementary protein–protein interactions to the ACP, covalent crosslinking will occur between the electrophilic 4′-PPant arm and the nucleophilic cysteine of the KS, resulting in a crosslinked complex. PDB IDs: DEBS2 ACP, 2JU2; DEBS3 KS-AT, 2QO3.

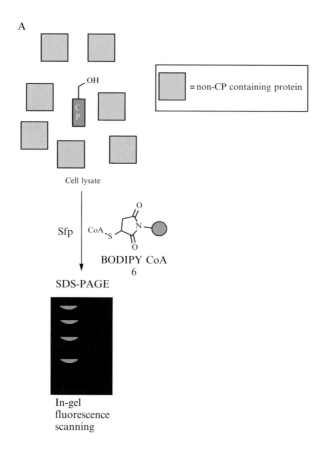

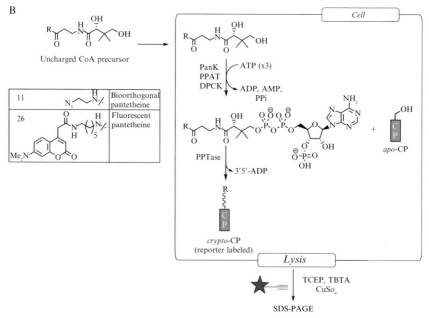

Jordan L. Meier and Michael D. Burkart, Figure 9.8 (Continued)

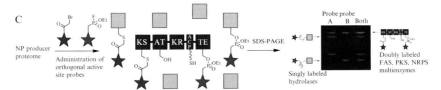

Jordan L. Meier and Michael D. Burkart, Figure 9.8 Strategies for proteomic labeling of PKS and NRPS enzymes. (A) *Apo*-CP domains in crude cell lysate can be directly labeled by use of reporter-labeled CoA analogues (6) and Sfp. This strategy is most useful in organisms in which the secondary PPTase has been inactivated, such as *B. subtilis* 168. (B) *In vivo* labeling of CP domains by bioorthogonal (11) or fluorescent (26) CoA precursors. Cellular uptake and processing of 11 by the endogenous CoA biosynthetic pathway results in formation of reporter-labeled CoA *in vivo*, which can be utilized by endogenous PPTases to form labeled CP domains. Labeled carrier proteins are visualized after cell lysis by chemoselective ligation to a bioorthogonal reporter molecule followed by SDS-PAGE. (C) Multienzymatic labeling of PKS and NRPS megasynthases. The treatment of natural product producer proteomes with reporter-labeled electrophiles can be used to distinguish PKS and NRPS multienzymes from monofunctional hydrolases, provided the inhibitors target orthogonal PKS active sites (pictured are inhibitors targeting KS and TE domains).

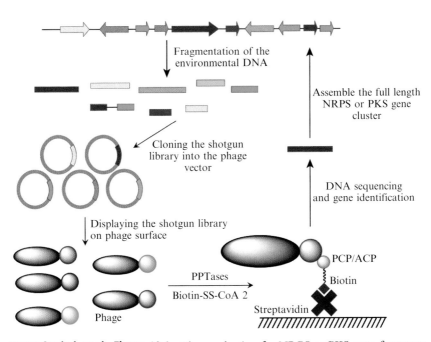

Murat Sunbul *et al.*, Figure 10.4 Phage selection for NRPS or PKS gene fragments encoded in a bacterial genome.

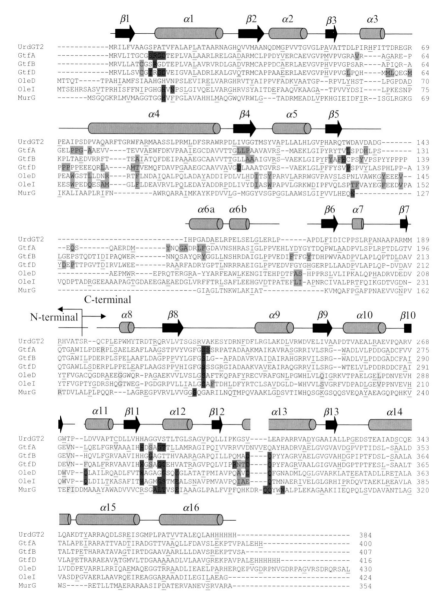

Johannes Härle and Andreas Bechthold, Figure 12.2 Alignment of the amino acid sequences of GT-B superfamily members. The predicted secondary structure of UrdGT2 (Prediction by PSIPRED Protein Structure Prediction Server http://bioinf.cs.ucl.ac.uk/psipred/psiform.html) is shown. The amino acids involved in substrate binding based on X-ray structure determination are highlighted in colors. Red: hexose-binding-site; pink: phosphat-binding-site; yellow: base-binding-site; green: aglycone acceptor-binding-site; turquoise: both hexose- and aglycone acceptor-binding-site; blue: proposed catalytic base. UrdGT2 is from *S. fradiae* Tü2717; GtfA, GtfB, and GtfD are from *Amycolatopsis orientalis*; OleD and OleI are from *S. antibioticus*; MurG is from *E. coli*.

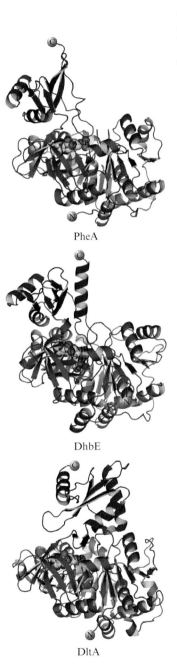

M. A. Marahiel and L.-O. Essen, Figure 13.2 Domain organization observed for the adenylate forming enzymes activating L-phenylalanine (PheA), 2,3-dihydroxybenzoate (DhbE), and D-alanine (DltA), depicted in equivalent orientations for the large N-terminal (red) subdomain. The small C-terminal subdomains (blue) show different orientations during the catalytic cycle. The active sites are tightly clamped between the two subdomains and accommodate either ATP (PheA and DltA) or 2,3-dihydroxybenzoic-AMP (DhbE).

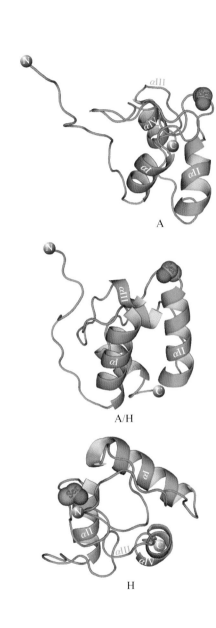

M. A. Marahiel and L.-O. Essen, Figure 13.3 The NMR structures of the TycC$_3$-PCP conformers for apo- (A, A/H) and holo- (H, A/H) PCP. The location of the active site serine, to which the ppan cofactor in the holo-PCP form is attached, is shown in red.

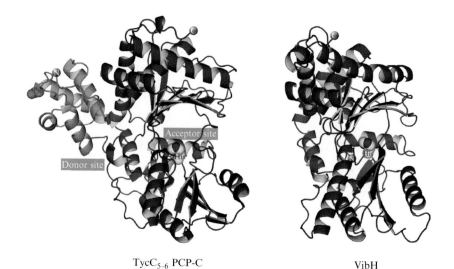

TycC$_{5-6}$ PCP-C   VibH

M. A. Marahiel and L.-O. Essen, Figure 13.4 Structures of the C-domains in PCP-C and for the stand-alone VibH C-domain are shown in grey, whereas the PCP carrier domain located on the donor site of the C-domain in PCP-C is shown in green. PCP adopts an A/H form and its active-site serine (red) is located some 50 Å from the active-site histidine, located at the bottom of the C-domain canyon.

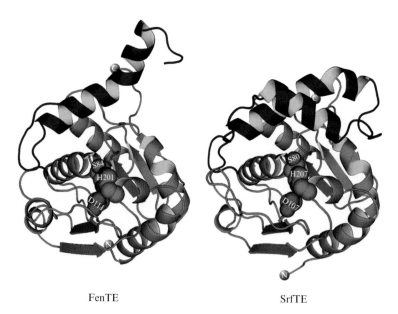

FenTE   SrfTE

M. A. Marahiel and L.-O. Essen, Figure 13.5 The structures of the FenTE and SrfTE domains reveal a core comprising an α/β-hydrolyase fold (gold) connected to an α-helical lid-region (blue). The catalytic triad residues (Ser, His, Asp) are located at the bottom of the canyon-like substrate-binding site.

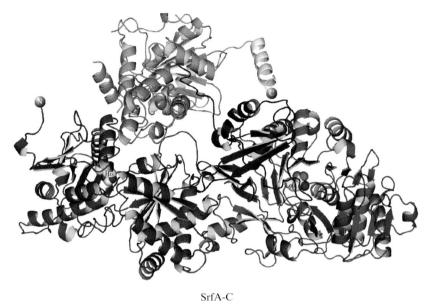

M. A. Marahiel and L.-O. Essen, Figure 13.6 Overall structure of the SrfA-C termination module (C-A-PCP-TE) showing the relative orientation of the four essential domains. Shown are a leucine residue in the active site of the A-domain and the active site histidine in the C-domain. Linkers connecting the domains are shown in blue and all domains have the same color code as in other figures. The peptide stretch arising from parts of an affinity tag is shown in yellow.

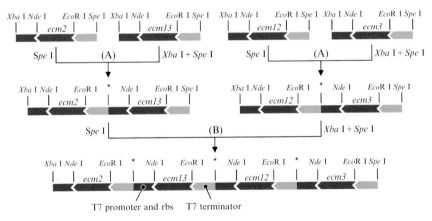

Kenji Watanabe et al., Figure 15.5 Strategy employed for constructing a multiple monocistronic expression system. (A) The first round of ligation joins the cohesive ends of an XbaI–SpeI restricted promoter–gene–terminator cassette with the compatible ends created by either SpeI- or XbaI-digestion of an acceptor plasmid (SpeI-digestion is shown in the diagram). This ligation creates a DNA sequence resistant to both XbaI and SpeI digestion (as indicated by ★), leading to the formation of a dual promoter–gene–terminator cassette that can be isolated as an intact single fragment upon XbaI–SpeI double digestion. (B) A second round of ligation fuses two dual promoter–gene–terminator cassettes to give a single fragment consisting of a quadruple promoter–gene–terminator cassette. T7 promoter and ribosome binding site are shown in blue, while T7 terminator is in green. Abbreviations are: rbs, ribosome binding site; and ★, XbaI- and SpeI-resistant ligation product between cohesive overhangs generated by XbaI and SpeI.

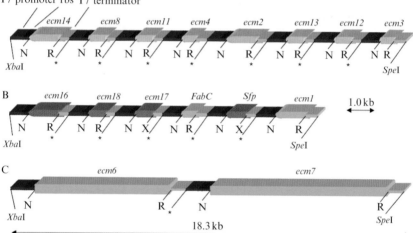

Kenji Watanabe et al., Figure 15.6 Organization of the plasmid-borne echinomycin biosynthetic gene cluster on three plasmids. T7 promoters, T7 terminators, and ribosome binding sites are shown in dark blue, light blue, and red, respectively. Abbreviation: rbs, ribosome binding site. Restriction endonuclease recognition sites are: N, NdeI; E, EcoRI; X, Xho I; and *, XbaI- and SpeI-resistant ligation product between cohesive overhangs generated by XbaI and SpeI. (A) Map of plasmid pKW532 for production of the quinoxaline-2-carboxylic acid biosynthetic proteins, carrying ecm2–4, ecm8, and ecm11–14 (yellow), a carbenicillin resistance gene (not shown) and a pBR322 origin of replication (not shown). (B) Map of pKW538 for production of the peptide-forming proteins coded by ecm1 and fabC (green), peptide-modifying proteins coded by ecm17 and ecm18 (pink), and auxiliary proteins coded by ecm16 and sfp (orange). This plasmid also carries a streptomycin resistance gene (not shown) and a CDF origin of replication (not shown). (C) Map of pKW541 for the production of the NRPSs, carrying ecm6 and ecm7 (green), a kanamycin resistance gene (not shown), and a RSF origin of replication (not shown).

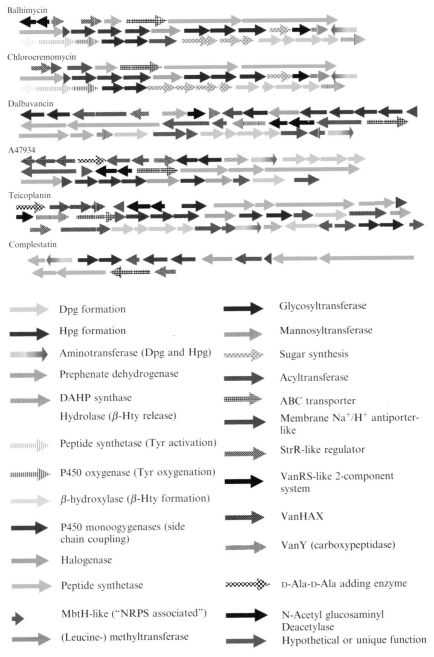

Wolfgang Wohlleben et al., Figure 18.3 Comparison of the organization of biosynthetic gene clusters for balhimycin, chloroeremomycin, dalbavancin (A40296), A47934, teicoplanin, and complestatin, modified according to Donadio et al. (2005); arrows are not in scale.

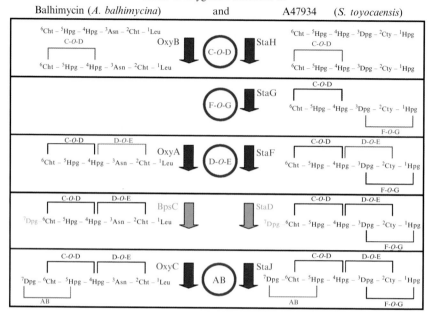

Wolfgang Wohlleben et al., Figure 18.4 Order of the oxygenase reactions (in red) in vancomycin-like glycopeptides and in lipoglycopeptides. Before the final cross-link reaction takes place it is hypothesized that the peptide synthetase, BpsC or StaD, respectively, catalyzes the addition of the seventh amino acid (in green).

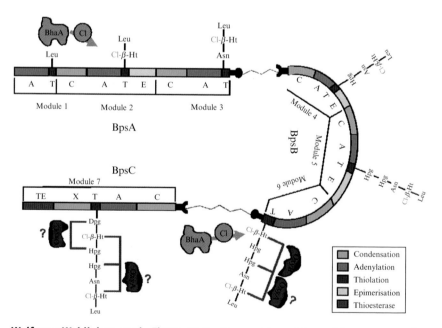

Wolfgang Wohlleben et al., Figure 18.5 Composition of the multienzyme complex catalyzing peptide backbone formation and its simultaneously cross-linking by the oxygenases (OxyA, B, C) and its chlorination by BhaA. The supposed time points of cross-linking by OxyA and OxyC have to be definitively proven.

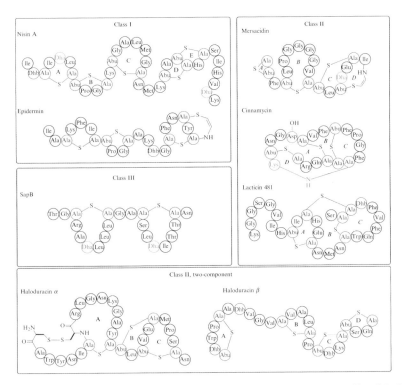

Bo Li et al., Figure 21.2 Representative examples of the three classes of lantibiotics. The same color-coding and shorthand notation is used as defined in Figs. 21.1 and 21.4. For Lan and MeLan structures, the segments derived from Ser/Thr are in red whereas those derived from Cys are in blue. The ring lettering is shown for some members and is typically alphabetical from the N- to C-terminus.

Eric W. Schmidt and Mohamed S. Donia, Figure 23.2 *Prochloron*-containing didemnid ascidians from a tropical reef. The green color in these animals is due to symbiotic cyanobacteria, *Prochloron* spp. Three species of didemnid ascidians are shown, including a bluish encrusting form, a plate-like green form, and the large animal in the center of the frame, *Lissoclinum patella*. This picture was taken in the Solomon Islands by Chris Ireland (University of Utah).